THE ROUTLEDGE HANDBOOK OF RELIGIOUS AND SPIRITUAL TOURISM

The Routledge Handbook of Religious and Spiritual Tourism provides a robust and comprehensive state-of-the-art review of the literature in this growing sub-field of tourism.

This handbook is split into five distinct sections. The first section covers past and present debates regarding definitions, theories, and concepts related to religious and spiritual tourism. Subsequent sections focus on the supply and demand aspects of religious and spiritual tourism markets, and examine issues related to the management side of these markets around the world. Areas under examination include religious theme parks, the UNESCO branding of religious heritage, gender and performance, popular culture, pilgrimage, environmental impacts, and fear and terrorism, among many others. The final section explores emerging and future directions in religious and spiritual tourism, and proposes an agenda for further research.

Interdisciplinary in coverage and international in scope through its authorship and content, this book will be essential reading for all students, researchers, and academics interested in Tourism, Religion, Cultural Studies, and Heritage Studies.

Daniel H. Olsen is a Professor in the Department of Geography at Brigham Young University, USA.

Dallen J. Timothy is a Professor in the School of Community Resources and Development at Arizona State University, USA.

THE ROUTLEDGE HANDBOOK OF RELIGIOUS AND SPIRITUAL TOURISM

Edited by Daniel H. Olsen and Dallen J. Timothy

LONDON AND NEW YORK

First published 2022
by Routledge
2 Park Square, Milton Park, Abingdon, Oxon OX14 4RN

and by Routledge
605 Third Avenue, New York, NY 10158

Routledge is an imprint of the Taylor & Francis Group, an informa business

British Library Cataloguing-in-Publication Data
A catalogue record for this book is available from the British Library

Library of Congress Cataloging-in-Publication Data
A catalog record has been requested for this book

ISBN: 978-0-367-19195-5 (hbk)
ISBN: 978-1-032-02077-8 (pbk)
ISBN: 978-0-429-20101-1 (ebk)

Typeset in Bembo
by codeMantra

CONTENTS

FIGURES

TABLES

CONTRIBUTORS

Suzanne Amaro, Management Department, School of Technology and Management, Polytechnic Institute of Viseu, Portugal.

Silvia Aulet, Faculty of Tourism, University of Girona, Spain.

Sharenda H. Barlar, Department of Modern and Classical Languages, Wheaton College, Illinois, USA.

Cristina Barroco, Management Department, School of Technology and Management, Polytechnic Institute of Viseu, Portugal.

Dino Bozonelos, Department of Political Science and Department of Global Studies, California State University, USA.

Thomas S. Bremer, Religious Studies Department, Rhodes College, Tennessee, USA.

Chadwick Co Sy Su, Department of Arts and Communication, University of the Philippines Manila.

Simon Coleman, Department for the Study of Religion, University of Toronto, Canada.

Carole M. Cusack, Department of Studies in Religion A20, University of Sydney, Australia.

Michael A. Di Giovine, Department of Anthropology and Sociology, West Chester University of Pennsylvania, USA.

Rev. Ruth Dowson, UK Centre for Events Management, Leeds Beckett University, UK.

Tomasz Duda, Institute of Spatial Management and Socio-Economic Geography, University of Szczecin, Szczecin, Poland.

Carlos Fernandes, Department of Tourism Studies, Polytechnic Institute of Viana do Castelo, Portugal.

Paula Fonseca, Management Department, School of Technology and Management, Polytechnic Institute of Viseu, Portugal.

Babu George, Department of Management, Fort Hayes State University, Kansas, USA.

Maureen Griffiths, Department of Marketing, Monash University, Melbourne, Australia.

Farooq Haq, Faculty of Management, Canadian University Dubai, Dubai, UAE.

Paul Heintzman, Leisure Studies, University of Ottawa, Canada.

Hillary Kaell, Department of Anthropology and School of Religious Studies, McGill University, Montreal, Canada.

Maximiliano E. Korstanje, Department of Economics, University of Palermo, Buenos Aires, Argentina.

Darius Liutikas, Lithuanian Centre for Social Sciences, Vilnius, Lithuania.

Avril Maddrell, Department of Geography and Environmental Science, University of Reading, UK.

Alison J. McIntosh, School of Hospitality & Tourism, Auckland University of Technology, New Zealand.

Ian S. McIntosh, Office of International Affairs, Indiana University Purdue University Indianapolis, USA.

Daniel H. Olsen, Department of Geography, Brigham Young University, Utah, USA.

Pravin S. Rana, Tourism Management, Faculty of Arts, Banaras Hindu University, Varanasi, India.

Lena Rose, Centre for Socio-Legal Studies, University of Oxford, UK.

Purna Roy, Department of Religions and Cultures, Concordia University, Montreal, Canada.

Brooke Schedneck, Department of Religious Studies, Rhodes College, Memphis, USA.

Tony Seaton, Kemmy Business School, University of Limerick, Ireland.

Richard Sharpley, Lancashire School of Business and Enterprise, University of Central Lancashire, UK.

Kiran A. Shinde, Department of Social Inquiry, La Trobe University, Australia.

Rana P.B. Singh, Former Professor, Department of Geography, Banaras Hindu University, Varanasi, India.

Melanie Kay Smith, School of Tourism, Leisure, and Hospitality, Budapest Metropolitan University, Hungary.

Dallen J. Timothy, School of Community Resources and Development, Arizona State University, USA.

Anna Trono, Department of Cultural Heritage, University of Salento, Italy.

Gregory Willson, School of Business and Law, Edith Cowan University, Australia.

Kaori Yanata, Graduate School of Tourism, Wakayama University, Japan.

1

INVESTIGATING THE INTERSECTIONS BETWEEN RELIGION, SPIRITUALITY, AND TOURISM

Daniel H. Olsen and Dallen J. Timothy

Introduction

Travel for religious and/or spiritual reasons has long been a motivator for human mobility. Reaching back possibly as long as humanity has existed (Timothy & Olsen 2006a; Butler & Suntikul 2018), people have traveled to sacred sites to worship gods, participate in initiatory, cleansing, and fertility rituals, for educational purposes, or because of curiosity (Morinis 1992). Some of the oldest human structures, such as the Göbekli Tepe in Turkey and the Ħal Saflieni Hypogeum in Malta, are related to religious and spiritual travel. Travel to these and other sacred sites has not only led to the creation of regional networks of pilgrimage paths to sacred sites and cities, but also led to entire pilgrimage economies that catered to religious travelers on the move—paths and economies that undergird many modern transportation and economic systems (Greif 2006; Coleman & Eade 2018; Timothy & Olsen 2018; Olsen 2019a).

The distance to faraway pilgrimage destinations and the dangers associated with such travels have historically inhibited the religious and spiritual mobilities of most people. However, in modern times, advances in transportation and telecommunication technologies, as well as the rise of better travel amenities, a growing middle class, and increasing religiosity and spirituality in post-secular societies, have led millions of people to travel to sacred sites around the world. While good, reliable statistics regarding religious and spiritual travel are difficult to come by (Olsen 2019b), the World Tourism Organization (UNWTO) estimates that presently, approximately 600 million people travel internationally for religious reasons let alone domestically and for spiritual reasons (UNWTO 2011).[1] If this number is accurate, this means that almost half of the 1.46 billion pre-COVID-19 international trips were motivated because of religion, making religious tourism the largest tourism niche market in the world![2] While increasingly efficient and affordable travel and access to religious and spiritual sites have removed, in many cases, the traditionally sacrificial aspects of religious pilgrimage, this surge in visitation to sacred sites has led to several countries, destinations, and tourism companies and entrepreneurs to invest in religious and spiritual tourism niche markets (Olsen 2003). In addition, several scholars have investigated the economic and social potential of these niche markets in various developing regions of the world (Horák et al. 2015; Kartal, Tepeci & Atlı 2015; Okonkwo 2015; Bayih 2018; Giuşcă, Gheorghilaş & Dumitrache 2018; Heydari Chianeh, Del Chiappa & Ghasemi 2018; Ozcan, Bişkin & Şimşek 2019). There

Table 1.1 A list of recent conferences on religious and spiritual tourism sponsored by the UNWTO

Date	*Name*	*Location*	*Declaration*
29–31 October 2007	International Conference on Tourism, Religions and Dialogue of Cultures	Cordoba, Spain	
21–22 November 2013	First UNWTO International Conference on Spiritual Tourism for Sustainable Development	Ninh Bihn, Viet Nam	*The Ninh Binh Declaration on Spiritual Tourism*
17–20 September 2014	First UNWTO International Congress on Tourism and Pilgrimage	Santiago de Compostella, Spain	*Santiago de Compostela Declaration on Tourism and Pilgrimages*
27–28 November 2014	Religious Heritage and Tourism: Types, Trends and Challenges	Elche, Spain	
15–16 June 2015	International Conference on Religious Tourism: Fostering Sustainable Socioeconomic Development for Host Communities	Bethlehem, Palestine	*Bethlehem Declaration on Religious Tourism as a Means of Fostering Socio-Economic Development of Host Communities*
5–7 October 2016	Religious Heritage and Tourism: How to Increase Religious Heritage Tourism in a Changing Society	Utrecht, The Netherlands	
22–23 November 2017	International Congress on Religious Tourism and Pilgrimage: The Potential of Sacred Places as a Tool for Sustainable Tourism Development	Fatima, Portugal	

have also been attempts to create organizations to better structure and represent the religious tourism industry, such as the Faith Travel Association[3] and the presently defunct World Religious Travel Association (2007–2011). Religious and spiritual tourism has also played an important role at tourism trade shows (Schmude 2009). Interest in these niche markets can also be seen in several UNWTO-sponsored conferences related to these tourism markets (Table 1.1) and in the publication of two UNWTO books related to religious tourism (UNWTO 2011, 2020).

However, when compared to other tourism niche markets, such as ecotourism, heritage tourism, and cultural tourism, religious and spiritual tourism have been relatively understudied until recently. Academically, the first attempt to understand the phenomenon of religious tourism was a special issue of *Annals of Tourism Research* in 1992. Boris Vukonić's (1996) *Tourism and Religion* was the first book published specifically on the intersections between tourism and religion. Several books by scholars of different disciplinary backgrounds followed these publications (Shackley 2001; Swatos Jr. & Tomasi 2002; Badone & Roseman 2004; Collins-Kreiner et al. 2006; Timothy & Olsen 2006a; Raj & Morpeth 2007; Oakes & Sutton 2010; Stausberg 2011). Rashid (2018) notes that since 2006, there has been an increasing academic interest in the religious and now spiritual tourism markets, as manifested in the plethora of journal articles and books being published on these topics, as shown in a non-exhaustive search of Google Scholar (Figures 1.1 and 1.2). These publications have been facilitated by the creation of two book series related to the intersections between pilgrimage, religion, and tourism (Routledge's Studies in Pilgrimage, Religious Travel and Tourism[4] and CABI's Religious Tourism and Pilgrimage Book Series[5]) and two specialized journals on the same topics (*International Journal of Religious and Spiritual Tourism*[6] and *International Journal of Tourism & Spirituality*[7]).

Another indicator of the growing interest in religious tourism has been the publication of several bibliographic reviews of the religious tourism literature over the past three years

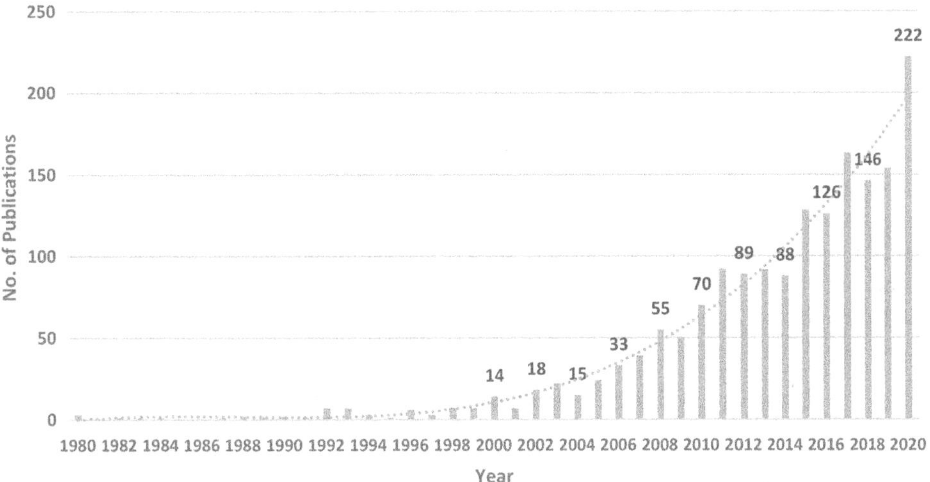

Figure 1.1 A graph showing the growth of publications related to religious tourism
Source: Google Scholar, "allintitle: religious tourism" – as of November 2020; sum = 1702.

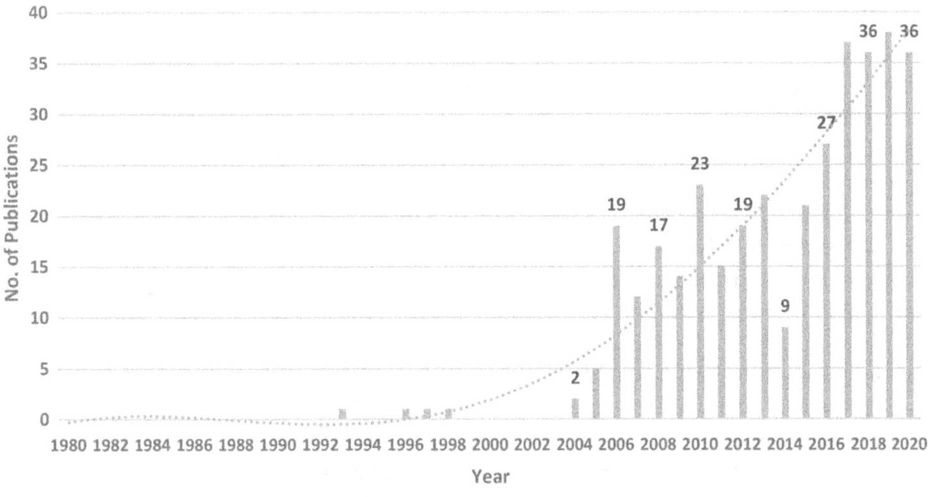

Figure 1.2 A graph showing the growth of publications related to spiritual tourism
Source: Google Scholar, "allintitle: religious tourism" – as of November 2020; sum = 356.

(Durán-Sánchez et al. 2018; Heidari et al. 2018; Rashid 2018; Durán-Sánchez et al. 2019; Iliev 2020; Kim, Kim & King 2020; Collins-Kreiner 2020), which have attempted to summarize the key research themes and suggest avenues of future research (Table 1.2). While the methods used in these bibliographic reviews are extremely limited in scope (compare Table 1.2 with Figures 1.1 and 1.2), these reviews may be seen as a sign of some level of maturity in this research field. In the context of spiritual tourism, because of the newness and lack of literature on this niche market, there have been no attempts to summarize the nascent literature in any meaningful manner.

The purpose of this handbook is to examine many of the key themes both identified and not identified by the bibliographic reviews (Norman & Cusack 2015). As noted in the conclusion chapter of this handbook (see Chapter 33), there are several themes and topics that are not covered herein which can be potentially fruitful avenues for further study. As such, this book is not meant to be the definitive word on religious and spiritual tourism. Indeed, most of the chapters focus on religious tourism rather than on spiritual tourism per se, which may in part reflect the maturity of the literature on religious tourism (Figures 1.1 and 1.2). At the same time, even though there are separate religious and spiritual tourism markets, the religious and the spiritual are intimately intertwined, with considerable overlap between them. As such, in many cases, discussions on religious tourism in this handbook will be equally applicable to spiritual tourism. To set the context for the remaining chapters, the authors discuss the relationships between religious and spiritual tourism before introducing the handbook chapters.

Defining and differentiating religious and spiritual tourism

As is common in academic circles, it is difficult coming to a consensus regarding defining topics or phenomena of interest. This is the case in the context of religious and spiritual tourism, where there are no definitive definitions of the two terms (Olsen 2015b). While there are several reasons for this, four reasons are listed here. First, there has been much ink spilt over defining and differentiating religion and spirituality. As Norman (2011: 17) notes, the terms 'religion' and 'spirituality' are problematic within the study of religion, let alone tourism. They are 'fuzzy terms', in that people know what they are, but their definitions are difficult to articulate (Pals 1996). For example, religion can be found in every culture. However, the term 'religion' itself is a Western conceptual label used to compare Western understandings of religion to similar sociocultural constructs in non-Western societies (Smith 1998; Braun 2000). As Horri (2018) notes, while many Western cultures differentiate between the religious and secular spheres, in other cultures this distinction does not exist, with religion being intimately intertwined with politics, culture, heritage, and social norms and values, which only adds to the difficulties in defining religion. Spirituality has also been a difficult term to define, in part because while spirituality is generally viewed as an individual's search or quest for meaning, this quest is also found within institutionalized religion. As Stausberg (2014: 355) puts it, 'spiritual/spirituality is semantically situated within and beyond the social realm of religion'. Presently, spirituality is generally used to describe this quest outside of institutionalized religion as people 'explore—and deeply and meaningfully connect one's inner self—to the known world and Beyond' (Kale 2004: 93).

Second, difficulties in developing definitive definitions of religious and spiritual tourism stem from whether definitions come from a supply-side perspective, which focuses on the destinations and activities that religious and spiritual tourists visit, or a demand-side definition, which focuses on motivations for travel. As Timothy and Olsen (2006b: 272) argue, 'the

Table 1.2 A list of recent attempts to summarize the academic research on religious tourism

Author(s)/Date	*Literature examined*	*Current research themes*	*Avenues of future research*
Collins-Kreiner (2020)	40 journal articles published in tourism journals between 2015 and 2019	Pilgrim-tourist dichotomy; religious tourism as 'exponomy' and transformative travel; religion as a 'product'; religion and sustainability, over-tourism, and community-based tourism	Post-secular tourists and postmodernist conceptualizations; the politics of religion and the rise of identity politics through tourism
Durán-Sánchez et al. (2018)	1999 documents from Scopus database—articles, books, and book chapters	A focus on author productivity and academic journals that published the most religious tourism articles	None
Durán-Sánchez et al. (2019)	103 documents from the Web of Science database prior to 2017	A focus on most cited articles, country affiliation of authors, and academic journals that published the most religious tourism articles	None
Heidari et al. (2018)	122 journal articles from five different databases published between 1990 and 2017	A focus mainly on bibliometrics, such as research methods used, research type (e.g., literature review, philosophical paper, applied paper, case study), geographical areas of research, and publication venues; only two themes mentioned: marketing and demand	Expansion of meta-data analysis of academic journals excluded from this paper
Iliev (2020)	124 'relevant documents' from the Web of Science, Scopus, Google Scholar, ResearchGate, and Academia.edu databases prior to 2018—articles, books, and book chapters	Pilgrimage-tourism dichotomy; conceptualizing religious tourism; segmenting the religious tourism market; route-based religious tourism and pilgrimage; supply-side analysis of religious tourism; religious tourism as mass or alternative tourism	Role of religion in motivating people to visit religious sites; religious tourist experiences; post-journey experiences; neglected supply-side elements of religious tourism; postmodern approaches to religious tourism
Kim, Kim, and King (2020)	84 journal articles in 12 leading tourism journals	Religious tourist perceptions; pilgrim-tourist dichotomy; religious commercialism; religious tourism destinations; religious motivations; authenticity; economic impacts; religious tourism infrastructure	Hospitality and religion; research beyond Asia and Europe; research beyond Islam, Buddhism, Hinduism, and Christianity; visitor experience at sacred sites; sustainability issues; stakeholder perspectives; accommodations at sacred sites; religious tourist experiences
Rashid (2018)	All journal articles published in Emerald Insight journals between January 2006 to December 2017 (total number of articles not given)	Who is a 'religious tourist' dark tourism; locations linked to religious tourism; earthly gains vs. heavenly aspirations; development of religious amenities; role of technology; supply-side angst; demand-side motivations; marketing to specific religious groups	None

term [religious tourism] is generally used by tourism promoters and researchers to describe the phenomenon in two different ways: those whose impetus to travel combines both religious (dominant) and secular (secondary) motives and people who visit sacred sites during their journeys to other attractions and destinations'. Timothy and Olsen continue:

> On the one hand, from a religious perspective, religious tourism is separate from other forms of tourism because it is characterized by its aims, motivations, and destinations…, as equating pilgrimage with tourism or even religious tourism would make pilgrimage co-equal with more hedonic types of undertakings, such as wine tourism or sex tourism….From an industry or academic perspective, definitions of tourist types are typically not based on motivations but rather on the activities tourists engage in while traveling….From the perspective of tourism planners, promoters, and scholars, pilgrims are simply tourists, for tourists are not defined by motives, rather by the simple fact that they travel away from home.
>
> *(p. 272)*

This leads to the third difficulty in defining religious and spiritual tourism, in that postmodernists argue that creating definitive definitions is too constraining to the study of religious and spiritual tourism (Collins-Kreiner 2020). As such, rather than developing absolute definitions for these specialized niches, many scholars instead create 'stipulative' definitions, or definitions that reflect the way in which individual scholars understand and use these terms within a particular research context. In many cases, these stipulative definitions are also based on supply- and demand-side definitions. For example, Olsen (2008: 22) defines religious tourism as 'travel by tourists to religious destinations, cultures and sites regardless of motivation, whether the visits to these sites are of primary or secondary interest'. This definition is heavily skewed toward a supply-side definition of religious tourism and ignores the demand-side or motivational aspects of why people chose to travel to religious destinations. However, this definition was stipulative in nature, considering that Olsen's research was on the theological aspects of sacred site management, a part of the supply side of religious tourism. El-Gohary (2016), in studying both the supply and demand sides of halal tourism, uses a more balanced definition found in the *FICCI Religious Tourism Report* (2012: 2): 'travel with the core motive of experiencing religious forms, or the products they induce, like art, culture, traditions and architecture.' Alternatively, Blackwell (2007: 37) defines religious tourism as 'encompass[ing] all kinds of travel that is motivated by religion and where the destination is a religious site'. While the supply-side aspect of religious tourism is present in this definition, religious motivations are the central focus of this definition. The use of a demand-side definition, however, was probably purposeful, considering that the focus of Blackwell's writing was on the motivations for religious tourism, pilgrimage, festivals, and events.

In the context of spiritual tourism, Norman (2012: 23) focuses on developing a 'working definition' of spiritual tourism because of the difficulty in defining the phenomenon. He does note, however, that 'What confusion that exists over the term "spiritual tourism" seems to be, in fact, largely a scholarly creation, as the emic voice is quite clear in its focus on individual self-discovery and wellbeing maintenance'. In the same vein, Willson, McIntosh and Zahra (2013: 152) define spiritual tourism as a phenomenon undertaken by an 'individual in the quest for personal meaning from and through travel'. Indeed, spiritual tourism is generally defined based on motivational or demand-side aspects of travel, possibly because spiritual tourism is often compared to a 'quest', where the search for meaning does not necessarily mean travel to specific destinations or physical locations—life itself, as well as everyday living, can be seen as a spiritual quest (Kujawa 2017).

Fourth, as Olsen (2015a; forthcoming) argues, the religious tourism market has become increasingly fragmented into other separate yet overlapping tourism niche markets: religious tourism, pilgrimage or faith tourism, spiritual tourism, and New Age tourism. The religious tourism market, which is usually situated within the broader cultural and heritage tourism marketplace, refers to tourism that involves visitation to both natural and human-built religious sites. This tourism type focuses on supply-side aspects of religious travel, with religious sites, ceremonies, and cultures being included as a part of broader cultural or heritage package tours. The 'pilgrimage tourism' or 'faith-based tourism' market focuses specifically on the religious motivations of people—or 'believer[s] as tourist[s]' (Terzidou, Scarles & Saunders 2018). Travels by people in this category focus more on engaging in religious rituals and educational pursuits at religious sites with little attention paid to leisure activities (Seyer & Muller 2011; Fang 2020). The spiritual tourism market is closely related to the religious pilgrimage/faith tourism markets, because of, as mentioned above, the overlap between religion and spirituality (Lopez, González & Fernández 2017). However, spiritual tourism is 'tourism characterised by a self-conscious project of spiritual betterment' (Norman 2012: 20) outside of religious constraints and is tied more closely to wellness tourism (Bushell & Sheldon 2009; Bowers & Cheer 2017; Cheer, Belhassen & Kujawa 2017). The New Age tourism market overlaps with the spiritual tourism market. What makes the New Age tourism market different is that New Agers draw upon and experiment with a wide variety of religious, spiritual, and indigenous worldviews, rituals, and practices within a growing 'spiritual marketplace' (Aupers & Houtman, 2006; Pernecky & Johnson 2015; Timothy & Conover 2006).

This fragmentation of the religious tourism market into these four segments has come about in part because globalization and its neoliberal market consumerist logistics have given rise to secularizing and individualistic forces have led to rising 'cultural deregulation' (Gauthier, Martikainen & Woodhead 2013: 270). This cultural deregulation has led to both a metaphorical and literal expansion of what constitutes religion (Olsen 2014), with individual experiences and worldviews being the ultimate arbiter of truth and understanding. As such, there has been a proliferation of new forms of religion and spirituality, each of which has the potential for commodification (Olsen 2019a). Indeed, as Weidenfield and Ron (2008: 358) argue, 'the tourism and hospitality industries are constantly searching for new customer segments'. As such, these niche markets have been segmented even further, with the marketplace catering to an ever-growing list of faith communities, socioeconomic classes, and destinations. For example, each of these market segments offers meetings, incentives, conferences, and exhibitions; events; themed cruises, safaris, adventure travel, weekend getaways, leisure vacations, and retreats/guesthouses; volunteer vacations; and visiting pertinent religious/spiritual sites (Seyer & Muller 2011; Fang 2020). Many tour companies also offer services to specific faith communities, such as Buddhists, Muslims, Christians, and Jews, based on their specific supply and demand needs (Olsen forthcoming; Ron & Timothy 2019). This commodification, however, has been criticized by scholars and theologians, who argue that the (in)compatibility of merging the material and immaterial or spiritual aspects of religion leads to issues related to authenticity and the development of kitsch (Gauthier, Martikainen & Woodhead 2013; McIntyre 2014; Olsen forthcoming).

With the current COVID-19 pandemic, it has become clear that not just economic forces, but also broader political forces, shape these tourism market segments. The shutting down of religious mass gatherings by government and health authorities in 2020 led to different reactions by faith communities, some positive and some negative (Nhamo, Dube & Chikodzi

2020; Olsen & Timothy 2020; see Olsen 2020). While the short- and long-term future of tourism more generally is uncertain, religious and spiritual tourism are thought to be one of the first tourism markets to rebound post COVID-19, particularly regionally, considering the importance of sacred places within the context of salvation and the search for meaning in an uncertain world. It is hoped that this handbook will both highlight the importance of these two travel niches not just economically but also socioculturally and help provide a context for the future development of these market segments in an ethical and sustainable manner.

Chapter contributions

This book is comprised of 33 chapters written by 38 scholars from 16 countries and is divided into six sections. Section 1 focuses on definitions, theories, and concepts related to religious and spiritual tourism. In Chapter 2, Tony Seaton discusses different types of journeys within historical religious practice that are not generally acknowledged in the academic literature. Drawing on the work of Max Weber and research in the field of mobilities, Seaton takes a sociological approach to examine eight ideal role types of religious travelers within the context of Christianity. Dino Bozonelos looks at the intersection of political economy and religious and spiritual tourism in Chapter 3, an area of research that has been relatively untouched. Bozonelos discusses the field of political economy, as well as the small but growing literature related to political economy and religious tourism, before presenting a framework to elucidate how political, economic, and religious institutions are linked together. In Chapter 4, Paul Heintzman examines the scholarship on the links between religion, spirituality, and leisure and provides a preliminary framework to organize and discuss empirical research that examines the spiritual and religious outcomes of leisure travel.

In Chapter 5, Melanie Smith investigates the intersections between religion, spirituality, and wellness; the purpose of her chapter is to examine the relationship between religion and spirituality in the context of wellness tourism, focusing on holistic meditation at yoga retreat centers and the importance of landscape and nature in the development of spiritual wellbeing. In Chapter 6, Carole Cusack examines the 'spiritual marketplace' in the context of new religious movements (NRMs) and spiritual and New Age tourism, arguing that sites favored by spiritual tourists with 'New Age' interests are very different from sites valued by members of NRMs which are more aligned to traditional pilgrimage practices. In Chapter 7, Daniel Olsen discusses the intersections between fan pilgrimage, religion, and spirituality, examining the ways in which religion is related to fandom and how some fandoms act as quasi-religions. In Chapter 8, Chadwick Co Sy Su attempts to trace the developments in nonreligious spiritualism, alternatively called 'secular pilgrimage', with a view to indicate areas of agreement, disagreement, and subsequent convergence in both literature and individual scholarship. Ian McIntosh tackles difficult questions related to pilgrimage, tourism, and peacebuilding in Chapter 9. In particular, McIntosh asks whether tourism can be an avenue for de-escalating tensions or even peacebuilding, and if the religious and spiritual tourism industries can help break down political, cultural, and ethnic barriers within and between states.

Section 2 examines aspects of spaces and places related to the supply side of religious and spiritual tourism. In Chapter 10, Rana Singh, Pravin Rana, and Daniel Olsen look at religious tourism in the context of Hinduism, particularly focusing on how the natural environment is tied to Hindu theology and cosmology, how pilgrimages can lead to better human-nature relationships, and the greening of Hindu pilgrimage practices. Drawing on concepts of space-place, Richard Sharpley suggests in Chapter 11 that the greater the sense of place, the greater the sense of spirituality. He discusses this idea by presenting a continuum of place-related spirituality through examining blue spaces, green spaces, and dark (outer) spaces. Thomas Bremer discusses national parks as treasured sacred places in the United States in Chapter 12 with an emphasis on three elements that have made American national parks attractive pilgrimage destinations for generations of tourist visitors. In Chapter 13, Lena Rose focuses on the role of religious theme parks as powerful tools for proselytizing, making money, promoting tourism, and engaging in nation-building. She sets religious theme parks in the context of regular theme parks and then explores their pedagogies, performances, experiences, and the themes of authenticity, representation, power, and orthodoxy. Brooke Schedneck explores the common themes and practices within spiritual and religious retreats within the traditions of Christianity, Buddhism, Hinduism, and New Age institutions in Chapter 14. In Chapter 15, Michael Di Giovine conceptualizes the interrelatedness between spiritual tourism, pilgrimage, and UNESCO's World Heritage program, examining both the way traditional pilgrimage destinations have become World Heritage Sites and how travel to modern World Heritage Sites are akin to pilgrimage and sacred travel.

Section 3 looks at the topics of motivation, experiences, and performances, or the demand side of religious and spiritual tourism. In Chapter 16, Darius Liutikas discusses the main motivating factors that lead pilgrims, religious tourists, and spirituality seekers to travel. While reviewing the literature on motivation, he highlights the fact that most people travel with a mix of sacred-secular motivations—motivations that are related to people's individual values and identities. In Chapter 17, Gregory Willson and Alison McIntosh examine whether volunteer tourism experiences are religious or spiritual in nature. They examine how volunteer tourism is linked to the search for meaning, personal growth, and connectedness with the self, the environment, and others. Sharenda Barlar looks at Protestant experiences along the Camino de Santiago de Compostela in Chapter 18. Barlar does this by presenting the experiences of evangelical students from Wheaton College who walked the French Way of the Camino in 2019. In Chapter 19, Avril Maddrell focuses on the issue of gender in religious and spiritual tourism studies, highlighting the importance of embodied participant-centered experiences and meaning-making across organized and individual pilgrimages. In Chapter 20, Maureen Griffiths and Maximiliano Korstanje focus on the concept of, and issues related to, authenticity in the context of religious and spiritual tourism. They focus particular attention on several questions, including: Is religious tourism an activity that alienates pilgrims or simply a modern manifestation of sacred pilgrimage? What are the motivations of this emerging sub-segment? Is secularization theory enough to explain these types of complex issues? What are the main challenges religious tourism will face in the future? Purna Roy and Hillary Kaell consider how pilgrims negotiate the aftermath of their journeys in Chapter 21. The authors consider questions as such: Are the pilgrims changed post-pilgrimage? Did they attain their goal?' 'How do scholars evaluate such outcomes?

Section 4 of this handbook focuses on management issues related to religious and spiritual tourism. In Chapter 22, Dallen Timothy examines several of the sociocultural and economic implications of religious and spiritual tourism and how they relate to other tourism impacts or how they are unique to this context. In Chapter 23, Kiran Shinde discusses the environmental impacts of religious and spiritual tourism, highlighting four different themes in the academic literature, including differences and commonalities of impacts across different sacrosanct environmental settings; the peculiarities of environmental behavior in religious tourism that influence impacts on sacred sites; the nature of environmental management in religious tourism; and the distinguishing characteristics of spiritual tourism–environment interactions. In Chapter 24, Farooq Haq examines the marketing of religious and spiritual tourism experiences, suggesting ways in which marketers and businesses can better capitalized on these specialized tourism markets. Managing sacred sites is the topic of Chapter 25, where Simon Coleman and Daniel Olsen focus on a wide range of strategies that are utilized in human-built and natural religious and spiritual sites. In Chapter 26, Ruth Dowson reviews current research on the organization and management of religious and spiritual events, festivals, and celebrations.

In Chapter 27, Suzanne Amaro, Cristina Barroco, and Paula Fonseca provide an overview on the use of information and communication technologies (ICTs) in the religious tourism industry, demonstrating how ICTs can be useful in managing religious sites, events, and pilgrimages more effectively in challenging environments. The authors suggest six areas where ICTs can enhance the religious tourism market: tourists' experience, religious tourism promotion, facilitating accessible tourism, ensuring more sustainable religious tourism, safety, and advancing religious tourism research. In Chapter 28, Tomaz Duda discusses issues related to the interpretation of religious or spiritual heritage with the broader context of heritage tourism interpretation. Duda walks readers through how to better interpret and present narratives at religious tourism sites to help facilitate deeper understandings and experiences for visitors. Kaori Yanata and Richard Sharpley (Chapter 29) look at how residents around temple-stays deal with the commodification of their religious practices and their co-existence with tourists. Silvia Aulet, Carlos Fernandes, and Dallen Timothy (Chapter 30) examine the relationships between food, religion, and tourism, emphasizing the ways in which different religions use food for ritual and communal purposes as well as at broader cultural and heritage attractions. In Chapter 31, Maximiliano Korstanje and Babu George look at the interface of security, safety, and risk in the context of religious tourism, while in Chapter 32, Anna Trono reflects on religious and spiritual tourism in the context of sustainable development.

In Section 5, Dallen Timothy and Daniel Olsen conclude this handbook by discussing several emerging themes related to religious and spiritual tourism that were not discussed in depth in this handbook, including unusual spaces and places of religious tourism, dark tourism and pilgrimage, slow and religious tourism, religious hospitality, emerging travel markets and motives, and the effects of globalization on religious tourism.

Notes

1 The UNWTO (2011) suggests that most religious tourism takes place in Asian and European countries.
2 https://www.unwto.org/global-and-regional-tourism-performance.
3 https://ntaonline.com/markets/faith-travel-association/.
4 https://www.routledge.com/Routledge-Studies-in-Pilgrimage-Religious-Travel-and-Tourism/book-series/RSRTT.
5 https://www.cabi.org/products-and-services/about-cabi-books/cabi-religious-tourism-and-pilgrimage-series/.
6 https://arrow.tudublin.ie/ijrtp/.
7 http://ijts.usc.ac.ir/.

References

Aupers, S., & Houtman, D. (2006) 'Beyond the spiritual supermarket: The social and public significance of New Age spirituality', *Journal of Contemporary Religion*, 21(2): 201–222.

Badone, E., & Roseman, S.R. (2004) *Intersecting Journeys: The Anthropology of Pilgrimage and Tourism*. Chicago: University of Illinois Press.

Bayih, B.E. (2018) 'Potentials and challenges of religious tourism development in Lalibela, Ethiopia', *African Journal of Hospitality, Tourism and Leisure*, 7(4): 1–17.

Blackwell, R. (2007) 'Motivations for religious tourism, pilgrimage, festivals and events', in Raj, R., & N.D. Morpeth (eds.) *Religious Tourism and Pilgrimage Festivals Management: An International Perspective* (pp. 35–47). Wallingford: CABI.

Bowers, H., & Cheer, J.M. (2017) 'Yoga tourism: Commodification and western embracement of eastern spiritual practice', *Tourism Management Perspectives*, 24: 208–216.

Braun, W. (2000) 'Religion', in Braun, W., & McCutcheon, R.T. (eds.) *Guide to the Study of Religion* (pp. 3–18). New York: Cassell.

Bushell, R., & Sheldon, P.J. (eds.) (2009) *Wellness and Tourism: Mind, Body, Spirit, Place.* New York: Cognizant Communication.

Butler, R., & Suntikul, W. (2018) *Tourism and Religion: Issues and Implications.* Bristol: Channel View Publications.

Cheer, J.M., Belhassen, Y., & Kujawa, J. (2017) 'The search for spirituality in tourism: Toward a conceptual framework for spiritual tourism', *Tourism Management Perspectives*, 24: 252–256.

Coleman, S., & Eade, J. (2018) 'Pilgrimage and political economy: Introduction to a research agenda', in Coleman, S., & Eade, J. (eds.) *Pilgrimage and Political Economy: Translating the Sacred* (pp. 1–20). New York: Berghahan Books.

Collins-Kreiner, N. (2020) 'Religion and tourism: A diverse and fragmented field in need of a holistic agenda', *Annals of Tourism Research*, 82: 102892.

Collins-Kreiner, N., Kliot, N., Mansfeld, Y., & Sagi, K. (2006) *Christian Tourism to the Holy Land: Pilgrimage during Security Crisis.* Aldershot: Ashgate.

Durán-Sánchez, A., Álvarez-García, J., Río-Rama, D., de la Cruz, M., & Oliveira, C. (2018) 'Religious tourism and pilgrimage: Bibliometric overview', *Religions*, 9(9): 249.

Durán-Sánchez, A., Del Río, M.D.L.C., Oliveira, C., & Álvarez-García, J. (2019) 'Religious tourism and pilgrimage: Study of academic publications in scopus', in Álvarez-García, J., del Río Rama, M.D.L., & Gómez-Ullate, M. (eds.) *Handbook of Research on Socio-Economic Impacts of Religious Tourism and Pilgrimage* (pp. 1–18). Hershey, PA: IGI Global.

El-Gohary, H. (2016) 'Halal tourism, is it really Halal?', *Tourism Management Perspectives*, 19: 124–130.

Fang, W.-T. (2020) *Tourism in Emerging Economies: The Way We Green, Sustainable, and Healthy.* Singapore: Springer.

FICCI Religious Tourism Report (2012) 'Diverse beliefs: Tourism of faith religious tourism gains ground', available at: http://ficci.in/spdocument/20207/Diverse-Beliefs-Tourism-of-Faith.pdf (accessed 10 November 2020).

Gauthier, F., Martikainen, T., & Woodhead, L. (2013) 'Acknowledging a global shift: A primer for thinking about religion in consumer societies', *Implicit Religion*, 16(3): 261–276.

Giuşcă, M.C., Gheorghilaş, A., & Dumitrache, L. (2018) 'Assessment of the religious-tourism potential in Romania', *Human Geographies*, 12(2): 225–237.

Greif, A. (2006) *Institutions and the Path to the Modern Economy: Lessons from Medieval Trade.* Cambridge: Cambridge University Press.

Heidari, A., Yazdani, H.R., Saghafi, F., & Jalilvand, M.R. (2018) 'The perspective of religious and spiritual tourism research: A systematic mapping study', *Journal of Islamic Marketing*, 9(4): 747–798. DOI: 10.1108/JIMA-02-2017-0015.

Heydari Chianeh, R., Del Chiappa, G., & Ghasemi, V. (2018) 'Cultural and religious tourism development in Iran: Prospects and challenges', *Anatolia*, 29(2): 204–214.

Horák, M., Kozumplíková, A., Somerlíková, K., Lorencová, H. and Lampartová, I. (2015) 'Religious tourism in the south-Moravian and Zlín regions: Proposal for three new pilgrimage routes', *European Countryside*, 7(3): 167–178.

Horri, M. (2018) *The Category of 'Religion' in Contemporary Japan.* Cham: Palgrave Macmillian.

Iliev, D. (2020) 'The evolution of religious tourism: Concept, segmentation and development of new identities', *Journal of Hospitality and Tourism Management*, 45: 131–140.

Kale, S.H. (2004) 'Spirituality, religion, and globalization', *Journal of Macromarketing*, 24(2): 92–107.

Kartal, B., Tepeci, M., & Atlı, H. (2015) 'Examining the religious tourism potential of Manisa, Turkey with a marketing perspective', *Tourism Review*, 70(3): 214–231.

Kim, B., Kim, S., & King, B. (2020) 'Religious tourism studies: Evolution, progress, and future prospects', *Tourism Recreation Research*, 45(2): 185–203.

Kujawa, J. (2017) 'Spiritual tourism as a quest', *Tourism Management Perspectives*, 24: 193–200.

Lopez, L., González, R.C.L., & Fernández, B.M.C. (2017) 'Spiritual tourism on the way of Saint James the current situation', *Tourism Management Perspectives*, 24: 225–234.

McIntyre, E. (2014) 'Rescuing God from bad taste: Religious kitsch in theory and practice', *Literature & Aesthetics*, 24(2): 83–108.

Morinis, A. (1992) 'Introduction: The territory of the anthropology of pilgrimage', in Morinis, A. (ed.) *Sacred Journeys: The Anthropology of Pilgrimage* (pp. 1–28). Westport, CT: Greenwood Press.

Nhamo, G., Dube, K., & Chikodzi, D. (2020) *Counting the Cost of COVID-19 on the Global Tourism Industry*. Cham: Springer.

Norman, A. (2011) *Spiritual Tourism: Travel and Religious Practice in Western Society*. London: Continuum.

Norman, A. (2012) 'The varieties of the spiritual tourist experience', *Literature & Aesthetics*, 22(1): 20–37.

Norman, A., & Cusack, C. (eds.) (2015) *Religion, Pilgrimage, and Tourism, Volume 1*. London: Routledge.

Oakes, T., & Sutton, D.S. (eds.) (2010) *Faiths on Display: Religion, Tourism, and the Chinese State*. Lanham, MD: Rowman & Littlefield Publishers.

Okonkwo, E. (2015) 'Religious activities and their tourism potential in Sukur Kingdom, Nigeria', *International Journal of Religious Tourism and Pilgrimage*, 3(1): 1–11.

Olsen, D.H. (2003) 'Heritage, tourism, and the commodification of religion', *Tourism Recreation Research*, 28(3): 99–104.

Olsen, D.H. (2008) *Contesting Identity, Space and Sacred Site Management at Temple Square in Salt Lake City, Utah*. PhD dissertation, University of Waterloo, Ontario, Canada.

Olsen, D.H. (2014) 'Metaphors, typologies, secularization, and pilgrim as hedonist: A Response', *Tourism Recreation Research*, 39(2): 248–258.

Olsen, D.H. (2015a) 'Definitions, motivations and sustainability: The case of spiritual tourism', in UNWTO (ed.) *First UNWTO International Conference on Spiritual Tourism for Sustainable Development* (pp. 35–46). Madrid: World Tourism Organization.

Olsen, D.H. (2015b) 'Religion, tourism', in Jafari, J., & Xiao, H. (eds.) *Encyclopedia of Tourism* (pp. 784–787). New York: Springer.

Olsen, D.H. (2019a) 'Religion, spirituality, and pilgrimage in a globalizing world', in Timothy, D.J. (ed.) *Handbook of Globalisation and Tourism* (pp. 270–283). London: Edward Elgar.

Olsen, D.H. (2019b) 'Religion, pilgrimage, and tourism in the MENA region', in Timothy, D.J. (ed.) *Routledge Handbook on Tourism in the Middle East and North Africa* (pp. 109–124). London: Routledge.

Olsen, D.H. (2020) 'Disease and health risks at mass religious gatherings', in Shinde, K.A., & Olsen, D.H. (eds.) *Religious Tourism and the Environment* (pp. 116–132). Wallingford: CABI.

Olsen, D.H. Forthcoming. 'Faith, New Age spirituality, and religion: Negotiating the religious tourism niche market', in Novelli, M., Cheer, J., Dolezal, C., Jones, A., & Milano, C. (eds.) *Handbook of Niche Tourism*. Cheltenham: Edward Elgar.

Olsen, D.H., & Timothy, D.J. (2020) 'COVID-19 and religious travel: Present and future directions', *International Journal of Religious Tourism and Pilgrimage*, 8(7): 170–188.

Ozcan, C.C., Bişkin, F., & Şimşek, Ç. (2019) 'Regional economic effects and marketing of religious tourism: The case of Konya', in de la Cruz del Río Rama, M., & Gómez-Ullate García de León, M. (eds.) *Handbook of Research on Socio-Economic Impacts of Religious Tourism and Pilgrimage* (pp. 250–274). Hershey, PA: IGI Global.

Pals, D.L. (1996) *Seven Theories of Religion*. New York: Oxford University Press.

Pernecky, T., & Poulston, J. (2015) 'Prospects and challenges in the study of New Age tourism: A critical commentary', *Tourism Analysis*, 20(6): 705–717.

Raj, R., & Morpeth, N.D. (2007) *Religious Tourism and Pilgrimage Festivals Management: An International Perspective*. Wallingford: CABI.

Rashid, A.G. (2018) 'Religious tourism – A review of the literature', *Journal of Hospitality and Tourism Insights*, 1(2): 150–167.

Ron, A.S., & Timothy, D.J. (2019) *Contemporary Christian Travel: Pilgrimage, Practice and Place*. Bristol: Channel View Publications.

Schmude, J. (2009) 'Religion and the tourism market – An analysis of products offered at ITB 2009', in Trono, A. (ed.) *Tourism, Religion & Culture: Regional Development through Meaningful Tourism Experiences* (pp. 365–373). Lecce: Mario Congedo.

Seyer, F., & Muller, D. (2011) 'Religious tourism: Niche or mainstream?', in Papathanassis, A. (ed.) *The Long Tail of Tourism: Holiday Niches and their Impact on Mainstream Tourism* (pp. 45–56). Heidelberg: Springer.

Shackley, M. (2001) *Managing Sacred Sites: Service Provision and Visitor Experience.* London: Continuum.
Smith, J. (1998) 'Religion, religions, religious', in Taylor, M. (ed.) *Critical Terms for Religious Studies* (pp. 269–284). Chicago: University of Chicago Press.
Stausberg, M. (2011) *Religion and Tourism: Crossroads, Destinations and Encounters.* London: Routledge.
Stausberg, M. (2014) 'Religion and spirituality in tourism', in Lew, A.A., Hall, C.M., & Williams, A.W. (eds.) *The Wiley Blackwell Companion to Tourism* (pp. 349–360). Malden, MA: Wiley Blackwell.
Swatos, Jr., W.H., & Tomasi, L. (2002) *From Medieval Pilgrimage to Religious Tourism: The Social and Cultural Economics of Piety.* London: Praeger.
Terzidou, M., Scarles, C., & Saunders, M.N. (2018) 'The complexities of religious tourism motivations: Sacred places, vows and visions', *Annals of Tourism Research*, 70: 54–65.
Timothy, D.J., & Conover, P.J. (2006) 'Nature religion, self-spirituality and New Age tourism', in Timothy, D.J., & Olsen, D.H. (eds.) *Tourism, Religion and Spiritual Journeys* (pp. 139–155). London: Routledge.
Timothy, D.J., & Olsen, D.H. (eds.) (2006a) *Tourism, Religion and Spiritual Journeys.* London: Routledge.
Timothy, D.J., & Olsen, D.H. (2006b) 'Conclusion: Wither religious tourism?', in Timothy, D.J., & Olsen, D.H. (eds.) *Tourism, Religion and Spiritual Journeys* (pp. 271–278). London: Routledge.
Timothy, D.J., & Olsen, D.H. (2018) 'Religious routes, pilgrim trails: Spiritual pathways as tourism resources', in Butler, R., & Suntikul, W. (eds.) *Tourism and Religion: Issues, Trends and Implications* (pp. 220–235). Bristol: Channel View Publications.
UNWTO (2011) *Religious Tourism in Asia and the Pacific.* Madrid: UNWTO.
UNWTO (2020) *Buddhist Tourism in Asia: Towards Sustainable Development.* Madrid: UNWTO.
Vukonić, B. (1996) *Tourism and Religion.* Oxford: Pergamon.
Weidenfeld, A., & Ron, A.S. (2008) 'Religious needs in the tourism industry', *Anatolia*, 19(2): 357–361.
Willson, G.B., McIntosh, A.J., & Zahra, A.L. (2013) 'Tourism and spirituality: A phenomenological analysis', *Annals of Tourism Research*, 42: 150–168.

SECTION I

Definitions, theories, and concepts

2

PILGRIMAGES, JOURNEYS, AND OUTINGS

The historical mobilities of religious praxis

Tony Seaton

Introduction

What is "Religious Tourism"? To some it may seem like a strange oxymoron—a conflation of the sacred and the profane that incongruously links spiritual travel to a more trivial, hedonistic pursuit. For historians and social scientists, religious tourism has held no such contradictions, being considered a reality of modern, industrialised societies in which commerce now provides the facilities, including inclusive, conducted tours for those wishing to make journeys to holy places (Vukonić 1996; Olsen 2016). Indeed, both historians and social scientists view religious tourism as a sacred journey, whether undertaken in pre-industrial societies or more recently as a part of commercially organised and packaged tours. As such, religious tourism represents a type of pilgrimage—an implicit, two-way relationship between reverenced, holy figures and individuals who wish to travel to places or events associated with them to worship and pay homage or for more secular reasons such as education and leisure.

However, has religious travel and tourism always been about sacred journeys, which this 'pilgrimage model' suggests? There are many types of journeys made within historical religious practice that may not necessarily fit the category of a sacred journey. Such instances include:

- Papal delegates in the Middle Ages attending "summits" in Europe with secular leaders from the Holy Roman Empire on European politics;
- Islamic religious leaders bargaining with architects and builders for new hotel developments near Mecca;
- Church officers in Victorian England making and taking commercial trips in negotiating church furnishings, electric lighting, or catering facilities; and
- Edwardian churchmen acting as tour brokers with travel agencies marketing holidays for Anglicans.

These instances suggest that there needs to be an accounting for a *variety* of different types of religious travel throughout history that extend beyond 'pilgrimage' within any sociological appraisal of religious tourism. Indeed, sacred texts say little regarding travel *per se* as a dependent or independent variable in religious behaviour, except when it comes to pilgrimage, which has long featured as an institutional practice in Buddhist, Jainist, Judaic, Islamic, and Christian discourse. Non-pilgrimage religious travel has rarely been featured in social science approaches to religion (Braun 2000; Lambek 2002; Hinnells 2010) until recently (Stausberg 2011). Yet travel, journeying, and the role of hosts and guests have played a major part in both the *narratives* and *social practices* of religion, not least in their foundation myths. Christianity began with the journey of a man and his pregnant wife to Bethlehem where all the inns were full, followed by the visitation of three wise men from the East, and later a hurried refugee escape out of town to avoid persecution. The foundation myths of Judaic mythology narrate the struggles to establish a homeland and journeys of exodus out of Egypt to a promised land. Buddhism began with a young man rejecting and abandoning his privileged home to wander the world seeking truth and spiritual enlightenment.

This chapter, therefore, seeks to rectify this situation by highlighting several less obvious kinds of journeys that constitute unmarked pragmatics and praxis in the religious life of the past as well as the social experiences of its followers. The chapter draws on three areas of Max Weber's works, including his ideas on religion (Weber 1964), bureaucratic organisations (Weber 1946: 196–266), and his methodological use of 'ideal-type' analysis (Weber 1949: 59–61), which has already been usefully adapted in segmenting religious travellers (Olsen 2010). Weber viewed religion as a corporate entity that runs on embodied and represented practices, managed and enacted by participants who serve and discharge several *functional roles* within the life of a specific religion. Unfortunately, Weber nowhere singles out travel or tourism as discrete elements for separate consideration. However, religious travel is immanently embedded in the pragmatics of establishing and promoting relationships between the different significant 'Others', whose collaborative efforts are directed to promoting religious ideas to a community of believers.

In addition, this chapter draws on perspectives from the newer field of *mobilities*, first named and theorised within sociological thought in the 1990s (Sheller and Urry 2006, Urry 2007). While still a somewhat nebulous and diffuse discursive field, mobilities is a multi-disciplinary field that brings together studies in different subjects such as geography, technology, economics, sociology, psychology, transportation history, media studies, management, and travel and tourism and examines their conjoined impact on the movement of people, goods, and ideas. Radical converts to the mobilities agenda suggest that its concerns address issues that constitute the underlying fabric of modern living and the power relations that structure them, including the impact *of deliberate* human decisions and their *unanticipated consequences*. Accordingly, this chapter approaches the study of religion historically as the movement of ideas and people and the networks of influence and power produced in the genesis, dissemination, celebration, and reproduction of religious faiths, including the production and consumption of travel and tourism among believers and secular associates. As such, this chapter attempts to identify some of the relationships of what may be called the *historical* mobilities of religious life.

For the purposes of this chapter, 'religious tourism history' is provisionally defined as

> The study of historical journeys made by religious individuals and groups in the pursuit and practice of religious goals, and those organizations and individuals that support and advance these journeys, together creating travel patterns, network connections, and configurations of power and influence that would not otherwise have existed, and which have been constantly subject to change.

In seeking to operationalise this definition, the chapter appraises different types of journeys, activities, and performances that make up the totality of religious travel. Eight ideal *role types of religious travellers* are presented, each with its own distinctive travel and tourism patterns. These functional roles are particularly well exemplified in Christianity, the main, though not exclusive, focus of this historical analysis. These eight historical roles include the Founder-Prophet; Organisational Functionaries; Spiritual Virtuosi; the local Pastoral Carer; the Reforming Revivalist; the Missionary; the Pilgrim; and the Church Member/ Believer.

Travel and tourism takers and makers in religious history

The charismatic founder/prophet

For Weber, the beginning of religion starts with a charismatic prophet/leader who embodies and articulates, through the power of his personal presence and teaching, the foundation narratives of prophecy on which churches are built (Weber 1946: 245–266; 1964: 46–60). Belief systems are thus top-down articulations that then pass through lower discursive levels of interpreters, organisational functionaries, and crusading mediators, before reaching a broader public that will constitute the community of believers (Roth & Schluchter 1979: 144–165).

Time, place, and travel have typically played a relatively small part in studying belief systems, though a more prominent one in the organisation of rituals. Charismatic, religious leaders have often been peripatetic figures appearing during times of political and philosophical upheaval. Charles Eliot (1921/1971: xix) notes that the sixth century BC

> ...was a time of intellectual ferment in many countries. In China it produced Lao-tzu and Confucius: in Greece, Parmenides, Empedocles, and the sophists were only a little later. In all these regions we have the same phenomenon of restless, *wandering teachers*, ready to give advice on politics, religion or philosophy, to any who would hear them Eliot.

Eliot might have also added to the list the life and times of Moses, Jesus, and Mohammed as religious leaders who had a great impact on Judaism, Christianity, and Islam, respectively, in the Middle East, creating sacred spaces celebrated in travelogues about all three faiths. The spatial and political context was more than just background colour or historical actualité. In Christianity, the locations of Christ's life and physical movements around the Holy Land have provided authenticating topographical details of his ministry; contributed to its symbolic resonance (e.g., the devil tempts in the desert from a high place; Christ's humility symbolised by his birth in a Bethlehem stable); provided the context for his parables (e.g., the Good Samaritan, the woman of Samaria); and underscored and the geo-religious irony of his death (a local, Jewish prophet, spared by a Roman occupier, sacrificed by the Jewish clerisy to death in Jerusalem at Passover time). This mapping of 'celestial geographies' (Aiken 2010) helped make the 'Holy Land' a must-see destination across European and later North American Christendom through package tours, as offered during medieval times up to the 1860s by Thomas Cook, the pioneering travel agent (Rae 1891; Pudney 1953; Brendon 1992) to the present.

The organisational functionary

Once a charismatic founder and his inner-circle of disciples are dead, a religion confronts the problems of perpetuation, which includes agreeing a successor and then the dissemination, maintenance, reproduction, and control of religious doctrine and practices. These steps involve the creation of institutional structures and organisational functionaries to manage them. The size and complexity of a religious organisation and the number of functionaries to maintain it will depend upon the numerical and geographical range of the membership and the communication channels necessary and available to reach all believers. This communication may vary from era to era, from oral transmission and face-to-face contact for small, sectarian, or denominational belief systems, to travel and the use of multi-media technologies when a faith community is national or international in scale. Religious organisations and their functionaries may also have the problem of identifying and adapting to social and technological changes that they do not have control over but which may have profound consequences for their religion, such as printing, TV, and the Internet (Weber 1946: 209–214).

Here Weber's work on religion and bureaucracy come together to demonstrate how charismatic leadership gives way to corporate direction. Specifically, Weber saw Latin Christendom as the classic case of a religion that began as a close-knit, informal group, which morphed into a multi-national concern with the Pope at its head in Rome, the nerve centre of a complex web of inter-communication across European borders. Though the Catholic Church was conceived as "one foundation" in spiritual terms, managerially it was a hierarchically ordered pyramid of organisational functionaries that ranged from the Pope enthroned in the Vatican, to the poorest priest in an obscure European village, each occupying different roles that came with their own jurisdictional travel and communication networks. These networks varied from great Papal 'summits' for high-ranking churchmen from across Europe, to smaller, regional meetings, such as an abbess might have with a bishop making a diocesan visitation to check out discipline among her nuns in a convent.

Organisational functionaries come and go as the roles they occupy are created, change, or disappear. Some functionaries may occupy several roles and continue to serve when one of those roles becomes redundant, while others may have to take on new roles due to social and/or organisational change. In all cases, travel and tourism demands and commitments will continue, and in the case of functionaries of high rank, may at times look less like sacred journeys and more like personal travel or even business tourism.

Spiritual virtuosi: monastic hosts and travellers

Some religions have marked out certain individuals or groups as ones embodying exemplary holiness, making them exempt from the everyday life of 'getting and spending'. They were permitted to live their lives apart within or outside the main religious community as seers, wise men/women, oracles, or monks. Their specific roles included prayer, intercession, prophecy, meditation, and advice to their religious community and on occasion to secular leaders in times of crisis.

In Western Christianity, the majority of these special individuals of groups were two different kinds of monks: 'anchorites' who were hermits living in ascetic isolation, and the religious orders, who were monks and nuns living in enclosed communal quarters, separated from the main population with their lives regulated by rules of poverty, chastity, and obedience. For both these groups, travel and tourism was nominally out of bounds since they were confined to base by their vows of 'stability'. However, they influenced tourism in several ways. First, though monks and nuns were in principle removed from contact with the outside world and unable to travel, the degree of their separation from the 'real' world varied. In Venice, for example, where there were as many prostitutes as nuns, convents were 'hotbeds' of political scheming for travellers and visiting dignitaries (Laven 2002). Moreover, although monks and nuns could not themselves travel, unless senior figures such as abbots and abbesses agreed, they welcomed travellers to their monasteries and convents, offering them food and shelter, particularly if they were pilgrims. This duty of hospitality to strangers was a religious obligation that led to the creation of specialist roles within a monastery or convent, in which the 'cellarer', 'cook,' and 'guest master' would work as a team to host visitors. The residual memory of monastic traditions of open-house hospitality survives today in the nominal missions of the corporate food and lodging industry who engage in hospitality on a strictly commercial basis.

A different monastic influence on the imaging of travel and tourism came from the one group of monks who did not take vows of stability and could officially travel. These were the friars, otherwise called the 'Mendicant Orders', licensed by the Papacy in the thirteenth century as a kind of peripatetic, monastic vigilante group who were allowed to preach, teach, study, and dispense charity in the outside world, begging or working for sustenance among the general population (Knowles 1948–1959; Rowlands 1999; Andrews 2006). Though many of these friars were devout and learned scholars, some acquired a reputation in popular culture for immorality, worldliness, and gluttony that the hospitality industry has intermittently used as branding device, such as 'The Jolly Monk' pubs and restaurants in Dublin, Swindon, UK, and New York where the caricature of a jolly, overweight monk with a love of food and drink is used to draw customers.

There were also a small group of monks between 600 and 800 AD that adopted freelance travel as a way of life, making sea voyages to places as far away as North America and converting non-believers to Christianity. These monks were all Irish, and the voyages and travels they took were part of cultural rituals variously named *Navigatio* and/or *Peregrinatio*. Those who took such voyages were called *Peregrini*, and a few of them became saints, including St Brendan, St Columba, and St Patrick, who are regarded as early Christian world-builders. While evidence regarding their lives and work is fragmentary—part literary myth and part historic chronicle (see Wooding 2000)—their legendary travels survive in the word 'peregrination' as a synonym for travelling in a roaming manner.

The most potent legacy of monasticism in modern religious tourism was the reclamation of monasteries, particularly ruined ones, in late eighteenth-century Europe for tourism purposes in Protestant countries, especially Great Britain where the formal practice of monasticism had been abolished. This change came about due to the impact of Romanticism and the Gothic Revival in Britain, and this use of former monasteries has since been officially reinforced in 'heritage' promotion by governments and tourism authorities.

The Pastoral Carer

Religious groups have different organisational structures based on scale (e.g., international, national, regional, and local) and, using a Christian example, hierarchal administration (e.g., archbishopric, deanery, and diocese). Typically, the smallest, most basic unit is a local church community based on a parish which is managed by a local religious leader (e.g., vicar, priest, or minister) with pastoral responsibilities for the congregation of his or her church. The role of this local leader is to provide a visible church presence in the community and to support its members through providing opportunities for routine worship/observance, with cyclic celebrations of special dates in the church calendar, including seasonal festivals, holy days, and national holidays. Pastoral carers also typically officiate at special *rites of passage* in their parishioner's lives, including performing blessings and baptisms at birth, presiding over marriage ceremonies, officiating in death rituals, and offering support as counsellors in times of crisis (e.g., economic hardship, war, and epidemics).

These localised responsibilities meant that travel and tourism played a limited role in pastoral care and parish life. However, local leaders were expected to know their parish socially and geographically, even if only on a 'show-your-face' level. The relationship between church members and their appointed pastoral leader—whether rabbi, priest, or vicar—was the closest the laity came to contact with the church as organisation. This made the local leader the most knowable of church officers and one who has often featured in European literature from the Middle Ages to modern times in many different guises as absentee vicar, slum vicar, reforming vicar, drunkard priest, the priest losing his faith, the marriageable vicar, and philandering priest. Among the many writers who have made the local priest a main character include Jane Austen, Stendhal, Balzac, Graham Greene, and Evelyn Waugh.

While this travel was geographically limited in scope, the familiarity local religious leaders developed within their assigned area led them to become both social and natural historians for the region, with many of them writing natural history books. One such book, Gilbert White's (1789/1836) *History of Selborne*, became a classic work of natural history that was written in an obscure Hampshire parish in England. Elsewhere vicars wrote major county histories, sometimes at great cost. In the English county of Buckinghamshire, the two most important historians, Browne Willis and Lipscombe, were both vicars who impoverished themselves by publishing histories at their own expense. Since the nineteenth century, numerous, lesser-known clerics have published more modest histories of their church or community that have attracted visitors to their areas and/or acted as guides while they were there.

The pilgrim

Although this chapter seeks to broaden the idea of religious tourism beyond the confines of pilgrimage, it will always retain its historical importance, even in the West where pilgrimage has not been a significant feature of Protestant Christian behaviour for 500 years and where Christian beliefs are presently believed to be in decline. The enduring interest in pilgrimage can be inferred from the volume of research works on the subject since the 1970s and the varieties of approaches they represent.

One category of broad cross-cultural research related to pilgrimage has been the study of *the history of pilgrimage* in the main world religions, including Christianity, Judaism, Islam, Buddhism, Hinduism, Sikhism, and Jainism (Finucane 1977; Barber 1991; Coleman & Elsner 1995; Webb 2001; Barnes & Branfoot 2006; Ron & Timothy 2018). There has also been a rise in research related to specific religious sites by academics and guidebook publishers, including Christian sites in Ireland, Britain, Spain, and Italy (Loxton 1978; Dunne 1989; Confraternity of St. James 1992; Carroll 1999; Vatican City 1999; Gitlitz & Davidson 2000). Other research on pilgrimage has been *by time period*, such as pilgrimages to the Holy land during the first millennium (Wilkinson 1977: viii), Roman times (Hunt 1982), prior to the Crusades (Wilkinson 1977), and during the nineteenth century (Ben-Arieh Jehoshua 1979). Pilgrimage during medieval times is probably the most written-about time period (Kendall 1972; Sumption 1975; Birch 1998; Webb 2000, 2002). *Biographical studies* of individual pilgrims or groups of pilgrims have also flourished including those on Muslim travellers (Farahani, Daniel, & Farmayan 1990), the Christian nun Egeria (Wilkinson 1981), and Holy Land pilgrims (Osband 1989).

While these texts are generally descriptive studies, the first and arguably the most enduring attempt to theorise and synthesise pilgrimage traditions was Turner and Turner's (1978) introduction of 'liminality' and 'communitas' into tourism discourse (Turner & Turner 1978), which, they argued, were the quintessential characteristics found in all pilgrimage traditions. 'Liminality' referred to the threshold that pilgrims cross during their travels from their mundane reality into a heightened state of being as they approach their destination, while 'communitas' was the feeling pilgrims could develop of a shared oneness with other pilgrims, as they proceeded together along their journey.

In contrast to these studies, a revisionist notion that has periodically been debated regarding the term pilgrimage is that of 'secular pilgrimage'—the idea that a pilgrimage could be a journey to a site other than one sacralised in an established religion. Secular pilgrimage is related to the idea of 'civil religion', a term introduced by Bellah (1967; see Parsons 2002), which held that certain kinds of nationalism in the United States could be considered akin to religion as expressed by visits or pilgrimages to war memorials and graves and historic battlefields. In 1997, a variant on secular pilgrimage arose in Britain where thousands of visitors came to grieve at Buckingham Palace after the death of Lady Diana (Grünhagen 2010).

While visiting war memorials, battlefields, or Lady Diana's grave would not meet the Turners' criteria of 'liminality' and 'communitas' as features of pilgrimages that emerge over prolonged journeys, the fact that people seek to invest personal journeys with sacred significance says something about tourism as an important investment in self-definition and identity. Indeed, tourism as self-definition is an ideal the tourism industry and the media have heartily endorsed, with terms like 'quest', 'odyssey', and 'pilgrimage' becoming common descriptors for recreational journeys. The desire for pilgrimage to places that are not orthodox religious sites has generated the search for new kinds of sacred (Olsen 2016: 786–787) as well as pilgrimage manuals based on pilgrimage-as-art and autobiographical journeys of people to places they considered sacred (Cousineau 1998; Egan 2019). One effect has been the valorisation of travelling distance and remoteness in *its own right* as sacred journeys (Herrero & Roseman 2015). Another type of journey that can be considered a pilgrimage is one related to 'metempsychosis' or 'metensomatosis'—forms of tourism where people retrace the steps of significant historical people or groups (Seaton 2001, 2002, 2013). For example, H.V. Morton, one of the twentieth century's most successful travel writers made his name following in the footsteps of Jesus, St Paul, and other biblical persons (Morton 1934, 1936, 1938), and many secular figures have been successfully stalked to literary advantage since (Seaton 2013). Metempsychosis has also been a form of memorial pilgrimage by family members seeking to trace the final paths of dead soldier relatives (Seaton 2014).

The missionary

Missionary travels, along with pilgrim travels, have the most obvious claims to being forms of religious travel. They are most prevalent in the domain of Buddhism, Islam, and Christianity, the religions that most actively seek converts to their faith. In Christianity, missionary travel started shortly after Christ's death with the evangelising of St Paul and the Apostles. Much of the text in the New Testament is comprised of epistolary exhortation for spreading Christianity through the Mediterranean world. Most Christian missionary efforts were concentrated in Europe until the seventeenth century when European exploration led to the discovery of the Americas and greater geographical knowledge of Africa and Asia. Most missions to these world regions were, under papal direction, spearheaded by the Jesuits, an order convened as evangelical expeditionaries (Chadwick & Neill 1986). Protestants' missions to these regions began in the late seventeenth and early eighteenth centuries by missionaries from Britain, Denmark, and France. English missionary efforts were concentrated in their new American colonies, especially the Carolinas, Pennsylvania, New York, and New Jersey during the eighteenth century (Stock 1933: 421).

A comparison between missionary journeys in the days of St Paul with later Protestant ones suggests functional variations in the nature of religious travel and the moral ambiguities inherent in the notion of missionaries as sacred travellers. St Paul's travels had no official sponsor or political goal. However, his itinerary was purposeful in that where he chose to travel was based on which groups of people he thought were most likely to hear, understand, and accept the gospel. As such, he travelled to places that were simultaneously 'centres of Roman administration... of Hellenic civilisation...(and) of Jewish influence, as well as being on great trading routes' where he could 'plant...churches' (Allen 2011: 6). In contrast, the *Society for the Propagation of the Gospel in Foreign Parts*, a missionary body founded in Britain in 1701, was established by a powerful group of ecclesiastics and English gentry, one of them the great scientist Robert Boyle, to facilitate Anglican ideology and principles in America and British overseas landholdings. While a non-governmental organisation, it gained a royal charter from William III (Humphreys 1730: 1–29). To be a missionary for the Society, missionaries had to undergo a rigorous selection process, including supplying character references from church leaders and go through an intensive interview process where Society leaders vetted potential candidates to ensure strict adherence to Anglican orthodoxy. Once in tenure, these missionaries had to report on their efforts to the Society through monthly letters. Society leaders could remove missionaries at any time without appeal if they believed missionaries were not adequately overcoming the problems of having to preach to many dissenting groups, including Independents, Anabaptists, Presbyterians, and Quakers, as well as indigenous tribal groups and African slave groups that did not always take kindly to the Anglican gospel message.

The travel and touring conditions Society missionaries faced to discharge their duties were formidable. In addition to difficulties preaching to the dissenting and cultural groups noted above, the potential for shipwrecks as they crossed the Atlantic Ocean was a strong reality. Missionaries also encountered challenges related to frontier life in the semi-wilderness areas of settler America. On arrival, they frequently found small, scattered settlements that had a newly built community church, but no preacher or religious literature. Out of necessity, they were pressed into becoming vicars of embryonic parishes which they knew little about. The most common description of these settlements in their mission reports was of sparsely populated parishes that could be as much 40 × 30 miles in size, with journeys around them taking as long as 70 miles a day by horse. The constant travelling produced "frequent and severe illness" (Humphreys 1730: 141). There was also the threat of slave rebellions and Indian Wars, one of which impelled the Society to award some of its missionaries an extra half-year's pay as danger money (Humphreys 1730: 94–102, 128–143).

In contrast to Anglican missionaries in America, the Moravian Brethren, formerly known as the *Unitas Fratrum* (Latin: Unity of the Brethren), were a pietist church with no imperial or political motivations. Dating back to the fifteenth century, the religion revived at Herrnhuth in Saxony under Count Zinzendorf in the 1730s. The Brethren became famous for missionary work in remote regions of the world that were traditionally neglected by other Christian missionaries, the most famous being Greenland. Though patronising in their opinion of the indigenous population they taught, and often having their teachings rebuffed, the Moravians established two missions in Greenland over a 30-year period. The missionary work of the Brethren resulted in the publication of a book entitled *The History of Greenland* by David Crantz (1767), a Moravian missionary serving in Greenland, which, like other missionary accounts from far-flung parts, contributed to the developing discipline of anthropology (Walls 1996).

There were also to be more, widespread popular, trickle-down effects of missionary travel in late Victorian England and America. Returning missionaries often reported back to the Society and the church communities whose donations had funded their travel, not just in written reports, but by giving talks, magic lantern presentations, and stereoscopic view illustrations, narrating their virtual travels. These became popular, parish entertainment forms for churchgoers, and part of the recreational agenda increasingly mounted by churches in the nineteenth century.

Reformers/revivalists

Religious belief systems and practices are rarely immune to internal and external criticisms. Internal conflict may arise for many reasons, including choosing the successor to a charismatic leader who has died, as was the case with Islam after the Prophet Mohammed's death, which led to permanent schism between Sunnis and Shiites. More commonly, challenges arise from reformist groups believing that religious leaders have strayed from the original intent of the faith's founders—that spiritual values, doctrinal orthodoxies, or sacred rituals are not being performed or maintained in the proper or orthodox way, and thus urging changes that church leaders may regard as heresies.

These conflicts emerge in particular places and always have geographical consequences that affect religious travel patterns and reconfigure the mobilities of the religion. This is acutely observed in Christianity, where, for example, in the sixth century a schism led to the separation of Eastern and Western Christendom. In addition, the Reformation in sixteenth-century Europe redrew the map of Western Europe as nationalistic states based on religion (i.e., Catholic or Protestant).

The first Christian reformists to make travel an evangelical tool of protest and then pioneer new religious tourism and leisure initiatives were the Quakers and the Methodists in Britain and America in the eighteenth and nineteenth centuries. The rise of the Quakers and Methodists developed out of religious controversies triggered by the Reformation in England in the 1530s and 1540s, which led in the 1640s to a Civil War. This war, in which Protestant Parliamentary forces defeated Pro-Catholic Royalists, ended in the execution of King Charles and was followed by years of struggles for power between different reformist Protestant groups and sects. The struggles were followed by a 20-year republic under the "Protectorate" of Oliver Cromwell. After Cromwell's death, the English monarchy was restored in 1665 under the rule of Charles II, along with the precarious re-establishment of a Protestant Anglican Church that was only partially supported by Charles II. Long conflicts regarding what post-Reformation Church of England would look like was settled in favour of Protestantism with the accession to the English throne in 1689 of William III and Mary II, the former being a Protestant from Holland.

The Anglican Church then entered a period of peace, during which there were negligence and inertia among sections of the clergy tasked with pastoral care for their parishes. Some clergymen were 'pluralists', holding several 'livings', and thus were incapable of giving full attention to any one parish. Socially, clergymen tended to be the sons of gentry and aristocracy and therefore more closely linked with the local squire than their parishioners were. They enjoyed the hunting, shooting, and fishing recreations of the upper classes and vacation time away at Bath and other fashionable spa towns. As such, these clergymen would pay poor curates to perform their religious duties for them. The social divide between Anglican clergy and their congregations was memorably drawn in Hogarth's caricature *The Sleeping Congregation*, in which parishioners are depicted ignoring a parson droning on at the pulpit, leading to an upper-class woman falling asleep. This divide and the lack of religious attention given to parishioners by the clergy led to a situation that invited reform.

The Quakers had been an early thorn in the side of established religion since its inception in the 1640s. This group was led by the charismatic preacher-traveller George Fox, the son of a weaver, who left home at the age of 19 and began preaching a stripped-down version of Christianity without hierarchy, appointed ministers, or structured church services. Taking the view that God was in everyone, he taught pacifism and that church meetings should be silent to encourage prayerful reflection that could be broken only when an individual felt moved to declare his/her thoughts to the community. Fox was an eccentric mystic who regularly kept journals about his revelations and encounters he had with witches (Fox 1911, Vol 1: 104–105, 110; Vol 2: 26–27, 387), and miraculous deliverances he experienced from dangers, including one where an owl hooted a warning that warned him of men lying in wait to do him harm (Fox 1911, Vol 1: 168). Fox was a fearless preacher who denounced corruption and backsliding in the English church, a critique that provoked violent persecution, with Fox being assaulted, and imprisoned by local justices in appalling gaol conditions that caused the death of 400 of his followers. Addressing audiences of thousands of people in urban spaces or in farmer's fields, Fox revived a tradition of itinerant, open-air preaching that had not been seen since the days of Catholic Friars and Protestant martyrs like Hugh Latimer, a famous Protestant martyr, a century before. This open-air preaching would later be reconfigured and greatly extended by the Methodists in the century following Fox's death.

Methodism was provoked through dissatisfactions with the Anglican Church as voiced by three young men in their 20s: John Wesley, his brother Charles, and George Whitehead, who were students together at Oxford in the early 1730s. Their intention was not to replace, but rather reform, Anglicanism, which they felt had become lethargic and corrupt. While they made little headway convincing people at Oxford regarding their reforms, they took their reforming ideas directly to the people, preaching throughout Britain, America, and other British colonies and recording their teachings and speeches in personal journals. Their voluminous memoirs, which date between 1735 and 1790, became the foundation of Methodism and were referred to by one commentator as the 'incurable scriptomania' of Methodists (Johann Wilhem; quoted in Piette 1937: 210). John Wesley's travel journals ran to seven volumes (Wesley, ed. Ward and Heitzenrater 1988–2003) and have a unique place as travel literature, being full of descriptions of hair-raising incidents in England and America, including a near shipwreck, encounters with violent mobs, being threatened by highwaymen, and the daily uncertainty of where to find food and shelter.

As a lover of travel stories himself, Wesley periodically issued his stories in cheap, serial instalments to expose his experiences to a wider audience. Wesley's book editors wrote regarding the importance of Wesley's travels: 'To accompany Wesley on his pilgrimage through two-thirds of the eighteenth century on journeys which extended from Georgia to Upper Lusatia, from Derry to Deal, and from Aberdeen to Land's End, is taxing for an editor who cannot count on the life-span of his subject; for the reader it should be a liberal education' (Wesley, ed. Ward and Heitzenrater 1988 Vol 1: 105). The picaresque appeal of Wesley's narratives and the emotional resonance of his preaching were at least as important as theological arguments. The message to his listeners was less about 'scrupulosity of doctrine' and more about 'right dispositions': '…all that was required of members of the Methodist Society was a real desire to flee from the wrath to come; their *opinions doctrinal…or liturgical…ecclesiastical might be what they would*' (Ward and Heitzenrater, in Wesley 1988, Vol 3: 54, italics added).

Whitefield was the equal of John Wesley as a charismatic preacher in England and America. He described how, after finishing an address at Blackheath near London, he retired to an inn where many came 'drowned in tears to take a last farewell' (Whitefield 1965: 285). At Gloucester he wrote that he was received with 'inexpressible joy', adding that 'late report of my being dead...only served to make my present visits more welcome' (Whitefield 1965: 295). In 1740, Whitefield preached to 6,000 people in a friend's house in Boston with great numbers standing at the doors, and afterwards on a public common to another 8,000 persons (pp. 459–460). The next day he went back to the common and addressed another 15,000 persons before becoming hoarse and retiring to his lodgings. At another meetinghouse, the crush of people led to disaster, as the overcrowded conditions caused panic, leading to people jumping from the gallery, throwing themselves out of windows, and trampling each other, leading to the death of five people with many others being injured (Whitefield 1965: 461).

Fox, Wesley, and Whitehead made travel and touring part of the mythology of dissenting religion, with Wesley and Whitehead in particular having something of rock star glamour in their power to draw crowds during their preaching tours of one-night stands in what modern hospitality observers would consider bed and breakfast stops. They spoke from any public platform they could find, including inns, private houses, in fields, at village crossroads, and outside churches. They also slept in inns, hay lofts, or spare rooms shared with charitable strangers. Their meetings were theatrical occasions in places where public performances were rare. They brought not only spiritual inspiration but also tales of adventure in places at home and abroad that most of their audiences hardly knew. This charismatic preaching stood in stark contrast to the audiences nodding off in the pews of Anglicanism, and in the last decade of Wesley's life Methodist membership grew from 43,380 to 71,568 (Dimond 1926: 48).

Members of religious communities

The types of travel considered so far have been associated with either individuals or small groups of people—the 'makers' and 'movers' of religion. But what of the rank and file believers for whom most of this travel was directed? What are the forms of religious travel and tourism in which religious adherents have engaged? In historical and sociological terms, the answer is that outside of upper-class travellers, religious travel has been limited to the groups and individuals mentioned above. While some intrepid lay believers did participate in pilgrimage travel in the Middle Ages, pilgrimage was abolished among English Protestants by the end of the 1530s. This limited the religious mobility of lay Christian believers in England, causing them to stay put within their own parishes and conform to the norms of the state religion, whether Anglicanism or, briefly under Queen Mary, Catholicism. This conformity included church attendance by law and for many the obligation not to move outside their own parish to another. Up until the late eighteenth century, the only religious travel most of the population engaged in was journeying to and from church.

By the last decade of the eighteenth century, both Quakers and Methodists were facing difficulties. With their original founders long dead, the novelty of Quaker and Methodist preaching tours was wearing off. In addition, major social and political developments in Europe, such as the French Revolution in the early 1790s which Methodists were believed to support (even though Wesleyans repeatedly declared their loyalty to the Crown (Wearmouth 1937)), cast a shadow over dissenting religious groups. As well, once laws discriminating against Quakers and Methodists were abolished in the first quarter of the nineteenth century, they ceased to be persecuted minorities, thus losing something of their marginalised 'otherness'. As well, the change in laws regarding religious freedom also led to the rise of other religious groups, including the Moravian Brethren, Swedenborgians, Unitarians,

Baptists, and Plymouth Brethren (Bogue & Bennett 1810), all of whom competed with the Quakers and Methodists for congregants. The most important development, however, was the advent of the Industrial Revolution which both increased the population and changed the social geography of England. In the first half of the eighteenth century, the population only increased from five to six million; the population increased by three million people during the second half of the century. The Methodist response to this population growth was to focus its proselytising efforts on the big, new industrialising cities, paying less attention to former rural areas, where pockets of industrial development in coal and iron mining were also springing up (e.g., in parts of Staffordshire and Shropshire where a proletariat of iron workers, coal miners, and pottery workers was increasing). Working for these resource extractive industries, however, was punishing work, and workers found relief in 'binge' recreational activities such as drinking, brutal spectator sports (among them bear baiting and cock fighting), as well as opportunistic sexual encounters, particularly, when seasonal fairs, local 'wakes' and festivals came around.

Within this context arose a variant of Methodism that would have a profound and unanticipated influence on the relationship between travel, tourism, and religion for more than a century. Primitive Methodism was established between 1800 and 1807 by Hugh Bourne and his associate William Clowes in Tunstall, England (Kendall n.d.: 7–86 1919: 560–586; Kendall n.d. for a full account of their work and association). Bourne, who was brought up on a remote farm near the Staffordshire moors showed a precocious interest in faith matters at an early age. In 1788, his family moved to Bemersley near the growing pottery town of Burslem which had a Methodist community, which Bourne subsequently joined, learning about the life and travels of Fox, the itineraries of John Wesley, and the evangelical work of John Fletcher, the Minister of the town of Madeley where miners and labourers from the ironworks at Coalbrookdale came to worship. In 1799 Bourne set up as a carpenter and timber merchant, occupations which involved travelling to the surrounding communities. During his travels, Bourne was shocked to find that many of the people had little or no religion as evidenced in 'the pagan plight' of the neighbouring mining villages of Harrishead and Mow Cop (PMMCCP 1957: 21–35). Bourne concluded that this 'pagan flight' was due to Methodism's remoteness from the lives of workers, its top-down evangelical reliance on preachers arriving and sermonising at length in places they hardly knew before moving on, and the lack of religious participation for people in places where there were few churches.

Bourne's initial solution was to spend two years (1801–1803) building a church in Harrishead, one of the offending 'pagan' villages. He also came up with a novel idea from America, that of the 'Camp Meeting'—an open-air gathering for prayer, communal worship, hymn singing, and public faith declarations that harked back to the field preaching of Fox and the Wesleyans (Graham 1936: 6–8). The camp meeting also drew on the 'Love Feast'—a Methodist ritual where church members came together in Christian fellowship and worship over a meal (Farndale 1950: 24–33). For Bourne, the Camp Meeting model could bring widely scattered communities together in great, open-air conventions that lasted several days. Borne had read about Camp Meetings in a Methodist magazine by American promoter Lorenzo Dow. In May 1807, Dow came to England where he met Bourne, and the two collaborated to stage the first English Camp Meeting on 31 May 1807 on Mow Hill, a spectacular peak 1,100 ft above sea level, that was located near the other 'pagan' settlement at Mow Cop (Kendall n.d., Vol 1: 561–571). It lasted from 6 am to 8 pm and was a great success, with people streaming up the hill all day to the tented show ground and stands where invited and impromptu speakers stepped up to tell their stories of faith and conversion. At intervals, additional preachers on wagons came and formed 'prayer rings' into which sinners were invited (Graham 1936: 5). A few months later, Bourne organised a second, improved version at Mow Cop with 'shorter speechifying' to counter the 'longwinded system' of preaching that he had always opposed. It was an even greater success than the first and led Bourne to stage a total of 17 camp meetings during his lifetime.

For the next 150 years, Camp Meetings became identifying features of the Primitive Methodists. They were typically staged in locations with elevated positions metaphorically symbolising spiritual spaces above the dark lowlands of manufacture and mining. The lure of the outdoors at Camp Meetings was not tranquil nature worship, but rather wind-in-your-face release from the spatial confinement of crowded workplaces and cramped homes. Bailey (1978), in a study of working-class pleasures among urban populations in the 1860s, observed that 'Recreation out of doors was generally brisker than the domesticities of domestic leisure' (p. 60). Camp Meetings combined religion, travel, and leisure that were carnivalesque occasions for factory hands working long hours and miners in dark jeopardy underground, where all could come together freely with neighbours and fellow believers.

Primitive Methodists staged Camp Meetings at every opportunity. They were the mainstay of Annual General Meetings that always took place in late May or early June to maximise the likelihood of good weather. Anniversaries were fervently recorded and celebrated with events that lasted days. For example, the 100th Anniversary of Mow Cop Camp Meeting took place in 1907, beginning with a great mass meeting on the first night. Next day there were processions to Mow Cop at 7.00 am, followed by two Camp Meetings lasting two hours each and then a Love Feast at Mow Cop Chapel. On the following day, two more Camp Meetings dominated the morning and afternoon events (PMMPCP 1907). In 1929, the 110th anniversary of Tunstall, the town where Primitive Methodism began, staged a ten-day event with three Camp Meetings at different local venues, and two large open-air services (PMMPCP 1929). Twenty-eight years later, the 150th Anniversary of the First Camp Meeting at Mow Cop was celebrated, including a love feast, processions, youth rallies, and services in nine different churches around the Tunstall region ((PMMPCP 1957).

The importance of these Camp Meetings was highlighted by one lifelong member of the Primitive Methodists who described them as 'the occasions of the year' (Patterson 1909: 104–105). Another commentator suggested that Mow Cop, as the site of the first Camp Meeting, was 'sacred ground, the foundation myth of Primitive Methodism' (PMMCCP 1907: 96). Like Quakers and Wesleyans before them, Primitive Methodists encountered violent opposition on occasion from those who objected to their beliefs and their intrusive expression of them. Spoiling tactics used against them included being pelted with mud, stones, and rotten eggs at Camp Meetings, and having Anglican vicars ringing their church bells to drown or disrupt their noisy assemblies (Kendall 1919: 43; Graham 1936: 12). However, Primitive Methodism continued to grow into the early twentieth century, particularly in the Midlands and the North of England among working and lower middle-class populations (see Wickham 1957: 131–134). In its first seven years of existence there were approximately 8,000 believers. By 1850, there were 104,762, by 1880 182,681, and by 1900 196,408 (PMMCCP 1907: 54). The Camp Meetings have also been characterised by historians as templates for other social movements, including socialist and trade union crusades, whose leadership comprised many Primitive Methodists (Wearmouth 1937; Farndale 1950: 62–64; Wickham 1957: 131–134). Conversely, other historians have seen Primitive Methodism with its demonstrative populism as a conservative force that diverted working people from political activism (Thompson 1963; Cunningham 1980).

Camp Meetings, however, were not to only leisure innovation by the Primitive Methodists. Another was the 'Outing'—a day trip to the countryside, seaside, or to a special event, mentored by a religious leader. The Outing was first recorded in Quaker history, when in 1801 at Coalbrookdale Richard Reynolds, a Quaker philanthropist who made his fortune off the iron works founded by his father-in-law, Abraham Darby, organised a picnic for his family and workers to the Wrekin, a prominent hill and local landmark (Rathbone 1852: 38–41). However, the Outing only took off as a national institution with the advent of the railway age. The pioneering first religious event was in 1843, when Thomas Cook, a pious Unitarian, organised his first excursion by negotiating a cheap, day-return, party rate for a trip from Leicester to Loughborough. His purpose was not profit but to distract working class people from drinking to excess and causing public disorder at local wakes and fairs in his hometown of Leicester (Rae 1891; Pudney 1953; Brendon 1992). These types of excursions were initially frowned upon by Lord's Day observers, who thought trips on Sunday profaned the Sabbath, and by rail companies who were reluctant to lower their prices for groups. However, this changed during the Great Exhibition in London in 1851, the biggest tourism event of its age. Demand for group excursions became so great that the railways had to run additional trains. Thereafter, railway outings grew in popularity, and religious organisations fully embraced them. For example, one Methodist excursion from Diss to Great Yarmouth, started in 1853, ran annually for 59 years, and in 1912 attracted 1,500 passengers (Jordan & Jordan 1991: 147). In 1863, Primitive Methodists held a great temperance event at Cleethorpes that attracted 12,000 (Ibid: 47). In the North of England, rail excursionists from Newcastle to Carlisle, some travelling with church groups and others independently, visited famous religious sites, including Lanercost Priory, Finchale Priory, and the hermitage at Warkworth, each fully described in cheap guide pamphlets (Robson 1852). Outings were also staged by Primitive Methodist as part of their anniversary programmes. In the twentieth century, these Outings included 'motor omnibus' excursions to Mow Cop, to Bourne's home at Bemersley, and to Bourne's gravesite (PMMPCP 1929).

The success of the Methodists' open-air forms of leisure and tourism acted as a wake-up call to the Church of England. Though Anglicans often looked down on non-conformists, they were forced to take notice of their combination of religious and leisure practices in planning their own parish efforts. Tourism and leisure thus became a battle front for souls, with Anglicans developing their own outdoor outings and events. Vicars in industrial cities became active in their parishes and increasingly appeared in family photograph albums as expedition leaders, seated among parties of parishioners in their Sunday best (Wickham 1957). The Anglican Church became so complicit with tourism that it set up as a tour operator by launching the *Free Church Touring Guild* in 1906 (F.C.T.G. 1928), formed in partnership with the Lunn travel agency. The Guild offered holidays, vetted by a committee of 15 vicars, at reduced terms for clerics. Its annual catalogue ran to 160 pages of cruises, and holidays in destinations that included the Riviera, Palestine and Egypt with tour guides from Mansfield College, Oxford. It also brokered holidays at home using accommodation with University Halls of Residence and approved hotels in England and Scotland. Profits were paid to the National Free Church Council for funded 'summer chaplaincies', and contributions to various religious and philanthropic groups (F.C.T.G. 1928: 169).

Despite the popularity of Camp Meetings and Outings, the middle classes were not as interested in participating in these activities, taking fewer Sunday and Bank Holiday excursions because they had the money and vacation time to take longer holidays. Indeed, it was the middle classes who made up the prime market for costly, packaged tours in Europe with Thomas Cook and other travel companies during the second half of the nineteenth century. However, middle-class tourists were actually more directed to religious sites than travellers from other socio-economic classes in the nineteenth century. This was not because of greater religious belief (although church going was more common among the upper than the lower classes), but due more to aesthetic tastes promulgated during the Gothic Revival movement by cultural critics like Gilbert Scott, Augustus Pugin, and John Ruskin, the movement focused on all things medieval and ecclesiastical in art, design, architecture, and furnishings. Medievalism ruled as a tourism attraction in pocket guidebooks and in expensive, coffee-table volumes of engraved plates with descriptions of cathedrals, churches or monasteries in Britain and Europe. John Britton, a talented engraver and a specialist in gothic illustrations, published a four-volume work on *Cathedral Antiquities*, another four-volume set on *Ecclesiastical Antiquities*, and many similar works that reached upper- and middle-class audiences (Britton 1850).

The first four decades of the twentieth century saw a great expansion of religious tourism and outdoor recreation, particularly for children that were legacies of Camp Meetings and Outings. One major outdoor institution was the Boy Scouts Movement (and later, the Girl Guides) in England, which was founded by Sir Robert Baden-Powell, a distinguished soldier who had fought in the Boer War and was a hero at the siege of Mafeking. In 1908 he published *Scouting for Boys*, a book that went through 31 editions in 30 years and was said to rival Sherlock Holmes' stories in popularity (Baden-Powell 1908). A year later, Baden-Powell retired from the army to lead the scouting movement, which focused on teaching young men the imperial values of honour and patriotism, along with survivalist skills in woodcraft, field crafts and cookery, military drilling, marching, and parading in uniform, and Christian values. One of the adverts for the scouting program read: 'If you believe in Christianity you should support scouting'. Within two decades of its inception the scouting program had a world-wide membership of 2,000,000 boys and a presence in 20 European countries including Latvia, Lithuania, Iceland, and Liechtenstein (Groom 1938). The movement also coined the term *the Jamboree* to describe their yearly Camp Meetings.

The United States had its own scout movement, including camps where boys engaged in horseback riding, swimming, learning rope tricks, and dressing like cowboys and Indians. Summer camps and year-long scout trooping led to summer camps more generally to become a part of the American way of life. For example, Frank Higgins, who was born on Toronto, Canada, began preaching to lumberjacks, and by 1912 had established logging missions in Washington, Oregon, Arkansas, and the Adirondacks. Part of his success was to combine religion and recreation in the form of what he called *Trailblazer* camps. One admirer wrote of Higgins: 'In work and play, in camp and city, in palace and shack, Higgins was always the same – natural, interesting, zealous for his "boys", a man among men, were they hoboes or millionaires'.

Through the twentieth century, the United States saw an expansion in many forms of camps in tents, cabins, and hostels for children, adults, and families. These staycations included those with religious affiliations among them: Namaschaug (1920), a Catholic Boys Camp for summer activities; Camp Lenape Pennsylvania (1927) on the banks of the Fairview Lake which was home to a number of YMCAs; the Pennsylvania State Sabbath School Camp in 1923; and Camp Carolina in Brevard, North Carolina which started in 1924 (see Messenger 1999). These camps are part of the history of religious travel and tourism and yet do not attract systematic study within the sociology religion.

Conclusion

This chapter has adopted a role-based, sociological appraisal of religious tourism history that, while recognising the central role of pilgrimage, has been an attempt to incorporate the variety of other travel and tourism functionally undertaken and generated within the domain of religion with a focus on Christian journeys. These journeys have been differentiated in relation to eight historical 'ideal types' of role occupancy within religious behaviour, each type generating its own distinctive mobilities as illustrated by select historical case histories. This chapter is a tentative first attempt to deconstruct religious travel and tourism into differentiated categories. The number of ideal travel types and the differentiations between them must be refined and extended. There may also be overlap between the different types as well as differences in the travel and tourism repertoires within these types. Moreover, not all the types discussed here will exist in every religion, and the relative importance of each type may vary between different religions. These types, however, constitute a start in the sociological exploration of aspects of religious tourism that are difficult to place as pilgrimages or sacred journeys.

As suggested in this chapter, there are many aspects of religious tourism that have been marginalised by historians and sociologists. One is the relatively minor part pilgrimage has played historically in the life of most Western, religious believers, as 'rank and file' church members have had little access to travel, let alone religious travel, until the nineteenth and twentieth centuries as compared to those in elite socio-economic classes. Another less obvious aspect is how regularly the lives of charismatic founders, reformers, and revivalists in religious history have been shaped and affected by travel (e.g., Christ, St Paul, the Buddha, Wesley, and others), and thereafter integrated into popular pilgrimage and tourism itineraries in the modern era. A final aspect is that travel by lay communities in the past was confined to *leisure time* which, in Christian Europe during the Industrial Revolution, meant Sundays, half-holidays, and Bank Holidays for the working classes. It was the Methodists who first recognised that combining religion and leisure was an opportunity to embed church life into the new industrial urban calendars of work and leisure (Bailey 1978; Cunningham 1980). Outings, excursions, camp meetings and lesser outdoor and indoor church events such as bazaars, fetes and Christmas Fairs replaced older, rural cultures with their customary practices of Saints days, feasts and fasts, wakes, and fairs. In mining villages, iron towns, textile valleys, and mid-West settlements, churches or chapels became community hubs that for many were the centre of organised leisure and tourism, as well as spiritual life. The gradual progression from rural/traditional, to urban/modern, patterns of leisure was a key leitmotif in the evolution of religious travel and tourism in Britain into the mid-twentieth century, when church attendance and membership began to decrease, and with them the community recreational life religion had helped to shape.

References

Aiken, E.J. (2010) *Scriptural Geography: Portraying the Holy Land.* London: I.B. Tauris.

Allen, R. (2011) *Missionary Methods: St Paul's of Ours, A Study of the Church in the Four Provinces.* Mansfield Centre, CT: Martino Publishing.

Andrews, F. (2006) *The Other Friars Carmelite, Augustinian, Sack and Pied Friars in the Middkle Ages.* Woodbridge: Boydell Press.

Baden-Powell, Sir Robert (1908) *Scouting for Boys.* London: Arthur Pearson.

Bailey, P. (1978) *Leisure and Class in Victorian England: Rational Recreations and the Contest for Control, 1830–1885.* London: Routledge and Kegan Paul.

Barber, R. (1991) *Pilgrimages.* Woodbridge, UK: The Boydell Press.

Barnes, R., & Branfoot, C. (2006) *Pilgrimage: The Sacred Journey.* Oxford: Ashmolean Museum.

Bellah, R. (1967) Civil religion in America, *Daedalus*, 96(1): 1–21.

Ben-Arieh, Y. (1979) *The Rediscovery of the Holy Land in the Nineteenth Century.* Jerusalem: Magnes Press/The Hebrew University Israel Exploration Society.

Birch, D.J. (1998) *Pilgrimage in the Middle Ages.* Woodbridge, UK: The Boydell Press.

Bogue, D., & Bennett, J. (1810) *History of Dissenters from the Revolution in 1688, to the year 1808*. London: Frederick Westley and A.H Davis.

Braun, W. (2000) *Guide to the Study of Religion*. London: Continuum.

Brendon, P. (1992) *Thomas Cooks 150 years of Popular Tourism*. London: Martin Secker and Warburg.

Britton, J. (1850) *The Auto-Biography of John Britton*. London: The Author.

Carroll, M.P. (1999) *Irish Pilgrimage Holy Wells and Popular Catholic Devotion*. Baltimore, MD: Johns Hopkins University Press.

Chadwick, O., & Neill, S. (1986) *A History of Christian Missions*, 2nd Edition. Harmonssworth, UK: Penguin.

Coleman, S., & Elsner, J. (1995) *Pilgrimage: Past and Present in the World Religions*. London: British Museum Press.

Confraternity of St. James. (1992) *The Pilgrim's Guide. A 12th Century Guide for the Pilgrim to St. James of Compostella*, translated from the Latin by James Gogarth. London: Confraternity of St. James, London.

Cousineau, P. (1998) *The Art of Pilgrimage: The Seeker's Guide to Making Travel Sacred*. San Francisco: Conari Press.

Crantz, D. (1767) *The History of Greenland containing a Description of the Country and Its Inhabitants, and Particularly, A Relation of the Mission, Carried On for above These Thirty Years by the Unitas Fratrum at New Herrnhuth and Lichtenfels, in that Country*. London: Printed for the Bethren's Society for the Furtherance of the Gospel among the Heathen and sold by J. Dodsley in Pall Mall....and at all the Brethren's Chapels.

Cunningham, H. (1980) *Leisure in the Industrial Revolution*. London: Croom Helm.

Dimond, S. (1926) *The Psychology of the Methodist Revival*. Oxford: Oxford University Press.

Dunne, J.J. (1989) *Shrines of Ireland*. Dublin: Veritas.

Egan, T. (2019) *A Pilgrimage to Eternity: From Canterbury to Rome in Search of a Faith*. New York: Viking.

Eliot, Sir C. (1921, reprinted 1971) *Hinduism and Buddhism: A Historical Sketch*. London: Routledge and Kegan Paul.

Farahani, M.M.H., Daniel, E.L., & Farmayan, H. (1990) *A Shi'ite Pilgrimage to Mecca 1885–1886. The Safarnameh of Mirza Mohammed Hosayn Farahani*. London: Saqi Books.

Farndale, W.E. (1950) *The Secret of Mow Cop. A New Appraisal of Primitive Methodist Origins*. London: Epworth Press.

Finucane, R.C. (1977) *Miracles and Pilgrims*. London: J.M. Dent.

Fox, G. (1911) *The Journal of George Fox*. Edited from the MSS. by Norman Penney, F.S.A. With an introduction by T. Edmund Harvey. Cambridge: Cambridge University Press.

F.C.T.G. (1928) *Free Church Tourism Guild Prospectus*. London.

Gitlitz, D.M., & Davidson, L.K. (2000) *The Pilgrimage Road to Santiago*. New York: St. Martin's Griffin.

Graham, T. (1936) *Mow Cop – and After*. London: Epworth Press.

Groom, A. (1938) *Scouting in Europe*. No Place of Publication: The Scout Book Club.

Grünhagen, C. (2010) "Our queen of hearts'—The glorification of Lady Diana Spencer: A critical appraisal of the glorification of celebrities and new pilgrimage', *Scripta Instituti Donneriani Aboensis*, 22: 71–86.

Herrero, N., & Roseman, S.R. (2015) *The Tourist Imaginary and Pilgrimages to the Edges of the World*. Bristol, UK: Channel View Publications.

Hinnells, J.R. (2010) *The Routledge Companion to the Study of Religion*. London and New York: Routledge.

Humphreys, D. (1730) *An Historical Account of the Incorporated Society for the Propagation of the Gospel in Foreign Parts, Containing their Foundation, Proceedings and the Successes of the Missionaries in the British Colonies, to the Year 1728*. London: John Downing.

Hunt, E.D. (1982) *Holy Land Pilgrimage in the Later Roman Empire*. Oxford: Clarendon Press.

Jordan, A., & Jordan, E. (1991) *Away for the Day: The Railway Excursion in Britain, 1830 to the Present Day*. Kettering, UK: Silver Link Publishing.

Kendall, A. (1972) *Medieval Pilgrims*. London: Wayland Publishers.

Kendall, H.B. (n.d.) *The Origin and History of the Primitive Methodist Church*, 2 vols. London: Edwin Dalton.

Kendall, H.B. (1919) 'The Primitive Methodist Church and the independent Methodist churches', in W.J. Townsend, H.B. Workman, & G. Eayrs (eds), *A New History of Methodism* (pp. 553–598). London: Hodder and Stoughton.

Knowles, D. (1948–1959) *The Religious Orders in England*, 3 vols. Cambridge: Cambridge University Press.
Lambek, M. (2002) *A Reader in the Anthropology of Religion*. Oxford: Blackwell.
Laven, M. (2002) *Virgins of Venice: Broken Vows and Cloistered Lives in the Renaissance Convent*. London: Penguin Books.
Loxton, H. (1978) *Pilgrimage to Canterbury*. Newton Abbot, UK: David and Charles.
Messenger, T. (1999) *Holy Leisure: Recreation and Religion in God's Square Mile*. Minneapolis: University of Minnesota Press.
Morton, H.V. (1934) *In the Steps of the Master*. London: Rich and Cowan.
Morton, H.V. (1936) *In the Steps of St. Paul*. London: Rich and Cowan.
Morton, H.V. (1938) *Through the Lands of the Bible*. London: Methuen.
Olsen, D.H. (2010) 'Pilgrims, tourists and Max Weber's 'ideal types'', *Annals of Tourism Research*, 37: 848–851.
Olsen, D.H. (2016) 'Religion, tourism', in J. Jafari and H. Xiao (eds), *Encyclopedia of Tourism* (pp. 784–787). New York: Springer.
Osband, L. (1989) *Famous Travellers to the Holy Land. Their Personal Impressions and Reflections*. London: Prion.
Parsons, G. (2002) *Perspectives on Civil Religion*. Aldershot: Open University and Ashgate.
Patterson, W.M. (1909) *Northern Primitive Methodism*. London: Dalton.
Piette, M. (1937, trans by Rev. J.B. Howard) *John Wesley in the Evolution of Protestantism*. London: Sheed and Ward.
PMMCCP = Primitive Methodist Church Commemorative Programmes:
PMMCCP (1907) The Primitive Methodist Centenary Souvenir Programme. Mow Cop, May
PMMCCP (1929) 100th Annual Conference Programme, Tunstall.
PMMCCP (1957) The Mow Cop Story. The 150th Anniversary of the First Camp Meeting. Stoke-on-Trent
Pudney, J. (1953) *The Thomas Cook Story*. London: Michael Joseph.
Rae, R.F. (1891) *The Business of Travel, a Fifty Year's record of Progress*. London: Thomas Cook and Son.
Rathbone, H.M. (1852) *Letters of Richard Reynolds with a memoir of his life*. London: Charles Gilpin.
Robson, J.P. (c. 1852) *Summer Excursions in the North of England; Including a Trip to Warkworth; A Ramble to Marsden Rocks; Picnicings at Finchale Priory; A Week at Gilsland, with Visits to Naworth Castle and Lanercost Priory, and a Ride, by Rail, from Carlisle to Newcastle*. Newcastle, UK: Robert Ward.
Ron, A., & Timothy, D.J. (2018) *Contemporary Christina Travel: Pilgrimage, Practice, and Place*. Bristol, UK: Channel View Publications.
Roth, G., & Schluchter, W. (1979) *Weber's Vision of History Ethics and Methods*. Berkeley: University of California Press.
Rowlands, K. (1999) *The Friars: A History of the Medieval Friars*. Sussex, UK: The Book Guild.
Seaton, A.V. (2001) 'In the footsteps of Acerbi: Metempsychosis and the repeated journey', *Tutkimusmatkalla Pohjoisseen, Acta Universitatis Oulensis Humaniora, B*, 40: 121–138.
Seaton, A.V. (2002) 'Tourism as metempsychosis and metensomatosis: The personae of eternal recurrence', in G.S. Dann (ed), *The Tourist as a Metaphor of the Social World* (pp. 135–168). Wallingford, UK: CABI.
Seaton A.V. (2013) 'Cultivated pursuits: Cultural tourism as metempsychosis and metensomatosis', in M. Smith & G. Richards (eds), *Routledge Handbook of Cultural Tourism* (pp. 19–27). London and New York: Routledge.
Seaton, A.V. (2014) 'The unknown mother: Thanatourism and metempsychotic remembrance and rituals after World War I', in L. Sikorska (ed), *Of What is Past, or Passing, or to Come: Travelling in Time and Space in Literature in English* (pp. 143–170). New York: Peter Lang.
Sheller, M., & Urry, J. (2006) 'The new mobilities paradigm', *Environment and Planning A*, 38: 207–226.
Stausberg, M. (2011) *Religion and Tourism: Crossroads, Destinations, and Encounters*. London and New York: Routledge.
Stock, E. (1933) 'Protestant missions', in C.H.H. Wright, C.S. Carter, & G.E.A. Weeks (eds), *The Protestant Dictionary, Containing Articles on the History, & Doctrines, & Practices of the Christian Church* (pp. 420–427). London: The Harrison Trust.
Sumption, J. (1975) *Pilgrimage: An Image of Mediaeval Religion*. London: Faber and Faber.
Thompson, E.P. (1963) *The Making of the English Working Class*. London: Gollancz.
Turner, V., & Turner, E. (1978) *Image and Pilgrimage in Christian Culture*. New York: Columbia University Press.

Urry, J. (2007) *Mobilities*. Cambridge: Polity Press.
Vatican City. (1999) *Pilgrims in Rome: The Official Vatican Guide for the Jubilee Year 2000*. London: Chapman.
Vukonić, B. (1996) *Tourism and Religion*. New York: Pergamon.
Walls, A.F. (1996) 'The nineteenth-century missionary as scholar', in A.F. Walls (ed), *The Missionary Movement in Christian History* (pp. 187–198). Edinburgh: T. & T. Clark.
Wearmouth, R.F. (1937) *Methodism and the Working-Class Movements of England, 1800–1850*. London: Epworth Press.
Webb, D. (2000) *Pilgrimage in Medieval England*. London and New York: Hambledon and London.
Webb, D. (2001) *Pilgrims and Pilgrimage*. London: I.B.Tauris.
Webb, D. (2002) *Medieval European Pilgrimage*. Basingstoke, UK: Palgrave.
Weber, M. (1946) *From Max Weber: Essays in Sociology*. trans. and ed. H.H. Gerth and C Wright Mills. New York: Oxford University Press.
Weber, M. (1949) *The Methodology of the Social Sciences*. New York: The Free Press.
Weber, M. (1964) *The Sociology of Religion*. Boston: Beacon Press.
Ward, E.R., & Heinzenrater, R.P. (eds) (1998–2003). *The Works of John Wesley*. Vols. 18–24. *Journals and Diaries* 1–7, Nashville, TN: Abingdon Press.
White, G. (1789/1836) *The Natural History of Selborne; With Observations on Various Parts of Nature, and the Naturalist's Calendar*, 6th ed. Edinburgh: Fraser.
Whitefield, G. (1965) *George Whitefield's Journals*. London: The Banner of Truth Trust.
Wickham, E.R. (1957) *Church and People in an Industrial City*. London: Lutterworth Press.
Wilkinson, J. (1977) *Jerusalem Pilgrims before the Crusades*. Jerusalem: Ariel Publishing House.
Wilkinson, J. (1981) *Egeria's Travels to the Holy Land*. Warminster, UK: Aris and Philips.
Wooding, J.M. (2000) *The Otherworld Voyage in Early Irish Literature*. Dublin: Four Courts Press.

3

THE POLITICAL ECONOMY OF RELIGIOUS AND SPIRITUAL TOURISM

Dino Bozonelos

Introduction

The search for religious and spiritual experiences has been an inspiration for travel from the earliest parts of human history (Butler & Suntikul 2018). Indeed, there is a vast literature on pilgrimage and its associated institutions, with much of the earliest writings rooted in the late Middle Ages (Di Giovine & Choe 2019). Yet the academic study of religion and tourism as separate from the field of pilgrimage studies has a very short history, beginning in the late 1990s (Vukonić 1996; Badone & Roseman 2004; Timothy & Olsen 2006; Raj & Morpeth 2007; Stausberg 2011; Raj & Griffin 2015). Often dichotomized into the disciplines of religious studies or management, scholars have brought different disciplinary perspectives to understanding this complicated field, in which sacred places often double as tourist destinations (Bremer 2006; Norman 2011).

One area of productive research has been on the management of the sacred sites, which includes the study of the organizations that control and manage these sites (Olsen 2009, 2019), the contrasting motivations of the tourists and pilgrims that visit these sites (Blackwell 2007; Drule et al. 2012; Abbate & Di Nuovo 2013; Bond, Packer & Ballantyne 2015; Nyaupane, Timothy & Poudel 2015), and the management practices that are used to provide services and experiences to visitors (Shackley 2001a, 2001b; Olsen 2006; Wong, McIntosh & Ryan 2016; Dowson, Yaqub & Raj 2019). The social sciences have also contributed to the study of religion and tourism. For example, geographers, anthropologists, and sociologists have developed theories and case studies related to religion, spirituality, pilgrimage, and tourism through employing interdisciplinary research and ethnographic approaches (Badone & Roseman 2004; Coleman & Eade 2004, 2018; Collins-Kreiner 2016; Lopez, Lois González & Fernández 2017; Farias et al. 2019; Sanagustín-Fons, Gregory & Martínez-Quintana 2019).

One of the academic fields that is now beginning to examine the relationships between religion and tourism is political science. While there exists small literature on terrorism, political crises, and political instability as it relates to religious tourism (Shackley 2001a; Collins-Kreiner et al. 2006; Suleiman & Mohamed 2011; Chowdhury et al. 2017), there are many more avenues of research that political scientists can explore. These include theories on political attitudes, preferences, and public opinion formation that could help refine research on religious destination choice; institutionalist theories regarding how religious tourist destinations are governed; and the management decision-making process at sacred sites. Newer research on comparative political economy can also help researchers better understand how religious organizations and their efforts to promote religious tourism and pilgrimage are affected by differences in national political economies that help shape the development of macro and micro religious institutions within a country. As Hall and Soskice (2001) notes, the arrangements between the state and the market strongly impact how firms (like religious organizations) solve coordination problems in the market regarding the development of existing employment relations. For example, in liberal market economies, a free market approach allows for tourist destinations to develop without much state interference, whereas, social democratic economies seek to ensure equal access to a country's resources. Thus, the state will often intervene at times and regulate tourist destinations for the benefit of society overall. As such, political economy can affect the path development of many types of institutions, including tourism and religious institutions.

Yet, an additional layer of institutional complexity exists for religious tourist destinations that do not exist for their secular counterparts. Religious arrangements in a society, such as concordats, established state religions, or tax and regulatory arrangements, can additionally affect how religious tourist destinations are developed and managed. Thus, the intersection of political economy types and religious arrangements could help explain how religious tourist institutions have developed. As 'carriers of history', religious tourist institutions thus carry two types of histories—political-religious and political-economic (David 1994).

Added to this complexity are the newer wrinkles of neoliberalization reforms, where market competition is often proposed as the model to solve coordination problems. From an international political economy perspective, the global push for free trade and liberalization has transformed the global tourism industry (Giampiccoli 2007). More importantly, it has impacted the management of religious tourist destinations regarding the development of transnational tourism corporations (Bianchi 2002). The opening of domestic economies to foreign direct investment and the deregulation of foreign ownership of land and access to resources, such as seaports, has allowed multinational hotel chains, mass transit carriers, and specialist tour operators to capture larger shares of the tourist market within a given country. However, their presence is less pronounced in the religious tourism segment. Clearly then, political economy as a discipline has much to offer to the study of religious tourism.

The purpose of this chapter is to provide a better understanding of the political economy of religious tourism. First, political economy is defined and the historical trajectory of the field is presented. Then, the small but growing literature on the political economy of religious tourism is discussed, after which the author brings in research from comparative political economy to help map the intersection of political, economic, and religious institutions, and provides a framework from which to better understand how these institutions are linked before concluding.

Defining political economy

Political economy as a disciplinary field of study focuses on the relationship between the marketplace and the powerful actors that work within it, with an emphasis on the political, economic, and societal institutions that shape the distributional outcomes between and within countries as well as the results that arise from these interactions (Gilpin 2001). Political economy is often considered as a subfield of political science that provides a deeper understanding of individual institutional arrangements. Even though the field often finds itself touching upon the areas where economics and political science intersect, the study of institutions and their effect on society is often a study of politics and the compromises made through political decisions (Menz 2017). Often, the development of said institutions is key in understanding why certain arrangements exist. Political economy, however, is not just an exercise in historical analysis but also attempts to explain contemporary phenomena through acknowledging that the rationality behind current behavioral decisions is often bounded or constrained by the institutions that condition decision-making. As such, political economists continually focus on the 'ever-evolving nexus of relationships between state governance and economic transactions' (Coleman & Eade 2018: 2).

Historically, Adam Smith's (1977) book, *The Wealth of Nations*, is considered the foundation for the study of political economy. Smith was followed by seminal nineteenth-century scholars such as Smith, Ricardo, List, Marx, and Engels. These philosophers saw no distinction between politics and economics. They contended that the endogenous relationships between institutions made it difficult to disentangle cause and effect mechanisms from each other. Classical scholars understood the field holistically. In the early twentieth century, economics began to separate itself formally from politics by focusing on theories of economic behavior as they related to human behavior. Employing rational discourse to individual decision-making, institutions took on less of a role in both understanding and ultimately predicting behavior.

However, advanced mathematical modeling of economic behavior has struggled at times to account for what could be deemed 'less rational' decisions made by individual actors. Scholars will often assign the dearth of rationality to the lack of proper information or a misinterpretation of signals or actions taken by a more informed actor to convey information to less informed actors. The basic assumption in rational-choice modeling is the maximization of utility, however, defined (Green & Shapiro 1994). Organizations and institutions are seen simply as an aggregation of interests—where structures are created to reduce transaction costs, thereby allowing for wealth maximization for economic actors.

This singular focus on rationalization, and the locus of the individual as the unit of analysis was not accepted by all. Referring to themselves as neoinstitutionalists, these scholars argued that collective action should be at the center of analysis. The norms and values of institutions could be more relevant than achieving one's desires as explanations for decision-making. March and Olsen (2011) explain this logic as decision-making driven by rules. Rules are followed by actors because 'they are seen as natural, rightful, expected, and legitimate' (p. 1). The obligations of upholding the norms and values of an institution is what justifies an actor's place within the institution. This then reinforces rules of appropriate behavior and can justify the norms that undergird the institutions themselves (Peters 2019).

These neoinstitutionalist approaches gave space for political economy to reemerge as a discipline. Beginning in the 1970s, political scientists reintroduced the study of political and economic institutions to explain behavioral differences between countries and within countries. These scholars highlighted that the differences in political institutions, both global and national, were the essential variable in comparative studies. The neoinstitutionalist approach is explained best by North (1991), who reminded behavioralists that the market itself is an institution guided by certain norms and expectations, and that markets undergo evolutionary change as well. This approach has led to a dramatic upsurge in neoinstitutionalist publications focusing on how institutions shape economic decision-making at different levels of analysis and in different segments of society (March & Olsen 1984; Pierson & Skocpol 2002; Orren & Skowronek 2004).

An important division in the study of political economy revolves around the level of institutional analysis. The first level of analysis is the global or international level, which involves the study of *international* institutions that govern global political-economic behavior (Frieden 2017). The second level is the domestic level of analysis, which analyzes the *internal* institutions that govern domestic political-economic behavior within a country. A focus on these two levels of analysis has led to the development of two major subfields in political economy: International Political Economy (IPE) and Comparative Political Economy (CPE). Menz (2017) frames this distinction as being about understanding the outcomes of institutional arrangements as either top-down (IPE) or bottom-up (CPE). However, Menz notes that there is a consistent blurring of the lines exists between the two subfields, and that this 'interchange and interaction in the domestic-international nexus' (p. 47) has become more pronounced as state governments have forcibly liberalized their economic systems to participate in global trade. This liberalization has led Oatley (2019) to describe IPE as the 'political battle between the winners and the losers of global economic exchange...[that] shapes the economic policies government adopt' (pp. 25–26).

Similarly, the study of political institutions has also led academics to develop several approaches to understand how institutions are created, rationalized, and maintained. For example, rational-choice institutionalism involves the study of rules and incentives as a system, which system is forever negotiated by political actors. From this viewpoint, institutions are products of agreements that under pareto-optimal conditions could distribute goods in an equitable manner, where one party is made better but no parties are made worse-off (Rhodes 2008). However, rarely are such conditions optimal, and actors in an unequal relationship often struggle to roll back the hegemony of rules and incentives that have been institutionalized (Zey 1998).

Historical institutionalism, on the other hand, focuses on the concept of path dependency—a methodological inquiry that investigates 'historical sequences in which contingent events set in motion institutional patterns or event chains that have deterministic properties' (Mahoney 2000: 57). Institutions do not just spontaneously appear (David 1994); rather, they are often the codification of preexisting socially established 'conventions' or social norms for negotiating within a society. Because there are high costs associated with the formalization of social norms and the transaction costs required for reforming institutions can be expensive, changes in institutions are uncommon. Rhodes (2008) referred to institutions as 'dried cement' where 'cement can be uprooted when it has dried, but the effort to do so is substantial' (xv). As such, the actions and decisions made by institutional leaders can be fairly predictable, helping to explain why institutions are long-lasting. It is only when actions and decisions by institutional leaders or negotiations between institutions go against the norm that radical change takes place.

Path dependency and its accompanying methodology of process-tracing—'an analytic tool for drawing descriptive and causal inferences from diagnostic pieces of evidence—often understood as part of a temporal sequence of events or phenomena' (Collier 2011: 823)—have become part of the methodological foundation for political economy, especially in the smaller field of comparative political economy. While often used in case study work, path dependency as a method is also quite useful in conjunction with explanatory typologies. For example, Hall and Soskice (2001) employ path dependency as a part of their Varieties of Capitalism (VoC) theory, which attempts to understand the similarities and differences of institutions within two types of capitalism found in developed economies: liberal market economies (LMEs) and coordinated market economies (CMEs; also referred to as corporatist economies). LMEs are found in countries where firms depend more heavily on market mechanisms to solve coordination problems,[1] whereas CMEs are found in counties where firms rely more on strategic cooperation between important industry-specific constituencies.[2]

The political economy of religious tourism

There is significant literature on the political economy of tourism development (Bianchi 2002; Kütting 2010; Bramwell 2011; Mosedale 2011, 2016; Nunkoo & Smith 2013; Müller 2019; Bianchi & de Man 2021). In his seminal work *Tourism and Politics*, Hall (1994) noted that there was already over 30 years of work on the political economy of tourism development from which he could draw upon. Focusing on the role of the state or the domestic level of analysis, Hall provided context regarding how tourism became both an economic sector and a driver of industrialization policies for many countries, stating that 'there is almost universal acceptance by governments around the world, regardless of ideology, that tourism is a good thing' (p. 28). Hall also notes that it is through national tourism organizations that developing countries have adopted policies that aggressively promoted tourism as a development tool.

In their edited volume *Tourism and Politics: Global Frameworks and Local Realities*, Burns and Novelli (2007) also discuss tourism development, but more so from a policy evaluation perspective. Suggesting that 'tourism is a powerful mix of cultural, economic and political phenomena' (p. 1), the book's contributors examine how tourism development has impacted other institutions. For example, in one of the chapters, Richter (2007) notes how little democratic debate there has been regarding national tourism policies, noting that while national governments set visa requirements and formulate policies on transportation and health, there is very little participation by the public in tourism development. Mosedale's (2011) edited volume, *Political Economy of Tourism: A Critical Perspective*, critiqued tourism development through the use of neomarxist critiques and IPE and CPE frameworks to discuss issues such as equitable distribution of power, decision-making, labor rights, class inequality of access to tourism, the gentrification of tourist destinations, and the effect on neoliberalism on industrial relations.

An analysis of the chapters in the books by Burns and Novelli (2007) and Mosedale (2011) shows that there are almost an even number of chapters dedicated to the two levels of institutional analysis noted above, where a total of 15 chapters focused on international and/or supranational actors, while another 15 of the chapters centered on the actions taken by domestic actors, usually within the context of responding to changes at a global level. Interestingly, the international chapters were more theoretical in nature, while domestic chapters were almost exclusively case studies that documented the development or traced the process of the particular case in question. Only one chapter did not fit the two levels of institutional analysis. In this chapter, the authors used a comparative political economy framework from which to understand tourism development (Webster, Ivanov & Illum 2010).

As a subfield within tourism studies, religious tourism is ripe for analysis using political-economic frameworks and methodologies, given the importance of existing religious institutions as 'carriers of history'—representing the codification of preexisting socially established 'conventions' or the use of social norms for understanding the presence and practice of religion within a society. The high costs on formalizing religion in different societies, often the result of centuries-long negotiations involving frank discussions regarding secularism, religious minorities, religious education, and the management of religious destinations, helps to explain why religious institutions are long-lasting. Indeed, the transaction costs required for reforming or removing religious institutions from both the public and private sphere would be prohibitively expensive socially, economically, culturally, and morally, and any trade-offs that may be involved often make reform not worth the societal costs. As such, religious institutions persevere in a way that might be best understood through the path development approach often used in political economy.

However, even though there is a small but growing literature on the political economy of *religious* tourism, with a heavy focus, as will be shown below, on the interactions between institutions that support pilgrimage activities and on how governments use tourism policies to impact the management of pilgrimage sites and routes, these authors do not couch their work in the political economy literature or approach religious tourism from a distinctly political economy perspective. For example, so far nothing has been published on applying IPE or CPE approaches or using political economy types to understand the relations between the state and the market and their effect on religious tourism destination sites. Given this, it may be better to understand this body of work as *religious tourism policy studies.*

A good example is the discussion by Trono (2015), who discusses how state and private enterprises, in conjunction with religious bodies functioning as corporate entities, can lead to better management of religious tourism through an analysis of tourism development around the new Church of Padre Pio in San Giovanni Rotondo in Italy. Her work is indicative of newer research that seeks to examine and better understand the relationships between state institutions, market actors, and religious authorities. Another good example is Shinde's (2018, 2020) examination of examples in India of institutional vacuums that exist regarding the impacts of religious tourism. Shinde notes that during the transition of a destination from a pilgrimage site to a religious tourism site, there is a lack of both foresight and cooperation between religious and governmental institutions in terms of who is to take care of the social and environmental impacts of pilgrimage and religious tourism. This is particularly acute in places where there is already a lack of sufficient sanitation and waste management systems and accommodations. In these cases, government officials are ill-equipped to handle the demands of religious tourists on top of the demands of pilgrims, with religious organizations are not interested in doing the job of government.

These two case studies illustrate the difference between religious tourism management and religious tourism policy studies. In religious tourism management, the variable of analysis is the pilgrim or religious tourist. The focus in religious tourism management is often on the motivations of the traveler, whether the traveler should be defined as a pilgrim or a secular tourist, and what this designation means for the religious destination and/or sacred site. Nolan and Nolan (1993) were one of the first to discuss the types of visitors who visit Christian shrines in Europe, explaining how these shrines have evolved in their management practices to meet the needs of these visitors. Other scholars have continued this tradition (Winter & Gasson 1996; Shackley 2001a, 2001b, 2003; Rössler 2003; Woodward 2004; Olsen 2006; Raj & Griffin 2015; Griffiths & Wiltshier 2019). This research, however, tends to take both the presence and the actions of institutions as a priori or self-evident and also focus on understanding the relationship between visitors and religious institutions from a phenomenological perspective.

More recently, there has been a focus on religious tourism policy studies where the variable of analysis is the institutional actions that support pilgrimage activities. For example, Reader (2014) refers to a 'pilgrimage market' where efforts are made by different religious, state, and economic actors to publicize pilgrimages and religious tourism destinations in order to increase the number of visitors. As Reader notes, 'pilgrimages cannot rely on a continual flow of pilgrims and that those who oversee them need to engage in publicity activities in order to ensure that their sites continue to attract pilgrims and flourish' (p. 6). This highlights the fact that it is difficult to separate out the profane or secular from the sacred; that attempts to study pilgrimage and religious journeys set apart from 'normative everyday routines and the mundane world' (p. 14) are fraught with limitations. The dynamics of the economic marketplace are essential to pilgrimage activities, as are the roles of the state (e.g., responding to local, national, and/or international tourism industry pressures), marketplace actors (e.g., local merchants, tourism agencies, or transportation companies), and religious authorities (e.g., priests seeking to promote their religious agendas or managers opening their sacred sites for visitors). Reader argues that these forces of consumption and commodification of religious destination sites should not be considered as caustic to religion, but instead as important for the maintenance and possibly survival of pilgrimage sites.

A more comprehensive approach to religious tourism policy studies is the edited volume by Coleman and Eade (2018) entitled *Pilgrimage and Political Economy: Translation of the Sacred.* This edited volume focused on examining political economy approaches in the context of 'pilgrimage studies'. Albera (2018: 174) suggested that there are two different approaches in pilgrimage studies: the study of culture or society *through* the lens of pilgrimage and the study *of* pilgrimage to construct theories of pilgrimage. The chapters in Coleman and Eade's (2018) book are mostly orientated on the study *of* pilgrimage, with several case studies focusing on different aspects of pilgrimage in different faith traditions, such as Hindu pilgrimages in Mauritius, the promotion of Buddhist pilgrimages to Japanese tourists, a description of pilgrimages to shrines of popular Islam in Uzbekistan, and New Age Energy pilgrims visiting Catholic shrines. Each chapter incorporates a discussion of the existing relations between institutions regarding how resources are allocated to pilgrimage sites and how these same pilgrimage sites are resources themselves for state and other political entities.

Two of the chapters in Coleman and Eade's volume in particular are excellent examples of the dynamic relationships that exist between political, economic, and religious actors. Delage (2018), for example, analyzes the role that South Asian Sufi saints have in the development of economic activities. Delage describes in detail how local institutions organize and promote the religious fairs associated with the tombs of Sufi saints, facilitate trade between tribes, and determine how resources are distributed in the local context. Katić (2018) uses the term 'pilgrimage capital' to explain how different civil society associations, through interactions with political and economic actors, use social, economic, and political resources for their own particular benefits. Using post-war Bosnia i Hercegovina as the case study, Katić contends that it took strong institutional energies to promote pilgrimage sites in Bosnia to the displaced Bosnian Croat communities living in Croatia proper, noting how these efforts have paid off for Croatian politicians who have benefited tremendously from the political capital generated from the reconstruction of Catholic Churches and the construction of new shrines in the Croat parts of Hercegovina.

Applying comparative political economy to religious tourism

To date, there has been no attempt to apply CPE frameworks to religious tourism. Similarly, little has been written regarding CPE frameworks in the political economy of tourism, with Webster et al.'s (2010) as the only work that hypothesizes that political economy types (referred to as paradigms by the authors) are correlated with the relative weakness and/or strength of national tourism organizations (NTOs). In doing so, Webster et al. appropriate O'Neil's (2018) framework of comparative political economy paradigms, using his typology of liberalism, social democracy, communism, and mercantilism and show that the liberalist political economies are associated with weak NTOs, whereas social democracy political economies are associated with stronger NTOs. As well, the literature review above highlights some of the deficiencies that still exist in the application of political economy theories and frameworks to religious tourism. The focus on pilgrimages at the expense of religious tourism, case study work on sacred sites and destinations, and the lack of frameworks to understand cross-national patterns, leave opportunities for scholars to incorporate political economy concepts into religious tourism studies. Coleman and Eade (2018: 16) are correct in their claim that pilgrimage perspectives should be broadened and that the complex economic and political relations between actors should be better understood. Indeed, the relationships between economic, political, and religious actors are best understood through institutional analyses, such as the frameworks developed by comparative political economists.

Here I create a model to demonstrate how political-economic and political-religious institutions are related to each other. To understand political-economic institutions, I follow the VoC approach noted above, which has become quite popular in Comparative Political Economy (CPE) circles, with no fewer than a dozen books have been published since Hall and Soskice's (2001) work on the VoC approach (Amable 2003; Schmidt 2009; Becker 2014). However, rather than using Hall and Soskice's (2001) work where the firm is at the center of analysis, I use Amable's (2003) broader work on five models of capitalism that roughly group into particular world regions: Market (LME); Social Democratic (CME), which encompasses the Nordic countries; Asia (Meso-Communitarian); Mediterranean; and Continental European (see Table 3.1). This is because in Amable's work firm behavior is less rational and conditioned by the institutions that surround them. Also, Amable's analysis of political economy ideal-types examines five fundamental institutional areas which are also important to consider: product-market competition; wage-labor nexus and labor-market institutions; the financial intermediation sector and corporate governance; social protection and the welfare state; and the education sector. In particular, Amable's descriptions of product-market competition are important, as religious tourism is often described as a process of consumption (Aulet & Vidal 2018). I also only use the 21 OECD countries that Amable included in his initial analysis. This is because, as Menz (2017) argues, there are enormous cultural differences and divergent historical trajectories between these countries and other OECD member countries, and as such any comparison between all the countries would become unwieldy and analytically flawed. As such, it is easier to provide in-depth coverage of a smaller selection of countries that in many ways are representative cases of larger clusters of countries in this model.[3]

Table 3.1 Models of capitalism: product market regulation indicators

Market (LME)	Deregulated product markets. Low barriers to entrepreneurship; low administrative regulation; low state control and public ownership. (USA, UK, Canada, and Australia)
Social Democratic (CME)	Regulated product markets. Coordination between the state, firms, and unions. (Denmark, Finland, Sweden)
Asia (Meso-Communitarian)	'Governed' rather than regulated product-market competition. State-firm coordination is the norm for product-markets. (Japan and South Korea)
Mediterranean	Regulated product markets. Administrative burdens for corporations; barriers to entrepreneurship; strong public sector. (Greece, Italy, Portugal, and Spain)
Continental European	Competitive to mildly regulated markets. Some markets are more heavily regulated. (Switzerland, Netherlands, Ireland, Belgium, Norway, Germany, France, and Austria)

After Amable 2003.

In terms of including political-religious institutions into this model, I focus on the political-religious arrangements that exist within countries. A simple way to identify these arrangements is through models of Church-State relations. For example, Riedel (2008) identifies four possible models of church-state relations. The first model is the established state church, in which she includes countries such as Greece, whose Constitution references a 'predominant church'. The second is the cooperationist model, where the state works with traditional religious institutions and may provide support, such as in collecting a 'church tax'. Reidel's third and fourth models are variants of laicism or secularism. The first variant is authoritarian laïcité, where a strict interpretation on the separation of the state and the church is understood. The government is prohibited from recognizing or subsidizing any religion. Often laicic (secular) governments are accused of promoting secularism at the expense of religion, with the bulk of criticism coming from religious minorities feeling pressure to not practice their faith (Riedel 2008). The second variant is modern secular or neutral laïcité. This variant centers on state neutrality in religious affairs and is characterized by the absence of state influence within religious institutions, neither promoting nor demoting religious practice.

In addition to these four models of church-state relations, I add one additional model which, while listed under Riedel's cooperationalist variant, needs its own category: the Concordat model of church-state relations. The Concordat is a contract between the Vatican (Holy See) and other countries regarding the governance of religious affairs, institutions, and in some cases religious destinations within those countries. In essence, the Concordat replaces the Catholic Church as the legal established church in the constitution, ostensibly separating church from state. In practice, countries or provinces/states that have a Concordat are often majority Catholic, and the Concordats carry much weight legally and can act as if they are the established church. As such, the Concordat category needs to be separated from the cooperational category, as the actions of the Catholic Church within these countries are less cooperational and more obligations under international law (Table 3.2).

Table 3.2 Models of church-state relations

Concordat	Agreement between the Holy See of the Catholic Church and a sovereign state on religious matters. Legally, it is an international treaty. (Italy, Portugal, Spain, Bavaria[4])
Established State Church	Gives the church an official status & is part of the state structures. (Denmark, Finland, Norway, Iceland and England[5])
State-Provided Support (Cooperationist)	Official separation of church and state, however, state support for the church exists. Also includes countries with a 'church tax'. (Greece, Germany (excluding Bavaria), Austria, Sweden,[6] Japan)
Liberal Neutrality (Secular Laïcité)	Government is absent from religion and vice versa. None to little relations exist at all between the state and religion. Free exercise of religion and no establishment of religion. (the USA, rest of the UK, Canada,[7] Ireland,[8] Australia, the Netherlands, South Korea, and Switzerland[9])
Laïcité (Authoritarian)	No religion is to receive any legal establishment, state must be neutral in religious matters, and churches are part of the private sector and may not receive any state funding. Principle of secularism in public spheres is to be respected as well, though the degree varies by country and era. (France, Belgium, Quebec)

The mapping of Amable's political economy types onto models of church-state relations reveals several clusters (see Table 3.3). First, there appears to be a tighter link between market-based political economies and countries that have adopted a policy ofliberal neutrality. In liberal market political economies, firms experience deregulated product markets with low barriers to entrepreneurship, low administrative regulation, low state control, and low levels of public ownership. In this system, religious institutions are also often non-hierarchical and are often characterized as firms in a market. Using neoclassical economic theory, Iannaccone (1998) highlights how religious leaders are quite entrepreneurial in a deregulated market, especially in the creation of new religious dominations and sects, or in some cases new religions altogether. His research helps explain why the liberal market political economies, such as the United States, are more religiously diverse and why rates of religiosity might be higher. The exception appears to England, where Anglicanism is the state church operating in a liberal market political economy. However, in the case of the Church of England, one can make an argument that the religious climate is more similar to that of the United States.

A second cluster appears in those countries with Mediterranean political-economic systems that have a Concordat with the Vatican. As mentioned in Table 3.1, these countries experience regulated product markets with administrative burdens for corporations, barriers to entrepreneurship, and a strong public sector. As Concordats are legally binding documents regarding management of religious institutions, this makes sense for countries with strong Catholic traditions. A third cluster involves Continental European political economies with countries that use a liberal neutral model of church-state relations. As Amable (2003) notes, these countries experience competitive to mildly regulated markets. Interestingly, these countries exhibit features of a liberal market and have instituted a wide range of neoliberalization reforms. An argument could be made that liberal market reforms have possibly influenced church-state relations, especially in Ireland where voters have sought to decrease the influence of the Catholic Church in societal affairs.

A final cluster involves the strongly secular countries of France and Belgium and the Canadian province of Quebec. Amable's description of a competitive to mildly regulated market is evident in the management of religious institutions in France. The 1905 law of Laïcité, which led to a strong separation of church and state institutions in French society, led to several developments. On the one hand, the state privatized the ownership of local churches

Table 3.3 Mapping models of political economies and models of church-state relations

	Market	*Social Democratic*	*Asia*	*Mediterranean*	*Continental European*
Established State Church	England	Denmark, Finland			Norway
Concordat				Italy, Portugal, Spain	Bavaria
State-Provided Support		Sweden	Japan	Greece	Germany, Austria
Liberal Neutrality	USA, Canada, Australia, rest of UK		South Korea		Switzerland, Netherlands, Ireland,
Laïcité					Belgium, France, Quebec

to local parishes. Unfortunately, over time the lack of state support of local churches has led to the deterioration of some church buildings, forcing some Catholic dioceses to close entire parishes. On the other hand, cathedrals are still state-owned, with these cathedrals, which are often grandiose, being major tourist destination sites in France. This leads to a peculiar situation where the state regulates the use of these cathedrals as cultural heritage sites. This often leads the resident priest to ask local authorities for permission to conduct liturgical practices (Bernard 2019).

What does the above discussion mean for pilgrimage and religious tourism? The clustering of countries in Table 3.3 highlights how closely tied institutions are within particular OCED countries. Those countries with liberal market political economies tend to adopt liberal neutral models of church-state relations and allow the market to serve as the mechanism for solving problems between political, economic, and religious institutions. This is true for religious tourist destinations as well. In liberal market economies, the government permits religious organizations to manage their own sacred sites with little to no government regulation unless it is related to the medicalization of religious sites and mass gatherings (Olsen 2020). If anything, religious tourist destination sites often reach out to local authorities for help and guidance (Tilson 2001, 2005).

This can be contrasted with countries that have continental political economies as well as liberal neutral models of church-state relations. In these countries, such as Switzerland, The Netherlands, and Ireland, there exists competitive to mildly regulated markets where the government intervenes at times in the religious market but only in support of local efforts. This is illustrated well in the case of Failte (Welcome) Ireland, where the national tourism development authority commissioned a report to establish an integrated strategy to position the West and Northwest regions of Ireland for increased religious and spiritual tourism (McGettigan, Griffin & Candon 2011). In the United States, local religious organizations generally have to reach out to the governing authorities and major economic actors to engage in joint stakeholder projects. As such, even with the same model of church-state relations, the differences in the political-economy structures can help us better understand how institutions can guide behavior.

This is not the case with the Mediterranean political economies that have a Concordat with the Vatican. Religious tourist destinations are an important part of the cultural heritage of any country, but particularly so with Roman Catholic heritage that is millennia old. As Beltramo (2015: 89) notes in a study of hospitality among pilgrimage routes,

> The reflection on the meaning and ways of receiving the faithful in sacred places, such as sanctuaries convents, monasteries and churches, is part of the pastoral care in religious tourism, in regard to which the importance of the relationship between the [Catholic] Church and contemporary society has recently been reaffirmed.

Indeed, the Catholic Church plays a critical role in the management of sacred sites, and private groups that desire to profit from these destinations need to partner or accede to the demands made by the local religious elites. Thus, religious tourist destinations are a regulated product market, where the concordat, not state laws, govern religious tourism development (see Trono 2015).

A final cluster to consider is the Continental European political economies of Belgium, France, and the Canadian province of Quebec. Again, these countries have also adopted policies of laïcité, where a separation of church and state is the legal norm and the state does not interfere with religious affairs. In Belgium, the laïcité policy is referred to as the

Neutrality of the State, defined by Velaers and Foblets (2015) as 'The neutrality of the State and of the public authorities implies that the State itself has no particular religion or belief, and does not pass judgment on the religions or beliefs held by its citizens' (p. 101). However, non-interference can mean a number of things, and does not necessarily mean a sense of a-religiousness, or a strict rigid sense of secularism, nor does it mean hostility. Rather, this non-intervention can range from the relegation of religious observance to the private sphere in France to actual state assistance to recognized religions in Belgium (Chelini-Pont & Ferchiche 2015; Velaers and Foblets 2015).

Even with their official policies of laïcité, Belgium, France, and Quebec recognize the value of religious institutions, particularly regarding their cultural and historical significance. Often, local governments will work with important religious tourist destination sites, such as the Lourdes in France, to help manage tourist flows and build up tourist infrastructure (Kaufmann 2005). This is also the case in Quebec, where Catholicism is closely tied to French heritage and what sets them apart from the rest of Canada (Tourisme Montreal 2010). Clearly, there is a recognition of the economic impact that religious tourism can have, even in laïcité countries. Still, any help from the state is often seen strictly as a cultural or economic contribution and *not* privileging one religion over another.

Future research regarding the political economy of religious tourism

Table 3.3 shows the promise of using institutional frameworks to better understand the institutional arrangements and relationships within countries. These clusters provide evidence that correlations exist between political economy types and models of church-state relations—that there is a nexus not only between political governance and economic relations but also with religious institutions. While some scholars might contend that these arrangements have always been studied in disciplines such as religious studies, sociology, and economics, the trend in the literature of religious tourism and pilgrimage has been to ignore the role of the institution. To paraphrase Shinde (2018), an 'institutional vacuum' has developed in this scholarship, something that this chapter has attempted to rectify.

As noted in the introduction to this chapter, every country, regardless of its political-economic system, has had to accommodate the pressures of globalization, particularly regarding neoliberal policies adopted at the state level. Pressures by the World Bank, International Monetary Fund, and the World Trade Organization (often referred to as the Washington Consensus) have forced countries to open domestic markets and allow foreign entities to do business within a country. This move toward neoliberalization has also clearly impacted the tourism marketplace and by extension the religious marketplace as well. The effect of privatization regarding capital and production on sacred sites has been profound. In addition, the retreat of the state in the hotel and transportation industries and the weakening of labor and employment rights has led to a new understanding of how to best manage them (Bähre 2007). This is especially more acute during periods of political instability, where the state may have to step in and coordinate efforts in the nest interests of society. This could have an effect in religious tourism, where the state has a vested role in maintaining access to the site.

Understanding these changes to the religious marketplace from an institutional perspective is important, considering how strongly correlated political, economic, and religious institutions are within a country. For example, Dreher (2020) writes how the neoliberalization reforms promoted by right-wing parties went hand-in-hand with a concomitant rise in religiosity. Both movements fought against secular elites who often employed statist policies to promote religious equality in society. This global intertwining of religious conservatives and economic elites can be seen with evangelical Christians in the Republican Party in the United States, in the Hindutva movement in the Bharatiya Janata Party in India, and with Islamists within the Justice and Development Party in Turkey. How neoliberalization can be used to advance the cause of religious movements, and how that shapes religious institutions within a country, will have important consequences for religious tourist destinations.

In the future, the application of political-economic theories and frameworks to religious and spiritual tourism will lead to more research in religious tourism. Neoinstitutionalism is rich in theory. Varieties of institutionalisms have the potential to help scholars. So long as churches, individuals, and businesses find value in institutions, better theory is needed to explain stakeholder actions. The obligations of upholding the norms and values of an institution is what justifies an actor's place within the institution. This then reinforces rules of appropriate behavior and can justify the norms that undergird the institutions themselves. The key mechanism in institutionalism theory will be in understanding how institutions connect micro-level behavior of actors within the institution, with the macro-level constraints that are determined by the institution. Doing so will provide more theoretical context to a growing and quite relevant industry.

Notes

1 Hall and Soskice (2001) list the United States, the UK, Australia, Canada, New Zealand, and Ireland as LME countries.
2 Hall and Soskice (2001) list Germany, Japan, Switzerland, the Netherlands, Belgium, Sweden, Norway, Denmark, Finland, and Austria as CME countries.
3 Becker's (2014) work on the BRICs would disagree with Menz's work. He argues that change in institutions can be measured and should eventually be expanded to include non-OECD countries.
4 The majority Catholic German state of Bavaria has a Concordat with the Vatican, separate from the rest of the Republic of Germany.
5 Church of England is the established church in England, but not in the rest of the UK (Riedel, 2008).
6 The Church of Sweden was the established state church until 2000. Since then, the Swedish government provides support, mostly through the collection of fees, previously referred to as the "church tax" (Riedel, 2008).
7 Excludes the province of Quebec, where the policy of Laïcité is predominant.
8 Until recently, one could make the argument that the Republic of Ireland provided state support the Catholic church and pro-Catholic policies. The recent push for secularization in Irish society pushes it closer to the Liberal Neutrality model.
9 In theory, Switzerland maintains a policy of Liberal Neutrality. However, there is variance among the cantons and occasionally religious institutions can develop a special relationship within a canton.

References

Abbate, C.S., & Di Nuovo, S. (2013) 'Motivation and personality traits for choosing religious tourism. A research on the case of Medjugorje', *Current Issues in Tourism*, 16(5), 501–506.

Albera, D. (2018) 'Going beyond the elusive nature of pilgrimage', in S. Coleman and J. Eade (eds), *Pilgrimage and Political Economy: Translation of the Sacred* (pp. 173–190). New York: Berghahn Books.

Amable, B. (2003) *The Diversity of Modern Capitalism*. Oxford: Oxford University Press.

Aulet, A., & Vidal, D. (2018) 'Tourism and religion: Sacred spaces as transmitters of heritage values', *Church, Communication and Culture*, (3)3, 237–259.

Badone, E., & Roseman, S.R. (2004) *Intersecting Journeys: The Anthropology of Pilgrimage and Tourism*. Urbana and Chicago: University of Illinois Press.

Bähre, H. (2007) 'Privatisation during market economy transformation as a motor of development', in P.M. Burns & M. Novelli (eds), *Tourism and Politics: Global Frameworks and Local Realities* (pp. 33–58). Oxford, UK: Elsevier Ltd.

Becker, U. (2014) *The BRICs and Emerging Economies in Comparative Perspective: Political Economy, Liberalisation and Institutional Change*. New York: Routledge.

Beltramo, S. (2015) 'Medieval architectures for religious tourism and hospitality along the pilgrimage routes of Northern Italy', *International Journal of Religious Tourism and Pilgrimage*, 3(1): 79–94.

Bernard, J. (2019, April 16) 'Church and state disagree over management of religious heritage in France', *The Art Newspaper*. Available at: https://www.theartnewspaper.com/analysis/church-and-state-disagree-over-management-of-religious-heritage-in-france (accessed 1 October 2019).

Bianchi, R.V. (2002) 'Towards a new political economy of global tourism', in R. Sharpley and D. Telfer (eds), *Tourism and Development: Concepts and Issues* (pp. 265–299). Clevedon, UK: Channel View Publications.

Bianchi, R.V., & de Man, F. (2021) 'Tourism, inclusive growth and decent work: A political economy critique', *Journal of Sustainable Tourism*, 29(2–3): 353–371.

Blackwell, R. (2007) 'Motivations for religious tourism, pilgrimage, festivals and events', in R. Raj and N.D. Morpeth (eds), *Religious Tourism and Pilgrimage Festivals Management: An International Perspective* (pp. 35–47). Wallingford, UK: CABI.

Bond, N., Packer, J., & Ballantyne, R. (2015) 'Exploring visitor experiences, activities and benefits at three religious tourism sites', *International Journal of Tourism Research*, 17(5): 471–481.

Bramwell, B. (2011) 'Governance, the state and sustainable tourism: A political economy approach', *Journal of Sustainable Tourism*, 19(4/5): 459–477.

Bremer, T.S. (2006) 'Sacred spaces and tourist places', in D.J. Timothy & D.H. Olsen (eds), *Tourism, Religion and Spiritual Journeys* (pp. 25–35). London and New York: Routledge.

Burns, P.M, & Novelli, M. (eds) (2007) *Tourism and Politics: Global Frameworks and Local Realities*. Oxford, UK: Elsevier.

Butler, R. & Suntikul, W. (2018) *Tourism and Religion: Issues and Implications*. Bristol, UK: Channel View Publications.

Chelini-Pont, B. & Ferchiche, N. (2015) 'Religion and the secular state: Rapport Français,' in J. Martinez-Torron & W.C. Durham (eds), *Religion and the Secular State: National Reports*, (pp. 299–318). Madrid: Servicio de Publicaciones de la Facultad de Derecho de la Universidad Complutense de Madrid.

Chowdhury, A., Razaq, R., Griffin, K.A., & Clarke, A. (2017) 'Terrorism, tourism and religious travellers', *International Journal of Religious Tourism and Pilgrimage*, 5(1): 1–19.

Coleman, S., & Eade, J. (eds) (2004) *Reframing Pilgrimage: Cultures in Motion*. London and New York: Routledge.

Coleman, S., & Eade, J. (eds) (2018) *Pilgrimage and the Political Economy: Translation of the Sacred*. New York: Berghahn Books.

Collier, D. (2011) 'Understanding process tracing', *PS: Political Science and Politics*, 44(4): 823–830.

Collins-Kreiner, N. (2016) 'The lifecycle of concepts: The case of 'Pilgrimage Tourism'', *Tourism Geographies*, 18(3): 322–334.

Collins-Kreiner, N., Kilot, N., Mansfeld, Y., & Saig, K. (2006) *Christian Tourism to the Holy Land: Pilgrimage during Security Crisis*. Aldershot, UK: Ashgate.

David, P.A. (1994) 'Why are institutions the 'carriers of history'? Path dependence and the evolution of conventions, organizations and institutions', *Structural Change and Economic Dynamics*, 5(2): 205–220.

Delage, R. (2018) 'Sufism and the pilgrimage market: A political economy of a shrine in Southern Pakistan', in S. Coleman & J. Eade (eds), *Pilgrimage and Political Eocnomy: Translating the Sacred* (pp. 59–76). New York: Berghahn Books.

Di Giovine, M.A., & Choe, J. (2019) 'Geographies of religion and spirituality: Pilgrimage beyond the 'officially' sacred', *Tourism Geographies*, 21(3): 361–383.

Dowson, R., Yaqub, J., & Raj, R. (eds) (2019) *Spiritual and Religious Tourism: Motivations and Management*. Wallingford, UK: CABI.

Dreher, S. (2020) *Religions in International Political Economy*. Cham, Switzerland: Palgrave Macmillian.

Drule, A.M., Chiş, A., Băcilă, M.F., & Ciornea, R. (2012) 'A new perspective of non-religious motivations of visitors to sacred sites: Evidence from Romania', *Procedia-Social and Behavioral Sciences*, 62: 431–435.

Farias, M., Coleman III, T.J., Bartlett, J.E., Oviedo, L., Soares, P., Santos, T., & Bas, M.D.C. (2019) 'Atheists on the Santiago Way: Examining motivations to go on pilgrimage', *Sociology of Religion*, 80(1): 28–44.

Frieden, J.D. (2017) *International Political Economy: Perspectives on Global Power and Wealth*, 6th Edition. New York: W. W. Norton & Company.

Giampiccoli, A. (2007) 'Hegemony, globalisation and tourism policies in developing countries', in P.M. Novelli (ed), *Tourism and Politics: Global Framework and Local Realities* (pp. 175–191). Oxford, UK: Elsevier.

Gilpin, R. (2001) *Global Political Eocnomy: Understanding the international Economic Order*. Princeton, NJ: Princeton University Press.

Griffiths, M. & Wiltshier, P. (2019) *Managing Religious Tourism*. Wallingford, UK: CABI.

Green, D., & Shapiro, I. (1994) *Pathologies of Rational Choice: A Critique of Applications in Political Science* (pp. 13–17). New Haven, CT: Yale University Press.

Hall, C.M. (1994) *Tourism and Politics: Policy, Power and Place*. New York: John Wiley & Sons.

Hall, P.A., & Soskice, D. (2001) *Varieties of Capitalism: The Institutional Foundations of Comparative Advantage*. New York: Oxford University Press.

Iannaccone, L.R. (1998) 'Introduction to the economics of religion', *Journal of Economic Literature*, 36(3): 1465–1495.

Katić, M. (2018) '"Pilgrimage capital" and Bosnian Croat pilgrimage places: Bosnian Croat pilgrimage and transnational ties through time and space', in S. Coleman & J. Eade (eds), *Pilgrimage and Political Economy: Translations of the Sacred* (pp. 93–111). New York: Berghahn Books.

Kaufman, S. (2005) *Consuming Visions: Mass Culture and the Lourdes Shrine*. Cornell: Cornell University Press.

Kütting, G. (2010) *The Global Political Economy of the Environment and Tourism*. New York: Palgrave Macmillan.

Lopez, L., Lois González, R.C., & Fernández, B.M.C. (2017) 'Spiritual tourism on the way of Saint James the current situation', *Tourism Management Perspectives*, 24: 225–234.

Mahoney, J. (2000) 'Path dependence in historical sociology', *Theory and Society*, 29(4): 507–548.

March, J.G., & Olsen, J.P. (1984) 'The new institutionalism: Organizational factors in political life', *American Political Science Review*, 78(3): 734–749.

March, J.G., & Olsen, J.P. (2011) 'The logic of appropriateness', in R.E. Goodin (ed), *The Oxford Handbook of Political Science* (pp. 479–497). Oxford: Oxford University Press.

McGettigan, F., Griffin, C., & Candon, F. (2011) 'The role of a religious tourism strategy for the West and North West of Ireland in furthering the development of tourism in the region', *International Journal of Business and Globalisation*, 7(1): 78–92.

Menz, G. (2017) *Comparative Political Economy: Contours of a Subfield*. Oxford: Oxford University Press.

Mosedale, J. (2011) *Political Economy of Tourism: A Critical Perspective*. London and New York: Routledge.

Mosedale, J. (ed.) (2016) *Neoliberalism and the Political Economy of Tourism*. London and New York: Routledge.

Müller, D.K. (ed.) (2019) *A Research Agenda for Tourism Geographies*. Cheltenham, UK: Edward Elgar Publishing.

Nolan, M.L., & Nolan, S. (1993) 'Religious sites as tourism attractions in Europe', *Annals of Tourism Research*, 19(1): 68–78.

Norman, A. (2011) *Spiritual Tourism: Travel and Religious Practice in Western Society*. New York: Bloomsbury Academic.

North, D.C. (1991) 'Institutions', *The Journal of Economic Perspectives*, 5(1): 97–112.

Nunkoo, R., & Smith, S.L. (2013) 'Political economy of tourism: Trust in government actors, political support, and their determinants', *Tourism Management*, 36: 120–132.

Nyaupane, G.P., Timothy, D.J., & Poudel, S. (2015) 'Understanding tourists in religious destinations: A social distance perspective', *Tourism Management*, 48: 343–353.

Oatley, T. (2019) *International Political Economy*. New York: Routledge.

Olsen, D.H. (2006) Management issues for religious heritage attractions. In D.J. Timothy & D.H. Olsen (eds), *Tourism, Religion and Spiritual Journeys* (pp. 104–118). London and New York: Routledge.

Olsen, D.H. (2009) '"The strangers within our gates": Managing visitors at Temple Square', *Journal of Management, Spirituality & Religion*, 6(2): 121–139.

Olsen, D.H. (2019) 'Best practice and sacred site management: The case of Temple Square in Salt Lake City, Utah', in M. Griffins & P. Wiltshier (eds), *Managing Religious Tourism* (pp. 65–78). Wallingford, UK: CABI.

Olsen, D.H (2020) 'Disease and health-related issues at mass religious gatherings', in K.A. Shinde & D.H. Olsen (eds), *Religious Tourism and the Environment* (pp. 116–132). Wallingford, UK: CABI.

O'Neil, P.H. (2018) *Essentials of Comparative Politics*, 6th Edition. New York: W.W. Norton & Company.
Orren, K., & Skowronek, S. (2004) *The Search for American Political Development*. Cambridge: Cambridge University Press.
Peters, B.G. (2019) *Institutional Theory in Political Science: The New Institutionalism*, 4th Edition. Cheltenham, UK: Edward Elgar Publishing.
Pierson, P., & Skocpol, T. (2002) 'Historical institutionalism in contemporary political science', in I. Katznelson & H.V. Miller (eds), *Political Science: State of the Discipline* (pp. 693–721). New York: Norton.
Raj, R., & Griffin K.A. (2015) *Religious Tourism and Pilgrimage Management: An International Perspective*, 2nd Edition. Wallingford, UK: CABI.
Raj, R., & Morpeth N.D. (2007) *Religious Tourism and Pilgrimage Festivals Management: An International Perspective*. Wallingford, UK: CABI.
Reader, I. (2014) *Pilgrimage in the Marketplace*. New York: Routledge.
Rhodes, R.A. (2008) *The Oxford Handbook of Political Institutions*. Oxford: Oxford University Press.
Richter, L.K. (2007) 'Democracy and tourism: Exploring the nature of an inconsistent relationship', in P.M. Burns (ed), *Tourism and Politics: Global Frameworks and Local Realities* (pp. 5–16). Oxford, UK: Elsevier.
Riedel, S. (2008) 'Models of church-state relations in European democracies', *Journal of Religion in Europe*, 1(3): 251–272.
Rössler, M. (2003) 'Managing world heritage cultural landscapes and sacred sites', *World Heritage Papers*, 13: 45–48.
Sanagustín-Fons, M.V., Gregory, R.B., & Martínez-Quintana, V. (2019) Holy Grail route: A sociological analysis of a spiritual and religious tourist route', in J. Álvarez-García, M. de la Cruz del Río Rama, M., & Gómez-Ullate (eds), *Handbook of Research on Socio-Economic Impacts of Religious Tourism and Pilgrimage* (pp. 38–53). Hershey PA: IGI Global.
Schmidt, V.A. (2009) 'Putting the political back into political economy by bringing the state back in yet again' *World Politics*, 61(3): 516–546.
Shackley, M. (2001a) *Managing Sacred Sites: Service Provision and Visitor Experience*. London: Continuum.
Shackley, M. (2001b) 'Sacred world heritage sites: Balancing meaning with management', *Tourism Recreation Research*, 26(1): 5–10.
Shackley, M. (2003) 'Management challenges for religion-based attractions', in A. Fyall, B. Garrod, & A. Leask (eds), *Managing Visitor Attractions: New Directions* (pp. 159–170). Oxford, UK: Butterworth-Heinemann.
Shinde, K. (2018) 'Governance and management of religious tourism in India', *International Journal of Religious Tourism and Pilgrimage*, 6(1): 58–71.
Shinde, K.A. (2020) 'Managing the environment in religious tourism destinations: A conceptual model', in K.A. Shinde & D.H. Olsen (eds), *Religious Tourism and the Environment* (pp. 42–59). Wallingford, UK: CABI.
Smith, A. (1977) *The Wealth of Nations*. London: Dent.
Stausberg, M. (2011) *Religion and Tourism: Crossroads, Destinations and Encounters*. New York: Routledge.
Suleiman, J.S.H., & Mohamed, B. (2011) 'Factors impact on religious tourism market: The case of the Palestinian territories', *International Journal of Business and Management*, 6(7), 254–260.
Tilson, D.J. (2001) 'Religious tourism, public relations and church-state partnerships', *Public Relations Quarterly*, 46(3): 35–39.
Tilson, D.J. (2005) 'Religious-spiritual tourism and promotional campaigning: A church-state partnership for St. James and Spain', *Journal of Hospitality and Leisure Marketing*, 12(1/2): 9–40.
Timothy, D.J., & Olsen, D.H (eds) (2006) *Tourism, Religion and Spiritual Journeys*. London and New York: Routledge.
Tourisme Montreal (2010) *Religious and Spiritual Tourism in Québec: An Action Plan for Marketing the Four National Sanctuaries of Québec*. Available at: https://www.newswire.ca/news-releases/religious-and-spiritual-tourism-in-quebec-an-action-plan-for-marketing-thefour-national-sanctuaries-of-quebec-546327742.html (accessed 27 July 2020)
Trono, A. (2015) 'Politics, policy and the practice of religious tourism', in R. Raj & K.A. Griffin (eds), *Religious Tourism and Pilgrimage Management: An International Perspective*, 2nd Edition (pp. 16–36). Boston, MA: Elsevier.
Velaers, J., & Foblets, M. (2015) 'Religion and the State in Belgian Law', in J. Martinez-Torron and W.C. Durham (eds), *Religion and the Secular State: National Reports* (pp. 99–122). Madrid: Servicio de Publicaciones de la Facultad de Derecho de la Universidad Complutense de Madrid.

Vukonić, B. (1996) *Tourism and Religion*. Oxford, UK: Elsevier.
Webster, C.S., Ivanov, S., & Illum, S.F. (2010) 'The paradigms of political economy and tourism policy: National tourism organizations and state policy', in J. Mosedale (ed), *Political Economy of Tourism: A Critical Perspective* (pp. 55–73). London and New York: Routledge.
Winter, M., & Gasson, R. (1996) 'Pilgrimage and tourism: Cathedral visiting in contemporary England,' *International Journal of Heritage Studies*, 2(3): 172–182.
Wong, C.U.I., McIntosh, A., & Ryan, C. (2016) 'Visitor management at a Buddhist sacred site,' *Journal of Travel Research*, 55(5): 675–687.
Woodward, S.C. (2004) 'Faith and tourism: Planning tourism in relation to places of worship', *Tourism and Hospitality Planning & Development*, 1(2): 173–186.
Zey, M. (1998) *Rational Choice Theory and Organizational Theory: A Critique* (pp. 55–58). Thousand Oaks, CA: Sage Publications.

4

THE RELIGIOUS AND SPIRITUAL DIMENSIONS OF LEISURE TRAVEL

Paul Heintzman

Introduction

Twenty-two years ago, I (Heintzman 1998) reviewed one of the first books published on the intersections between religion and tourism, Boris Vukonić's (1996) *Tourism and Religion*. In the first chapter of his book, Vukonić discussed the relationships between spiritual life and leisure time and stated that the book's emphasis was on "those characteristics of free time and leisure time that determine or promote people's relationship toward their spiritual needs" (p. 5). Although alluding to Dumazedier's (1974) understanding of leisure as the activities in which a person participates, Pieper's (1968) classical understanding of leisure as contemplation, and Veblen's (1899/1953) work on leisure as a function of social class characterized by conspicuous consumption, Vukonić primarily focused on an understanding of leisure as free time, suggesting that "people will find spiritual fulfillment mostly in the time that is free from organized work…the effect of free time is to intensify the various forms of our spiritual life, and in a certain sense to encourage and develop it" (p. 4). Later, he explained that "spiritual life is being increasingly transferred to, and manifested, in free time. Free time has thus become a space for the contemplative and the creative, a unity of thought and action" (p. 8). Contextualizing this idea within the Christian spiritual tradition, he argued that

> …free time and leisure are a unique and unified time given to people by God, which should therefore be used to serve God. Leisure time, the part of free time in which people will express their most intimate inclinations and devote themselves only to that which satisfies them completely, is the ideal time for people to find the peace they need to give themselves to God and receive Him.
>
> *(p. 9)*

Thus, he concluded, "Leisure is also our way to God" (p. 10).

While Vukonić (1996) viewed leisure as free time, he also recognized that free time and leisure time are not synonymous. Rather, he viewed leisure as the positive use of free time:

> Although it is quite difficult to distinguish between the concepts of free time and leisure time, because these two concepts overlap and complement each other, their conceptual

> features are clear. This is not Veblen's idleness....This is leisure which should continually make more room for true human living. This undeniably includes everything that enriches people, first of all their spiritual life.
>
> *(p. 11)*

Vukonić then moved to connect free time and leisure to tourism

> ...in its origins and duration. Numerous theorists tend to go so far as to claim that tourism is a classical product of free time or leisure. When free time became the property of the masses, modern (mass) tourism appeared as a form of using that time. That is why the analysis of tourism as a phenomenon is often approached from the aspect of free time and leisure time.
>
> *(p. 14)*

Vukonić later noted that both leisure and tourism are beneficial

> ...we see tourism as a productive part of leisure time, as a phenomenon that will have a positive and productive effect on the total life of humankind. This is because tourism provides people with the conditions for a constant search for the spiritual enrichment of the individual and his or her constant self-improvement as a personality.
>
> *(p. 18)*

Since the publication of Vukonić's (1996) book, there has been a dramatic increase in scholarship on the intersections between leisure, tourism, religion, and spirituality. The purpose of this chapter is to examine this scholarship, first by discussing the prevalent concepts in the domain of leisure research and then connecting these concepts with religious and spiritual tourism. The chapter then provides a preliminary framework within which to organize empirical research that examines the spiritual and religious outcomes of leisure travel.

Concepts of leisure

Within the leisure studies field, the view of leisure as free time as promoted by Vukonić (1996) is only one of many conceptualizations of leisure. Other common understandings of leisure include classical leisure, leisure as activity, leisure as a function of social class, the psychological understanding of leisure as subjective experience, feminist perspectives of leisure, and holistic leisure (Heintzman 2013a). Each of these understandings of leisure, as well as the view of leisure as free time, is examined here in turn.

Classical leisure. Classical views of leisure that have their roots in ancient Greece emphasize "a spiritual and mental attitude, a state of inward calm, contemplation, serenity, and openness" (Kraus 1984: 42). This classical view, which focuses on a state of being as well as moral choices and conduct, has evolved through the centuries, from Aristotle to early Christian writers such as Augustine, and then to Thomas Aquinas and monasticism during the Middle Ages. More recently, this tradition has continued in the Roman Catholic philosopher and theologian Pieper (1968) who, in his book *Leisure: The Basis of Culture*, defined leisure as "a mental and spiritual attitude...a condition of the soul...a receptive attitude of mind, a contemplative attitude" (pp. 40–41).

Leisure as activity. This view of leisure focuses on the activities that are undertaken during free time. Murphy (1974) defined leisure as "non-work activity in which people engage during their free time—apart from obligations of work, family and society" (p. 4). Historically, the activity view of leisure was usually utilitarian in nature—that is, activities engaged in during leisure time were done in order to achieve some sort of benefit or to meet targeted outcomes, such as becoming healthier or to improve social relationships. From this perspective, leisure has often been viewed as subservient to work-related activities and associated with a life rhythm of work and recreation. More recently, however, the leisure as activity concept has not necessarily been a utilitarian view. The French sociologist Joffre Dumazedier (1967), for example, has argued that "Leisure is activity—apart from the obligations of work, family and society—to which the individual turns at will, for relaxation, diversion, or broadening his knowledge and his spontaneous social participation, the free exercise of his creative capacity" (pp. 16–17). For Dumazedier, leisure had three main functions: relaxation, entertainment, and personality development. Based on the activity view of leisure, Stebbins (1999) developed the concepts of "serious leisure" and "casual leisure." Serious leisure is "the systematic pursuit of... an activity that participants find so substantial and interesting that...they launch themselves on a career centred on acquiring and expressing its special skills, knowledge, and experience" (p. 69). In contrast, casual leisure is an "immediately, intrinsically rewarding, relatively short-lived pleasurable activity requiring little or no special training to enjoy it" (Stebbins 1997: 18). This pleasure-oriented activity tends to be non-utilitarian.

Leisure as free time. This view of leisure, as introduced earlier in the discussion of Vukonić's (1996) work, is a quantitative understanding that defines leisure as "that portion of time which remains when time for work and basic requirements for existence have been satisfied" (Murphy 1974: 3). From this viewpoint, life is divided into existence (i.e., attending to biological needs such as eating and sleeping), subsistence (i.e., work), and leisure (i.e., non-obligated or discretionary time). This view assumes that the more free time one has, the more leisure time one has, reducing leisure to a unit of time with no focus on the quality of that period of time.

Leisure as a function of social class. This view of leisure can be defined as "a way of life for the rich elite" (Murphy 1974: 92). In his classic book, *The Theory of the Leisure Class*, the American economist and sociologist Thorstein Veblen (1899/1953) questioned the intrinsic character of leisure activities and hypothesized that leisure behavior was influenced by the desire to impress others and distinguish oneself from other people. Veblen therefore defined leisure as "non-productive consumption of time" (p. 46). From his perspective, "Time is consumed non-productively (1) from a sense of the unworthiness of productive work, and (2) as an evidence of pecuniary ability to afford a life of idleness" (p. 46). Veblen coined the terms "conspicuous leisure" and "conspicuous consumption" to explain how the visible display of leisure and its consumption was more important than actually engaging in leisure activities for their own sake. Therefore, according to Veblen, leisure was symbolic, with wealthy classes of citizens throughout time being identified by both their use of available leisure and their possessions while lower classes of citizens tried to imitate or emulate the wealthy classes.

Leisure as a psychological, subjective experience. The study of leisure as subjective or psychological experience, also known as the state of mind view which gained prominence within leisure studies in the 1980s, focuses on the human "experience that results from recreation engagements" (Driver & Tocher 1970: 10). This view of leisure is founded upon psychological concepts such as William James' (1890) notion of "stream of consciousness," in which conscious states or mental experiences are perceived as continuous and ever-changing, and Abraham Maslow's (1968) idea of "peak experience," which refers to "moments of highest happiness and fulfillment" (p. 73). Another psychological concept frequently associated with this view of leisure is Csikszentmihalyi's (1975) concept of "flow," where participation in some activities leads to intensely absorbing experiences, in that the challenge of an activity matches the skill level of an individual so that the person loses track of both time and awareness of self. Leisure as psychological experience includes properties such as moods and emotions; levels of activation and arousal; cognitions such as images, ideas, thoughts, and beliefs; the perception of how quickly time is passing; levels of attention and concentration; self-awareness and self-consciousness; sense of competence and mastery; sense of autonomy and freedom; and sense of interpersonal relationships (Walker, Kleiber & Mannell 2019; cf. Graef et al. 1983; Mannell & Iso-Ahola 1987).

Feminist views of leisure. Feminist scholars have long been critical of the traditional views of leisure as free time or a set of activities that differ from work because these views are built on premises that do not always take into consideration the constraints on women's leisure time and leisure activities (Henderson 1991). Feminist research and theorizing on leisure have led to an enhanced understanding of leisure, not as time or activities, but as a meaningful experience characterized by enjoyment (Henderson et al. 1996). These meaningful experiences may be found in many aspects of life. Often, the meaningful experience is associated with time for one's self to relax and do nothing rather than participating which is often the case for males. Feminists have developed the concept of affiliative leisure that suggests relationships with other people such as friends and family are more important than the specific type of leisure they engage in (Henderson et al. 1996). Feminists emphasize agentic leisure characterized by autonomy where one can express oneself through self-determined (rather than determined by other people) activities and experiences (Freysinger & Flannery 1992; Henderson et al. 1996). Because the emphasis is on meaningful experience, the activity, social setting, or physical location is seen as a leisure container in which the experience of leisure may take place (Henderson et al. 1996; Freysinger & Kelly 2004). Feminists also speak of leisure enablers, the opposite of leisure constraints, that allow and facilitate leisure experiences. An example of a leisure enabler is a sense of entitlement to leisure (Henderson et al. 1996; Gibson, Ashton-Schaeffer, Green & Autry 2003).

Holistic views of leisure. From this perspective, leisure is seen as a total way of life—that there is very little that differentiates work and leisure. Leisure may be experienced within the various contexts of life such as family, education, religion, or work. Leisure is fused with satisfying work and is continuous instead of fragmented. Therefore, the holistic concept unites leisure as an end, as in the classical view, with leisure as a means, as in the activity view. It combines a focus on "being" with a focus on "doing" and therefore reflects a return to a more traditional way of life (Kaplan 1974).

According to Kraus (2001), an additional "way of conceptualizing leisure…[is] in terms of its contribution to spiritual expression or religious values" (p. 36). However, spirituality has been linked to many of the existing definitions of leisure (Heintzman 2002). For example, the classical leisure perspective, with its emphasis on contemplation, has had spiritual overtones for centuries. Defined as free time, leisure time can be used for spiritual growth. Defined as activity, spiritual activities can be included. Peak experience, optimal experience, and flow, which are associated with the state-of-mind view of leisure, also describe spiritual experiences. Holistic leisure integrates spirituality into all of life, including leisure.

Connecting leisure concepts with religious and spiritual tourism

Now that we have reviewed the main concepts of leisure, how might they be related to religious and spiritual tourism? In regard to *classical leisure*, Voigt, Brown and Howat (2011) found that spiritual retreat visitors, in contrast to lifestyle resort visitors and beauty spa visitors, placed more value on transcendent experiences that involved self-awareness at the spiritual level, a sense of spiritual renewal, experiencing peace and calmness, contemplating one's life, and meditative practices. Such a description has similarities with classical leisure. Likewise, in another study, reflection was an important part of spiritual retreatants' experience that deepened their relationship with God (Gill, Packer & Ballantyne 2018; see Chapter 14 this volume).

Regarding *leisure as activity* and its connection to spiritual and religious tourism, in his dissertation on spiritual tourism, Alex Norman (2004) defined tourism as "a leisure activity that can range from a convalescent-type "recharge" away from the normal working world to an existential search for meaning and truth, or a quest for the sacred that bears many of the marks of pilgrimage" (p. 12). Following Stebbins' (1999) idea of serious versus casual leisure, Haq and Jackson (2006) made a distinction between the purposeful spiritual tourist with deep intention, for whom "personal spiritual growth is the main reason for visiting" and the casual spiritual tourist, for whom "personal spiritual growth is a casual motivation for the visit" and who has a lower level of spiritual experience (p. 2). Moufakkir and Selmi (2018) suggested that participants in Sharpley and Jepson's (2011) study of tourists to England's Lake District should be categorized as casual spiritual tourists, as their travel was not purposefully to seek spiritual fulfillment, whereas Moufakkir and Selmi considered participants in their own study to be purposeful spiritual tourists because their motivation was "to intentionally seek spiritual fulfilment" (p. 109). Another example related to casual and serious leisure is Voigt, Howat, and Brown's (2010) study where they found that beauty spa tourists focused on hedonistic outcomes, whereas spiritual retreat visitors progressed through various stages of obtaining new skills, knowledge, and training in religious or philosophical teachings and meditation techniques, much in the same way that someone develops skills during a work career. These retreat visitors described their experiences which did not necessarily exclude negative emotions, as deeply fulfilling, which in some cases led to increased self-knowledge and self-identity. Yoga tourism has also been categorized by tourism researchers as a form of serious leisure (Patterson, Getz & Gubb 2016; Bowers & Cheer 2017; Dillette, Douglas & Andrzejewski 2019).

As noted above, Vukonić's (1996) work on tourism and religion used the *leisure as free time* perspective to discuss religious and spiritual travel during leisure time. Indeed, tourism, religion, and spirituality have many structural similarities, in addition to the fact that for the most part, they take place during people's leisure time. Thus, activities related to tourism, religion, and spirituality compete with each other as well as with other leisure activities from which people can choose to participate in during their free time. These activities may also complement each other if the religious and spiritual tourist combines them within their free time activities (Weidenfeld & Ron 2008).

The possibility that tourism might be linked to the concept of *leisure as a function of social class* is alluded to in MacCannell's (1976/2013) classic book *The Tourist: A New Theory of the Leisure Class*, where he argued that the expansion of modern society was "intimately linked in diverse ways to modern mass leisure, especially to international tourism and sightseeing" (p. 3). However, this view of leisure does not seem very relevant to religious and spiritual tourism, as according to this view, people often engage in leisure for extrinsic reasons—to impress others with their conspicuous leisure and their conspicuous consumption (Veblen 1899/1953). This leisure perspective is most likely to be reflected in Yiannakis and Gibson's (1992) "high-class tourists" and "jetsetters" tourist types, while those who engage in religious and spiritual tourism are more likely to be intrinsically motivated and aligned with Yiannakis and Gibson's "seeker" tourist type, who seek "spiritual and/or personal knowledge to better understand self and meaning of life (p. 291); and "who, through travel, seek to learn more about themselves, and ultimately, the meaning of existence. Seekers are clearly on some type of quest" (pp. 297–298).

A good example of the *leisure as psychological/subjective experience* perspective is Little and Schmidt's (2006) study of the spiritual dimensions of experiences people have during leisure travel. Lamenting that tourism research focuses more on defining and objectifying tourists and understanding supply and demand determinants for travel rather than the subjective tourist experience, Little and Schmidt argued that "understanding personal meanings, structural frameworks and the inner worlds of participating individuals, is core to further unpacking the experience itself" (p. 108). As such, the authors investigated the spiritual dimension of the tourist experience for ten independent leisure travelers. They found that leisure travel was a multifaceted experience that had spiritual effects and meaning for the participants—where participants obtained a greater awareness of self, others or God; experienced an enhanced sense of relationship with something greater than the self; and had an intensely spiritual leisure travel experience, that was characterized by release, fear, awe, and wonder.

Peak experiences and "flow," as discussed above, are important aspects of both subjective leisure experiences and tourist experiences. Cohen (2006), for example, suggested that experiences during pilgrimage travel reflect "a peak experience (a special moment in Time), where the sacred and profane meet" (p. 80). In a study of Christian pilgrimages to the Holy Land, participants identified "peak moments" when they felt connected to their faith's history and to Jesus (Belhassen, Caton & Stewart 2008), in part because "the tour was steeped in moments in which toured objects, tourist performances, socialization, and faith merged to produce powerful spiritual experiences" (p. 682). Flow experiences have also been discussed in the context of heritage tourism where heritage resources and sites have the potential to create flow-like experiences characterized by reverence and transcendence (e.g., Powell et al. 2015). It is important to note, however, that spiritual experiences may become conflated with peak and flow experiences in which certain states of mind are achieved through engaging in certain travel experiences. Some empirical studies have documented this confusion. For example, Jepson and Sharpley (2015) argued that for many tourists, "the distinction between emotional and spiritual experiences was not always clear" (p. 1165). Likewise, Jarratt and Sharpley (2017) recognized that some of the experiences that seaside tourists have may fit better with wellness and psychological concepts rather than spirituality (see Chapter 11, this volume). However, as Esfahani, Musa, and Khoo (2014) argued, "spirituality is a more stable state of mind, achieved through perception and feedback from people and the environment" (p. 4).

As mentioned above, *feminists* have critiqued traditional concepts of leisure and have placed an emphasis on meaningful experience characterized by enjoyment. This focus on meaningful experience and enjoyment is seen in a number of tourism studies. Tourism memories have been found to be more psychologically important for women compared to men (Anderson & Littrell 1995). Likewise, the moderating effects of memorable tourist experiences upon subjective well-being have been discovered to be greater for women than men (Sthapit & Coudounaris 2017). Experience has been found to be very important to Goddess pilgrims, who were not mere sightseers, but rather intimately experienced the sacred landscape at a multisensory level that resulted in a strong embodied connection (Rountree 2002). In a study of contemporary women's travel narratives of Paris, experience was a significant theme with one of the over-arching themes being emotional experience of place: "Paris arouses extremely meaningful and emotional experiences for the women travel writers in this study....They are re-inventing themselves as women who confidently embrace the sensuous and emotional experiences of everyday life..." (McClinchey 2017: 7).

The feminist concept of affiliative leisure is also evident in tourism research. Both quantitative and qualitative research on the holiday travel of single women established that being together, sharing experiences, and bonding with friends during holiday travel was important to single women of all ages (Heimtun & Morgan 2012). Similarly, a study of Goddess pilgrims found that they experienced community and a high level of connection with other women with whom they shared their pilgrimage (Rountree 2002).

The idea of *holistic leisure* and its connection to religious and spiritual tourism, clearly visible in the title of Smith's (2003) paper, "Holistic holidays: Tourism and the reconciliation of body, mind and spirit," is to some extent, tied to the understanding of tourism as a form of modern pilgrimage (MacCannell 1973; Cohen 2006). Through travel, people who find their lives as fragmented or who search for meaning may experience a more holistic and authentic way of life in which the fragmented parts of their lives are brought together. Such travel may involve a search for personal authenticity through an inner journey of significant self-transformation. Cohen (2006; see Kelner 2012) gives the example of Jewish youth who travel to Israel, with the purpose of bringing together in a holistic way their fractured identities by experiencing different dimensions of Jewish life (social, political, religious, and spiritual) and to engage in significant actions that express national and ethnic identity. Cohen (2006) noted that these Jewish youth may experience transformations that are behavioral (i.e., participation in the local Jewish community, engaging in religious rituals), affective (i.e., greater feelings of commitment and attachment), and/or cognitive (i.e., attitudinal adjustment, growth in knowledge) in nature. For retreat tourism participants, retreat centers offer a holistic retreat experience of a home away from home, healthy foods, spiritual practices, and transformational learning (Bone 2013). Another example is that of health holidays or holistic tourism (Smith 2003; Smith & Kelly 2006; Voigt, Brown & Howat 2011; Bowers & Cheer 2017). Many people travel to accomplish goals related to holistic living and self-transformation. In contrast to mass tourism that tends to focus on escaping one's everyday setting, holistic tourism helps people to engage their inner being and resolve internal conflict through spiritual and deeply personal activities. This type of tourism, often focused on healing and fitness, is based on the holistic premise that each dimension of health (e.g., spiritual and physical) is interrelated with all the others.

Spiritual and religious outcomes of leisure travel: a framework for synthesizing empirical research

In recent decades, there has been a dramatic increase in empirical research on the spiritual outcomes of leisure travel and tourism; however, there are few frameworks or models that synthesize this empirical research. Some researchers (Bond, Packer & Ballantyne 2015; Gill, Packer & Ballantyne 2018) have applied Beeho and Prentice's (1997) Activity, Setting, Experience, Benefits (ASEB) framework to their research on the spiritual outcomes of tourism. More recently, Cheer, Belhassen, and Kujawa (2017) developed a conceptual framework for spiritual tourism, whereas Chhabra (2020) created a conceptual model of slow spiritual tourism. However, none of these frameworks are based on an extensive synthesis of empirical research. In this paper, I provide a framework to organize empirical research that has examined the spiritual outcomes of leisure travel and tourism. This framework is based on a slight modification to an existing framework of outdoor activities and spirituality that includes antecedent conditions, setting, and recreation components, which together lead to short- and long-term spiritual outcomes (Heintzman 2016a; cf. Heintzman 2012, 2013c, 2016b). In the tourism context, the recreation component will be renamed the tourism activity component.

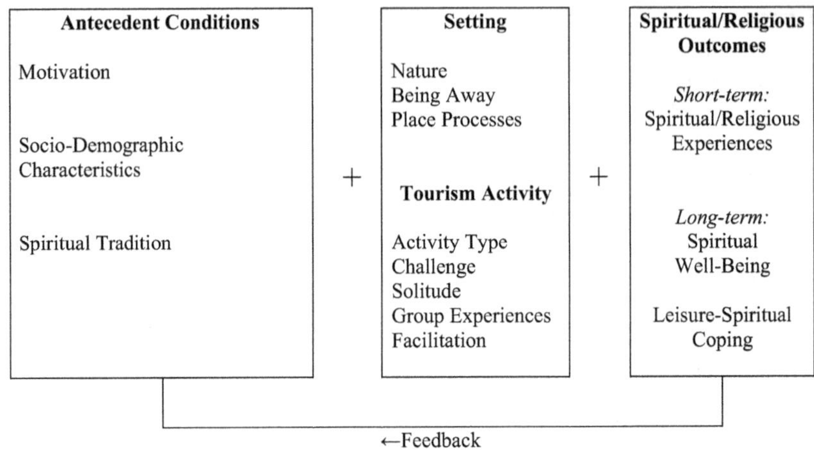

Figure 4.1 Spiritual/religious outcomes of leisure travel.

The following preliminary but not exhaustive review of empirical studies on the spiritual dimensions of leisure travel and tourism reveals that most of the elements of this framework are present (Figure 4.1).

Antecedent conditions

The presence and type of spiritual outcomes that result from tourism activities may be influenced by antecedent conditions, which refer to people's characteristics prior to traveling, such as their motivations, social demographic characteristics, and religious and/or spiritual traditions. There is some evidence that *tourist motivations* may be related to spiritual outcomes. For example, Bond, Packer, and Ballantyne (2015) discovered that not only do Christian pilgrimage festivals and pilgrimage shrines appeal to those with a focus on spiritual interests, while grand cathedrals attract those with an interest in cultural heritage and religious history, but that shrine and festival visitors rated spiritual growth as the second highest of five tourist benefits whereas cathedral visitors had the lowest spiritual growth levels. Some research has found that there is not necessarily a clear relationship between tourist motivations and spiritual outcomes. In a study of visitors to Chichester Cathedral, even visitors who did not consider spiritual outcomes as their main motivation for visiting the cathedral tended to obtain some degree of spiritual blessing from their cathedral visit (Gutic, Caie & Clegg 2010: 757). Research by Sharpley and Sundaram (2005) on the motivations and experiences of western tourists to an Ashram in India discovered that although there were a diversity of motivations ranging from purposeful spiritual need satisfaction to secular knowledge-driven curiosity, different degrees of spiritual outcomes were experienced even if it was unintentional. Significantly, they found that spiritual benefits resulted from a desire to learn and a curiosity rather than spiritual motivations, a discovery that is consistent with the finding that intellectual motivations, along with stimulus-avoidance motivations, were correlated with spiritual well-being in a study of leisure, including travel and tourism activities, and spiritual well-being (Heintzman 1999).

In terms of *social demographic characteristics*, spirituality has been recognized as a determining factor in the travel motivations of senior tourists (Moal-Ulvoas 2014), and the resulting travel produces spiritual benefits for them (Moal-Ulvoas & Taylor 2014). In a study of visitors to English cathedrals, both the spiritual feelings evoked by the visit and the religious significance of the cathedral visit were dramatically higher for the oldest age group compared to the youngest age group (Jackson & Hudman 1995). Furthermore, more females (30%) than males (20%) noted that religious feelings accompanied their cathedral visit. These findings are consistent with studies of leisure and spirituality, which suggest that spiritual outcomes are more likely with older populations and with women in comparison to men (Heintzman 2016a, 2016b).

Spiritual tradition also influences religious and spiritual motivations and subsequently religious and spiritual outcomes. Andriotis (2009), for example, found that many of the visitors to Mount Athos were motivated by a strong commitment to their Orthodox Christian faith, and identified spiritual reasons for visiting Mount Athos more so than other types of visitors to the site. These "proskinites," as they are referred to, stated that their inner journey was more important than their outdoor journey, and that the pull motives for their journey included "to pray," "to venerate," "to meditate," "to get closer to God," "to be in a sacred shrine," "to strengthen their belief," and "to improve their religious faith." They also tended to not be as interested in secular or touristic activities and elements of the site as other visitors. A similar study of visitors to Mount Athos and Meterora in Greece found that these Orthodox "proskinites" found spiritual experiences through participation in religious rituals, while non-Orthodox and non-religious visitors experienced a different type of spiritual experience through stillness and aesthetic contemplation (della Dora 2012). In another study, Williams et al. (2007) discovered that tourists who attended weekly church gatherings tended to report higher rates of experiencing the presence of God during cathedral visits than tourists who only rarely or occasionally attended church gatherings.

Setting

Factors related to tourist settings, such as *being in nature*, may influence spiritual outcomes. Whether it be hills and lakes (Sharpley & Jepson 2011; Jepson & Sharpley 2015), mountains (Andriotis 2009; Huang et al. 2020), the seaside (Jarratt & Sharpley 2017), the ocean (Jirásek & Hurych 2019), a rainforest (Bidder 2018), or the desert (Moufakkir & Selmi 2018), *being in nature* has been found to be conducive to spiritual outcomes. For example, as one participant in Andriotis' (2009) study explained, "Sitting on the wall overlooking the forest and the sea beyond, gazing at distant peaks and trekking to the summit, offer me a glimpse of the sacred. I feel as though I'm connecting with God" (p. 77; see also Chapter 11 this volume).

Being away from home and being immersed in a different environment or place is also conducive to spiritual outcomes. For example, leaving urban areas and travelling to other locations may allow people to escape into a simpler way of living, which can produce or trigger spiritual experiences (Jepson & Sharpley 2015). For outdoor adventure tourists, the isolated and remote location of a Field Centre in Borneo, along with the absence of telephones and internet, contributed to the participants being present in the moment and thus encountering the timeless dimension of spiritual experiences (Bidder 2018). Jarratt and Sharpley (2017) discovered that seaside visitors felt as if they were a part of something bigger than what they experienced in their everyday life. At Mount Athos, pilgrims escaped the distractions, cares, and pace of their everyday lives in order to receive spiritual sustenance (Andriotis 2009), and pilgrimage trips from a Greek community in California to a Greek Orthodox monastery in Arizona helped participants to leave behind their everyday concerns and focus on doing spiritual work (Klimova 2011). Similarly, for retreat tourism participants in New Zealand, the retreat centers offered a place to be away from the city, work, and home as well as distractions such as technology, in order to participate in a new way of life that included spiritual practices and transformational learning (Bone 2013).

Spiritual outcomes may be more readily found in *places* that are labeled or considered as being sacred or set apart from the mundane world. Indeed, descriptions of tourist's experiences in England's Lake District suggest an interrelationship between place attachment and spiritual experiences that are dependent upon or enhanced by a sense of place (Jepson & Sharpley 2015). Elsewhere, Preston (1992) found that Mount Athos exhibits a "spiritual magnetism" that draws visitors to it. Nearly half of the participants in a study of visitors at Apostle Islands National Seashore identified sacred sites as spiritual places (Salk, Schneider & McAvoy 2010). Likewise, tourists driving ATVs in the Australian desert stated that they discovered a spiritual dimension to their activities due to the journey through desert-like areas being perceived as a pilgrimage to a sacred space (Narayanan & Macbeth 2009).

Tourism activity factors

Spiritual outcomes derive from *diverse tourism activities* which range from the more traditional—pilgrimages to the Holy Land (Belhassen et al. 2008), visits to Greek Orthodox monasteries (Andriotis 2009; Klimova 2011) and spiritual retreat centers (Voigt et al. 2010, 2011; Schedneck Chapter 14 this volume)—to the novel—independent leisure travels (Little & Schmidt 2006), rural tourism activities (Sharpley & Jepson 2011; Jepson & Sharpley 2015); seaside tourism (Jarratt & Sharpley 2017); tourism at desert camps (Moufakkir & Selmi 2017), four-wheel drive tourism in the desert (Narayanan & Macbeth 2011), sailing tourism (Jirásek & Hurych 2019), outdoor adventure tourism (Bidder 2018), and dark tourism (Zheng, Zhang, Qui, Guo & Zhang 2020). Thus, religious and spiritual outcomes are not limited to one type of religious or spiritually oriented tourism activity.

Tourism activities with *challenge* can facilitate spiritual outcomes (Sharpley & Jepson 2011). The physical effort of climbing mountains combined with the scenic view from the mountain-top made the tourist experience explicitly spiritual for some rural tourism participants (Jepson & Sharpley 2015). Although there were variations among participants in this study, the challenge of physical activity and physical achievement led to spiritual fulfillment and spiritual feelings. For outdoor adventure, tourism participants in Borneo, overcoming physical, emotional, and mental challenges, including long and difficult treks during hot and humid weather conditions, was one dimension of their spiritual experience (Bidder 2018).

For many tourists, spiritual outcomes can stem from tourism activities that encourage *silence and quietness*, as these aesthetic attributes can lead to introspection and therefore spiritual experiences (della Dora 2012). This silence and quietness has been found to be helpful not only for visitors to monasteries (della Dora) and retreat centers (Bone 2013) but also in natural settings especially combined with physical activity (Jepson & Sharpley 2015).

In some cases, a balance of both solitude and interactive group *experiences* was helpful to spirituality (Sharpley & Jepson 2011). The importance of others and inspiration from others were also spiritual themes in a study of travelers to Buddhist mountain destinations (Huang et al. 2020). Community was an important theme in research on New Zealand retreat tourism. Community included friendship, group safety, a sense of "being at home," a sense of belonging, caring for others, spiritual companionship, unity of goal and camaraderie, which all together were viewed as fostering the spiritual dimension of interconnectedness (Bone 2013).

While *facilitation* may play an important role in spiritual outcomes for outdoor activities that occur in a group or as part of a program (Heintzman 2016a), there is less research on the facilitation of spiritual outcomes within the tourism context. Parsons, Houge Mackenzie and

Filep (2019) discovered that spiritual tourism guides in addition to facilitating in ways common to all tourism such as providing access to sites, encounters in and outside of the travel group, understanding, empathy and self-development, also facilitated a five-stage chronological process more directly related to self-development and spiritual outcomes: preparation, enclave development, mentoring, reflection, and integration of the spiritual lessons learned. This process began before the travel and continued after the travel.

Spiritual outcomes

One spiritual outcome of tourism documented by many studies, is *spiritual/religious experience* which is characterized by affective dimensions, cognitive processes, feelings of transcendence, and a high level of emotional intensity (e.g., Little & Schmidt 2006; Belhassen et al. 2008; Narayanan & Macbeth 2009; Andriotis 2009; Klimova 2011; Sharpley & Jepson 2011; della Dora 2012; Jepson & Sharpley 2015; Jarratt & Sharpley 2017; Bidder 2018; Moufakkir & Selmi 2018). For example, the spiritual experience outcomes of independent leisure travelers (Little & Schmidt 2006) and Christian pilgrims to the Holy Land (Belhassen et al. 2008) were described earlier in this chapter when leisure as psychological/subjective experience was connected to tourism. Another example is provided by Bidder (2018) who found that outdoor adventure tourism settings contributed significantly to spiritual experiences of a new sense of meaning, timelessness, overcoming challenges, connectedness, and ineffability. Haluza-Delay (2000) criticized these types of studies that focus on pleasant emotional states and urged investigation of whether these experiences lead to life transformation.

While religious and spiritual tourism can lead to the transformation of the quality of life of people (Reisinger 2013; Dillette, Douglas & Andrzejewski 2019), there has been little research in terms of whether these experiences lead to life-long transformation and *spiritual well-being*, which can be defined as

> A high level of faith, hope, and commitment in relation to a well-defined worldview or belief system that provides a sense of meaning and purpose to existence in general, and that offers an ethical path to personal fulfillment which includes connectedness with self, others, and a higher power or larger reality.
>
> *(Hawks 1994: 6)*

Although little has been researched in terms of tourism and the concept of spiritual well-being, one study of note is Reis' (2007) analysis of Brazilian pilgrims on the Way of St. James of Compostela. In this study, Reis discovered that pilgrims exhibited four key dimensions of spiritual well-being: transcendence, a life of significance, a community of shared values and support, and intrinsic values (Reis 2007). Sailing tourism has been found to be associated with five factors of spiritual health: relationship to oneself, relationship with others, relationship with nature, meaning in life, and transcendence (Jirásek & Hurych 2019). A study of New Zealand retreat tourism noted that participating in a retreat contributed to the participant's spiritual well-being (Bone 2013). Heintzman (1999, 2013b; Heintzman & Mannell 1999) investigated the relationships between several leisure activities, including travel and tourism activities, and spiritual well-being. Heintzman found that while participating in most travel and tourism activities, such as travelling within foreign countries, going on boat cruises, or visiting resorts, did not increase behavioral or subjective spiritual well-being, traveling to religious and spiritual retreats did increase these types of well-bring. Interestingly, participation in adventure trekking had a significant negative correlation with both behavioral and subjective well-being possibly because (1) adventure trekking is significantly correlated with competence-mastery leisure motivations, which makes it difficult to focus on spirituality (Heintzman 2013b); (2) adventure recreationists tend to view nature as something to be conquered rather that something in which to be immersed and seek a relationship with (Morgan, 1994); and (3) adventure trekkers seem to be akin to an "explorer tourist" who "prefers adventure travel, exploring out of the way places and enjoys challenges involved in getting there" (Yiannakis & Gibson 1992: 291). As such, they seek adventure, newness, and challenge, whereas a "seeker tourist," who engages in more passive types of recreational activities in nature, is "a seeker of spiritual and/or personal knowledge to better understand self and meaning in life" (Yiannakis & Gibson 1992: 291).

Another spiritual outcome is *leisure-spiritual coping*, which refers to the ways in which people seek out and receive help in the context of their leisure from spiritual resources (e.g., higher power, spiritual practices, and faith community) during periods of life stress (Heintzman 2008). For example, many people travel to religious, spiritual, New Age, or secular retreat centers to deal with and overcome negative life events, such as death of a loved one, a divorce, or a serious illness (Voigt et al. 2010). Another example is desire for personal transformation by visitors to Mount Athos. This desire was the case with one middle-aged man who stated "…my main objective is to strengthen my faith in a way that will enable me to continue my life back home with new energy and a feeling of purpose" (as quoted in Andriotis 2009: 74).

Conclusion

Over 20 years ago, Vukonić made preliminary connections between leisure as free time, spirituality, religion, and tourism. This chapter has documented some of the complexity of these relationships both in terms of the ways that leisure is conceptualized beyond the notion of free time and the various factors that link leisure travel with religious and spiritual outcomes. Many factors such as antecedent conditions, setting factors, and tourism activity components influence whether there are spiritual or religious outcomes of leisure travel and whether these outcomes are short-term or long-term. The components of this framework have some overlap with the activity, setting, experience, and benefit components of the ASEB framework of visitor experience (Beeho & Prentice 1997); however, unlike other frameworks (Beeho & Prentice 1997; Cheer, Belhassen & Kujawa 2017; Chhabra 2020), the one presented in this paper is based upon a synthesis of empirical research. Future research on the spiritual and religious outcomes of leisure travel and tourism may serve to strengthen and modify this framework so that we have a better understanding of the processes that link leisure travel with spiritual outcomes. While much research exists on the spiritual outcome of spiritual experience, further research is needed on how leisure travel may bring about spiritual well-being as well as assist with leisure-spiritual coping.

References

Anderson, L., & Littrell, M. (1995) 'Souvenir-purchase behavior of women tourists', *Annals of Tourism Research*, 22(2): 328–348.

Andriotis, K. (2009) 'Sacred site experience: A phenomenological study', *Annals of Tourism Research*, 36(1): 64–84.

Beeho, A.J., & Prentice, R.C. (1997) 'Conceptualizing the experiences of heritage tourists: A case study of New Lanark World Heritage Village', *Tourism Management*, 18(2): 75–87.

Belhassen, Y., Caton, K., & Steward, W.P. (2008) 'The search for authenticity in the pilgrim experience', *Annals of Tourism Research*, 35(3): 668–689.

Bidder, C. (2018) 'Outdoor adventure tourism: Exploring the spiritual dimension of wellness', *International Journal of Academic Research in Business and Social Sciences*, 8(16): 199–217.

Bond, N., Packer, J., & Ballantyne, R. (2015) 'Exploring visitor experiences, activities and benefits at three religious tourism sites', *International Journal of Tourism Research*, 17: 471–481.

Bone, K. (2013) 'Spiritual retreat tourism in New Zealand', *Tourism Recreation Research*, 39(2): 295–309.

Bowers, H., & Cheer, J.M. (2017) 'Yoga tourism: Commodification and western embracement of eastern spiritual practice', *Tourism Management Perspectives*, 24: 208–216.

Cheer, J.M., Belhassen, Y., & Kujawa, J. (2017) 'The search for spirituality in tourism: Toward a conceptual framework for spiritual tourism', *Tourism Management Perspectives*, 24: 252–256.

Chhabra, D. (2020) 'A conceptual paradigm to determine behavior of slow spiritual tourists', in L. Cantoni, S. De Ascaniis, & K. Elgin-Nijhuis (eds), *Proceedings of the Heritage, Tourism and Hospitality International Conference 2020* (pp. 115–124). Lugano, Italy: Università della Svizzera Italiana.

Cohen, E.H. (2006) 'Religious tourism as an education experience', in D.J. Timothy & D.H. Olsen (eds), *Tourism, Religion and Spiritual Journeys* (pp. 139–155). London and New York: Routledge.

Csikszentmihalyi, M. (1975) *Beyond Boredom and Anxiety: The Experience of Play in Work and Games*. San Francisco, CA: Jossey-Bass.

della Dora, V. (2012) 'Setting and blurring boundaries: Pilgrims, tourists, and landscape in Mount Athos and Meterora', *Annals of Tourism Research*, 39(2): 951–974.

Dillette, A.K., Douglas, A.C., & Andrzejewski, C. (2019) 'Yoga tourism–A catalyst for transformation?', *Annals of Leisure Research*, 22(1): 22–41.

Driver, B.L., & Tocher, S.R. (1970) 'Toward a behavioral interpretation of recreational engagements, with implications for planning', in B.L. Driver (ed), *Elements of Outdoor Recreation Planning* (pp. 9–31). Ann Arbor: The University of Michigan Press.

Dumazedier, J. (1967) *Toward a Society of Leisure*. New York, NY: The Free Press.

Dumazedier, J. (1974) *Sociology of Leisure*. Amsterdam: Elsevier.

Esfahani, M., Musa, G., & Khoo, S. (2014) 'The influence of spirituality and physical activity level on responsible behaviour and mountaineering satisfaction on Mount Kinabalu, Borneo', *Current Issues in Tourism*, 20(11): 1162–1185.

Freysinger, V.J., & Flannery, D. (1992) 'Women's leisure: Affiliation, self-determination, empowerment and resistance?', *Loisir et société/Society and Leisure*, 15(1): 303–321.

Freysinger, V.J., & Kelly, J.R. (2004) *21st Century Leisure: Current Issues*. State College, PA: Venture.

Gibson, H., Ashton-Shaeffer, C., Green, J., & Autry, C. (2003) 'Leisure in the lives of retirement-aged women: Conversations about leisure and life', *Leisure/Loisir*, 28(3–4): 203–230.

Gill, C., Packer, J., & Ballantyne, R. (2018) 'Exploring the restorative benefits of spiritual retreats: The case of clergy retreats in Australia', *Tourism Recreation Research*, 43(2): 235–249.

Graef, R., Csikszentmihalyi, M., & McManama Gianinno, S. (1983) 'Measuring intrinsic motivation in everyday life'. *Leisure Studies*, 2(2): 155–168.

Gutic M, J., Caie, E., & Clegg, A. (2010) 'In search of heterotopia? Visitor motivations to an English Cathedral', *International Journal of Tourism Research*, 12(6): 750–760.

Haluza-Delay, R. (2000) 'Green fire and religious spirit', *The Journal of Experiential Education*, 23: 143–149.

Haq, F., & Jackson, J. (2006) 'Exploring consumer segments and typologies of relevance to spiritual tourism', Paper presented at the Australia and New Zealand Academy of Marketing Conference. Brisbane, Australia: Central Queensland University.

Hawks, S. (1994) 'Spiritual health: Definition and theory', *Wellness Perspectives*, 10: 3–13.

Heimtun, B., & Morgan, N. (2012) 'Proposing paradigm peace: Mixed methods in feminist tourism research', *Tourist Studies*, 12(3), 287–304.

Heintzman, P. (1998) Book Review of: B. Vukonić (1996). *Tourism and Religion*. Tarrytown, NY: Elsevier. (Trans. S. Matešić). In *Annals of Tourism Research*, 25(2), 531–535.

Heintzman, P. (1999) *Leisure and Spiritual Well-Being: A Social Scientific Exploration* (Unpublished doctoral dissertation). University of Waterloo, Waterloo, ON.

Heintzman, P. (2002) 'Leisure and spirituality: The re-emergence of a historical relationship', *Parks and Recreation Canada*, 60(1): 30–31.

Heintzman, P. (2008) 'Leisure-spiritual coping: A model for therapeutic recreation and leisure services', *Therapeutic Recreation Journal*, 42(1): 56–73.

Heintzman, P. (2012) 'Spiritual outcomes of wilderness experience: A synthesis of social science research', *Park Science*, 28(3): 89–92, 102.

Heintzman, P. (2013a) 'Defining leisure', in R. McCarville & K. MacKay (eds), *Leisure for Canadians* (2nd ed., pp. 3–14). State College, PA: Venture.
Heintzman, P. (2013b) 'Retreat tourism as a form of transformational tourism', in Y. Reisinger (ed), *Transformational Tourism* (pp. 68–81). Cambridge, MA: CABI.
Heintzman, P. (2013c) 'Spiritual outcomes of park experience: A synthesis of social science research', *The George Wright Forum*, 30(3): 273–279.
Heintzman, P. (2016a) 'Outdoor studies and spirituality', in H. Prince, K. Henderson, & B. Humberstone (eds), *International Handbook of Outdoor Studies* (pp. 388–397). New York, NY: Routledge.
Heintzman, P. (2016b) 'Religion, spirituality and leisure', in G. Walker, D. Scott, & M. Stodolska (eds), *Leisure Matters: The State and Future of Leisure Studies* (pp. 67–75). State College, PA: Venture.
Heintzman, P., & Mannell, R. (1999) 'Leisure style and spiritual well-being', in W. Stewart & D. Samdahl (eds), *Abstracts from the 1999 Symposium on Leisure Research* (p. 68). Ashburn, VA: National Recreation and Park Association.
Henderson, K.A. (1991) 'The contribution of feminism to an understanding of leisure constraints', *Journal of Leisure Research*, 23(4): 363–377.
Henderson, K.A., Bialeschki, D.M., Shaw, S.M., & Freysinger, V.J. (1996) *Both Gaps and Gains: Feminist Perspectives on Women's Leisure*. State College, PA: Venture.
Huang, K., Pearce, P., Guo, Q., & Shen, S. (2020) 'Visitors' spiritual values and relevant influencing factors in religious tourism destinations', *International Journal of Tourism Research*, 22(3): 314–324.
Jackson, R.H., & Hudman, L. (1995) 'Pilgrimage tourism and English cathedrals: The role of religion in travel', *The Tourist Review*, 50(4): 40–48.
James, W. (1890) *The Principles of Psychology*. New York, NY: Henry Holt.
Jarratt, D., & Sharpley, R. (2017) 'Tourists at the seaside: Exploring the spiritual dimension', *Tourism Studies*, 17(4): 349–368.
Jepson, D., & Sharpley, R. (2015) 'More than sense of place? Exploring the emotional dimension of rural tourism experiences', *Journal of Sustainable Tourism*, 23(8–9): 1157–1178.
Jirásek, I., & Hurych, E. (2019) 'Experience of long-term transoceanic sailing: Cape Horn example', *Journal of Outdoor Recreation and Tourism*, 28, Article 100221.
Kaplan, M. (1974) 'New concepts of leisure today', in J.F. Murphy (ed), *Concepts of Leisure: Philosophical Implications* (pp. 229–236). Englewood Cliffs, NJ: Prentice-Hall.
Kelner, S. (2012) *Tours that Bind: Diaspora, Pilgrimage, and Israeli Birthright Tourism*. New York: New York University Press.
Klimova, J. (2011) 'Pilgrimages of Russian Orthodox Christians to the Greek Orthodox monastery in Arizona', *Tourism*, 59(3): 305–318.
Kraus, R. (1984) *Recreation and Leisure in Modern Society*. 3rd ed. Glenview, IL: Foresman.
Kraus, R. (2001) *Recreation and Leisure in Modern Society*, 6th ed. Toronto, ON: Jones & Bartlett.
Little, D.E., & Schmidt, C. (2006) 'Self, wonder and God! The spiritual dimensions of travel experiences', *Tourism*, 54(2): 107–116.
MacCannell, D. (1973) 'Staged authenticity: Arrangements of social space in tourist settings', *The American Journal of Sociology*, 79: 589–603.
MacCannell, D. (1976/2013) *The Tourist: A New Theory of the Leisure Class*. New York, NY: Schocken Books.
Mannell, R.C., & Iso-Ahola, S.E. (1987) 'Psychological nature of leisure and tourism experience', *Annals of Tourism Research*, 14(3): 314–331.
Maslow, A.H. (1968) *Toward a Psychology of Being* (2nd ed). Toronto, ON: Van Nos Reinhold.
McClinchey, K.A. (2017) 'Paris, je t'aime: (Post) feminist identities, emotional geographies and women's travel narratives of Paris', *Travel and Tourism Research Association: Advancing Tourism Research Globally*. 7. https://scholarworks.umass.edu/ttra/2017/Academic_Papers_Oral/7.
Moal-Ulvoas, G. (2014) 'Retired adults' motivations to travel: The influence of aging studied through the theory of gerotranscendence', *Decisions Marketing*, 76: 29–45.
Moal-Ulvoas, G., & Taylor, V.A. (2014) 'The spiritual benefits of travel for senior tourists', *Journal of Consumer Behaviour*, 13(6): 453–462.
Morgan, G. (1994) 'The mythologies of outdoor and adventure recreation and the environmental ethos', *Pathways: The Ontario Journal of Outdoor Education*, 6(6): 11–16.
Moufakkir, O., & Selmi, N. (2018) 'Examining the spirituality of spiritual tourists: A Sahara desert experience', *Annals of Tourism Research*, 70: 108–119.
Murphy, J.F. (1974) *Concepts of Leisure: Philosophical Implications*. Englewood Cliffs, NJ: Prentice Hall.

Narayanan, Y., & Macbeth, J. (2009) 'Deep in the desert: Merging the desert and the spiritual through 4WD tourism', *Tourism Geographies*, 11(3): 369–389.

Norman, A. (2004) *Spiritual Tourism: Religion and Spirituality in Contemporary Travel.* (Unpublished B.A. Honours thesis). University of Sydney, Sydney, Australia.

Parsons, H., Houge Mackenzie, S., & Filep, S. (2019) 'Facilitating self-development: How tour guides broker spiritual tourist experiences', *Tourism Recreation Research*, 44(2): 141–152.

Patterson, I., Getz, D., & Gubb, K. (2016) 'The social world and event travel career of the serious yoga devotee', *Leisure Studies*, 35(3): 296–313.

Pieper, J. (1968) *Leisure: The Basis of Culture.* London: Faber & Faber.

Powell, R.B., Ramshaw, G.P., Ogletree, S.S., & Krafte, K.E. (2015) 'Can heritage resources highlight changes to the natural environment caused by climate change? Evidence from the Antarctic tourism experience', *Journal of Heritage Tourism*, 11(1): 71–87.

Preston, J. (1992) 'Spiritual magnetism: an organizing principle for the study of pilgrimage,' in A. Morinis (ed.) *Sacred Journeys: The Anthropology of Pilgrimage* (pp. 31–46). Westport, CT: Greenwood Press.

Reis, G.G. (2007) 'Spiritual well-being and tourism: Analysis of reports from pilgrims of the Way of St James of Compostela', *Turismo - Visão e Ação*, 9(2): 233–248.

Reisinger, Y. (ed.) (2013) *Transformational Tourism: Tourist Perspectives.* Wallingford, UK: CABI.

Rountree, K. (2002) 'Goddess pilgrims as tourists: Inscribing the body through sacred travel', *Sociology of Religion*, 63(4): 475–496.

Salk, R., Schneider, I.E., & McAvoy, L.H. (2010) 'Perspectives of sacred sites on Lake Superior: The case of the Apostle Islands', *Tourism in Marine Environments*, 6(2/3): 89–99.

Sharpley, R., & Jepson, D. (2011) 'Rural tourism: A spiritual experience', *Annals of Tourism Research*, 38(1): 52–71.

Sharpley, R., & Sundaram, P. (2005) 'Tourism: A sacred journey? The case of ashram tourism, India', *International Journal of Tourism Research*, 7(3): 161–171.

Smith, M. (2003) 'Holistic holidays: Tourism and the reconciliation of body, mind and spirit', *Tourism Recreation Research*, 28(1): 103–108.

Smith, M., & Kelly, C. (2006) 'Wellness tourism', *Tourism Recreation Research*, 21(1): 1–4.

Stebbins, R.A. (1997) 'Casual leisure: A conceptual statement', *Leisure Studies*, 16(1): 17–25.

Stebbins, R.A. (1999) 'Serious leisure', in E.L. Jackson & T.L. Burton (eds), *Leisure Studies: Prospects for the Twenty-First Century* (pp. 69–79). State College, PA: Venture.

Sthapit, E., & Coudounaris, D.N. (2017) 'Memorable tourism experiences: Antecedents and outcomes', *Scandinavian Journal of Hospitality and Tourism*, 18(1): 72–94.

Veblen, T. (1899/1953) *The Theory of the Leisure Class.* New York, NY: The New American Library.

Voigt, C., Brown, G., & Howat, G. (2011) 'Wellness tourists: In search of transformation', *Tourism Review*, 66(1/2): 16–30.

Voigt, C., Howat, G., & Brown, G. (2010) 'Hedonic and eudaimonic experiences among wellness tourists: An exploratory enquiry', *Annals of Leisure Research*, 13(3): 541–562.

Vukonić, B. (1996) *Tourism and Religion.* Oxford: Pergamon.

Walker, G.J., Kleiber, D.A., & Mannell, R.C. (2019) *A Social Psychology of Leisure.* Champaign, IL: Sagamore-Venture.

Weidenfeld, A., & Ron, A. (2008) 'Religious needs in the tourism industry', *Anatolia: An International Journal of Tourism and Hospitality Research*, 19(2): 357–361.

Williams E., Francis L.J., Robbins, M., & Annis J. (2007) 'Visitor experiences of St Davids Catherdal: The two worlds of pilgrims and secular tourists', *Rural* Theology, 5(2): 111–123.

Yiannakis, A., & Gibson, H. (1992) 'Roles tourists play', *Annals of Tourism Research*, 19: 287–303.

Zheng, C., Zhang, J., Qui, M., Guo, Y., & Zhang, H. (2020) 'From mixed emotional experience to spiritual meaning: Learning in dark places', *Tourism Geographies*, 22(1): 105–126.

5

RELIGION, SPIRITUALITY, AND WELLNESS TOURISM

Melanie Kay Smith

Introduction

Spirituality and wellness are arguably inextricably connected. Wellness can be defined as the path to achieving well-being (Nahrstedt, 2008), a path that includes physical, mental, and spiritual health, self-responsibility, social harmony, environmental sensitivity, intellectual development, emotional well-being, and occupational satisfaction (Smith & Puczkó, 2013), whereas spirituality "includes experiencing oneness with nature and beauty and a sense of connectedness with self, others and a higher power or larger reality, concern for and commitment to something greater than self" (Hawks, 1994: 1). Moal-Ulvoas and Taylor (2014: 454) define spirituality as being "concerned with understanding reality in a broad sense and includes understanding one's 'self', other human beings or alterity, and the sacred". Some authors have argued that spirituality is at the core of wellness (Myers, Sweeney & Witmer, 2000; Steiner & Reisinger, 2006), and that travel, whether in the form of spiritual tourism or of wellness tourism based body-mind-spirit activities are closely connected.

However, questions remain regarding the separation of spirituality from organized or institutional religion and how far spiritual wellness is connected to secular wellness practices which focus more on self-development. While the seminal work by Timothy and Olsen (2006) explored the relationship between tourism, religion, and spirituality, paving the way for a robust scholarship on the subject, the meaning of spirituality in the context of wellness tourism has not been defined very clearly. Therefore, the purpose of this chapter is to examine the relationship between religion and spirituality in the context of wellness tourism, including those activities that offer tourists greater meaning in their lives and better connections to the world around them. Often, this takes place in holistic meditation or yoga retreat centres, but the importance of landscape and nature in the development of spiritual well-being is also explored.

Spiritual tourism: a personal quest beyond religion?

The quest for spirituality is in some ways based on the increasing secularization of society. This is partly a result of the rise of consumerism as a global ethos, which has led to the growth of consumer-based forms of spirituality outside of religion (Gauthier et al. 2013). However, the relationship between religion and spirituality is still being debated. Kujawa (2017), for example, argues that the quest for spirituality represents dissatisfaction with or alienation from the dogmatism of institutionalized religion, which is why many people travel in search of or to regain or rediscover spiritual traditions in order to fill a spiritual vacuum that religion cannot fill. Kujawa further suggests that it is a person's internal experience, rather than the external expression of faith, which is the driving force for spiritual tourism. Kato and Progano (2017) also suggest that religiousness and spirituality have not declined, but rather that the way in which religion is practised has changed in the modern world. As the influence of traditional religious institutions has declined, spirituality has been 'de-regulated' from religious institutions, with individuals having the freedom to create their own spiritual lives and choose which spiritual traditions to practice (Kujawa, 2017; Gauthier et al., 2013).

Stausberg (2014) suggests that any definition of spirituality belongs both within and beyond religion. Norman (2012) argues that while religion describes a shared system of beliefs and participation, spirituality is connected to a personal or individual focus. This focus, he suggests, leads spiritual tourists to engage in a "variety of practices or behaviours that are self-consciously seen as contributory to meaning and identity, and/or beneficial for the individual's health and wellbeing" (p. 21). Within these practices or behaviours, Norman lists five categories of experiences that spiritual tourists seek when travelling:

- **Healing**: which includes the search spiritual, emotional, and psychological healing;
- **Experiment**: where spiritual tourists experiment with different cultures and religious philosophies or immerse themselves in different religious traditions and practices;
- **Quest**: travel to search for meaning and personal discovery;
- **Retreat**: where people seek socio-geographical escape, meaning that people seek experiences away from their home and home culture and use secular attractions for spiritual purposes; and
- **Collective**: this category includes people who travel to "spiritual hubs" for spiritual holidays.

Wilson, McIntosh and Zahra (2013), in their phenomenological analysis of the meaning of spirituality, suggest that spirituality concerns a human being's individual search for meaning in life; that people engage in activities that result in experiencing transcendence, connection, and/or deep personal meaning which also promotes growth; and that people seek harmonious relationships or 'oneness' with the 'self', the 'other' (including other people, animals, the earth, and nature) and/or God/Higher Power. Although the individual and personal quest is given emphasis in their understanding of spirituality, connections with nature, people around them, and what lies beyond them are also clearly important for the spiritual tourist. Likewise, Kujawa (2017), in describing spiritual tourism as a transformative experience that often takes place in a collective setting with like-minded people, suggests that spiritual tourism involves experiences related to transcendence, connectivity, and transformation. However, Kujawa also refers to the "peculiarity of spiritual tourism that external travel is needed to achieve an internal experience of spiritual meaning" (p. 194).

However, some authors have defined spiritual tourism even more broadly. For example, Kato and Progano (2017: 245) state that

> Religious tourism is defined as travelling for spiritual purposes usually carried out according to the interpretation of sanctioned religious leaders and institutions and to institutionally sanctioned destinations, whereas spiritual tourism is characterized as a more subjective and individual travel for spiritual betterment and self-discovery and may include secular sites such as war memorials, natural landscapes or places related to celebrities.

Spirituality and personal wellness journeys

Travel has been viewed as a way of enhancing and transforming individual self-development and even changing world views. For example, Reisinger (2015: 5) states that "Travel can offer physical, psychological, cognitive affective and spiritual experiences that can change one's assumptions, expectations, world views and fundamental structures of the self". There are strong connections to personal well-being. Indeed, Norman and Pokorny (2017: 203) refer to spiritual tourism as a practice of subjective well-being work, arguing that it is a "reflexive well-being intervention" that leads to spiritual betterment and can include both religious and non-religious activities. They argue that spiritual tourists are being proactive in addressing a problem in their lives.

Smith and Puczkó (2012) highlighted the importance of spirituality in the relationship between tourism and quality of life by adding a 'spiritual well-being' category to the domains of well-being that are affected by tourism. This means that a person might adhere to

> A high level of faith, hope, and commitment in relation to a well-defined worldview or belief system that provides a sense of meaning and purpose to existence in general, and that offers an ethical path to personal fulfillment which includes connectedness with 'self', others, and a higher power or larger reality.
>
> *(Hawks, 1994: 6)*

In their spectrum of well-being and types of tourism, Smith and Diekmann (2017) categorised 'retreat tourism' and 'spiritual pilgrimage' as types of tourism that contribute to long-term, eudaimonic well-being and enhance existential authenticity. Religious and spiritual retreats have become more commonplace in the tourism landscape in recent years (Sharpley & Sundaram, 2005; Heintzman, 2013; Chapter 14, this volume). Smith and Diekmann (2017) also stated that tourists are more likely to move towards a sense of self-transcendence through forms of spiritual tourism of extended duration, such as through engaging in long pilgrimages or ashram-stays. Voigt, Howat, and Brown (2011) also proposed that in the context of wellness tourism, those eudaimonic experiences are more likely to be gained from spiritual retreats, whereas more hedonic well-being experiences might take place in, for example, a beauty spa.

Bandyopadhyaya and Nairb (2019) suggest that spiritual tourism is prominent within the rapidly expanding wellness industry. For example, they describe how Western tourists embrace spirituality in India because it represents a panacea to hectic modern lifestyles and helps them to attain self-fulfilment, inner harmony, and bliss. Cheer, Belhassen, and Kujawa (2017) describe spiritual travel experiences as enabling people to renew connections with others, life in general, and most importantly with themselves. Smith (2003: 104) described spiritual tourists as those searching for "an authentic sense of self" in which "[t]he tourist's own self thus becomes the object of the tourist gaze, rather than any external attractions or activities". Smith also suggests that travel becomes transformative when it reveals a person's true or authentic 'self'. Although many spiritual tourists tend to be young backpackers who are attempting to 'find themselves' and explore the meaning of life through travel, some research has suggested that tourists tend to become more spiritual as they get older. For example, Moal-Ulvoas and Taylor (2014) have shown that spirituality can motivate older adults to travel and can result in knowing the 'self' better, giving greater meaning to their lives, gaining a clearer understanding of others and connecting more closely to nature.

Spiritual and wellness destinations and experiences

The role of the destinations or locations in the creation of spiritual and wellness experiences plays a somewhat ambivalent role in spiritual and wellness tourism, as the internal journey is often given greater emphasis than any external setting. However, Norman and Pokorny (2017) discuss how spiritual tourism activities tend to take place in spaces that are 'alternate' to the spaces of the tourists' everyday life, such as spaces that are quiet, slow, rural, or religious. Kato and Progano (2017; see Howard, 2012) suggest that pilgrimage in the form of walking pilgrimages is a type of slow tourism that utilizes the healing qualities of the natural environment and affords moments of spiritual engagement.

Although many retreat visitors prefer to stay within retreats themselves rather than exploring the surrounding area (Kelly, 2010), it is nevertheless important that any retreat is located within a peaceful and natural landscape (Smith & Kelly, 2006). At the same time, many spiritual seekers tend to visit so-called sacred landscapes and destinations, which represent "relations between bodily existence, felt practice and faith in something immanent but not manifest" (Dewsbury & Cloke, 2009: 695). Tourists may be drawn to a country that offers different spiritual traditions from their own (e.g. Western tourists visiting Indian ashrams, Korean temples, or Thai meditation centres). Heelas and Woodhead (2005: 150) metaphorically refer to this as "tectonic shifts in the sacred landscape" to describe the shift away from "denominational religions" to the adoption of different forms of spirituality Jiang, Ryan, and Zhang (2018) suggest that rising interest in meditation retreats represent such a shift which retreats promote ancillary components such as beautiful natural landscapes, physical and mental well-being, and temple foods in addition to the religious and spiritual practices. There are also increasing overlaps between more hedonic (e.g. spas and beauty) and more eudaimonic forms of wellness (e.g. self-development and spirituality). For example, Ashton (2018) describes that while the spiritual retreat is a relatively new concept in Thailand, several of the retreats combine experiences that are often offered at specific types of retreats, such as experiences catering to spa and beauty, self-awareness, inner peacefulness or meditation, mind and body wellness, and spiritual renewal.

Holistic and spiritual retreats

Several authors have argued that spiritual retreat tourism is part of wellness tourism (Voigt et al., 2010; Heintzman, 2013; Ashton, 2018). Voigt et al. (2010) separated wellness tourism into three types of retreats: spa and beauty, lifestyle resort, and spiritual. Spiritual retreat tourism can be religious or non-religious in nature (Ashton, 2018). It tends to offer accommodation for people seeking peace, quiet and spiritual nurture, and sometimes combines religion and wellness in some form (Heintzman, 2013). Bone (2013) describes spiritual retreat tourism as a form of contemporary pilgrimage and suggests that the spiritual experience of being in a retreat is pervasive, regardless of whether participants gain the most benefits from escapism, forming part of a community, or being in a therapeutic landscape. In this vein, Kelly and Smith (2017) produced a typology of retreats:

- *Religious retreats*: spaces owned by or run by religious communities, e.g. temples and monasteries;
- *Spiritual retreats*: spiritually informed spaces (e.g. ashrams and meditation centres);
- *Yoga retreats*: retreats focusing specifically or partially on yoga;
- *Health retreats*: offering lifestyle improvement workshops (e.g. nutrition and weight loss);
- *Fitness retreats*: offering scheduled fitness classes and courses (e.g. bootcamps and outdoor adventure sports);
- *Mind-based retreats*: focusing on relaxation, reflection, mindfulness, and meditation;
- *Body-Mind-Spirit*: aiming to balance the physical, mental, emotional, and spiritual through a programme of carefully selected workshops (e.g. holistic retreat centres); and
- *Miscellaneous*: place-based retreats where special kinds of landscapes play a role (e.g. silent retreats in deserts and eco-retreats in jungles).

Spirituality is the key element in many of these retreats, but it is by no means central. As Kelly (2012) notes, the rest/relaxation factor is the main motivator why people visit these types of retreats, followed to a lesser extent by social and spiritual reasons.

Many tourists seeking spiritual retreat experiences might search online sites such as The Retreat Company (https://www.theretreatcompany.com/) which allows potential spiritual tourists to search retreats related to 'Spiritual Awareness'. Within this category, approximately 50 different options are available located in a wide range of countries around the world. The spectrum of activities is rather broad, ranging from meditation and mindfulness retreats in Bhutan to 'healing and adventure' retreats in Peru; 'dolphin therapy' in Hawaii; 'grey whale' experiences in California; and acoustic sound retreats in Finnish natural landscapes. This particular retreat company makes very little reference to religion or faith communities of any kind, preferring to use the word 'spiritual'. Another company, Retreat Finder (https://www.retreatfinder.com/), offers over 200 retreats, which are organized by religious grouping, including Christian, Catholic, Protestant, Quaker, Buddhist, Zen, Hindu, interfaith, and 'open to all'. Another popular website is Book Meditation (https://www.bookmediation.com/home), which lists more than 2,400 meditation retreats, of which 241 are labelled as 'spiritual retreats'. Most of these retreats are based on meditation and mindfulness. According to the Book Meditation website,

> Spiritual retreats grasp their essence from ancient religions. Whether the core teachings are Buddhist, Christian, Hindu, or Tibetan, one thing is certain. A spiritual holiday will help you regain presence and peace in your life. You will be given the opportunity to deepen your spiritual focus and understand yourself and your loved ones better.
>
> *(n.p.)*

Yoga tourism and retreats

Heelas and Woodward (2005) have noted the growing range of holistic and physical practices or activities like yoga to help facilitate the convergence of the personal and the spiritual path. Smith and Sziva (2017) describe yoga as being comprised of a series of physical, mental, and spiritual practices. While yoga initially focuses on the physical body through a series of movements or postures, mental and spiritual transformation can follow with regular practice. This transformation can be intensified if people decide to pursue their interest in yoga through retreat holidays. Ponder and Holladay (2013) therefore suggest that yoga tourism is a form of transformational tourism which can lead to self-actualization and spiritual renewal. Reisinger (2013) concurred and included yoga in a list of practices that leads to transformational experience that helps people find their authentic selves and enhance their psychological, emotional and spiritual well-being. Smith and Sziva (2017) note that for those who participate in yoga retreats, having a spiritual experience is not as important as other motivations (e.g. stress relief, self-understanding, and health) but also note that those participants who equated yoga with spirituality are more likely to travel to yoga retreats. Garrett (2001) also suggests that some yoga practitioners might call themselves spiritual, whereas other practitioners say they are 'working on myself'. Park, Braun, and Siegel (2015) concur, arguing that yoga practitioners are reported to be more spiritual, but not necessarily more religious. Bowers and Cheer (2017) undertook a more extensive study which analysed the extent to which yoga tourism is connected to spirituality. They suggest that individuals may well gain wellness benefits or have spiritual experiences while practicing yoga, but that it very much depends on their personal and individual perspectives and experiences. They also view yoga tourism and spiritual travel as being innately connected in that they both focus on inquiry and self-discovery. However, the extent to which yoga tourism can be considered spiritual is still debatable, partly because of increasing commodification and the fact that self-improvement and career enhancement as motivations overshadow spiritual ones for many participants.

Many of the above-mentioned websites list yoga activities at some spiritual or meditation retreats. However, because of their popularity, specific yoga-based retreats are usually given their own category. For example, The Retreat Company lists yoga retreats separately from spiritual awareness retreats. One company, Book Yoga Retreats (https://www.bookyogaretreats.com/) describes itself as 'the world's largest yoga site' and includes a separate category for 'Spirituality & Chanting'. Presently there are 3,962 'Spiritual Yoga Retreats' in addition to almost 5,000 'Yoga Meditation Retreats'. Interestingly, there are only six retreats that are described as 'Christian Yoga Retreats', and most yoga-based retreats do not identify with a faith tradition.

Temples and monasteries

Yiannakis and Gibson (1992) suggest a special category of tourist—'seekers'—who are on a spiritual or personal quest to understand themselves and their lives better. These tourists travel to many places in this quest for meaning, including monasteries and temple stays where guests participate in monastic life, ceremonies, meditation, and cooperative activities. Ouellette et al.'s (2005) study of visitors in a monastery suggests that the expected outcomes such as deepening faith, devoting time to prayer and feeling closer to the God are strong motivators for staying in monasteries and temples. Jiang, Ryan, and Zhang (2018) estimate that there are over 100 Chinese temples that offer meditation programmes. They suggest that even though many of the tourists who visit these retreats are not motivated by religious or spiritual reasons, they often have some sort of 'sacred-spiritual' experience. The authors sub-divide experiences into 'Search for Meaning' (focusing on one's inner being) and 'Search for Escape' (creating connections and relationships with the world and others). The authors also argue that although aspects of these retreats, such as Zen meditation may lose some of their sacred or spiritual nature when presented and consumed as a mediated tourism product, they nevertheless form an integral part of the overall experience and the search for the real 'self' and self-realization. However, participants in temple stays and monasteries tend to have either a secular experience that is enhanced by the natural landscape and religious culture or a sacred experience through reaching a sense of the divine and having a clear view of the purpose of their lives.

(Spiritual) landscapes and nature

While spiritual seeking revolves around connecting to a higher being or power and searching for the meaning of 'self' and purpose of life, Moufakkir and Selmi (2018) also note that a connection to nature is another aspect of spirituality. Taylor (2010) suggested that in some ways the natural environment has become a spiritual substitute for the search for meaning in traditional religious and spiritual groups. de Botton (2003) argued that humans began to become attracted to sublime landscapes at a time when traditional belief in the God was declining, and that rural landscapes have acquired sacred meaning in contrast to the urban environments seemingly devoid of spiritual spaces. As such, there has been a rise in what Taylor (2010) describes as 'nature religion', which has been connected to paganism, animism, the New Age movement, the romanticization of the rural, deep ecology, and environmental sustainability (Timothy & Conover, 2006; Olsen, 2020; Chapter 12, this volume).

Vallés-Planells et al. (2014) suggest that natural landscape can contribute to people's experiences through enjoyment (e.g. recreation and aesthetics), personal fulfillment (e.g. education, inspiration, and spiritual benefits), health (e.g. escapism and calm) and social fulfillment (e.g. social relations, cultural heritage, and sense of place). Maller et al. (2009) similarly summarized the benefits of parks and protected areas for human health and well-being, which included the following benefits:

- *Physical*: settings for recreation, sport, and other leisure activities;
- *Mental*: restoration from fatigue, peace and solitude, artistic inspiration, and education;
- *Spiritual*: reflection and contemplation, feeling a sense of place, and connecting to something greater than oneself;
- *Social*: including couples, families, networks, and associations' recreational activities and events; and
- *Environmental*: preservations and conservation of ecosystems.

Vallés-Planells et al. (2014) argued that spirituality is perceived as an important part of personal fulfilment, whereas Maller et al. (2009) suggested that spirituality is connected to reflection, connection, and transcendence. Sharpley and Jepson (2011) concluded that spirituality was mainly connected to respondents' subjective experiences in terms of meaning, harmony, and connectedness, and that most transcendental experiences tend to take place in areas like mountains, lakes, and coastlines where there is solitude, quietness, and remoteness.

In their development of a Customer Well-being Index for visitor satisfaction at national parks, Lee et al. (2014) found that spiritual experiences were rated much higher than relaxation or socialising in terms of what constituted visitor satisfaction. Ram and Smith (2019) suggest that the benefits derived from tourist interactions with different types of natural landscapes can be divided into four distinct types: spiritual interaction, physical-emotional interaction, intellectual interaction, and aesthetic interaction. The authors found that spiritual and physical-emotional factors have the most impact in influencing tourists' revisit intentions. As such, the emotional and spiritual benefits of natural landscapes are more important for visitors than intellectual and aesthetic benefits. In their study, Ram and Smith found that seaside landscapes were rated superior in all four factors. However, the landscape that rated the highest in terms of spiritual interaction was the desert. This parallels research by Moufakkir and Selmi (2018), who found that visitors to deserts view the landscape as a space that affords them feelings of awe because of its immenseness, peace, silence, simplicity, and emptiness. Participants in their study made a metaphorical connection between the physical emptiness of the desert and their inner/spiritual emptiness. Paradoxically, this empty landscape filled their spiritual emptiness, bringing them a sense of comfort and security and, in some cases, closeness to the God.

Concluding thoughts

Quests for spiritual enlightenment or meaning in the context of wellness tend to be based on a search outside of institutionalized religion in a tourist's home country. This does not mean, however, that the tourist's quest is secular or devoid of religious meaning. On the contrary, spiritual tourists tend to engage with other religious traditions in different geographical contexts and expose themselves to a wider variety of beliefs. However, this quest is not necessarily focused on religious sites or sacred landscapes. In fact, the type of destination travelled to is not as important as the inner journey which is facilitated through travel away from home. At the same time, the importance of a quiet, peaceful, and simple setting has been recognized as an important part of this quest. As such, most retreats are located in rural or wilderness landscapes, whether in small villages, mountains, forests, or near water. Hectic and stressful day-to-day living in urban environments clearly requires an antidote in slower forms of tourism in calm, natural settings (Norman & Pokorny, 2017; Bandyopadhyaya & Nairb, 2019). This quest is also typically an individual or personal quest for meaning and the development of the 'self'. However, the quest often allows people to engage with others in communal settings with like-minded people in retreats, monasteries, temples, and yoga centres. The presence of inspiring spiritual individuals, such as well-known gurus or famous yoga teachers, can also facilitate this search for meaning.

Some authors have argued that spiritual tourism can be defined even more broadly to include adventure tourism which challenges the individual and leads to greater self-fulfilment (Cheer, Belhassen & Kujawa, 2017; Gezon, 2018). Indeed, the individual search for the 'self' includes transcendence and connections to a higher power (whether God or nature), as well as relationships with others. It also includes growth in the form of self-development or transformation, which is why many retreat centres focus on activities that enhance personal growth and change. Spirituality is also a way to solve problems in people's lives (Norman & Pokorny, 2017). However, spirituality should not be defined too broadly, especially in the context of the wellness tourism industry, as there are very few tourism experiences that do *not* improve wellness in some way. Rather, it should be argued that the search for spiritual experiences as a proactive component of travel is important—and this is where wellness and spirituality overlap.

The quest for spiritual experiences has become an integral part of tourism and well-being (Smith & Puczkó, 2012). The types of tourism that include spiritual experiences, such as retreats or pilgrimages, are described as offering more eudaimonic benefits and the enhancement of existential authenticity than other forms of travel (Smith & Diekmann, 2017). This stands in contrast to the more hedonic and relaxation-based experiences offered by other aspects of the wellness tourism industry (Voigt, Howat & Brown, 2011). Wellness usually involves trying to balance one's body, mind, and spirit, yet spiritual health has often been a neglected component in Western culture. This, however, has not been the case in Eastern cultures. As many of the spiritual practices that are now practised by wellness tourists, such as yoga and meditation are derived from Asian health systems or lifestyles, spiritual tourists often go to Asia as a part of their spiritual quest (Smith & Puczkó, 2013).

Although wellness tourists choose to get away from their everyday lives to seek or practice spirituality, not all wellness tourists visit sacred or religious destinations. While they may choose to spend their time in a temple, monastery, or ashram, more commonly they visit holistic retreat centres that have no religious or spiritual affiliations. As well, certain landscapes influence feelings of spirituality more so than others. As noted above, deserts and seasides afford the most wellness benefits to visitors, many of which are spiritual. Many tourists who visit temples or monasteries clearly also want to benefit from the surrounding natural landscape as well (Jiang, Ryan & Zhang, 2018). The concept of the healing or therapeutic landscape is by no means a new one, but growing concerns about the planet and environment are influencing the rise of sustainable destinations, eco-friendly retreats, and a renewed emphasis on nature stewardship and preservation on the part of tourists. Among many communities of the world, the spiritual connections to nature are very strong, and the natural environment can even become a spiritual substitute for religion (Taylor, 2010).

Although many wellness practices may be undertaken more for rest, relaxation, or health than for spiritual reasons (Kelly, 2012), tourists may choose to engage in spiritual practices like yoga that can later take them on a more intense spiritual journey in the future. Yoga practitioners tend to see themselves as spiritual rather than religious, but true spiritual practitioners also tend to travel for wellness reasons rather than just doing yoga at home (Smith & Sziva, 2017). Meditation tourism is like yoga-based tourism in the sense that tourists may go to retreats with a non-religious motivation but still have sacred or spiritual experiences (Jiang, Ryan & Zhang, 2018). As such, seeking spirituality is not always a prior motivation for wellness tourists, but it nevertheless can become a significant outcome or benefit.

References

Ashton, A.S. (2018) 'Spiritual retreat tourism development in the Asia Pacific region: Investigating the impact of tourist satisfaction and intention to revisit: A Chiang Mai, Thailand case study', *Asia Pacific Journal of Tourism Research*, 23(11): 1098–1114.

Bandyopadhyaya, R., & Nairb, B.B. (2019) 'Marketing Kerala in India as *God's Own Country!* for tourists' spiritual transformation, rejuvenation and well-being', *Journal of Destination Marketing & Management*, 14. doi: 10.1016/j.jdmm.2019.100369.

Bone, K. (2013) 'Spiritual retreat tourism in New Zealand', *Tourism Recreation Research*, 38(3): 295–309.

Bowers, H., & Cheer, J.M. (2017) 'Yoga tourism: Commodification and western embracement of eastern spiritual practice', *Tourism Management Perspectives*, 24: 208–216.

Cheer, J.M., Belhassen, Y., & Kujawa, J. (2017) 'The search for spirituality in tourism: Toward a conceptual framework for spiritual tourism', *Tourism Management Perspectives*, 24: 252–256.

de Botton, A. (2003) *The Art of Travel*. London: Penguin Books.

Dewsbury, J.D., & Cloke, P. (2009) 'Spiritual landscapes: Existence, performance and immanence', *Social & Cultural Geography*, 10(6): 695–711.

Garrett, C. (2001) 'Transcendental meditation, Reiki and yoga: Suffering, ritual and self-transformation', *Journal of Contemporary Religion*, 16(3): 329–342.

Gauthier, F., Martikainen, T., & Woodhead, L. (2013) 'Acknowledging a global shift: A primer for thinking about religion in consumer societies', *Implicit Religion*, 16(3): 261–276.

Gezon, L.L. (2018) 'Global scouts: Youth engagement with spirituality and wellness through travel, Lake Atitlán, Guatemala', *Journal of Tourism and Cultural Change*, 16(4): 365–378.

Hawks, S. (1994) 'Spiritual health: Definition and theory', *Wellness Perspectives*, 10(4): 3–13.

Heelas, P., & Woodhead, L. (2005) *The Spiritual Revolution: Why Religion Is Giving Way to Spirituality*. Malden, MA: Blackwell Publishing.

Heintzman, P. (2013) 'Retreat tourism as a form of transformational tourism', in Y. Reisinger (ed) *Transformational Tourism Tourist Perspectives* (pp. 68–81). Wallingford, UK: CABI.

Howard, C. (2012) 'Speeding up and slowing down: Pilgrimage and slow travel through time', In S. Fullagar, K. Markwell, & E. Wilson (Eds.), *Slow Tourism: Experiences and Mobilities* (pp. 11–24). Bristol, UK: Channel View Publications.

Jiang, T., Ryan, C., & Zhang, C. (2018) 'The spiritual or secular tourist? The experience of Zen meditation in Chinese temples', *Tourism Management*, 65: 187–199.

Kato, K., & Progano, R.N. (2017) 'Spiritual (walking) tourism as a foundation for sustainable destination development: Kumano-kodo pilgrimage, Wakayama, Japan', *Tourism Management Perspectives*, 24: 243–251.

Kelly, C. (2010) 'Analysing wellness tourism provision: A retreat operator's study', *Journal of Hospitality and Tourism Management*, 17: 108–116.

Kelly, C. (2012) 'Wellness tourism: Retreat visitor motivations and experiences', *Tourism Recreation Research*, 37(3): 205–213.

Kelly, C., & Smith, M.K. (2017) 'Journeys of the self: The need to retreat', in M.K. Smith & L. Puczkó (eds) *The Routledge Handbook of Health Tourism* (pp. 138–151). London: Routledge.

Kujawa, J. (2017) 'Spiritual tourism as quest', *Tourism Management Perspectives*, 24: 193–200.

Lee, D.J., Kruger, S., Whang, M.-J., Uysal, M., & Sirgy, M.J. (2014) 'Validating a customer well-being index related to natural wildlife tourism', *Tourism Management*, 45: 171–180.

Maller, C., Townsend, M., St Leger, L., Henderson-Wilson, C., Pryor, A., Prosser, L., & Moore, M. (2009) 'The health benefits of contact with nature in a park context: A review of current literature', *The George Wright Forum*, 26(2): 51–83.

Moal-Ulvoas, G., & Taylor, V.A. (2014) 'The spiritual benefits of travel for senior tourists', *Journal of Consumer Behaviour*, 13: 453–462.

Moufakkir, O., & Selmi, N. (2018) 'Examining the spirituality of spiritual tourists: A Sahara desert experience', *Annals of Tourism Research*, 70: 108–119.

Myers, J.E., Sweeney, T.J., & Witmer, J.M. (2000) 'The Wheel of Wellness counseling for wellness: A holistic model for treatment planning', *Journal of Counseling & Development*, 78(3): 251–266.

Nahrstedt, W. (2008) *Wellnessbildung: Gesundheitssteigerung in der Wohlfühlgesellschaft*. Berlin: Erich Schmidt Verlag.

Norman, A. (2012) 'The varieties of the spiritual tourist experience', *Literature & Aesthetics*, 22(1): 20–37.

Norman, A., & Pokorny, J.J. (2017) 'Meditation retreats: Spiritual tourism and well-being interventions', *Tourism Management Perspectives*, 24: 201–207.

Olsen, D.H. (2020) 'Pilgrimage, religious tourism, biodiversity, and natural sacred sites', in K.A. Shinde & D.H. Olsen (eds) *Religious Tourism and the Environment* (pp. 23–41). Wallingford, UK: CABI.

Ouellette, P., Kaplan, R., & Kaplan, S. (2005) 'The monastery as a restorative environment', *Journal of Environmental Psychology*, 25(2): 175–188.

Park, C.L, Braun, T., & Siegel, T. (2015) 'Who practices yoga? A systematic review of demographic, health-related, and psychosocial factors associated with yoga practice', *Journal of Behavioral Medicine*, 38: 460–471.

Ponder, L.M., & Holladay, P.J. (2013) 'The transformative power of yoga tourism', in Y. Reisinger (ed) *Transformational Tourism: Tourist Perspectives* (pp. 98–108). Wallingford: CABI.

Ram, Y., & Smith, M.K. (2019) 'An assessment of visited landscapes using a Cultural Ecosystem Services framework', *Tourism Geographies*. doi: 10.1080/14616688.2018.1522545.

Reisinger, Y. (ed.) (2013). *Transformational Tourism: Tourist Perspectives*. Wallingford, UK: CABI.

Reisinger, Y. (2015) *Transformational Tourism: Host Perspectives*. Wallingford, UK: CABI.

Sharpley, R., & Jepson, D. (2011) 'Rural tourism: A spiritual experience?', *Annals of Tourism Research*, 38(1): 52–71.

Sharpley, R., & Sundaram, P. (2005) 'Tourism: A sacred journey? The case of ashram tourism, India', *International Journal of Tourism Research*, 7(3): 161–171.

Smith, M K. (2003) 'Holistic holidays: Tourism and the reconciliation of body, mind and spirit', *Tourism Recreation Research*, 28(1): 103–108.

Smith, M.K., & Diekmann, A. (2017) 'Tourism and Wellbeing', *Annals of Tourism Research*, 66: 1–13.

Smith, M.K., & Kelly, C. (2006) 'Holistic tourism: Journeys of the self', *Tourism Recreation Research* 31(1): 15–24.

Smith, M.K., & Puczkó, L. (2012) 'An analysis of TQoL domains from the demand Side', in M. Uysal, R.R. Perdue, & M.J. Sirgy (eds) *Handbook of Tourism and Quality-of-Life (QOL) Research: The Missing Links* (pp. 263–277). Cham, The Netherlands, Springer.

Smith, M.K., & Puczkó, L. (2013) *Health, Tourism and Hospitality: Spas, Wellness and Medical Travel*. London, Routledge.

Smith, M.K., & Sziva, I. (2017) 'Yoga, transformation and tourism', in M.K. Smith & L. Puczkó (eds) *The Routledge Handbook of Health Tourism* (pp. 168–180). London: Routledge.

Stausberg, M. (2014) 'Religion and spirituality in tourism', in A.A. Lew, C.M. Hall, & A.M. Williams (eds) *The Wiley Blackwell Companion to Tourism* (pp. 349–360). Chichester, UK: Wiley.

Steiner, C., & Reisinger, Y. (2006) 'Ringing the fourfold: A philosophical framework for thinking about wellness tourism', *Journal of Tourism Recreation Research*, 31(1): 5–14.

Taylor, B. (2010) *Dark Green Religion: Nature, Spirituality and the Planetary Future*. Berkeley and Los Angeles: University of California Press.

Timothy, D.J., & Conover, P.J (2006) 'Nature religion, self-spirituality, and New Age tourism', in D.J. Timothy & D.H. Olsen (eds) *Tourism, Religion, and Spiritual Journeys* (pp. 139–155). London and New York: Routledge.

Timothy, D.J., & Olsen, D.H. (2006) 'Conclusion: Whither religious tourism?' in D.J. Timothy, & D.H. Olsen (eds) *Tourism, Religion and Spiritual Journeys* (pp. 271–276). London and New York: Routledge.

Vallés-Planells, M., Galiana, F., & Van Eetvelde, V. (2014) 'A classification of landscape services to support local landscape planning', *Ecology and Society*, 19(1). doi:10.5751/ES-06251-190144.

Voigt, C., Brown, G., & Howat, G. (2011) Wellness tourists: in search of transformation. *Tourism Review* 66(1/2): 16–30.

Voigt, C., Laing, J., Wray, M., Brown, G., Howat, G., Weiler, B., & Trembath, R. (2010) 'Health tourism in Australia: Supply, demand and opportunities'. https://sustain.pata.org/wp-content/uploads/2015/02/120002-Health-Tourism-In-Australia-WEB.pdf (accessed 22 July 2020).

Wilson, G.B., McIntosh, A.J., & Zahra, A.L. (2013) 'Tourism and spirituality: A phenomenological analysis', *Annals of Tourism Research*, 42: 150–168.

Yiannakis, A., & Gibson, H. (1992) 'Roles tourists play', *Annals of Tourism Research*, 19: 287–303.

6
A NEW SPIRITUAL MARKETPLACE
Comparing new age and new religious movements in an age of spiritual and religious tourism

Carole M. Cusack

Introduction

New religious movements (NRMs) are described as religions that have emerged from the nineteenth century up to the present (Ashcraft 2018). There are scholarly disagreements about whether the defining characteristic of such groups is their 'newness' (Barker 2014), or whether, given no new religion is completely original, historical links with 'parent' traditions offer a more accurate way to classify NRMs (Melton 2004, 76). Early scholars of the New Age movement argued that these movements were more fluid and eclectic than many NRMs, which tend to have strong organizational boundaries. However, presently New Age movements are often considered a subset of NRMs because they are a part of what is considered the 'cultic milieu' (Campbell 1972)—both having alternative and non-mainstream beliefs and practices that are rejected by both Enlightenment science and Western Christianity. The secularization of Western culture, which led to the retreat of Christian churches from public life and with it the loss of both membership and public and personal relevance for a sizeable portion of western populations, became the enabling context within which NMRs have flourished. Increasingly secularized public spaces and dialogues encouraged those who were dissatisfied with both traditional Christianity and modern science—'seekers' as Colin Campbell termed them (Campbell 1972)—to experiment with the 'spiritual marketplace' as facilitated by late capitalism which focuses on individualism and the decline of communal relations (Roof 1999; Gauthier, Martikainen & Woodhead 2013).

This 'spiritual marketplace' is a powerful lens through which to view contemporary spiritual seekers, many of whom reject the 'New Age' label while still engaging in similar practices. The spiritual marketplace also drives many of these activities, including 'secular pilgrimage' (Digance 2006) and 'spiritual tourism' (Norman 2011, 2012), which types of tourism overlap with 'religious pilgrimage' and secular forms of tourism. While these forms or niche markets of travel have the capacity to be life-transforming (Cohen 1979), the desire for spiritual experiences—encounters with the 'other' and journeys of the 'self' (Smith & Kelly 2006)—rather than for recreational experiences is what motivates them to travel. In the same way that NRMs often build upon broader parent traditions, a wide variety of experiences offered in the ever-increasing spiritual marketplace have also evolved from older beliefs and groups. Hanegraaff (1996) situated the popularity of New Age movements with in a secular, open marketplace that has made previously esoteric, occult, and cultural more mainstream through commodification (Olsen 2019).

While spiritual tourists and members of New Age movements travel to a wide variety of sacred sites (as noted below), NRMs tend to have fewer, more focused sacred sites that appeal mainly to members of those movements. This is in part because NRMs focus more on institutional religious belonging rather than opening their sites to multiple interpretations. As well, because NRMs are comparatively recent in origin means that they do not have the historical development of sacred spaces like other, older religious faiths. The Church of Jesus Christ of Latter-day Saints (the case study used in this chapter; hereafter the Latter-day Saint Church or the Church) is a case in point. Established in 1830 by Joseph Smith in upper state New York, the idea of pilgrimage and sacred sites is a more recent phenomenon. Sites of interest to Mormons include the 150-acre Sacred Grove outside of Palmyra, New York, where the founder Smith's epiphany, or 'First Vision', occurred (Brown 2018), and the reconstruction of the Nauvoo Temple, which had been destroyed in 1850 (Madsen 2006), among other heritage and religious places (Olsen & Pierce 2021).

This chapter is divided into two sections. The first section focuses on the types of sites visited by members of NRMs and New Age movements or spiritual seekers. Second, this chapter examines the motivations, meanings, and benefits religious and spiritual tourists place on destinations (Cohen 1979; Norman 2011; Cheer, Belhassen & Kujawa 2017), highlighting how sites favoured by spiritual tourists with 'New Age' interests and very different from sites valued by members of NRMs, which are more aligned to traditional pilgrimage practices. These two sections are intertwined with major themes within the contemporary academic study of religion, including conspiracy culture and conspirituality (Voas & Ward 2011; Asprem & Dyrendal 2015); the physical and material culture of tourism (Stausberg 2011); eclecticism and *bricolage* in the construction of meaning (Redden 2016); and the commodification and marketing of both religion and spirituality (Carrette & King 2005).

Tourism destinations for members of new age and NRMs

Modern western tourism has its roots in the English upper-class tradition of the 'Grand Tour,'—a type of travel from the seventeenth to the early nineteenth centuries where young aristocratic young men would travel to study the great cities and monuments of western culture. As Norman (2011) notes, the Grand Tour, which was secular, educational, and cultural in nature, was like "a finishing school; an essential part of a 'gentleman's' education that gave invaluable experience of the world" (p. 81) and marked their passage from youth to adulthood. In contrast, religious travel or pilgrimage was also a popular form of travel during the Middle Ages. However, the number of people taking a pilgrimage declined after the Protestant Reformation, only seeing a revival post World War II with the advent of better transportation and communication technologies (Olsen 2019). As Kaelber (2006) rightly argues, in the medieval and early modern west the motivations of pilgrims and tourists were clearly different, with pilgrims seeking forgiveness of sins and religious benefits rather than educational or social benefits. Pilgrims were focused on destinations that were meaningful in the context of institutional Christianity, such as the tombs of saints and sites where sacred relics were housed (Norman 2011, 165), whereas while secular tourists might visit great cathedrals and other ecclesiastical sites, they did so in a spirit of cultural appreciation, aesthetics, and educational value. In contemporary travel, the distinctions between religious pilgrimage, spiritual tourism, and tourism *simpliciter* have become blurred (Cohen 1979; Digance 2006; Kaelber 2006; Collins-Kreiner 2010; Olsen 2010; Norman 2011).

New Age practitioners tend to visit five types of sites. The first type relates to unique sites of great natural beauty. Some of these sites include mountains like Mount Shasta in California, United States; rock formations like Uluru in Australia; and lakes and waterfalls such as Lake Titicaca on the border of Bolivia and Peru and Niagara Falls in Ontario, Canada. One defining characteristic of the New Age (and some NRMs) is suspicion of modernity, which they characterize as consisting of meaningless jobs, the alienation and subjugation of nature, and 'patriarchal religions and materialistic forms of healing' (Dubisch 2015: 145). This suspicion fuels the conviction that modern urban life is at best less authentic than life in the past and at worst is a toxic illusion (Coats 2011: 205). From a New Age perspective, nature and one's authentic self are interconnected, leading many seekers inspiration to seek for this authenticity in Indigenous spiritualities and traditions that emphasize 'the oneness of humanity, nature and the cosmos and [are] essentially animist in perceiving that the earth and the cosmos are alive and conscious' (Timothy & Conover 2006: 142). This spirituality is also related to social trends such as environmentalism and vegetarianism/veganism, which emphasizes animal rights, and engaging in activities such as hiking, camping, and visiting "power sites" where "earth energies" can boost wellness and effect personal transformation, in wilderness settings (Attix 2002: 53).

The second type of site are prehistoric monuments, which, like natural wonders, have a strong aesthetic appeal (Ezzy 2016). Stonehenge, arguably the most famous megalithic site, has drawn people to marvel at it since at least the Middle Ages if not longer (Cusack 2012). Lesser-known circles, such as the Rollright Stones in Oxfordshire, UK, Avebury, UK, and the Ring of Brodgar in Orkney, Scotland (Cope 1998; Cusack 2012; Cusack 2018a) similarly attract spiritual tourists because of their age, mysterious origins, pre-Christian antiquity, and its setting in an aesthetic landscape. For many, encountering stones is a 'numinous' experience (Cusack 2018a, 72), particularly as they are usually associated with ley lines or considered 'portals' between worlds—beliefs that have become influential motifs in popular culture and media such as novels, film, and television, which examples include the BBC television series *Children of the Stones* (1976); the Dr Who series 'The Stones of Blood' (1978); Penelope Lively's children's novel *The Whispering Knights* (1971); and Robin Hardy's classic pagan horror film, *The Wicker Man* (1973) (Parker 2009).

The third group of sites are historic buildings from the ancient and medieval worlds, such as Egyptian pyramids, Inca and Mayan cities, and Greek and Roman temples (Dubisch 2015). These buildings both share the aesthetic appeal of natural and prehistoric sites and are linked to them by esoteric ideas such as including hidden traditions about mainstream religions; visitations by aliens and UFOs; mystical beliefs about landscapes such as geomancy, planetary alignments; and collective human endeavours to raise the consciousness of the earth or the universe. For example, while Greek and Roman temples and Gothic cathedrals are remarkable and beautiful structures, but when they are connected to ley lines and occult legends about the bloodline of Christ, as for example the church of Saint-Sulpice in the Latin Quarter, Paris is in Dan Brown's novel *The Da Vinci Code* (Brown 2004 [2003]), their appeal is intensified.[1] Beliefs about alien visitations and planetary alignments are also linked to natural sites, but human-made structures are these sites may also be significant. For example, the Bradshaw Ranch in the National Forest near Sedona, Arizona (a renowned New Age destination) is said to "have been confiscated by the U.S. Government because it housed one of the most powerful inter-dimensional portals on the planet" (Dobson 2018, n.p.). People also hypothesize that the Inca city of Manchu Picchu and the Pyramids of Giza are sites of UFO and alien influence (Ivakhiv 2007).

The fourth group of sites are places related to wellness and healing. These sites include ashrams in India, yoga retreats in Bali, and meditation retreats (Timothy & Conover 2006; Bowers & Cheer 2017; Norman & Pokorny 2017) and alternative spirituality meccas such as Sedona, Arizona, United States (Ivakhiv 2001) and Glastonbury in the United Kingdom (Bowman 2005). Being eclectic in beliefs and practices spiritual tourists visit these towns because they offer a range of spiritual experiences, ranging from UFOlogy to tarot, yoga and meditation, astrology, and body-based methods such as Feldenkrais and Rolfing.[2] These sites tend to be the focus of a marketplace that combines beliefs, practices, and spaces from a wide variety of Indigenous and religious communities into a complex economy with a vast range of products on offer, including crystals, vegan and organic food, and spiritual and alternative services such as massage, tarot readings, astrology consultations, seminars, and workshops. This new bricolage of spiritual products, which "emphasis[es] the self, secular drivers for spiritual tourism[,] are consumptive by nature" (Cheer, Belhassen & Kujawa 2017: 254), and has energized and expanded this eclectic spiritual marketplace in recent decades (Redden 2016).

The final category of spiritual tourist destinations includes sites that are personally meaningful to individuals or groups but detached from traditional religious and spiritual beliefs and activities. Digance (2006) defines secular pilgrimage as "undertaking a journey that is redolent with meaning" (p. 36). The twenty-first century has seen the rise of intense fandoms that act as a middle ground between popular cultural interests and passionate devotion to a book series, musician or musical group, actor, or a film or television series among other media. Elvis Presley (1935–1977) attracted such devoted fans, and his home Graceland has been a site of secular pilgrimage since his death (King 1993; Rigby 2001). Hollywood or Disneyland would be sacred destinations for film buffs, and Bayreuth, Germany for aficionados of Richard Wagner's *Ring Cycle*. This type of elective journey is very diverse and can encompass so-called 'dark tourism', which includes prisons, death camps, cemeteries, massacre sites, and battlefields (Timothy 2018), to anime tourism and fan conventions (Buljan 2017; Asimos 2019).

These sites of interest to members of New Age movements or eclectic spiritual groups can be contrasted with the sacred sites of institutionalized NRMs. Arguably, very few if any NRMs have developed a full theology of pilgrimage to rival that of long-established religions like Buddhism, Islam, or Catholic Christianity. However, there are some common pilgrimage themes between them. For example, The Church of Jesus Christ of Latter-day Saints has engaged in the restoration and marketing of several of their historical sites as a form of historical remembrance, a proselytizing tool, and a way to construct and maintain the religious identity of church members (Madsen 2006; Olsen 2013). While the Church does not have a formal theology of pilgrimage, where sacred travel is mandated or related to ideas or rituals of salvation, it does have an informal theology of pilgrimage. For example, each year thousands of church members travel the Mormon Trail, which marks the route early church members took as they travelled from the United States into Indian Territory in search of a 'promised land', through vehicle (Olsen & Hill 2018). During the 1997 sesquicentenary of the original journey, a grass roots wagon train re-enacted the pioneer trek trip from Omaha to Salt Lake City as a part of that year's commemoration of this event. While only 250 people completed the 93-day trip, almost "10,000 participants walked, pulled a handcart, rode a horse, or took a spot in a mule-pulled wagon for one or more days" (Olsen & Hill 2018: 240). Other church historical sites like the 'Sacred Grove' in New York also act as sites of memory related to church's founding and resultant historical development (Madsen 2006; Olsen 2006, 2013, 2016). Unlike many sacred sites that appeal to those who identify with the New Age movement and spiritual tourism, the Church does not offer souvenirs or charge an entry fee at their historical sites and as such offer little in the way of a commodified spiritual marketplace.

Meanings of sites and the motivations to visit in the contemporary spiritual milieu

Norman (2012) has defined spiritual tourism as "tourism characterised by a self-conscious project of spiritual betterment" (p. 20). Norman has further suggested that spiritual tourists, whose principal spiritual task is self-actualization, have five main motivations for travel: the search for healing, to experiment with new ways of understanding reality and creating meaning, questing for new wisdom, retreating from the world to contemplate and reflect, and bonding with a community of like-minded others. These motivational categories are useful in examining how meanings are attached to particular spiritual tourism destinations. For the purposes of this article, two spiritual tourist 'meccas'—Glastonbury and Sedona—will be used to illustrate the meanings that are attached to these similar places. These two sites manifest the four phenomena that are important research areas in the academic study of religion: esoteric or occult elements and conspirituality; the material culture and physical infrastructure of spiritual tourism; eclectic meaning-making and *bricolage* activities on the part of visitors who sample and combine many religions and spiritualities; and the commodification and marketization of both religion and spirituality.

Glastonbury

The attachment of spiritual meanings to Glastonbury began in the Middle Ages with William of Malmesbury's *De antiquitate Glastoniensis ecclesie* (c. 1135 CE), Geoffrey of Monmouth's *History of the Kings of Britain* (c. 1136), Gerald of Wales' *Speculum Ecclesiae* (c. 1216), and Richard Pynson's *Life of Joseph of Arimathea* (1520), which linked the town with miraculous events in Christian history. Richard Pynson cemented the tale of Joseph of Arimathea's visit with the boy Jesus to Glastonbury, and then later returning to Glastonbury with the Holy Grail after the death and resurrection of Jesus. Legend holds that when Joseph set his staff into the ground that it burst into flower, known as the Glastonbury Thorn—a visible sign of life conquering death (Digance & Cusack 2002: 269). William of Malmesbury stated that some early Celtic saints, including St. Patrick and a disciple, Benignus, had lived and died in the monastic community of Glastonbury (Cusack 2018b). Gerald of Wales popularized the town as the burial site of King Arthur after the bodies of Arthur and Guinevere were found in 1191, and Geoffrey of Monmouth added to this legend the wise man Merlin and the Isle of Avalon where Arthur was taken after his defeat at the battle of Camlann. Avalon became identified with the former island of Glastonbury Tor and contributed a druidic and Celtic pagan element to the mythos.

These and other similar traditions attached to Glastonbury accentuate the town's appeal for multiple types of spiritual seekers. As Bowman (2004) observed, the 'official' Christian presence in the town is fractured between Anglicans and Catholics, with their annual summer pilgrimage and processions being on different dates. In addition, Goddess worshippers have convened conferences in the town since 1996 and have instituted a Goddess procession, honoring the Goddess in all her forms, that 'sometimes includes a procession of the Goddess in a cart through the streets to Chalice Well' (Bowman 2004: 282). Goddess-oriented spiritual tourists read Glastonbury's Christian heritage, in which the Virgin Mary and Saint Brigit are prominent, as merely a Christianized, sanitized Goddess cult. In addition to mainstream Christian groups, esoteric Christian groups also feature in the tale of Jesus's visit with Joseph of Arimathea and the 'Somerset tradition' of Britain as a 'holy land'.[3] Indeed, the myriad of Arthurian legends are read as either a part of esoteric Christianity, in which Arthur is a great

Christian warrior entombed in the Abbey, or as a part of pagan traditions, with the Isle of Avalon as a portal to the Celtic 'otherworld' (Bowman 2005: 161). The claims related to the Celtic otherworld are strengthened by the assertion that Glastonbury hosted a great druidic university in antiquity (Bowman 2004) and by situating the Celts as the 'indigenous' people of Britain. In addition, New Age groups have made their presence known in Glastonbury since the 1970s, deeming the place to be the "Heart Chakra of the Planet" on a confluence of ley lines—"a centre of earth energies … where spiritual energy comes into the physical plane" (Bowman 2005, 163), with UFO sightings being relatively common and attributed to this energy. As such, Glastonbury is a site where Indigenous, Christian, Pagan, New Age, and esoteric wisdom meet and at times merge into a broad spiritual marketplace.

Sedona

Sedona is a similar magnet for spiritual tourists, combining healing through earth energies at a "vortex site," Indigenous wisdom connected to the Hopi and Navajo First Nations, shamanism (both Native American and entheogen- or drug-based), and ecological and Goddess spirituality within a range of retreats, seminars, attractions, and activities (Attix 2002: 54). The attribution of esoteric and spiritual meanings to Sedona has its roots in the artists' community founded by Max Ernst (1891–1976) and his wife Dorothea Tanning (1910–2012), who lived in Sedona from 1943 to 1957 (Ivakhiv 2001: 158). Prior to this time, the town was a popular setting for Hollywood westerns and was also a naturalistic tourist attraction on account of its dramatic red rock formations. In the 1950s, Ernst and Tanning's 'metaphysical community' was established, and over time groups and retreats related to Hatha Yoga, Eckankar, the Sri Aurobindo Center, the Aquarian Educational Center, the Sedona Church of Light, and many other New Age establishments were founded (Ivakhiv 2001: 173–174). In the 1970s, Dick Sutphen and Page Bryant identified Sedona as a place with multiple power spots or vortices (Ivakhiv 2007: 272), which made Sedona a destination for UFOlogical seekers and those who sought places that served as portals to other worlds or dimensions. In 1987, a large medicine wheel was constructed on Schnebly Hill which became a key ritual site for New Agers, and in August of that year, Sedona hosted the first Harmonic Convergence—a synchronized global peace meditation, which event brought between five and ten thousand spiritual tourists to the town (Ivakhiv 2001: 174).

Spiritual tourists to Sedona often claim that they were 'drawn' there, and like Glastonbury, the site is replete with power spots and vortices that enhance the energy flow at Sedona, leading to enhanced 'spiritual awareness as well as healing experiences' (Coats 2009: 385). This is the most strongly articulated meaning linked to Sedona, though its Native American past is important. While Native American-derived practices, including sweat lodges, medicine wheel rituals, drumming circles, and smudging are popular, most teachers are not Native American, and tensions exist between spiritual tourists and Native Americans. Several First Nations have connections to Sedona: for example, the Yavapai-Apache view Sedona as the centre of the world and home to a lake from which the first humans emerged; whereas the Hopi regard the Sinagua culture of Sedona as ancestral to them. Ivakhiv notes that First Nations also connect Sedona with more recent historical events; 'the Yavapai-Apaches gather every February at Boynton Canyon to mark Exodus Day, a day of remembrance of forced exodus and the March of Tears in 1875' (Ivakhiv 2001: 152). The New Age practice of leaving offerings, such as crystals and candles, at local Native American sites, is also regarded negatively "as desecration of sacred space" (Timothy & Conover 2006, 150), and the Hopi vigorously deny any Hopi involvement with or connection to the Harmonic Convergence, which was claimed by its founder José Argüelles (Ivakhiv 2001: 194–196).

Glastonbury, Sedona, and the spiritual marketplace

Spiritual tourists believe that Glastonbury and Sedona are important spiritual and power locations in a global network of energy lines—places where spiritual healing and self-actualizing quests can be actively pursued, and where transformation is possible. Both sites draw upon a range of meanings, combining natural landscapes with Indigenous pasts and an esoteric present (Coats 2011: 198). Glastonbury foregrounds an Indigenous Celtic strand, occult Christianity, Arthurian traditions, and also accommodates a range of NRMs, including Bahais, Sufis, UFO spiritualists, and followers of Sai Baba and Krishna Consciousness (Bowman 2005: 166). At Sedona, Native American and UFOlogical belief systems are most prominent (Ivakhiv 2001: 176). The combination of spiritual beliefs that are not susceptible to falsification is basic to what Voas and Ward (2011) have termed "conspirituality"—a phenomenon in which conspiracy theories and alternative spirituality are merged. This conspiracy has three core foundational beliefs; that "a) nothing happens by accident, b) nothing is as it seems, c) everything is connected" (Voas & Ward 2011: 104; see Asprem & Dyrendal 2015). Bricolage and eclecticism are thus normalized as the basis for beliefs and practices; unrelated traditions are drawn into relationship by seekers who are experimental and ask only that experiences resonate with them as a testimony of their authenticity (Ivakhiv 2001: 136).

For many spiritual tourists, travel to these sacred sites is meant to help them '[attain]... some spiritual benefit, such as getting in touch with one's inner self or achieving an altered state of consciousness' (Cheer, Belhassen & Kujawa 2017: 254). As the self and the cosmos are viewed as interconnected, healing is considered a powerful driver for many spiritual tourists. As Dubisch (2015) notes, hope for transformational change operates 'at the personal, social, planetary and cosmic levels' (p. 147). This quest to be healed—whether from physical and psychological ailments or in terms of achieving an optimal self—opens spiritual tourists to engaging in experimental modes of tourism and personal being. They are considered 'deep' tourists who desire to become 'part of the destination' (Timothy & Conover 2006: 144) in order to 'retreat' from everyday life and the banal secular, seeking to leave as a different, more spiritually aware, person. Also, being with other like-minded individuals at these sites creates a sense of community; a collectivity that has ongoing relevance for the spiritual tourist after his or her journey home. This collectivity may result in return trips to the destination to maintain a connection to both the location of spiritual power and also fellow travellers (Norman 2012: 32–33). Interestingly, this interest in personal development has led to some places requiring that visitors undergo purification or 'clearing' rituals, as well as a self-evaluation or 'assessment of one's motivation[s]' regarding whether they have the requisite 'sense of humility and respect' for the site and whether they are willing to give 'some expression or token of gratitude before departing' (Ivakhiv 2007: 274).

The motivations of spiritual tourists in visiting powerful spiritual destinations differ from people who belong to NRMs who journey to sites of theological or historical significance for their faith (Esplin 2019). As noted above, NRMs possess slightly different views of pilgrimage as compared to older, more established religious traditions and spiritual and New Age movements. Members of The Church of Jesus Christ of Latter-day Saints do not expect that visiting their religious heritage sites will result in forgiveness of sins in the same way medieval pilgrimage conferred. However, travel to sites in the Holy Land have a strong appeal to church members, as do locations 'associated with...the Book of Mormon [and] church history' (Hudman & Jackson 1992: 111; Olsen 2006).[4] Church temples are considered the holiest places in the church, particularly the Salt Lake City Temple that serves as 'the symbolic centrepiece of the Mormon world' (ibid: 115; see Jackson & Henrie 1983). In Nauvoo, where the church's founder Joseph Smith is buried, there has been some tension between the Latter-day Saint Church and the Community of Christ, the latter also claiming to be the true successor organization of Joseph Smith after his martyrdom in Carthage, Illinois. Both groups own historical property related to the founding of the church and interpret their sites to visitors, albeit from different theological and historical perspectives (Olsen & Timothy 2002. This had led the city to resemble older Christian sites that are used by multiple congregations (Esplin 2019).

Conclusion

Spiritual tourists and members of New religious movements (NRMs) have very different attitudes towards spiritual and religious tourism. The former are inheritors of New Age spiritual interests which are both eclectic and universalist, seeking to combine disparate traditions such as Indigenous wisdom, esoteric traditions, diffuse new religious practices such as yoga and meditation, and a conspiracist interest in UFOlogy and related topics, into one harmonious tradition (Ivakhiv 2001: 136). This combination of traditions has led to a multiplication of potential sacred destinations, including natural sites, prehistoric monuments, historic locations, places of wellness and healing, and sites of personal significance. The most eclectic of these destinations are Sedona and Glastonbury, as noted in this chapter (Bowman 2005; Timothy & Conover 2006; Ivakhiv 2007), which are in many cases are overwhelmingly white, middle class, and affluent, participants who engage in activities directed towards spiritual betterment within a heavily commercialized spiritual marketplace (Cheer, Belhassen & Kujawa 2017: 254).

In contrast, the Latter-day Saint historical sites and pilgrimage practices are examples of the types of quasi-religious pilgrimage or spiritual tourism favoured by members of NRMs. First, sacred sites are specifically related to Church history that exemplifies important events or achievements in the development of the faith (Olsen 2006, 2013). These sites appeal mostly to church members, although others may visit on for heritage reasons or curiosity) and are therefore oriented towards the religious identity of adherents. There is very little development of any commercial development, let alone an eclectic spiritual marketplace, and there is no evidence of combining church traditions with the tradition of other faith communities apart from sites related to the Bible (Hudman & Jackson 1992).

Spiritual tourism and non-traditional religious travel are research areas that have become prominent relatively recently, and there are many new religious movement pilgrimages and spiritual tourism destinations that are unresearched, presenting opportunities for scholars to further map this emergent field. This brief survey of spiritual tourism among adherents of New Age spirituality and members of new religious movements reveals that the former are engaged in a broad-ranging, commercial, eclectic praxis at a vast range of sites, whereas the latter are concerned exclusively with verified sites of historical significance for their particular faith.

Notes

1 *The Da Vinci Code*, based substantially on *The Holy Blood and the Holy Grail*, an esoteric bestseller by Michael Baigent, Richard Leigh, and Henry Lincoln (2006 [1982]), claims that Jesus and Mary Magdalene were married and that their descendants have been protected throughout history by the Templars, the Masons, and the mysterious Priory of Sion. This mélange of fact and fiction posits that the mythical Holy Grail is actually the bloodline of Christ. This storyline also involves the alleged treasure of Rennes-le-Château, itself a magnet for New Age tourists and conspiracy

theorists (Radford 2019), and foregrounds the fictional Rose Line, a ley line on which Saint-Sulpice and Rosslyn Chapel, near Edinburgh (and the culminating destination *The Da Vinci Code*), are said to be sited (Cusack 2020).

2 Moshe Feldenkrais (1904–1984) developed a method of exercise therapy called the Feldenkrais Method, which aimed to improve health through slow, repeated movements to improve brain-body connections and thereby improve a person's psychological state. Ida Rolf (1896–1979) developed structural Integration (popularly known as Rolfing) as a technique to optimize human biomechanical functioning through realigning a person's energy field with the earth's gravitational field (Hanegraaff 1996: 54).

3 The 'Somerset Tradition' states that Joseph of Arimathea and the boy Jesus built the first church at Glastonbury in honour of the Virgin Mary, thus making England a 'holy land' and Glastonbury an English Jerusalem (Digance & Cusack 2002: 269).

4 Members of The Church of Jesus Christ of Latter-day Saints members consider the Book of Mormon a holy book of scripture akin to the Bible.

References

Ashcraft, W.M. (2018) *A Historical Introduction to the Study of New Religious Movements*. London and New York: Routledge.

Asimos, V. (2019) 'Navigating through space butterflies: CoxCon 2017 and fieldwork presentation of contemporary movements', *Fieldwork in Religion*, 14(2): 181–194.

Asprem, E., & Dyrendal, A. (2015) 'Conspirituality reconsidered: How surprising and how new is the confluence of spirituality and conspiracy theory?', *Journal of Contemporary Religion*, 30(3): 367–382.

Attix, S. (2002) 'New-age oriented special interest travel: An exploratory study', *Tourism Recreation Research*, 27(2): 51–58.

Baigent, M., Leigh, R., & Lincoln, H. (2006 [1982]) *The Holy Blood and the Holy Grail*. New York: Random House.

Barker, E. (2014) 'The not-so-new religious movements: Changes in the 'cult scene' over the past forty years', *Temenos: Nordic Journal of Comparative Religion*, 50(2): 235–256.

Bowers, H., & Cheer, J.M. (2017) 'Yoga tourism: Commodification and western embracement of eastern spiritual practice', *Tourism Management Perspectives*, 24: 208–216.

Bowman, M. (2004) 'Procession and possession in Glastonbury: Continuity, change and the manipulation of tradition', *Folklore*, 115: 273–285.

Bowman, M. (2005) 'Ancient Avalon, New Jerusalem, Heart Chakra of Planet Earth: The local and the global in Glastonbury', *Numen*, 52(2): 157–190.

Brown, D. (2004 [2003]) *The Da Vinci Code*. London: Corgi.

Brown, J.M. (2018) 'Managing for the spirit: Valuing the Mormon Sacred Grove', *Journal for the Study of Religion, Nature and Culture*, 12(3): 285–306.

Buljan, K. (2017) 'Spirituality-struck: Anime and religio-spiritual devotional practices', in C.M. Cusack & Kosnáč, P. (eds), *Fiction, Invention, and Hyper-reality: From Popular Culture to Religion* (pp. 101–118). London and New York: Routledge.

Campbell, C. (1972) 'The cult, the cultic milieu, and secularization', *A Sociological Yearbook of Religion in Britain*, 5: 119–136.

Carrette, J., & King, R. (2005) *Selling Spirituality: The Silent Takeover of Religion*. London and New York: Routledge.

Cheer, J.M., Belhassen, Y., & Kujawa, J. (2017) 'The search for spirituality in tourism: Toward a conceptual framework for spiritual tourism', *Tourism Management Perspectives*, 24: 252–256.

Coats, C. (2009) 'Sedona, Arizona: New Age pilgrim-tourist destination', *CrossCurrents*, 59(3): 383–389.

Coats, C. (2011) 'The melodramatic structure of New Age tourist desire', *Tourist Studies*, 11(3): 197–213.

Cohen, E. (1979) 'A phenomenology of tourist experiences', *Sociology*, 13(2): 179–201.

Collins-Kreiner, N. (2010) 'Researching pilgrimage: Continuity and transformations', *Annals of Tourism Research*, 37(2): 440–456.

Cope, J. (1998) *The Modern Antiquarian: A Pre-Millennial Odyssey Through Megalithic Britain*. London: Thorsons.

Cusack, C.M. (2012) 'Charmed circle: Stonehenge, contemporary paganism, and alternative archaeology', *Numen*, 59(2): 138–155.

Cusack, C.M. (2018a) 'Prehistoric monuments in Britain as numinous sites of spiritual tourism: The Rollright Stones', *Fieldwork in Religion*, 13(1): 61–80.

Cusack, C.M. (2018b) 'The Glastonbury Thorn in Christian traditions and popular culture', *Journal for the Study of Religion, Nature and Culture*, 12(3): 307–326.

Cusack, C.M. (2020) 'Esoteric tourism in Scotland: Rosslyn Chapel, *The Da Vinci Code*, and the appeal of the 'New Age'', in J.M. Wooding & L. Barrow (eds), *Prophecy, Fate and Memory in the Early and Medieval Celtic World* (pp. 247–270). Sydney: Sydney University Press.

Digance, J. (2006) 'Religious and secular pilgrimage: Journeys redolent with meaning', in D.J. Timothy & Olsen, D.H. (eds), *Tourism, Religion and Spiritual Journeys* (pp. 36–48). London and New York: Routledge.

Digance, J., & Cusack, C.M. (2002) 'Glastonbury: A tourist town for all seasons', in G. Dann (ed.), *The Tourist as a Metaphor of the Social World* (pp. 263–280). Wallingford, UK: CABI.

Dobson, J. (2018) 'The world's 8 best places to hunt for extraterrestrials and search for UFOs', *Forbes*, 28 January. At: https://www.forbes.com/sites/jimdobson/2018/01/28/the-worlds-8-best-places-to-hunt-for-extraterrestrials-and-search-for-ufos/#6f7aa0182ddb. Accessed 20 July 2020.

Dubisch, J. (2015) 'The seduction of the past in New Age pilgrimage', in M.A. Di Giovine, & D. Picard (eds), *The Seductions of Pilgrimage: Sacred Journeys Afar and Astray in the Western Religious Tradition* (pp. 145–168). Farnham and Burlington, VT: Ashgate.

Esplin, S.C. (2019) 'Creating and contesting Latter-day Saint pilgrimage to Nauvoo, Illinois', *International Journal of Religious Tourism and Pilgrimage*, 7(4): 11–17.

Ezzy, D. (2016) 'Religion, aesthetics and moral ontology', *Journal of Sociology*, 52(2): 266–279.

Gauthier, F., Martikainen, T., & Woodhead, L. (2013) 'Acknowledging a global shift: A primer for thinking about religion in consumer societies,' *Implicit Religion*, 16(3): 261–276.

Hanegraaff, W.J. (1996) *New Age Religion and Western Culture: Esotericism in the Mirror of Secular Thought.* Leiden and Boston, MA: Brill.

Hudman, L.E., & Jackson, R.H. (1992) 'Mormon pilgrimage and tourism', *Annals of Tourism Research*, 19: 107–121.

Ivakhiv, A. (2001) *Claiming Sacred Ground: Pilgrims and Politics at Glastonbury and Sedona.* Bloomington: Indiana University Press.

Ivakhiv, A. (2007) 'Power trips: Making sacred space through New Age pilgrimage', in D. Kemp & J.R. Lewis (eds), *Handbook of New Age* (pp. 263–286). Leiden and Boston, MA: Brill.

Jackson, R.H., & Henrie, R. (1983) 'Perception of sacred space', *Journal of Cultural Geography*, 3(2): 94–107.

Kaelber, L. (2006) 'Paradigms for travel: From medieval pilgrimage to the postmodern virtual tour', in D.J. Timothy & Olsen, D.H. (eds), *Tourism, Religion and Spiritual Journeys* (pp. 49–63). London and New York: Routledge.

King, C. (1993) 'His truth goes marching on: Elvis Presley and the pilgrimage to Graceland', in I. Reader & T. Walter (eds), *Pilgrimage in Popular Culture* (pp. 92–104). London: The Macmillan Press Ltd.

Madsen, M.H. (2006) 'The sanctification of Mormonism's historical geography', *Geographies of Religions and Belief Systems*, 1(1): 51–73.

Melton, J.G. (2004) 'Perspective: Toward a definition of 'new religion'', *Nova Religio: The Journal of Alternative and Emergent Religions*, 8(1): 73–87.

Norman, A. (2011) *Spiritual Tourism: Travel and Religious Practice in Western Society.* London and New York: Continuum.

Norman, A. (2012) 'The varieties of the spiritual tourist experience', *Literature & Aesthetics*, 22(1): 20–37.

Norman, A., & Pokorny, J.J. (2017). Meditation retreats: Spiritual tourism well-being interventions. *Tourism Management Perspectives, 24*: 201–207.

Olsen, D.H. (2006) 'Tourism and informal pilgrimage among the Latter-Day Saints', in D.J. Timothy & D.H. Olsen (eds), *Tourism, Religion and Spiritual Journeys* (pp. 256–270). London and New York: Routledge.

Olsen, D.H. (2010) 'Pilgrims, tourists, and Weber's "ideal types"', *Annals of Tourism Research*. 37(3): 848–851.

Olsen, D.H. (2013) 'Touring sacred history: The Latter-day Saints and their historical sites', in J.M. Hunter (ed), *Mormons and American Popular Culture: The Global Influence of an American Phenomenon*, Vol. 2, (pp. 255–242). Santa Barbara, CA: Praeger Publishers.

Olsen, D.H. (2016) 'The Church of Jesus Christ of Latter-day Saints, their "three-fold mission," and practical and pastoral theology', *Practical Matters: A Journal of Religious Practices and Practical Theology*, 9: 27–51.

Olsen, D.H. (2019) 'Religion, spirituality, and pilgrimage in a globalizing world', in D.J. Timothy (ed), *Handbook of Globalisation and Tourism* (pp. 270–283). London: Edward Elgar.

Olsen, D.H., & Hill, B.J. (2018) 'Pilgrimage and identity along the Mormon Trail', in D.H. Olsen & A. Trono (eds), *Religious Pilgrimage Routes and Trails: Sustainable Development and Management* (pp. 234–246). Wallingford, UK: CABI.

Olsen, D.H., & Pierce, G. (2021) 'The Latter-day Saints, the Bible, and tourism', in J. Bielo & J. Wijnia (eds), *The Bible and Global Tourism* (pp. 83–103). New York: Bloomsbury Press.

Olsen, D.H., & Timothy, D.J. (2002) 'Contested religious heritage: Differing views of Mormon heritage', *Tourism Recreation Research*, 27(2): 7–15.

Parker, J. (2009) *Written on Stone: The Cultural Receptions of British Prehistoric Monuments*. Newcastle on Tyne: Cambridge Scholars Publishing.

Radford, R. (2019) 'Psychogeography: An (old) new method for viewing the religious in the urban and the sacred', *Fieldwork in Religion*, 14(2): 195–215.

Redden, G. (2016) 'Revisiting the spiritual supermarket: Does the commodification of spirituality necessarily devalue it?', *Culture and Religion*, 17(2): 231–249.

Rigby, M. (2001) 'Graceland: A sacred place in a secular world?', in C.M. Cusack & P. Oldmeadow (eds), *The End of Religions? Religion in an Age of Globalisation* (pp. 155–165). Sydney: Sydney Studies in Religion.

Roof, W.C. (1999) *Spiritual Marketplace: Baby Boomers and the Remaking of American Religion*. Princeton, NJ: Princeton University Press.

Smith, M., & Kelly, C. (2006) 'Holistic tourism: Journeys of the self?', *Tourism Recreation Research*, 31(1): 15–24.

Stausberg, M. (2011) *Religion and Tourism: Crossroads, Destinations and Encounters*. London and New York: Routledge.

Timothy, D.J. (2018) 'Sites of suffering, tourism, and the heritage of darkness: Illustrations from the United States', in P.R. Stone, R. Hartmann, T. Seaton, R. Sharpley, & L. White (eds), *The Palgrave Handbook of Dark Tourism Studies* (pp. 381–398). London: Macmillan.

Timothy, D.J., & Conover, P.J. (2006) 'Nature religion, self-spirituality and New Age tourism', in D.J. Timothy & D.H. Olsen (eds), *Tourism, Religion and Spiritual Journeys* (pp. 139–155). London and New York: Routledge.

Voas, D., & Ward, C. (2011) 'The emergence of conspirituality', *Journal of Contemporary Religion*, 26(1): 103–121.

7

FAN PILGRIMAGE, RELIGION, AND SPIRITUALITY

Daniel H. Olsen

Introduction

In recent years, there has been a growing interest in the connections between popular culture, religion, and spirituality. However, these connections have tended to be marginalized by several scholars, in part because of their tendency to "categorize some religions as normal and some as deviant" (Pike 2009: 67). As Orsi (2005: 188) puts it, within the study of religion, there are some "ways of living between heaven and earth" that should be excluded, marginalized, and given the status of the "Other"—that do not fit nicely into the more general definitions of religion, and as such, act as foils for what experts consider to be religion. This marginalization has occurred not just because of the difficulty in defining religion (Harrison 2006; Oman 2013; Neville 2018) but also in part because of the difficulty in defining what exactly constitutes "popular culture". Indeed, popular culture is not a "thing" that can be studied (Mitchell 1995). As Storey (2006: 1) notes, the study of "popular culture" for many scholars is the study of "otherness"—where popular culture is compared and contrasted to other conceptual categories, such as "folk culture, mass culture, dominant culture, working-class culture", as well as "high" and "avant-garde" culture.

This division between high and popular culture, however, stands in stark contrast to post-modern views that attempt to dissolve the seemingly restrictive boundaries between pilgrimage and tourism as well as the sacred/secular divide (Kaelber 2006; Collins-Kreiner 2010; Olsen 2010). From this perspective, modern tourism takes on some of the traditional characteristics of pilgrimage, with modern pilgrims being considered a sub-type of tourist (MacCannell 1976; Graburn 1989; Timothy & Olsen 2006). This view has been accompanied by a reconceptualization of what constitutes sacred and profane space, what is meant by "transcendence", and how experiences formerly in the realm of religion are translated into non-religious contexts (Evans 2003; Knoblauch 2008; Knox & Hannam 2014; Shilling & Mellor 2014). These dissolvings of boundaries have led to travel to and activities at sporting events and stadiums, war memorials and other sites related to nationalism, and, indeed, any type of activity, being considered a pilgrimage and/or a religio-spiritual experience.

However, it is now impossible to ignore the influence that popular culture, religion, and spirituality have on each other. The purpose of this chapter is to briefly examine some of the ways in which fandoms have become interwoven with religion and spirituality. In doing so, this chapter examines both "religion = fandom" (Zubernis & Larsen 2018: 154), or how religious practices and terminology are used to explain the development of narratives and identities that connect to people to something they consider to be sacred, and "fandom = religion" (Hills 2000: 13), or how fandoms have led to the development of "invented religion[s]" (Cusack 2013; Taira 2013; Lužný 2020) and "fiction-based religion[s]" (Davidsen 2013; Cusack & Kosnáč 2016). Both angles of inquiry focus on certain symbolic aspects of popular culture that are used in ways that "go…beyond culturally marked religiosity" (Knoblauch 2008: 148). The chapter begins with a discussion of what constitutes "fandom" and highlights the "waves" or topics of research on fandoms more generally. Then, the chapter examines the interconnections between fan pilgrimage, religion, and spirituality, focusing first on "religion as fandom" or some of the similarities between fandom and religion, and then on "fandom as religion", or how some fandoms have become in part fiction-based religions, before concluding.

What is fandom?

Like most attempts by academics to define an object or subject of interest, a search for a definitive definition of "fandom" can be daunting. While the growth and influence in modern society over the past 30 years (Fuschillo 2020) of "fandoms" and "fans" have led to their taken-for-granted status in present lexicons, fandoms entail "multiple and ongoing cultural processes" (Nikunen 2007: 113) related to the intersection of ever evolving ideas regarding culture, consumption, and community. In addition, there is a wide range of ways in which fans can engage in fandoms (Williams 2011). As Sandvoss (2005: 10) argues, the "true core" of some fans is the "communal context of their fandom", while for other fans, "their fandom is driven more by an idiosyncratic bond with their object of fandom".

However, this has not stopped scholars from attempting to define this area of research. For example, according to Fiske (1992: 30), fandom is a common feature in industrialized societies where aspects of mass-produced and -distributed forms of entertainment are taken and "reworked into an intensely pleasurable, intensely signifying popular culture that is both similar to, yet significantly different from, the culture of more 'normal' popular audiences". Sandvoss (2005: 8) suggests that fandoms involve the "regular, emotionally involved consumption of a given popular narrative or text". For Davis (2015: 423), "fandom is a term used to refer to a subculture composed of fans characterized by a feeling of sympathy and camaraderie with others who share a common interest".

In this context, a "fan" is a person who is a "communal participant" in a fandom related to a sport, form of media, activity, or a famous person or group (Kirby-Diaz 2009: 147). Kirby-Diaz (2009: 147) argues that to be a fan, one needs to be more than consuming an activity, such as movie watching, book reading, or watching sports on TV. Rather, to be a fan means to be heavily engaged in "text" or a particular "team", to be an active participant with that text or team (Jenkins 2018). For fans, belonging to a fandom represents "forging a shared identity and consciousness by providing members with a sense of collective belonging based on strong interpersonal bonds similar to family-like ties" (Fuschillo 2020: 349). In this vein, Jenson (1992: 27) suggests that for many people, "Fandom is an aspect of how [people] make sense of the world, in relation to mass media, and in relation to [their] historical, social, [and] cultural location". For Gray, Sandvoss, and Harrington (2007: 10), fandoms are an important part of how people "form emotional bonds with [themselves] and others in a modern, mediated world".

Sandvoss, Gray, and Harrington (2017) suggest that historically there have been three different waves of research related to "fan studies" or "fandom studies". The first research wave was treating fandom as a form of activism—a reaction by regular people against high cultural elites, media producers, and industries that denigrated certain aspects of mass or lower culture, such as pop music, comic books, romance novels, and movies that have mass appeal (Fiske 1992). From this perspective, people who were a part of fandoms were characterized as groups of disenfranchised persons who, being subaltern, powerless, and marginalized within their societies, were deviant in their popular culture interests and expressions, leading fans to be branded as pathologically fanatic in their devotion to their fandoms (Jenson 1992; see Fuschillo 2020). However, as fandoms moved from being a marginalized segment of society into the cultural mainstream, a second research avenue began to focus on the sociology of consumption, examining how broader cultural and social hierarchies were replicated in fan communities and subcultures. While still interested in the broader concerns regarding power and inequality within society, scholars focused more on developing typologies of individual and collective practices. However, as Sandvoss, Gray, and Harrington (2017: 6) note, this research "had little to say about the individual motivations, enjoyment, and pleasure of fans". This led to a third wave of research on fandom, which considers the growing number of fandoms in the face of what they call fandom's "growing cultural currency", where "being a fan [has become] an ever more common mode of cultural engagement" (p. 6). Instead of fandoms being relegated to specialized and themed fan conventions, digital technologies and social media have led to the development of broader fandoms and the closer integration of fandoms with the everyday lives of fans everywhere. In addition, this wave of research focuses on "*intra*personal pleasures and motivations among fans, focusing on the relationship between fans" selves and their fan objects' (p. 6).

Another, newer, research avenue involves "post-fandoms" or "post-object fandoms". Williams (2011: 269) defines "post-object fandom" as a "fandom of any object which can no longer produce new texts". For example, the ending or cancellation of TV serials such as *Firefly*, *Buffy the Vampire Slayer*, and *The West Wing*, or book series like *Harry Potter*—each of which had almost cult-like followings—has led to "fan practices and interactions inevitably chang[ing]" (Williams 2011: 269) as these fan objects move from "ongoing to dormant" (Williams 2015: 14). When this occurs, questions arise regarding how fans shift from one object of fandom to another, how fans attempt to extend their fandoms or revive/rejuvenate dormant texts or objects, and why some fans grow "stagnant" or "retire" from fandoms (Que 2014; Lee 2015).

Academic research on specific types of fandoms has grown dramatically in the past two decades. These fandoms, as noted above, range from books and comics to TV serials, movies, sports teams, and music groups. As Jenkins (2018) notes, fandoms are generally categorized by single texts or celebrities, with fans either being given or creating their own nicknames to describe their affinity and associated bonds to these texts and celebrities. Some examples include:

- *Musicians/groups*: "ARMY" (K-pop group BTS) (Lee, Oh & An 2019; McLaren & Jin 2020); "Beliebers" (Justin Bieber) (Sherbine 2013); "BeyHive" (Beyoncé) (Blyth 2017); and "Directioners" (One Direction) (Korobkova 2014; Arvidsson et al. 2016);
- *TV serials*: "Browncoats" (*Firefly*) (Wilcox 2015; McCormick 2018); "Sherlockians" (*Sherlock Holmes*) (Stein & Busse 2014; Hills 2017); and "Whovians" (*Doctor Who*) (Wright & Wright 2015);

- *Movies*: "Trekkies" (Star Trek) (Jindra 1994; Coppa 2008) and "Warsies" (*Star Wars*) (Kruger 2015); and
- *Books*: "Potterheads" (*Harry Potter*) (Lee 2015) and Janeites (*Jane Austen*) (Yaffe 2013; Glosson 2020; Seaton 2020).

At the same time, fandoms are also "better understood as a more expansive subculture... whose members engage with a broad array of different media objects but who share traditions and practices built up over many years" (Jenkins 2018: 16). Because of this, several articles and books, along with the *Journal of Fandom Studies*, have looked at fandoms from a broader perspective, including:

- "Anime fandom" (Leonard 2005; Napier 2006; Ito, Okabe & Tsuji 2012);
- "Media fandom" (Hills 2000; Coppa 2006);
- "Music fandom" (Duffett 2013a; Giuffre 2014);
- "Sports fandom" (Brown 1998; Wann & James 2018);
- "Video game fandom" (Swalwell, Ndalianis & Stuckey 2017; Brown et al. 2018; Moreno 2020); and
- Fandom more generally (Pearson 2010; Duffett 2013b; Gray, Sandvoss & Harrington 2007; Booth 2018).

Intersections between fan pilgrimage, religion, and spirituality

Duffett (2013a: 18) argues that attempts to define "fandom" lead to a reification of this object of study, "stopping the process of fandom and artificially trying to pin it down". Instead, he suggests, following Cavicchi (1998), that instead of focusing on what fandom *is*, the emphasis should be on what fandom *does*. By extension, this should lead to discussions regarding what fandoms do *to* fans. This section focuses on this last part through investigating ways in which fandoms, religion, and spirituality are interrelated. As noted above, two ways to understand these relationships include "religion as fandom" and "fandom as religion".

Religion as fandom

Many scholars have turned to religious terminology to metaphorically describe, for example, tourism as a "sacred journey" or a form of modern pilgrimage (MacCannell 1976; Graburn 1989; Knox & Hannam 2014; Olsen 2014) or people's quasi-religious ties to nationalistic dogmas, ideals, and rites (i.e., "civil religion"; see Bellah 1975; Haberski Jr. 2018). This metaphorical extension also extends to research on fandoms in attempts to understand how fan communities are maintained and how fandom activities affect or transform fans. Within this context, several scholars suggest that there are strong ties between popular culture, fandoms, religion, and spirituality (Jindra 1994; Istoft 2010; Davidsen 2013; Buljan & Cusack 2015). For example, Hills (2002: 118) notes that both religion and fandom "are both centered around acts of devotion, which may create similarities", leading Hills to suggest that fandoms exhibit a type of "neoreligiosity" or neospirituality. Aden (1999: 152) also suggested that like the pilgrim's journey, fandom travel consists of "separation" or the beginning of the journey, the "liminal stage", which includes the journey, experiences at the shrine, encountering the "sacred", and "reaggregation" or the return home. Because of space, this chapter highlights only a few interrelated ways in which fandoms are akin to religion, including visiting sacred places, liminality and communitas, ritual and performance, affective experiences, and politicization.

Visiting sacred places

Travel to sacred places has long been an important part of religion and integral in the creation and maintenance of both individual and group religious identities and sacred geographies (Olsen 2012b, 2019). Religious pilgrimage has also long influenced politics, economic development, and societal cohesiveness at local, regional, national, and transnational scales (Coleman & Eade 2018). Presently, an estimated 600 million people visit religious sites to perform religious rituals or for educational purposes (UNWTO 2011). Pilgrimage sites, which range from entire holy cities to natural sacred sites to the burial spots of holy men and women (Shackley 2001), draw religious people for a variety of reasons, including curiosity, worship, initiatory and/or cleansing rituals, healing, to be educated, and to maintain religious identities (Morinis 1992; Olsen 2012b). When these sites are deemed to be sacred, whether for mystic-religious or historical reasons (Jackson & Henrie 1983), they are demarcated from the surrounding profane space and are maintained as sacred space—which sacredness is reinforced by pilgrims and tourists who travel to this site because it is sacred (Bremer 2004). Both pilgrims and tourists who visit these sites want these sites to be authentic. As Bremer (2004: 3–7) argues, many tourists are generally concerned with aesthetic (primarily authentic) experiences as they journey, which frame their experiences of religion, while religious travelers seek authentic religious experiences which can be enhanced through aesthetics.

Like many religions, fandoms are intimately tied to places related to aspects of their particular fandom, whether literary, media, sports, or virtual in nature. As Zubernis and Larsen (2018: 151) argue, "[fans do not] experience the things [they] love—films, television shows, books, music—as just a text, but as a site of intense emotional engagement". While for some fans watching their favorite TV show or reading their favorite book might be considered a "symbolic pilgrimage" (Brooker 2017), Yamamura (2020) notes that fandoms eventually transition from being fans of the objects of fandoms to being fans of the places associated with fandoms. These places or sites are sometimes referred to as "power spots", "sacred sites" (Okamoto 2019), or "venerated fan spaces" (Geraghty 2019). As noted by Zubernis and Larsen (2018: 149), "When fans love a movie, book, or television show, they often want to take an active role in connecting with that world and the characters (or actors) in it", which connectivity comes in part through traveling to places that are an integral part of the establishment of these fandoms. For example, many fans travel to cities in which a movie or TV serial was filmed (e.g., Vancouver and Melbourne) (Brooker 2007). These cities are "multiply coded" (Brooker 2007: 430), meaning that because they are settings for multiple fandoms, the importance of the urban landscape will differ for each fan. Another type of power spot includes specific locations where the filming of a movie or TV serial took place (Beeton 2005). Some examples include:

- The spot where Sherlock Holmes jumped to his death outside St. Bartholomew's Hospital in *Sherlock* (Toy 2017; Zubernis & Larsen 2018);
- Graceland, home of Elvis Presley (Duffett 2003; Brooker 2007);
- The town of Forks, WA, the location for the *Twilight* book and movie series (Crowe 2013; Larson, Lundberg & Lexhagen 2013);
- Iceland and Northern Ireland for *Game of Thrones* enthusiasts in search for the fictitious land of Westeros (Brooker 2007; Murray 2017a, 2017b); and
- "Hobbiton", the film set for scenes in *The Lord of the Rings* movies (Singh & Best 2004; Davis et al. 2014).

Cemeteries and the individual burial plots of famous celebrities are also a common power spot for fans (Barron 2014; Levitt 2018), as are fandom-themed conventions like Comic-Con and theme parks that are based on popular anime and manga publications (Dunn & Herrmann 2020), such as "Naruto World" and "J-World" in Japan. Whether multi- or single-fandom in nature, conventions in particular serve as "liminal spaces" (see below) or "sites of performance, play, veneration, and community" for like-minded "believers" to connect with objects and communities of veneration (Zubernis & Larsen 2018: 145). Conventions serve as "decentered" places, where it is the fandom, not the place, which "represents the true center of the convention pilgrimage process" (Porter 2004: 168).

These fandom power spots sometimes overlap with religious sites. For example, fans of the anime *Lucky Star* visit the Washinomiya Shrine in Kiku, Japan, because the shrine was used as inspiration for some of the illustrated backgrounds in the anime. Once fans made the connection between the shrine and the anime, visitation to the shrine increased from 90,000 to 300,000 in one year (Okamoto 2019: 150). As well, many Buddhist shrines in Japan utilize elements of popular culture and fandom, such as Hello Kitty, as mascots to brand themselves to younger generations based on contemporary socioeconomic circumstances (Porcu 2014; Maud 2017).[1] Sugawa-Shimada (2015) discusses the *rekjio* phenomenon in Japan—where young women have taken a strong interest in Japanese history. As part of this phenomenon, these "history fan girls" also visit Buddhist temples and Shinto shrines, which are considered "power spots", to gain spiritual power. There are also examples of Christian groups using particular fandoms, such as *My Little Pony*, as a way to promote Christian doctrine (Crome 2014).

Many scholars have referred to travel to these power spot sites by fans as a form of pilgrimage (King 1993; Alderman 2002; Gammon 2004; Porter 2004; Brooker 2005, 2007; McCarron 2006; Margry 2008; Williams 2012; Norris 2013; Andrews 2014; Larsen 2015; Okamoto 2015; Erdely & Breede 2017; Geraghty 2018). Many fans also refer to their travel to these places as "pilgrimages" (Porter 2004; Zubernis & Larsen 2018). This use of the term "pilgrimage" to describe visits to places fans find important or meaningful follows Morinis' (1992) view that any journey "undertaken by a person in quest of a place or a state that he or she believes to embody a *valued ideal*"(4; emphasis added). Indeed, the continued secularization of society has led to the decline of the symbolic, mystical, and religious qualities of pilgrimage (Cohen 1992a; Margry 2014; Olsen 2019), which qualities have been co-opted by secular journeys of discovery. As such, the term pilgrimage" is used to frame how fans, as "pilgrim-tourists" (Cohen 1992a), view their travel to sites related to their preferred fandom—to "center[s] out there" (Turner 1973; Cohen 1992b) or a person's sociocultural center—and as a way of "making meaning and establishing identity" (Zubernis & Larsen 2018: 145) and finding their "multiple selves" (Erdely & Breede 2017: 43). At many of these sites, fans take pictures of themselves in "religious-looking poses" and then post these pictures to social media. In doing so, they both sacralize and reinforce the sacred nature of these sites (Jang 2020a: 119).

Liminality and communitas

Another way in which fandom and religion are related is their relation to what Van Gennep (1960) called "liminality" and Turner (1969) termed "communitas". Liminality refers to the temporary, transitional, and immersive state that people enter while traveling to or during their time at sacred sites. During this transitional state, a person is "in-between situations and conditions where established structures are dis-located, hierarchies reversed, and traditional settings of authority possibly endangered" (Mälksoo 2012: 481). This liminal, "betwixt and between" state is thought to lead to "a reconstruction of identity (in which the sense of self is significantly disrupted) in such a way that the new identity is meaningful for the individual and their community" (Beech 2011: 287). As Turner and Turner (1978: 3) suggested, liminality is "not only [about] transition but also potentiality, not only 'going to be' but also 'what may be'". This temporary, transitional, and immersive state has been compared to a "third space" (Bhabha 1994; Soja 1996), where "people's religious identities can be reflexively remade through the encounter and interaction with both the physical and the meta-physical aspects of [sacred] site[s]" (Olsen 2012a: 233; see Collins-Kreiner 2010).

Communitas refers to the bonds that pilgrims form with each other when like-minded people are in physical proximity to each other in a liminal space (Turner 1969; Wu et al. 2020). For Turner (1969), these bonds take place because pilgrimage transcends ordinary socioeconomic structures (i.e., anti-structure), and through worship and ritual performances people find commonalities that make them a part of the same socioreligious community (Turner & Turner 1978). According to Turner, there are three types of communitas: "spontaneous" (i.e., where communitas spontaneously happens), "normative" (i.e., where space and experience are structured in such a way to foster and maintain communitas), or "ideological" (i.e., the ideological and utopian tenets that drive normative communitas) (Di Giovine 2011; Higgins & Hamilton 2020). Cox (2018) discusses another type of communitas—"tangible communitas", which refers to imagined communities that coalesce around tangible items or material culture. While all these forms of communitas do not mean that tension or contestation within or between groups at sacred sites are non-existent (Eade & Sallnow 1991; Coleman & Elsner 1995), communitas as a concept is merely meant to describe how people from different backgrounds can come together, feel a sense of togetherness, and form a cohesive group. At the same time, Collins and Murphy (2014), following Carse (1986, 2008), expand upon communitas to distinguish between "civitas" (producers) and "communitas" (community of consumers) to understand the relationships between churches and their adherents and, in this case, fandoms and their fans.

Both liminality and communitas have been discussed briefly within fandom studies, particularly in the context of conventions. Both multi-fandom and single fandom conventions are opportunities for fans to enter into a liminal space—a "temporary space with different norms for behavior and self-expression"—with like-minded "believers" to connect with objects and become a part of a community of veneration (Zubernis & Larsen 2018: 148). As Zubernis and Larsen (2018: 147) argue, "Conventions can be viewed as liminal spaces in the sense that they take place on the border between fans and fannish objects"—which "space is not used to bring the real and fictional world together, but to bridge the gap between the fan and the object of fandom". As such, convention space "is often a transitional space where fans experiment with and perform their identities....a transitional space where a temporary transgression of gender, sexual, racial, and ethnic norms is allowed and where boundaries between the self and the world are negotiated and reconstructed" (p. 149). In other words, a "third space".

In terms of communitas, while conventions are generally spaces of "communal celebration", Zubernis and Larsen (2018: 152) note that "While fan pilgrimage often replicates many of the dynamics of conventions such as ritual, performance, and entrance into a liminal space, it is often a solitary undertaking. Where conventions are spaces of communal celebration, pilgrimage sites are often places for quiet reflection and solitary communing". At the same time, entering the liminal space of conventions allows individuals to participate in "the communal experience of meeting fellow enthusiasts", which can strengthen their fandom. This in turn leads to a better understanding of "how collective experience can add to the self-defined notion of the fans as a family" (Phillips 2011: 478). For many fans, one way in which communitas occurs is through cosplay, in which fans "transgress the boundary between the real world and the fictional world and find pleasure in straddling these two worlds to explore aspects of identity" (Zubernis & Larsen 2018: 149). Outside of fan conventions, Guschwan (2011: 1996) suggests that communitas is achieved for football fans in Europe when they "momentarily comes together as one" through singing team football songs or cheering for their team in unison. At the same time, because for most fans engagement with the text or object of their fandom is solitary, "[t]he feeling of connection is not [actually] with other fans, but with the fiction" (Brooker 2007: 159). As such, "there are times when the presence of others actually detracts from and impedes full immersion", such as when die-hard fans interact with casual fans, which may cause a disconnect between convention attendees, inhibiting the prospects for true communitas to occur (Zubernis & Larsen 2018: 153).

Ritual/performance

While the written and spoken word is used to express theological approaches toward the faith, belief, and values of a religion, rituals and performances are also important forms of religious expression. Rituals are special social events that take place in special places that have been "bracketed off" from the ordinary world (Cavicchi 1998: 89; Beeman 2015: 40). While the ordinary world represents the "way things are", rituals, according to Smith (1980: 125), idealize the "way things ought to be". As Smith explained, "Ritual is a means of performing the way things ought to be in conscious tension to the way things are in such a way that this ritualised perfection is recollected in the ordinary, uncontrolled, course of things" (*ibid*). Rituals generally involve the use and/or manipulation of material objects and sacred embodied actions, whether through bowing, hand gestures, or other forms of bodily movement. While Smith believed that ritual by definition was "sacral" and therefore in the domain of religion, Grimes (1999: 269) argued that "not all ritual is religious", and that "sacred reenactment is a subcategory of ritual". As such, what is considered ritualistic is, much like pilgrimage, determined by the people who label certain actions as such.

According to Dionísio, Leal, and Moutinho (2008: 23), "Fandom experience is expressed through a formal series of public and private rituals requiring a symbolic language and space deemed sacred by its worshipers". For example, cosplayers who dress up as their favorite character perform the ritual of "striking a particular pose on stage that typifies that character" (Buljan 2016: 104). Another form of ritual at conventions is standing in line to wait for an encounter with actors who either voice or play characters within different fandoms—a form of "celebrity worship" (Zubernis & Larsen 2018: 155). Another ritual in which many fans participate is inscribing their names in the form of graffiti at fandom sites, such as Beatles fans writing their names outside of Abbey Road Studios or fans of the TV series *Sherlock* doing the same outside of St. Bartholomew's Hospital. In doing so, "It allows the fan to leave a bit of herself in the fictional world.... It also allows fans to connect to the real-world objects of fandom via the performance itself" (Zubernis & Larsen 2018: 153–154). Jang (2020a, 2020b) also gives the example of fans of *Love Live! School Idol Project*—a multimedia project revolving around a fictional story about a group of high school girls who start an idol group—engaging in several religious-like or ritualistic behaviors. For example, "armament", as a form of cosplay, refers to "the act of wearing a large and often chaotic array of *Love Live!* merchandise, such as badges and soft toys, when fans [travel]" (Jang 2020a: 121). Another ritualistic behavior within this fan group is "kowtowing" (i.e., kneeling and bowing very low) to *Love Live!* advertising. A further behavior is referred to as "birthday pilgrimage", where fans climb Mount Umi in South Korea on March 15 because one of the characters of *Love Live!* is named Umi and her birthday is on that date. At the top of Mount Umi, fans have created an altar where they leave "offerings", such as stuffed toys and Umi-related merchandise, and then "kowtow" to the altar before returning home.

Affective experiences

Liminality, communitas, and ritual/performance can lead to fans having affective experiences (Zubernis & Larsen 2018). The idea of "affect" is not an easily understood concept, being poorly defined in the academic literature. At a very basic level, affect can be defined as a person's "ability to affect or be affected" in some way (Massumi 2015: 48). From a psychological and neurological perspective, affect refers to the ways in which people react to objects or situations they encounter (Duncan & Barrett 2007). Affect, however, is considered as separate from moods or emotions, in that affect is "a background state that continually changes in response to a host of events, most beyond conscious monitoring" (Russell 2015: 196). In the same way that a person feels their body temperature rising or falling prior to cognitively determining whether they are "hot" or "cold" or making any determination why they are feeling this way (Russell 2003: 148), affect is a neurophysiological "non-reflective" state that determines the intrinsic pleasure or displeasure (i.e., hedonic valence) and interest or lack of interest (i.e., arousal or activation) a person has toward an object or situation. The status of these two psychological properties at any given time informs the precognitive attitudes and feelings a person has toward an object or situation. Affect also determines how the brain "privileges incoming information based on one's history of experience of what is motivationally relevant in a given context, thereby generating a predisposition to attend to certain categories of stimuli over others" (Todd et al. 2012: 367). As Pykett (2018: 161) puts, "human action precedes cognition; put simply, we act before we think".

While much of the research on affect has been in the realm of neuroscience and psychology, geographers have also examined affective or emotional geographies as part of "neural", "bio-social", and "affective" turns in the discipline (Pykett 2018). Affective or emotional geographies involve understanding how people produce geographical knowledge and experience life through embodied, sensuous, and "energetic outcome[s] of encounters between bodies in particular places" (Conradson & Latham 2007: 232). For Conradson and Latham (2007: 232, emphasis added), "Affect...emerges through engagement and interaction: it is an outcome of *emplaced* encounters. Human emotions then reflect our recognition and perceptions of affective states". These emplaced encounters can occur either during the "performative practices of everyday life" (Jones 2011: 876) or through travel to places that have special meaning. However, these emplaced encounters are often fixed or framed by economic, political, social, and cultural representations and interpretations of how a space should be experienced and the type of bodily performances that are appropriate for these encounters (Lorimer 2005). For example, religious site managers take care to create an environment or "sense of place" (Shackley 2001) to promote an authentic religious or spiritual atmosphere in order to elicit certain emotional responses, thereby offering affirmations of religious identity while also shifting the identities of tourists to pilgrims (Olsen 2012a, 2012b). As such, much of the work on affective geographies comes from the perspective of "non-representational theories" that go beyond how space is represented and focus on the "pre-cognitive aspects of embodied life" (Simpson 2011: 344).

Zubernis and Larsen (2018: 151, emphasis added) argue that "Fandom, much like religion, involves an intensely *affective experience*". As noted above, the objects or texts of fandom are "sites of intense emotional engagement" (*ibid*). In this context, Zubernis and Larsen are using "affect" as being synonymous with "emotion", which is in line with how many academics understand and utilize affect—to describe experiences and activities that are both meaningful and transformative. From this perspective, Zubernis and Larsen suggest that "it is the tangible physicality of the pilgrimage site that allows fans to immerse themselves in the fictional world" (p. 153). In other words, immersion into a fandom through embodied encounters with its (im)material worlds is what leads to "affective experiences". These embodied encounters not only take place during travel to and interaction with a site related to a fandom, but also through what Collins (2004) labels as "interaction rituals", where interactions between, in this case, fans, "generates positive emotional energies" which "lift[s] their spirits and creat[es] social solidarity among them" (Lim 2016: 687). This "positive emotional energ[y]" is also important in the formation of people's religiosity and spirituality (Saroglou, Buxant & Tilquin 2008). Indeed, "Affective experience serves as an important source of information when people are making global judgments about the quality of their lives" (Joshanloo 2018: 629). Positive affective experiences also lead many fans to describe their fandom experiences as a "religious" or a "spiritual experience" (Cavicchi 1998; Jang 2020a), which means that their emotional attachments to their fandom have increased in a positive matter. This can potentially create "affective shifts" to both cognitive and pre-cognitive processes which affect both a person's unconscious and personality (Hill & Hood 1999: 1016) and more fundamentally how they "act before they think" (Pykett 2018: 161).

Politicization

Any type of movement can be utilized for political purposes. This has long been the case with religion, which is often used to justify nationalistic movements and terrorist ideologies and to politically argue for or against certain social issues (Stump 2000; Brubaker 2012; Bloomer, Pierson & Claudio 2019). In many cases, fringe elements of a faith community can distort religious beliefs, leading to an extreme religious fanaticism or radicalism that advocates for political action, and in some cases violence, to promote a certain religious ideology, reclaim religious homelands, or to fight against globalizing influences within a religious society (Kundnani 2012). Religious pilgrimage can also be viewed through a political lens, where different religious groups or differences within one religious group can lead to contestation over the ownership, management, representation, and interpretation of religious sites (Olsen 2017). Pilgrimage can also have strong religious undertones, such as American "Zionist tourism" to Israel by evangelical Christians to express support and solidarity with the state of Israel (Belhaussen 2009; Belhaussen & Ebel 2009; Ron & Timothy 2019).

Fandoms are also occasionally utilized for political or activist purposes. According to Jenkins (2014: 65), "fan activism" refers to "forms of civic engagement and political participation that emerge from within fan culture itself". In many cases, fan activism consists of lobbying for fan-related outcomes (Brough & Shresthova 2012), such as engaging in complex campaigns to either protest the cancellation of TV shows, as in the case of *Star Trek* in the 1960s, or to show support for a potential film project (Jenkins 2015). In other cases, fan activism revolves around attempts to have more say in the development of fan content, such as the inclusion of sexual and racial minorities, to resist forms of censorship copyright infringements, and pushing back against the sexualization of women (Jenkins 2015, 2018). Even though most forms of activism are oriented toward nonpolitical ends (Earl & Kimport 2009), van Zoonen (2005: 63) notes that these fandom efforts at activist engagement and organization involve "customs that have been laid out as essential for democratic politics:

information, discussion, and activism". As such, "fandom may represent a particularly powerful training ground for future activists and community organizers" (Jenkins 2015: 211). However, there are other forms of fan activism that aim to make a political or social difference. One example is the Harry Potter Alliance, a US-based non-profit organization that advocates for social justice issues, such as fair trade and marriage equality because that is what Harry and Dumbledore would do (Brough & Shresthova 2012; Jenkins 2015). Many celebrities also seek to utilize their fans to promote activist and philanthropic causes, such as Lady Gaga, who uses social media to encourage her fans to contribute to and actively petition for change regarding social causes such as gay rights, HIVAIDS awareness, youth homeless shelters, and disaster relief (Bennett 2014).

Fandom as religion

As noted earlier, the idea of "fandom as religion" refers to how some fandoms can become quasi-religious in nature, and in some cases lead to the development of "invented religion[s]" (Cusack 2013; Taira 2013; Lužný 2020) or "fiction-based religion[s]" (Davidsen 2013; Cusack & Kosnáč 2016). For many fandoms, religion serves as a template for fan practice, with "the discourse of religious conversion...provid[ing] fans with a model for... describing the experience of becoming a fan" (Cavicchi 1998: 51; see Hills 2002) and the development of individual and group identity and meaning. Many fandoms revolve around what Davidsen (2013: 378–379) calls "speculative fiction"—where authors and producers take motifs or aspects of magic, pagan mythologies, science fiction, and the "dark side of the supernatural" and "disperses these ideas to a wide audience and enhance...their plausibility by inviting people to identify with protagonists who inhabit worlds in which the supernatural is notoriously real, and by investing the alternative supernatural with symbolic capital".

There are several examples of fandoms tied to a fiction-based religion. For example, Jindra (1994) argues that some aspects of *Star Trek* fandom are similar to religious movements, including the sacralization of certain aspects of modern culture, engagement in ritualized practices, and the development of a "canon" and a particular hierarchy within *Star Trek* fandom groups—all of which adds to the "alternative universe" of *Star Trek*. As McLaren and Porter (1999) note, many "Trekkies" combine different aspects of new age spirituality in part because the ties between science and spirituality are explored by various characters in the *Star Trek: The Next Generation*, *Star Trek: Deep Space Nine*, and *Star Trek: Voyager* series. Another example is "Snapeism", which comes from the Harry Potter fandom. "Snapeism" involves women who believe that one of the main characters, Snape, "exists as a being with thoughts and feelings independent of [the] author". Because Snape exists on an astral plane, he can be channeled, particularly by women, to help them in their daily lives (Alderton 2014: 221). Jediism is another fandom that has been compared to a quasi-religion. Based on the *Star Wars* franchise, fans that follow Jediism identify as Jedi knights and "really" believe that the "Force" is real, even though the franchise does not constitute real history (Possamai 2012; Davidsen 2013; Taira 2013). As Williams, Miller and Kitchen (2017: 119) note, these Jedi "take inspiration from the ideas and ideals of the fiction and its source materials to create a pragmatically spiritual way of life", based in part on Zen Buddhism (Bainbridge 2017), and live as a Jedi Knight "always and everywhere" (Davidsen 2016, 2017: 8). Davidsen (2013: 391) explains how Jediism has succeeded in becoming a quasi-religion where other fandoms have failed:

> It seems that fictional narratives can be used as authoritative texts for religion if they (a) tell of superhuman agents or powers that are real within the fictional universe, but

> supernatural from the perspective of the reader/viewer and if (b) these agents and powers are not obvious analogical references to one particular existing religion....It furthermore seems to boost the religious potential of fiction when it includes (c) an explicit and institutionalised 'narrative religion' (such as the Force religion of the Jedi Knights in Star Wars) whose (d) main ideas are presented by an authoritative teacher figure to a 'disciple' with whom the reader/viewer is invited to identify.

The use of fandoms as substitutes for religion is a vestige of the broader processes of secularization, which refers to "the process whereby sectors of society and culture are removed from the domination of religious instructions and symbols" (Berger 1967: 107). Rooted in early capitalism, this "secularization thesis" posited that religion would play a lesser role in the public sphere and in the daily lives of individuals and eventually disappear. While secularization has manifested itself at a more geographically limited scale than originally predicted, the weakening of religion in many modern societies and the rise of consumerism as a "social structuring vector" or logic has led to the development of consumer-based forms of spirituality (Gauthier et al. 2013), displacing religious authorities as the creators and containers of meaning and identity. This is because of the move toward "cultural deregulation" in consumer societies (Beyer 2007), which deregulation has led to individuals being "less institutionally bound and more personalized manifestations of religion that also offer access to transcendental human concerns" (Olsen 2019: 273). This may also explain in part why fandoms and their accompanying consumeristic logics are replacing religious symbols and institutions as transmitters of tradition and identity, acting as secular religious outlets for individuals who seek personal meaning and symbols outside of the realm of religion that fit their personalities and interests.

These examples of fan-based religions, however, raise questions regarding the validity and seriousness with which scholars should take these religions. As Jindra (1994) observed, these fandoms do not fit the conventional definition of religion with its belief in deity and the supernatural. For many people, "Fandom is an aspect of how [fans] make sense of the world, in relation to mass media, and in relation to [their] historical, social, [and] cultural location" (Jenson (1992: 27), which includes implicitly combining religion and spirituality within fandoms. As noted above, there are several commonalities between these fandoms and religion, even if only on a structural level. However, several scholars have questioned whether comparisons to religion and fandom are valid. As Porter (2009: 271) notes, many scholars tend to "pathologize the implicit religions dimensions" of fandom. In some cases, scholars view these fiction-based religions as a form of fanaticism. Fuschillo (2020: 357) defines fanaticism as "an intense emotional commitment toward a set of values and takes place as an individual process and/or a collective movement". As Kelly (2004: 1) suggests, "To many nonfans, especially among the professional and upper middle classes, [fans] are either the "obsessed individuals" or the 'hysterical crowd'", and as Alderton (2014: 220) notes, many fiction-based religions, such as Snapeism, are "usually interpreted as a ludicrous—and therefore invalid—religion". As such, the intersections between pop culture, pilgrimage, and spirituality should really fall under "secular pilgrimage" or "secular" or "civil" religion, in part because the object of travel is not religious per se—at least not in the historical and etymological understandings of religion and what constitutes a pilgrimage (Bickerdike 2016).

At the same time, Davidsen (2013) suggests that most fiction-based religions recognize that their "faith" is couched in an alternate world—a simulacrum of religion (Possamai 2012)—as compared to traditional or "history-based" religions that are founded on divine revelation (e.g., Judaism, Islam, Christianity) (Cusack 2010). As Porter (2009: 271) notes,

"Star Trek conventions are "like" pilgrimage[,]...the [National Hockey League] is "like" an ecclesia[, and]...Elvis is 'like' a saint". As noted above, Sherlockians who inscribe their names outside of St. Bartholomew's Hospital "require...a *performance of the belief* in the fictional world of Sherlock (as opposed to an actual belief in that world). It also allows fans to connect to the real-world objects of fandom via the performance itself" (Zubernis & Larsen 2018: 153–154, italics in original). This follows Falque's (2016: 88, emphasis in original) view that

> If a belief is a mental state and this state means recognizing that a certain representation is true, we could certainly doubt the adequation of the representation to the thing ("I believe *that* the sky is blue"), but not the *belief in the representation* ("I believe *in* the blue sky")—in particular when we entrust ourselves to beings that are dear to us, albeit imaginary ("I believe in Santa Claus").

As such, as noted above, for many fans, religion serves as a template for fan practice—at least regarding the use of religiously-oriented terminology to describe how one "answers spiritual needs in an age when established religions have lost their relevance" (Duffett 2003: 519). While some critics of "fandom as religion" suggest that "religion requires a degree of faith that fandom does not...fandom delivers a kind of individualistic satisfaction that religion discourages" (Cusack & Robertson 2019: 3). At the same time, as Doss (1999: 75) notes, Elvis fans do not necessarily consider their visits and actions at Graceland as pilgrimage or ritual. They also do not consider their actions as elevating Elvis to a "cult religious figure". Doss suggests that this is a confusing admission for scholars who examine fandom through the lens of religion. As such, it is the outside scholar that more often than not that imposes the use of religious terminology to describe the travel and actions of fans. As such, both fans and scholars of fandom at times use religion as a frame-of-reference to understand better both the structure and the experience of fans and their fandoms.

At the heart of approaching fiction-based religions from this perspective are questions related to "authenticity", which "is used to undermine and devalue the 'sacred' spaces that fans create for themselves" (Porter 2009: 273). Doing so, however, "diminishes the fan and trivializes the object of fandom" (Zubernis & Larsen 2018: 154). For many fans, fandom acts as a secular substitution for religion, in the same way that dark tourism is considered by some a substitute for religious pilgrimage in a post-modern world (Korstanje & Olsen 2020). As Adler (1992: 408) suggests, present-day travel is intimately tied to the "world and self-construction", and that travel itself can be considered a form of "religious ritual and quest". As Buljan (2016: 101) argues, "One of the primary human activities is to seek, find and construct meaning". Indeed, many fans believe that the objects of their fandom can offer both "an emotive outlet for those in search of a secure identity" and an avenue for them to "express their identity, affiliation, and how they chose to be seen by others" (King 2010: 1). If "religion and spirituality are understood as a search for or experience of the sacred, as defined by the individual", then "what is sacred is what the individual finds to be sacred in that he or she attributes qualities to the object that are sacred-like" (Hill & Hood 1999: 1019).

Conclusion

While scholars have historically criticized popular culture and fandoms, there has been an increasing academic interest in the organization of fandoms and their relationship to religion and spirituality in the past two decades. This is in part because fandoms are one way in which people seek to create meaning in their lives, particularly with the seeming decline of religion in a secularizing world. As Knoblauch (2008) argues, postmodern culture dissolves the boundaries between the religious and the non-religious, the sacred and secular divide, and the boundaries between the private and the public. As such, what constitutes religion, the sacred, and transcendence has become negotiable. This is the case with popular culture and its many fandoms, many of which can be seen as modern forms of popular religion. Indeed, in a time when people are becoming disenchanted by religion, popular culture seems to be a key reenchanting factor, particularly in the context of the search for meaning.

The purpose of this chapter was to summarize part of the rapidly growing literature on the intersections between popular culture, religion, and spirituality. This was done by examining both "religion as fandom" and "fandom as religion". Indeed, there are structural similarities between religion and fandom, whether based on the shared ideas of sacred space, liminality and communitas, ritual and performance, affective experiences, and politicization. As noted in this chapter, there are also instances where fandoms take on a quasi-religion function in the lives of fans that brings meaning to their lives outside of the realm of institutionalized religion. While there is much work to be done parsing out additional intersections between popular culture, religion, and spirituality, one fruitful area might be in the realm of fan experiences. While there has been work done on fan experiences (Karpovich 2008; Porat 2010; Phillips 2011; Lee 2015), unfortunately, there has been no work done in the area of affect and fandom from an affective-cognitive perspective. This might prove a fruitful field of research for better understanding how experience and emotion a person has regarding their fandom sets the "affective tone of the person's spiritual life" (Fallot 2001: 113).

Note

1 Following the Washinomiya Shine example above, Moy and Phongpanichanan (2018: 217) highlight several examples of TV and film locations that saw a large growth in visitor numbers after these TV dramas and films were released. Some examples include the Wallace Monument in Scotland, which saw a 300% increase in visitors a year after the release of *Braveheart*; Lyme Park in Cheshire, UK, which saw a 150% increase in visitors after the release of *Pride and Prejudice*; and Southfork Ranch in Dallas, USA, saw a 500,000 increase in visitors after the TV serial *Dallas* became popular.

References

Aden, R.C. (1999) *Popular Stories and Promised Lands: Fan Cultures and Symbolic Pilgrimages*. Tuscaloosa, AL: University of Alabama Press.

Adler, J. (1992) 'Mobility and the creation of the subject: Theorizing movement and the self in early Christian monasticism', in J.P. Jardel (ed.) *Le tourisme international: Entre tradition et modernite* (pp. 407–415). Nice: Centre d'Etudes Tourism et Civilisation.

Alderman, D.H. (2002) 'Writing on the Graceland wall: On the importance of authorship in pilgrimage landscapes', *Tourism Recreation Research*, 27(2): 27–33.

Alderton, Z. (2014) '"Snapewives" and "Snapeism": A fiction-based religion within the Harry Potter Fandom', *Religions*, 5: 219–267.

Andrews, D.K. (2014) 'Genesis at the shrine: The votive art of an anime pilgrimage', *Mechademia*, 9: 217–233.

Arvidsson, A., Caliandro, A., Airoldi, M., & Barina, S. (2016) 'Crowds and value. Italian directioners on Twitter', *Information, Communication & Society*, 19(7): 921–939.

Bainbridge W.S. (2017) 'Jediism: The most popular online virtual religion', in *Dynamic Secularization: Information Technology and the Tension between Religion and Science* (pp. 121–149). Cham, Switzerland: Springer.

Barron, L. (2014) *Celebrity Cultures: An Introduction*. Thousand Oaks, CA: SAGE.

Beech, N. (2011) 'Liminality and the practices of identity reconstruction', *Human Relations*, 64(2): 285–302.

Beeman, W.O. (2015) 'Religion and ritual performance', in J.-M. Pradier (ed.) *La croyance et le corps. Esthétique corporéité des croyances et identités* (pp. 35–58). Bordeaux: Presses Universitaires de Bordeaux.

Beeton, S. (2005) *Film-induced Tourism*. Bristol, UK: Channel View Publications.

Belhaussen, Y. (2009) 'Fundamentalist Christian pilgrimages as a political and cultural force', *Journal of Heritage Tourism*, 4(2): 131–144.

Belhaussen, Y., & Ebel, J. (2009) 'Tourism, faith and politics in the Holy Land: An ideological analysis of evangelical pilgrimage', *Current Issues in Tourism*, 12(4): 359–378.

Bellah, R.N. (1975) *The Broken Covenant: American Civil Religion in Time of Trial*. New York: Seabury.

Bennett, L. (2014) 'If we stick together we can do anything': Lady Gaga fandom, philanthropy and activism through social media', *Celebrity Studies*, 5(1–2): 138–152.

Berger, P. (1967) *The Social Reality of Religion*. London: Faber and Faber.

Beyer, P. (2007) 'Religion and globalization', in G. Ritzer (ed.) *The Blackwell Companion to Globalization* (pp. 444–460). Malden, MA: Blackwell.

Bhabha, H.K. (1994) *The Location of Culture*. New York: Routledge.

Bickerdike, J.O. (2016) *The Secular Religion of Fandom*. Thousand Oaks, CA: SAGE.

Bloomer, F., Pierson, C., & Claudio, S.E. (2019) *Reimagining Global Abortion Politics: A Social Justice Perspective*. Bristol, UK: Policy Press.

Blyth, C. (2017) 'In the Name of Our Lord Beysus Christ', *Auckland Theology & Religious Studies*. Available at: https://aucklandtheology.wordpress.com/category/theorel-101g/page/2/ (accessed 16 October 2020).

Booth, P. (ed.) (2018) *A Companion to Media Fandom and Fan Studies*. Hoboken, NJ: John Wiley & Sons, Inc.

Bremer, T.S. (2004) *Blessed with Tourists: The Borderlands of Religion and Tourism in San Antonio*. Chapel Hill, NC: The University of North Carolina Press.

Brooker, W. (2005) 'The Blade Runner experience: Pilgrimage and liminal space', in W. Brooker (ed.) *The Blade Runner Experience* (pp. 11–30). New York: Columbia University Press.

Brooker, W. (2007) 'Everywhere and nowhere: Vancouver, fan pilgrimage and the urban imaginary', *International Journal of Cultural Studies*, 10(4): 423–444.

Brooker, W. (2017) 'A sort of homecoming: fan viewing and symbolic pilgrimage', in J. Gray, C. Sandvoss, & C.L. Harrington (eds.) *Fandom: Identities and Communities in a Mediated World*, 2nd edition (pp. 149–164). New York: New York University Press.

Brough, M.M., & Shresthova, S. (2012) 'Fandom meets activism: Rethinking civic and political participation', *Transformative Works and Cultures*, 10. DOI: 10.3983/twc.2012.0303.

Brown, A. (ed.) (1998) *Fanatics!: Power, Identity, and Fandom in Football*. London and New York: Routledge.

Brown, K.A., Billings, A.C., Murphy, B., & Puesan, L. (2018) 'Intersections of fandom in the age of interactive media: eSports fandom as a predictor of traditional sport fandom', *Communication & Sport*, 6(4): 418–435.

Brubaker, R. (2012) 'Religion and nationalism: Four approaches', *Nations and Nationalism*, 18(1): 2–20.

Buljan, K. (2016) 'Spirituality-struck: Anime and religio-spiritual devotional practices', in C.M. Cusack & P. Kosnáč (eds.) *Fiction, Invention and Hyper-Reality* (pp. 115–132). London and New York: Routledge.

Buljan, K., & Cusack, C.M. (2015) *Anime, Religion and Spirituality: Profane and Sacred Worlds in Contemporary Japan*. Bristol, CT: Equinox.

Carse, J. (1986) *Finite and Infinite Games*. New York: Random House.

Carse, J. (2008) *The Religious Case Against Belief*. New York: Penguin Press.

Cavicchi, D. (1998) *Tramps Like Us: Music and Meaning among Springsteen Fans*. New York: Oxford University Press.

Cohen, E. (1992a) 'Pilgrimage and tourism: Convergence and divergence', in A. Morinis (ed.) *Sacred Journeys: The Anthropology of Pilgrimage* (pp. 47–61). Westport, CT: Greenwood Press.

Cohen, E. (1992b) 'Pilgrimage centers: concentric and excentric', *Annals of Tourism Research*, 19(1): 33–50.

Coleman, S., & Eade, J. (eds.) (2018) *Pilgrimage and Political Economy: Translating the Sacred*. New York: Berghahn Books.

Coleman, S., & Elsner, J. (1995) *Pilgrimage Past and Present in the World Religions*. Cambridge, MA: Harvard University Press.

Collins, N., & Murphy, J. (2014) 'Communitas and civitas: An idiographic model of consumer collectives', *Journal of Global Scholars of Marketing Science*, 24(3): 279–294. DOI: 10.1080/21639159.2014.911495.

Collins, R. (2004) *Interaction Ritual Chains*. Princeton, NJ: Princeton University Press.
Collins-Kreiner, N. (2010) 'Researching pilgrimage: Continuity and transformations', *Annals of Tourism Research*, 37(2): 440–456.
Conradson, D., & Latham, A. (2007) 'The affective possibilities of London: Antipodean transnationals and the overseas experience', *Mobilities*, 2(2), 231–254. DOI: 10.1080/17450100701381573.
Coppa, F. (2006) 'A brief history of media fandom', in K. Hellekson & K. Buuse (eds.) *Fan Fiction and Fan Communities in the Age of the Internet* (pp. 41–59). Jefferson, NC: McFarland.
Coppa, F. (2008) 'Women, Star Trek, and the early development of fannish vidding', *Transformative Works and Cultures*, 1. DOI: 10.3983/twc.2008.044.
Cox, N. (2018) 'Tangible communitas: The Los Angeles wisdom tree, folklore and non-religious pilgrimage', *Western Folklore*, 77: 29–55.
Crome, A. (2014) 'Reconsidering religion and fandom: Christian fan works in *My Little Pony* fandom', *Culture and Religion*, 15(4): 399–418, DOI: 10.1080/14755610.2014.984234.
Crowe, J. (2013) 'The Twilight of Forks? The effect of social infrastructure on film tourism and community development in Forks, WA', *Journal of Rural Social Sciences*, 28(1): 1–25.
Cusack, C.M. (2010) *Invented Religions: Imagination, Fiction and Faith*. Burlington, VT: Ashgate.
Cusack, C.M. (2013) Play, narrative and the creation of religion: Extending the theoretical base of 'invented religions', *Culture and Religion*, 14(4): 362–377.
Cusack, C.M., & Kosnáč, P. (eds.) (2016) *Fiction, Invention and Hyper-reality: From popular culture to religion*. Abingdon and New York: Routledge.
Cusack, C.M., & Robertson, V.L.D. (2019) 'Introduction: The study of fandom and religion', in C.M. Cusack, J.W. Morehead, & V.L.D. Robertson (eds.) *The Sacred in Fantastic Fandom: Essays on the Intersection of Religion and Pop Culture* (pp. 1–14). Jefferson, NC: McFarland.
Davidsen, M.A. (2013) 'Fiction-based religion: Conceptualising a new category against history-based religion and fandom', *Culture and Religion*, 14(4): 378–395.
Davidsen, M.A. (2016) 'From Star Wars to Jediism: The emergence of fiction-based religion', in E. van den Hemel & A. Szafraniec (eds), *Words: Religious Language Matters* (pp. 376–389). New York: Fordham.
Davidsen, M.A. (2017) 'The Jedi community: History and folklore of a fiction-based religion', *New Directions in Folklore*, 15(1/2): 7–49.
Davis, C.H., Michelle, C., Hardy, A., & Hight, C. (2014) 'Framing audience prefigurations of The Hobbit: An unexpected journey: The roles of fandom, politics and idealised intertexts', *Participations: Journal of Audience & Reception Studies*, 11(1): 50–87.
Davis, L. (2015) 'Football fandom and authenticity: A critical discussion of historical and contemporary perspectives', *Soccer & Society*, 16(2–3): 422–436. DOI: 10.1080/14660970.2014.961381.
Di Giovine M (2011) 'Pilgrimage: Communitas and contestation, unity and difference - An introduction', *Tourism*, 59(3): 247–269.
Dionísio, P., Leal, C., & Moutinho, L. (2008) 'Fandom affiliation and tribal behaviour: A sports marketing application', *Qualitative Market Research: An International Journal*, 11(1): 17–39.
Doss, E. (1999) *Elvis Culture: Fans, Faith, &* Image. Lawrence, KS: University Press of Kansas.
Dunn, R.A., & Herrmann, A.F. (2020) 'Comic Con communion: Gender, cosplay, and media fandom', in R.A. Dunn (ed.) *Multidisciplinary Perspectives on Media Fandom* (pp. 37–52). Hershey, PA: IGI Global.
Duffett, M. (2003) 'False faith or false comparison? A critique of the religious interpretation of Elvis fan culture', *Popular Music and Society*, 26(4): 513–522.
Duffett, M. (2013a) *Popular Music Fandom: Identities Roles and Practices*. London and New York: Routledge.
Duffett, M. (2013b) *Understanding Fandom: An Introduction to the Study of Media Fan Culture*. London and New York: Bloomsbury Publishing Inc.
Duncan, S., & Barrett, L.F. (2007) 'Affect is a form of cognition: A neurobiological analysis', *Cognition and Emotion*, 21(6): 1184–1211. DOI: 10.1080/02699930701437931.
Eade, J, & Sallnow, M.J. (eds.) (1991) *Contesting the Sacred: The Anthropology of Pilgrimage*. London and New York: Routledge.
Earl, J., & Kimport, K. (2009) 'Movement societies and digital protest: Fan activism and other non-political protest online', *Sociological Theory*, 27: 220–243. DOI: 10.1111/j.14679558.2009.01346.x.
Erdely, J.L., & Breede, D.C. (2017) 'Tales from the tailgate: The influence of fandom, musical tourism and pilgrimage on identity transformations', *The Journal of Fandom Studies*, 5(1): 43–62.

Evans, M.T. (2003) 'The sacred: Differentiating, clarifying and extending concepts', *Review of Religious Research*, 45(1): 32–47.

Fallot, R.D. (2001) 'Spirituality and religion in psychiatric rehabilitation and recovery from mental illness', *International Review of Psychiatry*, 13(2): 110–116.

Falque, E. (2016) *Crossing the Rubicon: The Borderlands of Philosophy and Theology*. Trans. Reuben Shank. New York: Fordham University Press.

Fiske, J. (1992) 'The cultural economy of fandom', in L.A. Lewis (ed.) *The Adoring Audience: Fan Culture and Popular Media* (pp. 30–49). London and New York: Routledge.

Fuschillo, G. (2020) 'Fans, fandoms, or fanaticism?', *Journal of Consumer Culture*, 20(3): 347–365.Gammon, S. (2004) 'Secular pilgrimage and sport tourism', in B.W. Ritchie & D. Adair (eds.) *Sport Tourism: Interrelationships, Impacts and Issues* (pp. 30–45). Clevedon, UK: Channel View Publications.

Gauthier, F., Martikainen, T., & Woodhead, L. (2013) 'Acknowledging a global shift: A primer for thinking about religion in consumer societies', *Implicit Religion*, 16(3): 261–276.

Geraghty, L. (2018) 'Hallowed place, toxic space: "celebrating" Steve Bartman and Chicago Cubs' fan pilgrimage', *Participations: Journal of Audience & Reception Studies*, 15(1): 348–365.

Geraghty, L. (2019) '"Everybody needs good neighbours": Transcultural capital, fan pilgrimage and the official Neighbours tour', in C. Lam, & J. Raphael (eds.) *Aussie Fans: Uniquely Placed in Global Popular Culture* (pp. 89–104). Iowa City: University of Iowa Press.

Giuffre, L. (2014) 'Music for (something other than) pleasure: Antifans and the other side of popular music appeal', in L. Duits, K. Zwaan, & S. Reijnders (eds). *The Ashgate Research Companion to Fan Cultures* (pp. 49–62). Burlington, VT: Ashgate.

Glosson, S. (2020) *Performing Jane: A Cultural History of Jane Austen Fandom*. Baton Rouge, LA: Louisiana State University Press.

Graburn, N.H.H. (1989) 'Tourism: The sacred journey', in V.L. Smith (ed.) *Hosts and Guests: The Anthropology of Tourism* (pp. 21–36). Philadelphia: University of Pennsylvania Press.

Gray, J., Sandvoss, C., & Harrington, C.L. (2007) 'Introduction: Why study fans?', in J. Gray, C. Sandvoss, & C.L. Harrington (eds.) *Fandom: Identities and Communities in a Mediated World* (pp. 1–16). New York: New York University Press.

Grimes, R.L. (1999) 'Jonathan Z. Smith's theory of ritual space', *Religion*, 29(3): 261–273. DOI: 10.1006/reli.1998.0162.

Guschwan, M. (2011) 'Fans, Romans, countrymen: Soccer fandom and civic identity in contemporary Rome', *International Journal of Communication*, 5: 1990–2013.

Haberski, Jr., R. (2018) 'Civil religion in America', *Oxford Research Encyclopedia of Religion*. Available at: https://oxfordre.com/religion/view/10.1093/acrefore/9780199340378.001.0001/acrefore-9780199340378-e-441 (accessed 16 October 2020).

Harrison, V.S. (2006) 'The pragmatics of defining religion in a multi-cultural world', *International Journal for Philosophy of Religion*, 59(3): 133–152.

Higgins, L., & Hamilton, K. (2020) 'Pilgrimage, material objects and spontaneous communitas', *Annals of Tourism Research*, 81: 102855. DOI: 10.1016/j.annals.2019.102855.

Hill, P.C., & Hood, Jr., R.W. (1999) 'Affect, religion, and unconscious processes', *Journal of Personality*, 67(6): 1015–1046.

Hills, M. (2000) 'Media fandom, neoreligiosity, and cult(ural) studies', *The Velvet Light Trap*, 46: 73–84.

Hills, M. (2002) *Fan Cultures*. London and New York: Routledge.

Hills, M. (2017) 'Sherlock "content" onscreen: Digital Holmes and the fannish imagination', *Journal of Popular Film and Television*, 45(2): 68–78.

Istoft, B. (2010) 'Avatar fandom as nature-religious expression?', *Journal for the Study of Religion, Nature & Culture*, 4(4): 394–413.

Ito, M., Okabe, D., & Tsuji, I. (eds.) (2012) *Fandom Unbound: Otaku Culture in a Connected World*. New Haven, CT: Yale University Press.

Jackson, R.H., & Henrie, R. (1983) 'Perception of Sacred Space', *Journal of Cultural Geography*, 3(2): 94–107. DOI: 10.1080/08873638309478598.

Jang, K. (2020a) 'Contents tourism and religious imagination', in T. Yamamura & P. Seaton (eds.) *Contents Tourism and Pop Culture Fandom: Transnational Tourists Experiences* (116–127). Bristol, UK: Channel View.

Jang, K. (2020b) 'Creating the sacred places of pop culture in the age of mobility: Fan pilgrimages and authenticity through performance', *Journal of Tourism and Cultural Change*, 18(1): 42–57.

Jenkins, H. (2014) in M. Ratto, & M. Boler (eds.) *DYI Citizenship: Critical making and Social Media* (pp. 65–73). Cambridge, MA: MIT Press.

Jenkins, H. (2015) '"Cultural acupuncture": Fan activism and the Harry Potter alliance', in L. Geraghty (ed.) *Popular Media Cultures* (pp. 206–229). New York: Palgrave Macmillan.

Jenkins, H. (2018) 'Fandom, negotiation, and participatory culture', in P. Booth (ed.) *A Companion to Media Fandom and Fan Studies* (pp. 13–26). Hoboken, NJ: John Wiley & Sons, Inc.

Jenson, J. (1992) 'Fandom as pathology: The consequences of characterization', in L.A. Lewis (ed.) *The Adoring Audience: Fan Culture and Popular Media* (pp. 9–29). London and New York: Routledge.

Jindra, M. (1994) 'Star Trek fandom as a religious phenomenon', *Sociology of Religion*, 55(1): 27–51.

Jones, O. (2011) 'Geography, memory and non-representational geographies', *Geography Compass*, 5(12): 875–885.

Joshanloo, M. (2018) 'Cultural religiosity as the moderator of the relationship between affective experience and life satisfaction: A study in 147 countries', *Emotion*, 19(4): 629–636. DOI: 10.1037/emo0000469.

Kaelber, L. (2006) 'Paradigms of travel: From medieval pilgrimage to the postmodem virtual tour', in D.J. Timothy & D.H. Olsen (eds.) *Tourism, Religion and Spiritual Journeys* (pp. 49–63). London and New York: Routledge.

Karpovich, A.I. (2008) 'Locating the Star Trek experience', in L. Geraghty (ed.) *The Influence of Star Trek on Television, Film and Culture* (pp. 199–217). Jefferson, NC: McFarland.

Kelly, W.W. (2004) 'Introduction: Locating the fans', in W.W. Kelly (ed.) *Fanning the Flames: Fans and Consumer Culture in Contemporary Japan* (pp. 1–16). Albany: State University of New York.

King, C. (1993) 'His truth goes marching on: Elvis Presley and the pilgrimage to Graceland', in I. Reader and T. Walter (eds.) *Pilgrimage in Popular Culture* (pp. 92–104). London: Palgrave Macmillan.

King, E.F. (2010) *Material Religion and Popular Culture*. London and New York: Routledge.

Kirby-Diaz, M. (2009) *Buffy and Angel Conquer the Internet: Essays on Online Fandom*. Jefferson, NC: McFarland & Company.

Knoblauch, H. (2008) 'Spiritualty and popular culture in Europe', *Social Compass*, 55(2): 140–153.

Knox, D., & Hannam, K. (2014) 'The secular pilgrim: Are we flogging a dead metaphor?', *Tourism Recreation Research*, 29(2): 236–242.

Korobkova, K.A. (2014) *Schooling the Directioners: Connected Learning and Identity-Making in the One Direction Fandom*. Irvine, CA: Digital Media and Learning Research Hub.

Korstanje, M., & Olsen, D.H. (2020) 'Negotiating the intersections between dark tourism and pilgrimage', in D.H. Olsen & M.E. Korstanje (eds.) *Dark Tourism and Pilgrimage* (pp. 1–15). Wallingford, UK: CABI.

Kruger, R. (2015) 'Stormtrooper: Intellectual property', *Without Prejudice*, 15(11): 59–61.

Kundnani, A. (2012) 'Radicalisation: The journey of a concept', *Race & Class*, 54(2): 3–25.

Larsen, K. (2015) '(Re)claiming Harry Potter fan pilgrimage sites', in L.S. Brenner (ed.) *Playing Harry Potter: Essays and Interviews on Fandom and Performance* (pp. 38–54). Jefferson, NC: McFarland.

Larson, M., Lundberg, C., & Lexhagen, M. (2013) 'Thirsting for vampire tourism: Developing pop culture destinations', *Journal of Destination Marketing & Management*, 2(2): 74–84.

Lee, C.T. (2015) 'Keeping the magic alive: The fandom and "Harry Potter experience" after the franchise', in C.E. Bell (ed.) *From Here to Hogwarts: Essays on Harry Potter Fandom and Fiction* (pp. 54–77). Jefferson, NC: McFarland.

Lee, Y.J., Oh, H.J., & An, S.K. (2019) 'Characteristics analysis and utilization plans of K-POP fandom records for popular music archives: Focused on the case of BTS fandom', ARMY', *The Korean Journal of Archival Studies*, 60) 161–194.

Leonard, S. (2005) 'Progress against the law: Anime and fandom, with the key to the globalization of culture', *International Journal of Cultural Studies*, 8(3): 281–305.

Levitt, L. (2018) 'Fandom and its afterlife: Celebrity cemetery tourism', in C. Lundberg & V. Ziakas (eds.) *The Routledge Handbook of Popular Culture and Tourism* (pp. 195–200). London and New York: Routledge.

Lim, C. (2016) 'Religion, time use, and affective well-being', *Sociological Science*, 3: 685–709.

Lorimer, H. (2005) 'Cultural geography: The busyness of being 'more-than-representational'', *Progress in Human Geography*, 29(1): 83–94.

Lužný, D. (2020) 'Invented religions and the conceptualization of religion in a highly secular society: The Jedi religion and the Church of Beer in the Czech context', *European Journal of Cultural Studies*, DOI: 1367549420919876.

MacCannell, D. (1976) *The Tourist: A New Theory of the Leisure Class*. New York: Schocken.

Mälksoo, M. (2012) 'The challenge of liminality for international relations theory', *Review of International Studies*, 38: 481–494.

Margry, P.J. (2008) 'The pilgrimage to Jim Morrison's grave at Père Lachaise Cemetery: The social construction of sacred space', in P.J. Margry (ed.) *Shrines and Pilgrimage in the Modern World: New Itineraries into the Sacred*. Amsterdam, The Netherlands: Amsterdam University Press.

Margry, P.J. (2014) 'Whiskey and pilgrimage: Clearing up commonalities', *Tourism Recreation Research*, 39(2): 243–247.

Massumi, B. (2015) *Politics of Affect*. Cambridge, UK: John Wiley & Sons.

Maud, J. (2017) 'Buddhist relics and pilgrimage', in M. Jerryson (ed.) *The Oxford Handbook of Contemporary Buddhism* (pp. 421–435) New York: Oxford University Press.

McCarron, K. (2006) 'A universal childhood: Tourism, pilgrimage and the Beatles', in K. Womack & T.F. Davis (eds.) *Reading the Beatles* (pp. 169–182). Albany: State University of New York Press.

McCormick, C.J. (2018) 'Active fandom: Labor and love in the Whedonverse', in P. Booth (ed.) *A Companion to Media Fandom and Fan Studies* (pp. 369–384). Hoboken, NJ: John Wiley & Sons, Inc.

McLaren, C., & Jin, D.Y. (2020) '"You can't help but love them": BTS, transcultural fandom, and affective identities', *Korea Journal*, 60(1): 100–127.

McLaren, D.L., & Porter, J.E. (1999) '(Re)covering sacred ground: New age spirituality in Star Trek: Voyager', in J.E. Porter & D.L. McLaren (eds.) *Star Trek and Sacred Ground: Explorations of Star Trek, Religion, and American Culture* (pp. 101–115). Albany: State University of New York.

Mitchell, D. (1995) 'There's no such thing as culture: Towards a reconceptualization of the idea of culture in geography', *Transactions of the Institute of British Geographers*, 20(1): 102–116.

Moreno, J.A. (2020) 'Game rules vs. fandom. How Nintendo's Animal Crossing fan-made content negotiates the videogame meanings', *Ámbitos: Revista Internacional de Comunicación*, 47: 212–237.

Morinis, A. (1992) 'Introduction: The territory of the anthropology of pilgrimage', in A. Morinis (ed.) *Sacred Journeys: The Anthropology of Pilgrimage* (pp. 1–28). Westport, CT: Greenwood Press.

Moy, L.Y.Y., & Phongpanichanan, C. (2018) 'Thai star's appeal to Chinese fans and its impact on Thailand popular culture tourism', in C. Lundberg & V. Ziakas (eds.) *The Routledge Handbook of Popular Culture and Tourism* (pp. 214–225). London and New York: Routledge.

Murray, N. (2017a) 'Northern Ireland Tourist Board and HBO: A critical evaluation of a digital media marketing alliance', in M. Stieler (ed.) *Creating Marketing Magic and Innovative Future Marketing Trends. Developments in Marketing Science: Proceedings of the Academy of Marketing Science* (pp. 253–263). Cham, Switzerland: Springer.

Murray, N.M. (2017b) 'GoT Belfast?: How a television epic about a war-torn land was employed to rebrand Northern Ireland', in A. Bayraktar & C. Uslay (eds.) *Global Place Branding Campaigns across Cities, Regions, and Nations* (pp. 1–24). Hershey, PA: IGI Global.

Napier, S. (2006) 'The world of anime fandom in America' *Mechademia*, 1(1): 47–63.

Neville, R.C. (2018) *Defining religion: Essays in philosophy of religion*. New York: SUNY Press.

Nikunen, K. (2007) 'The intermedial practises of fandom', *Nordicom Review*, 28(2): 111–128.

Norris, C.J. (2013) 'A Japanese media pilgrimage to a Tasmanian bakery', *Transformative Works and Cultures*, 14: 1–16.

Okamoto, R. (2019) *Pilgrimages in the Secular Age: From El Camino to Anime*. Tokyo, Japan: Japan Publishing Industry Foundation for Culture.

Okamoto, T. (2015) 'Otaku tourism and the anime pilgrimage phenomenon in Japan', *Japan Forum*, 27(1): 12–36.

Olsen, D.H. (2010) 'Pilgrims, tourists, and Max Weber's "ideal types"', *Annals of Tourism Research*, 37(3): 848–851.

Olsen, D.H. (2012a) 'Teaching truth in 'third space': The use of history as a pedagogical instrument at Temple Square in Salt Lake City, Utah', *Tourism Recreation Research*, 37(3): 227–237.

Olsen, D.H. (2012b) 'Negotiating religious identity at sacred sites: A management perspective', *Journal of Heritage Tourism*, 7(4): 359–366.

Olsen, D.H. (2014) 'Metaphors, typologies, secularization, and pilgrim as hedonist: A response', *Tourism Recreation Research*, 39(2): 248–258.

Olsen, D.H. (2017) 'Social politics on the move: The case of the Marian Ocean to Ocean Pilgrimage', in M.S.C. Mariani, & A. Trono (eds.) *The Ways of Mercy: Arts, Culture and Marian Routes between Eastand West* (pp. 405–430). Galatina, Italy: Mario Congedo.

Olsen, D.H. (2019) 'Religion, spirituality, and pilgrimage in a globalizing world', in D.J. Timothy (ed.) *Handbook of Globalisation and Tourism* (pp. 270–283). London: Edward Elgar.

Oman, D. (2013) 'Defining religion and spirituality', in R.F. Paloutzian & C.L. Park (eds.) *Handbook of the Psychology of Religion and Spirituality* (pp. 23–47). New York: Guilford Press.
Orsi, R. (2005) *Between Heaven and Earth: The Religious Worlds People Make and the Scholars Who Study Them*. Princeton, NJ: Princeton University Press.
Pearson, R. (2010) 'Fandom in the digital era', *Popular Communication*, 8(1): 84–95.
Phillips, T. (2011) 'When film fans become fan family: Kevin Smith fandom and communal experience'. *Participations*, 8(2): 478–496.
Pike, S.M. (2009) '"Why Prince Charles instead of "Princess Mononoke?" The absence of children and popular culture in The Encyclopedia of Religion and Nature', *Journal of the American Academy of Religion*, 77(1): 66–72.
Porat, A.B. (2010) 'Football fandom: A bounded identification', *Soccer & Society*, 11(3): 277–290.
Porcu, E. (2014) 'Pop religion in Japan: Buddhist temples, icons, and branding', *The Journal of Religion and Popular Culture*, 26(2): 157–172.
Porter, J. (2009) 'Implicit religion in popular culture: The religious dimensions of fan communities', *Implicit Religion*, 12(3): 271–280.
Porter, J.E. (2004) 'Pilgrimage and the IDIC ethic: Exploring Star Trek convention attendance as pilgrimage', in E. Badone & S.R. Roseman (eds.) *Intersecting Journeys: The Anthropology of Pilgrimage and Tourism* (pp. 160–179). Urbana and Chicago: University of Illinois Press.
Possamai, A. (2012) 'Yoda goes to Glastonbury: An introduction to hyper-real', in A. Possamai (ed.) *Handbook of Hyper-Real Religions* (pp. 1–21). Leiden, The Netherlands: Brill.
Pykett, J. (2018) 'Geography and neuroscience: Critical engagements with geography's "neural turn"', *Transactions of the Institute of British Geographers*, 43: 154–169.
Que, G.J. (2014) 'Stagnant fans and retired fans: The other side of Hallyu fandom in the Philippines', in A.U. Gueverra (ed.) *The Hallyu Mosaic in the Philippines: Framing Perceptions and Practice* (pp. 48–62). Manila: Ateneo de Manila University.
Ron, A.S., & Timothy, D.J. (2019) *Contemporary Christian Travel: Pilgrimage, Practice and Place*. Bristol, UK: Channel View Publications.
Russell, J.A. (2003) 'Core affect and the psychological construction of emotion', *Psychological Review*, 110(1): 145–172.
Russell, J.A. (2015) 'My psychological Constructionist perspective, with a focus on conscious affective experience', in L.F. Barrett and J.A. Russell (eds.) *The Psychological Construction of Emotion* (pp. 183–208). New York: The Guilford Press.
Sandvoss, C. (2005) *Fans: The Mirror of Consumption*. Cambridge, UK: Polity Press.
Sandvoss, C., Gray, J., & Harrington, C.L. (2017) 'Why still study fans?', in J. Gray, C. Sandvoss, & C.L. Harrington (eds.) *Fandom: Identities and Communities in a Mediated World* (2nd ed., pp. 1–26). New York: New York University Press.
Saroglou, V., Buxant, C., & Tilquin, J. (2008) 'Positive emotions as leading to religion and spirituality', *The Journal of Positive Psychology*, 3(3): 165–173.
Seaton, P. (2020) 'The contents tourism of Jane Austin's American fans', in T. Yamamura & P. Seaton (eds.) *Contents Tourism and Pop Culture Fandom: Transnational Tourists Experiences* (19–33). Bristol, UK: Channel View.
Seo, M.G., Barrett, L.F., & Bartunek, J.M. (2004) 'The role of affective experience in work motivation', *Academy of Management Review*, 29(3): 423–439.
Shackley, M. (2001) *Managing Sacred Sites: Service Provision and Visitor Experience*. London: Continuum.
Sherbine, K. (2013) 'Becoming-Belieber: Girls' passionate encounters with Bieber culture', *Bank Street Occasional Paper Series*, 30: 36–41.
Shilling, C., & Mellor, P.A. (2014) 'Re-conceptualizing sport as a sacred phenomenon', *Sociology of Sport Journal*, 31(3): 349–376.
Simpson, P. (2011) '"So, as you can see...": Some reflections on the utility of video methodologies in the study of embodied practices', *Area*, 43(3): 343–352.
Singh, K., & Best, G. (2004) 'Film-induced tourism: Motivations of visitors to the Hobbiton movie set as featured in The Lord of the Rings', in W. Frost, G. Croy, & S Beeton (eds.) *International Tourism and Media Conference Proceedings*, Vol. 24 (pp. 98–111). Melbourne: Tourism Research Unit, Monash University.
Smith, J.Z. (1980) 'The bare facts of ritual', *History of Religions*, 20: 112–127.
Soja, E.W. (1996) *Thirdspace: Journeys to Los Angeles and Other Real-and-Imagined Places*. Malden, MA: Blackwell.

Stein, L.E., & Busse, K. (eds.) (2014) *Sherlock and Transmedia Fandom: Essays on the BBC Series*. Jefferson, NC: McFarland.

Storey, J. (2006) *Cultural Theory and Popular Culture: An Introduction* (4th ed). Athens, GA: The University of Georgia Press.

Stump, R.W. (2000) *Boundaries of Faith: Geographical Perspectives on Religious Fundamentalism*. Plymouth, UK: Rowman & Littlefield.

Sugawa-Shimada, A. (2015) 'Rekijo, pilgrimage and "pop-spiritualism": Pop-culture-induced heritage tourism of/for young women'. *Japan Forum*, 27(1): 37–58.

Swalwell, M., Ndalianis, A., & Stuckey, H. (eds.) (2017) *Fans and Videogames: Histories, Fandom, Archives*. London and New York: Routledge.

Taira, T. (2013) 'The category of 'invented religion': A new opportunity for studying discourses on 'religion', *Culture and Religion*, 14(4): 477–493.

Timothy, D.J., & Olsen, D.H. (2006) 'Conclusion—Whither religion and tourism?', in D.J. Timothy & D.H. Olsen (eds.) *Tourism, Religion and Spiritual Journeys* (pp. 271–278). London and New York: Routledge.

Todd, R.M., Cunningham, W.A., Anderson, A.K., & Thompson, E. (2012) 'Affect-biased attention as emotion regulation', *Trends in Cognitive Sciences*, 16(7): 365–372.

Toy, J.C. (2017) 'Constructing the fannish place: Ritual and sacred space in a *Sherlock* fan pilgrimage', *The Journal of Fandom Studies*, 5(3): 255–266.

Turner, V. (1969) *The Ritual Process: Structure and Anti-Structure*. London: Routledge.

Turner, V. (1973) 'The center out there: Pilgrim's goal', *History of Religion*, 12(3): 191–230.

Turner, V., & Turner, E.L.B. (1978) *Image and Pilgrimage in Christian Culture*. New York: Columbia University Press.

UNWTO (2011) *Religious Tourism in Asia and the Pacific*. Madrid: World Tourism Organization.

van Gennep, A.V. (1960) *The Rites of Passage*. Chicago: University of Chicago Press.

van Zoonen, L. (2005) *Entertaining the Citizen: When Politics and Popular Culture Converge*. Lanham, MD: Rowman and Littlefield.

Wann, D.L., & James, J.D. (2018) *Sport Fans: The Psychology and Social Impact of Fandom*. London and New York: Routledge.

Wilcox, R.V. (2015) 'Whedon, Browncoats, and the big damn narrative: The unified meta-myth of Firefly and Serenity', in J.P. Telotte & G. Duchovnay (ed.) *Science Fiction Double Feature: The Science Fiction Film as Cult Text* (pp. 98–114). Liverpool, UK: Liverpool University Press.

Williams, A., Miller, B.A., & Kitchen, M. (2017) 'Jediism and the Temple of the Jedi Order', in C.M. Cusack & P. Kosnáč (eds.) *Fiction, Invention and Hyper-Reality: From Popular Culture to Religion* (pp. 119–133). New York: Routledge.

Williams, J. (2012) 'The Indianapolis 500: Making the pilgrimage to the 'Yard of Bricks'', in J. Hill, K. Moore, & J. Wood (eds.) *Sport, History and Heritage: An Investigation into The Public Representation of Sport* (pp. 247–262). Woodbridge, UK: Boydell and Brewer.

Williams, R. (2011) '"This is the night TV died": Television, post-object fandom and the demise of *The West Wing*', *Popular Communication*, 9(4): 266–279. DOI: 10.1080/15405702.2011.605311.

Williams, R. (2015) *Post-Object Fandom: Television, Identity and Self-Narrative*. London and New York: Bloomsbury Publishing Inc.

Wright, R.R., & Wright, G.L. (2015) 'Doctor who fandom, critical engagement, and transmedia storytelling: The public pedagogy of the doctor', in K. Jubas, N. Taber, & T. Brown (eds.) *Popular Culture as Pedagogy* (pp. 11–30). Rotterdam, The Netherlands: Sense Publications.

Wu, S., Li, Y., Wood, E.H., Senaux, B., & Dai, G. (2020) 'Liminality and festivals—Insights from the East', *Annals of Tourism Research*, 80: 102810. DOI: 10.1016/j.annals.2019.102810.

Yaffe, D. (2013) *Among the Janeites: A Journey through the World of Jane Austen Fandom*. Boston, MA and New York: Houghton Mifflin Harcourt.

Yamamura, T. (2020) 'Travelling *Heidi*: International contents tourism induced by Japanese anime', in T. Yamamura & P. Seaton (eds.) *Contents Tourism and Pop Culture Fandom: Transnational Tourists Experiences* (pp. 62–81). Bristol, UK: Channel View.

Zubernis, L., & Larsen, K. (2018) 'Make space for us! Fandom in the real world', in P. Booth (ed.) *A Companion to Media Fandom and Fan Studies* (pp. 145–159). Hoboken, NJ: John Wiley & Sons, Inc.

8

SECULAR PILGRIMAGES IN A POST-SECULAR WORLD?

Experiential journeys and hope for the future

Chadwick Co Sy Su

Introduction

This chapter discusses the phenomenon of secular pilgrimages. In particular, I focus here on areas of agreement, disagreement, and subsequent convergence in the literature and individual scholarship regarding religious and secular pilgrimage. In doing so, I ask several questions. First, is "secular pilgrimage" a worthwhile label to be used for the long term? Second, what qualifies and does not qualify as secular pilgrimage? Third, what directions can be taken to ensure that pilgrimage sites, whether religious or secular ones, remain sustainable? Fourth, what possibilities exist for the development of countries and sites both in the Global North and Global South as pilgrimage locales? Fifth, how can religious privilege be replaced with a sense of shared humanity? Several assertions and recommendations will be made for the reader to consider.

To answer these questions and make reasonable recommendations, I draw from some of my previous research on this topic (Sy Su 2017, 2018b) as well as others who have wrestled with defining and negotiating secular forms of pilgrimage in a post-secular world. I approach this topic also from my life experiences as a "militant atheist". I have no compunction about being public about my atheism, and I am willing and able to point out inconsistencies with religion—indeed, this atheism is part and parcel of my thought process while writing this chapter. However, unlike other militant atheists who are intolerant of religion and "see themselves as riding to the defence of a world besieged by threatening nonsense" (Kitcher 2011: 2), the militancy of my atheism is tempered by a willingness to find common ground, as tolerance in other beliefs and actions can be "a pragmatic [and effective] tool for avoiding a clash of fundamentalisms and for ending wars about truth and justification" (Fiala 2009: 142).

Secular pilgrimage?

Defining secular pilgrimage may seem simple enough. Often used as a foil to the "sacred" and implying a disassociation with religion, "secular" pilgrimage can be defined as any journey filled with meaning devoid of religious or spiritual content or experiences. However, pilgrimage is generally considered a term with religious parallels, describing journeys motivated by religion or spirituality. Margry (2008) argues that the term "secular pilgrimage" is troublesome at best—a forced intermixing of seemingly incompatible or oxymoronic terms. Indeed, "it is contra-productive to use the concept of pilgrimage as a combination term for both secular and religious phenomena, thereby turning it into much too broad a concept" (p. 14). This was exemplified by Morinis (1992), who broadly, or probably too broadly, defined pilgrimage as a "journey undertaken by a person in quest of a place or a state that he or she believes to embody a valued ideal" (pp. 4–5)—diluting the term to imply through suggesting that any travel that is meaningful is a pilgrimage. As an example, Knox and Hannam (2014) have playfully extended the pilgrimage as a metaphor to describe what they label "hedonistic" types of tourism that encourage the fulfilling of sensual pleasures and self-indulgent lifestyles. They argue that this is a valid use of the term "pilgrimage" because historically religious pilgrimages were partly hedonistic in nature, and as such any journey in search for authenticity and pleasure can be termed a pilgrimage.

In this context, I have always had a certain discomfort with the "pilgrim" label. Before entering into pilgrimage studies, I considered pilgrimages to be solidly in the domain of the religious. As I have studied travel and pilgrimage, I have come to consider the journeys dealing with non-religious locations and motivations to be also pilgrimages, albeit with the qualifying adjective "secular". When I travel, I consider myself a "traveler", a term which to me is probably the most neutral term one can use in order to forego having to justify whether or not I am a pilgrim and why I feel this way. As I have written elsewhere (Sy Su 2017), I think that pilgrimage and travel are one and the same, and I continue to do so. Religious people may have had pride of place at the table of pilgrimage centuries ago. While they may still have that pride of place, non-religious people now have a seat, or even several ones, at the table. In the same manner that virtue was previously thought of as the exclusive province of the theist (Hickson 2020), the atheist now has land in the provinces of both pilgrimage and virtue.

For my purposes here, I am interested in the use of technology, and how through technology one can undertake a pilgrimage without ever leaving the comforts of home. A concrete example of people this is watching a live feed of people climbing Mount Everest or participating in rituals at religious sites virtually because of the COVID-19 pandemic. However, while virtual travel is better for the natural environment and can further democratize (virtual) travel to places that hold special meaning to people around the world, can this really be considered pilgrimage? What about pilgrimages of the "self" that are mediated by books and meditation, where one can sit in their homes and find themselves through inner journeys rather than taking an arduous journey to some "center out there" (Cohen 1992)? Or what about walking labyrinths, in which a person walks through a labyrinth-like design while meditating and finding themselves (Griffith 2002; McGettigan & Voronkova 2016)? Without the act of real movement outside of a home environment, can these activities be considered pilgrimages?

As such, I find myself in a continual need for conceptual clarity regarding what exactly constitutes a pilgrimage. My views of what can be considered a pilgrimage have been shaped immensely by the writings of Goodnow and Bloom (2017) and Greenia (2018). Goodnow and Bloom (2017) suggest that research and philosophical debates regarding the dichotomy between sacred and profane has not had much change since the 1960s. They go on to define one of the features of pilgrimage as it being time and space set apart from ordinary life. Goodnow and Bloom also outline 12 properties of the sacred. Of these 12, commitment defined as a focused attachment to the sacred, stands out to me. Also of the seven components of pilgrimage proposed by Greenia (2018), the idea of pilgrimage being a body-centered enterprise is singularly important, in that traveling itself is a *sine qua non* for something to be called a pilgrimage. Indeed, as Collins-Kreiner (2010) points out, pilgrimage unavoidably calls for movement across space. I thus believe that these two factors (commitment and body-centric movement from the familiar to the unfamiliar) must be in place before a journey can be called a pilgrimage.

What destinations and activities qualify as secular pilgrimages?

Having negotiated some of the difficulties in using the term "secular pilgrimage", here I examine different categorizations that have been claimed to belong under the umbrella of pilgrimage. For example, walking as therapy has been described as pilgrimage by Warfield et al. (2014), who assert that pilgrimage is a therapeutic activity for many people, not only physically but also psychologically, socially, and spiritually. In addition to thru-hiking, which is defined as walking long distances of up to thousands of kilometers along the Appalachian Trail over several months, and having the same motivations to walk as those who walk with religious intent on the Camino de Santiago (Bader 2018), there are micro-pilgrimages, operationalized by Goodnow and Bloom (2017) as lasting four days or less to a place reachable by a four- to five-hour drive. Bader (2018: n.p.) differentiates the micro-pilgrimage from the traditional, in that the former is closer to home, shorter, and more economical than the latter.

National parks, such as Yellowstone, also carry touristic, pilgrimage, and pecuniary valuations (Ross-Bryant 2013; Bremer 2016; see also Chapter 12 this volume). Yoder (2018) uses Monte Verde in Costa Rica as an example of how ecotourism sites are "made sacred." As noted by Sharpley (Chapter 11, this volume), people can find different geographical locales such as wilderness and rural areas, seascapes and oceans, and gazing into the night sky as sacred or spiritual locations, even when they are engaging with these spaces for secular or more hedonistic reasons. As such, in the same way that manifestations of the "hierophanies" or manifestations of the sacred can be found at religious sites, "kratophanies" or the appearance of power is also a likely outcome in aesthetic landscapes and places that inspire wonder and awe.

Visiting war memorials and sites of remembrance are also generally viewed as secular pilgrimages (Seaton 2002), even though they have religious connotations, in part because of their bent toward peace education. The works of Tamashiro (2018) on bearing witness to inhumanity, McIntosh (2015) on the Palestinian conflict, and Blankenship (2018) on Jewish visitors to Holocaust memorials in Berlin are but a few examples in this regard. These works are reminiscent of recent research on dark tourism or thanatourism, which involves travel to places relating to death or other types of human tragedy (Stone 2013; Stone et al. 2018; Martini & Buda 2020). Indeed, my own studies on thanathology and the process of plastination have led me to the conclusion that visits to places of death and the processing of these sites by visitors can be both construed as sacred and secular pilgrimages of and toward the self (Sy Su 2018; see Olsen & Korstanje 2020).

In the same way that visits to see religious figures and leaders in person or to visit their birthplaces are common in religious pilgrimage (Becker 2006; Navarro 2015; Kim & Chen 2020), people also take secular pilgrimages to places associated with the lives, accomplishments, and deaths of celebrities and philanthropists (Wesolowski 2019; Soligo & Dickens 2020), and pop culture locations such as where certain films, books, and songs are set (Brown 2016; Chen & Mele 2017; Salamone 2018) may also be considered secular pilgrimage. Places such as Strawberry Fields in Central Park, New York City, with its association with The Beatles (Kruse 2013), Graceland in Memphis, Tennessee (King 1993; Alderman 2002), and the graves of Ludwig von Wittgenstein and Oscar Wilde (Middleton 2009) most readily come to mind, being reminiscent of the discussion in the previous paragraph on dark tourism and death.

Study abroad programs also fit the definition of secular pilgrimage, with DeGraaf et al. (2013) arguing that at least a semester spent overseas is equivalent to a pilgrimage. Sienkewicz (2018) presents a cogent argument that an integrated studies curriculum enhanced by the opportunity to participate in a one-day walk is a form of pilgrimage. Smith (2019) describes a short-term study abroad program as a chance for academics to impart their enthusiasm for pilgrimage, while at the same time allowing students to make their own discoveries. The opportunities to reflect on different systems of belief may be seen as a journey to the self, a point that I myself make in describing my travels and teaching Intercultural Communication courses in my home university (2017, 2018b).

Of course, there is a long history of comparing tourism to pilgrimage or at least a form of secular pilgrimage (Ambrosio 2007) with tourists as secular pilgrims (Knox & Hannam 2014). Indeed, as my cogitations on the matter have evolved with my increasing involvement in pilgrimage studies, I am reminded of Bauman's (1996; see Tidball 2004) argument that the successors of pilgrims are the vagabond, the stroller, the player, and the tourist, which parallels with Reader's (2014) descriptions of department stores and airport malls as pilgrimage sites. Because of this the sacred and profane, and by implication, pilgrimage and tourism, can never be examined without considering its dichotomous other.

Sustainability

With increases in travel technologies and both full-service and budget airlines advertising discounts on fares, air travel has become accessible to people across a much broader spectrum of socioeconomic classes than in the past. At the same time, technology has also facilitated the growth of remote work, a concept that was close to nonexistent three decades ago. Indeed, the "gig economy" has increased economic flexibility and has contributed in part to an increased propensity to travel. However, this increased propensity to travel has resulted in threats to environmental sustainability, including in some cases threats to the very existence of pilgrimage sites (Shinde 2007). I witnessed this firsthand when I made a journey to Tai Shan (Mount Tai) in Tai'an in China's Shandong Province, the most famous of the five holy mountains of China. In part because the journey to the mountain was not difficult, I encountered hordes of journeyers, even though the temperatures were in the low single digits before wind chill. What made the journey a challenge to me was seeing how this UNESCO Heritage Site's existence was being threatened by overtourism. While I would describe Mount Tai and its footpaths as reasonably clean, there was a lack of waste disposal facilities, as attested to by the water bottles and polystyrene food containers strewn around the site. Every quarter hour or so, our walk was interrupted by the sound of a fellow traveler hocking up phlegm and spitting it onto the ground. The nadir was, unfortunately, in the queue to the restrooms, if one can call those as such. The restrooms were unpartitioned, in that people could see each other defecating while squatting on a toilet. These infrastructure threats certainly took away from any sense of holiness. If Tai Shan was to have inspired emperors and artists to write poetry and songs, my experiences at the site inspired in me, a far inferior person, revulsion, leading me to commit to the discipline of pilgrimage studies.

With increasing populations and disposable income (Allan et al. 2017), it can be reasonably inferred that travel to sites both holy and secular will also increase. There is thus a need for approaches that ensure that pilgrimage locales are responsibly visited by present journeyers and will be preserved for generations to come. Something as simple as conspicuous guideposts or brochures suggesting proper behavior within a site may lessen instances of improper waste disposal. To another extreme, alternative sites for pilgrimage and tourist activity may be developed. Charyn Canyon, located in the east of Almaty in Kazakhstan, is a Central Asian alternative to national parks in the United States, as is the Ala-Archa National Park in the outskirts of Bishkek in Kyrgyzstan.

From a Western perspective, Herrero (2008) describes the old route to the Santiago de Compostela as having nearly been destroyed by roadbuilding. The plenary indulgence awarded by the Catholic Church to those who confess and take communion at the cathedral can be inferred to have led to pilgrims arriving *en masse* on motorized transport, at least until the 1980s (Lois González 2013). The passage of time has inexorably resulted in damage to the images of the Pórtico de la Gloria or the Santiago de Compostela Cathedral's main gate (De los Rios Murillo & Montero Delgado 2019). Even if we were to take away human traffic from the equation time is a most formidable enemy when it comes to the erosion of both human-built and natural sacred sites—humans just speed up this inevitable process. Fortunately, cooperation between the public and private sectors, along with academics and scientists, has resulted in a Santiago de Compostela that inspires awe instead of ruminations about a past long gone. It has been restored, quite simply, to a walking city (Lois González 2013). In a similar vein, there is also a growing literature examining how natural sacred sites can be preserved for both present religious rituals and use by future generations (Wild & McLeod 2008; Verschuuren et al. 2010; Pungetti, Oviedo & Hooke 2012).

Replacing religious privilege with shared humanity

Even with views starting in the 1960s that religion was on the decline (Luckmann 1963; Berger 1967; Gorski 2003), religion still occupies a place of privilege in post-secular societies. This is not surprising, given that around the world an estimated 85–93% of people claim some sort of religious affiliation (Pew Forum 2012). At the same time, different government types view and treat religion differently based upon whether they view religion as truth, danger, utility, or identity, the last which calls for so-called respect for religion (Modood 2010). Yet regardless of how governments interact with religion, religion still undergirds much of the world's sociopolitical structures, including policymaking and views of what constitutes morality, ethics, and values. The dominance of religion also undergirds the debate regarding the privileging of religious forms of pilgrimage over their secular counterpart. An example of this is the Way or Camino of St. James to Santiago de Compostela in Spain, where recent studies have shown that today, many of the people who walk the Camino are not Catholic, but rather either of different religious faiths or no faith at all (Frey 2004; Doi 2011; Egan 2011). Many of the people who travel to Santiago via bicycles or, worse, take buses or drive cars, are not considered "true" or "authentic" pilgrims by those who walk the Camino, as are those who are not religiously affiliated. While those that cycle or drive the Camino may consider themselves "pilgrims", as those for whose motivations are not religious or spiritual, they are considered by those walking the Camino to be nothing more than "tourists" who have no desire for spiritual improvement or who chose to eschew the physical travail of the Camino (Graham & Murray 1997).

At least from my experience as an academic in this part of the Global South, secular pilgrimage is certainly the proverbial poor cousin in the pilgrimage family. In the Philippines where I live, pilgrimage trips are almost exclusively described as being religious in nature, with secular pilgrimages being labeled as educational trips (Moncawe 2017). Given that "pilgrimage" has a religious overtone and is popularly held to be the province of the religious, adding the word "secular" to it as a modifier appears to be an unnecessary deference. Maybe it would be better if instead of using the term "secular pilgrimage" the term "experiential journeys" might be a better descriptor. Using this term would remove the religious overtones and at the same time maintain the good faith accorded to individual travelers where their unique, personal experiences are recognized with no attempt to either denigrate or romanticize their perceived motives. Indeed, the term "experiential journeys" recognizes the shared human experience where everyone, irrespective of religion, is attempting to derive meaning from life and searching for occasional relief from the sufferings or humdrum of daily living. I actually prefer this term to any usage of "pilgrimage", in part because if the point of going on pilgrimage is to negotiate with a god for favors such as fertility, healing, or deliverance from harm, not only does the act of pilgrimage question the supposed omniscience of god—a telling inconsistency that removes any claim to sacredness—but also converts pilgrimage into a business transaction between pilgrims and god(s).

Conversations with theist colleagues and the resulting introspection from these conversations have led me to recognize that there are more similarities than differences within humanity, and as such, I would now much rather look at religious pilgrimages as journeys that allow people to process their experiences and come to terms with their circumstances. Whether they believe that these circumstances are given to them by a god is irrelevant to me; what is relevant to me is that they are doing what they can to understand these experiences. On that point, we are all the same. Even as I was raised Catholic, transitioned to agnosticism in my late teens, became atheist in my early 20s, and am unlikely to be religious ever again, I recognize that living is a chore made easier by the presence of other people who are able to recognize the difference and in spite of such recognition, lessen disagreement.

Looking to the future with optimism

In the end, there will always be theists and atheists. What makes the future bright, at least for me, is that in spite of fundamental differences in belief, people from both sides of the aisle have the ability to work together to improve the quality of life for everyone. While I have presented alternative terminology in this chapter with the use of "experiential journey," I do so first in an attempt at inclusiveness before looking at it as a clean break from the religious fetters of "pilgrimage." Whatever viewpoints one has about morality, god, and the afterlife, common ground can be found in the decision to preserve sites of human and natural value, regardless of whether they are labeled heritage, pilgrimage, or tourist sites, for future generations to enjoy. While it is usually unsafe to generalize, I daresay that every person is likely to end their stay on this planet without being able to visit even a tenth of its attractions. Indeed, there remains a multitude of places to be developed, commoditized, and marketed to broaden human experience and alleviate the burden of overtourism on more popular locales. Such an ambitious goal will enable governments, whether local or national, to improve the facilities and better manage and preserve these well-trodden places.

Even though I have visited 84 countries, I will probably never see the entirety of this world. However, I have seen enough to make certain reasonable conclusions, including the fact that beauty can be found everywhere, whether from the polluted and crowded railways of Dhaka, Bangladesh to the Gullfoss in Iceland. I wish to help in the preservation of all sites of human and natural value so that many more people in the years ahead can see the places I have seen and from which I have derived immense pleasure and equanimity. As such, humanity needs to better educate students, as future travelers, to recognize the stories behind these places of beauty and inspiration and how they can help people find common ground with others. In breaking down religious and cultural barriers, people can learn to live and let live, and in so doing help with seeing the difference with openness of mind and a willingness to question assumptions, as I am willing to do.

References

Alderman, D.H. (2002) 'Writing on the Graceland wall: On the importance of authorship in pilgrimage landscapes', *Tourism Recreation Research*, 27(2): 27–33.

Allan, J.R., Venter, O., Maxwell, S., Bertzky, B., Jones, K., Shi, Y., Watson, J.E.M. (2017) 'Recent increases in human pressure and forest loss threaten many Natural World Heritage Sites', *Biological Conservation*, 206: 47–55. https://doi.org/10.1016/j.biocon.2016.12.011.

Ambrosio, V. (2007) 'Sacred pilgrimage and tourism as secular pilgrimage', in R. Raj & N.D. Morpeth (eds), *Religious Tourism and Pilgrimage Festivals Management: An International Perspective* (pp. 78–88). Wallingford, UK: CABI.

Bader, G. (2018) 'Thru-hiking and pilgrimage: The invention of "secular" pilgrimage routes', Paper presented in the 5th Sacred Journeys Global Conference, Indiana University Gateway, Berlin, Germany, July 5 and 6, 2018.

Bauman, Z. (1996) 'From pilgrim to tourist–or a short history of identity', in S. Hall & P. du Gay (eds), *Questions of Cultural Identity* (pp. 18–36). London: SAGE.

Becker, C. (2006) 'Gandhi's body and further representations of war and peace', *Art Journal*, Winter 2006, 78–95.

Berger, P.L. (1967) *The Sacred Canopy: Elements of a Sociological Theory of Religion*. New York: Anchor Books.

Blankenship, A.M. (2018) 'Jewish tourism in Berlin and Germany's public repentance for the Holocaust', *Academica Turistica*, 11(2), 117–126.

Bremer, T.S. (2016) 'Worshiping at nature's shrine', *Practical Matters*, 9: 1–12.

Brown, L. (2016) 'Tourism and pilgrimage: Paying homage to literary heroes', *International Journal of Tourism Research*, 18(2): 167–175.

Chen, F., & Mele, C. (2017) 'Film-induced pilgrimage and contested heritage space in Taipei City', *City, Culture and Society*, 9: 31–38.

Cohen, E. (1992) 'Pilgrimage centers: Concentric and excentric', *Annals of Tourism Research*, 19(1): 33–50.

Collins-Kreiner, N. (2010) 'Geographers and pilgrimages: Changing concepts in pilgrimage tourism research', *Journal of Economic and Social Geography*, 101(4): 437–448.

DeGraaf, D.G., Slagter, C., Larsen, K., & Ditta, E. (2013) 'The long-term personal and professional impacts of participating in a study abroad program', *Frontiers: The Interdisciplinary Journal of Study Abroad*, 23: 42–59.

De los Rios Murillo, A., & Montero Delgado, J. (2019) 'Estudio del biodeterioro en el Pórtico de la Gloria', *Rudesindus*, 12: 139–146.

Doi, K. (2011) 'Onto emerging ground: Anticlimatic movement on the Camino de Santiago de Compostela', *Tourism*, 9(3): 271–285.

Egan, K. (2011) '"I want to feel the Camino in my leg": Trajectories of walking on the Camino de Santiago', in A. Fedele & R.L. Blanes (eds), *Encounters of Body and Soul in Contemporary Religious Practices: Anthropological Reflections* (pp. 3–22). New York: Berghahn.

Fiala, A. (2009) 'Militant atheism, pragmatism, and the God-shaped hole', *International Journal for Philosophy of Religion*, 65(3): 139–151.

Frey, N.L. (2004) 'Pilgrimage and its aftermath', in E. Badone & S.R. Roseman (eds), *Intersecting Journeys: The Anthropology of Pilgrimage and Tourism* (pp. 89–109). Chicago: University of Illinois Press.

Goodnow, J., & Bloom, K.S. (2017) 'When is a journey sacred? Exploring twelve properties of the sacred,' *International Journal of Religious Tourism and Pilgrimage*, 5(2): 10–16.

Gorski, P.S. (2003) 'Historicizing the secularization debate: An agenda for research', in M. Dillon (ed), *Handbook of the Sociology of Religion* (pp. 110–122). Cambridge: Cambridge University Press.

Graham, B., & Murray, M. (1997) 'The spiritual and the profane: The pilgrimage to Santiago de Compostela', *Ecumene*, 4(4): 389–409.

Greenia, G. (2018) 'What is pilgrimage?', *International Journal of Religious Tourism and Pilgrimage*, 6(2): 7–15.

Griffith, J.S. (2002) 'Labyrinths: A pathway to reflection and contemplation', *Clinical Journal of Oncology Nursing*, 6(5): 295–296.

Herrero, N. (2008) 'Reaching "land's end": new social practices in the pilgrimage to Santiago de Compostela', *International Journal of Iberian Studies*, 21(2): 131–149.

Hickson, M.W. (2020) 'How a Huguenot philosopher realized that atheists could be virtuous'. Available at: https://getpocket.com/explore/item/how-a-huguenot-philosopher-realised-that-atheists-could-be-virtuous?utm_source=pocket-newtab/ (accessed 31 July 2020).

Kim, B., & Chen, Y. (2020) 'Effects of religious celebrity on destination experience: The case of Pope Francis's visit to Solmoe Shrine', *International Journal of Tourism Research*, 22(1): 1–14.

King, C. (1993) 'His truth goes marching on: Elvis Presley and the pilgrimage to Graceland', in I. Reader & Walter, T. (eds), *Pilgrimage in Popular Culture* (pp. 92–104). London: Palgrave Macmillan.

Kitcher, P. (2011) 'Militant modern atheism', *Journal of Applied Philosophy*, 28(1): 1–13.

Knox, D., & Hannam, K. (2014) 'The secular pilgrim: Are we flogging a dead metaphor?', *Tourism Recreation Research*, 39(2): 236–242.

Kruse, R.J. (2003). 'Imagining Strawberry Fields as a place of pilgrimage', *Area*, 35(2): 154–162.

Lois González, R.N.C. (2013) 'The Camino de Santiago and its contemporary renewal: Pilgrims, tourists and territorial identities', *Culture and Religion*, 14(1): 8–22.

Luckmann, T. (1963) *The Invisible Religion: The Problems of Religion in Modern Society.* New York: Macmillan.

Margry, P.J. 'Secular pilgrimage: A contradictionin terms', in P.J. Margry (ed.) *Shrines and Pilgrimage in the Modern World: New Itineraries into the Sacred* (pp. 13–46). Amsterdam, The Netherlands: Amsterdam University Press.

Martini, A., & Buda, D.M. (2020) 'Dark tourism and affect: Framing places of death and disaster', *Current Issues in Tourism*, 23(6): 679–692.

McGettigan, F., & Voronkova, L. (2016) 'Walking labyrinths: Spirituality, religion and wellness tourism', *International Journal of Religious Tourism and Pilgrimage*, 4(5): 37–50.

McIntosh, I. (2015) 'Gaza: Visioning peace in a place like hell', *Palestine-Israel Journal*, 20(2/3): 154–159.

Middleton, D.J.N. (2009) 'Dead serious: A theory of literary pilgrimage', *Cross Currents*, 59(3): 300–319.

Modood, T. (2010) 'Moderate secularism, religion as identity, and respect for religion', *The Political Quarterly*, 81(1): 4–14.

Moncawe, C.M. (2017) *A Study on the Marketing Communications Methods used by Travel Agencies in Promoting Pilgrimages.* Unpublished undergraduate thesis, University of the Philippines, Manila.

Morinis, A. (1992) 'Introduction: The territory of the anthropology of pilgrimage', in A. Morinis (ed), *Sacred Journeys: The Anthropology of Pilgrimage* (pp. 1–28). Westport, CT: Greenwood Press.

Olsen, D. H., & Korstanje, M. (eds.) (2020) *Dark Tourism and Pilgrimage.* Wallingford, UK: CABI.

Navarro, D. (2015) 'Tourist resources and tourist attractions: Conceptualization, classification and assessment', *Cuadernos de Turismo*, 35: 481–484.

Pew Forum on Religion. (2012) 'The global religious landscape'. Available at: https://www.pewforum.org/2012/12/18/global-religious-landscape-exec/ (accessed 21 July 2020).

Pungetti, G., Oviedo, G., & Hooke, D. (eds) (2012) *Sacred Species and Sites: Advances in Biocultural Conservation.* Cambridge: Cambridge University Press.

Reader, I. (2014) *Pilgrimage in the Marketplace.* New York: Routledge.

Ross-Bryant, L. (2013) *Pilgrimage to the National Parks: Religion and Nature in the United States.* London and New York: Routledge.

Salamone, F.A. (2018) 'Jazz pilgrimage', in F.A. Salamone & M.M. Snipes (eds), *The Intellectual Legacy of Victor and Edith Turner* (pp. 71–84). Lanham, MD: Lexington Books.

Seaton, A.V. (2002) 'Thanatourism's final frontiers? Visits to cemeteries, churchyards and funerary sites as sacred and secular pilgrimage', *Tourism Recreation Research*, 27(2): 73–82.

Shinde, K.N. (2007) 'Pilgrimage and the environment: Challenges in a pilgrimage centre', *Current Issues in Tourism*, 10(4): 343–365.

Sienkewicz, T. (2018) 'Experiencing and teaching pilgrimage in a sacred spaces course', *Religions*, 9(4): 102, https://doi.org/10.3390/rel9040102.

Smith, A. (2019) 'Micro-pilgrimage in France and Spain', *The Reflective Practitioner*, 4(1): 7–14. http://cas.upm.edu.ph/journals/index.php/the-reflective-practitioner/article/view/54

Soligo, M., & Dickens, D.R. (2020) 'Rest in fame: Celebrity tourism in Hollywood cemeteries', *Tourism Culture & Communication*, 20(2–3): 141–150.

Stone, P. (2013) 'Dark tourism scholarship: A critical review', *International Journal of Culture, Tourism and Hospitality Research*, 7(3): 307–318.

Stone, P.R., Hartmann, R., Seaton, A.V., Sharpley, R., & White, L. (eds) (2018) *The Palgrave Handbook of Dark Tourism Studies.* London: Palgrave Macmillan.

Sy Su, C.C. (2017) 'Travel and/or pilgrimage—both sacred journeys: An atheist's attempt at inquiry and introspection', in I.S. McIntosh & L.D. Harman (eds), *The Many Voices of Pilgrimage and Reconciliation* (pp. 173–182). Wallingford, UK: CABI.

Sy Su, C.C. (2018a) 'The crossroads of self and others: Plastination and pilgrimage', *Religions*, 9(4): 87. http://dx.doi.org/10.3390/rel9030087.

Sy Su, C.C. (2018b) 'Living and letting live, reframing atheist travel into understanding intercultural differences', in H.A. Warfield & Hetherington, K. (eds), *Pilgrimage as Transformative Process: The Movement from Fractured to Integrated* (pp. 57–64). Leiden, The Netherlands: Brill.

Tamashiro, R. (2018) 'Planetary consciousness, witnessing the inhuman, and transformative learning: Insights from peace pilgrimage oral histories and autoethnographies', *Religions*, 9(5): 148. https://doi.org/10.3390/rel9050148.

Tidball, D. (2004) 'The pilgrim and the tourist: Zygmunt Bauman and postmodern identity', in C. Bartholomew & F. Hughes (eds), *Explorations in a Christian Theology of Pilgrimage* (pp. 184–200). Aldershot, UK: Ashgate.

Verschuuren, B., Wild, R., McNeely, J.A., & Oviedo, G. (2010) *Sacred Natural Sites: Conserving Nature & Culture*. London: Earthscan.

Warfield, H.A., Baker, S.B., & Sejal, B.P.F. (2014) 'The therapeutic value of pilgrimage: A grounded theory study', *Mental Health, Religion & Culture*, 17(8): 860–875: https://doi.org/10.1080/13674676.2014.936845

Wesolowski, A. (2019) 'Circumnavigations of charity: the eighteenth century, pilgrimage, and philanthropic celebrity', *The Reflective Practitioner*, 4(1): 85–99. http://cas.upm.edu.ph/journals/index.php/the-reflective-practitioner/article/view/61

Wild, R., & McLeod, C. (2008) *Sacred Natural Sites: Guidelines for Protected Area Managers*. Gland, Switzerland and Paris: IUCN/UNESCO.

Yoder, S.D. (2018) 'Ecotourism, religious tourism, and religious naturalism', *Journal for the Study of Religion, Nature, and Culture*, 11(3): 291–314.

9

PILGRIMAGE, TOURISM, AND PEACE BUILDING

Ian S. McIntosh

Introduction

Rising international tensions during the Cold War saw peace become one of the driving ideals for global tourism scholars and practitioners in the management of tourism development. Could the tourism industry be an avenue for the de-escalation of tensions or even peace building? *The Declaration on World Tourism* (1980) and the *Tourism Bill of Rights and Tourist Code* (1983), for example, provided guidelines for how tourism could contribute to the development of international understanding and the promotion of friendship and peace. According to Moufakkir and Kelly (2010: xxiii), tourism is an activity capable of not only promoting economic development but also of breaking down political, cultural, and ethnic barriers within and between nations.

However, as Higgins-Desboilles (2006) argues, despite efforts to use tourism as a vehicle for peace-making, especially with the emergence of tourism subfields such as reconciliation tourism, pro-poor tourism, and justice tourism, the industry has not delivered to any significant degree on its promise. While organizations like the International Institute for Peace through Tourism—established in the 1980s by Louis D'Amore—focus on using the global tourism industry to "bring about peace for humanity and for nature," tangible results have been at best minimal (Hill et al. 1995: 709; see D'Amore 1988, 2009). As Higgins-Desboilles (2006) notes, in the present-day neoliberal era, the discourse of tourism as an "industry" and the extolling of its economic benefits have overshadowed tourism's potential as a vital social force for the greater common good. Yet as Jafari (1989: 154) notes, tourism—properly designed—has the potential to help bridge the psychological and cultural distance that separates people of diverse races, ethnicities, and religions. Through tourism, he says, people "...can come to appreciate the rich human, cultural and ecological diversity that our world mosaic offers and to evolve a mutual trust and respect for one another and the dignity of all life on earth".

Taking an interdisciplinary approach to this proposition, this chapter begins with definitions of "positive peace" and then investigates how certain aspects of the tourism industry might contribute to this goal. In determining that pilgrimage has the greatest potential in the sector for delivering the desired results, I revisit the standard definitions of this universal practice and argue that a paradigm shift is required in the way that scholars and practitioners define pilgrimage in order to better comprehend its potential for peace building. While not all pilgrims are peace-oriented, and not all pilgrimages have a peace dimension, there is evidence arising from intergroup contact theory to show that pilgrims exposed to a wide diversity of cultures and landscapes will return with a greater openness to other ways of being religious in the world and generally more tolerant of others. I also present a critique of the theory of antagonistic tolerance in an analysis of the growing trend toward interfaith pilgrimage on the global stage. Finally, the chapter highlights other examples of pilgrimage and peace building, including the journeys of inspired individuals, political pilgrimages geared toward peace, and the green pilgrimage movement.

What is peace?

The term "peace" can be difficult to define, as it is often associated with words such as justice, love, and happiness. Indeed, people often identify or recognize peace simply by the absence of open conflict (Webel 2007). Standard dictionaries, for example, define peace as a state of tranquility or quiet between two parties after formal negotiations and the signing of a truce or an accord. Johan Galtung (1988), a founder of the discipline of peace studies, defines peace not by the aforementioned absence of strife—which he calls "negative peace"—but by the presence of justice, a concern for human rights, and an assurance of security—or "positive peace". According to Haessly (2010: 4), the idea of peace is a holistic one, existing in all cultures and languages—*Shanti* (Sanskrit), *Mir* (Russian), *Ping* (Chinese), *Amani* (Swahili), *Hotep* (ancient Egyptian), *Shalom* (Hebrew), and *Salaam* (Arabic). The writings of Lao-Tse, the founder of Taoism, echo these holistic ideas of peace:

> If there is to be peace in the world, there must be peace in the nations.
> If there is to be peace in the nations, there must be peace in the cities.
> If there is to be peace in the cities, there must be peace between neighbors.
> If there is to be peace between neighbors, there must be peace in the home.
> If there is to be peace in the home, there must be peace in the heart.
>
> *(Ni 1979)*

Following Galtung's lead, the idea of negative and positive peace allows us to consider the distinction between individual journeys designed for personal benefit, including inner peace, and those journeys that are associated with more universal or collective benefits. When we focus our attention on this latter dimension, Indian theologian Deenabandhu Manchala sets the bar very high. He says that the traveler with justice and peace in mind must begin by confessing his or her complicity with structures, cultures, and systems that cause, nurture, and legitimize injustice and human aggression. More specifically, the onus is upon all those involved in the tourism and pilgrimage industries to commit to

> ...effecting transformation of structures and cultures that deny life and keep many in endless cycles of oppression and exploitation, poverty and misery [in particular the]... victims of racism, casteism, and patriarchy...and many others who remain nameless and faceless, existing only as categories.
>
> *(Manchala 2014: 141)*

While this may seem an insurmountable barrier, tourism and pilgrimage have major roles to play in the peace-building process, for in what other fields of human activity are peoples coming together in goodwill and where there is the potential for positive interaction and dialogue? Probing more deeply into this issue, Inayatullah (1995) calls for nothing less than a full societal transformation as a basic prerequisite for peace building. If "positive peace" is the desired outcome of sacred and secular journeys, there are many important things to consider, including how tourism affects the distribution of wealth, creates sustainable economic growth, reduces structural or systemic violence, and fosters the idea of cultural pluralism (Inayatullah 1995: 413).

Tourists as ambassadors of peace?

Tourism, as a dominant form of and reason for human mobility, is heavily involved in the facilitation of cross-cultural interaction. With approximately 1.4 billion international arrivals on the global stage in 2018 (UNWTO 2019), there are innumerable opportunities for person-to-person encounters. However, the outcome of these encounters is unclear, and it is unknown whether such first-hand experiences help to create any alternative sets of relationships that might gradually overcome cultural stereotypes or prevent conflict (Kim & Crompton 1990; te Kloeze 2014; Pratt & Liu 2016). Indeed, the very opposite may occur. Travel may lead to conflict both directly and indirectly. Conflict can occur when tourism stakeholders, for example, disagree over the details or the scale of a particular tourism project (Bramwell & Lane 2000; Uddhammar 2006; Dredge 2010). The ownership, representation, interpretation, and commodification of heritage sites can also be a source of rising tensions (Boniface & Robinson 1999; Olsen 2003; Porter & Salazar 2005; Walton 2005; Winter 2007; Poria & Ashworth 2009; Yang, Ryan & Zhang 2013; Olsen and Emmett 2020). Conflict can also occur between tourists and local residents when the negative impacts of tourism are seen to outweigh the positive impacts, when there is no prior informed consent to a project or when it is perceived that there is too much tourism at a destination (*i.e.*, overtourism) (Dodds & Butler 2019; Milano, Cheer & Novelli 2019).

On the other hand, tourism can lead to peaceful outcomes when managed in a way that it becomes a "social force" for peace building (Higgins-Desboilles 2006). As Haessly (2010: 14) argues, tourism can contribute to positive peace only when everyone involved in the industry (i.e., politicians, tourists, employees, providers) commits to the following guidelines of

- honoring spiritual traditions;
- acknowledging and protecting diverse cultural spaces and traditions
- reducing poverty and engaging local communities in the development of tourism;
- eliminating the conditions that lead to conflict and violence, and engage in conflict resolution;
- promoting sustainable development;
- promoting ecotourism; and
- promoting and preserving a culture of peace by supporting business and organizations that engage in socially and environmentally responsible business practices.

Tourism providers therefore need to ask themselves whether the experiences they are offering are building bridges of trust and understanding between different ethnic, racial, or cultural groups, fostering respect for human rights, and also promoting a vision of a just and equitable world (Moufakkir & Kelly 2010).

Yet as Pratt and Liu (2016) argue, tourism is more often than not a beneficiary of peace rather than a driver of justice and reconciliation. Tourism is certainly an integral part of two-track diplomacy, which consists of the official government channels between countries and the informal or unofficial personal encounters between citizens of those countries (Kim &

Crompton 1990). A powerful example of this is in Sri Lanka following the end of the civil war (McIntosh & Paramananda 2020) where increasing numbers of Tamils and Sinhalese are exploring their country, often for the first time in a generation. This presents the population with an opportunity for breaking down stereotypes by engaging in "natural dialogue" (Farra Haddad 2020). With an easing of the former surveillance and security measures, and with passes no longer required for Tamils in the north and east, movement about the country has accelerated. Intergroup contact theory (Vezzali 2016) predicts that this interaction will have the potential to bridge the psychological and cultural distances that separate people of diverse ethnicities, religions, and stages of social and economic development. As American writer and humorist Mark Twain (1869: 650) once remarked, travel is fatal to prejudice, bigotry, and narrow-mindedness. Closed societies, by contrast, are prone to suspicion and hostility and are conducive to nurturing fear and conflict. From this perspective, perhaps the greatest role that the tourism industry can play in peace building in Sri Lanka is simply providing opportunities for the parties in conflict to come together in a neutral territory to interact as they please without interference.

While the empirical evidence for tourism as a force for peace is either lacking or under review (Becken & Carmigani 2016), one can rightfully ask if certain forms of tourism have a greater potential for peace building than others. To this end, LeSueur (2018) has suggested a conceptual framework that can assist in determining the tourism markets that might be the best fit. Based on the dichotomistic relationship between Hardship on one side of the horizontal axis and Ease and Comfort on the other side, and Isolation and Immersion on different ends of the vertical axis, LeSeuer, in a somewhat oversimplified manner, labels the quadrants *Comfort and Rest*, *Adventure Tourism*, *Cultural Tourism*, and *Pilgrimage or Religious Tourism*. Each of these types of tourism niche markets/experiences can add value to a tourism destination through economically supporting local and mid-size enterprises, protecting the natural and built environment, providing employment to indigenous or other minority groups, and fostering pride in local and national cultures. However, pilgrimage or religious tourism is the category that might best help to achieve peace-related activities and outcomes.

Revisiting the role of pilgrimage

Definitions matter, in that the way that we define a phenomenon will strongly influence and create culturally prescribed ways in which people view, understand, and react to its specific qualities and properties (McIntosh, Farra Haddad & Munro 2020). For the purposes of this chapter, understanding what constitutes pilgrimage is culturally contingent. While the term is used in a cross-cultural comparative manner in Western and non-Western modes of religious travel, it is based on Western conceptions of religious mobility. As such, attention needs to be paid to how the term is understood and utilized in non-Western cultural contexts (Albera & Eade 2015; Eade & Albera 2016). What are viewed as essential truths from one standpoint may be incomprehensible in other social and cultural contexts (McIntosh, Farra Haddad & Munro 2020).

Many scholars have offered their personal definitions of pilgrimage. For example, for Margry (2008: 17) says that pilgrimage is a journey "to a place that is regarded as more sacred or salutary than the environment of everyday life, to seek a transcendental encounter with a specific cult object for the purpose of acquiring spiritual, emotional or physical healing or benefit." Adding more nuance, Haberman (1993: 7) states that during a pilgrimage,

> Pilgrims take to the road in search of some object, often quite vague, which promises to provide something to fill the painful holes in their lives. This object of yearning is

> difficult to pin down; it is experienced as that which is missing, some unnamed object lost long ago; it is that haunting lack which engenders the incessant flight from one thing to another. The promise of fulfillment, of wholeness, of perfection, of completion lures us out onto the road to begin a quest.

These definitions of what pilgrimage is, and what pilgrims desire to achieve, suggest that the practice is primarily an individual quest, wherein people travel within a culturally prescribed mode of travel to find individual meaning and reward. However, this individualistic approach raises an important question: Can a profoundly individual act of devotion have consequences beyond these specific desired personal outcomes? Pilgrimage, when considered as a transformative journey focused on the individual or solitary traveler, severely limits consideration of any broader, societal impacts that may come from the pilgrimage experience. As McIntosh, Farra Haddad, and Munro (2020) argue, emphasizing the role and experiences of the individual, while paying scant attention to the social and cultural context of the pilgrimage, blinds us to this deeper dimension.

As such, there needs to be a paradigm shift in how one understands pilgrimage, particularly when it comes to its social and cultural ramifications. Pilgrims and pilgrimage practices are embedded not just within religions but also within sociocultural, political, and economic systems. While individuals may be seeking personal and spiritual growth, healing, or specific blessings through the act of pilgrimage, the societies in which these pilgrimages are embedded have their own specific interests, needs, and agendas and these are always in flux. When only focusing on the narrow lens of individuality, it is difficult to understand the potential impacts at any other level.

There are many instances where pilgrimages are based on accomplishing or celebrating broader community-oriented goals. For example, the Dhammayietra "pilgrimage of truth" initiated by Buddhist Monk Maha Ghosnananda in Cambodia during the last stages of the Khmer Rouge oppression, was framed in terms of rebuilding the country after the devastation of the Pol Pot years, with each step considered both a prayer and a stage in bridge-building (Poethig 2002). Likewise, in the great Wari pilgrimage in Maharashtra in India, the major beneficiary is society itself, with each pilgrimage act performed by hundreds of thousands of devotees reinforcing the sacred values that lie at the heart of their pilgrimage such as social justice, equality, and dignity. Likewise, Palka (2014) describes how Mayan pilgrimage was an integral part of ancient Mayan culture, being centered around community economic and religious goals. Pilgrimage, he argues, not only reinforced social and political roles and Mayan identity but also perpetuated devotion to the gods in return for protection and prosperity. Important for maintaining cosmic balance and world order, pilgrimage had a significance that lay well beyond the level of individuals.

If the role of pilgrimage is to be considered in relation peace building, classic definitions therefore need to acknowledge the aspirations that a community has for itself and the role that sacred journeys play in fulfilling these aspirations. As Cohen (1979) and Vukonić (1996) have argued, this journey within a journey needs viewing from a macro perspective if it is to make any sense at all. Just as definitions of tourism tend to have two distinct aspects—the traveler's experience and the industry itself—definitions of pilgrimages should look beyond individual motives and outcomes and more to its role as part of living communities with their own plans and special interests. In the light of such a repositioning, scholars will then be in a better position to both appreciate and document the potential impact of religious and spiritual journeys as they either challenge or endorse the status quo in any given setting.

Paradox I: pilgrimage and peace building

According to Vukonić (1996: 127; see Timothy & Iverson 2006), pilgrims participating in the Hajj or the pilgrimage to Mecca seek the forgiveness of their sins, the attainment of wealth and prosperity, and/or good health. In the author's research, Hajj pilgrims have also stressed that the Hajj promotes a message of peace—or at the very least the building blocks of peace—like justice, reconciliation, tolerance, and compassion. As one pilgrim said to me: "The goal is peace within, peace with other Muslims, peace with non-Muslims, and peace with the environment." (McIntosh 2017). However, Social Identity Theory (SIT), which focuses on how individual identity is formed in part from belonging to such social groups (Stets & Burke 2000; Hornsey 2008), suggests that participation in mass rituals such as the Hajj will inevitably lead to an intensification of in-group identification to the exclusion of other groups.

Paradoxically, however, research shows that the very opposite occurs during the Hajj. In their study of Pakistani pilgrims to Mecca, Clingingsmith, Khwaja, and Kremer (2009) investigated changes in in-group orientation by comparing the viewpoints of successful and unsuccessful applicants for the Hajj lottery used by the Pakistan government to allocate pilgrim visas. What they found was that those who participated in the Hajj did exhibit an intensification in in-group global Islamic practices such as prayer and fasting. However, they also found that their Hajj experience led to an increased belief in equality and harmony among ethnic groups and Islamic sects, and that they developed more favorable attitudes toward women, including their inclusion in education and employment. Most interestingly, Hajj participants also showed an increased belief in the necessity of working toward peace and equality and being in harmony with adherents of other monotheistic religions. Alexseev and Zhemukhov (2017) completed a similar study with Muslim Circassians from southern Russia who had participated in the Hajj and found that the pilgrimage to Mecca re-personalizes the pilgrims toward greater openness for engaging in public life and accepting religious, ethnic, and national diversity.

The research noted above opens the way for further studies on other mass rituals. For example, the Sanctuary of Our Lady of Lourdes in France annually attracts over five million visitors. One notable event on the annual Lourdes calendar is a sacred ceremony for former war combatants from up to 40 countries who come together in a spirit of forgiveness (Warriors to Lourdes 2020). What lessons for peace building can be drawn from their experience of what appears to be a profound healing and renewed faith that can be replicated elsewhere? Does their forgiveness and acceptance of others extend to other religions?

The Arba'een pilgrimage in Iraq, which has strong human rights and justice focus, is another important potential case study of mass ritual and peace building. Does participation in this walking pilgrimage, and being in the presence of millions of devotees from across the Shi'a world, promote greater openness and respect for other faith and cultural groups?

Likewise, the Camino de Santiago attracts more than 250,000 pilgrims each year with as many nonreligious and spiritual pilgrims as Catholic pilgrims. Does this life-changing sacred journey, famous for promoting camaraderie among the walkers, also promote tolerance and understanding across social, cultural, and religious boundaries post-pilgrimage?

Paradox II: antagonistic tolerance?

There is, however, another paradox within discussions of pilgrimage and peace building. Hayden et al. (2016) note that there are instances where there appears to be a real spirit of acceptance and cooperation between faith groups in varied settings which have witnessed centuries of conquest and domination. However, when something occurs to upset the status quo, these same groups may turn upon one another with an unparalleled fury. The powerful influence for peace building of interfaith cooperation is completely undermined.

In examining the old Balkan city of Sarajevo, Hayden et al. suggest that just because there is a mosque, a synagogue, a Catholic church and an Orthodox church located in close proximity does not mean that interfaith cooperation is or has ever been a cherished value in that city. Even though some Muslims attend certain Christian pilgrimages and Christians repair old mosques that were former sites of pilgrimage, "antagonistic tolerance" or "competitive sharing" underlies the relationships between the faiths. According to this theory, interactions between religious groups will be generally peaceful but when circumstances change the balance of power between these groups, allowing one to encroach upon the rights or domain of the other, all tolerance is forsaken and violence erupts (Hayden et al. 2016).

For Hayden et al., antagonistic tolerance is the fulcrum upon which all religious relations revolve. Such antagonism is overridden only when a disinterested power, like the Ottomans under Mehmed 2 or the Communists of former Yugoslavia under Tito, or in cases of extreme poverty like in India where differences in faith had no bearing of your chances of survival, that peaceful relations between religious groups might occur for any extended period of time.

The problem with Hayden et al.'s focus on "antagonistic tolerance" is that it unjustifiably places the gaze squarely in the negative realm, where fear and distrust are the defining features of the relationships between different faith groups. In today's plural and multi-faith communities, however, there seems to be a renewed urgency for people of faith to publicly affirm and celebrate their shared values (Cornille 2013). There is ample evidence today, for example, of how crossing religious borders in a public and highly visible way is the norm rather than an exception, such as in Ethiopia and Lebanon where for centuries cooperative social activities have united communities (Dagnachew 2020; Farra Haddad 2020a, 2020b). Indeed, even in locations where there is a real potential for conflict between faith groups, there is often an undercurrent of deep respect that has the potential to alter the trajectory of relations. In Egypt, for example, each year hundreds of thousands of Muslims join with Christians at the Virgin Mary apparition sites in Egypt, such as Zaitoun. Despite the very real threats to their person by hardliners, Islam has the deepest regard for the Virgin Mary and the shared human need for spirituality, mysticism, and beauty sees many Muslims and Christians engaged in worship side by side (McIntosh 2017). In interfaith pilgrimages across India, both a Muslim and a Dalit ("Untouchable") can freely participate in the Hindu Wari "pilgrimage of joy", as can non-Muslims in the Sufi pilgrimage to Ajmer. While terrorist bombing has occurred in Ajmer, pilgrim participants continue this collaborative practice because they are focused on the positive realm of unity and harmony rather than the negative realm of antagonistic tolerance (McIntosh 2017).

Peace pilgrimages

Although more research is needed to determine the effectiveness of pilgrimage as a tool for peace building, there are many examples that can be showcased where pilgrimage does seem to lead to more peaceful relations between religious and ethnic groups. Peace pilgrimages, some of which have been described above, can be divided into four categories. The first category is pilgrimages led by inspired individuals that focus specifically on peace-related goals. Perhaps the best-known example is Mildred Lisette Norman (1908–1981), a nondenominational spiritual teacher, mystic, pacifist, vegetarian activist, and peace activist who, after a spiritual awakening in 1952, continuously walked across the United States speaking about nonviolence. Adopting the name "Peace Pilgrim", Norman took a vow to remain a wanderer until humankind learned the way of peace (Tamashiro 2018). Another example is the "Reconciliation Walk" created by U.S. pastor Lynn Green, who had a deep desire to confront the legacy of the Christian Crusades. In March 1996, he led over 3,000 people from many different denominations and nations from Cologne in Germany to Jerusalem to provide an opportunity for Christians to apologize face-to-face to Muslims and Jews for the crimes of the Crusaders. The three-year, two-thousand-mile pilgrimage across Europe and the Balkans ended in Jerusalem in July 1999, the nine-hundredth anniversary of the sacking of Jerusalem (Weyeneth 2001; Megoran 2010; McIntosh 2018). A third example is the "Journey of the Magi" to Bethlehem, organized by Robin and Nancy Wainwright, members of the Holy Land Trust, who led an international group of pilgrims in a re-creation of the journey of the Three Wise Men to Bethlehem. Their goal was to arrive on December 25 2000 to commemorate the birth of Jesus and to build ties of friendship and cooperation between Christians and Muslims. Although the Second Palestinian Intifada made progress through the West Bank difficult, the group arrived on the evening of the anniversary of Jesus' birth, with the Wainwrights leading a thousand-strong procession of Palestinians to Manger Square and Bethlehem's Church of the Nativity (Dyer 2013; McIntosh 2018).

The second category includes those journeys with a distinctly political nature and where the focus is peace, social justice, and reconciliation. One noteworthy example is the annual Selma to Montgomery pilgrimage that recreates the 1960s civil rights march of Dr. Martin Luther King (Raiford & Romano 2006). Another example is the 2000 "People's Walk for Reconciliation" in Australia where more than 300,000 non-Indigenous and Indigenous Australians crossed the Sydney Harbour Bridge together in support of Indigenous rights and reconciliation (Edmonds 2016). There is also a yearly social justice-oriented pilgrimage in memory of Monsignor Oscar Romero, a liberation theologist assassinated while delivering mass in 1980 because he was an outspoken critic of the El Salvadorian government and the human rights abuses carried out by its military. In 2015, Romero was declared a martyr and canonized (Brett 2017), and today, pilgrims visit the monument to "memory and truth" in downtown San Salvador, a wall with over 30,000 names (including Romero's) of those who disappeared or were murdered during the repression of the 1970s or during the civil war that followed from 1980 to 1992 (Villatoro 2016).

Olsen (2017) also gives the example of the "From Ocean to Ocean International Campaign in Defense of Life". This pilgrimage was instituted to both promote a pro-life/anti-abortion message and to strengthen the pro-life convictions and faith of church members and the local populace in the locations through which the pilgrimage procession traveled. Starting in 2012 in eastern Russia, pilgrims participating in this pilgrimage carried a painting of the Black Madonna of Częstochowa across Europe to North America and then into Mexico. While the pilgrimage itself was not designed to be political in nature, with pilgrims avoiding engagement with political figures, this pilgrimage might be considered by some as political in nature because of the politics surrounding pro-life messaging.

The third type of peace pilgrimage is interfaith pilgrimage, where members of two or more faiths engage in pilgrimage activities together, including shared rituals and experiences. These pilgrimages can be considered a part of the interfaith "dialogue of life" and the "dialogue of religious experience" (Belaj & Zvonko 2014), wherein shared experiences and rituals lead to greater inter-faith dialogue and understanding (Knitter 2013). While some examples of interfaith pilgrimages were mentioned earlier in this chapter, there are several other examples of interfaith pilgrimages given here. In Nigeria, a syncretic religion,

Chrislam, has emerged where adherents hold both the Bible and the Koran as holy texts. One of their unique practices is engaging in spiritual running, or "running deliverance," where members liken their pilgrimage to the biblical story of Joshua's army circling Jericho or the Muslim practice of circumambulating the Kaaba (McIntosh 2017). In Lebanon, Christians and Muslims both participate in shared pilgrimages and visit the shrines the other faith (Farra Haddad 2020a, 2020b). Christians in the Mindanao region of the Philippines, likewise, undertake a Solidarity Ramadan in alliance with their Muslim brothers and sisters in a profound display of interfaith unity (McIntosh 2017).

Scholars have noted interfaith pilgrimages at Mary's House in Ephesus, Turkey (Gallagher 2016; Öter & Çetinkaya 2016), a Marian shrine on the outskirts of Nîmes in southern France (Albera 2012), and at multiple religious sites in Ethiopia (Dagnachew 2020). Noteworthy is the Sri Pada, Sri Lanka pilgrimage (Scott 1995; McIntosh & Paramananda 2020), and certain one-off interfaith pilgrimages, such as the eight-month "Interfaith Pilgrimage for Peace and Life" organized by Japanese Buddhists in 1995 where participants traveled from Auschwitz to Hiroshima (Deats 1995; Schiel 1996).

The final category is green pilgrimages, which focus on the ecological problems of religious pilgrimages and the use of pilgrimage to promote broader global ecological sustainability and understanding (Ivakhiv 2016). While religious faiths have various views of the relationship between humans and the natural environment, pilgrimages continue to cause damage to sensitive environments, including groves, rivers, lakes, mountains, and (Olsen 2020). To this end, in 2011 the Green Pilgrimage Network (GPN) was founded in Assisi, Italy with the aim of encouraging pilgrimage travel in a way that minimizes the environmental impacts of pilgrims and the development of pilgrimage cities and communities that are more environmentally sustainable (Palmer & Hilliard 2011). To accomplish this, the GPN suggests that pilgrims and pilgrimage organizers need to commit to several guidelines and principles, including choosing sustainable tourist agencies, minimizing water use, utilizing "green" religious buildings, energy and infrastructure, safeguarding the natural landscape, wildlife and parks, and bringing greener ideas home with them, among others. The GPN also encourages pilgrimage sites and cities to focus on projects that improve energy efficiency, greener food, biodiversity, environmental awareness and education, and recycling, food, and waste management systems (Palmer & Hilliard 2011).

Conclusion

While pilgrimage is often viewed as a journey to help overcome a personal deficiency, with rituals specifically designed to fulfill an obligation, repay a debt, undertake penance, deal with a loss, or purify oneself in the presence of God, the more positive dimension of pilgrimage recognizes its potential for broader societal transformation and healing writ large. In this chapter, the focus has been on the purposes and impacts of pilgrimage beyond the needs and interests of the individual pilgrim to the needs and interests of the greater society. This is a precondition for better understanding the peace building potential of sacred journeys.

Pilgrimage at its core is as much a group activity as it is an individual one, but attempts to understand and measure the peace building dimension of mass rituals are hampered by definitions that focus on the aforementioned individual motives and desired outcomes. While not all pilgrimages are peace-focused, nor are all religious and spiritual tourists always driven by peaceful motives, it is evident that the practice of pilgrimage can have significant impacts on increasing tolerance and respect for diversity. In a world filled with conflict over race, religion, political ideologies, and finite resources, pilgrimage can help to address deep-seated conflicts, historical injustices, and social inequalities.

The most significant ways in which pilgrimage is currently having impacts on peace is through the growing interfaith movement and through the mobilization of pilgrims who share specific views regarding peace, justice, and reconciliation, often in troubled political climates. The "green pilgrimage" movement has also provided an opportunity for pilgrims to consider the impacts of their sacred journeys and to think globally.

A common feature of pilgrimages that facilitate peace is the way in which they embody a wish or hope for the future. The seemingly paradoxical case of the Hajj, with its vision of a united human family, best illustrates this point. As Haessly (2010: 5) argues, for peace to flourish, pilgrims must embrace such a shared vision and articulate it to others. Pilgrimage scholar George Greenia has described pilgrimage as the least violent human gatherings that humans have so far designed for themselves (McIntosh 2017: 8). It is clear that pilgrimages have a role to play in breaking down barriers between people and dispelling stereotypes. Not as clear is the full extent of its potential contribution to world peace. As such, further research is warranted and should be prioritized.

References

Albera, D. (2012) 'The Virgin Mary, the sanctuary and the mosque: Interfaith coexistence at a pilgrimage centre', in W. Jansen & C. Notermans (eds), *Gender, Nation and Religion in European Pilgrimage* (pp. 193–208). London and New York: Routledge.

Albera, D., & Eade, J. (2015) 'International perspectives on pilgrimage studies: Putting the anglophone contribution in its place', in D. Albera & J. Eade (eds), *International Perspectives on Pilgrimage Studies: Itineraries, Gaps and Obstacles* (pp. 1–22). London and New York: Routledge.

Alexseev, M.A., & Zhemukhov, S.N. (2017) *Mass Religious Ritual and Intergroup Tolerance: The Muslim Pilgrims' Paradox.* Cambridge: Cambridge University Press.

Becken, S., & Carmigani, F. (2016) 'Does tourism lead to peace?', *Annals of Tourism Research*, 61: 63–79.

Belaj, M., & Zvonko, M. (2014) 'Pilgrimage site beyond politics: experience of the sacred and interreligious dialog in Bosnia' in J. Eade and M. Katić (eds) *Pilgrimage, Politics and Place-Making in Eastern Europe: Crossing the Borders* (pp. 59–77). London and New York: Routledge.

Boniface, P., & Robinson, M. (eds) (1999) *Tourism and Cultural Conflicts.* Wallingford, UK: CABI.

Bramwell, B., & Lane, B. (2000) *Tourism Collaboration and Partnerships: Politics, Practice and Sustainability.* Clevedon, UK: Channel View Publications.

Brett, E.T. (2017) 'The beatification of Monsignor Romero: A historical perspective', *American Catholic Studies*, 128(2): 51–73.

Clingingsmith, D., Khwaja, A.I., & Kremer, M. (2009) 'Estimating the impact of the Hajj: Religion and tolerance in Islam's global gathering', *The Quarterly Journal of Economics*, 124(3): 1133–1170.

Cohen, E. (1979) 'Pilgrimage and tourism: Convergence and divergence', in E.A. Morinis (ed), *Sacred Journeys: The Anthropology of Pilgrimage* (pp. 18–35). Westport, CT: Greenwood Press.

Cornille, C. (ed.) (2013) *The Wiley-Blackwell Companion to Inter-Religious Dialogue.* Chichester, UK: Wiley-Blackwell.

Dagnachew, S. (2020) 'Interfaith tourism in Ethiopia: An opportunity for socio-economic development and peace-building?', in I.S. McIntosh, N. Farra Haddad, & D. Munro (eds), *Peace Journeys: A New Direction in Religious Tourism and Pilgrimage Research* (pp. 80–97). Newcastle, UK: Cambridge Scholars Publishing.

D'Amore, L. (1988) 'Tourism – a vital force for peace', *Tourism Management*, 27(1): 151–154.

D'Amore, L. (2009) 'Peace through tourism: The birthing of a new socio-economic order', *Journal of Business Ethics*, 89(4): 559–568.

Deats, R. (1995) 'On pilgrimage in 1995', *Fellowship*, 61(1–2): 3.

Dodds, R., & Butler, R. (eds) (2019) *Overtourism: Issues, Realities and Solutions.* Berlin and Boston, MA: De Gruyter.

Dredge, D. (2010) 'Place change and tourism development conflict: Evaluating public interest', *Tourism Management*, 31(1): 104–112.

Dyer, E. (2013) 'Hope through steadfastness: The journey of "Holy Land Trust"', *Quest: Issues in Contemporary Jewish History. Journal of Fondazione CDEC*, 5: 182–204.

Eade, J., & Albera, D. (2016) 'Pilgrimage studies in global perspective', in D. Albera & J. Eade (eds), *New Pathways in Pilgrimage Studies: Global Perspectives* (pp. 1–17). New York and Oxon, UK: Routledge.

Edmonds P. (2016) '"Walking together" for reconciliation: From the Sydney Harbour Bridge Walk to the Myall Creek Massacre commemorations', in P. Edmonds (ed), *Settler Colonialism and (Re) conciliation* (pp. 90–125). London: Palgrave Macmillan.

Farra Haddad, N. (2020a) 'Shared pilgrimages: The potential of natural dialogue for religious tourism in Lebanon', in I.S. McIntosh, N. Farra Haddad, & D. Munro (eds), *Peace Journeys: A New Direction in Religious Tourism and Pilgrimage Research* (pp. 46–63). Newcastle, UK: Cambridge Scholars Publishing.

Farra Haddad, N. (2020b). 'Interreligious dialogue: Trees, stones, water, and interfaith ritual experiences in Lebanon', in K.A. Shinde & D.H. Olsen (eds), *Religious Tourism and the Environment* (pp. 95–104). Wallingford, UK: CABI.

Gallagher, D.A. (2016) 'Mary's House in Ephesus, Turkey: Interfaith pilgrimage in the age of mass tourism', *Vincentian Heritage Journal*, 33(2): https://via.library.depaul.edu/vhj/vol33/iss2/2.

Galtung, J. (1988) *Peace and Social Structure: Essays in Peace Research*. Copenhagen: Christian Eljers.

Haberman, D.L. (1994) *Journey through the Twelve Forests*. Oxford, UK: Oxford University Press.

Haessly, J. (2010) 'Tourism and a culture of peace', in O. Moufakkir & I. Kelly (eds), *Tourism, Progress and Peace* (pp. 1–16). Wallingford, UK: CABI.

Hayden, R., Tanyeri-Erdemir, T., Rangachari, D., Aguilar-Moreno, M., Lopez-Hurtado, E., & Bakic-Hayden, M. (2016) *Antagonistic Tolerance: Competitive Sharing of Religious Sites and Spaces*. New York: Routledge.

Higgins-Desboilles, F. (2006) 'More than an "industry:" The forgotten power of tourism as a social force', *Tourism Management*, 27: 1192–1208.

Hill, B., Gibbons, D., Illum, S., & Var, T. (1995) 'International Institute for Peace through Tourism', *Annals of Tourism Research*, 3(22): 709.

Hornsey, M.J. (2008) 'Social identity theory and self-categorization theory: A historical review', *Social and Personality Psychology Compass*, 2(1): 204–222.

Inayatullah, S. (1995) 'Rethinking tourism: Unfamiliar histories and alternative futures', *Tourism Management*, 16(6): 411–415.

Ivakhiv, A. (2016) 'Green pilgrimage: Problems and prospects for ecology and peace-building', in A.M. Pazos (ed), *Pilgrims and Pilgrimages as Peacemakers in Christianity, Judaism and Islam* (pp. 85–104). London and New York: Routledge.

Jafari, J. (1989) 'Tourism and peace', *Annals of Tourism Research*, 16(3): 439–443.

Kim, Y.K., & Crompton, J.L. (1990) 'Role of tourism in unifying the two Koreas', *Annals of Tourism Research*, 17(3): 353–366.

Knitter, P.F. (2013) 'Inter-religious dialogue and social action', in C. Cornille (ed), *The Wiley-Blackwell Companion to Inter-Religious Dialogue* (pp. 133–148). Wallingford, UK, Wiley-Blackwell.

LeSueur (2018) 'Pilgrimage: A distinctive practice', in I.S. McIntosh, E. Moore-Quinn, & V. Keely (eds), *Pilgrimage in Practice. Narration, Reclamation and Healing* (pp. 16–25). Wallingford, UK: CABI.

Manchala, D. (2014) Theological reflections on pilgrimage. *The Ecumenical Review*, 66(2): 139–146.

Margry, P.J. 'Secular pilgrimage: A contradictionin terms', in P.J. Margry (ed.) *Shrines and Pilgrimage in the Modern World: New Itineraries into the Sacred* (pp. 13–46). Amsterdam, The Netherlands: Amsterdam University Press.

McIntosh, I.S. (2017) 'Pilgrimages and peace-building on the global stage', in I.S. McIntosh & L.D. Harman (eds), *The Many Voices of Pilgrimage and Reconciliation* (pp. 3–18). Wallingford, UK: CABI.

McIntosh, I.S. (2018) 'Dreaming of Al-Quds (Jerusalem): Pilgrimage and visioning', in I.S. McIntosh, E.M. Quinn, & V. Keely (eds), *Pilgrimage in Practice: Narration, Reclamation and Healing* (pp. 122–136). Wallingford, UK: CABI.

McIntosh, I.S., Farra Haddad, N., & Munro, D. (eds), (2020) *Peace Journeys: A New Direction in Religious Tourism and Pilgrimage Research*. Newcastle, UK: Cambridge Scholars Publishing.

McIntosh, I.S., & Paramananda, P. (2020) 'To the top together: Pilgrimage and peace-building on Sri Lanka's holy mountain', in I.S. McIntosh, N. Farra Haddad, & D. Munro (eds), *Peace Journeys: A New Direction in Religious Tourism and Pilgrimage Research* (pp. 2–24). Newcastle, UK: Cambridge Scholars Publishing.

Megoran, N. (2010) 'Towards a geography of peace: Pacific geopolitics and evangelical Christian Crusade apologies', *Transactions of the Institute of British Geographers*, 35(3): 382–398.

Milano, C., Cheer, J.M., & Novelli, M. (eds) (2019) *Overtourism: Excesses, Discontents and Measures in Travel and Tourism*. Wallingford, UK: CABI.

Moufakkir, O., & Kelly, I. (2010) 'Introduction: Peace and tourism—Friends not foes', in O. Moufakkir & I. Kelly (eds), *Tourism, Progress and Peace* (pp. xvi–xxxii). Wallingford, UK: CABI.

Ni, H-C. (1979) *The Complete Works of Lao Tzu: Tao Teh Ching & Hau Hu Ching*. Santa Monica: Tao of Wellness Press.

Olsen, D.H. (2003) 'Heritage, tourism, and the commodification of religion', *Tourism Recreation Research*, 28(3): 99–104.

Olsen, D.H. (2017) Social politics on the move: The case of the Marian Ocean to Ocean Pilgrimage', in M.S.C. Mariani & A. Trono (eds), *The Ways of Mercy: Arts, Culture and Marian Routes between East and West* (pp. 405–430). Galatina, Italy: Mario Congedo.

Olsen, D.H. (2020) 'Pilgrimage, religious tourism, biodiversity, and natural sacred sites', in K.A. Shinde & D.H. Olsen (eds), *Religious Tourism and the Environment* (pp. 23–41). Wallingford, UK: CABI.

Olsen, D.H., & Emmett, C.F. (2020) 'Contesting religious heritage in the MENA region', in C.M. Hall & S. Seyfi (eds), *Cultural and Heritage Tourism in the Middle East and North Africa* (pp. 54–71). London and New York: Routledge.

Öter, Z., & Çetinkaya, M.Y. (2016) 'Interfaith tourist behaviour at religious heritage sites: House of the Virgin Mary in Turkey', *International Journal of Religious Tourism and Pilgrimage*, 4(4): 1–18.

Palka, J.W. (2014) *Maya Pilgrimage to Ritual Landscapes: Insights from Archaeology, History, and Ethnography*. Albuquerque: University of New Mexico Press.

Palmer, M., & Hilliard, A. (2011) *Green Pilgrimage Network: A Handbook for Faith Leaders, Cities, Towns and Pilgrims*. Bath: The Alliance of Religions and Conservation.

Poethig, K. (2002) 'Moveable peace: Engaging the transnational in Cambodia's Dhammayietra', *Journal for the Scientific Study of Religion*, 41(1): 19–28.

Poria, Y., & Ashworth, G. (2009) 'Heritage tourism—Current resource for conflict', *Annals of Tourism Research*, 36(3): 522–525.

Porter, B.W., & Salazar, N.B. (2005) 'Heritage tourism, conflict, and the public interest: An introduction', *International Journal of Heritage Studies*, 11(5): 361–370.

Pratt, S., & Liu, A. (2016) 'Does tourism really lead to peace? A global view', *International Journal of Tourism Research*, 18(1): 82–90.

Raiford, L., & Romano, R.C. (2006) 'Introduction: The struggle over memory', in R.C. Romano & L. Raiford (eds), *The Civil Rights Movement in American Memory* (pp. xi–xxiv). Athens, GA: University of Georgia Press.

Schiel, S. (1996) 'The interfaith pilgrimage for peace and life: A pilgrim's reflection', *Fellowship*, 62(1–2): 20.

Scott, D. (1995) 'Buddhism and Islam: Past to present encounters and interfaith lessons', *Numen*, 42(2): 142–155.

Stets, J.E., & Burke, P.J. (2000) 'Identity theory and social identity theory', *Social Psychology Quarterly*, 62(3): 224–237.

Tamashiro, R. (2018) 'Planetary consciousness, Witnessing the inhuman, and transformative learning: Insights from peace pilgrimage oral histories and autoethnographies', *Religions*, 9(5), 148–157.

te Kloeze, J. (2014) 'Analyzing the peace through tourism concept: The challenge for educators', *Sociology and Anthropology*, 2(3): 63–70.

Timothy, D.J., & Iverson, T. (2006) 'Tourism and Islam: Considerations of culture and duty', in D.J. Timothy & D.H. Olsen (eds), *Tourism, Religion and Spiritual Journeys* (pp. 202–221). London and New York: Routledge.

Twain, M. (1869) *The Innocents Abroad, or, The New Pilgrim's Progress*. San Francisco: H.H. Bancroft and Company.

Uddhammar, E. (2006) 'Development, conservation and tourism: conflict or symbiosis?', *Review of International Political Economy*, 13(4): 656–678.

UNWTO. (2019) *United Nations World Tourism Organization World Tourism Barometer*. https://doi.org/10.18111/wtobarometereng.

Vezzali, L. (2016) *Intergroup Contact Theory: Recent Developments and Future Directions*. New York: Routledge.

Villatoro, C. (2016) 'A pilgrimage on behalf of Central American migrants', *The Episcopal Church Online*, https://episcopalchurch.org/library/article/pilgrimage-behalf-central-american-migrants.

Vukonić, B. (1996) *Tourism and Religion*. Oxford: Pergamon.

Walton, J.K. (ed) (2005) *Histories of Tourism: Representation, Identity and Conflict*. Clevedon, UK: Channel View Publications.

Warriors to Lourdes. (2020) https://www.warriorstolourdes.com/wtl/en/index.html.

Webel, C. (2007) 'Introduction: Toward a philosophy and metapsychology of peace', in C. Webel & J. Galtung (eds), *Handbook of Peace and Conflict Studies* (pp. 1–13). London and New York: Routledge.

Weyeneth, R.R. (2001) 'The power of apology and the process of historical reconciliation', *The Public Historian*, 23(3): 9–38.

Winter, T. (2007) *Post-Conflict Heritage, Postcolonial Tourism: Tourism, Politics and Development at Angkor*. London and New York: Routledge.

Yang, J., Ryan, C., & Zhang, L. (2013) 'Social conflict in communities impacted by tourism', *Tourism Management*, 35: 82–93.

SECTION II

Spaces and places

10

ENVIRONMENT AS A SACRED SPACE

Religious and spiritual tourism and environmental concerns in Hinduism

Rana P.B. Singh, Pravin S. Rana and Daniel H. Olsen

Introduction

People form a sense of themselves and their environmental surroundings at a variety of scales (*spatialities*), time frames (*sequentialities*), functions (*activities*), mobilities (e.g., *pilgrimages*), quests (*sacrality*), and mental states (*belief systems*) (Singh & Rana 2020a: 137). It is at the local scale that people first experience the spirit and power of places (*genius loci*) and from there proceed to experience this spirit and power at larger scales (Singh & Rana 2020b: 97). The idea of sacred space is a universal theme in both sacred and secular settings. This chapter focuses on the interface between the natural environment and sacrality in the context of Hinduism, which has a long history of creating cosmic landscapes and places (Baindur 2009). After discussing how the natural environment is tied to Hindu theology and cosmology, attention is turned to different pilgrimage traditions (*tīrtha-yātrā*) in India that lead to spiritual and religious advancement, followed by a broader discussion of the environmental issues in Hindu pilgrimage traditions and how pilgrimages can lead to better human–nature relationships and greener pilgrimage practices (see also Chapter 9, this volume).

Environment as a sacred repository in Hinduism

The natural environment and its various attributes have long attracted people to gaze at and appreciate the aesthetics and sense of place that comes from natural elements, such as rivers, mountains, and forests. This is also the case as eulogized and described in Hindu literature (cf. Timothy 2012: 40; Olsen 2020; Chapter 23, this volume). According to Hindu theology, sacred places are consecrated or 'illuminated by faith', which, because of their religious content, become the focus of religious travel. Such places are considered sublime (*deva sthāna* in Hindu tradition) and holy (*pavitratā*: Sanskrit for "pure") as a result of hierophanic events that took place there, sacred events or festivals that are hosted there because of the presence of sacred relics, and because of healings that take place in those places (cf. Vukonić 2006: 242). As well, both classical and modern Hindu literature describe the reverence people should have for 'Mother Nature' (*prakrīti*) and 'Mother Earth' (*Bhū Devī*), as the entire earth is considered a personified goddess. This image of earth as 'Mother' is conceptualized by relating all geographical features, such as mountains, hills, rivers, and caves as living as well as imagined or constructed sacred landscapes, and as such all places and landscapes become part of the sacred geography of India (see Eck 2012: 11). As such, landscapes in Hinduism have become sacred because humans perceive them to be sacred. While this is a socially constructed view of sacred space (see Olsen 2019a), this construction does not automatically mean that these places are not sacred, as "one way of knowing does not negate the validity of another" (Carmichael et al. 1994: 7). While many people turn to rationality and productive value to determine what is valid or real and to order space, spirituality, indigenous people's beliefs, taboos, and traditions that cannot directly be explained should also be seen as a valid way of both understanding and territorializing the world.

Hinduism therefore has a long history of endowing certain spaces and landscapes with sacred meaning and the inherent power of healing. This is why Hinduism has a long history of religious pilgrimage to these places and landscapes. Indeed, many Hindus travel to these places to have an 'auspicious glimpse' (*darshan*) at and interact with these spaces and place them as a part of their moral duty (*dharma*). Through pilgrimage (*tirtha-yātrā/dharma-yātrā*, 'religious tourism') and its associated deep feelings of faith and performance of rituals, Hindus transform the materialistic world into a cosmic one—where the physical (nature) and the spiritual (transcendental) worlds are viewed as two parallel dimensions of existence, and it is through interacting with these dimensions through the lenses of faith and revelation that a person can perceive their interconnectedness.

Even though the entire earth is considered sacred, there are several aspects of the natural environment that are deemed more sacred than others in Hinduism. For example, as Eliade (1958) notes, in Hinduism, as well as in other religious faiths, mountains are

> endowed with a twofold holiness: on the one hand they share in the spatial symbolism of transcendence—being "high", "vertical", and "supreme"—and on the other, they are the especial domain of all hierophanies of atmosphere, and therefore, the dwelling of the gods.
>
> *(p. 99)*

Moreover, mountains serve as cosmic pillars, acting as the *axis mundi* between heaven and earth where "one can pass from one cosmic zone to another" (Eliade 1958: 99–100). Because of their physical distinctness, their dense vegetal cover, being the source of many rivers, acting as territorial markers, their calm and quietness characteristics, and other such several corresponding cosmic and spiritual geographies, mountain are viewed in Hindu cosmology as places of sacred wisdom and enlightenment (Cooper 1997). In fact, "There is a certain worldview of mythical history associated with these geographical places—-a narrative of *being* sacred by creation, rather than being *made* sacred" (Baindur 2009: 47).

Many Hindu mythologies describe mountains as places possessing a deep power of sacrality—where heaven and earth meet—as well as temples and sacred towns which are viewed as Sacred Mountains and being as *axis mundi* regarded as the meeting point of heaven, earth, and hell (Eliade 1965: 12 and 16, see also Bernbaum 1997: 206–214). In Hindu cosmology (as well as Jain and Buddhist cosmology), Mount Meru (also referred to as Sumeru) represents the centre of all spiritual, physical, and metaphysical universes, and temples that are built on high hills are often referred to as Meru/Sumeru temples. In Hinduism, Mount Meru is considered as "the cosmic conduit along which, or by means of which, the different aspects of reality communicate with each other: heaven, earth, and the underworld are connected by it" (Kinsley 1995: 61). The Himalayan mountain range, especially those mountains that serve as the source region of the Gaṅgā and the Yamunā rivers, is full of temples and broader sacredscapes having hierophanic connotations and "holy territories" or *kshetras* where ancient pilgrimage sites, including the five sacred abodes of Shiva (*Pañcha-Kedāras*, i.e. Kedaranath, Madhyameshvara, Tunganatha, Rudranatha, and Kalpeshvara) and the seven abodes of Vishnu (*Sapta-Badrīs*, i.e. Badrinarayan, Adibadri-1, Vriddha, Bhavisya, Yoga, Adibadri-2, and Nrisimha; see Figure 10.1) can be found. Mount Kailāsh (in Tibet, China) is equally venerated both by the Buddhists and the Hindus, as the mountain serves as the abode of Lord Shiva and his consort Pārvatī. Additionally, there are seven sacred mountains as described in an ancient epic the *Mārkaṅḍeya Purāṇa* (57.10–11), a CE 7th-century text, which includes Mahendra (Orissa), Malaya (southern part of Western Ghats), Suktiman (the mountains in the eastern part), Riksha (Nilagiri and Deccan Plateau), Sahya/Sahyadri (northern part of Western Ghats), Vindhya, and Pariyātra (older part of Vindhya in central India) (Dave 1970, vol. IV: 151; see Figure 10.2). These sacred mountains are surrounded by sacred forests and are considered to be the places of veneration and meditation. Unfortunately, many of these sacred forests have begun to disappear as they are used for spiritual resorts under the patronage of religious monasteries. These mountains also face several environmental threats, including submergence, clear cutting forestry, mining, quarry, urban and agricultural encroachment,

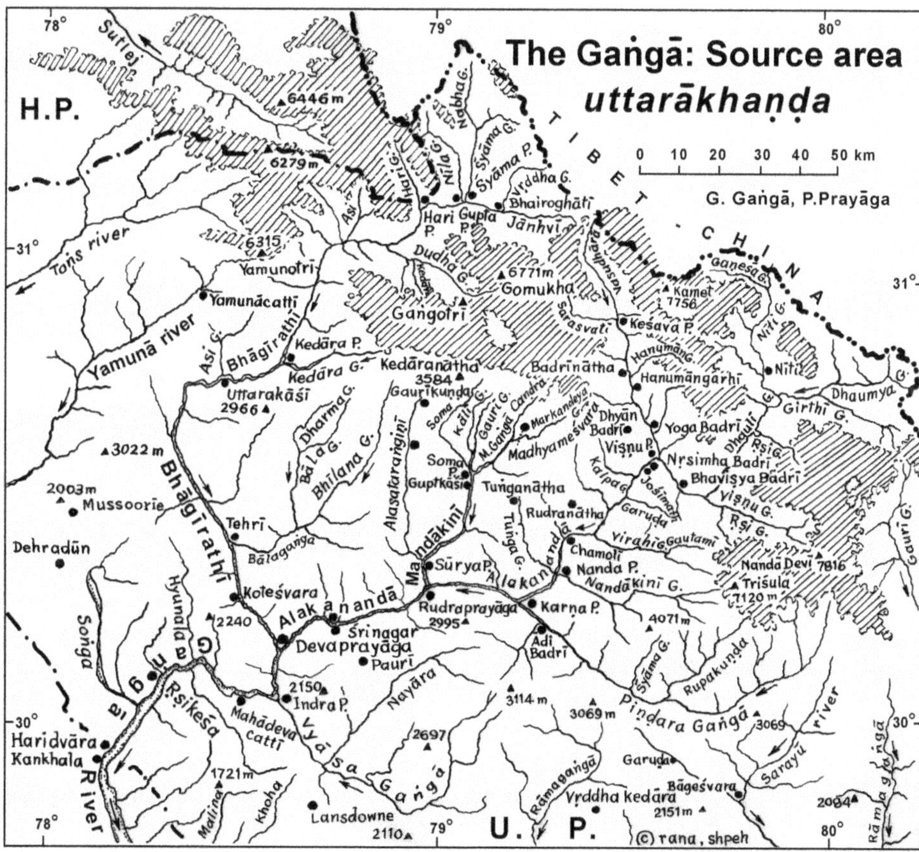

Figure 10.1 The source area of Ganga River and sacred places

Source: © Singh, Rana 2013: 164, represented with permission.

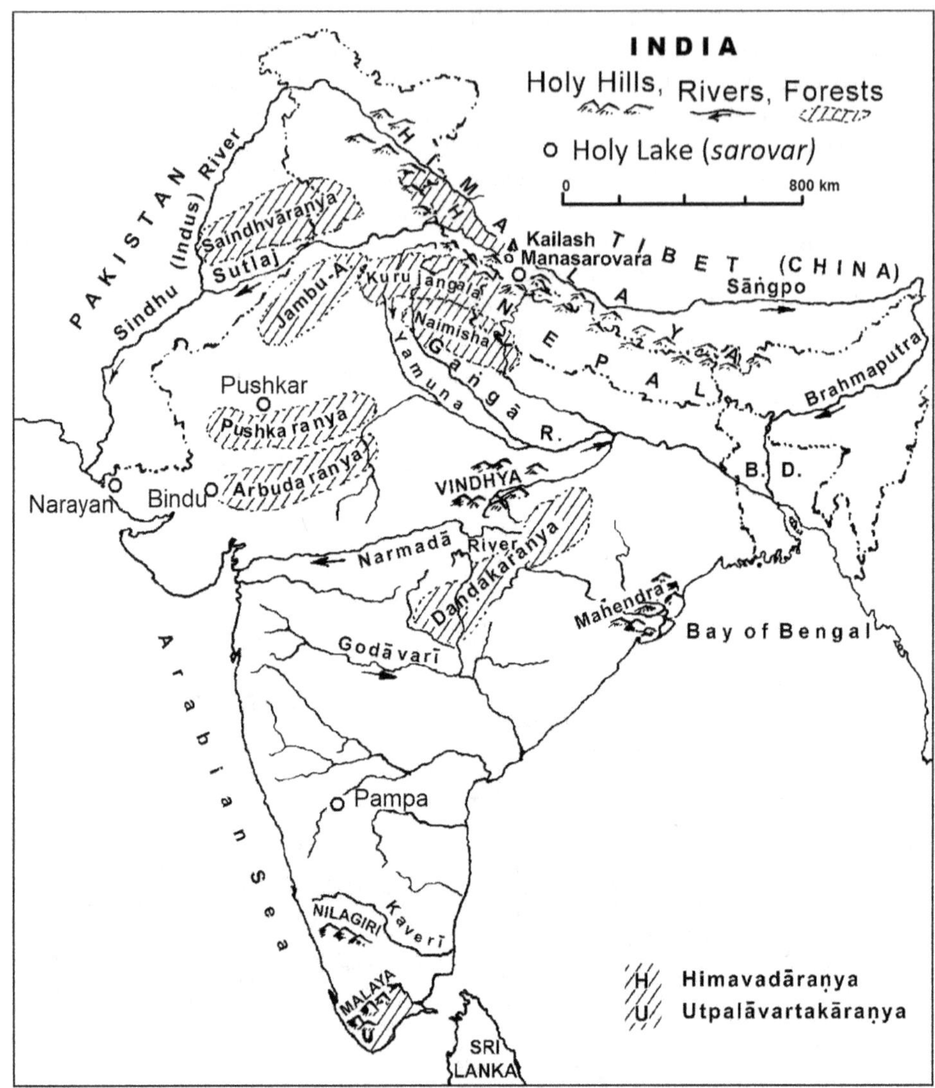

Figure 10.2 The holy hills, rivers, forests and Sarovars of India

Source: © Singh, Rana 2016: 49, reprinted with permission.

expanding built spaces, urbanization, over harvesting—all the consequence of materialistic attitudes and other depletive factors. There also appears to be a theological fallacy of religion (here Hinduism) which is generally perceived as separating the sacred or spiritual from the profane or utilitarian, which results in the lack of preservation, conservation, and sustainability efforts (Apffel-Marglin & Parajuli 2000: 309). The rejuvenation and adaptability of the ancient attitudes of the forest as a 'place of transcendence' and 'state of mind' would help to maintain the eco-spirituality of forests (Lutgendorf 2000: 280).

Water is also an important natural element that has significant impacts on Hindu religion and pilgrimage traditions (Alter 2001). The oldest text of Hinduism, the *Rig Veda* (RgV) (ca. 2500 BCE) describes water as the "greatest gross-element of life which pervades all, holding the embryo, and producing *agnih* [heat]" (RgV 10.121.7), and characterizes water as having a life-infusing power and motherly qualities (RgV 1.23.20, 6.50.7). Along with mountains, water is one of the most sacred symbols in Hinduism. Water sites, in the form of rivers (*tīrthas*) and the confluence of those rivers are considered sacred, and bathing in these water bodies allows a person to experience cosmic energy and be purified from sin (Agoramoorthy 2015; Singh 2020). Of all the rivers in India, the seven most holy or sacred rivers are the Gaṅgā, Yamunā, Godāvarī, Sarasvatī, Narmadā, Sindhu (Indus), and Kāverī

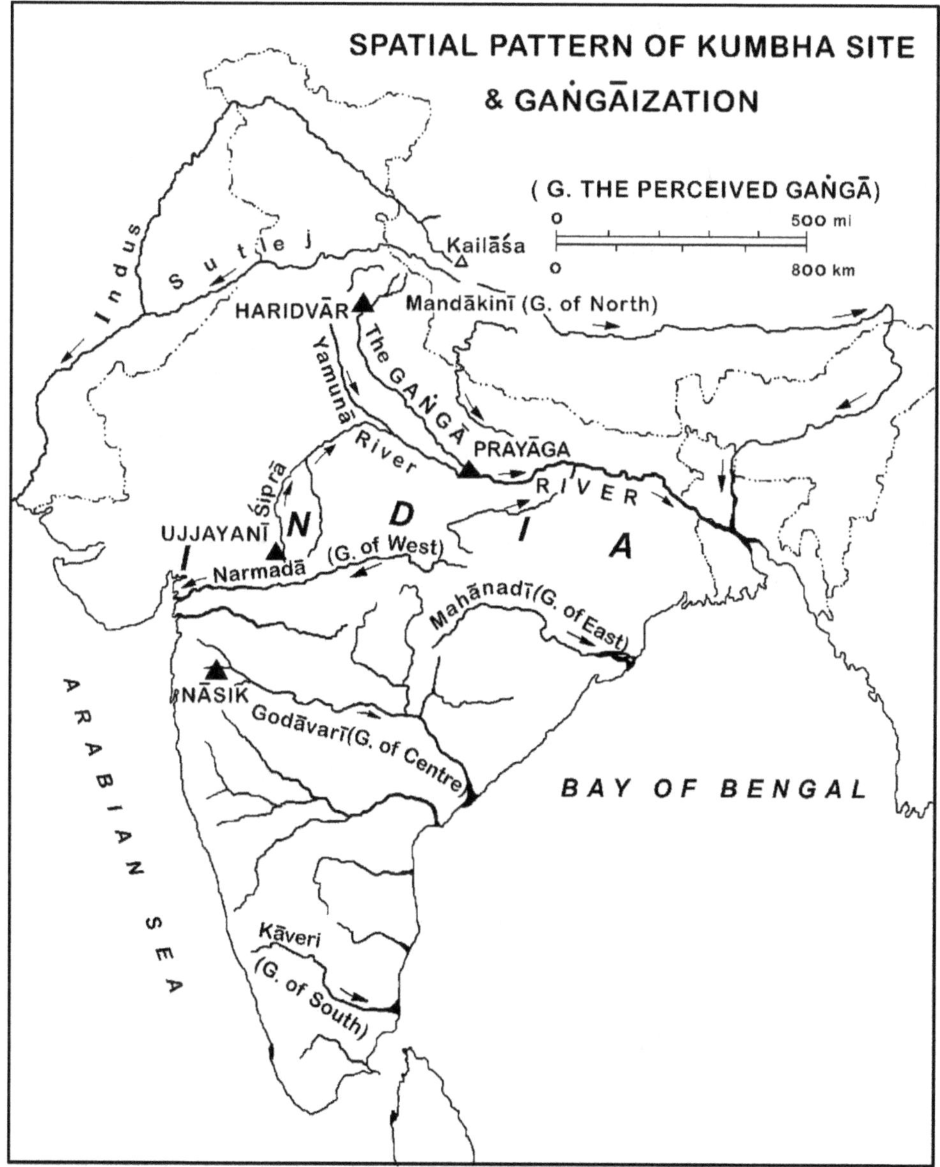

Figure 10.3 The Gaṅgā River, Gaṅgāization and Kuṁbha sites

Source: © Singh, Rana 2013: 168, reprinted with permission.

rivers (cf. Figure 10.3), making the whole of India intensely sacralized. This has powerfully promoted the development of a huge mass of sacred places along these rivers, in addition to replicating other such sacred places from various parts of India in the frame of spatial transposition. Small ponds (*kuṇḍas*) and lakes are also considered sacred. The *Bhāgavat Purāṇa* (6.5.3), dated ca. CE 6th century, eulogized the five holy lakes (*sarovars*), which are linked to sources of natural springs and also associated with the creation myth at different levels symbolizing primordial water. As such, these lakes, being eulogized as sacred and propitiated by the great sages in the ancient past, have become an important part of sacred bathing during pilgrimage rituals. The importance of bathing in these lakes and other rivers is narrated in ancient epics like the *Rāmāyaṇa* and the *Mahābhārata* (Kane 1975, vol. III: 1400–1471) (see Table 10.1).

Also, like in parts of Europe and other parts of Asia and as noted above, forests in Hinduism are been venerated as sacred, especially for those who wish to meditate for inner peace. The sacred groves in Hinduism functions for the appeasement of deity/spirits, ancestral spirits, and are also used for totems. Ormsby and Bhagwat (2010) suggest that there are over 100,000 sacred forests (including pockets of shrubs) in India, ranging in size from small groves located in between agricultural lands to large tracts of forests. These forests are believed to house a god or gods, and at times are named after certain deities (Ormsby & Bhagwat 2010). Forests are also often the abode of hermits and sannyasis (religious ascetics) (Robinson & Cush 1997). Forests are so sacred in Hindu cosmology that there is the belief that "cutting down the green tree is a sin" (Sivapriya 1989: 77–78; quoted in Robinson and Cush 1997: 26). Specific deities are also believed to manifest themselves in certain species of trees and plants, and to plant to a tree is believed to be an act of devotion and worship (Chandrakanth et al. 1990).

Table 10.1 Pancha *Sarovars*: characteristics

Se	*Sarovar, holy lake, and (location)*	*Deity/saint affiliation*	*Links to ancient tale*	*Religious merit*
1	Manasarovar (Tibet, China)	'Heart of the divine world': Buddhism, Hinduism, Jainism, and Bon Tibetan	A *Shakti Pītha*, where fell right palm of Devi; Fish incarnation of Vishnu held here.	Liberation from transmigration.
2	Bindu Sarovar (Siddhpur, Gujarat)	Sage Kardama and his son Kapila	Shiva filled this with holy water, known as Kapalamochan-tirtha.	Get blessings from (feminine) dead ancestors and release of sins.
3	Narayana Sarovar (Koteshvar temple, Kutchh, Gujarat)	Vishnu and Rama, saint Vallabhacharya	Vishnu rest place, symbolising Chhirasagar	Blessings of Vishnu, cleaning the sins.
4	Pampa Sarovar (Koppal, Karnataka)	Rama, and devotee Shabari (a female ascetic described in the *Rāmāyaṇa*)	Vishnu and seven ancient sages, nearby Anjani hill (Hanuman's mother's temple)	Blessing of Savitri, cleaning the sins.
5	Pushkar Sarovar (Ajmer, Rajasthan)	Brahma, the progeny in the Trinity; three *sarovars* linked together.	Savitri (Sarasvati), goddess of knowledge.	The first water progeny, cleansing the sins, place of ancestral rituals.

Source: based on Gita Press 1957: pp. 68, 533, 551, 410 and 385, see Figure 10.2.

In classical Hindu tradition, three types of forests are described: *Tapovana* (used for solace and penance), *Mahāvana* (natural forests with scenic beauty), and *Shrīvana* ('forest of prosperity', dense forest kept in wilderness). All these forest types are still used in various religious festivities, including salvific rituals. Most ancient temples or sacred sites are surrounded by the five sacred trees used in healing and ayurvedic medicines: Pañchavatī, viz. Pipal (*Ficus religiosa*), Vata (banyan, *Ficus benghalensis*), Gular (*Ficus glomerata*), Ashoka (*Saraca asoca*), and Nīma (*Azadirachta indica*). In some temple compounds, Nīma is replaced by Bel/Vilvā (*Aegle marmelos*). These trees archetypically represent the five gross elements of the universe (Space/Sky, Earth, Air, Water, and Fire). The five gross elements are represented with the five *tīrthas* associated with five forms of Shiva, all located in the southern part of India. These are used as part of religious and spiritual activities; and they are associated with different Hindu deities. In addition, there are four different types of temple forests or gardens in Hindu theology: (i) Star forest (representing a personified form of stars), (ii) Nine-planet forest (represented by sacred trees, shrubs, and grasses), (iii) Zodiac (27 constellations represented by sacred trees, shrubs and herbs) forest, and (iv) Pañchabhūta garden (the five sacred and holy trees, representing five gross elements, also symbolizing the five firms of Shiva) (Chandrakanth et al. 1990).

Hindu pilgrimages

An estimated 250–600 million people a year travel for religious purposes (Timothy 2011; UNWTO 2011). In the context of pilgrimage, pilgrims may leave home for religious purposes for stretches of time ranging from a few hours to days, weeks, months, or even years! Out of the 23 pilgrimage sites around the world that record above half a million pilgrims annually, nine are located in India: Sabarimala (34 mill), Tiupati-Tirumala (33 mill), Amritsar (30 mill), Shirdi (12 mill), Allahabad (10 mill; but every twelfth year celebrated as Kuṁbha Melā), Vrindavan (6 mill), Dvaraka (5 mill), Varanasi (1.5 mill), and Amarnath Cave Temple in Jammu and Kashmir (650,000) (see ARC 2014, also Singh & Haigh 2015).

Festivals (*melās*) at sacred sites in India are a vital part of Hindu pilgrimage traditions. These *melās* attract an estimated 450 million a year from all over India, meaning that most of the religious movement for pilgrimage worldwide takes place in India (cf. Singh & Haigh 2015: 783). The largest—and the oldest—of these festivals is the Kuṁbha Melā, a riverside festival that is held every three years and rotates between four different cities: Prayagraj (Allahabad), located at the confluence of the Gaṅgā and Yamunā Rivers (cf. Dubey 2001); Nasik on the Godāvarī River; Ujjain on the Shiprā River; and Haridwar on the Gaṅgā River (Figure 10.3). The Kuṁbha Melā is an important *melā* in Hinduism because bathing in these rivers during the Kuṁbha Melā is considered an endeavour of great merit, cleansing both body and spirit.

The Kuṁbha Melā has attracted the most attention in the academic literature because of the sheer number of pilgrims that attend these *melās* (Krasa 1965; Dubey 2003; Lochtefeld 2004; Maclean 2008). For example, the 2019 Prayagraj Melā received an estimated 150 million pilgrims and tourists, including one million foreign visitors in 2019 (Kashin 2019), making it the largest religious festival in history (Nanadoum 2019). The 2019 event was spread over an area of 45 square kilometres with bathing *ghāṭs* (stairways) stretching more than 8 kilometres along the Gaṅgā-Yamunā-mythical Sarasvatī rivers confluence. Twenty-two million people resided in a temporary tent city divided into 22 sectors, and infrastructural facilities included 122,500 toilets, 20,000 sanitation workers, 20,000 dustbins, 90 parking lots for 500,000 vehicles, 22 Pontoon bridges on the two rivers, 500 shuttle buses, and

22 hospitals (with 450 beds each) (PMA 2019). The forthcoming Kuṁbha Melā at Haridwar (in the state of Uttarakhand, north India), to be held during 14 January to 26 March 2021, is expected to match, if not exceed, the Prayagraj Kuṁbha Melā in terms of scale, pilgrim numbers, and needed infrastructure (TNN 2020).

Pilgrimage to one of the 51 *Śakti Pīthas*—a network of religious sites connected with the Mother Goddess—is also an important pilgrimage tradition. More specifically, these *Śakti Pīthas* are related to Shaktism, a goddess-focused Hindu tradition focused on the goddess Shakti or Sati (Figure 10.4). According to Hindu cosmology, the 51 *Śakti Pīthas* mark the location where a body part from Śakti's disembodied body fell to earth (Singh 2013: 136–137, Lochtefeld 2002). Most of the shrines occupy either hill or mountain tops or are associated

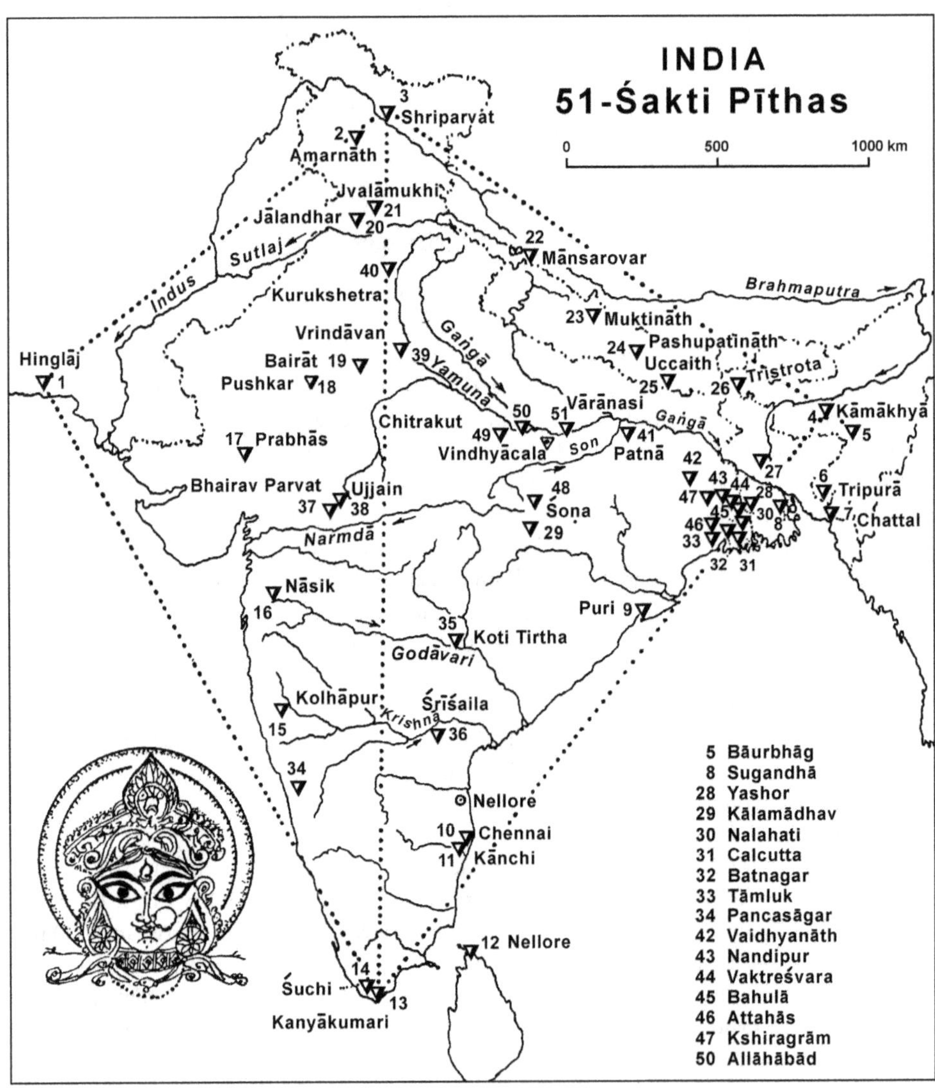

Figure 10.4 India: 51 Śakti Pīthas

Source: © Singh, Rana 2013: 135, reprinted with permission.

with holy springs and ancient groves and are believed to be places of cosmic power (Singh 2012; Bhandari 2013).

Similar to *Śakti Pīthas*, there also exist a series of Pan-India level sacred sites associated with the different groups of divinities and *sampradāyas* (religious groups/sects), which sites are distributed all over India and are part of religious journeys and pilgrimages of their groups (Gita Press 1957: 590–802). Some such examples include 275 forms of Shaiva Kshetra (holy segment), 108 forms of Shiva (forms of Shiva), 108 sites of Alvar saints, 108 tīrthas along the route of Lord Rāma's journey from Ayodhya to Sri Lanka when he was in exile, 51 Siddha Kshetra (areas of miraculous power and transcendence), 42 Shrāddha Tīrthas (holy sites for ancestral worship), 40 Digaṁbara Jain Tīrthas and 22 Svetāṁbara Jain Tīrthas, 21 abodes of Lord Gaṇesha's forms, 18 Niṁbārka (Vaishnavite) Tīrthas, 14 Gauḍīya Sampradāya (related to Chaitanya, and popularized by Bhaktivedāṅta Svāmi Prabhupāda and the ISKCON- International Society for Krishna Consciousness), 14 Prayāgas (confluence of the two sacred rivers; except for the one at Prayagraj, the rest are in the Uttarakhand Himalaya, see Figure 10.1), 12 Jyotir Lingams of Shiva (light manifested forms), 12 Svayaṁbhū Lingams of Shiva (self-manifested forms), 8 Svāmināšrāyaṇ Tīrthas, 7 Purīs (sacred abodes bestowing salvation), 7 Vallabha sites of Krishna devotion and 84 sub-sites (*bhaithakas*) associated to the life events of Vallabhāchārya, 5 Dādū Sampradāya Tīrthas, 5 Pañchabhūta Shiva Lingams (representing five gross elements), and several such groups. All the five Pañchabhūta Shiva Lingams exist in the southern part of India, viz. Ākāsha-Lingam ('sky/space/ether') in Chidaṁbaram, Appū-Lingam ('water') in the Jaṁbukeshvaram temple at Tiruchirappalli, Agni-Lingam ('fire') in the Arunachaleshvara temple at Tiruvannamalai, Prithvī-Lingam ('earth') in the Prithvīshvara temple at Kañchīpuram, Vāyu-Lingam ('air') at Kālahastī (see Howley 1996: 16–17).

Valuing cultural and spiritual perspectives in conserving natural sacred sites

In recent decades, cultural and spiritual values have been determined by many international conservation and non-government organizations to be crucial elements in the development of nature conservation priorities and plans (Verschuuren et al. 2010; Olsen 2020). Many institutional and indigenous spiritual traditions view the world as a multiple-level hierarchic reality (Singh 2013: 41), where the sacred or spiritual world manifests or irrupts into the human world as well as onto the natural world. The importance of a holistic view of the conservation and management of natural areas should therefore consider all the values and stakeholders involved, including religious institutions. This incorporation of religious and spiritual values into natural conservation efforts has led several international nature conservation agencies to work closely with representatives of many faith and spiritual traditions, such as the International Union for Conservation of Nature (IUCN), which has several commissions that focus on the cultural and spiritual values attributed to nature in collaboration with local and indigenous peoples on conservation issues in relation to sacred natural sites (SNS) (Verschuuren 2007, Verschuuren et al. 2010; Singh 2013), and the Alliance of Religions and Conservation (ARC), which works to help religious faiths develop environmental programmes around the preservation of natural sacred sites (cf. Palmer & Finlay 2003; Palmer 2012; Singh 2013). Other institutions that have focused on this religion–spirituality–conservation interface include various branches of the International Council on Monuments and Sites (ICOMOS) and the United Nations Educational, Scientific, and Cultural Organization UNESCO (Ivakhiv 2016).

Many meetings have taken place and much research has been published on this religion–spirituality–conservation interface. For example, at a symposium on conserving sacred natural sites in Tokyo, Japan in 2005, a "Declaration on the Role of Sacred Natural Sites and Cultural Landscapes in the Conservation of Biological and Cultural Diversity" was produced (UNESCO-MAB 2006), emphasizing the importance of faith perspectives and the development of sacred natural sites and cultural landscapes to help safeguard cultural and biological diversity for present and future generations and the spiritual well-being of pilgrims and local communities (Schaaf & Lee 2006). A World Bank report entitled *The Role of Indigenous Peoples in Biodiversity Conservation* noted the importance of including indigenous populations in the decision-making process related to natural sacred sites (Sobrevila 2008). As well, several other articles and books give examples of cooperation between religious faiths, spiritual groups, and government and non-government organizations at several different scales (e.g., Wild & McLeod 2008; Frisk 2015; Ivakhiv 2016; Verschuuren & Furuta 2016; Wexler 2016; Verschuuren et al. 2017; Olsen 2020).

Environmental concerns and greening Hindu pilgrimage

Within this context, while Hindu scripture and cosmology are replete with references to the importance of the natural environment and the relationship between the environment and pilgrimage, there seems to be a dissonance between the way pilgrims should interact with the natural environment and the way in which the natural environment is used for pilgrimage purposes. This occurs in part because Hindu pilgrimages tend to correspond with particular sacred times related to astronomical occurrences, meaning that Hindu pilgrimage sites see an inundation of pilgrimage at certain times of the year (Singh & Haigh 2015). This mass movement of people towards and around these sites at certain peak times of the year creates a greater intensity of environmental impacts than during other times of the year (Shinde 2020). With millions of people a year traveling to religious sites in India, many scholars and government officials have become concerned with the environmental damage that occurs because of these mass pilgrimages (van Horn 2006; Shinde 2007, 2011b, 2012, 2018; Gangwar & Joshi 2008; Sharma et al. 2012; Patange et al. 2013; Raj 2018; Luthy 2019).

As globalization accelerates, the expansion of pilgrimage has encouraged environmental cleanliness, and eco-development programmes have been created to protect the planet and awaken people's sense of environmental responsibility through deeper experiences—from realization to revelation—that ultimately that foster peace. Organizations, such as the Green Pilgrimage Network initiative (GPN), as discussed below, has made is easier (Singh 2016).

Water pollution in the context of Hindu pilgrimage has been a concern for decades. In fact, this issue is already sensitized long back in history, as the *Brahma Purāṇa*'s *Prāyashcittatattva* (2.535), a tenth century CE text warns that "one should not perform fourteen acts in and around the holy waters of the river Gaṅgā, to avoid any form of pollution" (Singh 2021: 90). Most of the major rivers in India are heavily polluted, and yet are the focus on pilgrims who bathe themselves in these waters in purification rituals, as flowing water is believed to wash away a person's sins. As Sinha (2019: 66) explains,

> Waste is interpreted within the Hindu concepts of purity and pollution, which form a unity in the experience of the sacred on the *ghāṭs*. The contrasting traditional religious and modern secular views of waste explain why visible pollution is ignored by stakeholders engaged in rituals who make a distinction between material and symbolic purity and assert that the River Gaṅgā is pure even when it is unclean.

Sinha also notes that

> The community of the faithful views the Gaṅgā through a different lens than those who see it as a polluted river, degraded by pathogens and requiring external aid in restoring its health. They ignore the very visible pollution in the river in the belief that the Gaṅgā will remove the waste and remain pure. The boundaries between purity and pollution, sacred and profane, waste and utility, appear to be blurred….

While rivers may therefore be unclean, rivers themselves, from a cosmological perspective, can never be unclean, and as such bathing in polluted waters will still wash away sins (Alley 2016; Sinha 2019).

The *creation of solid waste and its disposal* is another concern, whether it is from garbage and plastic water bottles that are left along pilgrimage routes and trails or from votive offerings to gods that need to be disposed of after religious rituals (Shinde 2018, 2020). In many cases, host communities either have a lack of or an inadequate waste infrastructure to accommodate the influx of solid waste brought into the region by pilgrims. The same can be said with *human waste disposal*, as many pilgrimage events lack of toilets and other types of basic sanitation, leading to the creation of open defecation fields by event organizers (Doron & Jeffrey 2014; Mozaffar 2014; Mara & Evans 2018). This open defecation, which has long been an issue in India, leads to concerns regarding *hygiene*, underground water contamination, site aesthetics, and the spread of disease at pilgrimage events (George 2008; Holman & Shayegan 2014; Vortmann et al. 2015; Coffey & Spears 2017). *Overcrowding* and concerns regarding other *health-related issues* also have been a concern (Griffin et al. 2018). These environmental issues arise in part because of both the lack of infrastructure and *governance* or oversight regarding the mitigation of the environmental impacts of pilgrimage—where governments at different scales and religious authorities argue over who is responsible for said mitigation, leading to an "institutional vacuum" regarding how to deal with the direct and indirect environmental impacts of pilgrimage (Shinde 2018, 2020) (Table 10.2). As Shinde (2011b: 453–459) notes, the environmental discourses that different social groups, including religious gurus and priests, local residents, government officials, and community leaders, adhere to regarding the environmental problems that arise from pilgrimage seem to fall under four categories:

- *Denial*: The view that environmental damage via pilgrimage does not matter—that pilgrimage deals with spiritual environments and salvation, and that damage done to the environment by pilgrims is not the concern of religious leaders. Indeed, it is believed that the gods will take care of any physical damage caused by pilgrims.
- *Indifference*: While some social groups may acknowledge the environmental damage caused by pilgrims, this damage is subsumed into the over poor environmental quality of destinations, and as such is overlooked by local residents.
- *Helplessness/Criticism*: While the environmental damage caused by pilgrimage is acknowledged and criticized, the lack of infrastructure and the fact that pilgrims are essential to the economic viability of many pilgrim destinations leads to locals feeling helpless regarding how to deal with this environmental damage, blaming other social groups for their lack of environmental awareness and help.
- *Stewardship*: There are some religious organizations and social groups that link theology with environmental ethics and management, focusing on strategies that minimize environmental damage and petition social groups to work together to this end.

Table 10.2 A table highlighting research on environmental issues related to Hindu pilgrimages

Environmental issue	*Examples of academic sources*
Water pollution	Alley (1994, 1998, 2002); Joshi et al. (2009); Sharma et al. (2012); Khan (2013); Shweta and Satyendra (2015); Vortmann et al. (2015); Chowdhury et al. (2019)
Solid waste disposal	Kuniyal et al. (1998); Gangwar and Joshi (2008); Kaushik and Joshi (2011); Singh and Gupta (2011); Chandrappa and Das (2012); Bashir and Goswami (2016); Abdulredha (2018)
Human waste disposal/sanitation	Hamner et al. (2006); Agoramoorthy and Hsu (2009); Doron and Jeffrey (2014); Mozaffar (2014); Mara and Evans (2018)
Congestion/health issues	Vortmann et al. (2015); Griffin et al. (2018)
Lack of governance	Shinde (2007, 2018, 2020)

Within the stewardship vein, governments, religious officials, and non-government organizations have in recent years begun to work individually and collectively to and create greener and environmentally responsible pilgrimages. The idea of "green pilgrimages" focuses mainly on the responsibilities that both pilgrimage organizers and pilgrims have as they travel to sacred sites. Green pilgrimage is in many ways connected to what Ivakhiv (2016) refers to as "global civil religion", dealing with broader environmental concerns related to "Mother Earth".

Concern over the environmental impacts of pilgrimage has led, for example, to the creation of the Green Pilgrimage Network in Europe (https://www.greenpilgrimageeurope.net/) that focuses on respecting natural environment through sustainable pilgrimage practices (Palmer & Hilliard 2011). Within the context of India, the Green Kuṁbha Foundation (http://greenkumbha.org/) engages in several "green campaigns" in efforts to promote environmental sustainability at the Kuṁbha Melās. The state government of the Kerala province in India has created the "Mission Green Sabarimala" initiative to distribute eco-friendly cloth bags in exchange for plastic bottles, and postcards reminding pilgrims to not pollute the river are distributed throughout the pilgrimage season at Sabarimala (Times of India 2019). Religious and government groups organizing the 2013 Kuṁbha Melā at Haridwar developed the moniker "Green Kuṁbha – Clean Kuṁbha". The organization Green Gaṅgā focuses on efforts to clean the Gaṅgā River both during and after pilgrimages to different sections of the river (Rashid 2013). However, questions arise as to what degree these organizations can be successful. With estimated pilgrim numbers attending the 2021 at the Haridwar Kuṁbha to be over 100 million (TNN 2020), the state government of Uttarakhand has allocated Rs 375 crore (approximately 53 million USD) to help with the building of infrastructure (see George 2020). As well, other international government and non-government agencies have also focused on maintaining and preserving the ecological sustainability several world heritage sites in India (e.g., Dudley et al. 2009; Lafortune-Bernard et al. 2020).

Conclusion

In the cosmogonic frame of a sanctified environment, the whole of India (Bhārata) is portrayed as one large sacredscape and is symbolized as Mother (Bhārata Mātā), consisting of holy spots (*pīthas*) that represent her body. However, there is a strong disconnect between this theological view and the practices of religious groups and pilgrims when it comes to the behaviours and practices of pilgrims as they relate to the treatment and use of the natural environment. Indeed, the strong ties between *dharma* (moral duty) and *karma* (phenomenal action) that was once the nucleus of Hinduism have now loosened its ties in the current era of globalization (Singh & Aktor 2015: 1928–1929). Jacobsen's (2013: 162) statement provokes the contemporary condition of pilgrimage sites:

> The polluted state of many *tīrthas* in India is due to several factors, such as tremendous population growth, urbanization, industrial production, poverty, mismanagement and corruption, and is not largely caused by religion. But it leads to a questioning of religion: How is it that a *tīrtha* site is able to purify invisible moral impurity, but is not able to handle the visible environmental pollution?

This leads to raise an ethical question:

> Are the pilgrims who take a bath in environmentally polluted water or drink it able to maintain the belief that the water is purifying? Will the pilgrims find it meaningful or even physically possible to continue to perform the ritual bathing in or drinking of contaminated water? Can the religion of *tīrthas* survive the environmental damage affecting pilgrimage places?
>
> *(ibid.)*

Solutions must be found if religious pilgrimage is to be a part of sustainable development (Olsen & Shinde 2020)!

Olsen (2019b: 279) argues that "as globalisation continues to intensify, religion and spirituality continue to be integral part of religious mobilities, tourism motivations, economic growth, and cultural conflict". In this vein, if Colin Turnbull (1992: 274) is correct in arguing that "the quest for society is one and the same thing as the quest for the self, which for some of us is also the quest for the Sacred that ultimately unites us all", then pilgrimage practices need to be attuned with the Green Pilgrimage paradigm and occur within the frame of environmental planning and ecological sensitivity (Singh 2016: 37). As Shinde correctly notes (2011a: 143),

> As cultural constructs, [pilgrimage landscapes] are ever evolving and therefore it is of little value to keep romanticising about the past landscapes or judge them within the binaries of sacred and profane. It is necessary to recognize that as traditional pilgrimage metamorphoses into religious tourism, the sacred landscape also tends to become a tourist landscape. This realisation is of utmost importance because of the negative ecological implications that accompany contemporary patterns of travel in sacred landscapes.

While efforts are underway to try to limit and change the environmental impacts of pilgrimage in India, it will take time for Hindu pilgrims, with their inherent virtues of tolerance, ethical values, and *dharma* linked to the four ends of life (economic gain, pleasure, religious duty, and liberation), accompanied with new infrastructural development, to present a model of green and sustainable world where spiritually and concerns for the natural environment will be at its core (Singh 2013: 363).

References

Abdulredha, M., Al Khaddar, R., Jordan, D., Kot, P., Abdulridha, A., & Hashim, K. (2018) 'Estimating solid waste generation by hospitality industry during major festivals: A quantification model based on multiple regression', *Waste Management*, 77: 388–400.

Agoramoorthy, G. (2015) 'Sacred rivers: Their spiritual significance in Hindu religion', *Journal of Religion and Health*, 54(3): 1080–1090.

Agoramoorthy, G., & Hsu, M.J. (2009) 'India needs sanitation policy reform to enhance public health', *Journal of Economic Policy Reform*, 12(4): 333–342.

Alley, K.D. (1994) 'Ganga and Gandagi: Interpretations of pollution and waste in Benaras', *Ethnology*, 33(2): 127–145.

Alley, K.D. (1998) 'Images of waste and purification on the banks of the Ganga', *City & Society*, 10(1): 167–182.

Alley, K.D. (2002) *On the Banks of the Gaṅgā: When Wastewater Meets a Sacred River.* Ann Arbor, MI: University of Michigan Press.

Alley, K.D. (2016) 'Rejuvenating Ganga: Challenges and opportunities in institutions, technologies and governance', *Tekton*, 3(1): 8–23.

Alter, S. (2001) *Sacred Waters: A Pilgrimage to the Many Source of the Ganga.* New Delhi: Penguin.

Apffel-Marglin, F., & Parajuli, P. (2000) '"Sacred Groves" and ecology: Ritual and science', in C.K. Chapple & M.E. Tucker (eds.), *Hinduism and Ecology: The intersection of Earth, Sky and Water* (291–316). Cambridge MA: CSWR, Harvard University Press.

ARC (2014) *Pilgrim Numbers.* Bath, UK: Alliance of Religions and Conservation. http://www.arcworld.org/projects.asp?projectID=500 <accessed: 15 December 2019>

Baindur, M. (2009) 'Nature as non-terrestrial: Sacred natural landscapes and place in Indian Vedic and Purāṇic Thought'. *Environmental Philosophy*, 6(2): 43–58.

Bashir, S., & Goswami, S. (2016) 'Tourism induced challenges in municipal solid waste management in hill towns: Case of Pahalgam', *Procedia Environmental Sciences*, 35: 77–89.

Bernbaum, E. (1997) *Sacred Mountains of the World.* Berkeley: University of California Press.

Bhandari, H. (2013) 'Sacred landscape: A documentation of the Chamunda Devi temple complex, Himachal Pradesh', *Context*, 10(1): 25–33.

Carmichael, D.L., Hubert, J., & Reeves, B. (1994) 'Introduction', in D.L. Carmichael, J. Hubert, B. Reeves, & A. Schanche (eds.), *Sacred Sites, Sacred Places* (pp. 1–8). London: Routledge.

Chandrakanth, M.G., Gilless, J.K., Gowramma, V., & Nagaraja, M.G. (1990) 'Temple forests in India's forest development,' *Agroforestry Systems*, 11(3): 199–211.

Chandrappa, R., & Das, D.B. (2012) *Solid Waste Management: Principles and Practice.* Cham: Springer.

Chowdhury, R., Bohara, S., Katuwal, A.K., Pagán, H., & Thacher, J.A. (2019) 'The impact of ritual bathing in a holy Hindu river on waterborne diseases', *The Developing Economies*, 57(1): 36–54.

Coffey, D, & Spears, D. 2017. *Where India Goes: Abandoned Toilets, Stunted Development and the Cost of Caste.* Noida, India: HarperCollins Publishers.

Cooper, A. 1997. *Sacred Mountains: Ancient Wisdom and Modern Meanings.* Edinburgh and London: Floris Books.

Dave, J.H. (1970), *Immortal India*; vol. IV. Bombay/Mumbai: Bharatiya Vidya Bhavan.

Doron, A., & Jeffrey, R. (2014) 'Open defecation in India', *Economic and Political Weekly*, 49: 72–78.

Dubey, D.P. (2001) *Prayāga, the Site of Kuṁbha Melā.* New Delhi: Aryan International Publisher.

Dubey, D.P. (2003) 'Kumbha Mela: The festival of eternal bliss', *The Oriental Anthropologist*, 3(2): 164–173.

Dudley, N., Higgins-Zogib, L., & Mansourian, S. (2009) 'The links between protected areas, faiths, and sacred natural sites', *Conservation Biology*, 23(3): 568–577.

Eck, D.L. (2012) *India: A Sacred Geography.* New York: Random House.

Eliade, M. (1958) *Patterns in Comparative Religion.* London: Sheed & Ward.

Eliade, M. (1965) *The Myth of the Eternal Return, or, Cosmos and History.* Princeton, NJ: Princeton University Press.

Frisk, K. (2015) 'Citizens of planet earth: The intertwinement of religion and environmentalism in a globalization perspective', *Journal for the Study of Religion, Nature & Culture*, 9(1): 68–86.

Gangwar, K.K., & Joshi, B.D. (2008) 'A preliminary study on solid waste generation at Har Ki Pauri, Haridwar, around the Ardh-Kumbh period of sacred bathing in the river Ganga in 2004', *The Environmentalist*, 28(3): 297–300.

George, J. (2020) 'Center to allocate Rs 375 crore for Kumbha Mela', *The Times of India.* https://malayalam.samayam.com/latest-news/india-news/narendra-modis-government-sanctions-375-crore-for-maha-kumbh-mela-in-haridwar/articleshow/74984631.cms.

George, R. (2008) *The Big Necessity: The Unmentionable World of Human Waste and Why It Matters.* New York, NY: Metropolitan Books.

Gita Press (1957), *Tīrthāṅka–Kalyāṇa*. Annual No. 31. Gorakhpur: Gita Press. [Popular essays on 1820 pilgrimage places of India; in Hindi].
Griffin, D., Radhakrishnan, Y.P., & Griffin, K. (2018) 'Risk management or the right to ritual: Attempting to control the masses at the Kumbh Mela', in M.E. Korstanje, R. Raj, & K. Griffin (eds.), *Risk and Safety Challenges for Religious Tourism and Events* (pp. 102–112). Wallingford, UK: CABI.
Hamner, S., Tripathi, A., Mishra, R.K., Bouskill, N., Broadaway, S.C., Pyle, B.H., & Ford, T.E. (2006) 'The role of water use patterns and sewage pollution in incidence of water-borne/enteric diseases along the Ganges River in Varanasi, India', *International Journal of Environmental Health Research*, 16(2): 113–132.
Holman, S.R., & Shayegan, L. (2014) 'Toilets and Sanitation at the Kumbh Mela', https://repository.gheli.harvard.edu/repository/10697/.
Howley, J. [Jata Bharata Dasa] (1996) *Holy Places and Temples in India*. Philadelphia: Krishna Center.
Ivakhiv, A. (2016) 'Green pilgrimage: Problems and prospects for ecology and peace-building', in A.M. Pazos (ed), *Pilgrims and Pilgrimages as Peacemakers in Christianity, Judaism and Islam* (pp. 85–104). Farnham, UK and Burlington, USA: Ashgate.
Jacobsen, K.A. (2013) *Pilgrimage in the Hindu Tradition: Salvific Space*. London and New York: Routledge.
Joshi, D.M., Kumar, A., & Agrawal, N. (2009) 'Studies on physicochemical parameters to assess the water quality of river Ganga for drinking purpose in Haridwar district', *Rasayan Journal of Chemistry*, 2(1): 195–203.
Kane, P.V. (1975) *History of Dharmashāstras*, vol. III. Lucknow: Hindi Samiti, Govt. of Uttar Pradesh. (in Hindi).
Khan, V.S. (2013) 'Hindu festivals: Hazards to environment and ecology', *Journal of Indian Research*, 1(1): 136–141.
Kinsley, D. (1995) *Ecology and Religion. Ecological Spirituality in Cross-Cultural Perspective*. Englewood, NJ: Prentice Hall Inc.
Kashin, V. (2019) 'At the Kumbh Mela in India', *Asia and Africa Today*, 7: 74–77.
Kaushik, S., & Joshi, B.D. (2011) 'A comparative study of solid waste generation at Mansa Devi and Chandi Devi temples in the Shiwalik Foothills, during the Kumbh Mela 2010', *Report and Opinion*, 4(8): 39–42.
Krasa, M. (1965) 'Kumbha Mela: The greatest pilgrimage in the world', *New Orient*, 6: 180–184.
Kuniyal, J.C., Jain, A.P., & Shannigrahi, A.S. (1998) 'Public involvement in solid waste management in Himalayan trails in and around the Valley of Flowers, India', *Resources, Conservation and Recycling*, 24(3–4): 299–322.
Lafortune-Bernard, A., Suwal, R.N., Weise, K., & Coningham, R.A.E. (2020) 'Religious tourism and environmental conservation in Lumbini, the birthplace of Lord Buddha, world heritage site, Nepal', in K.A. Shinde & D.H. Olsen (eds.), *Religious Tourism and the Environment* (pp. 83–94). Wallingford, UK: CABI.
Lochtefeld, J.G. (2002) *An Illustrated Encyclopedia of Hinduism*. New York: Rosen Publishing.
Lochtefeld, J.G. (2004) 'The construction of the Kumbha Mela', *South Asian Popular Culture*, 2(2): 103–126.
Lutgendorf, P. (2000) 'City, forest and cosmos: Ecological perspectives from the Sanskrit Epics', in C.K. Chapple & M.E. Tucker (eds.), *Hinduism and Ecology: The Intersection of Earth, Sky and Water* (269–289). Cambridge MA: CSWR, Harvard University Press.
Luthy, T. (2019) 'Bhajan on the banks of the Ganga: Increasing environmental awareness via devotional practice', *Journal of Dharma Studies*, 1(2): 229–240.
Maclean, K. (2008) *Pilgrimage and Power: The Kumbh Mela in Allahabad, 1765–1954*. Oxford and New York: Oxford University Press.
Mara, D., & Evans, B. (2018) 'The sanitation and hygiene targets of the sustainable development goals: Scope and challenges', *Journal of Water, Sanitation and Hygiene for Development*, 8(1): 1–16.
Mozaffar, P. (2014) *Open Defecation and the Human Waste Crisis in India*. Unpublished Doctoral dissertation. Lawrence: University of Kansas.
Nanadoum, C. (2019) 'Prayagraj Kumbh Mela surpasses total of 150 million pilgrims', *IR Insider*, https://www.irinsider.org/music-culture-1/2019/3/7/prayagraj-kumbh-mela-suprasses-total-of-150-million-pilgrims.
Olsen, D.H. (2019a) 'The symbolism of sacred space', in N. Crous-Costa, S. Aulet, & D. Vidal-Casellas (eds.), *Interpreting Sacred Stories: Religious Tourism, Pilgrimage and Intercultural Dialogue* (pp. 29–42). Wallingford, UK: CABI.

Olsen, D.H. (2019b) 'Religion, spirituality, and pilgrimage in a globalising world', in Dallen J. Timothy (ed.) *Handbook of Globalisation and Tourism* (pp. 270–283). Cheltenham, UK: Edward Elgar.

Olsen, D.H. (2020) 'Pilgrimage, religious tourism, biodiversity, and natural sacred sites', in K.A. Shinde & D.H. Olsen (eds.), *Religious Tourism and the Environment* (pp. 23–41). Wallingford, UK: CABI.

Olsen, D.H., & Shinde, K.A. (2020) 'Religion, environment, and sacred places: Lessons learned and future directions', in K.A. Shinde & D.H. Olsen (eds.), *Religious Tourism and the Environment* (pp. 162–168). Wallingford, UK: CABI.

Ormsby, A.A., & Bhagwat, S.A. (2010) 'Sacred forests of India: A strong tradition of community-based natural resource management', *Environmental Conservation*, 37(3): 320–326.

Palmer, M. (2012) *Sacred Land: Decoding Britain's Extraordinary Past Through its Towns, Villages and Countryside*. London: Brown Book Group.

Palmer, M., & Finlay, V. (2003) *Faith in Conservation*. Washington, DC: World Bank Publications.

Palmer, M., & Hilliard, H.A. (2011) *Green Pilgrimage Network: A Handbook for Faith Leaders, Cities, Towns and Pilgrims*. Bath, UK: The Alliance of Religions and Conservation.

Patange, P., Srinithivihahshini, N.D., & Mahajan, D.M. (2013) 'Pilgrimage and the environment: Challenges in a pilgrimage centre in Maharashtra, India', *International Journal of Environmental Sciences*, 3(6): 2269–2277.

PMA, Prayagraj Mela Authority (2019) *The Divine Glories of Kumbha 2019*. Lucknow: Govt. of Uttar Pradesh. [a booklet in Hindi].

Raj, N. (2018) 'Hindu pilgrimage: Issues and challenges in Kerala, India', in B. Varghese (ed.), *Evolving Paradigms in Tourism and Hospitality in Developing Countries* (pp. 445–463). Oakville: Apple Academic Press.

Rashid, O. (2013) 'A 'Green Kumbh Mela' initiative', *The Hindu*. https://www.gangaaction.org/a-green-kumbh-mela-initiative/.

Robinson, C., & Cush, D. (1997) 'The sacred cow: Hinduism and ecology', *Journal of Beliefs and Values*, 18(1): 25–37.

Schaaf, T., & Lee, C. (2006) *Conserving Cultural and Biological Diversity: The Role of Sacred Natural Sites and Cultural Landscapes*. Paris: UNESCO.

Sharma, V., Bhadula, S., & Joshi, B.D. (2012) 'Impact of mass bathing on water quality of Ganga river during Maha Kumbh-2010', *Nature and Science*, 10(6): 1–5.

Shinde, K.A. (2007) 'Pilgrimage and the environment: Challenges in a pilgrimage centre', *Current Issues in Tourism*, 10(4): 343–365.

Shinde, K.A. (2011a) 'Placing communitas: Spatiality and ritual performances in Indian religious tourism', *Tourism: An International Interdisciplinary Journal*, 59(3): 335–352.

Shinde, K.A. (2011b) '"This is religious environment": Sacred space, environmental discourses, and environmental behavior at a Hindu pilgrimage site in India', *Space and Culture*, 14(4): 448–463.

Shinde, K.A. (2012) 'Place-making and environmental change in a Hindu pilgrimage site in India', *Geoforum*, 43(1): 116–127.

Shinde, K.A. (2018) 'Governance and management of religious tourism in India', *International Journal of Religious Tourism and Pilgrimage*, 6(1): 58–71.

Shinde, K.A. (2020) 'Managing the environment in religious tourism destinations: A conceptual model', in K.A. Shinde & D. H. Olsen (eds.), *Religious Tourism and the Environment* (pp. 42–59). Wallingford, UK: CABI.

Shweta, S., & Satyendra, N. (2015) 'Water quality analysis of river Ganga and Yamuna during mass bathing, Allahabad, India', *Universal Journal of Environmental Research & Technology*, 5(5): 251–258.

Singh, A., & Gupta, G. (2011) 'Generated household and temple waste in Chitrakoot, a pilgrimage point in India: Their management and impact on river Mandakini', *Indian Journal of Science and Technology*, 4(7): 750–758.

Singh, R.P.B. (2012) 'The 51 Shakti pithas in South Asia: Spatiality and perspective for pilgrimage-tourism', *The Oriental Anthropologist*, 12(1): 1–37.

Singh, R.P.B. (2013) *Hindu Tradition of Pilgrimage: Sacred Space and System*. New Delhi: Dev Publishers.

Singh, R.P.B. (2016) 'Green pilgrimage initiatives for fostering peace and harmonising the world', in S. Mishra & B.B. Pradhan (eds.), *SOUVENIR, International Symposium on Fostering Tourism for Global Peace - Opportunities and Challenges: 1–3 February* (pp. 35–52). Bhubaneshwar, India: SOA University,

Singh, R.P.B. (2020) 'Sacrality and waterfront sacred places in India: Myths and the making of place', in C. Ray (ed.), *Sacred Waters: A Cross-Cultural Compendium of Hallowed Springs and Holy Wells* (pp. 80–94). London and New York: Routledge.

Singh, R.P.B. (2021) 'Heritagescapes of the Gaṅgā river Basin: Understanding global civilization', in N. Mishra (ed.), *Gaṅgā, the River of 'Sanātan' Civilization* (pp. 71–99). New Delhi: Research India Press.
Singh, R.P.B., & Aktor, M. (2015) 'Hinduism and globalization', in S.D. Brunn (ed.), *Changing World Religion Map: Sacred Places, Identities, Practices and Politics* (pp. 1917–1932). Dordrecht and New York: Springer Nature.
Singh R.P.B., & Haigh, M.J. (2015) 'Hindu pilgrimages: The contemporary scene', in S.D. Brunn (ed.), *The Changing World Religion Map: Sacred Places, Identities, Practices and Politics* (pp. 783–801). Dordrecht and New York: Springer Nature.
Singh, R.P.B., & Rana, P.S. (2020a) 'Contemporary perspectives of Hindu pilgrimage in India: The experiential exposition', in D. Liutikas (ed.) *Pilgrims: Values and Identity* (pp. 135–147). Wallingford, UK: CABI.
Singh, R.P.B., & Rana, P.S. (2020b) 'Faith and place: Hindu sacred landscapes of India', in T. Edensor, A. Kalandides, & U. Kothari (eds.), *The Routledge Handbook of Place* (pp. 15–108). London and New York: Routledge.
Sinha, A. (2019) 'Ghats on the Ganga in Varanasi, India: A sustainable model for waste management', *Landscape Journal*, 37: 65–78.
Sivapriya (1989) 'The Greatness of Hinduism', in J. Wolffe (ed.), *The Growth of Religious Diversity* (pp. 73–80). London: Hodder & Stoughton.
Sobrevila, C. (2008) *The Role of Indigenous Peoples in Biodiversity Conservation: The Natural but Often Forgotten Partners*. Washington, DC: The World Bank.
Timothy, D.J. (2011) *Cultural Heritage and Tourism: An Introduction*. Bristol: Channel View Publications.
Timothy, D.J. (2012) 'Religious views of the environment: Sanctification of nature and implications for tourism', in A. Holden & D.A. Fennell (eds.), *The Routledge Handbook of Tourism and the Environment* (pp. 31–42). London: Routledge.
Times of India (2019) 'Sabarimala pilgrimage to be a 'green' affair', http://timesofindia.indiatimes.com/articleshow/72363766.cms?utm_source=contentofinterest&utm_medium=text&utm_campaign=cppst.
TNN (2020) 'At least 15 crore pilgrims expected in Kumbh 2021: CM Triven', *The Times of India*. http://timesofindia.indiatimes.com/articleshow/73742668.cms?utm_source=contentofinterest&utm_medium=text&utm_campaign=cppst.
Turnbull, C.M. (1992) 'Postscript: Anthropology as pilgrimage, anthropologist as pilgrim', in A. Morinis (ed.), *Sacred Journeys: The Anthropology of Pilgrimage* (pp. 257–274). Westport, CT: Greenwood Press.
UNESCO-MAB (2006) *Conserving Cultural and Biological Diversity: The Role of Sacred Natural Sites and Cultural Landscapes*. Proceedings of the Tokyo Symposium: 30 May-2 June 2005. Paris: UNESCO.
UNWTO (2011) *Religious Tourism in Asia and the Pacific*. Madrid: World Tourism Organization.
van Horn, G. (2006) 'Hindu traditions and nature: Survey article', *Worldviews: Global Religions, Culture, and Ecology*, 10(1): 5–39.
Verschuuren, B. (2007) 'An overview of cultural and spiritual values in ecosystem management and conservation strategies', in B. Haverkort & S. Rist (eds.), *Endogenous Development and Bio-cultural Diversity: The Interplay of Worldviews, Globalisation and Locality* (pp. 299–325). Leusden, The Netherlands: Centre for Development and Environment.
Verschuuren, B., & Furuta, N., eds. (2016) *Asian Sacred Natural Sites: Philosophy and Practice in Protected Areas and Conservation*. London: Routledge.
Verschuuren, B., Wild, R., McNeely, J.A., & Oviedo, G., eds. (2010) *Sacred Natural Sites: Conserving Nature and Culture*. London: Earthscan.
Verschuuren, B., Wild, R., & Verschoor, G. (2017) 'Connecting policy and practice for the conservation of sacred natural sites', in B. Verschuuren & S. Brown (eds.), *Cultural and Spiritual Significance of Nature in Protected Areas: Governance, Management and Policy* (pp. 11–40). New York & London: Routledge.
Vortmann, M., Balsari, S., Holman, S.R., & Greenough, P.G. (2015) 'Water, sanitation, and hygiene at the world's largest mass gathering', *Current Infectious Disease Reports*, 17(2): 1–7.
Vukonić, B. (2006) 'Sacred spaces and tourism in the Roman Catholic tradition', in D.J. Timothy & D.H. Olsen (eds.), *Tourism, Religion, and Spiritual Journeys* (pp. 237–253). London: Routledge.
Wexler, J. (2016) *When God Isn't Green: A World-Wide Journey to Places Where Religious Practice and Environmentalism Collide*. Boston, MA: Beacon Press.
Wild, R., & McLeod, C. (2008) *Sacred Natural Sites: Guidelines for Protected Area Managers*. Gland, Switzerland and Paris: IUCN/UNESCO.

11
TOURISM AND SPIRITUALITY
Green places, blue spaces, and beyond

Richard Sharpley

Introduction

It has long been recognized that people's experiences with nature may elicit some sort of spiritual response (Szerszynski 2005). Equally, the positive effects of natural environments on both physical and mental well-being have also been understood since ancient times (Gesler 1993), although more general and recent awareness of their contribution to spiritual well-being coincides with the continuing decline in adherence to religious beliefs and practices since the early nineteenth century (Heelas & Woodhead 2005). As de Botton (2003: 171) observed,

> It is no coincidence that the Western attraction to sublime landscapes developed at precisely the moment when traditional beliefs in God began to wane. It is as if these landscapes allowed travelers to experience transcendent feelings that they no longer felt in cities and the cultivated countryside. The landscapes offered them an emotional connection to a greater power, even as they freed them of the need to subscribe to the more specific and now less plausible claims of biblical texts and organized religions.

More than a century ago, the sociologist Durkheim (2008) asserted that when institutionalized religion eventually succumbs to the advance of scientific knowledge and rational belief, people inevitably seek meaning in experiences and practices other than religion. Subsequently, it has been suggested that, unsurprisingly, tourism has become one such "secular substitute" for religion (Allcock 1988: 35) and is considered a potential source of emotional or spiritual fulfilment (Sharpley 2009). Indeed, tourism is now widely considered to be a contemporary practice that is "functionally and symbolically equivalent to other institutions that humans use to embellish and add meaning to their lives" (Graburn 1989: 22), which meaning may lie either in the journey in its entirety or in tourists' engagement with particular places they visit.

Significantly, however, understanding of this spiritual dimension of contemporary tourism remains relatively limited. Although the various manifestations of religious tourism, or tourism "whose participants are motivated either in part or exclusively for religious reasons" (Rinschede 1992: 52), are considered extensively in the burgeoning literature (Vukonić 1996;

Timothy & Olsen 2006; Stausberg 2011; Raj & Griffin 2015; Butler & Suntikil 2018), less attention has been paid to the phenomenon of spiritual experiences in "secular" tourism scenarios (see Chapter 8, this volume). This is not to say, however, that this topic has been overlooked completely. For example, Willson, McIntosh, and Zahra's (2013) study explored the concept of spirituality within the context of individual tourist's travel experiences, while other scholars have considered the secular spiritual dimension of the experience of traditional pilgrimage journeys, such as the Camino de Santiago (Devereux & Carnegie 2006; Norman 2011; Lois González 2013); at religious sites, including churches and cathedrals (Williams et al. 2007; Francis et al. 2010), ashrams (Sharpley & Sundaram 2005), and Buddhist temples (Hall 2006; Kaplan 2010); and in particular secular touristic settings, such as the countryside (Jepson & Sharpley 2015) and the seaside (Jarratt & Sharpley 2017). However, there continues to be a limited understanding of nature and extent of spiritual experiences among secular tourists, particularly with regard to tourists' experiences of and with spirituality in different places.

The purpose of this chapter is to address this gap in the literature. More specifically, this chapter considers whether different categories of place or, more precisely, the manner in which tourists engage with different places elicit different levels or intensities of spirituality, spirituality being defined here as a sense of connectedness or what de Souza (2012: 293) describes as "human relationality"—people are relational beings, and hence "living with an awareness of one's relationality is the essence of human spirituality" (de Souza 2012: 292). Focusing primarily on the broad categorizations of "green", "blue", and "dark" spaces, the argument put forward in this chapter is that tourists' relationality or sense of connectedness may become diluted the less physically engaged (or connected) they are with place, following a continuum from *being in* green spaces/places to *gazing on* blue space and, ultimately, *imagining beyond* into outer (dark) space. In so doing, this chapter seeks to contribute a more nuanced understanding regarding how secular tourists may experience a sense of spirituality in different spatial contexts. First, the concepts of space, place, and the creation of meaning are briefly reviewed before discussing spirituality-related issues in green places, blue spaces, and in the dark beyond.

Space, place, and meaning

Tourism and place are intimately related. Tourism involves the temporary movement from one place or centre of meaning—typically ordinary home life—to another place—a centre "out there" (Cohen 1979: 180) that offers the potential for new, meaningful experiences (Urry 2005). More simply stated, tourism is motivated by a desire to experience an "Other" (Wearing, McDonald & Ankor 2016). This, in turn, implies that tourists purposefully select not "anywhere" but "somewhere" to travel to; that is, they attribute meaning to the places they travel to, arguably in the expectation of desired meaningful experiences and, as such, this meaning represents the attraction or pull-factor of the destination (Dann 1981; Suvantola 2002). Hence, as Jepson and Sharpley (2015: 1159) observe, "tourism cannot be understood only in terms of geographic loci; it is also necessary to understand the social, cultural and psychological interactions that visitors have with the place". The questions that must then be addressed include what is "place", and how is meaning attached to place?

In addressing the first question, fundamental to understanding the concept of place is the recognition that "to exist at all...is to have a place – to be implaced....To "be" is to be in place" (Casey 1993: 13). In other words, human existence is defined by the relationships people have with place, such relationships being described by a variety of terms including "place attachment", "place identity", "place dependence", "place bonding", and "sense of place", the latter arguably being the most recognized and utilized term (Stedman 2003). Equally, it is suggested that the (post)modern world is increasingly defined by a condition of placelessness (Harvey 1990, 2012), whereby people's relationships with place and place-making are becoming diminished through technological and other transformations. Consequently, modern societies suffer uncertainty, instability, and a sense of alienation, precisely the reason why MacCannell (1976) first suggested that tourists are contemporary secular pilgrims seeking meaning or connectedness in other places.

Irrespective of terminology, a key factor in understanding the significance of place is the distinction between "space" and "place" (Cresswell 2004). According to Bremer (2006: 26), space is meaningless, whereas place is meaningful, with the distinction between the two lying in the "particularity of places [compared] to the homogeneity of space". In other words, undifferentiated space becomes place "when we endow it with value" (Tuan 1977: 6); or, as Gieryn (2000: 465) suggests, "place would revert to space if we vacuumed out the distinct collection of values, meanings and objects that created it". Significantly, space and place are not independent—that is, place cannot exist without space (Agnew 2011). Moreover, space is defined by movement (Tuan 1977)—specifically, movement from place to place; hence, the existence of place is dependent on space. This is the case in tourism, as the significance of two meaningful places (home and the destination) is defined by the movement (travel) through the space between them.

If place is indeed a "centre of meaning" (Relph 1976: 2) within existential space, the second question then logically follows: what factors determine that meaning? The social construction of place is considered widely in the literature and from a variety of disciplinary perspectives; hence, it is not possible to review fully this literature within the confines of this chapter (cf. Milligan 1998; Manzo 2003; Kyle & Chick 2007; Alkon & Traugot 2008). Nevertheless, there is a general consensus that there are three components that lead to a space becoming a place (referred to here as the "triad of place"). The first component is the physical/objective environment or, more broadly, the physical attributes of a particular place. Indeed, the very basis of a place is a set of tangible physical features that distinguish it from other places and spaces. In the context of this chapter, those tangible physical features offer tourists the opportunity to fulfil their particular desires or needs. For example, wilderness or mountainous areas may be sought out by those who wish to engage in physical activities, whereas tourists with an interest in history may be drawn to built heritage environments.

The second component of place is the social or cultural significance attached to place. Places are endowed with cultural meaning, or as Gieryn (2000: 473) puts it, "meanings that individuals and groups assign to places are more or less embedded in historical contingent and shared cultural understandings of the terrain". Consequently, countryside places, for example, may be understood through a cultural lens of "rurality", representing a bygone, arcadian yet imagined era (Short 1991). Equally, destinations in less developed countries may perhaps be viewed by Western tourists through culturally determined understandings of tradition and modernity. As such, any one place may be subject to different cultural interpretations. As Greider and Garkovich (1994: 2) note, "any physical place has the potential to embody multiple landscapes, each of which is grounded in the cultural definitions of those who encounter that place".

The third component of place is the significance or meaning ascribed to a particular place that lies in the manner in which people experience a place or interact with it, either in the past or in the present. For example, childhood memories of family holidays may contribute to a sense of belonging or connection to a particular place, while the opportunity to engage in what Urry (1992) describes as either a romantic (alone or with significant others) or collective (shared with many others) gaze may determine the significance attached to different places. Thus, the meaning of a wilderness area may lie in the opportunity to experience solitude and connection to nature. In contrast, popular tourist resorts are expected to be experienced collectively as part of a crowd. As such, "place social bonding" is increasingly considered a dominant factor in place attachment (Ramkissoon, Weiler & Smith 2012: 264), where the significance or sense of a place is reflected in the communal bonds developed between people living in or tourists who visit that place (Hammit, Blacklund & Bixler 2006).

These three components of place remain widely debated within the literature, not least with regard to their relative importance in defining place. Stedman (2003), for example, argues that the contribution of the physical environment to a sense of place remains underestimated, suggesting that it plays a more powerful role in determining a sense of place than most commentators acknowledge. Nevertheless, in the context of this chapter, it is the third component of place—people's experiences or behaviours—that is of most relevance, for without actually interacting with or being in a particular place, an individual will not be able to fully ascribe personal meaning to it, and consequently in that person's mind, that place may revert to space. Putting it another way, tangible, physical environments exist and almost invariably will be subject to shared cultural meanings. However, to be emotionally connected to or experiencing a sense of place, an individual must be or have been there. If not, any sense of emotional connection or place attachment/bonding will be limited. This in turn suggests that tourists are more likely to experience some sense of connectedness or spirituality, or feel it more intensely, when physically interacting with a place—that is, being in a place—as opposed to gazing upon places and spaces, such as landscapes and seascapes, or even imagining these places or viewing them virtually through the Internet. This argument is now explored in more detail by considering the degree of tourist interaction with different types of places and spaces and the extent to which spirituality/connectedness might subsequently be experienced.

Green place, tourism, and spiritual experiences

As noted earlier, there is evidence of a long-held belief in the restorative effects of the natural environment on people's physical and mental wellbeing (e.g., Hartig, Mang & Evans 1991; Thompson 2011; Shanahan et al. 2016). During the nineteenth century, the recognition that nature and access to green spaces was conducive to health and well-being was manifested first in the provision of public parks and other open spaces in burgeoning industrial towns and cities, and later with the emergence of organizations and the creation of policies dedicated to both preserving the natural environment and promoting access to it for recreation, leisure and, implicitly, well-being (Conway 1991). For example, commencing with the designation of Yellowstone in the US in 1872, national parks were subsequently established in many other countries during the late nineteenth and early twentieth centuries (Gissibl, Höhler & Kupper 2012), while in the UK, the National Trust, now boasting more than four million members, was founded in 1895 with the express initial purpose of preserving and maintaining access to natural places for the benefit of the nation (Weideger 1994). Presently, national parks represent a major tourism resource globally (Eagles & McCool 2002; Frost & Hall 2009), as do rural areas that cater to rural tourists (Sharpley & Sharpley 1997; Gartner 2004; Su 2011). For the purposes of this chapter, the term "green space" is understood more broadly to embrace any land-based, non-urban natural environment, whether or not it is "green" (i.e., irrespective of the presence of vegetation) that is accessible to and used by tourists. As noted above, green space has the potential to become "green place" through tourists' interaction with it, which interaction may elicit spiritual experiences.

While it is assumed that green spaces and places have a beneficial impact on people's health and well-being, several studies attempt to measure this assumption empirically. According to Pool (2016), the first such empirical study was undertaken by Ulrich (1979), who found that views of natural scenes led to lower levels of anxiety among people who were experiencing high levels of stress than scenes of urban environments. This finding was confirmed by MacKerron and Mourato (2013), who also found that people seem to have greater levels of happiness being in natural environments versus being in human-built environments. Since then, the relationship between natural places and well-being has attracted increasing academic attention, with numerous studies confirming the beneficial impacts of both viewing and physically engaging with green space on physical and mental health (e.g., Bowler et al. 2010; Thompson & Aspinall 2011; Pearson & Craig 2014; McMahan & Estes 2015).

Equally, other scholars have started to investigate the potential spiritual dimension of green spaces and places, whether from a public land management and environmental sustainability policies perspective (e.g., Fredrickson & Kerr 1998; Ashley 2007) or the extent to which spiritual or emotional outcomes may result from participating in tourism and leisure experiences in a variety of natural environments or green spaces. Some examples of research examining the spiritual outcomes of visiting natural environments include the connection between spirituality and wilderness experiences (Stringer & McAvoy 1992; Fredrickson & Anderson 1999; Heintzman 2003; Ashley 2007) or being in forest and desert environments (Williams & Harvey 2001; Narayanan & Macbeth 2009); the experiential (including spiritual) aspects of mountain tourism (Pomfret 2006); and the inherent spirituality of the outdoors more generally as a leisure and recreation resource (Allcock 2003; Schmidt & Little 2007; Ferguson & Tamburello 2015; Jepson 2015). These and other studies collectively suggest that a being in a variety of natural environments or green spaces can elicit spiritual or emotional responses although, perhaps unsurprisingly, these vary according to individual expectations, motivations, and experiences (Heintzman 2010). Nevertheless, experiencing green space is generally found to generate a sense of belonging and place and feelings of harmony and a deep connection with the world (Fredrickson & Anderson 1999; Heintzman 2010). In addition, according to Williams and Harvey (2001: 250), people in natural environments may experience a state of "flow", whereby "the usual distinctions between self and object are lost". Also, some studies reveal that being in and interacting with the natural environment is a significant factor in stimulating spiritual responses, as being there "heighten[s] one's level of sensory awareness" (Fredrickson & Anderson 1999: 34). In other words, being in green space offers opportunities to develop not only a "spiritual connectedness to the landscape" (Gelter 2000: 78), which transforms green space into green place or a space endowed with meaning), but "intimate contact with this environment leads to thoughts about spiritual meanings and eternal processes" (Kaplan & Talbot 1983: 178).

Although simply gazing upon green space or natural landscapes may elicit a sense of well-being or spirituality, physically being in and interacting with the natural environment not only completes the triad of "place-making" factors but also serves to enhance spiritual experiences. A study by Sharpley and Jepson (2011; cf. Jepson & Sharpley 2015) confirms this to be the case, their research focusing on the extent to which tourists engaging in a rural area (the English Lake District) sought and/or experienced spiritual meaning from their visits and, if so, what factors induced such meaning. Although it was found that the majority of those participating in the study did not consciously visit the Lake District in pursuit of spiritual refreshment—indeed, despite expressing a sense of connectedness, awe, peace, and solitude, few respondents considered their resultant experiences to be specifically spiritual—the research revealed three key factors that emphasized the significance of being "in place" to the intensity of spiritual (or emotional) touristic experiences. First, respondents noted that their emotional response to being in the Lake District was very much framed by their lived experiences of the area—that their expressions of a strong sense of place attachment were due to past experiences and childhood memories of the site. Second, the descriptions of the transcendental experiences of respondents were typically related to the physical characteristics of the area, specifically the mountains and the benefits they offered (e.g., solitude, being elevated, remoteness), experiences that came from being present in the landscape. Third, the physical interaction of respondents with the natural environment through trekking in the mountains and "conquering" them was the most powerful element in providing them with implicit spiritual experiences. As such, these outcomes collectively support the argument that when circumstances combine to give meaning to green spaces, tourists are likely to experience the most intense form of spiritual or emotional experience when in and interacting with green place. This in turn suggests that when there is no such physical connection and interaction with the natural environment, spiritual responses or experiences may be more limited or even lacking.

Blue space, tourism, and spirituality

In the preceding section, green space was defined as any land-based, non-urban natural environment. In contrast, blue spaces and places may be thought of comprising natural areas containing water. However, the concept of blue space becomes complex, as blue spaces and places can refer to both areas that are predominantly or entirely aquatic, such as oceans, seascapes, and, of particular relevance tourism, the seaside or coastal areas, as well as aquatic elements that are found in green spaces and places, such as lakes and rivers that run through national parks. In the latter case, identifying and assessing the restorative qualities or potential spiritualities of blue spaces and places in particular or distinguishing between the relative contributions of blue space and the green space that surrounds is problematic (Pool 2016). This perhaps explains why research into the impacts of blue space on health, well-being, and spirituality has been relatively limited (White et al. 2010; Völker & Kistemann 2011; Foley & Kistemann 2015), although this research does suggest that not only are people drawn more to water than to other landscape features but also that the presence of water is a predictor of the perceived attractiveness of a landscape (Pool 2016). More specifically, lakes and larger bodies of (inland) water have been rated highly in studies both for their restorative effects and for the sense of tranquility they generate (Whalley 1988; Herzog & Bosley 1992; Purcell, Peron & Berto 2001; Cooper 2006).

Given the evident difficulties in separating the potential emotional or spiritual responses to tourism derived from aquatic features in combined green/blue spaces and recognizing that the combination of lakes and mountains endows certain landscapes with particular aesthetic qualities (Walton & Wood 2013), for the purposes of this chapter the term "blue space" refers specifically to the sea and seascapes. While this is not to deny the potential of inland blue spaces in inducing a sense of spirituality, not only is the sea a major tourism resource in terms of activities both on the coast (e.g., sun-sea-sand tourism, coastal walks) and on the sea itself (e.g., sailing, cruise holidays), but also recent research has begun to reveal that the touristic experience of the sea embraces a spiritual or liminal dimension (Šunde 2008; Varley 2011). Moreover, it is primarily seascapes and gazing over blue space that has been found to induce feelings or emotions akin to a sense of spirituality, although, as discussed below, being in or interacting with the sea may also elicit such a response (Levine 1985; Brown & Humberstone 2015).

Like green space, there has long been a recognition that blue spaces/aquatic environments are beneficial to people's health, and this in turn led to the beginnings of the development of modern tourism in Europe and beyond. In the seventeenth and eighteenth centuries, spas in the UK and Europe became popular for the perceived benefits of their mineral waters, with many spa towns becoming nascent tourist resorts (Towner 1996). The subsequent development of seaside resorts followed a similar pattern, where in the mid-1700s the alleged benefits of bathing in and even drinking seawater were described in a medical treatise, and many resorts soon became fashionable among the leisured classes before transforming into major mass tourism resorts (Williams 2003: 23–25). Reasons for travelling to the seaside in the present also include reasons other than health, including leisure, adventure tourism, and cruising, that underpin the popularity of the sea as one of the most popular resources for tourism in all its related forms (Hall 2001; Jennings 2007; Orams & Lück 2014).

Given the significance of the sea to global tourism, it is surprising, that very few studies have explored the spiritual dimension of blue spaces and the extent to which sea-related tourism may offer spiritual outcomes, especially considering that many of the authors of British Romanticism, including Wordsworth, Keats, Coleridge and Byron, all "sensed in the vast organic entity of the sea the same amalgam of spirit that stirred in the depths of the human soul" (Lencek & Bosker 1998: 97), while Bachelard (1994) suggested that there exists a connection between the immensity and limitlessness of the seas and the depth of inner space within all people.

There are two studies, however, that are of particular relevance here. First, Bull (2006) speculates on the extent to which tourism to the seaside is undertaken in the expectation of or may result in some form of spiritual response or experience. Referring specifically to activities related to the sea, either in or on the water or utilizing the coastal margin, Bull suggests several factors that may influence motivations to travel to the seaside that either possess some "degree of natural spirituality" or may elicit a "set of human spiritual responses". Some of these factors relate to the physical coastal environment in terms of health and well-being, the natural rhythms of the tides, and its significance as a liminal place (cf. Shields 1991). Other factors include people's interactions with the water, including the potential for a spiritual connection with pre-terrestrial existence and surrendering to a greater power. Implicit in Bull's paper, however, is that spiritual experiences in blue spaces are most likely to result from being by the sea or gazing at the broader blue landscape as opposed to physically being in it.

In the second study, Jarratt and Sharpley (2017) also focus on the spiritual dimension of seaside visits, although their work is based on empirical research with tourists at a seaside resort in northern England whose motivations were to primarily enjoy the coastal environment without physically interacting with the water. Several themes emerged from their research, including the fact that as a tourist destination, the seaside was viewed by tourists more as a generic place. As one tourist commented, "I don't think it matters too much where you are, if you're at the seaside, you're at the seaside". While this attachment to the seaside might be based to an extent on cultural nostalgia (Jarratt & Gammon 2016), the sea itself is clearly a fundamental element in the experience of the seaside, with many respondents in Jarrat and Sharpley's study expressing an emotional response to being beside and gazing out over the sea. For example, some of the tourists spoke about how being by the seaside inspired them to contemplate and reflect deeply about themselves and their lives, while others said that by looking out towards the sea they felt a (re)connection to nature. Other tourists referred to their experiences at the seaside in a more explicitly spiritual way, as evidenced by their references to the timelessness of the seascape. As Jarratt and Sharpley (2017: 12) note, "As visitors look out to sea, they see the beautiful and the sublime, a timeless blank canvas without any evidence of the encroachment of mankind... [they were] encouraged to contemplate existence and creation". Equally, some tourists linked their perceived vastness and awesomeness of the seascape to their place in the world, while others suggested that the "nothingness" of the seascape gave rise within them some sense of spirituality.

From these two studies, it is evident that the blue space of the sea/seaside and its characteristics of liminality, vastness, timelessness, and nothingness provides a potential foundation for contemplation, for deep thought, and, perhaps, questioning. By gazing at and, in some circumstances, being in the sea, tourists may feel out of normal time and place, a sensation that may create the potential for, though not necessarily result in, spiritual experiences. Alternatively stated, the sea remains blue space; it lacks the cultural significance to be transformed into place (indeed, it would be illogical to refer to the vastness of the world's seas as "place") and the research referred to above suggests that emotional responses are generally elicited by viewing the seascape, not by being physically connected with the sea. This suggests that the potential spiritual experiences of tourists in blue spaces, where both a physical and cultural connection with the sea (as opposed to the seaside) is lacking, are more limited than for tourists who engage with green places.

Beyond: gazing at outer space

Having considered both green and blue spaces/places which, together, comprise the material basis of the Earth, this chapter extends this discussion to a third space which is increasingly becoming a literal focus of the tourist gaze—outer or dark space. Although outer space and the broader universe have fascinated humankind since time immemorial, and travel to gaze at the heavens or to witness astronomical events has a long history (Collison & Poe 2013), recent years have witnessed an increase in tourism related to looking up at and into space. Referred to variously as astronomical tourism (Collison & Poe 2013), astrotourism (Fayos-Solá, Marín & Jafari 2014; Soleimani et al. 2019), terrestrial astrotourism (Matos 2017), or even celestial ecotourism (Weaver 2011)—the latter term firmly establishing it as a variety of nature-based tourism with the emphasis on where it typically, but not always, occurs—tourism to dark spaces involves travel to and stays at places offering a view the night sky unimpeded by light pollution. As such, the astrotourism (the term used here) niche market promotes and develops places where astrotourists can travel to view space, although Matos (2017) distinguishes between specific (e.g., observatories, parks/reserves) and non-specific sites (e.g., grasslands) where it is possible to view dark skies.

People participate in astrotourism for a variety of reasons and motives. Matos (2017: 16) proposes a simple taxonomy of astrotourists based along a continuum from, at one end, scientists and professional astronomers to, at the other end, the general public who engage in astrotourism through casual or serendipitous visits to dark sky parks and other places offering the opportunity to gaze at the night sky. In between the two ends of the continuum are amateur astronomers. Not included in Matos' taxonomy are astrologers who also visually explore the night skies either for professional purposes or as a hobby and those who pay for trips to the International Space Station (Cater 2010). Matos' (2017) study is an example of the rise in recent years in the phenomenon of astrotourism from several perspectives (Fayos-Solá, Marín & Jafari 2014; Slater 2020; Soleimani et al. 2019; Jacobs, Du Preez & Fairer-Wessels 2020).

Given the evident link between gazing at space (that is, towards the heavens) and interpretations of spirituality that embrace both the individual search for meaning/connection and a belief in a higher or supreme power (Kale 2004), it is surprising that none of these studies consider explicitly the spiritual dimension of astrotourism, a potentially fruitful area for future research. Nevertheless, some offer some findings that contribute to the argument in this chapter. Matos (2017), for example, found that the two principal motives for engaging in astrotourism were education (i.e., learning about the stars /solar system) and meeting new friends or developing relationships with people having shared interest. However, the most satisfying aspect of the astrotourism experience was what respondents described as the opportunity to relax. Specifically, Matos' respondents (none of whom were astronomers) revealed that through participating in astrotourism, they were able to reflect on their lives, "to contemplate the universe and themselves, [which contributed] to their emotional well-being and making them forget momentarily their daily lives" (Matos 2017: 82). Indeed, the view was expressed that astrotourism provided a

> feeling of belonging, because it is so beautiful, it touches you, touches your emotional part...[a feeling of belonging] to the Earth, like we are on this site observing the stars and so we feel that we belong to this whole.
>
> *(p. 68)*

In a similar vein, in Slater's (2020) study of the experiences of astrotourists, one respondent claimed that "stargazing is one of the most profoundly human things you can do. When you look into infinity, you realise that there are more important things than what you do all day" (p. 218). Another respondent, perhaps experiencing a similar sensation to those discussed above who referred to the timelessness of blue space, stated, "I can see my future when I want... Looking at the light is looking into the past at a time gone by, a time to be treasured... having a handle on the future gives comfort..." (p. 222). Both studies mentioned here reveal the potential for spiritual experiences through gazing on dark space, at least among non-astronomers. Paradoxically, however, Slater's study revealed that, for many astrotourists, gazing in silence at the dark skies served to heighten their awareness of the sounds and smells around them. As such, they found that gazing into dark space (up there) resulted in a sense of connection with the Earth and nature (down here).

From the limited research to date, participation in astrotourism or looking into dark space (a space that is of undoubted significance as the counterpoint to physical green and blue space but that can only be seen or imagined from afar), appears to offer the opportunity to reflect, to question, to travel psychologically from the here and now to an undefinable and arguably non-physical space that cannot be entered into or connected with (at least until the advent of space tourism). As such, dark spaces can never be a "place", and consequently, it is in these spaces that spiritual experiences are least likely to occur. Indeed, gazing into dark space may raise more spiritual questions than it answers.

Conclusion

Although there exists a spiritual dimension to contemporary tourism, where secular (non-religious) tourists may experience something spiritually meaningful as a result of their journey and/or the places and people they encounter, understanding the relationship between tourism and spirituality remains limited. Despite the centrality of place to the tourist experience, few attempts have been made to consider how the tourist's relationship with place may determine subsequent spiritual or transcendental experiences. Hence, in addressing this gap in the literature, this chapter has attempted to demonstrate that the intensity of spiritual feelings or a sense of connectedness among tourists is dependent upon the extent to which people are attached to or "emplaced" in particular places. In other words, it is in places, not spaces, and when tourists are in and interacting with meaningful places, where and when spiritual experiences are most likely to occur. Thus, if one element of the "triad of place" mentioned earlier, such as physical interaction with place, is removed, then place reverts to space, leading to a diminishing of both the potential for and the intensity of spiritual experiences. Hence, it is suggested here that a continuum of tourism place-space spirituality exists, as conceptualized in Figure 11.1. In summarizing the argument in the chapter, Figure 11.1 suggests that tourists in green places, such as rural or wilderness destinations, may have the most intense spiritual experiences because they are more deeply immersed in place. In blue spaces, where tourists are beside or gazing at the sea, spiritual feelings and experiences will be less intense but of greater intensity than tourists who are looking out at dark spaces.

It is, however, important to emphasize three points here in conclusion. First, the model presented in Figure 11.1 inevitably over-simplifies complex and individualistic responses to green, blue, and dark spaces and places. As such, the intent of Figure 11.1 is principally to offer a basis for developing a more nuanced understanding of tourists' spiritual relationship with spaces and places. Second, as discussed elsewhere (Jepson & Sharpley 2015), not all tourists seek spiritual experiences or interpret them as such, as the distinction between spiritual and emotional responses is often blurred and may be dependent on both an individual's belief systems and how they interpret their experiences. Third, the discussion here has focused only on secular tourists and the extent of their potential spiritual experience in different (green, blue, dark) places/spaces. Other tourists, of course, may possess strong religious beliefs that act as a lens through which they interpret their experiences. Therefore, one might properly ask whether the argument made in this paper is reversed for tourists with a belief in a greater power. In other words, for religious tourists or tourists who are religious, material spaces—the countryside, the sea—are perhaps less spiritually significant than the manner in which they believe them to be created. Hence, might gazing into the beyond, into dark space towards a greater yet unseen power generate the most intense spiritual response?

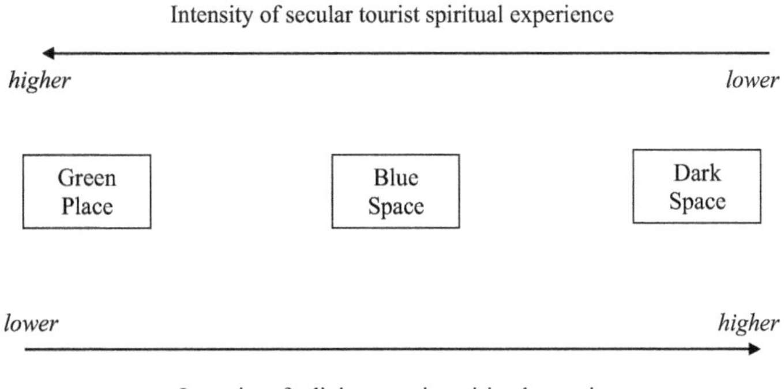

Figure 11.1 Continuum of spiritual experience

References

Agnew, J. (2011) 'Space and place', in J. Agnew & D. Livingstone (eds), *Handbook of Geographical Knowledge* (pp. 316–330). London: Sage Publications.
Alkon, A.H., & Traugot, M. (2008) 'Place matters, but how? Rural identity, environmental decision making, and the social construction of place', *City & Community*, 7(2): 97–112.
Allcock, D. (2003) 'From the plains to the peaks: The outdoors as a repository of spirituality', *Horizons*, 21: 10–12.
Allcock, J. (1988) 'Tourism as a sacred journey', *Loisir et Société,* 11: 33–48.
Ashley, P. (2007) 'Toward an understanding and definition of wilderness spirituality', *Australian Geographer,* 38(1): 53–69.
Bachelard, G. (1994) *The Poetics of Space*. Boston, MA: Beacon Press.
Bowler, D.E., Buyung-Ali, L.M., Knight, T.M., & Pullin, A.S. (2010) 'A systematic review of evidence for the added benefits to health of exposure to natural environments', *BMC Public Health*, 10(1): 456. https://doi.org/10.1186/1471-2458-10-456.
Bremer, T. (2006) 'Sacred spaces and tourist places', in D.J. Timothy & D.H. Olsen (eds), *Tourism, Religion and Spiritual Journeys* (pp. 25–35). London and New York: Routledge.
Brown, M., & Humberstone, B. (eds) (2015) *Seascapes: Shaped by the Sea*. Burlington, VT: Ashgate.
Bull, A. (2006) 'Is a trip to the seaside a spiritual journey?', Paper presented at the *Tourism: The Spiritual Dimension Conference*, University of Lincoln, UK (unpublished).
Butler, R., & Suntikil, W. (eds) (2018) *Tourism and Religion: Issues and Implications*. Bristol: Channel View Publications.
Casey, E. (1993) *Getting Back into Place: Toward a Renewed Understanding of the Place-World*. Bloomington: Indiana University Press.
Cater, C.I. (2010) 'Steps to space: Opportunities for astrotourism', *Tourism Management*, 31(6): 838–845.
Cohen, E. (1979) 'A phenomenology of tourist experiences', *Sociology*, 13(2): 179–201.
Collison, F, & Poe, K. (2013) "Astronomical tourism': The astronomy and dark sky program at Bryce Canyon National park', *Tourism Management Perspectives*, 7: 1–15.
Conway, H. (1991) *People's Parks: The Design and Development of Victorian Parks in Britain*. Cambridge: Cambridge University Press.
Cooper, C. (2006) 'Lakes as tourism destination resources', in C.M. Hall & T. Härkönen (eds), *Lake Tourism: An Integrated Approach to Lacustrine Tourism Systems* (pp. 27–42). Clevedon, UK: Channel View Publications.
Cresswell, T. (2004) *Place: A Short Introduction*. Oxford: Blackwell.
Dann, G. (1981) 'Tourist motivation: An appraisal', *Annals of Tourism Research*, 8(2): 187–219.
de Botton, A. (2003) *The Art of Travel*. London: Penguin Books.
de Souza, M. (2012) 'Connectedness and connectedness: The dark side of spirituality–implications for education', *International Journal of Children's Spirituality*, 17(4): 291–303.
Devereux, C., & Carnegie, E. (2006) Pilgrimage: Journeying beyond self. *Tourism Recreation Research*, 31(1): 47–56.
Durkheim, E. (2008) *The Elementary Forms of Religious Life* (trans. C. Cosman). Oxford: Oxford University Press.
Eagles, P.F., & McCool, S.F. (2002) *Tourism in National Parks and Protected Areas: Planning and Management*. Wallingford, UK: CABI.
Fayos-Solá, E., Marín, C., & Jafari, J. (2014) 'Astrotourism: No requiem for meaningful travel', *PASOS: Revista de Turismo y Patrimonio Cultural*, 12(4): 633–671.
Ferguson, T.W., & Tamburello, J.A. (2015) 'The natural environment as a spiritual resource: A theory of regional variation in religious adherence', *Sociology of Religion*, 76(3): 295–314.
Foley, R., & Kistemann, T. (2015) 'Blue space geographies: Enabling health in place', *Health & Place*, 35: 157–165.
Francis, L., Mansfield, S., Williams, E., & Village, A. (2010) 'Applying psychological type theory to cathedral visitors: A case study of two cathedrals in England and Wales', *Visitor Studies*, 13(2): 175–186.
Fredrickson, L., & Anderson, D. (1999) 'A qualitative exploration of the wilderness experience as a source of spiritual inspiration', *Journal of Environmental Psychology*, 19(1): 21–39.
Fredrickson, L., & Kerr, W. (1998) 'Spiritual values: Can they be incorporated into forest management and planning?', in H. Vogelsong (ed.), *Proceedings of the 1998 Northeastern Recreation Research Symposium* (pp. 239–245). Radnor, PA: US Department of Agriculture, Forest Service.

Frost, W., & Hall, C.M. (eds.) (2009) *Tourism and National Parks: International Perspectives on Development, Histories, and Change*. London and New York: Routledge.

Gartner, W.C. (2004) 'Rural tourism development in the USA', *International Journal of Tourism Research*, 6(3): 151–164.

Gelter, H. (2000) 'Friluftsliv: The Scandinavian philosophy of outdoor life', *Canadian Journal of Environmental Education*, 5(1): 77–92.

Gesler, W. (1993) 'Therapeutic landscapes: Theory and a case study of Epidauros, Greece', *Environment and Planning D: Society and Space*, 11(2): 171–189.

Gieryn, T. (2000) 'A space for place in sociology', *Annual Review of Sociology*, 26(1): 463–496.

Gissibl, B., Höhler, S., & Kupper, P. (eds) (2012) *Civilizing Nature: National Parks in Global Historical Perspective*. New York: Berghahn Books.

González, R.L. (2013) 'The Camino de Santiago and its contemporary renewal: Pilgrims, tourists and territorial identities', *Culture and Religion*, 14(1): 8–22.

Graburn, N. (1989) 'Tourism: The sacred journey', in V. Smith (ed.) *Hosts and Guests: The Anthropology of Tourism* (pp. 21–36). Philadelphia, PA: University of Pennsylvania Press.

Greider, T., & Garkovich, L. (1994) 'Landscapes: The social construction of nature and the environment', *Rural Sociology*, 59(1): 1–24.

Hall, C.M. (2001) 'Trends in ocean and coastal tourism: the end of the last frontier?', *Ocean and Coastal Management*, 44(9–10): 601–618.

Hall, C.M. (2006) 'Buddhism, tourism and the middle way', in D.J. Timothy & D.H. Olsen (eds), *Tourism, Religion and Spiritual Journeys* (pp. 172–185). Abingdon: Routledge.

Hammitt, W., Backlund, E., & Bixler, R. (2006) 'Place bonding for recreation place: Conceptual and empirical development', *Leisure Studies*, 25(1): 7–41.

Hartig, T., Mang, M., & Evans, G.W. (1991) 'Restorative effects of natural environment experiences', *Environment and Behavior*, 23(1): 3–26.

Harvey, D. (1990) *The Condition of Postmodernity*. Oxford: Blackwell.

Harvey, D. (2012) 'From space to place and back again: Reflections on the condition of postmodernity', in J. Bird, B. Curtis, T. Putnam, & L. Tickner (wds), *Mapping the Futures: Local Cultures, Global Change* (pp. 17–44). Abingdon: Routledge.

Heelas, P., & Woodhead, L. (2005) *The Spiritual Revolution: Why Religion Is Giving Way to Spirituality*. Oxford: Blackwell.

Heintzman, P. (2003) 'The wilderness experience and spirituality what recent research tells us', *Journal of Physical Education, Recreation & Dance*, 74(6): 27–32.

Heintzman, P. (2010) 'Nature-based recreation and spirituality: A complex relationship', *Leisure Sciences*, 32(1): 72–89.

Herzog, T., & Bosley, P. (1992) 'Tranquility and preference as affective qualities of natural environments', *Journal of Environmental Psychology*, 12(2): 115–127.

Jacobs, L., Du Preez, E.A., & Fairer-Wessels, F. (2020) 'To wish upon a star: Exploring Astro Tourism as vehicle for sustainable rural development', *Development Southern Africa*, 37(1): 87–104.

Jarratt, D., & Gammon, S. (2016) ''We had the most wonderful times': Seaside nostalgia at a British resort', *Tourism Recreation Research*, 41(2): 123–133.

Jarratt, D, & Sharpley, R. (2017) 'Tourists at the seaside: Exploring the spiritual dimension', *Tourist Studies*, 17(4): 349–368.

Jennings, G. (2007) *Water-Based Tourism, Sport, Leisure, and Recreation Experiences*. London and New York: Routledge.

Jepson, D. (2015) 'The lure of the countryside: The spiritual dimension of rural spaces of leisure', in S. Gammon & S. Elkington (eds.) *Landscapes of Leisure* (pp. 202–219). London: Palgrave Macmillan.

Jepson, D., & Sharpley, R. (2015) 'More than sense of place? Exploring the emotional dimension of rural tourism experiences', *Journal of Sustainable Tourism*, 23(8–9): 1157–1178.

Kale, S. (2004) 'Spirituality, religion, and globalization', *Journal of Macromarketing*, 24: 92–107.

Kaplan, R., & Talbot, J. (1983) 'Psychological benefits of a wilderness experience', in I. Altman & J. Wohlwill (eds), *Behavior and the Natural Environment* (pp. 163–203). Boston, MA: Springer.

Kaplan, U. (2010) 'Images of monasticism: The temple stay program and the re-branding of Korean Buddhist temples', *Korean Studies*, 34: 127–146.

Kyle, G., & Chick, G. (2007) 'The social construction of a sense of place', *Leisure Sciences*, 29(3): 209–225.

Lencek, L., & Bosker, G. (1998) *The Beach: The History of Paradise on Earth*. New York: Viking.

Levine, S.Z. (1985) 'Seascapes of the sublime: Vernet, Monet, and the oceanic feeling', *New Literary History*, 16(2): 377–400.

MacCannell, D. (1976) *The Tourist: A New Theory of the Leisure Class*. New York. Schocken Books.

MacKerron, G., & Mourato, S. (2013) 'Happiness is greater in natural environments', *Global Environmental Change*, 23(5): 992–1000.

Manzo, L. (2003) 'Beyond house and heaven: Towards a revisioning of emotional relationships with places', *Journal of Environmental Psychology*, 23(1): 47–61.

Matos, A.L. (2017) *Terrestrial Astrotourism: Motivation and Satisfaction of Travelling to Watch the Night Sky*. Master's Thesis, Aalborg University. Available at: https://projekter.aau.dk/projekter/files/260343239/THESIS_ASTROTOURISM_PDF.pdf (Accessed 10 October 2018).

McMahan, E.A., & Estes, D. (2015) 'The effect of contact with natural environments on positive and negative affect: A meta-analysis', *The Journal of Positive Psychology*, 10(6): 507–519.

Milligan, M.J. (1998) 'Interactional past and potential: The social construction of place attachment', *Symbolic Interaction*, 21(1): 1–33.

Narayanan, Y., & Macbeth, J. (2009) 'Deep in the desert: Merging the desert and the spiritual through 4WD tourism', *Tourism Geographies*, 11(3): 369–389.

Norman, A. (2011) *Spiritual Tourism: Travel and Religious Practice in Western Society*. London and New York: Bloomsbury Publishing.

Orams, M.B., & Lück, M. (2014) 'Coastal and marine tourism', in A.A. Lew, C.M. Hall, & A.M. Williams (eds), *The Wiley-Blackwell Companion to Tourism* (pp. 479–489). Chichester, UK: Wiley-Blackwell.

Pearson, D.G., & Craig, T. (2014) 'The great outdoors? Exploring the mental health benefits of natural environments', *Frontiers in Psychology*, 5: 1–4.

Pool, U. (2016) *The Impact of Water and Anthropogenic Objects on Implicit Evaluations of Natural Scenes: A Restorative Environments Perspective*. PhD Thesis: University of Central Lancashire. Available at: http://clok.uclan.ac.uk/17669/ (Accessed 6 October 2018).

Pomfret, G. (2006) 'Mountaineering adventure tourists: A conceptual framework for research', *Tourism Management*, 27(1): 113–123.

Purcell, T., Peron, E., & Berto, R. (2001) 'Why do preferences differ between scene types?', *Environment and Behavior*, 33(1): 93–106.

Raj, R., & Griffin, K. (eds) (2015) *Religious Tourism and Pilgrimage Management: An International Perspective*. Wallingford, UK: CABI.

Ramkissoon, H., Weiler, B., & Smith, L. (2012) 'Place attachment and pro-environmental behaviour in national parks: The development of a conceptual framework', *Journal of Sustainable Tourism*, 20(2): 257–276.

Relph, E. (1976) *Place and Placelessness*. Pion: London.

Rinschede, G. (1992) 'Forms of religious tourism', *Annals of Tourism Research*, 19(1): 51–67.

Schmidt, C., & Little, D.E. (2007) 'Qualitative insights into leisure as a spiritual experience', *Journal of Leisure Research*, 39(2): 222–247.

Shanahan, D.F., Franco, L., Lin, B.B., Gaston, K.J., & Fuller, R.A. (2016) 'The benefits of natural environments for physical activity', *Sports Medicine*, 46(7): 989–995.

Sharpley, R. (2009) 'Tourism, religion and spirituality', in T. Jamal & M. Robinson (eds), *The Sage Handbook of Tourism Studies* (pp. 237–253). London: Sage Publications.

Sharpley, R., & Jepson, D. (2011) 'Rural tourism: A spiritual experience?', *Annals of Tourism Research*, 38(1): 52–71.

Sharpley, R., & Sharpley, J. (1997) *Rural Tourism: An Introduction*. London: International Thomson Business Press.

Sharpley, R., & Sundaram, P. (2005) 'Tourism: A sacred journey? The case of ashram tourism, India', *International Journal of Tourism Research*, 7(3): 161–171.

Shields, R. (1991) *Places on the Margin: Alternative Geographies of Modernity*. London: Routledge.

Short, J. (1991) *Imagined Country: Society, Culture and Environment*. London: Routledge.

Slater, D. (2020) *Towards an Understanding of the Astro Tourist: A Conceptual and Empirical Study*. PhD Thesis, University of Central Lancashire.

Soleimani, S., Bruwer, J., Gross, M, & Lee, R. (2019) 'Astro-tourism conceptualisation as special-interest tourism (SIT) field: A phenomonological approach', *Current Issues in Tourism*, 22(18): 2299–2314.

Stausberg, M. (2011) *Religion and Tourism: Crossroads, Destinations and Encounters*. Abingdon: Routledge.

Stedman, R. (2003) 'Is it really just a social construction? The contribution of the physical environment to sense of place', *Society & Natural Resources*, 16(8): 671–685.
Stringer, L., & McAvoy, L. (1992) 'The need for something different: Spirituality and wilderness adventure', *Journal of Experiential Education*, 15(1): 13–20.
Su, B. (2011) 'Rural tourism in China', *Tourism Management*, 32(6): 1438–1441.
Šunde, C. (2008) 'The open horizon: Exploring spiritual and cultural values of the oceans and coasts', in M.G. Patterson & B.C. Glavovic (eds), *Ecological Economics of the Oceans and Coasts* (pp. 166–183). Cheltenham, UK: Edward Elgar.
Suvantola, J. (2002) *Tourist Experience of Place*. Aldershot: Ashgate.
Szerszynski, B. (2005) *Nature, Technology and the Sacred*. Oxford: Blackwell Publishing
Thompson, C.W. (2011) 'Linking landscape and health: The recurring theme', *Landscape and Urban Planning*, 99(3–4): 187–195.
Thompson, C.W., & Aspinall, P.A. (2011) atural environments and their impact on activity, health, and quality of life', *Applied Psychology: Health and Well-Being*, 3(3): 230–260.
Timothy, D.J., & Olsen, D.H. (eds) (2006) *Tourism, Religion and Spiritual Journeys*. Abingdon: Routledge.
Towner, J. (1996) *An Historical Geography of Recreation and Tourism in the Western World 1540–1940*. Chichester: John Wiley & Sons.
Tuan, Y. (1977) *Space and Place: The Perspective of Experience*. Minneapolis: University of Minnesota.
Ulrich, R. (1979) 'Visual landscapes and psychological well-being', *Landscape Research*, 4(1): 17–23.
Urry, J. (1992) 'The tourist gaze 'revisited'', *American Behavioral Scientist*, 36(2): 172–186.
Urry, J. (2005) *Consuming Places*. London: Routledge.
Varley, P.J. (2011) 'Sea kayakers at the margins: The liminoid character of contemporary adventures', *Leisure Studies*, 30(1): 85–98.
Völker, S., & Kistemann, T. (2011) 'The impact of blue space on human health and well-being–Salutogenetic health effects of inland surface waters: A review', *International Journal of Hygiene and Environmental Health*, 214(6): 449–460.
Vukonić, B. (1996) *Tourism and Religion*. Pergamon: Oxford.
Walton, J., & Wood, J. (eds) (2013) *The Making of a Cultural Landscape: The English Lake District as a Tourist Destination, 1750–2010*. Aldershot: Ashgate
Wearing, S., McDonald, M., & Ankor, J. (2016) 'Journeys of creation: Experiencing the unknown, the other and authenticity as an epiphany of the self', *Tourism Recreation Research*, 41(2): 157–167.
Weaver, D. (2011) 'Celestial ecotourism: New horizons in nature-based tourism', *Journal of Ecotourism*, 10(1): 38–45.
Weideger, P. (1994) *Gilding the Acorn: Behind the Façade of the National Trust*. London: Simon & Schuster Ltd.
Whalley, J.M. (1988) 'Water in the landscape', *Landscape and Urban Planning*, 16(1–2): 145–162.
White, M., Smith, A., Humphryes, K., Pahl, S., Snelling, D., & Depledge, M. (2010) 'Blue space: The importance of water for preference, affect, and restorativeness ratings of natural and built scenes', *Journal of Environmental Psychology*, 30(4): 482–493.
Williams, E., Francis, L., Robbins, M., & Annis, J. (2007) 'Visitor experiences of St Davids Cathedral: The two worlds of pilgrims and secular tourists', *Rural Theology*, 5(2): 111–123.
Williams, K., & Harvey, D. (2001) 'Transcendent experience in forest environments', *Journal of Environmental Psychology*, 21(3): 249–260.
Williams, S. (2003) *Tourism Geography*. London and New York: Routledge.
Willson, G., McIntosh, A., & Zahra, A. (2013) 'Tourism and spirituality: A phenomenological analysis', *Annals of Tourism Research*, 42: 150–168.

12

THE RELIGIOUS AND SPIRITUAL APPEAL OF NATIONAL PARKS

Thomas S. Bremer

Introduction

In the final decade of the nineteenth century, writer Olin Wheeler joined the growing number of tourists visiting Yellowstone National Park, where he experienced the spiritual powers that had made the park a premier pilgrimage destination for Gilded Age travelers. He later recalled a particular moment of spiritual awakening as he sat in the dawn light on the rim of the Grand Canyon of the Yellowstone River:

> Grand and glorious pageant, vision mighty and eternal; for unnumbered aeons thou hast been slowly, through the attritive powers given thee by Nature's God, working out thy destiny. With a perseverance sublime, by the power of torrent and beat of wave, the rush of the avalanche, the grinding of the glacier, the hot breath of the geyser, the subtle uplifting of the frost, the downpouring from the clouds; by all the powers of earth and sky, the wind and hail, the lightning's glare, the thunder's crash, hast thou worked onward, channeling the mountains, sculpturing the hills, painting the cliffs, that man, the noblest of God's creations, might stand before thee in awe and rapture, and feel himself uplifted to that spirit land from whence he came.
>
> *(Wheeler 1893: 82)*

Wheeler's response to Yellowstone, although with more Victorian poetic flourishes than most, is not unusual. Religious sentiments have been prominent in visitor experiences of American national parks at least since their founding with the establishment of Yellowstone in 1872.[1] Travel to the parks promised nineteenth-century visitors sublime experiences bordering on the spiritual and religious. Some of these tourist pilgrims brought overtly sectarian frames of religious reference for interpreting their experiences, while others saw the wonders of nature through an aesthetic lens that derived in large part from American Protestant spiritual traditions of the era. For nearly all national park visitors in nineteenth-century America, religion figured prominently in their experiences, interpretations, and memories of the nation's foremost natural attractions.

The religious aspect in visitors' experiences of American national parks has not lessened since nineteenth-century pilgrims first journeyed to Yellowstone and the other early parks.

Following a brief discussion about national parks in the scholarly literature of American religious history, this chapter will concentrate on three particular elements that have made American national parks attractive pilgrimage destinations for generations of tourist visitors. First, national parks have been places of healing both individually and collectively. The earliest place protected as a federal reserve, Hot Springs in Arkansas, was valued as a source of bodily healing where pilgrims with a variety of infirmities could find relief and even be cured in the steaming mineral waters. In addition, numerous national parks address the need for collective healing in memorializing atrocities that have occurred throughout American history.

The second element of appeal for pilgrim tourists to national parks has been a spiritual connection to nature. Especially in the large parks designated primarily for their scenic value and natural wonders, many visitors discover a mystical dimension in their experiences of American national parks. Divine encounters in the wild parklands are sometimes framed in the specific theological orientations of devotional traditions, but often they rely more on Romantic aesthetics popularized in nineteenth-century art and literature that regarded nature as the expression of divine presence. A good number of tourist pilgrims experience their gods most directly in the natural settings of national parks.

Finally, national parks serve as sacred sites of nationalist civil religion. Considering nationhood as a peculiarly modern form of religion, the notion of civil religion explains the fervent devotion that characterizes the patriotic sentiments of many citizens and the profound feelings experienced in the sacred places of a nation. In the United States, the National Park Service acts as a custodian in protecting and caring for the most meaningful sites of American history while also presenting the American story in official interpretations of these places of national significance.

Scholarship on national parks

The literature on national parks is extensive. Most of it aims for a general readership among those who love the parks, but a good number of more academic works have been published in recent decades (Runte 1987; Rothman 1989; Kaufman 1996; Sellars 1997; Heacox 2001; Mackintosh & McDonnell 2005; Duncan & Burns 2009; Miles 2009). Very few of these studies, though, even mention religion or spiritual practices, and fewer still are detailed critical studies regarding religious dimensions in the histories, management, and popular appeal of national parks. *Blessed with Tourists* (Bremer 2004), with its attention to the history, preservation, and interpretive challenges of the San Antonio Missions National Historical Park, may have been the first such published scholarly study in the field of American religious history. An impressive book on pilgrimage to the national parks (Ross-Bryant 2013) uses a religious studies approach that combines perspectives of historical, phenomenological, and cultural studies in an investigation of religious travel to popular nature parks. Through a different religious studies lens, another recent book (Mitchell 2016) analyzes the management and interpretive strategies of the National Park Service that encourage and facilitate spirituality for park visitors.

Though studies specifically of religion and spirituality in national parks have been few, several excellent works recount the more general history of American religious attitudes toward nature and the natural environment. Catherine Albanese's seminal work on "nature religion" (Albanese 1990) has little mention of parks, but it details a religious history of nature in American culture that has made parklands not only possible, but inevitable. A more recent work (Stoll 2015) seeks to contextualize the history of environmentalism in American religious history, and it convincingly shows that parks, including national parks, have their genesis in the influential impact of Calvinist Protestant traditions on the moral and aesthetic development of the American nation. Another work elucidating religious influences on American environmentalism (Berry 2015) concentrates more narrowly on the critical early decades of the twentieth century that, among other important environmental developments, produced the National Park Service as the agency responsible for America's national parks and other protected lands and structures.

In terms of their role in the patriotic devotions of American civil religion, national parks represent significant sites in the sacred landscape of the American nation. Although rarely mentioned in the scholarly literature on civil religion, the National Park Service has been responsible for many of the most important sites of national historical significance, including such iconic places as Independence Hall, the Statue of Liberty, the National Mall in Washington, D.C., Revolutionary and Civil War battlefields, as well as places of national tragedy such as the Sand Creek Massacre National Historic Site in Colorado, the Little Bighorn Battlefield National Monument in Montana, Manzanar National Historic Site in California, and the Flight 93 National Memorial in Pennsylvania.

Places of healing

One of the earliest attractions to places that would become national parks involved healing bodily infirmities. Yellowstone attracted visitors hoping to cure their ailing bodies in the hot mineral waters even before it became the world's first national park. As an expedition of curious tourists from Bozeman, Montana, made their way up the Yellowstone River in the summer of 1871, nearly a year before Congress set the area aside as a national park, they came upon a crudely constructed health spa called Chestnutville on Warm Spring Creek at Soda Mountain, later renamed Mammoth Hot Springs (Haines 1996: 144). Two years later, after Yellowstone had become a national park, Rev. Edwin Stanley found the health spa at Mammoth Hot Springs still flourishing (Stanley 1878: 57–58). In fact, Yellowstone's reputation for curing a variety of maladies continued throughout the nineteenth century. As late as 1890, a park visitor reported, "We have seen several consumptives who have come here for their health" (Harrison 1891: 71).

Yellowstone was not the first park that drew infirm pilgrims for healing their bodily ailments. In fact, a much earlier precedent for preserving natural areas was Hot Springs Reservation in Arkansas, created by the U.S. Congress in 1832 (Shugart 2009: 7). As early as 1809, Euroamerican settlers had recognized Arkansas's natural springs as a destination for ailing pilgrims, and by the 1830s, lodging and other accommodations were serving travelers from the United States and Europe (Shugart 2009: 3). An 1832 proposal to Congress for establishing a hospital at the springs was not included in the legislation establishing the reserve, but over the subsequent decades, numerous bathhouses and medical establishments have brought relief, comfort, and recovery for tourist pilgrims at Arkansas's Hot Springs. The reservation was placed under the jurisdiction of the newly established National Park Service in 1916, and when Hot Springs became a national park in 1932, it was stipulated that park superintendents were to be U.S. Public Health Service physicians (Shugart 2009: 33).

Seeking out health in national parks such as Hot Springs and Yellowstone derives from a long tradition of travel to auspicious destinations to heal bodily ailments. Ancient Romans frequented shrines of the god Asklepios to undergo healing rites, and Christians in late antiquity adapted the ancient practice of curative pilgrimage in their devotional travels to the *loca sancta* of Holy Land destinations and the shrines of holy personages (Talbot 2002: 153).

The practice continues today with popular pilgrimage traditions at places like Lourdes in southern France, the Bosnian shrine of Medjugorje, the English sites of Glastonbury and the Madron Well, the Basilica of Our Lady of Guadalupe in Mexico City, and Chimayó in New Mexico.

Although finding relief and even cures for bodily infirmities has been a motivation of some travelers to national parks, many more have found collective healing for the traumas of national history. Numerous national park units remember historical atrocities against marginalized groups and provide opportunities for visitors to mourn for those who suffered, to atone for past injustices, and to engage in national healing. The history of injustice, slaughter, and displacement of Native American peoples, for instance, is prominently featured at Sand Creek Massacre National Historic Site in Colorado. At Devils Tower National Monument in eastern Wyoming, a place that many native groups regard as Bear's Lodge (*Mato Tipila* in the Lakota language), a holy site for several tribal groups, contemporary Native Americans perform rituals of healing every year and leave offerings around the base of the monolithic tower (National Park Service 2016).

Other places of national healing include the former internment camp of Japanese Americans at Manzanar National Historic Site in California and the various civil rights locations of the American south. An annual pilgrimage of remembrance to Manzanar began in the 1960s and brought attention to the site. The National Park Service acquired it in 1992 and rebuilt a replica of the internment facility for visitors to experience the conditions that residents suffered during World War II (Hayashi 2003). The pilgrimage continues today and draws attention to parallel injustices in contemporary America (Manzanar Committee n.d.). Similarly, at places like the Ebenezer Baptist Church in Atlanta where the Rev. Dr. Martin Luther King, Jr., served as a minister and where he is buried, now part of the Martin Luther King, Jr. National Historical Park, the Park Service tells the story of racial divisions and the history of injustices and racial violence in America. In recent years, additional park units have been added, including the Selma to Montgomery National Historic Trail, which commemorates the events, people, and route of the 1965 Voting Rights March in Alabama. In total, the National Park Service lists 25 national park units that commemorate civil rights activists and leaders, plus places featuring African American, Native American, Latinx, and women's history (National Park Service 2018). At these and other sites, visitors participate in national healing by learning the history of marginalized groups in America.

The aesthetic experience of nature

Another sort of restorative ceremony took place in 2016 as part of the centennial celebration of Acadia National Park, located on Mount Desert Island and surrounding locations along the coast of Maine. The program involved a public arts and worship production staged by local clergy who exclaimed that "Mount Desert is a soul and spirit stretching place, and that is what we celebrate today" (Bremer 2019: 132). In short, religious leaders were commemorating Acadia National Park as a place of soulful regeneration.

Acadia is not alone among the soul and spirit stretching places in national parks. In fact, perhaps the most inviting appeal of the parks has been the aesthetic experience of nature. Whether tourist visitors are awed by majestic scenic vistas, seek adventures that challenge their souls, or find solace in the calm and quiet of natural settings, most all encounters with nature in the national parks involve aesthetic dimensions. For many park visitors in nineteenth-century America, the notion of "the sublime" was a common aesthetic frame for interpreting the human experience of the natural world.

Seeking the sublime in the natural world became a favorite pastime among America's elite genteel classes in the decades following the Civil War (Sears 1989: 10). Increased circulation of artworks and literature featuring the wonders of the American west inspired people to visit these places with expectations of transformative spiritual experiences. Travel accounts in newspapers, periodicals, and books, the burgeoning market for dime novel adventure stories, as well as the popularity of the Hudson River School of landscape painters and their successors brought widespread public attention to the wonders of natural landscapes of the American nation. Western destinations such as Yellowstone, Yosemite, and the Grand Canyon joined with images of Niagara Falls, the Hudson River Valley, the Adirondacks, and the White Mountains of the northeast to entice a burgeoning leisure class in Gilded Age America (Davidson 2006: 4–6).

Many of the travelers who endured the hardships of nineteenth-century journeys to the western parks struggled to put into words their experiences of these places. They repeatedly stressed their inability to convey what they had seen and felt in their tours of these natural wonderlands. For instance, Harry J. Norton, who toured Yellowstone in 1872, wrote of the newly established park, "The subject is beyond the conception of the most vivid imagination—language is inadequate to express the unapproachable picture presented—the eye only can photograph the gorgeous scene" (Norton 1873: 38). The impossibility of capturing the experience of national parks in words had a parallel challenge for artists. William O. Owen, who undertook a bicycle tour of Yellowstone in 1883, said of the Lower Falls of the Yellowstone River, "it forms a picture that… the most skillful artists could not hope to reproduce" (quoted in Hassrick 2002: 8). Likewise, a decade later artist Frederic Remington commented on the Golden Gate of Yellowstone: "It is one of those marvelous vistas of mountain scenery utterly beyond the pen or brush of any man. Paint cannot touch it, and words are wasted" (quoted in Hassrick 2002: 8). Nevertheless, despite the frustration of lacking words adequate to their experiences, visitors to the parks wrote effusively, usually with no hint of irony, and the national parks quickly became a popular topic in the burgeoning popular literary genre of travel writing.

Many of these earlier visitors resorted to religious language to describe what they experienced in the parks. This religious language took several forms. Christian-oriented writers often resorted to conventional biblical language, imagery, and themes in relating their experiences of the natural landscape. In Yellowstone, the frightening sights, sounds, and stench of gurgling hot springs and bubbling mud pots evoked Christian images of the hellish underworld, while the uplifting and emotionally moving mountain scenery and waterfalls sometimes evoked angelic references to the celestial paradise. This proclivity to describe natural features in familiar Christian terms translated to place names, some of which still bear Christian references in various national parks.[2]

Besides more traditional religious imagery for describing nature's wonders, Romanticism also influenced how many Americans responded to their experiences of national parks. As historian John Sears remarks,

> The eighteenth-century English tradition of landscape gardening; the aesthetics of the sublime, the beautiful, and the picturesque; and the work of Romantic writers, like William Wordsworth, Sir Walter Scott, and Lord Byron, had identified culture and landscape so closely with each other that they seemed almost identical.
>
> *(Sears 1989: 4)*

It was in this context that the American concept of parks developed, according to Katherine Early, "simultaneous with, and partly in response to, Romanticism" (Early 1984: 3). In what she describes as the "genteel romanticism" of nineteenth-century America, the notion of the "sublime" took precedence over the merely beautiful by inspiring deep emotional responses to nature (Early 1984: 23). For many of the earliest tourist visitors to national parks and other places of superlative natural attractions, the promise of sublime experience inspired their travels.

The popularity of sublime experience among American travelers can be traced in large part to the work of eighteenth-century British philosopher Edmund Burke who recognized the sublime in the immense power of exceptionally large or somewhat frightening natural objects. The German philosopher Immanuel Kant also recognized the power of the sublime in nature, but he reinterpreted it by locating sublime power not in objects but in the human psyche. He regarded the sublime as a state of mind obtained in the recognition of frightening and apparently unlimited aspects of nature, which led viewers to a realization of their own deepest being (McGreevy 1994: 10–11; Bremer 2020: 194). For nineteenth-century American travelers, this notion of the sublime experience of nature provided them with both an explanation for their reactions to wild scenery and a vocabulary for expressing those reactions.

Attention to the natural world and its effect in eliciting sublime experiences gained wider popularity through Transcendentalism, which introduced Romantic religious perspectives about nature for many Americans. Foremost were the essays and lectures of Ralph Waldo Emerson that contributed to a more aesthetically robust religious language for considering the human relationship to the natural world. Emerson's Transcendentalist philosophy helped to pivot the Christian perspective from regarding nature as an object separate from and inimical to human experience to one of subjective immersion in nature. His seminal 1836 essay *Nature* begins with a contemplation on the value of solitude and the need for individuals "to retire as much from [their] chamber as from society." He writes,

> But if a man would be alone, let him look at the stars. The rays that come from those heavenly worlds, will separate between him and what he touches. One might think the atmosphere was made transparent with this design, to give man, in the heavenly bodies, the perpetual presence of the sublime.
>
> *(Emerson 1850: 5)*

For Emerson and other Transcendentalists who followed his lead, especially Henry David Thoreau and poet Walt Whitman, nature revealed spiritual truths available in the immediacy of one's surroundings.

The growing influence of Romantic aesthetics introduced by Emerson and his literary heirs encouraged the American embrace of the sublime. Even scientifically minded observers sometimes indulged in a Romantic sense of impressive American landscapes. Geologist Clarence E. Dutton, who once exclaimed that "the geologist finds himself a poet," described the view from the promontory that he named Point Sublime overlooking Arizona's Grand Canyon as "the sublimest thing on earth" (quoted in Pyne 1998: 75). The Canyon, he remarked, "is not alone by virtue of its magnitude, but by virtue of its whole—its *ensemble*" (quoted in Pyne 1998: 82, emphasis in original). This ensemble of geological wonders coalesced for Dutton, like millions of visitors who followed in subsequent generations, as a sublime experience in what would eventually become America's fifteenth national park.

Besides its influence in literary tastes of nineteenth-century America, Romanticism contributed as well to the popularity of landscape painting, most famously with the rise of the Hudson River School painters, who "pioneered the quest for 'sublime' sights, seeking to convey on paper and canvas a divine presence in the marvels of nature" (Davidson 2006: 3). These artists concentrated on natural and pastoral landscape scenes, initially in the northeastern United States with later followers of this tradition depicting the wild scenery of the American west (Ferber 2009). Many of their romantic renderings of pastoral and wild landscapes were saturated with religious associations, especially for white Protestant audiences of the nineteenth century (Stoll 2015: 58). As their paintings gained acceptance and popularity, these artists contributed to an enduring religio-aesthetic that would figure decisively in designating scenic places as national parks.

One of the more inspirational places for these landscape painters was the island that would later be dominated by Acadia National Park. Thomas Cole and other artists in the Hudson River School, especially Cole's protégé Frederic E. Church, brought Mount Desert Island's spectacular scenery into public awareness in the 1840s and 1850s with their depictions of the shoreline, mountain vistas, and majestic sunsets and sunrises (Belanger & Sweeney 1999: 19; Little & Little 2016: 33–36). Similarly, Thomas Moran, a later artist in the Hudson River School tradition who accompanied the first official geological survey of the Yellowstone region in 1871, produced spectacular paintings that, along with the photographs of William Henry Jackson, also a participant in the 1871 geological survey, were decisive in convincing the U.S. Congress to set aside Yellowstone as the world's first national park. Their paintings and photographs confirmed the aesthetic appeal of Yellowstone's attractions for members of Congress as well as the public at large, especially Moran's famous masterwork "Grand Cañon of the Yellowstone," which helped to "fan the public wonder" about the new national park (Wilkins & Hinkley 1998: 5).

Perhaps the most influential work in establishing the aesthetic appeal of national parks as sacred destinations were the writings of naturalist John Muir. Raised in a strict Presbyterian household, Muir never lost a Calvinist regard for nature as a divine revelation in God's creation. But like many other environmentalists raised as Presbyterians, Muir abandoned the orthodoxy of Reformed theology in favor of a more spiritual natural theology (Stoll 2015: 148). He took this spiritual reverence for nature to the wilds of California, where he discovered the Sierra Nevada mountains as "the Range of Light, the most divinely beautiful mountain-chains I have ever seen" (Muir 1894: 3). Wandering these wild mountains, John Muir reveled in the aesthetic wonder of God's presence. He regarded the Yosemite Valley, a place of special affection where he passed many seasons, as "that sublime Sierra temple where every day one may see the grandest of sights" (Muir 1894: 43). "In the morning," he wrote of the mountain forests,

> everything is joyous and bright, the delicious purple of the dawn changes softly to daffodil yellow and white; while the sunbeams pouring through the passes between the peaks give a margin of gold to each of them.... every pulse beats high, every life-cell rejoices, the very rocks seem to tingle with life, and God is felt brooding over everything great and small.
>
> *(Muir 1894: 179)*

Besides his Scottish Presbyterian upbringing and college professors who "taught science as the study of God's creation" (Stoll 2015: 146), Muir also came under the spell of the Transcendentalists, especially Emerson, who visited him in Yosemite in 1871. Muir pressed the "sage of Concord" to spend with him "a month's worship with Nature in the high temples of the great Sierra Crown," but to Muir's great disappointment, Emerson declined (quoted in Worster 2008: 211). Their meeting, however, had a life-changing impact on Muir (Stoll 2015: 148) and initiated a friendship sustained through regular correspondence between the young naturalist and aging philosopher. Emerson sent to Muir a complete set of his essays, which Muir gratefully received and studied. He also studied Thoreau's essays, especially *Walden*, and he concluded that "the pure soul of Thoreau… would have been content with my log house" (quoted in Worster 2008: 212). Muir himself would follow the well-worn path of nineteenth-century romantically tinged nature essayists as he began his writing career in the 1870s.

By the opening decade of the twentieth century, John Muir had become one of the most widely read nature writers in America, with numerous books and articles extolling the beauty and power of the natural world. His 1901 book on the national parks inspired a generation of park visitors to seek spiritual rewards in places like Yellowstone, where

> a multitude of still, small voices may be heard directing you to look through all this transient, shifting show of things called 'substantial' into the truly substantial, spiritual world whose forms flesh and wood, rock and water, air and sunshine, only veil and conceal, and to learn that here is heaven and the dwelling place of angels.
>
> *(Muir 1901: 74)*

Among his many achievements, perhaps Muir's most significant influence has been as a prophetic voice in the spiritual value of wild places, especially national parks.

Equally influential in promoting a religio-aesthetic appreciation of American national parks were the photographs of Ansel Adams. Like John Muir, Adams adopted an Emersonian aesthetic of nature that profoundly influenced his artistic vision as a photographer in service of religious and moral purposes (Stoll 2015: 117). Adams professed in a 1961 commencement address at Occidental College in California "a somewhat mystical concept of nature," believing that "the world is incomprehensibly beautiful—an endless prospect of magic and wonder" (Stillman 2010: 11). He translated this mystical sense in photographs that have inspired millions of people who share his vision of nature to seek magic and wonder in national parks. With exquisite artistry, Adams framed the aesthetic value of parklands by rendering their sublime character in photographic images.

Adams' photographs of the national parks influenced a generation of visitors who came to the parks in the period following World War II. After more than a decade of waning visitation, American national parks in the postwar period experienced a marked increase in their popularity as travel destinations. Many of these visitors came as pilgrim tourists seeking an aesthetic experience of the natural world, inspired by the writings of such prophetic voices as John Muir and by Ansel Adams' striking images of the national parks.

American civil religion

Besides their aesthetic appeal, people also come to national parks out of a sense of patriotic devotion. As noted earlier, American national parks encompass some of the most sacred sites of nationhood, including Independence Hall in Philadelphia, the Lincoln Memorial in Washington, D.C., the Statue of Liberty in New York, Mount Rushmore in South Dakota, and the U.S.S. Arizona Memorial in Hawaii. In fact, all U.S. national parks contribute in some way or another to an American civil religion.

The place of national parks in American civil religion became more explicit beginning in the 1930s when the National Park Service became a prominent force in the debates over public memory. The agency joined other large bureaucratic organizations in attempting to impose and shape a nationalist framework for the public perception of the past. In particular, the park service promoted a middle-class narrative of progress and patriotism (Bodnar 1992: 170). This nationalist narrative and the values that it promotes serve to sanctify the iconic sacred places of American civil religion, many of which are under the protection and care of the National Park Service.

The notion of an American civil religion refers to nationalistic practices and traditions that legitimize state authority with reference to transcendent powers.[3] This idea gained attention in modern scholarship beginning in the 1960s when sociologist Robert Bellah claimed that a nationalistic religiosity flourishes in the United States alongside other religious traditions (Bellah 1967). Bellah's interpretation of American nationalist devotion utilizes the idea of civil religion first proposed by the eighteenth-century philosopher Jean-Jacques Rousseau. According to Bellah's definition, civil religion involves "a collection of beliefs, symbols, and rituals with respect to sacred things and institutionalized in a collectivity" (Bellah 1967: 4), specifically the institutions of a nation; he argues "that the civil religion at its best is a genuine apprehension of universal and transcendent religious reality as seen in or... revealed through the experience of the American people" (Bellah 1967: 12). American civil religion, in Bellah's analysis, "has its own prophets and its own martyrs, its own sacred events and sacred places, its own solemn rituals and symbols" (Bellah 1967: 18). Specifically, he discusses references to God in presidential addresses and characterizes the Declaration of Independence, the U.S. Constitution, and Abraham Lincoln's Gettysburg Address as "sacred scriptures" of the nation.

Following its initial introduction, Bellah's concept of civil religion was widely discussed and became a useful interpretive framework for regarding the religious nature of nationalism (Fenn 1977; Wilson 1979; Bellah & Hammond 1980; Handy 1980; Gehrig 1981; Mathisen 1989; Jones & Richey 1990; Bellah 1992; Castren 1993; Flowers 1994; Hammond et al. 1994; Davis 1998; Ellis 2005; Lippy 2006; Kao & Copulsky 2007; Chernus 2010). Scholars have used the concept in numerous studies of historical events and trends, national ceremonies, and important objects and sites of the American nation. At the same time, there has also been considerable criticism of Bellah's notion of an American civil religion. A common concern takes issue with designating "an incredibly rich and internally complex culture" in simple terms of "religion" (Wilson 1979: 148, 175). Other critics argue that Bellah's concept itself must be understood as a product of the peculiar cultural and religious circumstances of the Cold War period that allowed scholars like Bellah to imagine, in one critic's estimation, "a mythic past sustaining an American civil religion that gave coherence and cohesion to the common life of the American people" (Lippy 2006: 24). Much of the American population, several critics maintain, was excluded from this mythic past, "making the civil religion really the prerogative of white males of economic privilege" (Lippy 2006: 34).

Yet despite its drawbacks as an interpretive category, the notion of an American civil religion has proven useful for understanding nationalistic devotion and the popularity of destinations related to the history and heritage of the American nation, including national park sites. Perhaps the holiest ground of American civil religion is Philadelphia's Independence Hall, the birthplace of the United States, now the centerpiece of Independence National Historical Park. Both the Declaration of Independence and the U.S. Constitution, the two most sacred texts of the American nation, were authored and signed in this hallowed building by such legendary figures as George Washington, Thomas Jefferson, Benjamin Franklin, John Adams, James Madison, and Alexander Hamilton. Tourist pilgrims line up every day to enter the sanctified Assembly Room in Independence Hall that "became a shrine to the founding of the nation," according to the National Park Service (National Park Service 2017). Across the street in its own shrine visitors can see the Liberty Bell, an iconic symbol of the freedoms enshrined in the Constitution.

Besides Philadelphia, Washington, D.C., the nation's capitol, has the greatest concentration of American civil religion sites, most of them national park units, which attract millions of visitors each year. Jeffrey Meyer's study of the city's religious dimensions uses the metaphors of archaeology and pilgrimage to explore the historical contexts of Washington's sacred buildings, monuments, and memorials as well as contemporary visitors' experiences of these sites. Although he doubts that most people have what one scholar has described as "a blinding religious experience, a rite of communion" in their initial encounter with Washington's environs, Meyer acknowledges that the layout of the city lends itself to pilgrimage routes for exploring the sacred history of the nation (Meyer 2001: 9).

Battlefields also rank highly among national park sites that commemorate American civil religion, with over four million visitors to battlefield units of the National Park Service in 2018 (Ziesler 2019). Although the Park Service tends to emphasize "education over veneration," resisting "homage that approached worship" at these battlefield sites, Robert Utley, former assistant director of the National Park Service overseeing historical preservation, acknowledges that commemoration has always been a powerful impetus in preserving historical places (Linenthal 1993: x). Revolutionary War battlefields and Civil War locations are especially popular destinations for tourist pilgrims. Minute Man National Historical Park in Massachusetts, which highlights the opening battles of the Revolutionary War, hosts over a million visitors each year, while nearly a million annual visitors learn about the turning point of the Civil War at Gettysburg National Military Park in Pennsylvania (Ziesler 2019).

American national parks also bring attention to blemishes on the American historical narrative. The park system in recent decades has added numerous sites that tell a more complete and diverse story of the nation, which means including the mistreatment and injustices suffered by numerous marginalized groups. At sites such as Birmingham Civil Rights National Monument in Alabama, Manzanar National Historic Site in California, Sand Creek Massacre National Historic Site in Colorado, and Stonewall National Monument in New York, visitors confront the darker dimensions of the nation's historical legacy. As discussed above, these sites can be places of collective healing. But they also support the celebratory tone of American civil religion as evidence of national redemption in acknowledging the nation's transgressions.

The appeal of national parks

Tens of millions of visitors travel to American national parks each year, and a good number of them come with expectations of religious or spiritual experiences. Some of these tourist pilgrims find devotional fulfillment in the healing powers and aesthetic wonder according to their particular religious orientation. Many more experience a vague religio-aesthetic or spiritual connection to park attractions derived from artistic and literary traditions that translate the natural world as divine revelation. Millions of citizens also revel in a patriotic civil religion at the sacred sites of the American nation. Indeed, national parks have been destinations for the many varieties of spiritual tourism since their beginnings in nineteenth-century America.

Dividing various religious and spiritual motives into discrete categories may be useful for purposes of analysis and interpretation in understanding the experiences of national park visitors, but it also can be misleading. In actual practice, tourists' experiences are not always so easily separable into divisions of confessional faith traditions, aesthetic desires, and patriotic devotion. Often all of these categories play a simultaneous role in a particular individual's encounter with the features of a national park. This was especially true in the earlier years of American national parks as tourists began visiting the nature parks of the American west.

As an early example, when Rev. Edwin Stanley toured Yellowstone in 1873, his delight in the wonders of God's creation involved his own Methodist faith, an aesthetic appreciation of nature influenced by nineteenth-century American Romanticism, and a civil religion confidence in Manifest Destiny. At the Grand Canyon of the Yellowstone, for instance, the religious and aesthetic influences are apparent when he writes of "the strangely bewitching beauty and sublimity of this scene, the overpowering sense of the presence of Deity which it gives" (Stanley 1878: 77). At the same time, a shading of remorse colors his acknowledgment of the inevitable reality of Manifest Destiny that will sweep away "the untutored tribes of the forest" as "victims of savage life who must be destroyed because they will not be subdued and civilized" (Stanley 1878: 12–13). His journey to Yellowstone National Park brought a realization that the entire region, its wildlife and the people who had been there for many generations, "all must soon come under the influence of civilization, and yield to the scepter of the irrepressible white man" (Stanley 1878: 20). Rev. Stanley's experience of Yellowstone, like those of so many visitors to national parks across the continent in the nearly 150 years since, relied at once on his own faith tradition, a religio-aesthetic interpretation of the natural world, and elements of American civil religion.

The religious underpinnings of American national parks may be less obvious for many contemporary visitors. Their experiences, however, remain to some extent a product of the historical influence of devotional traditions that have exerted significant influence on American culture. Although most people today may not relate their travels to national parks in explicitly religious or spiritual terms, they are beneficiaries of the various traditions of religious travel and religio-aesthetic interpretations that have made national parks prized destinations for millions of visitors.

Notes

1 The term "national park" in this essay encompasses all units under the jurisdiction of the National Park Service of the United States, including national monuments, national historical parks, national recreational areas, and other sites managed by the Park Service. As a historian of American religions, I focus exclusively on national parks of the United States, but much of what is discussed in this essay would also apply to parks elsewhere, although with different histories and different cultural contexts.

2 Initially Yellowstone National Park had "no less than 56 devil, 6 hell, and 3 Satan place names," but most of these names disappeared in the efforts of United States Geological Survey official Arnold Hague, who was responsible for establishing Yellowstone's official place names in the 1880s and 1890s (Whittlesey 1988: xxxix). Zion National Park in Utah may be the park with the most surviving religious place names, including Angel's Landing, Cathedral Mountain, Tabernacle Dome, The Three Patriarchs, Towers of the Virgins, West Temple, the Virgin River, and the name of the park itself, Zion.

3 This discussion of American civil religion is adapted from Bremer (2015: 353–354).

Works cited

Albanese, C.L. (1990) *Nature Religion in America: From the Algonkian Indians to the New Age*. Chicago: University of Chicago Press.

Belanger, P.J., & Sweeney, J.G. (1999) *Inventing Acadia: Artists and Tourists at Mount Desert*. Rockland, ME: Farnsworth Art Museum.

Bellah, R.N. (1967) 'Civil religion in America', *Daedalus*, 96(1): 1–21.

Bellah, R.N. (1992) *The Broken Covenant: American Civil Religion in Time of Trial* (2nd ed.). Chicago: University of Chicago Press.

Bellah, R.N., & Hammond, P.E. (1980) *Varieties of Civil Religion*. San Francisco: Harper & Row.

Berry, E. (2015) *Devoted to Nature: The Religious Roots of American Environmentalism*. Oakland: University of California Press.

Bodnar, J.E. (1992) *Remaking America: Public Memory, Commemoration, and Patriotism in the Twentieth Century*. Princeton, NJ: Princeton University Press.

Bremer, T.S. (2004) *Blessed with Tourists: The Borderlands of Religion and Tourism in San Antonio*. Chapel Hill: University of North Carolina Press.

Bremer, T.S. (2015) *Formed from This Soil: An Introduction to the Diverse History of Religion in America*. Oxford, UK: Wiley Blackwell.

Bremer, T.S. (2019) 'Acadia National Park: A soul and spirit stretching place', *Chebacco: The Magazine of Mount Desert Island Historical Society*, XX: 131–141.

Bremer, T.S. (2020) 'Consider the tourist', in V. Narayanan (ed), *The Wiley Blackwell Companion to Religion and Materiality* (pp. 187–206): Newark, NJ: Wiley Blackwell.

Castren, A.M. (1993) 'Bellah's civil religion in America: New clothes for a classic--Emile Durkheim theology?', *Sociologia*, 30(4), 257–269.

Chernus, I. (2010) 'Civil religion', in P. Goff (ed), *The Blackwell Companion to Religion in America* (pp. 57–70). Malden, MA: Wiley-Blackwell.

Davidson, G.S. (2006) 'Landscape icons, tourism, and land development in the northeast', in G.S. Davidson, F.E. Church W. Holmer, & T. Moran (eds), *Frederic Church, Winslow Homer, and Thomas Moran: Tourism and the American Landscape* (pp. 3–73). New York: Bulfinch Press.

Davis, D. (1998) 'Civil religion as a judicial doctrine', *Journal of Church and State*, 40: 7–23.

Duncan, D., & Burns, K. (2009) *The National Parks: America's Best Idea: An Illustrated History*. New York: Alfred A Knopf.

Early, K.E. (1984) *"For the Benefit and Enjoyment of the People": Cultural Attitudes and the Establishment of Yellowstone National Park*. Washington, DC: Georgetown University Press.

Ellis, R. (2005. *To the Flag: The Unlikely History of the Pledge of Allegiance*. Lawrence: University Press of Kansas.

Emerson, R.W. (1850) *Nature, Addresses and Lectures*. Boston, MA: Phillips, Sampson.

Fenn, R.K. (1977) 'The relevance of Bellah's 'civil religion' thesis to a theory of secularization', *Social Science History*, 1(4): 502–517.

Ferber, L.S. (2009) *The Hudson River School: Nature and the American Vision*. New York: Skira Rizzoli.

Flowers, R.B. (1994) *That Godless Court? Supreme Court Decisions on Church-State Relationships*. Louisville: Westminster John Knox Press.

Gehrig, G. (1981) *American Civil religion: An Assessment*. Storrs: Society for the Scientific Study of Religion.

Haines, A.L. (1996) *The Yellowstone Story: A History of Our First National Park, Volume One* (rev. ed.). Niwot: University Press of Colorado.

Hammond, P.E., Porterfield, A., Moseley, J.G., & Sarna, J.D. (1994) 'American civil religion revisited', *Religion and American Culture*, 4(1): 1–23.

Handy, R.T. (1980) 'A decisive turn in the civil religion debate', *Theology Today*, 37: 342–350.

Harrison, C.H. (1891) *A Summer's Outing and the Old Man's Story*. Chicago: Dibble Publishing.

Hassrick, P.H. (2002) *Drawn to Yellowstone: Artists in America's First National Park*. Los Angeles: Autry Museum of Western Heritage in association with University of Washington Press, Seattle.

Hayashi, R.T. (2003) 'Transfigured patterns: Contesting memories at the Manzanar National Historic Site', *The Public Historian*, 25(4): 51–71.

Heacox, K. (2001) *An American Idea: The Making of the National Parks*. Washington, DC: National Geographic Society.

Jones, D.G., & Richey, R.E. (eds) (1990) *American Civil Religion*. San Francisco: Mellen Research University Press.

Kao, G.Y., & Copulsky, J.E. (2007) 'The pledge of allegiance and the meanings and limits of civil religion', *Journal of the American Academy of Religion*, 75(1): 121–149.

Kaufman, P.W. (1996) *National Parks and the Woman's Voice: A History*. Albuquerque: University of New Mexico Press.

Linenthal, E.T. (1993) *Sacred Ground: Americans and Their Battlefields* (2nd ed.). Urbana: University of Illinois Press.

Lippy, C.H. (2006) 'American civil religion: Myth, reality, and challenges', in C.H. Lippy (ed.), *Faith in America, vol. 2: Religious Issues Today* (pp. 19–36). Westport, CT: Praeger.

Little, D., & Little, C. (2016) *Art of Acadia: The Islands, The Mountains, The Main*. Camden, Maine: Down East Books.

Mackintosh, B., & McDonnell, J. (2005) *The National Parks: Shaping the System* (rev. ed.). Washington, DC: US Department of the Interior.

Manzanar Committee. n.d. Who We Are. https://manzanarcommittee.org/who-we-are/. Accessed September 19, 2019.

Mathisen, J.A. (1989) 'Twenty years after Bellah: Whatever happened to American civil religion?', *Sociological Analysis*, 50(2): 129–146.

McGreevy, P.V. (1994) *Imagining Niagara: The Meaning and Making of Niagara Falls*. Amherst: University of Massachusetts Press.

Meyer, J.F. (2001) *Myths in Stone: Religious Dimensions of Washington, D C*. Berkeley: University of California Press.

Miles, J.C. (2009) *Wilderness in National Parks: Playground or Preserve*. Seattle: University of Washington Press.

Mitchell, K. (2016) *Spirituality and the State: Managing Nature and Experience in America's National Parks*. New York: New York University Press.

Muir, J. (1894) *The Mountains of California*. New York: The Century Co.

Muir, J. (1901) *Our National Parks*. Boston, MA: Houghton Mifflin.

National Park Service. (2016) 'A sacred site to American Indians', *Devils Tower National Monument*. https://www.nps.gov/deto/learn/historyculture/sacredsite.htm. Accessed September 9, 2019.

National Park Service. (2017) 'Independence Hall', *Independence National Historical Park*. https://www.nps.gov/inde/learn/historyculture/places-independencehall.htm. Accessed September 20, 2019.

National Park Service. (2018) 'Visit Parks', *Civil Rights*. https://www.nps.gov/subjects/civilrights/parks.htm. Accessed September 19, 2019.

Norton, H.J. (1873) *Wonder-Land Illustrated; or, Horseback Rides Through the Yellowstone National Park*. Virginia City, Montana: Harry J. Norton.

Pyne, S.J. (1998) *How the Canyon Became Grand: A Short history*. New York: Viking.

Ross-Bryant, L. (2013) *Pilgrimage to the National Parks: Religion and Nature in the United States*. New York: Routledge.

Rothman, H. (1989) *America's National Monuments: The Politics of Preservation*. Lawrence: University Press of Kansas.

Runte, A. (1987) *National Parks: The American Experience* (2nd rev. ed.). Lincoln: University of Nebraska Press.

Sears, J.F. (1989) *Sacred Places: American Tourist Attractions in the Nineteenth Century*. New York: Oxford University Press.

Sellars, R.W. (1997) *Preserving Nature in the National Parks: A History*. New Haven, CT: Yale University Press.

Shugart, S. (2009) *When Did It happen? A Chronology of Events at the Hot Springs of Arkansas*. Hot Springs: Eastern National.

Stanley, E.J. (1878) *Rambles in Wonderland: or, Up the Yellowstone, and Among the Geysers and Other Curiosities of the National park*. New York: D. Appleton.

Stillman, A.G. (ed.) (2010) *Ansel Adams in the National Parks: Photographs from America's Wild Places*. New York: Little.

Stoll, M. (2015) *Inherit the Holy Mountain: Religion and the Rise of American Environmentalism*. New York: Oxford University Press.

Talbot, A-M. (2002) 'Pilgrimage to healing shrines: The evidence of miracle accounts. *Dumbarton Oaks Papers*, 56: 153–173.

Wheeler, O.D. (1893) *6,000 Miles through Wonderland: Being A Description of the Marvelous Region Traversed by the Northern Pacific Railroad*. Chicago: Rand, McNally, & Co.

Whittlesey, L.H. (1988) *Wonderland Nomenclature: A History of the Place Names of Yellowstone National Park, Being a Description of and Guidebook to Its Most Important Natural Features, Together with Appendices of Related Elements*. Helena: Montana Historical Society Press.

Wilkins, T., & Hinkley, C.L. (1998) *Thomas Moran: Artist of the Mountains* (2nd ed.). Norman: University of Oklahoma Press.

Wilson, J.F. (1979) *Public Religion in American Culture*. Philadelphia: Temple University Press.

Worster, D. (2008) *A Passion for Nature: The Life of John Muir*. New York: Oxford University Press.

Ziesler, P. (2019) *Statistical Abstract 2018*. Fort Collins, CO: National Park Service. Retrieved from https://www.nps.gov/subjects/socialscience/visitation.htm

13
RELIGIOUS THEME PARKS

Lena Rose

Introduction

Within the tourist industry, theme parks are particularly unique spaces. They consist of an enclosed or enclave space that usually features different rides, performances, shows, and amenities that are in line with a specific theme or a central story, which engulfs the visitor and unifies the different attractions. One example is the Huis Ten Bosch theme park in Sasebo, Nagasaki, Japan, also dubbed 'Little Europe,' which is a recreation of The Netherlands (Hendry 2000: 43). The most widely known and still most lucrative theme parks are the Disney theme parks (Rubin 2018). Started in the 1950s, Disney theme parks reinvented the largely 'un-themed' amusement parks that had been popular since the beginning of the twentieth century (Adams 1991; Davis 1996; Lukas 2013). Indeed, theme parks, as partly cultural displays, are an evolution of public exhibitions of cultural artifacts from the eighteenth century, which played an important role in forming a national citizenry amid the rise of nation-states (Dicks 2004). These later developed into world exhibitions and fairs (Rydell 1993; Greenhalgh 2011), which were designed as mini theme parks and therefore important precursors to contemporary theme parks (Mitchell 1988).

Since the mid-twentieth century, the genre of theme parks has spread widely from their origins in the United States across the world. In 2018, the top ten theme park groups worldwide drew an astounding number of over 500 million visitors, among which Disney parks lead by a share of more than 30% (Rubin 2018: 9). The number of visitors to theme parks rose by 5% between 2017 and 2018 alone (Rubin 2018). As the popularity of theme parks has grown, so have their size. For example, Walt Disney World Resort in Florida is twice the size of the New York Borough of Manhattan. This theme park has 25 million visitors yearly and employs approximately 30,000 people (Paine 2019). Disney Paris is over 2000 hectares in size, although only half of the land is developed (West 2018).

In addition to research into the history of theme parks (Botterill 1997; Young & Riley 2002; Silverman 2019), the theme park phenomenon has generated a large body of scholarship that analyses and critiques this form of entertainment as an example of the process of the 'globalization of Americanization of leisure' (Hendry 2000: 73). For example, scholars have emphasized how modern theme parks are a more 'clean' and 'safe' form of entertainment in contrast to the more raucous amusement parks of the past (Adams 1991; Fjellman 1992).

They create, like in the case of Disney theme parks, positive and idealized landscapes and promotion of American nationalism, such as Main Street USA at Disney's 'Magic Kingdom' (Bryman 1995: 11f). Other critical scholarship has examined the creation of childhood nostalgia based on Disney characters at Disney theme parks (Eco 1986; Hunt & Frankenberg 1997), the effects of commodified leisure in the form of theme parks on modern society (Moore 1980; Eco 1986), authenticity and theme parks (Lovell & Bull 2017; Weiler 2017; Waysdorf & Reijnders 2018), and how the Disney theme park phenomenon has been transferred to other countries and tourism contexts (Brannen 1992; Maanen 1992; Raz 1999; Hendry 2000; Srivastava 2014; McDaniel 2017; Wong 2017).

Religious theme parks

So how does religion fit into the world of theme parks? The idea of religious theme parks may seem odd to some, considering popular views of religion being related to piety, seriousness, and quiet contemplation rather than to fun-oriented, commercially driven, leisure. Yet, the market for religious theme parks is presently burgeoning, even though as a leisure genre they have not been well researched (Paine 2019). Religious theme parks share several key characteristics with theme parks in general—they are also enclaves set apart from everyday life that include various attractions, rides, shows, and amenities, and are also unified by a particular theme. They equally offer imaginary worlds to their visitors that are made coherent through landscaping, architecture, and theatrical effects that may also include costumed staff (Chidester 2005). In addition, however, their themes, rationale, or history are inspired by religion. According to one of the leading experts in the field, Paine (2019), there are only about two dozen 'pure' religious theme parks in the world, i.e., parks that closely reflect the theme park genre as defined above. The most thoroughly researched examples among religious theme parks are the Holy Land Experience in Orlando (e.g., Wharton 2006; Lukens-Bull & Fafard 2007; Dykins Callahan 2010; Stevenson 2013; Chmielewska-Szlajfer 2017), the Ark Encounter (e.g., Bloomfield 2017; Bielo 2018a) and the Creation Museum in Kentucky (e.g., Asma 2007; Butler 2010; Stevenson 2013; Scott 2014), Nazareth Village in Israel (e.g., Ron & Feldman 2009; Rose 2020), the giant Buddha in Hong Kong (Wong 2017), the Akshardham Temple in India (Srivastava 2014), and Anandsagar, a religious theme park in the pilgrim-town of Shegaon in Maharashtra, India (Shinde 2021). While there are a few Islamic theme parks in countries such as Iran, Lebanon, China, or the Qur'anic Park in Dubai (Lari Lyanna & Jabeen 2019; Paine 2019), they have not received much focused scholarly attention beyond their naming in the religious theme park literature.

Religious theme parks also share a similar history and rationale with their non-religious counterparts, albeit with important thematic differences. Religious theme parks have also been developed to cater to the needs of a burgeoning (religious) middle class. The technical advances that make theme parks in general so attractive are also used abundantly in religious theme parks. Where the Disney phenomenon provides a nostalgic rooting experience for their audiences who are affected by an uncertain world (Dicks 2004), religious theme parks too create 'safe' and orderly spaces for their visitors outside the anxieties of daily life (Paine 2016). These similarities have led some scholars in reverse to suggest that non-religious theme parks, especially Disney theme parks, can be considered as sacred space and contain elements of religiosity. For example, Caron (1993: 125) describes Disney's 'Magical Kingdom' as a mystical religious space that transforms the places where they are found. Moore (1980: 207ff) suggests that a visit to Disney theme parks can be compared to a visit to pilgrimage centers, where visitors escape into a magical or fantastical world made up of shared symbols and myths, willingly suspend belief in order to enter into the experience, and buy souvenirs to remember them. Here, the idea of play and ritual is intertwined as it might be in other pilgrimage centers, except it is offered to a secular, technologized audience. On the other hand, Bremer (2001) referred to Temple Square in Salt Lake City, Utah—the spiritual and ecclesiastical center of The Church of Jesus Christ of Latter-day Saints—'the Disneyland of Mormonism' because of attempts by church leaders to pattern their sacred site after Disneyland in order to create a more inviting atmosphere for visitors. It is only a small step from this analysis to viewing travel to religious theme parks as a kind of 'pseudo-pilgrimage,' particularly in cases like the Holy Land Experience that may even replace a pilgrimage to the original site (for example, a visit to Holy Land Experience, Florida, may substitute a visit to Israel-Palestine, see section on authenticity and representation), yet fulfill many functions of what an 'original' pilgrimage might entail.

Given the origin of the theme park genre, most religious theme parks are found in the United States and feature strong Christian messaging,[1] though there are several Buddhist and Hindu theme parks in Asia as well (Hendry 2000; Srivastava 2014; McDaniel 2017; Wong 2017; Paine 2019). In the United States, religious theme parks were developed as theme parks with religious elements and themes, whereas in Asia pre-existing religious structures such as Temples or monasteries were used as the basis to create religious theme parks (Graburn 1983; Srivastava 2014; Wong 2017; Paine 2019). However, in Asian theme parks that seek to represent 'exotic' foreign places, such as Japan's Huis Ten Bosch theme park in Nagasaki, religion enters in the form of the inclusion of churches, temples, and shrines as part of their display (Hendry 2000: 88), even though these buildings are usually not used for religious services. However, as Millar (1999) and Wong (2017) note, not all religious theme parks are purposefully built, meaning that over time, even religious sites that are not generally built and designed with tourism purposes in mind, may become tourist attractions and religious theme parks. This is the case of the Tian Tan Buddha theme park in Hong Kong. Located on Lantau Island, the site contains the big statue of the Buddha, a museum on Buddhism, a cable car, the century-old Po-Lin Buddhist monastery, and Ngong Ping Village. While these attractions are in many cases not connected to each other and were not designed for tourist consumption, over time development and tourist interest in the big Buddha led to the island turning into a themed religious park. Unlike at other religious theme parks, however, no fee is charged for entrance to the island, and the park is not under the management of a single committee making this park a deviation from how other religious theme parks are organized and managed.

Paine (2019) suggests that there are several main reasons why religious groups build and maintain religious theme parks. Four of the most important reasons include to make money; to proselytize; to promote tourism; and to engage in nation-building. While at first glance religious theme parks seem like a surprising combination of devotion and consumerism, the intermixing of these motivations is not new. International exhibitions and fairs—the forerunners of modern theme parks—often featured religious displays in a mixing of leisure and promotion. For example, The Church of Jesus Christ of Latter-day Saints operated religious-themed pavilions, including the Chicago World Fair (1893) and the New York World Fair (1964–1965) to promote religious messaging (Neilson 2011; Kogan 2013). The 1904 Louisiana World Fair included an exhibit on the Book of Genesis (Adams 1991) as well as a replica of Ottoman Jerusalem and other biblical sites that enabled visitors to go on 'pseudo-pilgrimages' to recreated locations from the New Testament (Long 2003). As Paine (2019) notes, even before the popularity of world exhibitions to display religion, some pilgrimage sites featured additional 'extras' that transformed them into a pseudo-theme park site that added value to the visitor experience. As Bielo (2018a: 21; see Olsen 2003) writes, 'the use of commerce and recreation to enhance religious ambitions has a long history.' Even before the rise of religious theme parks, in the United States visitors could enjoy educational attractions like the Chautauqua Institute's Palestine Park that were aimed at a Christian audience (Long 2003). Much like the Holy Land Experience today, Palestine Park was a representation of the Holy Land for those who could not afford to visit it in person.

Blurring the lines between religious theme parks and other attractions

The spectrum between religious theme parks and similar attractions with religious elements is fluid. For example, some biblical gardens (spaces planted with Biblical vegetation) have themed, experiential, and pedagogical elements that resemble those of religious theme parks (Bielo 2018b, 2018c). Even the distinction between 'living history museums' or re-creations of religious sites which offer an immersive experience of a particular element of a national or cultural past to its visitors and theme parks is blurry, as many museums today offer recreational and immersive elements such as role-playing, digital experiences, and cinematic shows. Theme parks based on, for example, Norse or Greek mythology also combine religion, myth, and leisure (Paine 2019). Entire cities can also be branded as religious holiday locations, such as Branson, Missouri, which overarching tourism plan provides for an overall religious emphasis on family-friendly entertainment, gospel shows, and natural beauty. Indeed, tourism in Branson is designed as a 'form of devotion expressed within theatrical and amusement-oriented venues which seek to impart everyday expressions of leisure with "sacred" value' (Ketchell 2007: xvi). While not distinctly branded as religious theme park, the Silver Dollar City amusement park, located in Branson, has a proselytizing ethos and requires its staff to adhere to 'Christian values' (Ketchell 2007: 96). Interestingly, many of the designers and operators of what would be obviously called religious theme parks prefer not to use the term 'religious theme park' themselves, especially when the inspiration for building the park was educationally based. As such, operators would rather consider these a religious 'ministry' or even a museum (Dykins Callahan 2010: 65; Chmielewska-Szlajfer 2017: 545).

While not theme parks in the truest sense of the word, the attractions mentioned here share common features of religious theme parks, such as a sense of immersion, efforts at pedagogy, a tension between authenticity and representation, and questions regarding religious orthodoxy. These attractions also use similar techniques and aesthetics that enhance the visitor experience, such as the use of panoramas, dioramas, cinemas, and other digital mechanisms, relics, statues, rides, models, water shows, religious architecture, volunteers, and guides. Through these techniques and aesthetical tools these attractions challenge, as Dykins Callahan (2010: 66) argues, 'the notion that entertainment, education, and religion have ever been mutually exclusive in American popular culture.'

Pedagogy and identity

One of the aims of many religious theme parks is to 'spread the word' and to teach their visitors about aspects of the religion as depicted in these parks. As Paine (2016: 402) puts it, one of the functions of religious theme parks is 'to missionize the world's burgeoning middle class through its new leisure time.' As such, religious theme parks, like other religious sacred sites (Olsen 2012), serve the purpose of attempting to foster particular religious identities and subjectivities similar to the way in which Disney theme parks contribute to the identity formation of their visitors by highlighting particular versions of the past (Bryman 1995: 11f).

For example, the Holy Land Experience in Orlando, Florida, which is set in a location that is saturated with world-renowned theme parks, offers a religious theme park experience with Christian visitors in mind. The park consists of a condensed array of reconstructed biblical sites, such as the Damascus Gate, a Jerusalem street market, Bedouin tents, the Garden Tomb, and the Qumran Caves, as well as various performance spaces that offer theatrical or cinematic performances of biblical stories, such as a daily re-enactment of Jesus' crucifixion and resurrection. The site is held together by what Dykins Callahan (2010: 26) identifies as a 'Salvation by Grace' narrative—a Christian doctrine centered on the death and resurrection of Jesus. While an amalgam of different time periods (from Old Testament times to Ottoman Jerusalem), the biblical past as depicted at the Holy Land Experience serves to illustrate what is considered by the site owners as biblical truth, with the aim to immerse the visitor in biblical landscape features that most visitors will know well from their scripture study.

A strong undercurrent in the pedagogical processes at the Holy Land Experience is the combination of a particular strand of American nationalism and cultural identity with religious messaging. One of the shows at the park is entitled 'Celebrate America,' which is held in the Shofar auditorium. This show does not contain a visibly religious message, but rather focuses on the 'sacrifice' of American soldiers in defense of the United States with the colors of the American flag dominating the stage. This inclusion of a deeply patriotic theme in the context of an otherwise faith-inspired biblical representation powerfully connects religion to a particular American national and cultural identity that was founded as 'one nation under God' (Chmielewska-Szlajfer 2017; cf. Stevenson 2013).

The promotion of nationalism and particular cultural religious identities is also a common theme in religious theme parks beyond the United States. Wong (2017: 170), for example, suggests that the unusual orientation of the Tian Tan Buddha (or Big Buddha) in Hong Kong toward China in the North, rather than toward the south as is typical for most Buddhas, is actually a political statement about the 'Chineseness' of Hong Kong (2017: 170). This seems counterintuitive, considering that the tourism slogan of Hong Kong is 'Hong Kong–Asia's World City,' an attempt to disassociate itself from China amid continuing tensions regarding Hong Kong's political status (Wong 2017: 170; cf. Zhang, Decosta, & Mckercher 2015). However, Srivastava's (2014) analysis of the Akshardham Temple complex in Delhi, India, shows that this site is also designed in part toward nationalistic ideas, in this case toward a particular construction of 'Indianness.' The Akshardham Temple complex was constructed by one of the major sub-sects within the Swaminarayan movement, the Bochasanwasi Shri Akshar Purushottam Swaminarayan Sanstha (BAPS), which is considered the 'the dominant form of transnational Gujarati Hinduism' (Dwyer 2004: 181), to remember the life of Swaminarayan, a teenage yogi who was believed to be a manifestation of God (Kim 2009). The site is approximately one hundred acres in size and is dominated visually by a very large Hindu temple. The site also includes a high-tech 'Hall of Values,' which highlights the life of Swaminarayan through robotics and dioramas; an IMAX cinema showing a film of Swaminarayan's early life; a 12-minute boat ride that depicts '10,000 Years' of heritage and life in Vedic India; and an impressive musical fountain. These attractions are designed specifically to promote the teachings of Hindutva (a form of Hindu nationalism), in part because the political party closest aligned to the teachings of the BAPS, the National Democratic Alliance, played a significant role in the contested allocation of the land on which the site sits, which allocation involved the clearing of several slums and their inhabitants (Srivastava 2014: 194).

However, unlike the Holy Land Experience that focuses the attention of visitors toward the idea of a 'nostalgic past,' the combination of religion, technology, and nationalism used at the Akshardham Temple complex is more oriented toward the present and future. As Srivastava notes, 'nostalgia has little appeal for an audience whose dominant memory of the immediate past might be of the license-permit regime of the Five-Year Plan state where material benefits were largely sequestered by an industrial-bureaucratic elite' (p. 206). As such, the interplay between nationalism, technology, leisure, and 'ancient' Hindu culture is set in the midst of the processes of contemporary modernity and serves to make 'new middle class identities that are in a dialogue between being a modern consuming citizen' while 'remaining "Indian" through adherence to religious practice' (p. 192). At the same time, the notion of consumption perhaps trumps the identity-shaping pedagogical experience, for as Srivastava shows, rather than sit through the shows and movies that teach about the life of notable members of the sect, the focus of visitors seems to be focused on grabbing the best seats for the next performance by leaving the current performance early and thereby missing the majority of its content, rather than immersing themselves in the experience and the pedagogical elements of the performances.

Experience and performance

An important part of the interpretative and experiential aspects of religious theme parks is the notion of immersion or the process of learning and being (in)formed by being at the park. A key to this immersion is the 'separateness' of these parks, which, as noted above, has caused some scholars to describe secular theme parks as 'sacred spaces' akin to pilgrimage centers (Hendry 2000: 88, 93). While this perception of sacredness in conjunction with theme parks and their offered experiences is determined by the perspectives of individual visitors (Paine 2019), the way in which religious theme parks are also bounded by walls, gates, and entry fees also sets them apart from ordinary, everyday spaces and provides the potential for spiritual experiences.

In his ethnography of the making of the Ark Encounter in Williamstown, Kentucky, Bielo (2018a) highlights the processes by which the religious pedagogy described above is actualized through immersion. The main message of the Ark Encounter, which is comprised of a giant reconstruction of Noah's Ark in the Old Testament and its corresponding story, revolves around a literalist reading of the Genesis account in the Bible, in which evolution is denied as a plausible explanation for human existence and God is the creator of the world and everything on it. This literalist view of the Earth's creation differs from the similarly inspired Creation Museum by trying to get the message across through entertainment and fun, rather than scientific debate (Asma 2007; Scott 2014; Trollinger & Trollinger 2016). Bielo (2018a) documents how the creative team behind the construction of the Ark Encounter professionally employs popular culture elements to achieve the greatest effect of compelling, immersive experience. For example, apart from the huge reconstruction of Noah's Ark based on the original biblical dimensions, there is an interactive graphic novel, exhibits of the living quarters, life-size reconstructions of animals that may have been on board the Ark, and 'pre-flood world' experience that includes auditory annotations, signage, dioramas, and murals. Interestingly, Bielo notes the lack of 'spiritualizing' in the branding and design of the site. Rather than focusing on whether concept art and design ideas were religiously attractive, team members at the park were more interested in whether they were 'cool' or engaging to visitors, with the aim of transporting 'the visitor away from the frame of everyday reality and to a frame defined by the creationist past' (Bielo 2018a: 67). This also functions to solidify the group identity of creationist believers through affect and sensory strategies against the majority of Americans who believe in evolution in a kind of culture war (Bielo 2018a).

Another technique that is often employed in religious theme parks to enhance a sense of immersion is the addition of (usually) costumed volunteers and staff with whom visitors can interact. In most religious theme parks, the volunteers and staff are lay adherents of the religion that maintains the site (Dykins Callahan 2010; Srivastava 2014; Paine 2019). Whether in first- or third-person persona, these costumed performers help create the image of visitors being in a particular time and place in the past and also being 'among kindred.' This is particularly the ace in religious theme parks, such as the Holy Land Experience, Ark Encounter, or Nazareth Village, that are based on historical reconstructions (Ron & Feldman 2009; Rose 2020), where volunteers and staff are clothed in representational costumes of the time periods in question. At Nazareth Village, volunteers and staff 'populate' the land to increase a sense of it being a more authentic working Jewish farm from the first century, despite its actual location being in the middle of a contemporary Arab Israeli city. Visitors can take pictures with the costumed characters, ask them questions about the processes of everyday life (even though the responses are often made up due to lack of historical knowledge of the volunteers), and can eat a first-century meal served by the characters to further 'transport' them back in time. While these reconstructions of sacred lands and times, particularly biblical territories and times, can be criticized for perpetuating orientalizing images of the population and not being forthcoming about contemporary situations in these places (e.g., Palestinian/Israeli conflict; see Rose 2020), this oriental gaze could be turned around in Asian attractions that feature Western volunteers, such as at *Tokyo Disneyland*, a non-religious theme park, where Japanese visitors can take their photos with presumably white foreign staff who play the role of characters in Disney films such as snow White and Peter Pan (Hendry 2000: 94). For example, while not a religious theme park per se, Japanese visitors can take their photos with presumably white foreign staff who impersonate 'natives' of Disney's stories. Unlike Japanese staff, foreign white staff members do not wear name tags, and are thus their cultural and historical identity is anonymized, much like the workers who take on biblical personas at Nazareth Village.

Sometimes religious theme parks are so successful in their creation of an imaginary world that visitors report having what they consider 'real' spiritual experiences. This is especially interesting when this takes place in a religious theme park that is not affiliated with a particular faith community or has clergy on site (Chmielewska-Szlajfer 2017). This phenomenon raises at least two important analytical questions: Are religious theme parks actual 'sacred space' like other sites of official worship, or do they merely *resemble* sacred space (Lukas 2007)? Is entertainment at religious theme parks a 'tool for spirituality,' or is spirituality 'yet another feature of entertainment' (Chmielewska-Szlajfer 2017: 548)? The ambiguity of these questions is perhaps a testament to the effectiveness of the techniques and mechanisms of immersion that are utilized at religious theme parks.

Authenticity and representation

While it is easy to raise questions regarding the authenticity of reconstructions of the past at religious theme parks, it is important to investigate both the distinction between authenticity and representation and how, like secular theme parks, religious theme parks intentionally blur the distinction between 'natural' and 'artificial.' Both secular and religious theme parks seek to help visitors suspend their disbelief in order to create scripted and desired responses and experiences. To make a visit to their park more 'authentic,' managers of the Holy Land

Experience offer snacks and products for sale that come directly from the Holy Land. At Nazareth Village in Israel, the play between authenticity and representation is even more acute since the site houses two 'genuine' archaeological sites, including a hole in the rock that served as a winepress in the first century (Pfann, Voss & Rapuano 2007). While the buildings at the park are all reconstructed, the emphasis on actual archaeological findings casts these reconstructed buildings in a more 'authentic' light. Yet, religious theme parks that are historical reconstructions do not necessarily want to be too authentic, as visitors still want access to modern amenities such as flushable toilets and water fountains. This follows work by Eco (1986) on entertainment spaces in the United States, where he argues that the hyperrealism offered at many tourism sites may eliminate the 'need for the original.' This is in agreement with other scholarship on tourism that identifies the desire of many tourists to encounter a 'knowable,' perhaps more palatable, Other, even at the expense of 'authentic' experiences (e.g., Boorstin 1987; MacCannell 1999, 2011; Urry & Larsen 2011). As Dykins Callahan (2010: 64) explains, in the case of the Holy Land Experience this translates into collapsing time and space to present a narrative that is uninterrupted and excludes the 'dead space' that is 'non-symbolic, ordinary, [and] unreadable' that might be encountered by a person traveling to Israel-Palestine.

This 'cleaning up' of unpleasant aspects of the past in religious theme parks inadvertently (or purposefully) can lead to the 'cleaning up of history' (Hendry 2000: 75), thereby creating a 'nostalgic need' (Fjellman 1992: 59f) within visitors. At Christian-themed sites, representations of biblical pasts help Christian visitors to go on an unmediated journey to the temporal, geographical, ethnic, and cultural origins of their faith (Bielo 2018a). This is the case at Nazareth Village, which claims that its site peels back 'twenty centuries of time, distance, and culture' (Kauffman 2005) so that the evangelical visitor who may otherwise be alienated by difficult and tiring experiences of Holy Land travel can enter into a seemingly 'authentic' experience of the first century and time of Jesus. In this focus on authenticity that is experienced, rather than accurate from a socio-historical perspective, visitors are active agents (Dicks 2004; Dykins Callahan 2010). For example, the Holy Land Experience is designed so that the visitor can choose their own route through the attraction, and decide how much time they want to spend at which part of the site, or if they want to return another day to complete their tour. Dykins Callahan (2010: 125) suggests that this lays the burden of achieving 'experiential authenticity' of the site on the visitor themselves. The journey through Nazareth Village is more restrictive, in that the site can only be accessed with a local guide, who determines how long each group spends at what attraction. Yet, how they experience the site also depends on how deeply visitors allow themselves to enter into the reconstruction—for example, when taking photos of the attractions, visitors can focus on a 'mid-range panorama' that blends out the surrounding (and conflicting) city-scape (Rose 2020).

Orthodoxy and power

Since in most cases managers of religious theme parks choose which doctrines and representations of the past are presented to their visitors, a final consideration here regards the role of power in the shaping of religious orthodoxy and identity at religious theme parks. As noted above, in many cases religious theme parks use religion to promote conservative, if not right-wing, agendas (Srivastava 2014; Wong 2017; Paine 2019). The question is who has a say in the design and implementation of the content of religious theme parks, and how is this content consumed by visitors? Dykins Callahan (2010), for example, documents the takeover of the Holy Land Experience by the wealthy and powerful Trinity Broadcast Network

TBN in 2007. The TBN brought in not just a particular prosperity gospel slant to the park, replacing more pietistic evangelical displays, but also introduced Christian Zionist doctrinal perspectives in the site's performances and attractions. This meant that the theatrical participation of Jews in the daily crucifixion of Jesus Christ was removed (Stevenson 2013) and that a push toward an emphasis for a Christian affinity for Israel was overtly introduced to other performances and souvenirs such as jewelry in the shape of a star of David (Beal 2005; Lukens-Bull & Fafard 2007; Dykens Callahan 2010).

Messaging related to a Christian affinity for Israel is also apparent in Nazareth Village in Israel-Palestine. The 'sensual indexicality' (in this context, the appeal to all senses aimed at pointing to and deepening religious truths; Bielo 2018b: 32) and performance of theologies at this site, ironically in the place that is at the heart of the theology of Christian Zionism itself, can have a profound effect on what is considered 'true' and dogmatically correct. As a representation of the 'Holy Land,' Nazareth Village is geared toward Western evangelical visitors who have struggled to connect the current realities of Israel-Palestine to their biblical imaginations of what the place should look like (cf. Kaell 2014). Nazareth Village is a pastoral landscaped area amid a bustling Palestinian Israeli town. The visitor experience is carefully guarded and monitored by the costumed staff who make sure to draw their attention away from the high-rise buildings surrounding the site and toward the examples of the materiality of Jesus' time presented within the themed site. The most politically contentious effect of Nazareth Village is the use of Palestinian staff who slip into the role of first-century Jews to portray an image of Israel to the Western visitor that is populated by Jews, suggesting a theological position that sees a return of the Jews to the territory of Israel as a pre-requisite of the end-times.

Similar strategies of inclusion and exclusion in the name of serving particular theological orthodoxies are also found at non-Christian sites. For example, at the Akshardham Temple complex Indian history is represented without the mention of Muslims, claiming that India is an entirely Hindu nation (Srivastava 2014).

Conclusion

In sum, religious theme parks are a powerful tool in efforts to proselytize, make money, promote tourism, and engage in nation-building. Religious theme parks aim to shape visitor's identity formation by offering immersive and compelling experiences in spaces set apart from the everyday world in which religious and spiritual connections can be made. These theme parks also provide rich material for studying material religion and the role of power in the formation of religious orthodoxy. While there are only a few examples of 'pure' religious theme parks, research into this phenomenon can throw into relief the connections between religion, leisure, nationalism, and technology.

A component that is largely missing from research on religious theme parks, with Dykins Callahan's work perhaps being an exception, is how successful they really are in influencing and shaping the (religious) identities of those who consume them. Who are the visitors that choose to make these sites their holiday destination? Are they already supportive of the messages conveyed, or are they being 'converted' to the religious institutions that own and run these sites as a result of the experience? How might visitors negotiate, resist, or subvert the intended messages of the theme park's designers? Do visitors revisit these sites, and how does their experience and perception of these sites change if visiting them multiple times? What is the influence of comparison or even alternative messaging if visitors travel to and engage with different religious parks? What might be the role of their companions (e.g., family, friends, and church groups) in making the attraction 'work'? How do online reviews or social media engagement with the sites shape a visitor's experience before or after their visit? This gap in knowledge is commensurate with the limited existence of scholarship on visitor's experiences of theme parks as a whole (Paine 2019) and would provide a fruitful avenue for future research. Furthermore, it would be fascinating to gain more academic insights into religious theme parks of other religious traditions, such as Islamic ones, and explore how the themes outlined above in relation compare to existing ones of the Christian, Buddhist, or Hindu traditions.

Note

1 See Bielo's www.materializingthebible.com project website for a useful list of religious theme parks.

Bibliography

Adams, J.A. (1991) *The American Amusement Park Industry: A History of Technology and Thrills*. Woodbridge: Twayne Publishers.

Anton Clavé, S. (2007) *The Global Theme Park Industry*. Wallingford: CABI.

Asma, S.T. (2007) 'Dinosaurs on the ark: The creation museum', *Chronicle of Higher Education*, 53(37): B 10.

Beal, T.K. (2005) *Roadside Religion: In Search of the Sacred, the Strange, and the Substance of Faith*. Boston, MA: Beacon Press.

Bielo, J.S. (2018a) *Ark Encounter: The Making of a Creationist Theme Park*. New York: New York University Press.

Bielo, J.S. (2018b) 'Biblical gardens and the sensuality of religious pedagogy', *Material Religion*, 14(1): 30–54. https://doi.org/10.1080/17432200.2017.1345099.

Bielo, J.S. (2018c) 'Flower, soil, water, stone: Biblical landscape items and Protestant materiality', *Journal of Material Culture*, 23(3): 368–387.

Bloomfield, E.F. (2017) 'Ark encounter as material apocalyptic rhetoric: Contemporary creationist strategies on board Noah's ark', *Southern Communication Journal*, 82(5): 263–277.

Boorstin, D.J. (1987) *The Image: A Guide to Pseudo-Events in America*. New York: Atheneum.

Botterill, J. (1997) *The 'Fairest' of the Fairs: A History of Fairs, Amusement Parks, and Theme Parks*. Doctoral dissertation, Simon Fraser University, Canada.

Brannen, M.Y. (1992) '"Bwana Mickey": Constructing cultural consumption at Tokyo Disney Land', in J.J. Tobin (ed), *Remade in Japan* (pp. 216–234). New Haven, CT: Yale University Press.

Bremer, T.S. (2001) 'Tourists and religion at Temple Square and Mission San Juan Capistrano', *Journal of American Folklore*, 113(450): 422–435.

Bryman, A. (1995) *Disney and His Worlds*. London: Routledge.

Butler, E. (2010) God is in the data: Epistemologies of knowledge at the Creation Museum, *Ethnos* 75(3): 229–251.

Caron, B. (1993) 'Magic kingdoms: Towards a post-modern ethnography of sacred places', *Kyoto Journal*, 25: 135–130.

Chidester, D. (2005) *Authentic Fakes: Religion and American Popular Culture*. Berkeley, CA: University of California Press.

Chmielewska-Szlajfer, H. (2017) '"Authentic Experience" and manufactured entertainment: Holy Land experience religious theme park', *Polish Sociological Review*, 200(4): 545–558.

Davis, S.G. (1996) 'The theme park: Global industry and cultural form', *Media, Culture & Society*, 18(3): 399–422. https://doi.org/10.1177/016344396018003003.

Dicks, B. (2004) *Culture on Display: The Production of Contemporary Visitability*. Berkshire, England: Open University Press.

Dwyer, R. (2004) 'The Swaminarayan movement', in K.A. Jacobsen & P.P. Kumar (eds), *South Asians in the Diaspora: Histories and Religious Traditions* (pp. 180–199). Leiden: Brill.

Dykins Callahan, S. (2010) *Where Christ dies daily: Performances of faith at Orlando's Holy Land Experience*. PhD Dissertation, University of South Florida, USA.

Eco, U. (1986) *Travels in Hyper Reality: Essays*. San Diego: Harcourt.
Fjellman, S.M. (1992) *Vinyl Leaves: Walt Disney World and America*. Boulder, CO: Westview Press.
Graburn, N.H.H. (1983) *To Pray, Pay and Play: The Cultural Structure of Japanese Domestic Tourism*. Aix-en-Provence: Centre des Hautes Etudes Touristiques.
Greenhalgh, P. (2011) *Fair World: A History of World's Fairs and Expositions, 1851–2010*. London: Papadakis.
Hendry, J. (2000) *The Orient Strikes Back: A Global View of Cultural Display*. Oxford: Berg.
Hunt, P., & Frankenberg, R. (1997) 'It's a small world: Disneyland, the family and the multiple re-representations of American childhood', in A. James & A. Prout (eds), *Constructing and Reconstructing Childhood: Contemporary Issues in the Sociological Study of Childhood* (pp. 105–123). London: Routledge.
Kaell, H. (2014) *Walking Where Jesus Walked: American Christians and Holy Land Pilgrimage*. New York: New York University Press.
Kauffman, J. (2005) *The Nazareth Jesus Knew*. Harrison, NY: Bch Fulfillment & Distribution.
Ketchell, A.K. (2007) *Holy Hills of the Ozarks: Religion and Tourism in Branson, Missouri*. Baltimore, MD: John Hopkins University Press.
Kim, H. (2009) 'Public engagement and personal desires: BAPS Swaminarayan temples and their contribution to the discourses on religion', *International Journal of Hindu Studies*, 13: 357–390.
Kogan, N.S. (2013) 'Mormons in the New York World's Fair, 1964–1965,' in J.M. Hunter (ed), *Mormons and Popular Culture: The Global Influence of an American Phenomenon*, Volume 2 (pp. 209–223). Santa Barbara, CA: ABC-CLIO.
Lari, L., Iyanna, S., and Jabeen, F. (2020) Islamic and Muslim tourism: service quality and theme parks in the UAE, *Tourism Review* 75(2): 402–413
Long, B.O. (2003) *Imagining the Holy Land: Maps, Models, and Fantasy Travels*. Bloomington: Indiana University Press.
Lovell, J., & Bull, C. (2017) *Authentic and Inauthentic Places in Tourism: From Heritage Sites to Theme Parks*. London and New York: Routledge.
Lukas, S.A. (2007) *The Themed Space: Locating Culture, Nation, and Self*. Lanham, MD: Rowman & Littlefield.
Lukas, S.A. (2013) *The Immersive Worlds Handbook: Designing Theme Parks and Consumer Spaces*. Oxford: Focal.
Lukens-Bull, R., & Fafard, M. (2007) 'Next year in Orlando: (Re)creating Israel in Christian zionism', *Journal of Religion & Society*, 9: 1–20.
Maanen, J. (1992) 'Displacing Disney: Some notes on the flow of culture', *Qualitative Sociology*, 15(1): 5–35. https://doi.org/10.1007/BF00989711
MacCannell, D. (1999) *The Tourist: A New Theory of the Leisure Class*. Berkeley, CA: University of California Press.
MacCannell, D. (2011) *The Ethics of Sightseeing*. Berkeley, CA: University of California Press.
McDaniel, J. (2017) *Architects of Buddhist Leisure: Socially Disengaged Buddhism in Asia's Museums, Monuments, and Amusement Parks*. Honolulu: University of Hawai'i Press.
Millar, S. (1999) 'An overview of the sector', in A. Leask & I. Yeoman (eds), *Heritage Visitor Attractions: An Operations Management Perspective* (pp. 1–21). New York: Cassell.
Mitchell, T. (1988) *Colonising Egypt*. Cambridge: Cambridge University Press.
Moore, A. (1980) 'Walt Disney World: Bounded ritual space and the playful pilgrimage center', *Anthropological Quarterly*, 53(4): 207–218. https://doi.org/10.2307/3318104.
Neilson, R. (2011) *Exhibiting Mormonism: The Latter-day Saints and the 1893 Chicago World's Fair*. New York: Oxford University Press.
Olsen, D.H. (2003) 'Heritage, tourism, and the commodification of religion', *Tourism Recreation Research*, 28(3): 99–104.
Olsen, D.H. (2012) 'Negotiating religious identity at sacred sites: A management perspective', *Journal of Heritage Tourism*, 7(4): 359–366.
Paine, C. (2016) 'Religious theme parks', *Material Religion: The Journal of Objects, Art and Belief*, 12(3): 402–403. https://doi.org/10.1080/17432200.2016.1192146
Paine, C. (2019) *Gods and Rollercoasters: Religion in Theme Parks Worldwide*. London and New York: Bloomberg.
Pfann, S., Voss, R., & Rapuano, Y. (2007) 'Surveys and excavations at the Nazareth Village farm (1997–2002): Final report', *Bulletin of the Anglo-Israel Archaeological Society*, 25: 19–79.

Raz, A.E. (1999) *Riding the Black Ship: Japan and Tokyo Disneyland*. Cambridge, MA: Harvard University Asia Center.

Ron, A.S., & Feldman, J. (2009) 'From spots to themed sites – The evolution of the Protestant Holy Land', *Journal of Heritage Tourism*, 4(3): 201–216. https://doi.org/10.1080/17438730802504108.

Rose, L. (2020) 'Nazareth Village and the creation of the "Holy Land" in Israel-Palestine: The Question of Evangelical Orthodoxy', *Current Anthropology*, 61(3): 335–355.

Rubin, J. (2018) 'Global attractions attendance report', *TEA/AECOM*. Available at: http://www.teaconnect.org/pdf/TEAAECOM2013.pdf (accessed 15 July 2020).

Rydell, R.W. (1993) *World of Fairs: The Century-of-Progress Expositions*. Chicago: University of Chicago Press.

Scott, D.W. (2014) 'Dinosaurs on Noah's Ark? Multi-media narratives and natural science museum discourse at the Creation Museum in Kentucky', *Journal of Media and Religion*, 13(4): 226–243. https://doi.org/10.1080/15348423.2014.971570.

Shinde, K.A. (2021) 'Religious theme parks as tourist attraction systems', *Journal of Heritage Tourism*, 16(3): 281–299. https://doi.org/10.1080/1743873X.2020.1791887.

Silverman, S.M. (2019) *The Amusement Park: 900 Years of Thrills and Spills, and the Dreamers and Schemers Who Built Them*. New York: Black Dog et Leventhal Publishers.

Srivastava, S. (2014) *Entangled Urbanism: Slum, Gated Community, and Shopping Mall in Delhi and Gurgaon*. New Delhi: Oxford University Press.

Stevenson, J. (2013. *Sensational Devotion: Evangelical Performance in Twenty-First-Century America*. Ann Arbor: University of Michigan Press.

Trollinger, S.L., & Trollinger, W.V. (2016) *Righting America at the Creation Museum*. Baltimore, MD: John Hopkins University Press.

Urry, J., & Larsen, J. (2011) *The Tourist Gaze 3.0*, 3rd Edition. London: SAGE.

Waysdorf, A., & Reijnders, S. (2018) 'Immersion, authenticity and the theme park as social space: Experiencing the Wizarding World of Harry Potter', *International Journal of Cultural Studies*, 21(2): 173–188.

Weiler K. (2017) 'Aspects of architectural authenticity in Chinese heritage theme parks', in K. Weiler & N. Gutschow (eds), *Authenticity in Architectural Heritage Conservation: Discourses, Opinions, Experiences in Europe, South and East Asia* (pp. 219–246). Cham, The Netherlands: Springer.

West, G. (2018) 'The world's 6 biggest theme parks', *Family Travel*. Available at: https://familytravel.com.au/the-6-biggest-theme-parks-in-the-world/ (accessed 14 July 2020).

Wharton, A.J. (2006) *Selling Jerusalem: Relics, Replicas, Theme Parks*. Chicago: University of Chicago Press.

Wong, C.U.I. (2017) 'The Big Buddha of Hong Kong: An accidental Buddhist theme park', *Tourism Geographies*, 19(2): 168–187. https://doi.org/10.1080/14616688.2016.1158204.

Young, T., & Riley, R.B. (eds) (2002) *Theme Park Landscapes: Antecedents and Variations*. Washington, DC: Dumbarton Oaks.

Zhang, C., Decosta, P., & Mckercher, B. (2015) 'Politics and tourism promotion: Hong Kong's myth making', *Annals of Tourism Research*, 54: 156–171. https://doi.org/10.1016/j.annals.2015.07.003.

14
RELIGIOUS AND SPIRITUAL RETREATS

Brooke Schedneck

Introduction

The idea of a religious retreat is a fairly modern phenomenon, with its roots in traditions of religious monasticism and pilgrimage. The desert hermit, forest monk, and ascetic who leave the householder life behind are tropes in early Christianity, Buddhism, and Hinduism. Fleeing cities, societies, and civilizations for the tranquil wilderness or desert is a common theme for the elite practitioners of these religions. This separation from society to an isolated quiet space remains an impulse inside and outside religious traditions. However, monastics were not always so distant from nearby cities and villages. The support from lay members of religious communities was needed to maintain monasteries. Most important, for those who remained in the world became an inner silence and external space that at least supported tranquility and contemplative practices (Adler 2006: 16–17). Ralph Waldo Emerson and Henry David Thoreau began the work of substituting nature for enclosed religious spaces and equating forests and caves with sacredness in North America (Adler 2006: 9). In Asia, the Buddhist monastic meditative traditions developed the possibility for intense lay practice beginning in the late 1950s (Schedneck 2015: 34). From here retreats outside of a formal monastery or religious tradition expanded so that today lay people without any religious affiliation participate in retreats of spiritual and religious origins.

A religious retreat involves travel to a space created and maintained by an institution where a person engages in "a limited period of isolation during which an individual, either alone or as part of a small group, withdraws from the regular routine of daily life, generally for religious reasons" (Lozano 1987: 7768). This isolation, in conjunction with certain religious and/or ascetic practices, is meant to create a time and space for people to connect with religious or spiritual phenomena beyond the ordinary routine of life, such as a divine figure, a more authentic sense of self, or a deeper understanding of the nature of reality. Although New Age retreats, which mix personal wellness with an eclectic hybrid of spiritual practices, exist and will be discussed below, for institutionalized religions, religious retreats are most common in Buddhism, Hinduism, and Christianity. At Buddhist meditation retreat centers and Hindu yoga retreats held in ashrams, non-adherents can participate with adherents because of the ritual and aesthetics focus on bodily practices. In contrast, Christian prayer retreats run by churches and monasteries are centered on belief and are usually attended by Christians hoping to deepen their relationship with the God.

Whether conducted in a group or individually, religious retreats are usually focused on intense religious and/or spiritual practices ranging from a period of a few days to a few months. Most retreats are held in a retreat facility (re)designed for this purpose. These retreat centers are found in both natural settings and in or close to an urban environment. Not everyone who attends or participates in a religious retreat does so purposefully, some people may accidentally stumble upon a retreat and decide to participate because they have time and space in their travel itinerary. At the same time, religious retreats tend to be more intense and reflective than casually visiting a religious site. However, experiences at religious retreats may not be as devotional as a pilgrimage undertaken by a religious practitioner.

Retreats are an alternative to a vacation because of certain perceived benefits, including therapeutic value, reduced stress, enhanced well-being, and possibly self-transformation (Kelly & Smith 2017: 140).[1] Newberg et al. (2018), in studying the neurophysiological effects of religious retreats, found that participants in a seven-day retreat had short-term positive impacts on their dopamine and serotonin functions. Tori (1999) found that Roman Catholic and Buddhist retreats had a positive influence on participants in terms of an increase in emotional maturity, sense of achievement, and sympathetic warmth. Heintzman (2013: 72) notes that the transformational benefits of retreat participation include restoration and spiritual transformation and well-being, among others. Fu et al. (2015: 86–87) surveyed guest experiences at wellness retreat centers through analyzing data from TripAdvisor and found that many people chose to participate in the retreat because the religious and spiritual exercises and experiences at the retreat would help them resolve a life challenge or hardship they were facing. As such, retreats offer safe spaces to undertake self-development, renew one's self spiritually, mentally, and physically, and to deal with negative life events through reflection and reassessment.

The purpose of this chapter is to examine how different religious faith traditions utilize religious retreats. After examining the major issues and questions raised in the present scholarship on religious retreats, attention is turned to the religious retreat phenomenon within Buddhism, Christianity, Hinduism, and New Age movements. The chapter concludes with a comparison of how religious retreats function within these individual traditions.

Orientations

While the idea of a retreat or an escape from the everyday world has traditionally been associated with religious institutions, in recent years retreat locations and experiences have tended to fall under the rubric of wellness (Stausberg 2011). Since the 1980s, wellness tourists have been using practices such as yoga and meditation along with eating organic food, which can also be found in some religious retreats. Stausberg (2011: 133) highlights the approaches the wellness industry uses to appropriate religious or spiritual vocabulary, such as transcendence, along with religious images and ritual forms. In particular, Asian religious iconography—usually from Buddhist and Hindu traditions—are often used at secularized wellness retreats as decorations.

The connection between religion and wellness, as Norman (2011) points out, is part of the rise of secularism and (post)modernity, where "cultural deregulation" (Beyer 2007) has led to people increasingly interacting with religion in an individualized fashion. Individual choice, experimentation, self-spirituality, and the "sacralization of the self" (Heelas 1996) have taken precedence over religious tradition and group belonging (Olsen 2019). As such, wellness tourism and spiritual tourism have become closely connected. Because of this, and the fact that many religious faiths open up their retreats to more people than just ones who identify with a particular religious affiliation, it can be difficult at times to differentiate between retreat practitioners that follow the norms of the faith tradition and practitioners who seek peace and relaxation independent of the religious structures in place. What makes these retreats attractive to wellness and spiritual tourists is that they are not designed to function as a missionary conversion tool. Instead, the focus is on being open and available to all people as a place of respite in the world.

Kelly and Smith (2017) have suggested that there are several categories of retreats, including spiritual, religious, yoga, and body–mind–spirit retreats. The spiritual category in this schema includes Buddhist retreat centers and Hindu ashrams and under the religious category are Christian settings such as monasteries and convents. Vipassana meditation retreats, as part of the Buddhist tradition, are included in the category of body–mind–spirit-based retreats. As useful as this typology is to understand the different types of retreats, it is very difficult to distinguish between spiritual and religious retreats. Buddhist retreat centers, Hindu ashrams, and yoga retreats can be considered religious by participants while an experience at a Christian monastic retreat can also be considered spiritual in nature. Indeed, this separation or distinction of the religious from the spiritual follows work by many scholars studying retreats. In trying to parse out the categories of individual tourist and religious practitioner, Heintzman (2013: 68) has classified tourists who visit retreat centers as spiritual travelers or seekers, asserting that this type of tourist can fit into both categories of religious and wellness tourism. Rather than categorizing retreats by participant type, Shackley (2004) notes that in the developed world, most religious retreat houses are either Buddhist, Hindu, or Christian, and offer accommodation for people seeking quiet, peace, and spiritual nurture. Whether these sites are spiritual, wellness, or religion oriented is difficult to classify as their characteristics, practices, and experiential offerings are so varied.

To better understand the different types of religious and spiritual retreats, the rest of this chapter focuses on comparing the practices, beliefs, and formats of religious retreats within the traditions of Buddhism, Christianity, and Hinduism, as well as the New Age movement. While there are retreats within other traditions, research within these four groups is most prevalent.

Buddhism

Buddhism is arguably the religion most compatible with the retreat format (Tori 1999), as vipassana meditation from Southeast Asia, Zen meditation from Japan, and Tibetan forms of meditation from the Himalayan regions all recommend various periods of time away from everyday life to focus on meditative practice in a deep, transformative, and intensive way. To facilitate this, Buddhist retreats have been developed to create the time and space needed to reach the soteriological goal of Enlightenment or *nirvana*. Historically, Buddhist retreats were more closely connected to monastic life. However, since the introduction of what scholars call "modern Buddhism" (Lopez 2002) or "Buddhist modernism" (McMahan 2008), Buddhist retreats have democratized, with non-Buddhists—who may have different motivations for participating in these retreats, such as solutions to personal problems, enhancing life experiences, and escaping daily pressures—being allowed to participate along with lay and monastic Buddhists.

More recently, interest in and the popularization of Buddhist meditative retreats for wellness purposes has increased due to the innovative use of Buddhist temples as accommodations during the 2002 FIFA World Cup, which South Korea co-hosted with Japan. With the expected influx of millions of people to attend the World Cup, the South Korean Ministry of Culture and Tourism was concerned the country might not have enough accommodations. Eventually the idea of housing them at Buddhist temples of the dominant Chogye order was proposed. Buddhist leaders decided that instead of just using their temples as temporary hotels, they would create a whole spiritual and cultural experience (Kaplan 2010: 131–132). Since then, the Korean "Temple Stay" program has grown in the number of temples and visitors they accommodate each year. This growth has resulted in several studies being conducted on different aspects of this program, ranging from Temple Stay marketing and branding (Kaplan 2010; Kim 2017) to the motivations, experiences, and satisfaction levels of domestic, foreign, non-Buddhist, and Buddhist participants (Shin, Jeon & Rha 2014; Ho-Sung 2015; Song et al. 2015; Yoon 2016; Chun, Roh & Spralls 2017; Chun et al. 2018; Bae, Lee & Chick 2019; Ross, Hur and Hoffman 2019); from the role of temple food and religious personalities in branding (Moon 2008; Ji et al. 2010; Son & Xu 2013; Park, Bonn & Cho 2020) to the environmental impacts of Temple Stays (Shin & Shin 2011).

Another area of interesting research has been the psychological and physical effects that occur in participants in Temple Stays. Emavardhana and Tori (1997: 203), for example, in measuring for the "development of a healthy and individuated self-concept" among Thai Buddhist participants at a seven-day vipassana meditation retreat in Bangkok, suggest that those who participated in the retreats experienced increased feelings of worth, benevolence, and self-acceptance after the retreat as well as a heightened belief in Buddhist precepts and less self-criticism (Emavardhana & Tori 1997, 201). Tori (1999: 126) notes that teenagers participating in a Buddhist religious retreat scored higher in measures of emotional maturity, achievement, and sympathy than those who participated in a Roman Catholic retreat, leading the author to surmise that "nontheistic Buddhist techniques may be particularly applicable in secular settings and for those who find faith in unknown entities unsatisfying." Yoon et al. (2019) found that meditation by participants at an intense four-day Buddhist meditation retreat had facilitated white matter myelination in brain regions that are important for cognitive functioning.[2] Jo et al. (2020) also found that over the course of a five-day temple stay the paraben levels in South Korean participants did not decrease, in part because of the temple stay dietary program which allows for the use of traditional condiments and seasonings that might have contained parabens.[3] Krygier et al. (2013) found that several measures of well-being improved among participants after a ten-day vipassana meditation retreat.

There has long been an interest in Buddhist retreats by North Americans and Europeans. Several memoirs about travel to Buddhist retreats in India and Nepal, as well as anthropological accounts of retreat experiences, add a depth of understanding to why people are drawn to this practice. In the 1970s, several Westerners traveled to South and Southeast Asia with an interest in Eastern spirituality. Their writings reveal a common theme, wherein they met a teacher or encountered a Buddhist retreat center, and after this, continued to practice and spread Buddhist meditation the rest of their lives. One of these travelers was Alison Murdoch, who at the age of 27 bought a ticket to New Delhi in order to find herself. In 1987, she happened across a Tibetan Buddhist temple and participated in a ten-day meditation course there taught by a Canadian monk. Later on in the same trip she did a month-long Tibetan Buddhist course in Nepal, and since then has remained committed to Buddhism (Mackenzie 2001: 29–31).

Another common theme in these ethnographic writings is some people who participated in these meditative retreat in South and Southeast Asia returning to North America or Europe and establishing a retreat center there, leading to an internationalization of Buddhist meditation. This happened in the case of Sharon Salzberg, who at the age of 18 turned to meditation to deal with confusion and unhappiness she was working through at the time. In 1970, she traveled to Dharamsala, India, to practice meditation with the Dalai Lama (b. 1935), and in 1971, Bodh Gaya, India, to participate in an intensive ten-day meditation retreat (Mackenzie 2001: 73).[4] After returning home in 1974, Salzberg became one of the founders of the Insight Meditation Society and created a now well-known meditation vipassana meditation center in Barre, Massachusetts.

Indeed, vipassana meditation retreats in particular have also become very popular in Thailand and Myanmar (Jordt 2007; Cook 2010), to the point where in 2004, the World Fellowship of Buddhists published *A Guide to Buddhist Monasteries and Meditation Centres in Thailand* (Sirikanchana 2004) in English. Vipassana meditation retreats and vipassana meditation in different social contexts have also become a popular topic among academics (Marlatt et al. 2004; Perelman et al. 2012; Pagis 2015; Vaccarino & Comrie 2015). For example, Schedneck (2015) has written a monograph on the phenomenon of international meditation retreats in Thailand, examining issues related of translation, commodification, and embodiment in order to compare the international retreat participants with Thai participants. Schedneck (2019) has also analyzed ten retreat memoirs that discuss the experiences and reflections in the vipassana retreat setting for non-Buddhists. These academic analyses and guides demonstrate the significance of meditation and retreat within Buddhism and the interest from non-Buddhist international audiences.

In Korea, Myanmar, and Thailand, Buddhist meditation retreats have become an integral part of national and religious heritage. However, while as noted above Buddhist retreat centers have been established in some non-Buddhist countries, they do not carry the same importance. Indeed, as Gilli and Ferrari (2017) note, people who participate in Buddhist retreat centers in Italy find that their expectations compared with the reality of the experience were very different. Participants were hoping for a more hotel-like atmosphere, not realizing that the centers were nonprofit organizations with limited beds and a monastic lifestyle. This may be because people in non-Buddhist countries are not familiar with the purpose of and regimented routines that are a part of retreat living. People in non-Asian countries often do not understand the differences between Asian religious retreats. Instead of seeing these retreats as embedded within contexts of lineage, cosmology, and faith, outsiders might choose to make their own meaning out of the experience (Palmer 2014; Schedneck 2015).

Christianity

Religious retreats, which also have a long history within Christianity, have also been the focus of scholarly treatment, albeit from a very different angle. Rather than focusing on the motivations, benefits, effects of Christian retreats, and the effects of transnationalism and secularism, scholars have focused more closely on how retreats are connected to Christian tradition itself. This may be because there is less non-Christian participation and decontextualizing Christian retreats for a broader audience. Rather, Christians tend to be the participants in Christian retreats, which primarily take place within the Christian monastic tradition. While some Christian retreats offer a scheduled program of prayer, most retreats are focused on Christianity's soteriological goal of attaining heaven through offering ways to enhance one's relationship and connection with the God.

Because of this, much of the research on Christian retreats has focused on how clergy benefit from retreats and the ways monasteries bring spiritual restoration to its Christian followers. For example, Gill, Packer, and Ballantyne (2018) investigated the role of Christian retreats in Australia in helping Christian clergy deal with mental fatigue. According to the authors, participants in the retreats experienced "a significant positive impact on participants' mental state" (p. 246), including social, spiritual, and cognitive benefits related to reflection, renewal, restoration, personal development, transformation, and healing. Another study by Ouellette et al. (2005) looked at the potential of a Canadian Benedictine Monastery to provide a restorative environment, analyzing the nature of this restorative environment as well as retreat participants' motivations. The authors found that beauty, spirituality, and spending time away from home were some of the main motivating factors for people staying at the monastery.

The Ignatian monastic tradition offers the most unique type of Christian retreat with its focus on what is referred to as the Ignatian Exercises, which involves solitude and silence. Tyers (2010), in his survey of retreat practices in the Church of England from 1858 onward, suggests that by the mid-1960s between twenty and thirty thousand people in the Anglican Church were making an Ignatian retreat each year. This popularity led to an expansion of Ignatian retreats in 1979, and by the 1980s training courses for retreat directors from all denominations were made available. This expansion of Ignatian retreats spurred Benedictine monastics as well as Celtic and Franciscan traditions to "increase their focus on silence and devotional practices for all, encouraging a less cerebral and a more contemplative approach in many retreats, some of which are totally given over to exploring ways of stillness" (Tyers 2010: 272). Other Christian traditions have begun to host retreats experimenting with different retreat programming such as walking retreats, workshops, and journaling practices (Tyers 2010).

The idea of stillness and silence continues to be an important theme on research related to Christian retreats. In his work on retreats at Benedictine monasteries in England, Conradsen (2007: 34) argues that "stillness has been subject to growing valorization in Western countries in recent years. With pressures on many configurations of work—life balance, the scarcity of stillness has contributed to heightened public interest in places of retreat." As noted above, places of retreat, whether they are Christian monasteries or Buddhist meditation centers, are meant for people looking for a place apart; a place of stillness, rest, and renewal, as many people have "difficulty achieving internal stillness in many domestic settings" (p. 38). Rather than being viewed as sign of laziness and lack of productivity, stillness is now viewed as a kind of productive non-activity. Benedictine monasteries are well suited and designed to facilitate this stillness and encounters with the divine because of their "long-standing rhythms of prayer and worship" (p. 43). The programming at the Benedictine retreats helps participants learn how to take the practices of stillness and reflection and apply them when they return home.[5] As such, guests at the Benedictine monasteries Conradsen studied were encouraged to not focus on "doing", but rather on "being" and engaging with the natural environment so as to create a connection with the God (p. 42).

Besides monasteries, retreat houses are another popular Christina retreat. Retreat houses are places that provide "accommodation and spiritual input for guests in search of peace and quiet, whether or not this is associated with a religious or monastic experience" (Shackley 2004: 228). The retreat houses usually have spartan accommodations, little to no choice of food, and have little of the modern conveniences of hotels (Shackley 2004; O'Gorman & Lynch 2008). While retreat houses usually have a worship component to the guest experience, this component is optional, and there is no set programming regarding what someone must do during their time. Besides taking time away from one's normal activities, one could join in the many activities offered, including individual or group spiritual practice, taking spiritually based courses, or just walking in nature.

Christian retreats are also found beyond Europe, including North and South America (Wright 2008; Ron & Timothy 2019), Africa (e.g., Ojo 1988; Harrison 2004; Okonkwo & Nzeh 2009), and Asia (Jesudass 2000). And it is not just Roman Catholicism and the Anglican Church that host Christian retreats—Baptist, Methodist, Episcopal, Lutheran, Orthodox, Presbyterian, Pentecostal, Quaker, and Ecumenical Christian churches also offer retreat experiences. While the Christian retreat tourism niche market is understudied as a whole, several websites have been established and news articles published to help interested participants find appropriate Christian retreats.[6]

Hinduism

Religious retreats also take place in a Hindu context, many of which are tied to the practice of yoga. The Indian Ministry of Tourism has proclaimed that "yoga is India's gift to the world," and that "India has the potential to brand itself as the land of yoga" (Bowers & Cheer 2017: 211). As a part of this push to promote yoga in India as a panacea for the ailments of modern Western societies, Rishikesh, India, has begun to market itself as the "Yoga Capital of the World," and India has created a special yoga visa for travelers to India (Bowers & Cheer 2017). At the same time, many travelers tend to conflate yoga, religious retreats, and wellness tourism in this context (Norman 2011; Schedneck 2015) and may engage in several different types of retreats when traveling in Asia. Norman (2011), for example, found that people who engaged in yoga retreats in Rishikesh also participated in Buddhist vipassana meditation retreats in other parts of India. Schedneck (2015, 82) found that spiritual tourists at a meditation retreat in Thailand saw this as a part of their broader wellness trip, with one informant having just arrived from a yoga retreat in Bali. Norman (2011) also found that many spiritual tourists in Rishikesh create their own religious retreats, combining yoga and *satsang* (spiritual or philosophical lectures offered by gurus) activities with many tourists spending on average four to six hours per day in these classes. As such, bricolage and hybridity are an important part of religious retreats in Hindu tradition, which often claims its openness to people of all religions and those who claim no religion.

Yoga retreats in India are seen by spiritual tourists to reset the mind and gain a fresh perspective after a difficult period in one's life. While vacations and other types of activities can lead to renewal, retreats are viewed by many spiritual tourists as the best way to receive the catharsis they need. As such, yoga retreats have been a focus of study, particularly by psychologists. Pandya (2018) found that participation in Indian guru-led yoga retreats led to increased well-being among participants. Sharpley and Sundaram (2005) described a retreat at the Sri Aurobindo Ashram, home of the controversial Sai Baba, and note that all visitors are welcome. There are also no compulsory activities, as everyone is free to follow their own practices and stay for as long as they like, as demonstrated by one respondent who has stayed at the ashram for 27 years with no plans to return home! Some of the participants in the study had traveled to India solely to visit several ashrams, seeking spiritual strength they could not find in their daily lives, while others visited this specific ashram to participate in a few meditation or yoga classes. Because these types of Hindu retreats have several different activities in which a person can participate and are not time-limited, there are a wide range of people who attend, ranging from beginning Hindu practitioners to long-time affiliated members.

Hindu and yoga retreats are also found in the United States, where many people spend their vacations. In their study of yoga retreats in Indiana, Lehto et al. (2006) note that yoga retreats in the Unites States are viewed in the same way as retreats in India—a place to focus on the self and receive spiritual nourishment instead of the demands of work. Other important reasons for attending a yoga retreat included renewing oneself, relaxing, being more flexible in mind and body, letting go of stress, and gaining a sense of balance in one's spiritual, mental, physical, and emotional dimensions. Many of the participants at this retreat felt that an hour or two of yoga at a local yoga studio was not enough to recharge or provide that sense of balance within the self, and as such wanted to spend a longer, more intense period of time at the retreat to fully restore their life balance.

New Age retreats

New Age retreats are a more recent phenomenon but fit well within the secularized and individualized nature of contemporary spirituality. Indeed, a major part of modern New Age practice is going on retreat. Attix (2002) has found that for New Age practitioners, retreat centers are not only places for practice and restoration, but often become hubs for independent travelers seeking information and directions to local sacred sites and special places. Attix, however, seems to take a critical stance against New Age practitioners, stating that "they also show little indication of awareness about how they may be externally impacting host populations and religious sensibilities" (p. 56). This criticism follows more general disapproval of the New Age movement by those who argue that while this movement provides an eclectic "spiritual marketplace" (Gauthier et al. 2013) for those who seek individualized or customized spiritual experiences and services, it is both heavily reliant on cultural appropriation and too closely tied to capitalism, thereby seen as a shallow form of spirituality (York 2001; Carrette & King 2004; Taira 2009; Wood 2016).

Regardless, New Age retreats are an important part of spiritual tourism and its emphasis on balance, wellness, and health. Bone (2013) found in her research on two New Age retreats in New Zealand that community, connection, and the possibility of escape, along with natural and spiritual landscapes, are the key features of spiritual retreat tourism. These retreat centers focused on a mixture of yoga, meditation, natural therapies, healing, and exercises meant to awaken a person's consciousness. Participants also found that companionship with others who have similar goals and desires to escape from their everyday life was an important component of the overall retreat experiences. As such, Bone suggests that because of these positive benefits, New Age "[r]etreat centres can be seen as antidotes to the alienation and disruption that is felt by the population on a global scale" (p. 299), and therefore "are becoming increasingly popular as niche tourism sites" (p. 307). For those seeking these experiences outside of a religious institution, New Age retreats have become a significant option (Ivakhiv 2003; Redden 2005; Pernecky & Johnston 2006), especially when tied to power places such as Glastonbury Tor, UK and Sedona, Arizona (Bowman 1993; Ivakhiv 1997; Digance & Cusack 2002; Coats 2009; MacLaran & Scott 2009).

New Age retreats are also a type of spiritual retreat, which Bone (2015: 123) defines as "a unique form of tourism, differentiated from wellness or spa tourism, due to its focus on spiritual ideals and the touristic experience whereby spiritual activities and ideologies are practiced and symbolically evident at retreat sites." Spiritual retreats generally package health, wellness, spiritual practices, and services together with healthy, often organic and vegetarian, food options and a calm and peaceful environment similar to religious retreats in order to give participants the feeling they are doing something good for their bodies and minds. Within the New Age movement, the label of "spiritual" retreats is appropriate because it implies a space outside of a single religious institution.

Conclusion

In the modern world, people seek healing and meaning through religion and spirituality. As such, many people choose to travel to and participate in retreats and temple stays, staying

> ...a few days of quiet and contemplation in secluded places of prayer and meditation. Some people retreat to a more primitive mode of living close to nature and away from the conveniences of modern civilization. People thereby wish to be free from the stresses of daily life and to restore themselves through quiet and contemplation to an original state of purity for renewed energy and readiness for life escaping for a few days in quiet solitude.
>
> *(Kim 2014: 21)*

And yet as important as religious and spiritual retreats are, they remain an understudied tourism niche market. Scholars in religious studies, anthropology, tourist studies, and psychological studies have conducted research on religious and spiritual retreats. While the focus of much of this research has been on creating typologies of religious and spiritual retreat participants and the outcomes of participation in these retreats, more ethnographic work would help better understand the perspectives of religious retreat participants and in particular the role of teachers at these retreats. Research at religious places, such as Buddha's birthplace of Bodh Gaya, India (Geary 2017) or with pilgrims traveling on the Santiago de Compostela in Spain (Norman 2011) is less challenging than studying a retreat. The structure and physical demands of a retreat offer unique struggles to researchers, as scholars must attend retreats themselves, become familiar with the participants and the retreat leaders, record notes and interviews, all while remaining in a silent or semi-silent environment.

What religious and spiritual retreats have in common is that time away from one's daily routine is the biggest reason why people travel to and participate in these retreats. People indicate their beliefs and values not only verbally but through the movement of their bodies in culturally conceived spaces. These removed and socially distant retreat spaces communicate to participants in retreats and to broader society, a performance of the self, which values contemplation and seeking a kind of truth (Adler 2002, 46). The chance to focus on one's self while undertaking religious or spiritual practices is the major benefit and motivation for undertaking a retreat. While retreats range from the austere to the luxurious accommodations, the point is to be away from one's home and community and to learn about and take care of oneself, physically and mentally. The popularity of retreats highlights the problems of modernity in many participants' minds—too fast paced with no time to relax and reflect. Religious and spiritual retreat therefore becomes spaces of anti-modernity, intended for those people who need a break from living modern lives.

Notes

1 Some of these benefits are the same as those found in other tourism niche markets.
2 Myelination serves to keep nerve impulses flowing to different parts of the brain.

3 Parabens are non-persistent preservatives that are used in cosmetics, personal care products and food items that break down quickly.
4 Bodh Gaya, in Bihar, eastern India, is known as the "seat of the Buddha's enlightenment." It is here that Buddhist pilgrims come to pay respect to the spot where the Buddha attained *nirvana*.
5 This relates to research by Gill, Packer, and Ballantyne (2019), who note that the restorative aspects of Christian retreats diminish among participants who have significant levels of stress or deal with an excessive workload immediately after returning home.
6 Website examples include https://bookretreats.com/s/other-retreats/religious-retreats; https://www.retreatfinder.com/Directory/Faith/Christian.aspx.

References

Adler, J. (2006) 'Cultivating wilderness: Environmentalism and legacies of early Christian asceticism'. *Comparative Studies in Society and History*, 48(1): 4–37.

Adler, J. (2002) 'The holy man as traveler and travel attraction: Early Christian asceticism and the moral problematic of modernity,' in W.H. Swatos & L. Tomasi (eds), *From Medieval Pilgrimage to Religious Tourism: The Social and Economics of Piety* (pp. 25–50). Westport, CT: Praeger.

Attix, S.A. (2002) 'New-age oriented special interest travel: An exploratory study', *Tourism Recreation Research*, 27(2): 51–58.

Bae, S.Y., Lee, C.K., & Chick, G. (2019) 'A multicultural retreat in exotic serenity: Interpreting temple stay experience using the Mandala of Health model', *Asia Pacific Journal of Tourism Research*, 24(8): 789–804.

Beyer, P. (2007) 'Religion and globalization', in G. Ritzer (ed), *The Blackwell Companion to Globalization* (pp. 444–460). Malden, MA: Blackwell.

Bone, K. (2015) 'Selling spirituality: Issues in tourism', *Tourism Review International*, 19: 123–132.

Bone, K. (2013) 'Spiritual retreat tourism in New Zealand', *Tourism Recreation Research*, 38(3): 295–309.

Bowers, H., & Cheer, J.M. (2017) 'Yoga tourism: Commodification and western embrace of eastern spiritual practice', *Tourism Management Perspectives*, 24: 208–216.

Bowman, M. (1993) 'Drawn to glastonbury,' in I. Reader & T. Walter (eds), *Pilgrimage in Popular Culture* (pp. 29–62). London: Palgrave Macmillan.

Carrette, J., & King, R. (2004) *Selling Spirituality: The Silent Takeover of Religion*. London: Routledge.

Chun, B., Roh, E.Y., & Spralls III, S.A. (2017) 'Living like a monk: Motivations and experiences of international participants in Templestay', *International Journal of Religious Tourism and Pilgrimage*, 5(1): 39–55.

Chun, B., Roh, E.Y., Spralls, S.A., & Kim, Y. (2018) 'Predictors of templestay satisfaction: A comparison between Korean and international participants', *Leisure Sciences*, 40(5): 423–441.

Coats, C. (2009) 'Sedona, Arizona: New age pilgrim-tourist destination', *CrossCurrents*, 59(3): 383–389.

Cook, J. (2010) *Meditation in Modern Buddhism: Renunciation and Change in Thai Monastic Life*. Cambridge, UK: Cambridge University Press.

Digance, J, & Cusack, C. (2002) 'Glastonbury: A tourist town for all seasons', in G.M.S. Dann (ed), *The Tourist as a Metaphor of the Social World* (pp. 263–280). Wallingford, UK: CABI.

Emavardhana, T., & Tori, C.D. (1997) 'Changes in self-concept, ego defense mechanisms, and religiosity following seven-day Vipassana meditation retreats', *Journal for the Scientific Study of Religion*, 36(2): 194–206.

Fu, X., Tanyatanaboon, M., & Lehto, X.Y. (2015) 'Conceptualizing transformative guest experience at retreat centers', *International Journal of Hospitality Management*, 49: 83–92.

Gauthier, F., Martikainen, T., & Woodhead, L. (2013) 'Acknowledging a global shift: A primer for thinking about religion in consumer societies', *Implicit Religion*, 16(3): 261–276.

Geary, D. (2017) *The Rebirth of Bodh Gaya: Buddhism and the Making of a World Heritage Site*. Seattle, WA: University of Washington Press.

Gill, C., Packer, J., & Ballantyne, R. (2019) 'Spiritual retreats as a restorative destination: Design factors facilitating restorative outcomes', *Annals of Tourism Research*, 79: https://doi.org/10.1016/j.annals.2019.102761.

Gill, C., Packer, J., & Ballantyne, R. (2018) 'Exploring the restorative benefits of spiritual retreats: The case of clergy retreats in Australia', *Tourism Recreation Research*, 43(2): 235–249.

Gilli, M., & Ferrari, S. (2017) 'Marginal places and tourism: the role of Buddhist centers in Italy', *Journal of Tourism and Cultural Change*, 15(5): 422–438.

Harrison, P. (2004) *South Africa's Top Sites: Spiritual*. Kenilworth, UK: New Africa Books.

Heelas, P. (1996) *The New Age Movement: The Celebration of the Self and the Sacralization of Modernity*. Oxford: Blackwell.

Heintzman, P. (2013) 'Retreat tourism as a form of transformational tourism', in Y. Reisinger (ed), *Transformational Tourism: Tourist Perspectives* (pp. 68–81). Wallingford, UK: CABI.

Ho-Sung, C. (2015) *Temple Stay: A Journey of Self-discovery*. Seoul: Seoul Selection.

Ivakhiv, A. (2003) 'Nature and self in New Age pilgrimage', *Culture and Religion*, 4(1): 93–118.

Ivakhiv, A. (1997) 'Red rocks, "vortexes" and the selling of Sedona: Environmental politics in the new age', *Social Compass*, 44(3): 367–384.

Jesudass, M. (2000) *Spiritual Awareness and Growth of Indian Adventist Youth through Retreats*. PhD dissertation, Andrews University, Berrins Springs, Michigan, USA.

Ji, K., Kho, Y.L., Park, Y., & Choi, K. (2010) 'Influence of a five-day vegetarian diet on urinary levels of antibiotics and phthalate metabolites: A pilot study with "Temple Stay" participants', *Environmental Research*, 110(4): 375–382.

Jiang, T., Ryan, C., & Zhang, C. (2018) 'The spiritual or secular tourist? The experience of Zen meditation in Chinese Temples', *Tourism Management*, 65: 187–199.

Jo, A., Kim, S., Ji, K., Kho, Y., & Choi, K. (2020) 'Influence of vegetarian dietary intervention on urinary paraben concentrations: A pilot study with 'Temple Stay' participants', *Toxics*, 8(1): http://doi:10.3390/toxics8010003.

Jordt, I. (2007) *Burma's Mass Lay Meditation Movement: Buddhism and the Cultural Construction on Power*. Athens, OH: Ohio University Press.

Kaplan, U. (2010) 'Images of monasticism: The Temple Stay program and the re-branding of Korean Buddhist Temples', *Korean Studies*, 34: 127–146.

Kelly, C., & Smith, M.K. (2017) 'Journeys of the self: The need to retreat', in M.K. Smith & L. Puczkó (eds), *The Routledge Handbook of Health Tourism* (pp. 138–151). London and New York: Routledge.

Kim, K.O. (2014) 'The human sciences and the healing of civilizations', in K.O. Kim (ed), *The Humanities and Healing* (pp. 17–36). Seoul: Korean National Commission for UNESCO.

Kim, S.S. (2017) 'Authenticity, brand culture, and Templestay in the digital era: The ambivalence and in-betweenness of Korean Buddhism', *Journal of Korean Religions*, 8(2): 117–146.

Krygier, J.R., Heathers, J.A., Shahrestani, S., Abbott, M., Gross, J.J., & Kemp, A.H. (2013) 'Mindfulness meditation, well-being, and heart rate variability: A preliminary investigation into the impact of intensive Vipassana meditation', *International Journal of Psychophysiology*, 89(3): 305–313.

Lehto, X.Y., Brown, S., Chen, Y., & Morrison, A.M. (2006) 'Yoga tourism as a niche within the wellness tourism market', *Tourism Recreation Research*, 31(1): 25–35.

Lopez, D. (2002) 'Introduction', in D.S. Lopez (ed), *A Modern Buddhist Bible* (pp. viii–xli). Boston, MA: Beacon Press.

Lozano, J.M. (1987) 'Retreat', in L. Jones (ed), *Encyclopedia of Religion*, 2nd edition (pp. 7768–7773). New York: MacMillan Reference USA.

Mackenzie, V. (2001) *Why Buddhism? Westerners in Search of Wisdom*. New South Wales, Australia: Allen & Unwin.

MacLaran, P., & Scott, L.M. (2009) 'Magic and merchandise: Spiritual shopping in Glastonbury,' *Advertising & Society Review*, 10(4): http://doi:10.1353/asr.0.0039.

Marlatt, G.A., Witkiewitz, K., Dillworth, T.M., Bowen, S.W., Parks, G.A., Macpherson, L.M., Lonczak, H.S., Larimer, M.E., Simpson, T., Blume, A.W., & Crutcher, R. (2004) 'Vipassana meditation as a treatment for alcohol and drug use disorders. Mindfulness and acceptance', in S.C. Hayes, V.M. Follett, & M.M. Lineham (eds), *Mindfulness and Acceptance: Expanding the Cognitive-Behavioral Tradition* (pp. 261–287). New York: The Guilford Press.

McMahan, D. (2008) *The Making of Buddhist Modernism*. Oxford: Oxford University Press.

Moon, S.S. (2008) 'Buddhist temple food in South Korea', *Korea Journal*, 48(4): 147–180.

Newberg, A.B., Wintering, N., Yaden, D.B., Zhong, L., Bowen, B., Averick, N., & Monti, D.A. (2018) 'Effect of a one-week spiritual retreat on dopamine and serotonin transporter binding: A preliminary study', *Religion, Brain, and Behavior*, 8(3): 265–278.

Norman, A. (2011) *Spiritual Tourism: Travel and Religious Practice in Western Society*. New York & London: Continuum.

O'Gorman, K.D., & Lynch, P.A. (2008) 'Monastic hospitality: Explorations', in S. Richardson, L. Fredline, L. Patiar, & M. Ternel (eds), *(CAUTHE) 17th Annual Conference 2008* (pp. 1162–1175). Gold Coast, Australia: Griffith University.

Ojo, M.A. (1988) 'Deeper Christian life ministry: A case study of the charismatic movements in western Nigeria', *Journal of Religion in Africa*, 18: 141–162.

Okonkwo, E.E., & Nzeh, C.A. (2009) 'Faith-based activities and their tourism potentials in Nigeria', *International Journal of Research in Arts and Social Sciences*, 1: 286–298.

Olsen, D.H. (2019) 'Religion, spirituality, and pilgrimage in a globalizing world', in D.J. Timothy (ed), *Handbook of Globalisation and Tourism* (pp. 270–283). London: Edward Elgar.

Ouellette, P., Kaplan, R., & Kaplan, S. (2005) 'The monastery as a restorative environment', *Journal of Environmental Psychology*, 25: 175–188.

Pagis, M. (2015) 'Evoking equanimity: Silent interaction rituals in Vipassana meditation retreats', *Qualitative Sociology*, 38(1): 39–56.

Palmer, D.A. (2014) 'Transnational sacralizations: When Daoist monks meet global spiritual tourists', *Ethnos*, 79(2): 169–192.

Pandya, S. (2018) 'The culture of yoga retreats, active followers and peripheral associates of new religious movements: Wellness enterprises promoting well-being', *Mental Health, Religion & Culture*, 21(5): 443–457.

Park, J., Bonn, M.A., & Cho, M. (2020) 'Sustainable and religion food consumer segmentation: Focusing on Korean Temple Food restaurants', *Sustainability*, 12(7): 3035.

Perelman, A.M., Miller, S.L., Clements, C.B., Rodriguez, A., Allen, K., & Cavanaugh, R. (2012) 'Meditation in a deep south prison: A longitudinal study of the effects of Vipassana', *Journal of Offender Rehabilitation*, 51(3): 176–198.

Pernecky, T., & Johnston, C. (2006) 'Voyage through numinous space: Applying the specialization concept to new age tourism', *Tourism Recreation Research*, 31(1): 37–46.

Redden, G. (2005) 'The new age: Towards a market model', *Journal of Contemporary Religion*, 20(2): 231–246.

Ron, A.S., & Timothy, D.J. (2019) *Contemporary Christian Travel: Pilgrimage, Practice and Place*. Bristol, UK: Channel View Publications.

Ross, S.L., Hur, J.C., & Hoffman, J. (2019) 'Temple Stay as transformative travel: An experience of the Buddhist Temple Stay program in Korea', *Journal of Tourism Insights*, 9(1): https://doi.org/10.9707/2328-0824.1090.

Schedneck, B. (2019) 'The promise of the universal: Non-Buddhists' accounts of their *Vipassana* meditation retreat experiences', *Religion*, 49(4): 636–660.

Schedneck, B. (2015) *Thailand's International Meditation Centers: Tourism and the Global Commodification of Religious Practices*. New York & London: Routledge.

Shackley, M. (2004) 'Accommodating the spiritual tourist: The case of religious retreat houses', in T. Shackley (ed), *Small Firms in Tourism: International Perspectives* (pp. 225–237). Amsterdam, Boston, MA: Elsevier.

Sharpley, R., & Sundaram, P. (2005) 'Tourism: A sacred journey? The case of ashram tourism, India', *International Journal of Tourism Research*, 7: 161–171.

Shin, H.K., & Shin, H.C. (2011) 'Activation programs of Temple Stay for low carbon green growth', *The Journal of the Korea Contents Association*, 11(8): 438–447.

Shin, K.Y., Jeon, H.W., & Rha, Y.A. (2014) 'The factor analysis of satisfaction with temple food, motivation for temple-stay and revisit intention to temple-stay in temple-stay tourism', *Culinary Science and Hospitality Research*, 20(1): 238–252.

Sirikanchana, P. (2004) *A Guide to Buddhist Monasteries and Meditation Centres in Thailand*. Bangkok: World Fellowship of Buddhists.

Son, A., & Xu, H. (2013) 'Religious food as a tourism attraction: The roles of Buddhist temple food in Western tourist experience', *Journal of Heritage Tourism*, 8(2–3): 248–258.

Song, H.J., Lee, C.K., Park, J.A., Hwang, Y.H., & Reisinger, Y. (2015) 'The influence of tourist experience on perceived value and satisfaction with temple stays: The experience economy theory', *Journal of Travel & Tourism Marketing*, 32(4): 401–415.

Stausberg, M. (2011) *Religion and Tourism: Crossroads, Destinations and Encounters*. London and New York: Routledge.

Taira, T. (2009) 'The problem of capitalism in the scholarship on contemporary spirituality', in T. Ahlbäck (ed), *Postmodern Spirituality* (pp. 230–44). Turku, Finland: Donner Institute for Research on Religious and Cultural History.
Tori, C.D. (1999) 'Change on psychological scales following Buddhist and Roman Catholic retreats', *Psychological Reports*, 84(1): 125–126.
Tyers, J. (2010) 'Ignatian and silent: A brief survey of the development of the practice of retreat in the Church of England, 1858–2008', *Theology*, 113(874): 267–275.
Vaccarino, F., & Comrie, M. (2015) 'An analysis of a New Zealand faith-based prison unit as a place of retreat,' *International Journal of Religion & Spirituality in Society*, 5(2): 1–9.
Wood, M. (2016) *Possession, Power and the New Age: Ambiguities of Authority in Neoliberal Societies*. London: Routledge.
Wright, K. (2008) *The Christian Travel Planner*. Nashville, TN: Thomas Nelson.
Yoon, S.Y. (2016) 'The effects of Temple-stay experienced tourism motivation of participation and choice factors on satisfaction and revisit intention', *Journal of Service Research and Studies*, 6(2): 1–21.
Yoon, Y.B., Bae, D., Kwak, S., Hwang, W.J., Cho, K.I.K., Lim, K.O., Park, H.Y., Lee, T.Y., Kim, S.N., & Kwon, J.S. (2019) 'Plastic changes in the white matter induced by Templestay, a 4-day intensive mindfulness meditation program', *Mindfulness*, 10(11): 2294–2301.
York, M. (2001) New Age commodification and appropriation of spirituality', *Journal of Contemporary Religion*, 16(3): 361–372.

15

RELIGIOUS AND SPIRITUAL WORLD HERITAGE SITES

Michael A. Di Giovine

Introduction: World heritage and sacred and secular travel

Sacred and secular travel seem to occupy opposite ends of the mobilities spectrum: While the former—especially pilgrimage—is often perceived as a "serious" voyage to hyper-meaningful destinations (Di Giovine & Choe 2020: 2), the latter is frequently thought of as a more superficial leisure-time activity, particularly when taking the form of mass tourism (Graburn 1977; Crick 1989). Pilgrimage sites may be considered places of transcendence and illumination, where the cosmological perfection of the sacred "irrupts" into the chaos of everyday life (Eliade 1959). In many cases, pilgrims themselves are often quick to note that their form of visitation is an anti-touristic practice and the antithesis of mass tourism (Di Giovine 2013, 2021). Yet both types of mobility move and inspire travelers, with pilgrimage, as one of the oldest and most significant forms of human mobility in history, sometimes moving millions at a time to participate in sacred rituals (Di Giovine & Elsner 2016), while according to the United Nations World Tourism Organization (UNWTO), tourism (including those of a religious nature) move over 1.2 billion people annually. Of this number, nearly half travel for cultural purposes, visiting destinations of historical, artistic, and scientific importance (UNWTO 2018: 9). Heritage sites are prime destinations for cultural tourists, as they are touchstones for communities' identities and values and connect members through time and space (Harrison 2013; Di Giovine & Cowie 2014), and in the UNWTO's survey nearly all respondents explicitly mentioned heritage sites in their definition of "cultural tourism" (UNWTO 2018: 9–10). The relatively recent phenomenon of World Heritage, which decouples sites of local importance and re-contextualizes them as places of "universal value" for the whole of humanity, has become particularly important destinations (UNWTO 2015: 26). At once sacralized (MacCannell 1976: 43–45) and heritagized (Harrison 2013: 79–84), they too are often understood to be hyper-meaningful, mobilizing not only millions of travelers per year but also ideas, imaginaries, and financial and intellectual resources for the purposes of preservation (Di Giovine 2009a). It is therefore little wonder that roughly a quarter of over one thousand unique cultural and natural World Heritage sites are designated for their spiritual or religious value (UNESCO n.d.), and many of the original properties on the World Heritage List were already well-established pilgrimage sites in Europe and Asia (UNESCO 1994a).

Thus, World Heritage sites occupy a liminal midpoint between the polarities of pilgrimage and tourism destinations: they are hyper-meaningful and hyper-valorized places that often assume a sacred air (in discourse if not in practice); they draw secular tourists and sacred visitors; and many are touchstones of identity formation irrespective of one's nationality. This dynamic, however, presents specific challenges to the identification, conservation, and utilization of sacred World Heritage sites, especially as diverse stakeholders imbue them with different meanings and engage in qualitatively different forms of interaction, including localized devotional practices, mass pilgrimage to a religion's sacred center to create solidarity, personal visitation to effect miraculous interventions, educational travel, missionary work, and cultural tourism (among other forms). It therefore becomes incumbent on scholars of spiritual and religious travel to understand World Heritage's processes, aims, stakeholder groups, and oft-conflicting meanings.

Serving as a reference guide for scholars and practitioners of both sacred tourism and heritage, this chapter explores the United Nations' Educational, Scientific and Cultural Organization's (UNESCO) "flagship" World Heritage program (Rössler 2006; Rössler & Cameron 2013). It begins with the history of the concept of World Heritage and the development of its short-, medium-, and long-term goals, which, it is argued, are intimately tied to the particular challenges faced by sacred sites and their communities during the violent upheavals of the twentieth century. Drawing on an increasingly more robust scholarly literature on World Heritage, as well as primary source documents produced by UNESCO and its affiliates in the half-century since the program's foundation, the chapter details the process of nomination and inscription of previously local sites of spiritual (and secular) import into sites of "universal value." It then explores the ways in which sacred sites are recognized and valorized as World Heritage, before presenting some of the challenges and pitfalls of effectively identifying these places of a spiritual nature from the larger pool of World Heritage sites. The chapter concludes with an examination of the common critiques of World Heritage and the challenges inherent in preserving and managing sacred World Heritage sites, and a discussion of the ways in which UNESCO and its affiliates are beginning to address them.

World heritage: origin and intended outcomes

Established through the 1972 *Convention Concerning the Protection of the World Cultural and Natural Heritage*, World Heritage sites are places primarily of local or national importance which have been designated by UNESCO as possessing "outstanding universal value" that transcends the interests of specific communities. Considered the "heritage of humanity," they are highly valued for the ways in which they reveal cultural interchange and diversity and have withstood the destructive forces of humans and nature. The *Revised Operational Guidelines*—which outlines the Convention's policies and procedures—defines "outstanding universal value" as

> cultural and/or natural significance, which is so exceptional as to transcend national boundaries and to be of common importance for present and future generations of all humanity. As such, the permanent protection of this heritage is of the highest importance to the international community as a whole.
>
> *(UNESCO 2005: 24)*

The logic that a site of universalized value leads to a universalized sense of stewardship that transcends national interest adheres to the Convention text, in which "the States Parties to this Convention recognize that such heritage constitutes a world heritage for whose protection it is the duty of the international community as a whole to co-operate" (UNESCO 1972: 1).

However, although these foundational documents seem to emphasize the safeguarding of tangible heritage, scholars (Di Giovine 2009a, 2010; Hall 2011; Jokilehto 2012; Meskell 2018) have argued that UNESCO's intent is more far-reaching: awareness of, active engagement with, and preservation of these universally valorized properties may foster utopian ideals of "peace in the minds of men [sic]," UNESCO's foundational principle (UNESCO 1945). Complementing the more formal, top-down diplomatic approaches of the United Nations, this is a grassroots method of cultivating peaceful intercultural communication.

UNESCO was born out of the horrors of the Second World War, which saw widescale destruction of important historic sites and landscapes. Many of these were monumental religious structures, such as synagogues in Berlin, cathedrals in Cologne and Dresden, and temples in Tokyo, Hiroshima, and Nagasaki. The looting of important artifacts (many of which were of religious subjects), both before and during the war, was also rampant despite several international Conventions concerning the illicit antiquities trade ratified in the early twentieth century. And importantly, the war brought the culmination of ethnoreligious persecution and scapegoating in the form of the Holocaust in Nazi Germany. Raphael Lemkin, the architect of the United Nations' *Convention on the Preservation and Punishment of the Crime of Genocide* (1948), specifically linked the destruction of tangible heritage and devotional sites as a common tactic of genocide—to erase a people requires erasing evidence of their existence through history—and, invoking not only the Jewish population of his native Poland but also the Ottoman Empire's genocide against Christian Armenians around the First World War, urged "protecting both the bodily and cultural integrity of persecuted groups" (Bevan 2016: 209–210; see also Power 2002: 43). Lemkin was ultimately unsuccessful in inserting the safeguarding of cultural heritage into the Genocide Convention, although The Hague Conventions link heritage protection with humanitarian law and the Geneva Conventions address the protection of civilians in warfare to include language limiting the unnecessary destruction of heritage properties in times of war (see Di Giovine 2020). The 1954 Hague Convention would later criminalize the destruction of cultural property, as would the 1996 Rome Statute of the International Criminal Court.

The decades following the Second World War saw more highly publicized destruction of cultural property. On the one hand, technological advances in warfare translated into even larger-scale damage to cultural (and religious) sites during the Cold War. For example, in 1970, U.S. President Richard Nixon was persuaded to halt the bombing of Mỹ Sơn Sanctuary [designated a World Heritage site in 1999[1]]—the sacred citadel of the ancient Hindu Cham people in central Vietnam—during the Vietnam War (Di Giovine 2009a: 229–230), though hostilities by both the Viet Cong and the Americans leveled much of the future World Heritage site of Huế [1993]. On the other hand, highly publicized natural and man-made environmental factors mobilized the international community to protect the mass cultural tourism sites of Abu Simbel, Egypt [1979] and Venice, Italy [1987]—both of which contain sacred elements. UNESCO successfully raised $80 million from 40 nations and private donors to relocate Pharaoh Ramses II's (1303–1213 BC) Nubian temples—brick by brick—to a man-made mountain in order to save them from destruction during the construction of the Aswan High Dam at the first cataract of the Nile. UNESCO called the process "a triumph of international solidarity" (UNESCO 1982) that revealed the strong emotional relationship that such heritage properties, and the prospect of their destruction, could exert on the international community.

Thanks in part to the efforts to save Abu Simbel in 1965, a Conference on International Cooperation, convened by US President Lyndon Johnson at the White House in Washington, DC, called for the creation of a World Heritage Trust to engage the international community in the preservation of exemplary sites "for the present and future of the entire world citizenry" (quoted in UNESCO 2008: 7; see Cameron and Rössler 2013: 4–8). The idea gained traction the following year, when UNESCO mobilized the international community to "Save Venice" after the historic flooding of Venice and Florence in 1966. In 1968, the International Union for the Conservation of Nature (IUCN) followed suit in adopting a similar framework for its membership. In 1970, US President Nixon, who was concerned with his environmental legacy, supported the adoption of an international plan of conservation that was similar to that of the US National Parks, in that land would be specially set aside and managed by a supra-local organization. These proposals were combined in 1972, when delegates to the United Nations Conference on Human Environment in Stockholm called for a new Convention that could better ensure the safeguarding and management of cultural and natural properties. This became the World Heritage Convention, which created both a World Heritage Committee to oversee the general body of State-Party delegates to the Convention as well as a small World Heritage Fund comprised of mandatory donations from each member state.

Although the Convention is explicitly concerned with preservation, it gives the World Heritage Committee little direct power to safeguard a site. Rather, the State-Party agrees to assume the responsibility for preserving the property. However, the Committee has limited resources to act should the responsibility not be upheld. In the case of a State-Party's dereliction of duty, the Committee's main recourse is to inscribe a property on the List of World Heritage in Danger and eventual de-listing. An example of this occurred in 2009 when the Dresden Elbe Valley (2004)—a cultural landscape that includes a number of religious structures including a reconstructed Dresden Cathedral (which was destroyed at the end of the Second World War)—was delisted after the civic government built a series of modern bridges across the picturesque Elbe River.

Implicit in the Convention text, and in subsequent publications sponsored by UNESCO and the World Heritage Centre, are several interlinked short-, medium-, and long-term objectives (Di Giovine 2018b). The development and publication of the World Heritage List and the World Heritage in Danger List are "fundamentally acts of knowledge creation and dissemination"—short-term "awareness-raising campaigns [that] will inspire international conservation and tourism efforts in the medium-term" (Di Giovine 2018b: 2). While visiting and preserving a site go relatively hand-in-hand—after all, a well-preserved and well-managed site is often a more desirable destination, and tourism development is often invoked (many times misguidedly) as a vehicle for generating income earmarked for preservation—UNESCO has traditionally held an ambivalent stance towards tourism. The World Heritage Convention does not mention "tourism" (or pilgrimage) at all, and the phrase "tourist development projects" appears only among a list of possible threats that could provide the basis for placing an already-inscribed property on the World Heritage List in Danger (see UNESCO 1972: 6). The Convention was ratified shortly after the foundation of the United Nations' World Tourism Organization (UNWTO), which emerged from the International Union of Tourism Officers and was (and continues to be) occupied with the promises of tourism development as a means of modernization and integration into global markets (see Jafari 1975). However, UNESCO's own research at the time (de Kadt 1976) found that many of their positive predictions did not materialize. For example, tourism was benefiting international tour operators and development companies more than locals, did not stave off locals' "flight from the land" into crowded cities or abroad, and funds generated through tourism were not staying in the destination's country. Meanwhile, early tourism theorists also argued that tourism creates inauthenticity, museumification, stress, and socio-cultural imbalance akin to imperialism (cf. MacCannell 1976; Nash 1977; Murphy 1985: 1, 3). This is particularly an issue when a living spiritual destination is heritagized—valued for its past historical properties without recognizing present communities' contemporary uses (see Isnart & Cerezales 2020)—and thus is required by UNESCO to maintain historic elements that are not adaptable to the contemporary needs of devotees, as Joy (2012) shows in her study of the Grand Mosque in Djenné, Mali [1988].

It was only at the turn of the millennium that the Convention found resonances in the UN's Millennium Development Project and the "sustainable tourism" turn of the mid-1990s (see Stronza 2001). In 2001, the World Heritage Committee founded what would become the World Heritage Sustainable Tourism Programme, whose aims would be to "engage in dialogue and actions with the tourism industry to determine how the industry may contribute to help safeguard these precious resources" (UNESCO 2001: 63, 2010). This led to more direct engagement with the UNWTO, and in 2015, representatives from the two organizations met in Siem Reap, Cambodia—the site of the great Angkor temple complexes [1992]—to discuss a "policy and governance framework necessary to foster a new collaboration model between tourism and culture" (Brooks et al. 2016: 12). This culminated in the joint issue of the *Siem Reap Declaration on Tourism and Culture* (UNESCO/UNWTO 2015), and a second joint meeting of the two organizations in Oman during the United Nation's Year of Sustainable Tourism for Development in 2017.

In line with some globalization theories' emphasis on individuality over that of collectivities, in which one's imagination negotiates between individuals and "globally defined fields of possibility" that transcend traditional geopolitical borders (Appadurai 1996: 31; cf. Robertson 1992: 8), UNESCO seems to posit that "peace in the minds of men" can emerge through the reordering of individuals' sense of place. Rather than identifying exclusively with sites of local or national interest, individuals may also entertain a "common recognition and identification with the world's shared cultural heritage" as exemplified in these monuments (Di Giovine 2009a: 34) and enacted through collective participation in their preservation and touristic visitation. This notion is made most explicit in UNESCO's 2001 *Universal Declaration of Cultural Diversity*, which, defining culture as the "distinctive spiritual, material, intellectual, and emotional features of a society or group," argued that "respect for the diversity of cultures…[and] an awareness of the unity of humankind, and of the development of intercultural exchanges…are among the best guarantees of international peace and security" (UNESCO 2002b: 1). Ratified shortly after the September 11 terrorist attacks on New York and Washington, DC, and the Taliban's destruction of the Bamiyan Buddhas earlier that year, the Declaration further acknowledges that UNESCO's "specific mandate" within the UN system was to "ensure the preservation and promotion of the fruitful diversity of cultures…as necessary for humankind as biodiversity is for nature" (2002b: 1–2). Acknowledging that "diversity is the essence of identity" (Matsura 2002a: 3), the Declaration "makes it clear that each individual must acknowledge not only otherness in all its forms but also the plurality of his or her own identity, within societies that are themselves plural" (Matsura 2002a: i). By simultaneously celebrating the differences that mark human life and yet positing some unanimously recognizable (and valued) universal culture, UNESCO proposes a peaceful world system based on the structural unity of difference, which Di Giovine (2009a, 2010, 2018a) terms a world "heritage-scape."

This ultimate objective is clearly utopian, and scholars with intimate connections to World Heritage decision-making at the bureaucratic level argue that UNESCO has "fallen short" of these goals, in part because of nationalistic and Eurocentric ideals (Smith 2006; Meskell 2018). Indeed, as early as 1994, UNESCO found that monumental religious sites, particularly in Europe, were over-represented, in part because they conformed more closely to Western aesthetic and political sensibilities. Indeed, the soaring cathedrals of Europe, filled with inestimably valued sculpture and paintings from Western-dictated art historical canons, were already embedded in national and international conceptualizations of monumentality with a majority of them already highly visited pilgrimage and tourism sites. Similar material religious and archaeological structures in the Near East, North Africa, and Asia also were represented, in part because they conformed to Western-accepted notions of "religion" and "art." However, countries that had structures and landscapes that were less monumental, or valued for their immaterial and spiritual qualities, had a difficult time applying based on the established World Heritage criteria. This remains an issue for UNESCO despite the passing of its *Global Strategy for Representative, Balanced, and More Credible World Heritage List* (UNESCO 1994a) that directs the World Heritage Committee to broaden its notion of universal value, monumentality, and authenticity. Indeed, critics continue to show how World Heritage marginalizes indigenous and descendent communities (Colwell-Chanthaphanh & Ferguson 2007; Carter 2010; Di Giovine 2017), both in how their cultural and spiritual sites are defined under UNESCO's rubric and in the ways through which they are (dis)engaged in decision-making, preservation, and tourism development processes.

Furthermore, although the *Global Strategy* also gives preference to serial and trans-boundary properties—those that span more than one geographically defined place or nation-states—in order to foster more international cooperation and peace, properties continue to be listed by country in which they are located geographically. This is the case as well for pilgrimage trails like the Routes of Santiago de Compostela—the *Camino Francés* and Routes of Northern Spain [1993], designated seven years after the endpoint, Santiago de Compostela [1985] was inscribed. Critics, however, point out that this continues to cause international competition and may exacerbate pre-existing political tensions. For example, the Khmer temple-mount of Preah Vihear, which straddles a contested border between Thailand and Cambodia, was listed as Cambodia's property in 2008, leading to uprisings in Thailand that helped topple Prime Minister Thaksin Shinawatra's regime and causing skirmishes between the two countries' armies at the border. Because of its extremely contested nature, the Old City of Jerusalem [1981], the center of the Abrahamic tradition that is claimed by both Israel and Palestine, was proposed by Jordan and is the only property listed without a State-Party. But for changing political reasons, The Birthplace of Jesus: Church of the Nativity and Bethlehem Pilgrimage [2012] in the still-contested West Bank of Jerusalem was inscribed under Palestine, providing the first formal international recognition of the country by UNESCO. This recognition caused the United States and Israel to withdraw from UNESCO and from contributing dues. Palestine has since inscribed two other properties, in part to show its autonomy and integration into the world stage.

Further arguing that the Convention is a toothless document that does not give UNESCO power over States-Parties (see O'Keefe 2004), critics also point to UNESCO's inability to stave off the recent destruction of a number of religious sites such as the Bamiyan Buddhas in Afghanistan; Sufi shrines in Mali; Biblical sites in Nineveh sacred to Yazidi, Christians, and Shiite Muslims; Muslim sites in India by Hindu nationalists; and the ancient city of Palmyra—all of which were committed precisely because their highly valued status with

UNESCO would demonstrate their power to contest Euro-American hegemony and draw followers to their radical regimes (see Di Giovine & Garcia-Fuentes 2014: 2).

Process and criteria of inscription

A place of local interest (religious or otherwise) is converted into a site of "universal value" through a complex, political, and bureaucratic process that involves national and supranational entities. With only a few notable exceptions, the process is initiated by a State-Party, a nation-state that is a signatory to the 1972 World Heritage Convention in good standing. Each State-Party compiles and consistently updates a Tentative List, an inventory of sites that it intends to nominate. Two exceptions to this process were highly vulnerable sacred sites in "failed states": the Angkor Archaeological Park [1992] was designated after the United Nations took over the post-genocide transition of Cambodia, and the remains of the Bamiyan Buddhas destroyed by the Taliban were inscribed during the U.S. war in Afghanistan [2003].

Once a "property" is on the country's Tentative List, a Nomination File can be opened. Compiled in collaboration with one or more of UNESCO's expert advisory bodies, the Nomination File is a comprehensive set of documents that makes the case for the place's outstanding universal value by showing how it meets one or more of the World Heritage Criteria established in the 1972 Convention text and identified in the first *Operational Guidelines*, passed in 1978. The World Heritage Criteria consists of a set of ten typologies, or idealized categories, which are used to evaluate, and offer the conditions for, a site's inscription. While originally separate for cultural and natural sites, these criteria were merged in 2003 and went into effect in the 2005 *Revised Operational Guidelines*. The criteria's wording is extremely important for assimilating disparate sites into a cohesive List and for crafting the universalizing narrative individual sites possess. Instead of indicating concrete categories, the criteria are purposefully vague, so as to be applicable to a wide variety of places across the globe. As Di Giovine (2018b: 3) argues,

> Rather than *being* something definite, these sites are intended to *represent*, *exhibit*, *bear testimony to*, *exemplify*, or be *tangibly associated with* generalized forms, cultural traditions, ideas, and human abilities that are loosely defined and easily interpreted in multiple ways. They are, in a way, empty signs whose values and meanings can be filled in by different parties throughout time.

The criteria are as follows:

1 Represent a masterpiece of human creative genius.
2 Exhibit an important interchange of human values, over a span of time or within a cultural area of the world, on developments in architecture or technology, monumental arts, town-planning, or landscape design.
3 Bear a unique or at least exceptional testimony to a cultural tradition or to a civilization that is living or which has disappeared.
4 Be an outstanding example of a type of building, architectural or technological ensemble, or landscape which illustrates (a) significant stage(s) in human history.
5 Be an outstanding example of a traditional human settlement, land use, or sea use which is representative of a culture (or cultures) or human interaction with the environment, especially when it has become vulnerable under the impact of irreversible change.

6 Be directly or tangibly associated with events or living traditions, with ideas, or with beliefs, with artistic and literary works of outstanding universal significance. (The Committee considers that this criterion should preferably be used in conjunction with other criteria).
7 Contain superlative natural phenomena or areas of exceptional natural beauty and aesthetic importance.
8 Be outstanding examples representing major stages of earth's history, including the record of life, significant ongoing geological processes in the development of landforms, or significant geomorphic or physiographic features.
9 Be outstanding examples representing significant on-going ecological and biological processes in the evolution and development of terrestrial, freshwater, coastal and marine ecosystems, and communities of plants and animals
10 Contain the most important and significant natural habitats for in situ conservation of biological diversity, including those containing threatened species of outstanding universal value from the point of view of science or conservation.

The Nomination File also includes the technical and managerial aspects of the property; it documents the site's history and development, outlines its specific geopolitical boundaries and "buffer zones"—which, in the case of religious tourism destinations, may or may not include the actual pilgrimage route—and details its authenticity and integrity. Although authenticity and integrity are slippery terms (see, for example, De Cesari 2010; Labadi 2013: 113–126), for UNESCO they are technical qualities that ensure that the site's value is "truthfully and credibly expressed" through its form, design, the materials of its construction, and its functionality. Outlined in the Venice Charter (1964) and subsequently modified in the Nara Declaration of Authenticity (UNESCO 1994b) and the Cairns Charter (UNESCO 2000b), these technical qualities refer to the physical completeness of the site, its preservation, and the level of human intervention subsequent to the time period that the site represents (UNESCO 1978: 4). These later charters broaden the Eurocentric notion of authenticity as "original as opposed to counterfeit" (Bendix 1992: 104; Jokilehto 1999: 296; Labadi 2013: 114; Winter 2014) to acknowledge the very fluid and socially constructed nature of authenticity. For example, much to the chagrin of preservationists, Buddhist devotees in Laos' Luang Prabhang [1995] allow statuary and artwork to naturally decay as a metonym for the impermanence of life, and will also dismantle and repurpose centuries-old temples as a means of magnifying and transferring merit to a new living devotional structure (Karlström 2015). Nevertheless, emerging nation-states have frequently adopted these Western, cosmopolitan meanings and practices. The nomination documents for Mt. Wutai [2009] in China, for example, make explicit that while the sacred landscape and Buddhist temples have been changed over time, many remain identical to the time when they were built (Labadi 2013: 114).

Yet there is fluidity to this as well. France's Notre Dame cathedral is inscribed in the World Heritage site of Paris, Banks of the Seine [1991], which broadly reveals "the evolution of Paris and its history." Although explicitly identified as an "architectural masterpiece of the Middle Ages," the cathedral's juxtaposition with the Renaissance Pont Neuf, French classical Louvre, and the modern Eiffel Tower seems to allows for its statement of value to encompass not only its original Gothic construction but also a modern spire constructed in 1859 (see http://whc.unesco.org/en/list/600). However, the spire and its medieval ceiling burned to the ground in the fire of 2019, and currently, authorities are debating whether it should be rebuilt in the same style or be radically different. These reconstruction decisions could affect its inscription. This is not to say that subsequent interventions cannot be made—especially if they are restorative—but rather that these modifications ensure truthfulness and intactness after a catastrophic event. The Basilica of St. Francis of Assisi [2000] is a case in point. Designated not only for its centrality to the cult of the founder of the Franciscan Order (and Patron Saint of Italy), the Basilica is also inscribed for Giotto's renowned Renaissance fresco cycle, *The Life of St. Francis*—which is generally understood to signal the birth of Renaissance art (UNESCO 2000b)—as well as other works by Cimabue and Simone Martini, among others. However, during the central Italian earthquake of 1997, the Upper Basilica containing Giotto's frescoes was significantly damaged, rendering the entire structure unstable. Italian authorities were unusually swift and exacting in its reconstruction. In just three years, Giotto's frescoes were painstakingly reconstructed using original paint flecks from the ruins, and the entire building was earthquake-proofed with cutting-edge technology used by NASA (Gayford 1999). UNESCO lauded the art restoration effort's authenticity and integrity despite such major damage (ICOMOS 2000).

Once the Nomination File is complete, the appropriate Advisory Body will provide its own assessment. Comprised of expert-members who are international scholars and professionals with expertise in conservation, these NGOs are independent from the nation-state, although they may provide technical assistance to the State-Party during the file's compilation and will sometimes provide reactive monitoring after the site is designated. The International Council of Monuments and Sites (ICOMOS) evaluates cultural heritage sites and their proposed management while the International Union for the Conservation of Nature (IUCN) evaluates proposed natural sites. Should a site's material conservation need to be assessed—as, for example, the early Buddhist cave-paintings of Ajanta [1983] and Ellora [1983] in India—the International Centre for the Study of the Preservation and Restoration of Cultural Property (ICCROM) will also be solicited for input. As Di Giovine argues (2009a: 199), their written report is less an act of evaluation as it is of idealization, for it effectively changes the property into an idealized material form of UNESCO's metanarrative of "unity in diversity."

Once the File is complete, it is presented at the annual meeting of the World Heritage Committee, which is composed of a rotating group of 21 representatives of state-parties elected from the General Assembly. The Committee examines each nomination based on the technical evaluations and votes to inscribe or reject a property. They may also vote to defer judgment on a site, requesting additional information from the State-Party. Along with a site's inscription, the Committee confirms the textual wording of the site's designation, which is often a slight modification of the Advisory Body's statement of the site's "outstanding universal value." The importance of this act should not be underestimated—it is often subject to politicking as other parties struggle to determine the site's specific narrative, which translates into the ways in which the site is conserved and packaged for touristic consumption (Harrison & Hitchcock 2005; Di Giovine 2009a, 2010; Harrison 2010). Finally, the World Heritage Committee inscribes the new World Heritage site's name and geographic location on the List, along with other similarly valorized places. Importantly, despite the recognition of their universal value, each site is listed under the country (or countries) which proposed its nomination and within whose geopolitical boundaries the sites physically lie.

Sacred world heritage sites and problems with religious categorization

According to UNESCO (n.d.), approximately 20% of the properties on the World Heritage List have "some sort of religious or spiritual connection." However, it is often a difficult and rather fruitless enterprise to definitively tease out these components, since the notion of the sacred itself is fluid, changeable, and "ambiguous" (Hobart & Zarcone 2017). Varying from group to group and even individual and individual, these sites gain and lose their devotional importance over time and do not necessarily need to be "authorized" by a religious authority to be a locus of devotion and hyper-meaningful travel (Di Giovine & Choe 2020). Furthermore, not all cultures perceive of religion as separate from daily life, or even separate from the natural environment. As such, not all spiritual sites conform to Western notions of monumentality. Finally, World Heritage sites in general may become loci of pilgrimage and hyper-meaningful travel even if not designated as spiritual or religious in nature. Roots or heritage travel to Ghana's coastal slave forts and castles [1979]; or visits to Robben Island [1999], the site of anti-Apartheid Nelson Mandela's imprisonment; or to the Nazi concentration camp at Auschwitz-Birkenau [1979] have all been considered pilgrimages by theorists and travelers themselves (see Ebron 1999, Shackley 2001, Feldman 2008).

Nevertheless, properties that are explicitly recognized by UNESCO for their spiritual or religious qualities most often fall under the first seven criteria reserved for cultural sites. Criterion i ("masterpiece of creative genius") is most often invoked, thanks in part to the impetus amongst many societies to produce monumental religious structures that have lasted the tests of time, such as the great Catholic and Orthodox Christian cathedrals of Europe; intricately carved Hindu, Jain, Shinto and Buddhist temples of Asia; mosques in the Middle East, Africa, and Asia; pre-Colombian temple-citadels of the Maya, Aztecs, and Inca in Latin America; ancient petroglyphs in Africa, Central Asia, and Northern Europe; or the archaeological remains of ancient devotional sites in the Middle East and North Africa (many archaeological sites may alternatively be categorized under criterion iii). Criterion iv, which focuses on the technological skill of construction—also is used to designate a number of religious or spiritual sites, especially if they respond to particular movements in history, or if they show notable integration with the landscape, such as Bangladesh's Historic Mosque City of Bagerhat [1985], which was created out of the "piety of Khan Jahan" and reveals a "perfect mastery of the techniques of planning" that allowed it to be situated in "the impenetrable mangrove swamps of the Sunderbans" (see https://whc.unesco.org/en/list/321). Another site, Bolivia's Jesuit missions of the Chiquitos [1993]—six mission settlements of Christianized natives designed by the Catholic Jesuit Order—were inscribed under this criterion for the ways in which the Jesuits were inspired by the "ideal cities" envisioned by Renaissance humanists and adapted for the landscape and indigenous practices (see https://whc.unesco.org/en/list/529). This may also be combined with criterion v, especially when they are threatened by culture change. Thus, the Jesuit Missions were also inscribed for the vulnerability of its religious art "under the impact of changes that threatened the Chiquitos populations following the agrarian reforms of 1953" (ICOMOS 1990: 61).

Criterion vi has also been applied to spiritual or religious sites, though as the second sentence in the criterion intimates, this is a rather complicated typology because living traditions, beliefs, or ideas may not be readily evident in built material culture, and therefore is usually combined with another criterion that speaks to some recognizable or metonymic aesthetic quality. One European example is The Luther Memorials in Wittenberg and Eisleben [1996], which, coupled with criterion iv, concretize the watershed moments of the Protestant Reformation. Criterion vi is invoked by UNESCO, in that the site bears "unique testimony to the Protestant Reformation, one of the most significant events in the religious and political history of the world, and constitute exceptional examples of 19th-century historicism" (see https://whc.unesco.org/en/list/783).

There are, however, other instances in which classifying a spiritual World Heritage site is difficult. Many properties are not singular constructions but "ensembles" (criterion ii), which include multiple spiritual or religious constructions mixed with civic structures. A case in point is Cidade Velha, Historic Centre of Ribeira Grande [2009], on the tiny island country of Cabo Verde off Western Africa, which was the first European colonial outpost in the tropics. Boasting two Catholic churches, among other structures, the site is not recognized for its spiritual qualities *per se*, but rather is recognized for its "considerable role in international trade associated with the development of European colonial domination towards Africa and America and the birth of Atlantic triangular trade" (see https://whc.unesco.org/en/list/1310). However, these ensembles also include the great urban centers of Europe, Asia, and the Americas, such as Venice and its Lagoon [1987], which includes the Byzantine St. Mark's Cathedral and the Doges' Palace, its many canals, and all of the other palazzi on some 118 islets; the imperial Vietnamese capital of Huế [1993], which includes the Perfume Pagoda and Buddhist temples on the picturesque Perfume River and also the Imperial Purple City, its military citadel, and the tombs of its emperors spread across a vast jungle; and the Historic Center of Morelia [1991], which was designated as "an outstanding example of urban development combining town planning theories of Spain and the Mesoamerican Experience." The Historic Center of Morelia includes over 200 distinct sites of importance, including 21 churches and 20 civic constructions (http://whc.unesco.org/en/list/585). Thematic ensembles do not necessarily have to be located in close proximity to each other; these are called "serial properties," such as The Twentieth Century Architecture of Frank Lloyd Wright [2019], which includes not only his famous Guggenheim Museum in New York and residential constructions such as Falling Water in Pennsylvania but also the Unity Temple—a Unitarian church constructed in Oak Park, a suburb of Chicago. While the Frank Lloyd Wright serial property was designated only under criterion iii ("bear exceptional testimony to a cultural tradition"), another similar one, The Architectural Work of Le Corbusier, an Outstanding Contribution to the Modern Movement [2016]—which encompasses sacred and secular buildings in six countries in three continents, including the modernist Chapel of Notre-Dame du Haut in Ronchamp, France—was categorized under criteria i, ii, and iv.

Another problem with identifying the spiritual nature of World Heritage sites is that many cultures do not distinguish between sacred and secular. "Religion" itself is a Western construct that was developed primarily in the colonial era to make sense of monumental places and spiritual practices in non-Western regions (Smith 1998: 270, 275–280). Even in the Western tradition, the religious and civic were often mixed. For example, the pan-Hellenic sanctuary of Delphi [1987], considered to be the "navel of the world," was the religious center for ancient Greece, and ancient Athenians would undergo a yearly civic pilgrimage called *theoria* to hear the oracle of Apollo speak (Rutherford 2000). Likewise, Vatican City [1984] was (and continues to be) a political entity as well as the center of Roman Catholicism.

Although unique landforms often hold great spiritual value for many sacred travelers today, and have traditionally been places of cult devotion for many cultures throughout history (see Eliade 1959: 37–41), UNESCO has more difficulty in applying the last four criteria—developed for "natural" sites—to sacred places. Indeed, the criteria regarding natural sites rests on the materiality of the environment—such as geomorphic features, rare biota, or aesthetically unique landscapes—and do not include language on intangible qualities. Rather, they are coupled with one or more cultural criteria as a "mixed" site, revealing both human and natural interface. Papahānaumokuākea [2010], a set of several low-lying atolls in Hawaii, is a case in point. Although designated as "a very significant testimony of hotspot volcanism," as with other Polynesian cultures, it is precisely these continuous volcanic processes that have provided the spiritual component to these peoples, and at Papahānaumokuākea, UNESCO specifies that the islets are dotted with the "well preserved heiau shrines" that are not only associated with 3,000-year-old Polynesian marae-ahu culture but are living sites of pilgrimage today (see https://whc.unesco.org/en/list/1326). UNESCO has also recognized the spirituality of sites that nevertheless may not contain material cult structures, such as the Palau's Rock Islands Southern Lagoon World Heritage site. In its inscription, UNESCO points out that the 445 uninhabited volcanic and limestone islands, with their "unique mushroom-like shapes in turquoise lagoons surrounded by coral reefs," played an important role in the spirituality of the indigenous peoples "through oral traditions that record in legends, myths, dances, and proverbs, and traditional place names the land- and seascape of their former homes" (see https://whc.unesco.org/en/list/1386).

However, it is often difficult to tease out cultural and spiritual qualities of a site primarily valued for its natural landscape, especially by technical experts in material preservation. The Tasmanian Wilderness [1982] is a case in point. When Australia submitted its Nomination File, it made the argument—supported by archaeological evidence—that the natural landscape has been an important cultural and spiritual feature for Aboriginal groups. IUCN evaluated quite positively its natural qualities, but ICOMOS did not do the same for its intangible cultural dynamics. In its original evaluation (ICOMOS 1982: 1–2), the experts acknowledged evidence of changing land-use by aboriginal peoples, but that they "do not, in themselves, offer sufficient reason to inscribe this cultural property on the World Heritage List but simply constitute new positive evidence which supplement the already favorable report by IUCN." ICOMOS subsequently re-evaluated the site twice more but continued to remain skeptical and urged managers to consult local Tasmanian aboriginal authorities for better documentation. It was nevertheless eventually inscribed under several cultural categories by UNESCO, testifying to the Committee's acknowledgment that for many indigenous groups, it is difficult to map intangible spiritual practices of the past onto the tangible environment.

But more recently UNESCO and its advisory bodies have come to acknowledge that nature/culture and tangible/intangible dichotomies are themselves Western constructs, and many societies, especially indigenous groups, do not make such distinctions. There are many natural sites with little human intervention that hold great cultural importance to communities, such as the spiritual places like Uluru-Kata Tjuta [1987] in Australia or Mount Taishan [1987] in China, while built cultural sites are also often part of a wider natural environment, such as Meteora [1988] and Mount Athos [1988] in Greece or the sanctuary of Machu Picchu [1983] in Peru. Consequently, around the turn of the millennium, UNESCO largely eschewed this mixed categorization, preferring to apply the more fluid distinction of a "cultural landscape" (Rössler 2006; Mitchel, Rössler & Tricaud 2009), as applied to Göreme National Park and the Rock Sites of Ancient Cappadocia [1985] in Turkey, where early Christians in Anatolia sought refuge amongst its unique "fairy chimneys" landforms (tall, thin spires of rock also known as hoodoo) and carved cave-churches and homes into the mountains.

Ultimately, however, since the process is a political one, similar properties can be interpreted differently. For example, the Maloti-Drakensberg Park [2000], which spans South Africa and Lesotho, was inscribed for both its natural and cultural qualities. Not only does this site harbor a number of endangered endemic wildlife, such as the Cape and bearded vultures, and the Maloti minnow (a "critically endangered fish species only found in this park"), but also contains "spectacular...caves and rock-shelters with the largest and most concentrated group of paintings in Africa south of the Sahara. They represent the spiritual life of the San [hunter-gatherer] people, who lived in this area over a period of 4,000 years" (see https://whc.unesco.org/en/list/985). This differs from a similarly described site, Botswana's Tsodilo [2001], that is also inhabited by the San people. Considered "the Louvre of the Desert" for its over 4,500 prehistoric rock-art images that are revered by local communities in the Kalahari Desert as an abode of the spirits, it is only classified as a cultural site despite the integral nature of the landscape, which is nevertheless vividly described in its documentation (see https://whc.unesco.org/en/list/1021).

A property's understood sacred nature and religious affiliation may also change over time and is often subject to political control. Angkor Wat, for example, was originally constructed as a Hindu temple to Vishnu but was converted into a Buddhist temple by Jayavarman VII, a convert, 30 years later. Turkey's Hagia Sofia is a more contemporary case in point. Originally constructed as the main cathedral of Orthodox Christian Byzantium, it was briefly a Roman Catholic church during the thirteenth-century Crusades before being re-appropriated as a mosque in the Ottoman era, wherein its golden mosaics were plastered over and its decorated marble floors covered with carpeting. In 1935, the secular Turkish government officially declared it a museum, uncovering the marble floor and removing the plaster over many Christian mosaics. In 1985, UNESCO inscribed it as the cornerstone of its Historic Areas of Istanbul World Heritage site, which fuses Byzantine, Ottoman, and other more modern urban features (religious and secular), and which is intended to reveal the interplay of various religious and cultural traditions over time. However, in 2020, the increasingly conservative Islamic government of President Recep Tayyip Erdoğan officially declared the Hagia Sofia a mosque once again amid great outcry from the international community, including UNESCO. Although the government assured UNESCO that the Christian elements would continue to be protected and visitation would be allowed outside of prayer times, UNESCO decried the unilateral decision and the potential that it would affect physical access to the site, the structure of the building, the site's movable property, and the site's management (Olsen & Emmett 2021).

The operational value of identifying spiritual and religious sites

Despite the difficulties in identifying spiritual or religious World Heritage sites, UNESCO nevertheless finds it operationally useful in addressing the very discrepancies enumerated above. Doing so orients the World Heritage Centre and its advisory bodies to be more sensitive to places that are hyper-meaningful to communities and to strive towards better inclusivity in designating properties that may not conform to Western ideals of authenticity, integrity, and monumentality. There have been several initiatives starting with the *Global Strategy* (UNESCO 1994a) and the Budapest Declaration (UNESCO 2002a), which encouraged under-represented groups to participate and promised UNESCO's greater empowerment of local communities. The *Nara Document on Authenticity* (UNESCO 1994b), furthermore, recognized that authenticity should encompass the intangible "spirit of the place"—a term that was enshrined in the *Quebec Declaration on the Preservation of the Spirit of Place* by ICOMOS in 2008. Indeed, advisory bodies have been active in identifying the special nature of sacred places as they are ultimately responsible for making technical assessments and interventions. In 2003, ICCROM held a Forum on the Conservation of Living Religious Heritage, and two years later ICOMOS' General Assembly passed a resolution calling for the "establishment of an International Thematic Programme for Religious Heritage" (UNESCO n.d.). ICOMOS followed with another resolution in 2011 on the *Protection and*

Enhancement of Sacred Heritage Sites, Buildings and Landscapes, and in 2008, IUCN issued the *UNESCO MAB/IUCN Guidelines for the Conservation and Management of Sacred Natural Sites*, which corresponds "to the areas of land or water having special spiritual significance to peoples and communities" (McLeod et al. 2008). This interest was elevated to the international level around the same time, when in 2010 the United Nations General Assembly brought together religious authorities, States-Parties, site managers, and experts at an international seminar on the role of religious communities in the management of World Heritage properties. The resulting *Statement on the Protection of Religious Properties within the Framework of the World Heritage Convention* (UNESCO 2011) recognized the value of spiritual sites for local communities, the role played by religious communities (in conjunction with national authorities and international experts) to create, maintain and use sacred places, and the importance of dialogue between religious authorities and all stakeholders. Furthermore, The United Nations Alliance of Civilizations itself has also issued its Plan of Action to Safeguard Religious Sites, many of which are World Heritage properties (see United Nations 2019). Finally, UNESCO has supported the creation of several international expert-networks to study the unique challenges of spiritual World Heritage sites. ICOMOS formally created the Scientific Committee for Places of Religion and Ritual (PRERICO); IUCN also created the IUCN-WCPA's Specialist Group on Cultural and Spiritual Values of Protected Areas; and UNESCO has a Steering Group on Heritage of Religious Interest made up of appointed members from both these advisory bodies (UNESCO n.d.).

Conclusion

This chapter has defined and outlined UNESCO's World Heritage program, paying particular attention to the roles in which religious and spiritual sites, broadly conceived, have been conceptualized, utilized, and disputed. Although nearly 300 World Heritage properties are formally classified as religious or spiritual sites, it is a difficult and ever-evolving process to adequately represent the breadth of sacred sites and their associated pilgrimage trails. On the one hand, the World Heritage Convention has been criticized for its innate political structure, as well as its Eurocentric nature, especially in how it dichotomizes sacred/secular, nature/culture, religion/spirituality, and materiality/immateriality. On the other hand, many of the most important and highly visited pilgrimage sites have been left off the List. This sometimes stems from political issues between state and local government, such as with India's Varanasi; or from conflict and warfare—such as in Karbala, Iraq, the sacred center for Shiite Muslims; or because of a lack of interest or incentive on the part of the nation-state, such as with Mecca and Medina in Saudi Arabia; or because of the lack of will on the part of the religious community, as with the sacred Shinto shrines of Ise in Japan.

Although UNESCO acknowledges the importance of visitation to these living spiritual sites (both through pilgrimage and international tourism), it has historically treated these practices as preservation challenges rather than opportunities. These opportunities may include economic development—though this should not be taken for granted—but also more intangible aspects of well-being, such as empowerment and valorization (Di Giovine 2009b, 2017). Indeed, as pilgrimage and religious travel is one of the oldest forms of mobility in the world (Di Giovine & Elsner 2016), it has shaped peoples, places, and the environment itself. Thus, as ICCROM points out, sites of living religious and spiritual heritage "are indeed the oldest protected areas of the planet" (qtd. UNESCO, n.d.), and, UNESCO believes, if harnessed properly, may foster peaceful coexistence through the safeguarding of cultural and natural diversity.

Yet there continue to be challenges within UNESCO's utopian project, and sacred sites pose some of the greatest obstacles. Through the political nature and structure of the World Heritage process, indigenous and religious communities may be marginalized, left out, or feel that they stand to lose more than they gain. It is incumbent to not only include these community stakeholders in decision-making processes before and after inscription but to allow them to present oral histories and alternative narratives that may run counter to those crafted by technocrats and professionals. Furthermore, technical emphasis on preservation and Eurocentric notions of authenticity may also prohibit living devotional sites from changing and adapting with the times, placing undue burdens on local communities. Finally, sacred sites are some of the most contested places in the world (see Eade & Sallnow 1991)—hyper-meaningful places that serve as touchstones of identity politics. On the one hand, multiple groups may compete for control over the same site, such as Jerusalem and Angkor (and Varanasi), while on the other hand, their highly valued status makes them vulnerable to looting and other destructive forces that enact violence not only on the property itself but also on those communities who lay claim to them.

Nevertheless, religious and spiritual World Heritage sites inspire millions of travelers each year, many of whom visit for cultural or historical reasons. Likewise, historical World Heritage sites not included for spiritual reasons may nonetheless be sacralized by travelers, being hyper-meaningful places of "secular pilgrimage" (Digance 2006, Margry 2008). This is a testament to the effectiveness of UNESCO's World Heritage Program, the value-laden imaginaries it promulgates, and its ability to adapt to changing sensibilities and geopolitical challenges, even if its ultimate goal of fostering peace remains elusive.

Note

1 As the years in which particular properties were designated can provide further contextualization of the evolution of the World Heritage process, particularly as it relates to changing conceptualizations of the nature and role of the spiritual in these heritage sites, I have henceforth included the dates of their inscription in brackets after each exemplary property mentioned in this text.

References

Appadurai, A. (1996) *Modernity at Large*. Minneapolis: University of Minnesota Press.

Bendix, R. (1992) Diverging paths in the scientific search for authenticity.' *Journal of Folklore Research*, 29(2): 103–132.

Bevan, R. (2016) *The Destruction of Memory*. London: Reakton Books.

Brooks, G. et al. (2016) *UNWTO/UNESCO World Conference on Tourism and Culture: Building a New Partnership. Siem Reap, Cambodia, 4–6 February 2015*. Madrid: UNWTO. Available at: www.e-unwto.org/doi/pdf/10.18111/9789284417360 (accessed on 16 January 2018).

Carter, J. (2010) 'Displacing indigenous cultural landscapes: The naturalistic gaze at Fraser Island World Heritage Area.' *Geographical Research*, 48(4): 398–410.

Colwell-Chanthaphanh, C., & Ferguson, T. (2007) *Collaboration in Archaeological Practice: Engaging with Descendent Communities*. Lanham: AltaMira Press.

Crick, M. (1989) 'Representations of international tourism in the social sciences: Sun, sex, sights, savings and servility.' *Annual Review of Anthropology*, 18: 307–344.

De Cesari, C. (2010) 'World heritage and mosaic universalism.' *Journal of Social Archaeology*, 10(3): 299–324.

De Kadt, E. (1976) *Tourism: Passport to Development? A Joint World Bank-UNESCO Study*. Oxford: Oxford University Press.

Di Giovine, M.A. (2009a) *The Heritage-scape: UNESCO, World Heritage, and Tourism*. Lanham: Lexington Books.

Di Giovine, M.A. (2009b) 'Revitalization and counterrevitalization: Tourism, heritage, and the Lantern Festival as catalysts for regeneration in Hoi An, Viet Nam.' *Journal of Policy Research in Tourism, Leisure and Events*, 1: 208–230.

Di Giovine, M.A. (2010) 'World heritage tourism: UNESCO's vehicle for peace?' *Anthropology News*, 8–9, November 2010.

Di Giovine, M.A. (2013) 'Sacred journeys as spaces for peace in Christianity' in A. Pazos (ed.), *Pilgrims and Pilgrimages as Peacemakers in Christianity, Judaism, and Islam* (pp. 1–38). Surrey: Ashgate.

Di Giovine, M.A. (2017) 'Anthropologists weigh in on the sustainability of tourism.' *Anthropology News*, 58(4): e180–e186. August 14, 2017.

Di Giovine, M.A. (2018a) 'The heritage-scape: Origins, theoretical interventions, and critical reception of a model for understanding UNESCO's World Heritage Program. *Via@ Tourism Review*, 13. https://doi.org/10.4000/viatourism.2017

Di Giovine, M.A. (2018b) 'World heritage objectives,' in C. Smith (ed.). *Encyclopedia of Global Archaeology*, 2nd edition. Springer. https://doi.org/10.1007/978-3-319-51726-1_1915-2

Di Giovine, M.A. (2020) 'Threats to destroy culture heritage harm us all' *Anthropology News*. February 19. Accessed from https://www.anthropology-news.org/index.php/2020/02/19/threats-to-destroy-cultural-heritage-harm-us-all/ on November 12, 2020.

Di Giovine, M.A. (2021) 'Between tourism and anti-tourism: Ethics and the study abroad experience,' in J. Bodinger de Uriarte & M.A. Di Giovine (eds.) *Study Abroad and the Quest for an Anti-Tourism Experience* (pp. 281–323). Lanham: Lexington Books.

Di Giovine, M.A., & Choe, J. (2020) *Pilgrimage beyond the Officially Sacred*. London: Routledge.

Di Giovine, M.A., & Cowie, S. (2014) 'The definitional problem of patrimony and the futures of cultural heritage.' *Anthropology News*, 55(3): e54–e55, August 22, 2014.

Di Giovine, M.A., & Elsner, J. (2016) 'Pilgrimage tourism,' in J. Jafari & H. Xiao (eds.) *The Encyclopedia of Tourism* (pp. 722–724). New York: Springer Publications.

Di Giovine, M.A., & Garcia-Fuentes, J.-M. (2014) 'Sites of pilgrimage, sites of heritage: An exploratory introduction', *International Journal of Tourism Anthropology*, 5: 1–23.

Digance, J. (2006) 'Religious and secular pilgrimage: Journeys redolent with meaning,' in D. Timothy & D. Olsen (eds.) *Tourism, Religion and Spiritual Journeys* (pp. 36–48). London: Routledge.

Eade, J., & Sallnow, M. (1991) *Contesting the Sacred*. Urbana-Champagne: University of Illinois Press.

Ebron, P. (1999) 'Tourists as pilgrims: Commercial fashioning of transatlantic politics', *American Ethnologist*, 26(4): 910–932.

Eliade, M. (1959) *The Sacred and the Profane*. New York: Harcourt.

Feldman, J. (2008) *Above the Death Pits, Beneath the Flag*. Oxford: Berghahn.

Fowler, P.J. (2003) *World Heritage Cultural Landscapes*. World Heritage Papers no. 6. Paris: UNESCO World Heritage Centre.

Gayford, M. (1999) 'The miraculous rebirth of Assisi', *The Telegraph*. 27 November. Available at: https://www.telegraph.co.uk/culture/4719139/The-miraculous-rebirth-of-Assisi.html (accessed 27 June 2019).

Graburn, N. (1977) 'Tourism: A sacred journey,' in V. Smith (ed.) *Hosts and Guests* (pp. 21–36). Philadelphia: University of Pennsylvania Press.

Hall, M. (2011) *Towards World Heritage: International Origins of the Preservation Movement, 1870–1930*. London: Routledge.

Harrison, D., & Hitchcock, M. (2005) *The Politics of World Heritage: Negotiating Tourism and Conservation*. Clevedon: Channel View Press.

Harrison, R. (2010) *Understanding the Politics of Heritage*. Manchester: Manchester University Press.

Harrison, R. (2013) *Heritage: Critical Approaches*. London: Routledge.

Hobart, A., & Zarcone, T. (2017) *Pilgrimage and Ambiguity*. Canon Pyon: Sean Kingston Publishing.

ICOMOS (1982) Western Tasmania Wilderness National Parks. Advisory Body Evaluation No. 181. November 13, 1981.

ICOMOS (1990) Jesuit Missions of Chiquitos, No. 529. Advisory Body Evaluation, August 7, 1989. Paris: ICOMOS.

ICOMOS (2000) Assisi, No. 990. Advisory Body Evaluation, June 30, 1999. Paris: ICOMOS.

ICOMOS (2008) *Quebec Declaration on the Preservation of the Spirit of Place*. Québec, Canada, October 4, 2008.

ICOMOS (2011) Resolution 17GA 2011/35 on the Protection and Enhancement of Sacred Heritage Sites, Buildings and Landscapes. 17th General Assembly of ICOMOS, 27 November–2 December.

Isnart, C., & Cerezales, N. (2020) *The Religious Heritage Complex*. New York: Bloomsbury.
Jafari, J. (1975) 'Creation of the inter-governmental World Tourism Organization.' *Annals of Tourism Research*, 11: 237–244.
Jokilehto, J. (1999) *A History of Architectural Conservation*. Oxford: Butterworth-Heinemann.
Jokilehto, J. (2012) 'Human rights and cultural heritage. Observations on the recognition of human rights in the international doctrine.' *International Journal of Heritage Studies*, 18(3): 226–230.
Joy, C. (2012) *The Politics of Heritage Management in Mali*. Walnut Creek, CA: Left Coast Press.
Karlström, A. (2015) 'Authenticity,' in K.L. Samuels & T. Rico (eds.) *Heritage Keywords* (pp. 29–46). Denver: University of Colorado Press.
Labadi, S. (2013) *UNESCO, Cultural Heritage, and Outstanding Universal Value*. Lanham: AltaMira.
MacCannell, D (1976) *The Tourist*. Berkeley: University of California Press.
Margry, P.J. (2008) 'Secular pilgrimage: A contradiction in terms?' in P.J. Margry (ed.), *Shrines and Pilgrimage in the Modern World: New Itineraries into the Sacred* (pp. 13–46), Amsterdam: Amsterdam University Press.
Matsuura, K. (2002a) 'The cultural wealth of the world is its diversity in dialogue,' in UNESCO, *Universal Declaration of Cultural Diversity*. Paris: UNESCO.
Matsuura, K. (2002b) 'Preface,' in UNESCO, *Cultural Diversity: Common Heritage, Plural Identities* (pp. 1–5). Paris: UNESCO.
McLeod, C.; Valentine, P., & Wild, R. (2008). *Sacred Natural Sites: Guidelines for Protected Area Managers*. Paris: UNESCO/IUCN.
Meskell, L. (2018) *A Future in Ruins*. Oxford: Oxford University Press.
Mitchel, N., Rössler, M. & Tricaud, P. (2009) *World Heritage Cultural Landscapes*. Paris: UNESCO.
Murphy, P.E. (1985) *Tourism: A Community Approach*. New York: Methuen.
Nash, D. (1977) 'Tourism as a form of imperialism,' in V. Smith (ed.) *Hosts and Guests* (pp. 37–52). Philadelphia: University of Pennsylvania Press.
O'Keefe, P.J. (2004) 'World cultural heritage: Obligations to the international community,' *International and Comparative Law Quarterly*, 53(1): 189–209.
Olsen, D.H., & Emmett, C.F. (2021) 'Contesting religious heritage in the Middle East,' in C.M. Hall & S. Seyfi (eds.) *Cultural and Heritage Tourism in the Middle East and North Africa: Complexities, Management and Practices* (pp. 54–71). London and New York: Routledge.
Power, S. (2002) *A Problem from Hell*. New York: Basic Books.
Robertson, R. 1992. *Globalization: Social Theory and Global Culture*. London: Sage.
Rössler, M. (2006) 'World heritage cultural landscapes: A UNESCO flagship programme 1992–2006.' *Landscape Research*, 31(4): 333–353.
Rössler, M., & Cameron, C. (2013) *Many Voices, One Vision: The Early Years of the World Heritage Convention*. London: Routledge.
Rutherford, I. (2000) 'Theoria and darśan: Pilgrimage and vision in Greece and India.' *The Classical Quarterly*, 50(1): 133–146.
Shackley, M. (2001) 'Potential futures for Robben Island: Shrine, museum or theme park?' *International Journal of Heritage Studies*, 7(4): 355–363.
Smith, J.Z. (1998) 'Religion, religions, religious,' in M.C. Taylor (ed.) *Critical Terms for Religious Studies* (pp. 269–284). Chicago: University of Chicago Press.
Smith, L. (2006) *Uses of Heritage*. London: Routledge.
Stronza, A. (2001) 'Anthropology of tourism: Forging new ground for ecotourism and other alternatives', *Annual Review of Anthropology*, 30: 261–283.
UNESCO (1945) *Constitution of the United Nations Educational, scientific and cultural organization*. London, November 16, 1945. Paris: UNESCO.
UNESCO (1972) *Convention Concerning the Protection of the World Cultural and Natural Heritage*. Paris, November 23, 1972. Paris: UNESCO.
UNESCO (1978) Operational guidelines for the implementation of the World Heritage Convention (adopted by the Committee at its first session and amended at its second session). Paris: UNESCO.
UNESCO (1982) *Nubia: A Triumph of International Solidarity*. Paris: UNESCO.
UNESCO (1994a) Expert meeting on the 'global strategy' and thematic studies for a representative world heritage list. 20–22 June 1994, WHC-94/CONF.003/INF.6. Paris: UNESCO.
UNESCO (1994b) *Nara Document on Authenticity*. Nara, Japan 1–6 November 1994.
UNESCO (2000a) 24COM.XC.1 Assisi, The Basilica of San Francesco and other Franciscan Sites. *Report of the 24th Session of the World Heritage Committee*, Cairns, Australia, 27 November–2 December.

UNESCO (2000b) CONF 24 VI. Cairns Decision. *Report of the 24th Session of the World Heritage Committee*, Cairns, Australia, 27 November–2 December.
UNESCO (2001) Report of the 25th session of the world heritage committee meeting, Helsinki, 16 December 2001. Paris: UNESCO.
UNESCO (2002a) Budapest declaration. (WHC-02/CONF.202/25). Records of the 26th World Heritage Committee meeting, Budapest, Hungary 24–29 June 2002.
UNESCO (2002b) *Universal Declaration of Cultural Diversity*. Paris: UNESCO.
UNESCO (2005) *Operational Guidelines for the Implementation of World Heritage*. WHC05/2 Paris: World Heritage Center.
UNESCO (2008) *World Heritage Information Kit*. Paris: World Heritage Centre.
UNESCO (2010) 'Report on the World Heritage thematic programs' 34 COM 5F.2 in *Report of the 34th session of the World Heritage Committee*. Brasilia, Brazil July 25–August 3.
UNESCO (2011) *Kyiv Statement on the Protection of Religious Properties within the Framework of the World Heritage Convention*. Paris: UNESCO.
UNESCO (n.d.) *Heritage of Religious Interest*. Website. Accessed from https://whc.unesco.org/en/religious-sacred-heritage/ on November 12, 2020.
UNESCO/UNWTO (2015) *Siem Reap Declaration on Tourism and Culture—Building a New Partnership*. Siem Reap, 5 February 2015. Available at: https://www.icomos.org/images/DOCUMENTS/18_April/2017/Siem_Reap_Declaration_2015.pdf (accessed on 16 January 2018).
United Nations (1948) *Convention on the Preservation and Punishment of the Crime of Genocide*. A/RES/3/260 Paris, December 9.
United Nations (2019) *The United Nations Plan of Action to Safeguard Religious Sites*. New York: United Nations Alliance of Civilizations.
UNWTO (2015) *Tourism at World Heritage Sites*. Madrid: United Nations World Tourism Organization.
UNWTO (2018) *Tourism and Culture Synergies*. Madrid: United Nations World Tourism Organization.
Winter, T. (2014) 'Beyond Eurocentrism? Heritage conservation and the politics of difference', *International Journal of Heritage Studies*, 20(2): 123–137.

SECTION III

Motivations, experiences, and performance

16

TRAVEL MOTIVATIONS OF PILGRIMS, RELIGIOUS TOURISTS, AND SPIRITUALITY SEEKERS

Darius Liutikas

Introduction

Pilgrims' motivations have long been of interest to scholars of religious tourism and pilgrimage. Many cultural influences cause people to undertake pilgrimages, and there are as many motives for pilgrimage as there are spiritual or religious needs (Davidson & Gitlitz 2002). There are many theories of motivation and social action that can help explore the phenomenon of pilgrimage. Different social and motivational theories and classifications of pilgrims are presented here to identify the main motivating factors behind people's decisions to undertake a religious journey. The term 'motivation' refers to factors that activate, direct, and sustain goal-directed behavior (Nevid 2013). Motivation is an inner state of an individual that helps satisfy both psychological and physiological needs. Motives are the wants or needs that drive behavior and explain why we humans do what we do (Nevid 2013).

Various forces can influence motivations and their strength. Cultural and social factors include pilgrims' cultural traditions, social classes, societal role and status, familial influence, and that of other reference groups. Personal factors include age and stage in the life cycle, financial situation, occupation, level of personal mobility, beliefs and attitudes, and political leanings (e.g. conservatism or liberalism), dependency or autonomy, risk avoidance or acceptance, adaptability, and ability to endure difficulties also play an important role. The desire to reinforce and express values and identity may cultivate strong and deep desires to undertake a religious journey.

This chapter has three purposes. First, it introduces concepts and approaches to understanding travelers' motivations. Second, it elucidates the motives associated with religious-oriented travel and the complex process of how spirituality and religion are integrated into the everyday lives of people. Third, it presents a different way of seeing travelers' motives in relation to sacred places and challenges the ordinary use of terms such as 'pilgrim' and 'religious tourist'. Such limited terminologies and binary conceptual assumptions about the crossover between religious sites and their visitors have so far been insufficient. The typology outlined herein will help remedy that problem.

Motivations

Raj, Griffin, and Blackwell (2015) divided motivational theories into content and process theories. The content perspective focuses on what motivates people and seeks to identify and explain relevant variables. Process theories highlight the actual process of motivation to identify the relationship between dynamic variables, such as values and expectations, which influence individual motivation and decision-making.

Researchers generally agree that people select their travel destinations according to different push-and-pull factors (Crompton 1979; Dann 1981; Iso-Ahola 1982; Mansfeld 1992; Yoon & Uysal 2005). As regards pilgrimage, travelers are pushed by internal and emotional motives and pulled by external factors, such as the importance of a certain pilgrimage destination. Push and pull factors might be both independent and interdependent. Mannell and Iso-Ahola (1987) and Krippendorf (1987) suggest that travel is motivated by the desire to escape from the normative environment and that travelers' motives and behavior are markedly self-oriented (Hudson 1999).

A destination's image and enticement develop through many means, including media, social media, stories, and cultural traditions. The image and desirability of a pilgrimage destination may derive from all of these, as well as through religious and family traditions, faith-filled stories, and notoriety of miraculous healings or spiritual manifestations. Together, these create dreams and expectations, ideas and perceptions that people hold of a particular place (Liutikas 2013). Thus, the destination image is built upon a set of functional and valuistic expectations. Certain narratives are created of this image before the journey through the internet and social media, books, brochures, newspaper and magazine articles, television, public presentations, and stories told by friends and family members. They also form during the journey through personal experiences, encounters and observations, stories told by local people and guides, brochures and booklets, postcards, interpretive markers at the sites, souvenirs, and interactions with other travelers. After the journey, the image continues to evolve through souvenirs, memorabilia, entrance tickets and other mementos, photographs, personal diaries, social media narratives, travel stories shared by co-travelers, and travel group discussions and meetings.

Scholars have analyzed the motives of religious travelers at great length. This has resulted in various motivation-based classifications of pilgrims. Some authors examine the religious and spiritual dimensions of pilgrimage, emphasizing people's desires to experience transcendence or to have a personal encounter with the divine or other life-changing experience (Cohen 1992; Digance 2003; Stausberg 2011; Turner 1973; Turner & Turner 1978; Vukonić 1996). In most cases, people go on pilgrimages with the expectation that their lives will be changed.

Numerous studies analyze pilgrimage characteristics in particular religions. These include, among others, motivations for pilgrimages among Christians (Eade & Sallnow 1991; Maddrell, Della Dora, Scafi & Walton 2015; Nolan & Nolan 1989; Ron & Timothy 2019; Timothy & Ron 2019), Muslims (Din 1989; Haq & Jackson 2009; Raj 2012; Zamani-Farahani, Carboni, Perelli & Torabi Farsani 2019), Jews (Cahaner & Mansfeld 2012; Collins-Kreiner 2007, 2010, 2019), Buddhists (de Silva 2016; Hall 2006a), and Hindus (Bhardwaj 1983; Reddy 2014; Singh 2006). There have also been studies that delve into the motives of travelers who visit specific religious attractions, such as monasteries and cathedrals, and participate in religious activities such as festivals (Blackwell 2007; Lupu, Brochado & Stoleriu 2019; Rodrigues & McIntosh 2014).

Despite worship and self-improvement being consistent underlying motives, faith-based motivations may vary between different religions. Roman Catholics undertake pilgrimage for penance, to seek forgiveness. Their pilgrimages also entail worship and spiritual renewal, pursuing blessings, and demonstrating gratitude to the God (Liutikas 2012; Ron & Timothy 2019). According to Raj (2012), one of the main reasons Muslims go on pilgrimage is to pray and participate in rituals at Islam's holiest sites in Mecca, Medina, and Jerusalem. The effectiveness and rewards of ritual worship (*salaah*) at these auspicious localities are magnified exponentially. Bhardwaj (1983) defines two broad categories of motives in Hindu pilgrimage. The first category relates to a commitment or vow to the deity whose blessing is essential for the solution of a traveler's problems. The second is learning religious merit. Collins-Kreiner (2010) suggests that Jewish pilgrimage often entails praying, swearing oaths, making requests, and placing notes (supplications) between the stones of the Western Wall in Jerusalem, as well as honoring the faith by visiting sites of historical value (Ioannides & Cohen Ioannides 2004).

As already noted, some pilgrims travel to gain a specific benefit, for example, forgiveness for sins, recovery from an illness or to find a solution to personal problems (Morinis 1992; Smith 1992; Tomasi 2002; Turner 1973). Davidson and Gitlitz (2002) identify additional benefits, including improved fertility, finding love, good fortune, or better grades in school. The power of healing and blessings in pilgrimage comes from the increased faith in times of spiritual intensity.

Pilgrimage motives are sometimes part of a social action or collective behavior (Locher 2001). In some cultures, pilgrimage is a cultural-religious obligation or a rite of passage (Turner & Turner 1978). Cultural or family traditions also encourage pilgrimage. Personal pilgrimages, which often have deep spiritual meaning, may include visiting ancestors' graves, familial heritage localities, and places related to personal identity (Timothy 2008).

Winkelman and Dubisch (2005) suggest that for some individuals, pilgrimage is a self-transforming experience or a tool for social healing particularly through emotional connections and broader community bonding, or even a means of empowerment against church hierarchy or dominant value orientations. Turner and Turner (1978) discuss the importance of *communitas* in pilgrimage—a sense of commonalty and solidarity with other pilgrims who, together, share a common purpose regardless of life's challenges and opportunities at home. Pilgrimage in Turnerian thinking is a liminal rite of passage; it is anti-structural and transformative (Turner & Turner 1978). However, Turnerian *communitas* has been challenged by many authors. Eade and Sallnow (1991) introduce pilgrimage as an arena for competing discourses and conflict rather than intragroup solidarity. Coleman and Elsner (1995) argue that *communitas* is overly idealistic and not very achievable because many pilgrimages involve intra-group and intergroup conflict and are liable to be controlled by temporal powers (Coleman & Elsner 1995: 202).

Olsen (2013) analyzes differences in religious tourists' motivations according to the spatial characteristics of pilgrim destinations. He outlines a scalar difference in the motivations and experiential expectations of people who travel to religious points (e.g. cathedrals), utilize religious lines (pilgrimage trails, e.g. the Camino de Santiago), and visit religious areas (e.g. the Holy Land).

Likewise, religious tourism and pilgrimage have essentially become enveloped within broader tourismscapes, tourism infrastructures, and tourism management regimes. Thus, pilgrimage nowadays resembles many other forms of tourism and may be motived by more traditional manifestations of tourism (Ron & Timothy 2019). Religious tourists of every kind visit other natural or cultural attractions en route or in the area of the sacred destination (Rinschede 1992). Motives in general depend on time, place, traditions, social structure, individual interests, and various other circumstances (Swatos & Tomasi 2002; Timothy & Olsen 2006).

Variations in the range of motivations depend on the types of pilgrimage: individual, family, or communal; short distance or long distance (Rinschede 1992); special time for traveling or any time of year; walking pilgrimages or public transportation-based pilgrimages.

Demographic variations can also influence people's motivations. Gender, age, education, social class, and denominational adherence create common variations in motivation. For example, women report desiring religious experiences more than men do (Beit-Hallahmi & Argyle 1997), and women constitute a disproportionately higher number of pilgrims in many faiths (Notermans 2016; Rinschede 1992). Youth tend to participate more in mass religious gatherings, such as World Youth Days in the Catholic Church (Ron & Timothy 2019). The implications of education vary in different countries. Some educated individuals are more active in some aspects of religion (Beit-Hallahmi & Argyle 1997), whereas some educated persons are less inclined to visit sacred sites for spiritual reasons (Rinschede 1992). Social class can affect a person's choice of shrines to visit, but some research shows social standing being irrelevant to the composition of the markets for some sacrosanct localities (Liutikas 2012). There is a lack of research on denominational membership and pilgrim motivations; however, church communities often organize pilgrimages for their members, so active adherents have more exposure to travel opportunities.

Finally, some 'pilgrimages' do not relate directly to religion per se. 'Secular pilgrims' see their journey as one of a cultural, educational or nostalgic quest: a sightseeing trip, an adventure, an escape, a learning opportunity, a chance to make friends or experience certain pleasures denied them at home (Ambrosio 2007; Clift & Clift 1996; Coleman & Elsner 1995; Digance 2003; Hall 2006b; Liutikas 2012; Tomasi 2002). Likewise, some religious tourists are motivated by interests in religious architecture and art rather than their faith or a sacred site's inherent spirit of place. Various kinds of visitors attend religious events (e.g. meeting the Pope in their own country or participating in an extraordinary religious celebration).

Dark tourism is associated with sites of death, disaster, and depravity (Lennon & Foley 2000). Pilgrimages to cemeteries, tombs of saints, or other places associated with death have existed since people began to travel. Seaton and Lennon (2004) identify two main motives for dark 'pilgrimage' visits: *Schadenfreude*, meaning pleasure in viewing the misfortune of others, and contemplating death, which means accepting or confronting death. Recent studies (e.g. Olsen & Korstanje 2019) examine dark tourism through the new lens of pilgrimage.

One current research trend is the impact of values on tourists' motivations. Many studies seek to understand the role of personal and societal values in people's motivations and travel decisions (Crick-Furman & Prentice 2000; Jewell & Crotts 2001; Mehmetoglu et al. 2010; Woosnam, McElroy & van Winckle 2009). General findings suggest that people travel to certain events and destinations, or avoid them entirely, because of the personal and social values they espouse (Liutikas 2014). The process of valuing involves three steps: choosing values, prizing values, and acting on values (Raths et al. 1978). The value must be chosen freely, and the choice must consider the consequences of the alternative. The person must be happy with the choice, and the choice enhances the emotional and spiritual development of the individual. Finally, values are the major priorities that a person chooses to act on. The person wants to act repeatedly to affirm the choice (Palispis 1995). This notion of values-based action is pertinent to religious travel motives.

The journey as an expression of values and identity

In recent years, scholars have come to realize that certain pilgrimage motives increase people's need for spiritual quests. The relationships between personal values and tourism have been assessed in many studies (e.g. Crick-Furman & Prentice 2000; Miller 2003; Pizam & Calantone 1987; Watkins & Gnoth 2005, 2011a, 2011b). Coleman and Elsner (1995: 214) suggest that if the sacred is in some respects an embodiment and representation of societal ideals, pilgrimage takes new forms that go far beyond standard religious practices. Morinis (1992: 4) defined pilgrimages as journeys undertaken to a place or state that embodies a valued ideal. According to Morinis (1992: ix), pilgrimage is a movement towards aspirational ideals that cannot be realized at home, and therefore, pilgrimage becomes a values-based endeavor that mingles the sacrum and the profanum.

The changing conception of spirituality also influences the notion of values-based journeys. The spirituality revolution (Tacey 2004) includes a complex labyrinth of ideas that might be understood broadly as 'spiritual'. Yamin (2008) defines social and personal identity as a consciousness of one's own and others' perceptions of individuality. From this perspective, identity is related to the representation of one's values, personal experiences, memories, and intelligence.

Values-rich journeys may be described variously, including pilgrimage, spiritual travel, personal heritage tourism, holistic tourism, and valuistic journeys (Liutikas 2012; Morinis 1992; Norman 2011; Smith 2003, Smith & Kelly 2006; Timothy 1997). However, in all of these conceptualizations, the motivations for traveling are related to the construction or manifestation of personal and/or social identity and self-fulfillment.

Liutikas (2012) defines a valuistic journey as an expression of valuistic ideals, as well as a confirmation and manifestation of identity. The main element of such journeys becomes a valuistic motive. These quests help develop or change one's personal or social identity. The inner disposition and motivation of the valuistic pilgrim become the main factor distinguishing this pilgrimage from a normative recreational journey. Valuistic travel can help people evaluate their existential selves and values, and their outlook on the world. Values-rich journeys may include both traditional religious notions of pilgrimage and secular (unrelated to religion) pilgrimage experiences. The values fostered by valuistic quests can embody national, cultural, or other collective ideals or the unique values of an individual.

Traditional religious pilgrimages played an important role in European religious and cultural life during the Middle Ages (Webb 2001, 2002). The officially declared motive of pilgrims was to venerate the relics of saints (e.g. the cult of St. James in Santiago de Compostela and St. Thomas Becket in Canterbury). Some pilgrims traveled to saints' tombs to heal their own or family members' health maladies, while others were compelled to undertake a pilgrimage as penance for their sins.

The primary 'official' motives of contemporary Catholic pilgrimage are seeking the God's grace and good health, expressing gratitude to Jesus or the Virgin Mary, and spiritual growth and renewal (Liutikas 2012). Confessions and penance are still typical elements of a religious pilgrimage. For example, most of the major pilgrimage places in Lithuania are visited during indulgence feasts—church festivals geared toward penance.

In addition to receiving spiritual benefits, many pilgrims also desire to leave something of themselves in sacred places (Ron & Timothy 2019). To leave their mark, add meaning, and become physically a part of a sacrosanct space, pilgrims sometimes leave written prayers, offerings and small gifts, votive amulets, symbols of devotion, and inscriptions of gratitude on tiles or in walls cracks. In Catholicism, this physical act becomes part of the repentance process as Catholics metaphorically leave their sins behind. Gratitude is expressed in other ways as well. In the simplest form, candles are lit, with the flames symbolizing the pilgrims' presence, their aspirations to improve their lives, and expressions of a desire to have their wishes come true. Donating money is also a common and simple ritual at sacred locales.

Secular pilgrimages also have spiritual and valuistic connotations (Clift & Clift 1996; Hall 2006b). Most destinations and activities associated with secular pilgrimage include veneration in the areas of nationalism (patriotism), music, sports, and individual self-deification (Margry 2008). For secular pilgrims, a journey may involve certain rituals, a search for something beyond oneself, a quest for identity, or spirituality in a very broad and abstract sense (Liutikas 2014). Most secular pilgrimage destinations include patriotic monuments (could be international, national, personal, symbolic) and certain events (e.g. sport events, concerts, and exhibitions). Sport fans flying national flags, wearing team paraphernalia, waving banners and slogans, and chanting praises for a team share many similarities with religious pilgrimage rituals and for some, these activities might be described as spiritually moving. Secular pilgrimages also include fan visits to the tombs or former homes of famous celebrities, communion with nature on hiking trails, personal identity-seeking quests, meeting with esteemed mentors, or attending events that are significant for personal or social identity-formation.

Valuistic journeys express personal values and the unique identity of an individual. According to Sharpley and Stone (2012), well-being and happiness are important manifestations of spirituality. Certain experiences during the journey may be a source of happiness (Sharpley & Stone 2012: 5). Manifestations of values and identity include socially engaged action (Liutikas 2017), which may enhance self-actualization that leads to an improved quality of life.

Visiting certain destinations often strengthens the values that drive pilgrims to travel. Changes in personal identity emerge out of the interplay between social circumstances and events during the journey and the way the individual responds to them. An argument with a fellow traveler or the unique experience at a sacrosanct destination can induce changes in a pilgrim's value orientation or identity. Other people learn to accept pilgrims as they wish to be accepted. During such journeys, social consciousness and self-identity are concretized as participants' self-esteem develops.

Overarching motives

Religious travelers' motives are diverse and complex. However, a core set of motives can be identified. Research into values-rich journeys suggests that participation in pilgrimages is based upon personal and social values and identities, as well as individual perceptions. Three different conceptualizations of religious travel can be identified: (1) a search for, or manifestation of, religious values and identity; (2) partly religious or non-religious motivations; and (3) practical motivations. These concepts provide a framework for understanding specific travel motives. This values-based framework allows us to distinguish seven groups of travel motives: (1) pure religious and spiritual motives, (2) ritual-oriented motives, (3) family or community traditions, (4) a desire for inward changes and new social relations, (5) a desire to understand religion, (6) holiday travel (leisure time, sightseeing), and (7) seeking other objectives (Table 16.1).

The first group of motives entails a desire to strengthen or renew a personal relationship with deity. These motives reflect pilgrims' wish to ask for the God's grace, health, or other blessings, or to express gratitude, to participate in religious feasts and rituals, to fulfill

Table 16.1 Comparison of pilgrims' and religious tourists' motives

Travel conception	*Motives*	*Pilgrims*	*Religious tourists*	*Description*
Pilgrimage journey based on seeking religious values and identity (2/3 of all travelers)	Pure religious and spiritual motives	Asking for divine grace, health, blessing, expressing gratitude, participation in religious feasts and rituals, fulfilling religious obligations, visiting sacred sites, spiritual search, and renewal		The objectives of journey are exclusively religious, which most of all expresses religious values
	Performing the ritual	Important values of the journey: carrying out specific rituals or ordinances, satisfaction after the objective has been achieved, penance manifesting through physical difficulties and weariness		The journey is understood as a ritual (e.g. foot pilgrimage), which requires physical effort. After the objective has been achieved, there is a sense of satisfaction
	Family's or community's tradition	Annual or other visiting of pilgrimage place within a set time frame. Nostalgic feelings and wish to meet friends or relatives		The journey is understood as a possibility to meet family members, friends, or community members who share the same values and identities
	Potential for inner changes and new social relations	The main motive of the journey is spiritual search and self-renewal. Possibility to change the self or community. Seeking changes in daily life after returning home		The journey aims to change one's personal life or social life, new social relations are important

(*Continued*)

Travel conception	Motives	Pilgrims	Religious tourists	Description
Pilgrimage journey motivated by religious and non-religious motivations (1/3 of all travelers)	Potential to understand religion, curiosity		Observation of religious feast and ritual, acquaintance with religious architecture and traditions, education about religious heritage	The objectives of the journey are religious education, curiosity to experience what happens at religious festivals or events, visiting religious conferences
	Pilgrimage journey as a new form of holiday travel – good leisure time, sightseeing		Motives include spending leisure time. Important values related to the journey: new impressions, new landscapes, new natural and cultural environments, relaxation, rest from mundane routines, and possibility to express one's self	The journey is understood in a similar way as a cultural tourist journey, a way of spending leisure time, seeking new impressions, new possibilities for expression
Practical motivations	Traveling as the instrument seeking other objectives			Motives for visiting a sacred site include financial incentives (trade, driving a bus, guiding a group, and pick-pocketing), accompanying pilgrims with serious illnesses. The journey lacks spirituality, motivated only by economic benefits or helping others

religious obligations (e.g. hajj), to visit sacred sites, or simply to seek spiritual renewal. Pilgrimage is a purely religious act in which the priority is communion with the God; prayer is the most common activity derived from this motivation.

The second group of motives—ritual-oriented pilgrimages—derives satisfaction from achieving a specific objective. Frey (1998) noted the desire of footpath-based pilgrims to Santiago de Compostela to earn special indulgences associated with Holy Years. She also identified such motives as the fulfillment of vows, prayer for others, reflection and meditation, expiation of one's own or others' sins, or demonstration of faith (Frey 1998: 32).

The third group of motives relates to satisfying family and cultural traditions of visiting pilgrimage sites within a normative timeframe (e.g. annually, every second year, every fifth year). Pilgrims may attend annual religious festivals as feasts of indulgence. Part of their motive may also be the chance to meet family members and friends or to enjoy time with other members of their religious community. A value that is shared by the whole family can also be a reason for undertaking such a journey. Some couples go on pilgrimages to celebrate an anniversary or honeymoon. Nostalgia inspires some visits as people may relive their pilgrimage experiences from their youth. Often, pilgrims travel to religious feasts that occur at their place of birth. Usually, this journey provides an opportunity to spend more time with family members.

The fourth group motivates pilgrims and non-pilgrim religious tourists. The potential for inward change and new social relations motivates many people. Frey (1998: 32) and others identify mid-life and faith crises as reasons for hiking the Camino de Santiago. Developing social relations and satisfying social needs are important both for pilgrims and other tourists. Such journeys help build a sense of *communitas*, thereby helping realize people's aspirations for social inclusion. Group pilgrimages renew and create social relationships and a sense of security. Immersion within a social network helps create an emotional link to the broader community (e.g. all Catholics or all Muslims).

Many religious tourists are motivated by a mix of religious and non-religious variables. Some people within this category seek to understand religion and its rituals. They want to learn about a specific faith and its practices (Cohen 2006; Swatos & Tomasi 2002). Cohen (2006) denoted that these kinds of journeys may be related to the development of a religious identity through intellectual and spiritual explorations by participating in religious rituals and following pilgrim trails.

The sixth set of motives is much less connected to faith and spirituality. Instead, these take the form of general tourism, including leisure tourism and cultural tourism. In this case, participants see pilgrimage as a form of holiday-making, having a good time, relaxing and sightseeing. These people choose a pilgrimage journey to diversify their holiday or weekend experiences. When participating in organized pilgrimages, these travelers could be defined as 'pseudo pilgrims' or 'self-styled pilgrims' (Liutikas 2012). Even if part of a pilgrimage group, pseudo pilgrims are not typically motivated by faith; instead, they don a tourist identity. The choice to participate in a pilgrimage might be determined by its lower price, which is often the case compared to other foreign package tours. Or, participants may be looking for social contacts and a safe group to travel with. Pilgrimage packages also often include visits to other famous cultural and natural attractions in pilgrimage destinations.

The last group of motives is fairly abstract but relates to visiting pilgrimage destinations for non-tourism and non-pilgrimage purposes. Even in the Middle Ages, traders, thieves, pickpockets, and prostitutes traveled pilgrimage routes. Likewise, the servants frequently accompanied nobility during their pilgrimages, even if the servants themselves were not believers. So, in addition to trade, driving a bus, guiding a group, and even stealing from attendees, motives in this category can include accompanying close friends and family members, such as the elderly or those with disabilities or illnesses.

Pilgrim motives are not fixed and stable. Frey (1998) indicates that some pilgrims formulate or change their motives after a journey begins. Rarely is there only one reason for a journey, even if one major motive drives the choice to travel. This section confirms that faith is not the only driver of religious tourism and pilgrimages.

Traveler types

There is no homogeneous pilgrim type. As already noted, a wide range of motives drive people to visit sacred places. Pilgrimage sites vary a great deal in importance, from small shrines that attract the faithful from the immediate area to world famous places visited by believers and non-believers from many countries. Pilgrimage also involves different sized groups, from individual travelers to mass gatherings. However, travelers' identities and the distinction between religious and secular motivations are commonly used to classify religious or spiritual travelers.

Personal or social identity is the foundation of values-rich journeys. Travel offers opportunities for the construction and manifestation of identity. Perceptions of oneself and others are fluid and dynamic. Therefore, where people go and what motivates them, reveals who they are to others and to themselves. The connection between identity and travel motives in the current period of rapid global change is important.

Traveling to pilgrimage destinations is an indicator of religious identity. Religious identity means people's ways of relating to religion, how strongly they feel about their faith, and how they choose to demonstrate their beliefs during their daily lives or during travel (Liutikas 2015). The main way of expressing religious identity is participating in religious services, events, or feasts, performing rituals, and being obedient to the words of gods, prophets, and religious leaders. Some believers participate in the activities of religious communities, whereas others exhibit their identity in public or private discussions, by reading and quoting scriptures, using religious symbols (e.g. crosses, religious souvenirs), or a combination of these.

Different types of pilgrims may be identified based on their identities and motivations (Collins-Kreiner 2010; Morinis 1992). Scholars have identified three types of faith-associated travelers: devout religious pilgrims on one end of the spectrum, secular tourists on the other hand, and religious tourists in between (Santos 2003; Stoddard 1997). Smith (1992) identified two additional subtypes: religious tourists who are more pilgrims and less tourists and religious tourists who are more tourists than pilgrims. Collins-Kreiner (2010) highlights how Jewish pilgrimage attractions appeal to diverse visitors, ranging from secular to spiritual. She identifies pure pilgrims, pilgrim-tourists (religious visitors), traditional believers, and secular visitors.

As involvement is a common variable to understand types of leisure activities (Funk & Bruun 2007), it is also an important variable in religious tourism (Edgell 2006). Religious involvement may be in public settings (e.g. frequency of church attendance or participation in public religious rituals, events, and pilgrimages) or in private (e.g. frequency of private prayer, frequency of reading scriptures, and other holy writs). Individuals who are more involved with religion demonstrate increased participation in pilgrimages.

From a faith perspective, considering motivations and values, 12 types of travelers may be identified, ranging from religious to secular, and from low levels of religious identity and involvement to high levels of religious identity and involvement (Table 16.2). Starting from

Table 16.2 Types of religion-related travelers

	Travelers types		
Religious motivation ↓	(1) Committed pilgrims	Intensely involved with religion. Has a deep religious experience. Religion is the primary reason for the journey.	**High religious identity and involvement** ↑
	(2) Pure pilgrims	Know the religious goal of the trip, manifest religious values and identity.	
	(3) Religious visitors	Take the opportunity to visit a sacred place while visiting a destination for other reasons. The holy place is related to their religious identity and values.	
	(4) Spiritual questors (seekers)	Seek the spiritual goal and religious experience. Is happy having a deep religious experience.	
	(5) Ritualists	Special reasons related to a particular ritual or foot pilgrimage. Ritual becomes more important than religious beliefs.	
	(6) Social pilgrims	*Communitas* seekers. Still the faith element is stronger than leisure motivation.	
	(7) Purposeful religious tourists	The need for cultural learning and accumulation of new experiences. Participates in some religious rituals and activities.	
	(8) Curiosity wanderers	Curiosity as a break from the routine of everyday life. Religious motivation and involvement are quite weak.	
	(9) Sightseeing cultural tourists	Religious attractions are only a part of the visit. Religion is used for recreational purposes. Sightseeing pilgrimage plays a part of entertainment and relaxation.	
	(10) Incidental travelers	Usually accompanies other travelers. Destination is determined by co-travelers.	
	(11) Beneficials	Provide services for pilgrims. Take the opportunity to receive tangible or intangible benefits not related to religion.	**Low religious identity and involvement**
Secular motivation	(12) Anti-pilgrims	Purposes oppose religious values. Motives include criminal activities, such as stealing, pickpocketing, robbing, cheating, selling drugs, prostituting themselves, or enacting acts of terror.	

the first, each type reflects a decreasing level of personal involvement and greater distance from religious values and identity (Liutikas 2021).

Committed pilgrims

Some individuals are intensely involved with religion, including clergy (priests and members of religious orders); recent converts, especially those who experienced sudden and dramatic conversions; fundamentalists, or people who demonstrate extreme commitment; people who have had intense religious experiences (e.g. mystical experiences, ecstasy) and those with psychological disorders or mental health issues (Beit-Hallahmi & Argyle 1997). For them, pilgrimage is an important component of their emotional well-being and intense religious life.

Pure pilgrims

These people's spiritual journeys are based on religious and faith-filled motives. Prayer and meditation, communication about religious topics, and questions about the meaning of life underscore their motives. They are typically involved in religious communities, regularly attend religious services, and participate in faith-related social events. They manifest strong religious values and identity during pilgrimages.

Religious visitors

These pilgrims take the opportunity to visit a pilgrimage place while visiting a destination for other purposes. Their primary goal of the trip might be business, participation in a conference or seminar, or attending a cultural event. They see such trips as an opportunity to visit a nearby sacred site, which is connected to their religious identity and values. Some religious visitors are repeat visitors from the neighboring area. They may have moderate or high levels of religious identity and involvement.

Spiritual questors (seekers)

Sometimes a pilgrim's journey may serve to renew a religious identity or to strengthen it. The rites performed at pilgrimage sites or the spirit of the place itself enables identity development. It may even bring about a real religious conversion. Spiritual seekers may be involved in a formal faith community; they are generally believers and practice a religious creed, but they seek to refresh their testimonies or desire to change their spiritual lives by undertaking a pilgrimage.

Ritualists

Schnell and Pali (2013) emphasize how some pilgrims to Santiago de Compostela perform personal rituals. The meaning-making potential of the ritual may reveal transformative experiences. Life-transforming experiences are the core elements of both traditional and more contemporary forms of pilgrimage (Winkelman & Dubisch 2005). Pilgrimage in this case is the ritual of transformation during which new insights are given and deeper understanding is attained.

Social pilgrims

For these pilgrims, the most important aspect of the journey is *communitas* and socialization with other pilgrims. Most of these travelers are traditional, largely cultural, believers. The journey provides an opportunity to spend time with family members, friends, or other like-minded people who share similar values and identities. These pilgrims seek security and conformity in a stable social environment. They sometimes choose their journey for non-faith reasons, such as social prestige or lifestyle affirmation.

Purposeful religious tourists

For this cohort, religious education or appreciating tangible and intangible heritage is the primary motive for visiting sacrosanct places. By visiting, they learn about religious traditions, architecture, and art, feasts, and rituals. They want knowledge and new experiences, which sometimes entails them participating in sacred rituals and activities. They may or may not be affiliated with any particular religion in their daily lives.

Curious wanderers

These visitors mainly want to see for themselves the happenings at a sacred site. They are curious and interested and may want to get away from the grinds of everyday life. The visit, therefore, may be a way of diversifying their leisure pursuits. The pilgrimage path or destination can stimulate their imagination and help in their search for self-development, although visiting a consecrated locality may be a spontaneous decision. The level of religious identity is low.

Sightseeing cultural tourists

Sacred sites are only part of the travel itinerary. Religious involvement and religious identity in relation to the visited locality are weak. Sightseeing, relaxation, recreational pursuits, and other secular interests are the primary activities. Sightseeing tourists do not travel for religious or pilgrimage reasons, but given that sacred sites appear on their itineraries, they may end up having a religious or spiritual experience.

Incidental travelers

These people visit under uncontrolled circumstances; the decision to visit a holy place was made by somebody else connected to the incidental guest. They usually accompany others—friends, business partners, or family members with serious illnesses or disabilities. A locality might be only a quick stopover on the way to a different destination, or part of a business trip. The religious identity and involvement of these travelers are very low in relation to the locality being visited, although they might choose to participate in certain activities at the destination.

Beneficials

These people visit pilgrimage locales as part of their job, rather than for purposes of spiritual or intellectual edification. These are service providers, which are especially important during religious festivals. They are drivers, guides, food service providers, security guards, and devotional leaders. They may also be politicians and government officials, who seek attention in the public eye. Advertisers and promoters of various products or religious dogmas also visit pilgrimage places.

Anti-pilgrims

This type of 'visitor' is usually unwelcome and a nuisance for pilgrims and site managers. Their purpose for going to a hallowed place is to oppose its religious values and take advantage of the 11 types of visitors described above. They may protest against religion or ridicule the faithful. Or, more commonly, their motives include criminal activities such as stealing, pickpocketing, robbing, cheating, selling drugs, prostituting themselves, or enacting acts of terror.

Conclusions

This chapter reviews the multitudinous motivations of faith-based or spiritually motivated travel. Many scholars and studies have emphasized a wide range of motivations, including but not limited to spirituality, life transformation, and community solidarity and social belongingness. The strength of the motivation in expressing values and identities, the level of religious identity and involvement, and attitude towards the destination are three key elements of socio-psychological motives for participating in pilgrimage journeys.

The chapter highlights two main motivational trends. First, pilgrims' attitudes and inner dispositions push them to visit a certain destination. Second, the destination's image pulls the visitor toward it. Collective behavior theories stress people's personal identification with pilgrimage places and events, and current thinking suggests a strong connection between people's values and their motivation to undertake a pilgrimage. The main element of values-rich (valuistic) journeys is a valuistic motive, which is related to one's personal and/or social identity and self-fulfillment. Values-induced journeys are prized, esteemed, desired and enjoyed by pilgrims.

Values influence a person's motives, behaviors, and attitudes. However, not all motives of religious site visitors are related to religion. Some forms of travel that crossover with faith are motivated by sightseeing, adventure, education, relaxation, or intragroup socialization. On the contrary, secular pilgrimages may be spiritual or valuistic. In the classic sense, visiting becomes a beacon of one's values system and what a person deems most meaningful.

Religious identity and religious involvement are key components in distinguishing different types of religion-associated travelers. This chapter presents 12 types of travelers based upon their closeness or distance from religious beliefs, practices and motives. These include committed pilgrims, pure pilgrims, religious visitors, spiritual seekers, ritualists, social pilgrims, purposeful religious tourists, curiosity seekers, sightseeing cultural tourists, incidental travelers, beneficials, and anti-pilgrims. This classification is based upon the wide array of motives that drive people to visit places of religious or spiritual importance.

References

Ambrosio, V. (2007) 'Sacred pilgrimage and tourism as secular pilgrimage', in R. Raj & N.D. Morpeth (Eds.), *Religious Tourism and Pilgrimage Festivals Management: An International Perspective* (pp. 78–88). Wallingford: CABI.

Beit-Hallahmi, B., & Argyle, M. (1997) *The Psychology of Religious Behaviour, Belief and Experience*. London: Routledge.

Bhardwaj, S. (1983) *Hindu Places of Pilgrimage in India: A Study in Cultural Geography*. Berkeley: University of California Press.

Blackwell, R. (2007) 'Motivation for religious tourism, pilgrimage, festivals and events', in R. Raj & N.D. Morpeth (Eds.), *Religious Tourism and Pilgrimage Festivals Management: An International Perspective* (pp. 35–47). Wallingford: CABI.

Cahaner, L., & Mansfeld, Y. (2012) 'A voyage from religiousness to secularity and back: A glimpse into 'Haredi' tourists. *Journal of Heritage Tourism*, 7(4): 301–321.

Clift, J.D., & Clift, B.W. (1996) *The Archetype of Pilgrimage: Outer Action with Inner Meaning*. Eugene, OR: Wipf & Stock.

Cohen, E. (1992) 'Pilgrimage and tourism: Convergence and divergence', in A. Morinis (Ed.), *Sacred Journeys: The Anthropology of Pilgrimage* (pp. 47–64). Westport, CT: Greenwood Press.

Cohen, E.H. (2006) 'Religious tourism as an educational experience', in D.J. Timothy & D.H. Olsen (Eds.), *Tourism, Religion and Spiritual Journeys* (pp. 78–93). London: Routledge.

Coleman, S., & Elsner, J. (Eds.) (1995) *Pilgrimage: Past and Present in the World Religions*. Cambridge, MA. Harvard University Press.

Collins-Kreiner, N. (2007) 'Graves as attractions: Pilgrimage-tourism to Jewish holy graves in Israel', *Journal of Cultural Geography*, 24(1): 67–89.

Collins-Kreiner, N. (2010) 'Current Jewish pilgrimage tourism: Modes and models of development', *Tourism*, 58(3): 259–270.

Collins-Kreiner, N. (2019) 'Contemporary Jewish tourism: Pilgrimage, religious heritage and educational tourism', in D.J. Timothy (Ed.), *Routledge Handbook of Tourism in the Middle East and North Africa* (pp. 137–146). London: Routledge.

Crick-Furman, D., & Prentice, R. (2000) 'Modeling tourists' multiple values', *Annals of Tourism Research*, 27: 69–92.

Crompton, J.L. (1979) 'Motivations for pleasure vacation', *Annals of Tourism Research*, 6: 408–424.

Dann, G.M. (1981) 'Tourism motivations: An appraisal', *Annals of Tourism Research*, 8: 189–219.

Davidson, K.L., & Gitlitz M.D. (2002) *Pilgrimage: From the Ganges to Graceland, an Encyclopedia*. Santa Barbara, CA: ABC-CLIO.

de Silva, D.A. (2016) 'Anthropological studies on South Asian pilgrimage: Case of Buddhist pilgrimage in Sri Lanka', *International Journal of Religious Tourism and Pilgrimage*, 4(1): 17–33.

Digance, J. (2003) 'Pilgrimage at contested sites', *Annals of Tourism Research*, 30: 143–159.

Din, K.H. (1989) 'Islam and tourism: Patterns, issues and options', *Annals of Tourism Research*, 16: 542–563.

Eade, J., & Sallnow, J.M. (1991) *Contesting the Sacred: The Anthropology of Christian Pilgrimage*. Chicago: Illinois University Press.

Edgell, P. (2006) *Religion and Family in a Changing Society*. Princeton, NJ: Princeton University Press.

Frey, N.L. (1998) *Pilgrim Stories on and off the Road to Santiago: Journeys along an Ancient Way in Modern Spain*. Berkeley: University of California Press.

Funk, C.D., & Bruun, J.T. (2007) 'The role of socio-psychological and culture-education motives in marketing international sport tourism: A cross-cultural perspective', *Tourism Management*, 28: 806–819.

Hall, C.M. (2006a) 'Buddhism, tourism and the middle way', in D.J. Timothy & D.H. Olsen (Eds.), *Tourism, Religion and Spiritual Journeys* (pp. 172–185). London: Routledge.

Hall, C.M. (2006b) 'Travel and journeying on the Sea of Faith: Perspectives from religious humanism', in D.J. Timothy & D.H. Olsen (Eds.), *Tourism, Religion and Spiritual Journeys* (pp. 64–77). London: Routledge.

Haq, F., & Jackson, J. (2009) 'Spiritual journey to Hajj: Australian and Pakistani experience and expectations', *Journal of Management, Spirituality & Religion*, 6(2): 141–156.

Hudson, S. (1999) 'Consumer behavior related to tourism', in A. Pizam & Y. Mansfeld (Eds.), *Consumer Behavior in Travel and Tourism* (pp. 7–32). New York: The Haworth Hospitality Press.

Ioannides, D., & Cohen Ioannides, M. (2004) 'Jewish past as a "foreign country": The travel experiences of American Jews', in T. Coles & D.J. Timothy (Eds.), *Tourism, Diasporas and Space* (pp. 95–110). London: Routledge.

Iso-Ahola, S. (1982) 'Toward a social psychology theory of tourism motivation', *Annals of Tourism Research*, 9: 256–262.

Jewell, B., & Crotts, J.C. (2001) 'Adding psychological value to heritage tourism experiences', *Journal of Travel & Tourism Marketing*, 11(4): 13–28.

Krippendorf, J. (1987) *The Holiday Makers*. London. Heinemann.

Lennon, J., & Foley, M. (2000) *Dark Tourism: The Attraction of Death and Disaster*. London: Continuum.

Liutikas, D. (2012) 'Experiences of valuistic journeys: Motivation and behaviour', in R. Sharpley & P. Stone (Eds.), *Contemporary Tourist Experience: Concepts and Consequences* (pp. 38–56). London: Routledge.

Liutikas, D. (2013) 'Experiences of pilgrimage in Lithuania: Expressions of values and identity at new destinations', *Journal of Tourism Consumption and Practice*, 5(2): 43–60.

Liutikas, D. (2014) 'Lithuanian valuistic journeys: Traditional and secular pilgrimage', *Journal of Heritage Tourism*, 9(4): 299–316.

Liutikas, D. (2015) 'Indulgence feasts: Manifestation of religious and communal identity', in A. Jepson, & A. Clarke (Eds.), *Managing and Developing Communities, Festivals and Events* (pp. 148–164). Basingstoke: Palgrave Macmillan.

Liutikas, D. (2017) 'The manifestation of values and identity in travelling: The social engagement of pilgrimage', *Tourism Management Perspectives*, 24: 217–224.

Liutikas, D. (2021) 'The expression of identities in pilgrim journeys', in D. Liutikas (Ed.), *Pilgrims: Values and Identities* (pp. 17–34). Wallingford, UK: CABI.

Locher, D. (2001) *Collective Behavior*. Upper Saddle River, NJ: Pearson.

Lupu, C., Brochado, A., & Stoleriu, O. (2019) 'Visitor experiences at UNESCO monasteries in northeast Romania', *Journal of Heritage Tourism*, 14(2): 150–165.

Maddrell, A., Della Dora, V., Scafi, A., & Walton, H. (2015) *Christian Pilgrimage, Landscape and Heritage: Journeying to the Sacred*. London: Routledge.

Mannell, R.C., & Iso-Ahola, S.E. (1987) 'Psychological nature of leisure and tourism experience', *Annals of Tourism Research*, 14: 314–331.

Mansfeld, Y. (1992) 'From motivation to actual travel', *Annals of Tourism Research*, 19: 399–419.

Margry, J.P. (Ed.) (2008) *Shrines and Pilgrimage in the Modern World: New Itineraries into the Sacred*. Amsterdam: Amsterdam University Press.

Mehmetoglu, M., Hines, K., Graumann, C., & Greibrokk, J. (2010) 'The relationship between personal values and tourism behavior: A segmentation approach', *Journal of Vacation Marketing*, 16(1): 17–27.

Miller, G. (2003) 'Consumerism in sustainable tourism: A survey of UK consumers', *Journal of Sustainable Tourism*, 11: 17–39.

Morinis, A. (1992) 'Introduction: The territory of the anthropology of pilgrimage', in A. Morinis (Ed.), *Sacred Journeys: The Anthropology of Pilgrimage* (pp. 1–28). Westport, CT: Greenwood Press.

Nevid, S.J. (2013) *Psychology: Concepts and Applications*. Belmont, CA: Wadsworth Cengage Learning.

Nolan, L.M., & Nolan, S. (1989) *Christian Pilgrimage in Modern Western Europe*. Chapel Hill: University of North Carolina Press.

Norman, A. (2011) *Spiritual Tourism: Travel and Religious Practice in Western Society*. London: Continuum.

Notermans, C. (2016) 'Interconnected and gendered mobilities: African migrants on pilgrimage to Our Lady of Lourdes in France', in W. Jansen & C. Notermans (Eds.), *Gender, Nation and Religion in European Pilgrimage* (pp. 19–36). London: Routledge.

Olsen, D.H. (2013) 'A scalar comparison of motivations and expectations of experience within the religious tourism market', *International Journal of Religious Tourism and Pilgrimage*, 1(1): 41–61.

Olsen, D.H., & Korstanje, E.M. (Eds.) (2019) *Dark Tourism and Pilgrimage*. Wallingford: CABI.

Palispis, S.E. (1995) *Introduction to Values Education*. Manila: Rex Book Store.

Pizam, A., & Calantone, R. (1987) 'Beyond psychographics – Values as determinants of tourist behavior', *International Journal of Hospitality Management*, 6(3): 177–181.

Raj, R. (2012) 'Religious tourist's motivation for visiting religious sites', *International Journal of Tourism Policy*, 4(2): 95–105.

Raj, R., Griffin, K., & Blackwell, R. (2015) 'Motivations for religious tourism, pilgrimage, festivals and events', in R. Raj & K. Griffin (Eds.), *Religious Tourism and Pilgrimage Management: An International Perspective* (2nd edn, pp. 103–117). Wallingford: CABI.

Raths, L.E., Harmin, M., & Simon, S.B. (1978) *Values and Teaching: Working with Values in the Classroom*. Columbus: Charles E. Merrill.

Reddy, C.P. (2014) *Hindu Pilgrimage: Shifting Patterns of Worldview of Shri Shailam in South India*. London: Routledge.

Rinschede, G. (1992) 'Forms of religious tourism', *Annals of Tourism Research*, 19: 51–67.

Rodrigues, S., & McIntosh, A. (2014) 'Motivations, experiences and perceived impacts of visitation at a Catholic monastery in New Zealand', *Journal of Heritage Tourism*, 9(4): 271–284.
Ron, A.S., & Timothy, D.J. (2019) *Contemporary Christian Travel: Pilgrimage, Practice and Place.* Bristol: Channel View Publications.
Santos, M. (2003) 'Religious tourism: Contributions towards a clarification of concepts', in C. Fernandes, F. Mcgettigan, & J. Edwards (Eds.), *Proceedings of the Atlas Special Interest Group Religious Tourism and Pilgrimage* (pp. 27–42). Fatima: Tourism board of Leiria.
Schnell, T., & Pali, S. (2013) 'Pilgrimage today: The meaning-making potential of ritual', *Mental Health, Religion and Culture*, 16(9): 887–902.
Seaton, A.V., & Lennon, J. (2004) 'Thanatourism in the early twenty-first century: Moral panics, ulterior motives and alterior desires', in T.V. Singh (Ed.), *New Horizons in Tourism: Strange Experiences and Stranger Practices* (pp. 63–82). Wallingford: CABI.
Sharpley, R., & Stone, P. (2012) 'Introduction: Experiencing tourism, experiencing happiness?', in R. Sharpley & P. Stone (Eds.), *Contemporary Tourist Experience: Concepts and Consequences* (pp. 1–8). London: Routledge.
Singh, R.P.B. (2006) 'Pilgrimage in Hinduism: Historical context and modern perspectives', in D.J. Timothy & D.H. Olsen (Eds.), *Tourism, Religion and Spiritual Journeys* (pp. 220–236). London: Routledge.
Smith, M.K. (2003) 'Holistic holidays: Tourism and the reconciliation of body, mind, spirit', *Tourism Recreation Research*, 28(1): 103–108.
Smith, M.K., & Kelly, C. (2006) 'Journeys of the self: The rise of holistic tourism', *Tourism Recreation Research*, 31(1): 15–24.
Smith, V. (1992) 'Introduction: The quest in guest', *Annals of Tourism Research*, 19: 1–17.
Stausberg, M. (2011) *Religion and Tourism: Crossroads, Destinations and Encounters.* London: Routledge.
Stoddard, R.H. (1997) 'Defining and classifying pilgrimages', in R.H. Stoddard & A. Morinis (Eds.), *Sacred Places, Sacred Spaces: The Geography of Pilgrimages* (pp. 41–60). Baton Rouge: Louisiana State University.
Swatos, H.W., & Tomasi, L. (Eds.) (2002) *From Medieval Pilgrimage to Religious Tourism: The Social and Cultural Economics of Piety.* Westport, CT: Praeger.
Tacey, D. (2004) *The Spirituality Revolution: The Emergence of Contemporary Spirituality.* London: Routledge.
Timothy, D.J. (1997) 'Tourism and the personal heritage experience', *Annals of Tourism Research*, 34: 751–754.
Timothy, D.J. (2008) 'Genealogical mobility: Tourism and the search for a personal past', in D.J. Timothy & J.K. Guelke (Eds.), *Geography and Genealogy: Locating Personal Pasts* (pp. 115–136). Aldershot: Ashgate.
Timothy, D.J., & Olsen, D.H. (Eds.) (2006) *Tourism, Religion and Spiritual Journeys.* London: Routledge.
Timothy, D.J., & Ron, A.S. (2019) 'Christian tourism in the Middle East: Holy Land and Mediterranean perspectives', in D.J. Timothy (Ed.), *Routledge Handbook of Tourism in the Middle East and North Africa* (pp. 147–159). London: Routledge.
Tomasi, L. (2002) 'Homo viator: From pilgrimage to religious tourism via the journey', in H.W. Swatos & L. Tomasi (Eds.), *From Medieval Pilgrimage to Religious Tourism* (pp. 1–24). Westport, CT: Praeger.
Turner, V. (1973) 'The center out there: Pilgrim's goal', *History of Religion*, 12(3): 191–230.
Turner, V., & Turner E. (1978) *Image and Pilgrimage in Christian Culture.* New York: Columbia University Press.
Vukonić, B. (1996) *Tourism and Religion.* Amsterdam: Elsevier.
Watkins, L.J., & Gnoth, J. (2005) 'Methodological issues in using Kahle's list of values scale for Japanese tourism behaviour', *Journal of Vacation Marketing*, 11: 225–233.
Watkins, L.J., & Gnoth, J. (2011a) 'The value orientation approach to understanding culture', *Annals of Tourism Research*, 38: 1274–1299.
Watkins, L.J., & Gnoth, J. (2011b) 'Japanese tourism values: A means-end investigation', *Journal of Travel Research*, 50(6): 654–668.
Webb, D. (2001) *Pilgrims and Pilgrimage in the Medieval West.* New York: I.B. Tauris.
Webb, D. (2002) *Medieval European Pilgrimage.* New York: Palgrave.
Winkelman, M., & Dubisch, J. (2005) 'Introduction: The anthropology of pilgrimage', in J. Dubisch & M. Winkelman (Eds.). *Pilgrimage and Healing* (pp. IX–XXXVI). Tucson: University of Arizona Press.

Woosnam, K.M., McElroy, K.E., & van Winckle, C.M. (2009) 'The role of personal values in determining tourist motivations: An application to the Winnipeg Fringe Theatre Festival, a cultural special event', *Journal of Hospitality Marketing & Management*, 18(5): 500–511.

Yamin, S. (2008) 'Understanding religious identity and the causes of religious violence', *Peace Prints: South Asian Journal of Peacebuilding*, 1(1): 1–21.

Yoon, Y., & Uysal, M. (2005) 'An examination of the effects of motivation and satisfaction on destination loyalty: A structural model', *Tourism Management*, 26(1): 45–56.

Zamani-Farahani, H., Carboni, M., Perelli, C., & Torabi Farsani, N. (2019) 'Islamic tourism in the Middle East', in D.J. Timothy (Ed.), *Routledge Handbook of Tourism in the Middle East and North Africa* (pp. 125–136). London: Routledge.

17

VOLUNTEER TOURISM

A spiritual and religious journey of meaning, transcendence, and connectedness

Gregory Willson and Alison J. McIntosh

Introduction

This chapter considers how volunteer tourism experiences may be religious or spiritual in nature. Tourism experiences are no longer considered to be shallow and peripheral to an individual's life. Indeed, it is widely acknowledged that individuals can derive deeply personal meaning from their travel experiences, meaning that they may be influenced and made sense of through a person's spiritual and/or religious foundations. For example, Dann and Cohen (1996) suggest that tourist journeys can come close in spirit to religious odysseys as they both involve movement and a search for meaning and understanding. Similarly, MacCannell (1973, p. 589) argues that tourism absorbs, "Some of the social functions of religion in the modern world". Participation in tourism is considered to occur in "extraordinary time". That is, it presents people a time and space that is non-ordinary—a space in which they can reflect, cultivate their spirituality, and experience the sacred (Graburn, 1989; Shar pley & Sundaram, 2005).

Vukonić (1996, p. 162) claims that tourism experiences may be similar in nature to religious experiences because, like religion, most travel provides people with leisure time and "A space for the contemplative and the creative, a unity of thought and action", and thus, "an opportunity for human beings to recognise and cultivate their spiritual needs". MacCannell (1973), Vukonić (1996), and Durkheim (1915) also observed the close parallels between tourism and religion by arguing that just as religion contains symbolic rituals, tourism itself is a social ritual that people often undertake at regular intervals, or utilise to mark a particular juncture in their lives. Graburn (1989, p. 22) argues that, "Tourism... is functionally and symbolically equivalent to other institutions that humans use to embellish and add meaning to their lives".

While arguably all tourism experiences may somehow be "spiritual" (Willson et al., 2013), volunteer tourism is considered a form of tourism that facilitates cultivating particularly extraordinary and spiritually or religiously fuelled experiences (Ron & Timothy, 2019; Willson, McIntosh & Zahra, 2013; Willson, 2010; Zahra, 2007). This is because the personal meaning derived from volunteer tourism is often magnified, particularly so because the motivation to engage in volunteering is generally driven by an individual's spiritual, religious or moral foundation; individuals generally feel a strong prompting or yearning to volunteer

(Willson, 2010). Volunteer tourism, or the more commodified "voluntourism" (Bargeman, Richards and Govers, 2018), is "broadly defined as an activity in which people pay to volunteer in development or conservation projects" (Conran, 2011, p. 1454), or it involves

> tourists who, for various reasons, volunteer in an organized way to undertake holidays that might involve aiding or alleviating the material poverty of some groups in society, the restoration of certain environments or research into aspects of society and environment.
>
> *(Wearing, 2001, p. 1)*

Recent years have seen a growth both in terms of the number of participants and the number of volunteer experiences being offered around the world (Freidus, 2017; Henry & Mostafanezhad, 2019; Kontogeorgopoulos, 2017). Motivations to engage in volunteer tourism and the personal meaning derived from volunteer experiences are often complicated and multi-faceted (Willson et al., 2013). There is therefore a need to further our understanding of this phenomenon. This chapter firstly contributes to the body of knowledge regarding volunteer tourism and its relationship with spiritual and religious experiences by considering how individuals' personal spiritual or religious foundation influences their volunteer tourism experiences. The chapter also explores how a perceived spiritual and religious deficit in the Western world has potentially led to a surge in demand for volunteer tourism. Next, the chapter explores how volunteer tourism experiences are intertwined with the three core constructs of spirituality and religion: a search for meaning and purpose in life, personal growth through transcendence, and connectedness with self, the environment, a Higher Power and/or other volunteers. The chapter concludes by providing recommendations for how tourism businesses may best facilitate spiritual and religious growth through volunteer tourism.

Motivations, religion, spirituality, and volunteer tourism

Spirituality and religion differ in that not all people hold religious beliefs; however, all are spiritual, even though they may not consider "spiritual" to be the correct label to describe the core constructs shared by both religion and spirituality. Spirituality entails a search for meaning and purpose in life, self-growth through experiences of transcendence, and connectedness with self or "other" (Meraviglia, 1999; Miner-Williams, 2006; Willson et al., 2013). Spirituality or religious belief is not something that can be "left at home" while an individual engages in tourism. Generally, tourists will seek out destinations that resonate with how they derive spiritual meaning. For example, many devout Christians view travel to Israel as a significant life goal (Ron & Timothy, 2019).

There is a particularly strong link between volunteer tourism and how individuals derive spiritual meaning within their lives. To illustrate, if one's spirituality is strongly tied to a Christian faith, an individual may seek to use travel to contribute to God's work by volunteering in orphanages or in travelling to feed the poor. Willson (2010) presented the personal story of Liz, a devout Christian. For Liz, volunteer tourism was not optional; it was a necessity driven by her religious beliefs and an understanding of "to whom much is given, much will be required".

Motivations to engage in volunteer tourism may also be driven by secular means. How one derives personal meaning in life is personal and varied. One can find meaning in music, animals, nature and/or painting, for example (Van Ness, 1996). A person's deepest meaning may then lead to, for example, travel to engage with restoring a natural habitat or to protect wildlife. Generally, though, there is an agreement in the literature that because much volunteer tourism is driven by how the participant derives spiritual meaning, volunteer tourism is often considered a particularly moralistic or value-led form of tourism. Thus, because morals and values are an outcome of people's spirituality (Willson, 2010; Willson et al., 2013), these must be closely considered when discussing motivations for engaging in volunteer tourism.

It is misleading to suggest that all volunteer tourism is driven by virtuous morals and values. Indeed, many volunteer tourists are said to be motivated by altruism, through a desire to give back or to help those perceived to be less fortunate than themselves (Coghlan, 2015; Kontogeorgopoulos, 2017; Mustonen, 2006; Stoddart & Rogerson, 2004). However, Coghlan and Fennell (2009) purport that some volunteer tourists are motivated by "social egoism", which may include a desire to use volunteer tourism as an avenue to advance over others and personal gratification. This may mirror motivations to engage, for example, in religious practices. For some people, the process of attending a holy place may be more for prestige or a photo opportunity than a divine calling (Williams, Francis, Robbins & Annis, 2007). The motive for engaging in volunteer tourism is often spurred by significant world or national events, with the media driving touristic movement across the globe. For example, the impact of Hurricane Katrina in New Orleans led to a surge in tourism to the area, much of which entailed volunteers being moved by images on their televisions and resulting in a desire to help (Pezzullo, 2009).

Similarly, it is not reasonable to suggest that all impacts and outcomes of volunteer tourism are positive and fuelled by spiritual or religious goodness. There are many examples of volunteer tourism having caused great harm, such as people seeking monetary gain over satisfying the needs of people and animals. For instance, Raes (2013) explores the impact of orphanage tourism in Cambodia and found that the orphan child is often commodified and objectified. In this case, harm is caused because children may bond emotionally with volunteers, who then disappear, leaving the children to mourn the volunteers' departure. Similarly, Freidus (2017) argues that voluntourists often leave their experiences with only superficial understandings of poverty and different cultures and leave orphans with unrealistic expectations for their futures.

As volunteer tourism and spirituality are closely intertwined, to understand why there has been a growth in volunteer tourism, it is pertinent to explore the reasons for the burgeoning rise in interest, both from scholarly and consumer perspectives in spirituality. Indeed, it is argued that the twenty-first century has spurred a "boom" in interest in spirituality, particularly in the Western world (Sharpley & Sundaram, 2005; Timothy & Conover, 2006; Willson, 2016). Reportedly, the predominant reason many people now seek to address their spirituality is their sense of discontent with modern society (Belk, 1985; Cushman, 1990; Hartmann, 1999; Lengfelder & Timothy, 2000; McIntosh & Mansfeld, 2006; Powers, Cramer & Grubka, 2007). Increasingly, individuals are said to be fuelled by feelings of emptiness with modern life that is characterised by high stress, a lack of personal time, isolation, uncertainty, rising fuel prices, global warming, and feelings of depression caused by rapidly advancing technology and civilisation growth (Faulkner, 2008; Garner, 2008; Lengfelder & Timothy, 2000; Sharpley & Sundaram, 2005; Timothy & Conover, 2006). Latham (2001, p. 42) suggests, "In this time, many are facing a deep abyss, with feelings of emptiness and longing. Technology as an organising force is pulling us away from the essence, the centre of meaning". Since Latham's (2001) assertion, technology has advanced considerably, in many cases distancing people away from real social life into a realm of virtual social networks that may exacerbate individuals' sense of emptiness.

As a result of the pressures of modern life, many people are failing to take time to relax, experience nature, and address their spiritual and mental health (Dugan & Barnes-Farrell, 2017; Lengfelder & Timothy, 2000). To mitigate the perceived negative aspects of society, significant numbers of people are said to be seeking outlets to foster their spirituality, as spirituality is said to offer people a source of enduring meaning in troubled times (Heintzman & Mannell, 2003; O'Sullivan, 2012). Thus, there appears to be a relationship between the rise in mental health concerns, and furthered interest in exploring one's spirituality.

The motivation to "escape" the pressures of modern life and experience a wholly new destination has been illustrated many times in the tourism literature. Mansfeld and McIntosh (2009), for example, observe that increasing numbers of Westerners are yearning to receive respite from their fast-paced world and move to natural environments, often in countries or settings where poverty and deprivation are the norm. Notably, certain volunteer tourists actively seek discomfort and experiences that are in direct contrast to their normal standards of living. To illustrate, Willson (2010) explores the experiences of Western volunteer tourists in Peru, Vanuatu and Vietnam. Many of his research participants described encountering illness during their travel; common ailments included vomiting, diarrhoea, headaches, fatigue, heatstroke, and altitude sickness. However, each research participant expected to become ill when travelling to their respective destinations, and some mentioned that overcoming illness was a source of personal growth and pride. To illustrate, Nyla, a grandmother from New Zealand, explained that she knew her volunteering activities in Vietnam would lead to discomfort. However, she had recently experienced a personal battle with breast cancer and explained that she made a vow to herself after her diagnosis to live life to its fullest and embrace uncomfortable situations. A similar example in Willson's (2010) study was provided by Claire, who actively told the organisers of her volunteer travel that she did not want to sleep in the bed provided; rather, she wanted to sleep on the floor with the locals.

The above discussion raises three considerations. First, it illustrates that travel is not always the hedonistic pursuit that it is sometimes portrayed to be (Boorstin, 1964; Hayward & Turner, 2019; Shakeela & Weaver, 2018). Second, it highlights the importance of understanding what is meaningful to an individual and facilitates travel experiences that are authentic. Third, it suggests that because volunteer tourism is often spiritually motivated, experiences that stimulate personal growth, discovery, and reflection are likely to be of high personal significance.

Further to the above, volunteer tourism is also said to be an avenue in which individuals can re-ground themselves and find solace from environments in which they feel their morals and values may be compromised. For instance, Lana in Willson's (2010) study, explained that for her, the United States' lifestyle had led her children to become "victims" of materialism. For Lana, the only way to "shock" her children into the values that were important to her, was to engage in volunteer tourism in Mexico, "to take them somewhere out of their comfort zone, no technology, no materialism" (Willson, 2010, p. 246). In the same study, Amber, a grandmother from New Zealand, believed that immersing her children in a volunteer travel experience in a deprived country was a moral duty. She explained, "I think it is the responsibility of every parent in the Western world to expose their children to this" (Willson, 2010, p. 246).

Additionally, volunteer tourism may facilitate experiences that alleviate personal unease and the distress of guilt that individuals might feel. Relieving guilt in one's life in particular is a growing consideration among individuals when choosing their volunteer travel opportunities (Sin, 2009). Sin (2009) argues that increasing numbers of individuals, particularly from developed countries, feel guilty that they have so much, while others are deprived. They seek to re-dress this imbalance through altruistic means such as volunteer tourism. Willson's (2010) research illustrates a relationship between a desire to cleanse guilt and volunteer tourism. Illustrative quotes from participants in his study include, "I have a feeling of being so privileged just because of where I was born and so I feel obligated in a way to help others through travel" (Willson, 2010, p. 257). In the same study, it was apparent that even individuals who may not have much by Western standards may feel a desire to alleviate guilt. Nyla from New Zealand explained, "When you travel overseas you are the wealthy foreigner – the fact that I work three jobs and clean dunnies at schools to pay for my trips doesn't mean a thing to them there, we are still the wealthy foreigners" (Willson, 2010, p. 237).

Volunteer tourism and the core constructs of spirituality and religion

A personal search for meaning and purpose

Philosophers have conceptualised human beings as "questing animals". That is, humans are forever striving for personal development and will, throughout their lives, undertake a search for meaning that is deeply entrenched in their biological, psychological, linguistic, and social nature (Hardy, 1979; Newberg & Waldman, 2006; Torrance, 1994). This quest is endless and occurs at different points in life, during which individuals will have more or less clarity about questions of fundamental importance regarding their perceived purpose in life (Torrance, 1994). As previously noted, time spent as a tourist may be considered sacred and an ideal avenue in which people may ask themselves questions related to life's meanings and purposes. Tourism may thus be overtly religious, or more spiritually subtle.

For many, questions of self-discovery are triggered or accelerated by a personal experience (Marques, 2006). An illness, loss of a close relationship, divorce, or other loss, for example, will often spur an individual to seek meaning and clarity (Marques, 2006; Ramanayake, McIntosh & Cockburn-Wootten, 2018; Schultz, 2005). Indeed, traumatic experiences may raise questions of mortality and force individuals to evaluate their life's direction (Tedeschi & Calhoun, 2006). For many people, these deeply personal experiences naturally lead to travel, and frequently, volunteer travel. To illustrate, Ramanayake, McIntosh and Cockburn-Wootten (2018) analysed the literature on the impact of personal experiences of loss (e.g. the death of a loved one, chronic illness, infertility, a miscarriage, and divorce) and found that travel is seen by many as part of the healing process and a bastion of hope. Specifically, individuals experiencing loss may view travel as an avenue for *optimism* by planning a pleasurable experience, as an *escape* by choosing a place for rest and respite from their everyday realities, and *new beginnings* by seeking space to create hope and build strategies for the future. As discussed previously, discontent with modern life can also lead people to question "what is my purpose?" and "what are the most important things in my life?", which again frequently lead to volunteer opportunities.

The ability of volunteer tourism experiences to facilitate exploring existential and personally meaningful questions, and even to heal, is widely acknowledged (Prayag, Mura, Hall & Fontaine, 2016; Singh, 2009; Willson et al., 2013). For example, Willson (2010) found that the meaning volunteer tourists derived from their travel experiences could, in certain circumstances, effect mental and physical healing. To illustrate, breast cancer survivor Nyla, explained that after she was cleared of breast cancer, she was mentally exhausted and "not herself". She explained that she had not listened to her favourite singer, Michael Bublé, for more than a year. However, after her volunteer experiences in Vietnam, Nyla explained, "I'm feeling good now and I'm listening to my music" (Willson, 2010, p. 191).

As Nyla's experience demonstrates, volunteer tourism may also lead to positive outcomes among participants. Often they return from a volunteer destination with greater clarity and meaning in their lives (McGehee & Santos, 2005; Scheyvens, 2007; Zahra & McIntosh, 2007). Positive mental and physical health are considered outcomes of a strong spiritual foundation (Voigt, Brown & Howat, 2011). For instance, Zahra and McIntosh (2007) interviewed young volunteers who had engaged with a partner non-government organisation (NGO) to work on welfare projects. Seeing how fortunate they were compared to others, many volunteers felt a deep, lasting impression that impacted their lives long after their journey ended. For example, interacting with people in the host community provided volunteers an opportunity to reflect on a range of issues related to culture and religion. For others, deep personal significance derived through volunteer tourism experiences, leading some individuals to embrace religion. Zahra and McIntosh's (2007) study of young volunteer tourists found that respondents had not valued religion as part of their lives prior to their experiences. Through encounters with host communities that talked freely about God and volunteering in an environment that had visible religious symbols and interactions, some respondents reflected upon the role of God in their own lives. One respondent commented,

> The Philippines was a country that was really poor... yet the people were happy and there had to be a reason for that, this reason was their faith. They had something that a lot of people in the West do not have and I wanted to share in what they had.
>
> *(Zahra & McIntosh, 2007, p. 117)*

Transcendence

Further to the above discussion, to achieve meaning or purpose in one's life, some people may have to experience transcendence. This transcendence, or self-transcendence/spiritual transcendence, is therefore seen by many scholars as an integral part of any conceptual discussion of spirituality (Abernethy & Kim, 2018; Emmons, 2000; Freeman, 1998; McCormick, 1994; Miner-Williams, 2006; Piedmont, 2001; Tanyi, 2002). Piedmont (1999, p. 988) suggests that spiritual transcendence concerns "The capacity of individuals to stand outside of their immediate sense of time and place to view life from a larger, more objective perspective". Similarly, Emmons (2000) purports that transcendence involves going past the ordinary confines of the body.

The transcendent dimension of a person can be seen through a focus on growth, such as that described by Torrance (1994), or through Abraham Maslow's self-actualisation wherein a person seeks to improve himself or herself and increase self-knowledge and awareness of people around them (Piedmont, 1999). Transcendence has also been similarly compared with the work of psychologist Mihaly Csikszentmihalyi, who described the optimal human experiences as "flow", whereby people are challenged, experience a loss of self-consciousness and time (Van Ness, 1996). Piedmont (1999, 2001) argues that people seek to transcend themselves because they are aware of their own mortality and that physical death is inevitable; individuals thus strive to find and construct meaning and purpose in their lives. According to Piedmont (1999, p. 5), answering existential questions "Help[s] us to weave the many diverse threads of our lives into a more meaningful coherence that gives us the will to live productively".

Often, volunteer tourism experiences occur within settings of natural beauty and awe-inspiring vistas. In these settings, transcendent experiences are common. Willson (2010) explains how experiences of sensing something greater than oneself often facilitate transcendence. One research participant, Amy, for example, explained how being close to a volcano in Vanuatu allowed her to be "Particularly close to God because I realised how insignificant we are as people" (Willson, 2010, p. 215). Another participant, Lana, similarly explained that, "Being in the jungle and realising there could be a tiger around anywhere makes you aware of how small you are" (Willson, 2010, p. 215).

Transcendence can also occur through encountering and overcoming a personal challenge. In many volunteer tourism situations, participants learn and apply new skills, or discover things about themselves they did not know. In Willson's (2010) study, Sharen, who rejected the opportunity to sleep in a bed, explained, "We knew it would be a total shock sleeping on the floor of the longhouse, and you couldn't just walk out! But we wanted to do it as a challenge" (Willson, 2010, p. 216). Lana similarly explained that in her view, living in America had made her soft. She explained, "I knew certain parts of my travel to India would be challenging but I needed to experience it to get myself away from the American way of thinking" (Willson, 2010, p. 216).

Of course, not all tourists will have transformative experiences in their volunteer travel encounters, and noticeable changes may take months or years to materialise. Amy, who volunteered in a community in Vanuatu, explained, "It was a great holiday but I wouldn't say it changed me at all really" (Willson, 2010, p. 227). In the same study, the delayed realisation of the personal meaning of volunteer tourism experiences was highlighted by Laura, who commented, "This travel really changed me and I'm sure I haven't realised yet quite how much it has changed me" (Willson, 2010, p. 245). Sharen similarly explained that "I think there's a processing time afterwards where you think, ok what does this all mean" (Willson, 2010, p. 245). These conceptualisations of volunteer experiences are in contrast to many traditional conceptualisations which view tourism experiences as beginning when one leaves home and finishing when one returns again (Ryan, 1997). That is, while tourism experiences may indeed be life-changing, they are often slow-burning, and their full impact may not be realised until later in life.

Connection

Scholars have provided much evidence that through tourism, individuals can find connectedness with themselves and discover "who they really are" (Beeho & Prentice, 1997; Breathnach, 2006; Bruner, 1991; Coghlan & Weiler, 2018; Daniel, 1996; Howe, 2001; Noy, 2004), connections to God, (Cohen, 1979; Harris, 1996; Rinschede, 1992; Vukonić, 1996; Zahra, 2006) or connections to "others" (Curtin, 2005; Harrison, 2003; McCain & Ray, 2003; Relph, 1976; Schanzel & McIntosh, 2000; Trauer & Ryan, 2005). Some authors have contended that tourism presents an ideal situation for connectedness to occur. Craib (1997, p. 160) argues that touristic connections can "reach across gender, age, race, class and other social realities if need be". Harrison (2003) suggests that tourists encounter much of what Simmel (1910) termed the "social ideal" while travelling. These are situations that meet the human need for connection, but the temporary nature of the association frees people from the need to seek ongoing attachment. Indeed, Harrison (2003) proposes that "social ideals", such as travelling on a bus with people, can in fact be deeply meaningful and memorable, meeting people's needs for belonging.

As volunteer tourism typically comprises intimate and meaningful encounters with other people, animals, or nature, opportunities for experiences of connectedness are plentiful. Indeed, because volunteer experiences are generally consumed in settings that are vastly different from an individual's daily life, it is likely that experiences shared among a group of tourists are likely to be unifying. To illustrate, Karen, a participant in Willson's (2010) study spoke at length about how her family's shared experiences volunteering in Mexico opened dialogue between her and her children, and ultimately, strengthened their relationship. She explained,

> It just answered so many questions and brought up thousands more and we would lie awake at night, like the whole family talking… you know, it led to questions that you don't normally hear from a 12 year old boy.
>
> *(Willson, 2010, p. 174)*

Conclusion

This chapter offers insights into how volunteer tourism experiences may be considered religious or spiritual in nature. Volunteer tourism experiences may be explicitly motivated by an individual's religious or spiritual foundation, the foundation may be influenced by the experience. In each of these contexts, volunteer tourism may constitute experiences that reaffirm personal meaning and involve transcendence and social connectedness. As such, volunteer tourism organisations may be able to further facilitate spiritual or religious growth by appealing to these motives in the programmes they provide. Indeed, as mentioned throughout this chapter, much previous research considers how volunteer tourism experiences "make a difference" in people's lives (Boluk, Kline & Stroobach, 2017; Wearing, 2001) and can be "cathartic", healing the soul (Coghlan & Weiler, 2018; Zahra & McIntosh, 2007). For many participants, volunteer tourism is an emotional journey that can put into perspective how individuals fit in the world. Given the close and shared interactions inherently part or the volunteer travel experience, the accompanying emotions have been described as "intimate" (Conran, 2011), "empathic pain" (Frazer & Waitt, 2016) and a sense of "feeling global" (Germann Molz, 2017)—emotions that can challenge volunteers' material privilege as they witness the hardships of others.

One can argue that the emotionally-charged nature of volunteer tourism may bring about new attitudes, behaviours and subjectivities. As such, volunteer tourism organisations ultimately have a moral and ethical obligation to facilitate experiences of personal meaning, connection and transcendence to encourage the spiritual and moral transformation of the self (Crossley, 2012; Everingham, 2016). In the context of modern western families, Germann Molz (2017) argues that volunteer tourism must equip children with the emotional skills they need to live in an uncertain, unequal world of neoliberal globalisation. Thus, in addition to the more widely reported arguments of volunteer tourism being a platform for education, advocacy, and cross-cultural understandings (McGehee, 2014), volunteer tourism projects must also be spaces for personal reflection, deep personal meaning, optimism, and spiritual transformation.

References

Abernethy, A. D., & Kim, S. H. (2018). The spiritual transcendence index: An item response theory analysis. *The International Journal for the Psychology of Religion, 28*(4), 240–256.

Bargeman, B., Richards, G., & Govers, E. (2018). Volunteer tourism impacts in Ghana: A practice approach. *Current Issues in Tourism, 21*(13), 1486–1501.

Beeho, A. J., & Prentice, R. C. (1997). Conceptualizing the experiences of heritage tourists: A case study of New Lanark World Heritage Village. *Tourism Management, 18*(2), 75–87.

Belk, R. (1985). Materialism: Trait aspects of living in the material world. *Journal of Consumer Research, 12*(3), 265–280.

Boluk, K., Kline, C., & Stroobach, A. (2017). Exploring the expectations and satisfaction derived from volunteer tourism experiences. *Tourism and Hospitality Research, 17*(3), 272–285.

Boorstin, D. (1964). *The image: A guide to pseudo-events in American society*. New York: Harper.

Breathnach, T. (2006). Looking for the real me: Locating the self in heritage tourism. *Journal of Heritage Tourism, 1*(2), 100–120.

Bruner, E. M. (1991). Transformation of self in tourism. *Annals of Tourism Research, 18*(2), 238–250.

Coghlan, A. (2015). Prosocial behaviour in volunteer tourism. *Annals of Tourism Research, 55*, 46–60.

Coghlan, A., & Fennell, D. (2009). Myth or substance: An examination of altruism as the basis of volunteer tourism. *Annals of Leisure Research, 12*(3–4), 377–402.

Coghlan, A., & Weiler, B. (2018). Examining transformative processes in volunteer tourism. *Current Issues in Tourism, 21*(5), 567–582.

Cohen, E. (1979). A phenomenology of tourist experiences. *Sociology, 13*(2), 179–201.

Conran, M. (2011). They really love me!: Intimacy in volunteer tourism. *Annals of Tourism Research, 38*(4), 1454–1473.

Craib, I. (1997). *Classical social theory: An introduction to the thought of Marx, Weber, Durkheim, and Simmel*. Oxford: Oxford University Press.

Crossley, É. (2012). Affect and moral transformations in young volunteer tourists. In D. Picard & M. Robinson (Eds.), *Emotion in motion: Tourism, affect and transformation* (pp. 85–98). Aldershot: Ashgate.

Curtin, S. (2005). Nature, wild animals and tourism: An experiential view. *Journal of Ecotourism, 4*(1), 1–15.

Cushman, P. (1990). Why the self is empty. *American Psychologist, 45*(5), 599–611.

Daniel, Y. P. (1996). Tourism dance performances: Authenticity and creativity. *Annals of Tourism Research, 23*(4), 780–797.

Dann, G., & Cohen, E. (1996). Sociology and tourism. In Y. Apostopoloulos, S. Leivadi & Yianakis (Eds.), *The sociology of tourism: Theoretical and empirical investigation* (pp. 301–314). London: Routledge.

Dugan, A. G., & Barnes-Farrell, J. L. (2017). Time for self-care: Downtime recovery as a buffer of work and home/family time pressures. *Journal of Occupational and Environmental Medicine, 59*(4), 46–56.

Durkheim, E. (1915). *The elementary forms of religious life*. New York: Free Press.

Emmons, R. A. (2000). Is spirituality an intelligence? Motivation, cognition, and the psychology of ultimate concern. *The International Journal for the Psychology of Religion, 10*(1), 3–26.

Everingham, P. (2016). Hopeful possibilities in spaces of 'the-not-yet-become': Relational encounters in volunteer tourism. *Tourism Geographies, 18*(5), 520–538.

Faulkner, F. (2008, May). *Tripping the fright fantastic? Prospects for a terror-free 21st century global leisure and tourism industry*. Paper presented at the 2nd International Colloquium on Tourism and Leisure, Chiang Mai, Thailand.

Frazer, R., & Waitt, G. (2016). Pain, politics and volunteering in tourism studies. *Annals of Tourism Research, 57*, 176–189.

Freeman, A. (1998). Spirituality, well-being and ministry. *The Journal of Pastoral Care, 52*(1), 7–17.

Freidus, A. L. (2017). Unanticipated outcomes of voluntourism among Malawi's orphans. *Journal of Sustainable Tourism, 25*(9), 1306–1321.

Garner, A. (2008, May). *Risk and reward: The (lost?) art of backpacking*. Paper presented at the 2nd International Colloquium on Tourism and Leisure, Chiang Mai, Thailand.

Germann Molz, J. (2017). Giving back, doing good, feeling global: The affective flows of family voluntourism. *Journal of Contemporary Ethnography, 46*(3), 334–360.

Graburn, N. (1989). Tourism: The sacred journey. In V. Smith (Ed.), *Hosts and guests: Theanthropology of tourism* (2nd ed., pp. 21–36). Philadelphia: University of Pennsylvania Press.

Hardy, A. (1979). *The spiritual nature of man*. Oxford: Clarendon Press.

Harris, M. (1996). An inner group of willing people: Volunteering in a religious context. *Social Policy and Administration, 30*(1), 54–68.

Harrison, J. D. (2003). *Being a tourist: Finding meaning in pleasure travel*. Vancouver, Canada: UBC Press.

Hartmann, T. (1999). *The last hours of ancient sunlight*. Milsons Point, NSW: Bantam.

Hayward, K., & Turner, T. (2019). 'Be More VIP': Deviant leisure and hedonistic excess in Ibiza's 'Disneyized' party spaces. In T. Raymen & O. Smith (Eds.), *Deviant leisure: criminological perspectives on leisure and harm* (pp. 105–134). Cham, Switzerland: Palgrave Macmillan.

Heintzman, P., & Mannell, R. C. (2003). Spiritual functions of leisure and spiritual well-being. Coping with time pressure. *Leisure Sciences, 25*(2&3), 207–230.

Henry, J., & Mostafanezhad, M. (2019). The geopolitics of volunteer tourism. In D. J. Timothy (Ed.), *Handbook of globalisation and tourism* (pp. 295–304). Cheltenham: Edward Elgar.

Howe, A. C. (2001). Queer pilgrimage: The San Francisco homeland and identity tourism. *Cultural Anthropology, 16*(1), 35–61.

Kontogeorgopoulos, N. (2017). Finding oneself while discovering others: An existential perspective on volunteer tourism in Thailand. *Annals of Tourism Research, 65*, 1–12.

Latham, G. (2001). A journey towards catching phenomenology. In R. Barnacle (Ed.), *Phenomenology*. Melbourne, Australia: RMIT University Press.

Lengfelder, J. R., & Timothy, D. J. (2000). Leisure time in the 1990s and beyond: Cherished friend or incessant foe? *Visions in Leisure and Business, 19*(1), 13–26.

MacCannell, D. (1973). Staged authenticity: Arrangements of social space in tourist settings. *American Journal of Sociology, 79*(3), 589–603.

Mansfeld, Y., & McIntosh, A. (2009). Wellness through spiritual tourism encounters. In R. Bushell & P. Sheldon (Eds.), *Wellness and tourism: Mind, body, spirit, place* (pp. 177–191). New York: Cognizant Communications.

Marques, J. F. (2006). The spiritual worker: An examination of the ripple effect that enhances quality of life in – and outside the work environment. *Journal of Management Development, 25*(9), 884–895.

McCain, G., & Ray, N. (2003). Legacy tourism: The search for personal meaning in heritage travel. *Tourism Management, 24*(6), 713–717.

McCormick, D. W. (1994). Spirituality and management. *Journal of Managerial Psychology, 9*(6), 5–9.

McGehee, N. G. (2014). Volunteer tourism: evolution, issues and futures. *Journal of Sustainable Tourism, 22*(6), 847–854.

McGehee, N. G., & Santos, C. A. (2005). Social change, discourse and volunteer tourism. *Annals of Tourism Research, 32*(3), 760–779.

McIntosh, A., & Mansfeld, Y. (2006). Spiritual hosting: An exploration of the interplay between spiritual identities and tourism. *Tourism: An Interdisciplinary Journal, 54*(2), 117–126.

Meraviglia, M. C. (1999). Critical analysis of spirituality and its empirical indicators. Prayer and meaning in life. *Journal of Holistic Nursing*, 17(1), 18–33.

Miner-Williams, D. (2006). Putting a puzzle together: Making spirituality meaningful for nursing using an evolving theoretical framework. *Journal of Clinical Nursing Science Quarterly*, 15(7), 811–821.

Mustonen, P. (2006). Volunteer tourism: Postmodern pilgrimage? *Journal of Tourism and Cultural Change, 3*(3), 160–177.

Newberg, A., & Waldman, M. R. (2006). *Why We Believe what We Believe: Uncovering our Biological Need for Meaning, Spirituality, and Truth*. New York: Simon and Schuster.

Noy, C. (2004). This trip really changed me: Backpackers' narratives of self-change. *Annals of Tourism Research, 31*(1), 78–102.

O'Sullivan, M. (2012). Spiritual capital and the turn to spirituality. In M. O'Sullivan & B. Flanagan (Eds.), *Spiritual capital: Spirituality in practice in Christian perspective* (pp. 43–59). Aldershot: Ashgate.

Pezzullo, P. C. (2009). "This is the only tour that sells": Tourism, disaster, and national identity in New Orleans. *Journal of Tourism and Cultural Change*, 7(2), 99–114.

Piedmont, R. L. (1999). Does spirituality represent the sixth factor of personality? Spiritual transcendence and the five-factor model. *Journal of Personality and Social Psychology, 67*(6), 985–1013.

Piedmont, R. L. (2001). Spiritual transcendence and the scientific study of spirituality. *The Journal of Rehabilitation, 67*(1), 4–14.

Powers, D. V., Cramer, R. J., & Grubka, J. M. (2007). Spirituality, life stress, and affective well-being. *Journal of Psychology and Theology, 35*(3), 235–243.

Prayag, G., Mura, P., Hall, C. M., & Fontaine, J. (2016), "Spirituality, drugs, and tourism: tourists' and shamans' experiences of ayahuasca in Iquitos, Peru", *Tourism Recreation Research*, 41(3), 314–325.

Raes, P. J. (2013). 'Boy, have we got a vacation for you': Orphanage tourism in Cambodia and the commodification and objectification of the orphaned child. *Thammasat Review, 16*(1), 121–139.

Ramanayake, U., McIntosh, A., & Cockburn-Wootten, C. (2018). Loss and travel: A review of literature and current understandings. *Anatolia, 29*(1), 74–83.

Relph, E. (1976). *Place and placelessness*. London: Pion Limited.

Rinschede, G. (1992). Forms of religious tourism. *Annals of Tourism Research, 19*(1), 51–62.

Ron, A. S., & Timothy, D. J. (2019). *Contemporary Christian travel: Pilgrimage, practice, and place*. Bristol: Channel View Publications.

Ryan, C. (1997). *The tourist experience: A new introduction*. London: Cassell.

Schanzel, H. A., & McIntosh, A. J. (2000). An insight into the personal and emotive context of wildlife viewing at the Penguin Place, Otago Peninsula. *Journal of Sustainable Tourism, 8*(1), 36–52.
Scheyvens, R. (2007). Exploring the tourism-poverty nexus. *Current Issues in Tourism, 10*(2–3), 231–254.
Schultz, E. K. (2005). The meaning of spirituality for individuals with disabilities. *Disability and Rehabilitation, 27*(21), 1283–1295.
Shakeela, A., & Weaver, D. (2018). "Managed evils" of hedonistic tourism in the Maldives: Islamic social representations and their mediation of local social exchange. *Annals of Tourism Research, 71*, 13–24.
Sharpley, R., & Sundaram, P. (2005). Tourism: A sacred journey? The case of ashram tourism in India. *International Journal of Tourism Research*, 7(3), 161–171.
Simmel, G. (1910). Sociability. In D. Levine (Ed.), *On individuality and social forms* (pp. 143–149). Chicago: University of Chicago Press.
Sin, H. L. (2009). Volunteer tourism – "involve me and I will learn?" *Annals of Tourism Research, 36*(3), 480–501.
Singh, S. (2009). Spirituality and tourism: An anthropologist's view. *Tourism Recreation Research, 34*(2), 143–155.
Stoddart, H., & Rogerson, C. M. (2004). Volunteer tourism: The case of habitat for humanity South Africa. *GeoJournal, 60*(3), 311–318.
Tanyi, R. A. (2002). Towards clarification of the meaning of spirituality. *Journal of Advanced Nursing, 39*(5), 500–509.
Tedeschi, R. G., & Calhoun, L. G. (2006). Time of change? The spiritual challenges of bereavement and loss. *OMEGA-Journal of Death and Dying, 53*(1), 105–116.
Timothy, D. J., & Conover, P. J. (2006). Nature, religion, self-spirituality and New Age tourism. In D. J. Timothy & D. H. Olsen (Eds.), *Tourism, religion and spiritual journeys* (pp. 139–155). London: Routledge.
Torrance, R. M. (1994). *The spiritual quest: Transcendence in myth, religion, and science.* Berkeley: University of California Press.
Trauer, B., & Ryan, C. (2005). Destination image, romance and place experience – An application of intimacy theory in tourism. *Tourism Management, 26*(4), 481–491.
Van Ness, P. H. (1996). Games. In P. H. Van Ness (Ed.), *Spirituality and the secular quest* (pp. 520–544). New York: The Crossroad Publishing Company.
Voigt, C., Brown, G., & Howat, G. (2011). Wellness tourists: In search of transformation. *Tourism Review, 66*(1/2), 16–30.
Vukonić, B. (1996). *Tourism and religion.* Oxford: Pergamon.
Wearing S. (2001). *Volunteer tourism: Experience that make a difference.* Wallingford: CABI.
Williams, E., Francis, L., Robbins, M., & Annis, J. (2007). Visitor experiences of St Davids Cathedral: The two worlds of pilgrims and secular tourists. *Rural Theology, 5*(2), 111–123.
Willson, G. B. (2010). Exploring travel and spirituality: The role of travel in facilitating life purpose and meaning within the lives of individuals (Thesis, Doctor of Philosophy (PhD)). University of Waikato, Hamilton, New Zealand.
Willson, G. B. (2016). Conceptualizing spiritual tourism: Cultural considerations and a comparison with religious tourism. *Tourism Culture & Communication, 16*(3), 161–168.
Willson, G. B., McIntosh, A. J., & Zahra, A. L. (2013). Tourism and spirituality: A phenomenological analysis. *Annals of Tourism Research, 42*, 150–168.
Zahra, A. (2006). The unexpected road to spirituality via volunteer tourism. *Tourism, 54*(2), 173–185.
Zahra, A. (2007). *Regional tourism organisations in New Zealand from 1980 to 2005: Process of transition and change.* Hamilton: University of Waikato.
Zahra, A., & McIntosh, A. (2007). Volunteer tourism: Evidence of cathartic tourist experiences. *Tourism Recreation Research*, 32(1), 115–119.

18

EXPERIENCES ALONG THE CAMINO DE SANTIAGO

Sharenda H. Barlar

Introduction

Historically, Protestants and Evangelicals have rejected the notion of pilgrimage to visit a saint or holy site, in part because of their "rejection of images as harbingers of divine presence" (Meyer 2017: 315n34). However, there has recently been an increasing acceptance of pilgrimage by Protestants and Evangelicals to both Protestant and non-Protestant destinations (Tweed 2000; Coleman 2014; Zimmer 2018). Indeed, there is a growing market for Protestant pilgrimage to different locations, particularly the Holy Land (Bajc 2007; Ron & Feldman 2009; Kaell 2014). Another popular Protestant pilgrimage is walking along the Camino de Santiago de Compostela pilgrimage trail in Spain for those who seek a prolonged period of meditation, retreat, and prayer.

Since one of the traditional motivations for pilgrimage is to pay penance for one's sins, how does the Protestant idea of grace and not works affect the Evangelical pilgrim? How do the experiences of Evangelical pilgrims differ from pilgrims from other religious faiths? What language do they use to describe their journey along the Camino? What biases do the students bring with them as Evangelicals, and are they willing to embrace other perspectives and views of the transformative nature of pilgrimage? How does the idea and actuality of pain and suffering affect the Evangelical pilgrim? Since suffering is not emphasized in Protestant traditions like it is in other religions, will students see pilgrimage as an opportunity for spiritual growth or something else?

The purpose of this chapter is to address these questions through examining how Evangelical students from Wheaton College, an Evangelical Protestant liberal arts school in Wheaton, Illinois, interpreted their experiences of walking the French Way of the Camino after walking it in the summer of 2019. As a part of a study abroad course, students were required to keep a reflective journal and invite pilgrims they met on the Camino to participate in an online Qualtrics survey. The author draws upon these student journals and some of the results of the survey to answer the questions above. Before doing so, a discussion on the evolution of Evangelical views toward pilgrimage is presented, followed by a brief history of pilgrimage along the Camino de Santiago de Compostela. Attention is then turned to discussing pilgrim's reflections on the transformative process of the Camino once they have returned to their homes.

Evolving evangelical pilgrimage

The idea of someone traveling for religious purposes—commonly referred to as a "pilgrim"— is not a uniquely Christian or medieval European idea. Indeed, people from all religious communities and faith traditions engage in travel to sacred sites for a myriad of purposes, including initiation, penance, worship, to petition for blessings, and curiosity (Morinis 1992). In fact, the word "pilgrimage" is a seventeenth-century reductionist, catch-all, and comparative term created by Western scholars and theologians to describe "all journeys motivated by religion to sacred locations" (Olsen 2019: 274; see Eade & Albera 2016).

From a Christian perspective, pilgrimage is as old as the Old and New Testaments, where Jews traveled to Jerusalem for specific festive holidays, such as the Feast of Tabernacles (Exodus 23:14–17; Deuteronomy 16:1–17) (see Janin 2002; Carr 2010). The Hebrew word for pilgrimage (magor/magur: מגָור) meant to wander (Josan 2009), as in being "the land of your sojournings" (see Genesis 17:8). The Greek words *paroikoi* (πάροικοι) and *paroikia* (παροικία) are also used in the Bible to identify those who are strangers, exiled, or sojourning or staying somewhere temporarily, and Christians, as pilgrims who are striving to complete the journey of life and return to their eternal home (Romans 13:14; Colossians 3:9–10), are encouraged to finish the journey in spite of "discouragement, distraction, desire, inattention, pride or other causes" (Laansma 2017: 243). St. Augustine seconded this notion, referring to Christians as "pilgrims through time" who are on a journey to the ultimate destination of heaven (Bourke 1958). Webb-Mitchell (2007: p. 1) also argued that "the length and breadth of the Christian life is a pilgrimage and that it is through the actual practice of pilgrimage that we may best understand the experience of growth and change in the Christian life".

Because of the hardships of wandering during Biblical times, specific codes of hospitality were given. According to Jipp (2017: p. 2), hospitality is

> ...the act or process whereby the identity of the stranger is transformed into that of guest. While hospitality often uses the basic necessities of life such as the protection of one's home and the offer of food, drink conversation, and clothing, the primary impulse of hospitality is to create a safe and welcoming place where a stranger can be converted into a friend.

Historically, hospitality was viewed in many cultures and religious communities as a sacred duty (see Olsen 2011). In Islam, "guests [were considered[guests of God, and hospitality ceased to be a choice and became a [religious] duty" (Aziz 2001: 152–153; see Din 1989: 552), while in the Old Testament, hospitality toward strangers demonstrated a person's practical commitment to God (Fields 1995: 459). In the early Christian church, hospitality (*philoxenia*) was meant to be an offering of kinship, brotherly love, and faith (*phileo*) to strangers (*xenos*). Hospitality was to include providing for all of the needs of a stranger regarding "food, shelter, and protection", thus providing "a recognition of their worth and common humanity" (Pohl 1999: 6), leading to the establishment of the Knights Templar and monasteries that doubled as pilgrim hospitals along pilgrimage routes throughout Europe (Howarth 1982; O'Gorman 2006; Brodman 2009).

The Protestant reformation brought about changes in the ways pilgrimage was viewed in Europe. Martin Luther was a major opponent of pilgrimages, especially as they were performed in the Catholic Church. Largely due to the association with the practice of indulgences, Luther believed that "all pilgrimages should be stopped. There is no good in them: no commandment enjoins them, no obedience attached to them. Rather do these pilgrimages give countless occasions to commit sin and to despise God's commandments" (Davies 1988: 98). Following Luther, other medieval and early modern literature, such as *The Canterbury Tales* (Chaucer 1985), *Don Quixote* (de Cervantes Saavedra 1998), and picaresque novels like *Lazarillo de Tormes* (Fiore 1984) and *La pícara Justina* (de Úbeda 1968), also sought to shed critical light on pilgrimage. These post-Reformation views of pilgrimage have led Evangelical Protestants to traditionally reject the notion of pilgrimage to visit a saint or a holy site.

To better understand the evolving definitions of pilgrimage from an Evangelical perspective, it is important to understanding the theological tenets of Evangelicalism. Larsen (2007) outlines five distinguishing characteristics of an Evangelical Christian. First, an evangelical is an Orthodox Protestant who believes in the Trinity. While some evangelicals align with the Apostle's Creed and the Nicene Creed, others reject these creeds and follow the doctrine of *sola Scriptura*, or that there should be "no creed but the Bible" (p. 4). Second, an Evangelical is a part of a Global Christian network that arose from eighteenth-century revival movements tied to reformers such as John Wesley and George Whitefield. Third, the Bible is the final authority on matters of faith and practice, and as such has a prominent place in the daily life of an Evangelical. As Larsen notes, "Devotional Bible reading is more foundational to evangelical piety than the rosary is to Roman Catholic piety" and "often see the sermon as the high point of corporate worship" (p. 7). Fourth, and perhaps most important characteristic of an evangelical, is the insistence that humans reconcile with God through the work of Jesus Christ alone on the cross. Fifth, evangelical Christians stress the importance of the Holy Spirit in the individual conversion process to Christ, which process can be a dramatic event that "leads to an on-going life of fellowship with God" (p. 11). As a part of this conversion process, it is the responsibility of an evangelical to share the gospel with others.

How do these five characteristics of evangelicals relate to the historical anathema toward pilgrimage travel? First, in contrast to the Medieval pilgrim who makes a pilgrimage to a sacred site as a penance for sins, Evangelicals are encouraged to instead seek God internally and rely on the Bible for guidance and to seek God's grace, as forgiveness of sin does not come through ritual travel. Second, because Evangelicals identify strongly with revival events, it is in these types of events and accompanying sermons that evangelicals feel an intensification of theological practice, faith development, and a continuation of religious tradition. This can be contrasted with pilgrims who travel to sacred sites that act as markers of tradition. Third, iconography, physical relics, and sites of miracles are often dismissed or minimized by evangelicals because *sola scriptura* is all that is needed to follow Christ. As well, the New Testament does not command evangelicals to visit sacred sites because the individual believer is a temple in which the Holy Spirit dwells (see I Cor. 6:19). Also, unlike the pilgrim who travels to sacred sites to increase their spirituality and access to God, Evangelicals instead identify specific seasons or times in their lives as opportunities for increased consecration and devotion. Fourth, since evangelical Christians are saved by grace and not works, they question why a person needs to suffer and offer penance when the price for their sins has already been paid by Christ. Fifth, although conversion to Christ comes through experiences with the Holy Spirit, this emotionalism does not fully depict the depth of an evangelical's faith. Instead of participating in pilgrimages, time is better spent sharing the Gospel with others.

Even though the above discussion clarifies why evangelical Christians have traditionally been adverse toward pilgrimage practices and rituals, in recent decades there has been an increased interest and acceptance of pilgrimage by Protestants and Evangelicals to both sites based on Protestant/Evangelical heritage (Tweed 2000; Goh 2021) and, more recently, sites based on the religious heritage of other faith traditions. For many Evangelicals, travel to religious heritage sites is seen as more of an educational experience, engaging with culture through immersion, language learning, and seeing how other religious communities attempt to transform the world around them.

The Camino de Santiago de Compostela pilgrimage in Spain has grown in popularity among Protestants and Evangelicals who seek a prolonged period of meditation, retreat, and prayer as well as forming new ways of interpreting and contributing to ongoing stories of the world. The Camino becomes a path where Evangelicals develop a new awareness of themselves and the world around them, with pilgrimage becoming a growth opportunity for the Evangelical. In many ways, participating in a pilgrimage along the Camino by Evangelicals makes them structurally akin to pilgrims of old. As Turner and Turner (1978) note, both medieval and modern pilgrims travel toward places where miracles have and continue to happen. In doing so, they leave their home environment and travel into the unknown, during which time they may encounter physical suffering. While a pilgrim may choose to travel alone, they will encounter others along the way who are seeking similar experiences and destinations, and while the pilgrim does not know what to expect in the course of their journey, in the end they will be "irrevocably changed" (Webb-Mitchell 2007: 19).

The History of the Camino de Santiago de Compostela

The Camino de Santiago is an ancient Christian pilgrimage in Spain. The Camino is comprised of several routes starting in different regions of Europe that eventually converge at Santiago de Compostela, the site of what are believed to be the remains of St. James. Before ascending into heaven, Jesus left instructions for this disciples to share the Gospel to every nation—to "go and make disciples of all nations, baptizing them in the name of the Father and of the Son and of the Holy Spirit, and teaching them to obey everything I have commanded you" (Matthew 28:19). This "Great Commission" led his disciples to travel long distances to preach the word. The apostle James was believed to be the first Christian missionary to the Iberian Peninsula, and later became the first apostolic martyr after being sentenced to die in 44 AD by Herod Agrippa.

At some point during the ninth century, tradition holds, the body of St. James was miraculously transported in a marble boat from Galilee to the region of Galicia (modern Spain). This was historically a very important event, as Spain was in the middle of the Reconquista ("reconquest") between the Moors and Christians, which had begun in 711 AD when the Umayyad empire began their conquest of Hispania. Under Alfonso II, king of Asturias and Galicia (792–842 AD), and Bishop Theodemir of Iria Flavia, the long-forgotten tomb of St. James was miraculously re-discovered in the northwest corner of the Iberian Peninsula, and a church was built on the tomb site in Santiago de Compostela in 899 AD. Soon people began to have visions of St. James (or Santiago as he was referred to in Spain), and he appeared in several battles to enable the Christian armies to eventually defeat the Umayyad empire. Santiago's assistance in the Holy Wars gave him the name "Matamoros" or "Moor-slayer". As the fame St. James grew, the number of legends and stories about him multiplied, and pilgrims visit Santiago de Compostela to pay homage to his relics. Devoted medieval Christian pilgrims, facing increasing dangers during the Crusades, shifted their destination from Jerusalem to Santiago de Compostela. Schmidt

(2012) suggests that "with the accessibility of Santiago for persons from all over Europe, it became a destination for tens of thousands of pilgrims annually for literally centuries and hence a profound cultural and spiritual unifier for all of Europe" (p. xiii). Pilgrimage to Santiago de Compostela became so popular that the Roman Catholic Church eventually recognized via Papal decree the authenticity of the remains of St. James in the twelfth century, with Santiago de Compostela becoming the third Holy City of the Western Christian world after Rome and Jerusalem. The *Codex Calixtinus*, attributed to Pope Callixtus II and written to coincide with this Papal decree, became an important guidebook to the Way of St. James (see Melczer 1993).

While the Camino saw centuries of neglect and eventual declines in the number of pilgrims traveling the Camino—in part because of the Protestant Reformation—in recent decades there has been a resurgence of pilgrimage along the Way, particularly when Santiago de Compostela was declared a World Heritage Site in 1985. This resurgence in pilgrimage travel has also coincided with what has been called the secularization of the Camino (Challenger 2014). While many of the pilgrims that walk the Camino today are Catholic, traveling the Way for penance, asking for otherworldly favors, or to express faith and thanks, a growing number of pilgrims are either from other faith traditions or claim no religion (Frey 1998). For these pilgrims, walking the Camino "provides a way to rediscover a sense of spirituality, global camaraderie, and general well-being otherwise difficult to encounter in our postmodern consumer society" (Gardner, Mentley & Signori 2016: 58–59). However, regardless of faith or non-faith tradition, traveling the Camino leads to a feeling of *communitas* or a oneness that comes with engaging in shared pilgrimage rituals (Turner 1969).

Preparing for the Camino

As noted earlier, during the summer of 2019, six Spanish-speaking students and two faculty members from Wheaton College (including the author) walked portions of the French Camino de Santiago. Since students were receiving academic credit for this study abroad course, they participated in a history, art, and culture class the Spring semester before their departure. During the semester, students were asked how a traditionally Catholic pilgrimage could be a transformative experience for Evangelical Christians based on Larsen's (2007) five characteristics of evangelicals. Interestingly, students argued that *sola scriptura* was a key reason they used to justify the value of going on a pilgrimage. This was because, as noted above, the faithful in the Old and New Testaments are referred to as sojourners and pilgrims in this world, and as such, sacred spaces, iconography in the form of art and images created over the last several centuries by different cultures grappling with Christianity, spiritual formation, a search for God, and the fact that Jesus walked from town to town during his ministry, all made their potential pilgrimage along the Camino a real and tangible act of humility and discipleship. During the pre-field course, students also created a set of devotionals that included topics like suffering, joy, solitude, patience, and community that would be shared while traveling the Camino to reflect on the group's daily experiences. These themes were grounded in the scriptures and were meant to help students to reflect on the experiences they would be having each day along the Way.

The students who decided to spend their summer semester walking the Camino did so for a variety of reasons. While they did not consider penance to be a primary motivator, traveling along the Camino would allow them a time and space to devote to prayer, devotion, and spiritual reflection. In addition, the students stated that they were looking forward to the physical challenge that the Camino would provide. Because walking the Camino can be an arduous and physically demanding journey, each weekend the students would walk with their pre-prepared backpacks and footgear in preparation for walking long distances along the Camino, strengthening their bodies so as to reduce any injuries they may receive along the Way. Although students walked an average of 10 kilometers along well-groomed suburban paths during these weekend excursions in preparation for their Camino pilgrimage, these walks did not really give them an accurate picture of the reality of walking the Camino on a variety of terrains and elevations for 20 or more kilometers a day. As well, this physical preparation was not adequate to help them avoid injury and pain during their pilgrimage. Putting an Evangelical spin on the idea of injury and pain, it was discussed how as in life, preparation cannot eliminate all forms of future suffering, and how the scriptures teach that suffering produces a maturity that comes from experience and perseverance, and in the process turning developing disciples into mature Christians.

Pain

As students found out firsthand, pain and suffering is one of the constants that is found along the Camino. Historically, the Camino de Santiago was often fraught with hardships and even death during the medieval and early modern periods. As Eade and Sallnow (1991) note,

> In the Catholic tradition, pilgrimage has always been seen as a form of penance, and indeed was imposed as a punishment for secular offences in the Middle Ages. The hardships and dangers of the journey and the bodily privations which pilgrims were obliged to undergo were thought to win the penitential pilgrim God's forgiveness and grace.... By voluntarily undergoing pain, the devotee hopes that his or her request for a material favour, for the devotee personally or for one of his or her family, will be granted by the shrine divinity.
>
> *(p. 22)*

Most, if not all the Wheaton students, developed blisters on their feet that ranged from mildly irritating to excruciating because of their size and/or location. Some students also struggled with calf, knee, Achilles heel, and back pain. The author, as faculty director, was responsible to remedy or reduce the suffering of the students during these trying times, efforts that included having the students mail home all the items in their backpacks that they had not used a week into their pilgrimage, thus lightening their backpacks; stretching and doing yoga in the morning before continuing the pilgrimage; inspecting, rubbing, taping, putting a special protective pad called "Compeed" on each other's blisters; and making a massage assembly line to work out the knots that formed from carrying backpacks for 8–10 hours a day.

As part of their academic credit, the students were required to keep a reflective journal in Spanish based on both free writing and pre-assigned topics. Some of the pre-assigned topics or reflections included open-ended prompts related to writing poetry, drawing, or writing songs describing their day. Several other prompts asked specific questions about pain and suffering. Being vulnerable and embracing a theology of suffering can be a new experience for Evangelical pilgrims. By reflecting on their personal struggles on the Camino, they were able to open themselves up for hospitality and deepen their personal faith. Below are two entries from students regarding the topic of pain and suffering while being a pilgrim:

> Blisters! What is rubbing you the wrong way? Where are you experiencing conflicts? How much is internal, and how much involves other people? What are you learning about how you interpret and handle difficulties? Is there anything you wish were different? What might you do to address the persistent blisters in your life?
>
> Baggage, burdens, necessities, and travelling light: Today is an invitation to reflect on your own background (what you bring with you). What are you glad or grateful you have with you? What has given you strength? Are there things that you brought that you wish to leave behind as your journey continues after the Camino?

The Wheaton students, however, were not the only people along the Camino had gone through physical pain and suffering. Many pilgrims who walk the Camino are dismayed to find out how physically challenging the Camino can be. This may be in part because twentieth-century books and films on the Camino, such as Coelho's (1998) *The Pilgrimage*, MacLaine's (2000) *The Camino*, and the film *The Way* (Estevez 2010) gloss over the physical hardships of the pilgrimage, instead focusing on the potential spiritual and therapeutic benefits of walking the Camino. As Norman (2009) found in his research on Camino pilgrims that many pilgrims contemplate quitting the Camino soon after they start because their expectations about having mystical experiences are not immediately met. One American that Norman interviewed had read MacLaine's (2000) book and wanted to have a similar spiritual awaking as the author had had. However, her experience quickly became blisters and aching legs, leading to an overall frustration with the journey. As Norman observed,

> Pain, although mentioned in both [Coelho and MacLean's] books, is also found to be much harsher and consuming in reality. Knees suffer under the weight of a pack and the many kilometers, and the blisters people often develop are nearly crippling. Dealing with the visceral reality of the body takes some getting used to and at first seems to take up almost all of the pilgrims' free time.
>
> *(p. 58)*

In contrast to the medieval pilgrimage who embraced the hardships of the pilgrimage journey, modern pilgrims seek to avoid injury and pain by any means necessary. Most pilgrims who plan on walking the Camino de Santiago begin to prepare for their journey months or even years in advance. There has in recent years been a proliferation of blogs and Facebook pages like the American Pilgrims Along the Camino that give advice to would-be Camino pilgrims regarding what gear, shoes, bag, and clothing should be purchased and worn; suggestions for proper footwear to eliminate blisters and orthopedic issues and dry wicking clothing to prevent chafing; and the best bags to alleviate back pain. While opinions can be diverse and subjective regarding the best shoes or footwear, one thing that everyone can agree upon is that pain is expected, no matter how prepared a person is.

Hospitality

Along with pain, hospitality is constant along the Camino. The *Codex Calixtinus* mentioned several hospices that were built at certain points of the Camino to provide shelter and to help pilgrims recover from sickness and suffering. The *Codex*, however, cautioned that these hospices were meant to only be used for one night except under extreme conditions (Melczer 1993). In addition, the *Codex* admonishes true pilgrims to only take what is on their backs as did the apostles who were sent out without money or footwear:

> What will become of those who go there with large and plump horses and mules with saddle bags of pleasing objects? If Blessed Peter went to Rome without footwear or money and, being crucified, finally went to the Lord, why do many pilgrims go to him, riding with much money and a second set of vestments, eating delicious foods, drinking very strong wine, and sharing nothing with their needy brethren?
>
> *(Coffey, Davidson & Dunn 1996: 29)*

This hospitality is important because of the dangers along the Camino. As the *Codex* warns, some false pilgrims will try to take advantage of the true and repentant sojourner through cunning means. Early modern Spain's picaresque literature is full of characters who use trickery to take advantage of naïve and unsuspecting travelers. For example, in *La pícara Justina* (de Úbeda 1968), the mischievous Justina spends her time walking the Camino participating in feast days and pilfering from intoxicated guests. Religious piety is certainly not the reason she walked the Camino! While there are still dangers along the Way, including occasional violent acts against pilgrims, most of the dangers have to do with health-related issues. Much of the Camino provides pilgrims a sense of isolation. Hospitality in the guise of a pilgrim's peer group, from amenity providers but more often from strangers along the Way, encourages weary pilgrims to continue along their journey.

As the students traveled the Camino, fellow pilgrims would offer advice about how to cope with the difficulties of the journey, and the students would share their essential oils with those who were experiencing physical ailments. Through this exchange of hospitable actions, not only did a form of *communitas* quickly take hold at the formed at the "albergues" or shelters in which the students stayed, as evenings were spent sharing mutual aches and pains with complete strangers, but the Wheaton students were inspired by the generosity and openness of the other pilgrims regarding their pain, often commenting on the kindness of strangers who had quickly become friends. In light of Christian theologies of hospitality, students had begun to model and were witnesses to true reflections of what Jesus meant when he declared in Matthew 25:34–46 that those who are welcoming and hospitable to strangers will inherit the kingdom of God. Sharing lodging, meals, and traveling together are mentioned frequently as hospitable acts in the New Testament (Luke 9: Acts 10:35). Similarly, pilgrims on the Camino share meals and lodging and walk with a community with strangers. In doing so, people change.

Resilience

The Qualtrics survey the students developed prior to their Camino pilgrimage focused on assessing the resilience of pilgrims as they were confronted with unexpected suffering, whether physical or emotional, while on their pilgrimage. The Qualtrics survey included 40 questions ranging from religious affiliation, age, amount of suffering, and distance walked. The survey was translated into Spanish, French, German, Italian, and Portuguese. Of the pilgrims that were invited to complete the survey, 281 pilgrims did so. During times of socialization along the Camino and after the three-week Camino journey through online platforms like the American Pilgrims on the Camino, we surveyed pilgrims to gauge their suffering, both physically and emotionally. For the purpose of this paper, a focus is placed on the last ten questions in the survey in which we asked pilgrims to describe the level and type of pain they encountered on the Camino and whether or not the pain increased their openness and compassion for others.

Researchers have found that physical or bodily stress can be a healthy way to regenerate the body and create resilience in the brain so humans can adapt better to society. McEwen, Gray and Nasca (2015) define resilience in this sense as

> ...achieving a positive outcome in the face of adversity. Even when the healthy brain and associated behavior appears to have recovered from a stressful challenge, studies of gene expression have revealed that the brain is not the same, just as the morphology after recovery appears to be somewhat different from what it was before stress.
>
> *(p. 8)*

Both in the survey and in speaking to pilgrims directly during the Camino, this idea of resilience was mentioned numerous times. Several pilgrims told us that would reach their chosen destination for a particular day, they would often feel that they could not continue their pilgrimage the following day. However, miraculously, the next day they would wake up feeling refreshed and ready to face another grueling day walking the Camino. In the survey results, two-thirds of respondents said that they suffered either physically or mentally. However, when asked to rank their level of physical and/or suffering on a scale from 1 to 10, there seemed to be no clear indication that one form of suffering was greater than another. Rather, respondents stated that both types of suffering equally encouraged them to increased spiritual contemplation during their pilgrimage. Respondents also mentioned that the hospitality that was given to them as they walked helped them to maintain their resilience—that sharing their personal tales of suffering with the broader Camino pilgrim community helped them to break down mental and pride-based barriers that were stopping them from experiencing true transformation. Indeed, prolonged time spent thinking and talking with others throughout the day while enduring pain often opened pilgrims' thoughts to prayer and mindful or contemplative practices. Some pilgrims commented:

> It taught me to lean on God and on my group. I had to ask for help, make me slow down, and took down my pride a few notches. It reminded me of how Jesus suffered and also reminded me of the holistic elements of the physical along with the emotional and spiritual.
>
> The suffering kept me grounded during the pilgrimage and allowed me to keep real with myself. I understood the toll this experience was taking on me, and I allowed myself to find the pain to help my self-discovery. I found out how I react to these situations I have never experienced before.
>
> I left on the Camino to help recover from traumas I'd faced but had never dealt with. The Camino forced me to process and taught me how to process my own suffering on my own, sans medication. As for when I had a bum ankle for a week, it helped give me confidence that I can honestly do anything I set my mind to. I climbed a mountain on a terribly swollen and bruised ankle, I can do all.
>
> It enabled me to grow and to finally look at the parts of myself that I had been ignoring for so long. It made me proud of parts of me that I'm typically ashamed of. It made me more confident, and it shaped my pilgrimage. I did not embark on this journey because I thought it would be a vacation. I knew I would suffer. But it all attributed (eventually) to the beauty of the Camino.

This suffering and resulting hospitality and resilience led again to a sense of *communitas*—the development of a spontaneous community that transcends markers of social structure (Di Giovine 2011). Indeed, as Schmidt (2012) notes, there is a "leveling action" that the Camino provides pilgrims, in that "we all end up hurting, aching, and limping, and seeking something...[The Camino] does away with so many of the role, class, religious, and other distinctions we place on our common humanity" (p. 135).

From an Evangelical perspective, Nouwen (1979) describes Christ as the "wounded healer," and that through suffering the wounds that people carry are truly revealed. As a person engages with and opens up to a Christian community these wounds can be healed, "not because wounds are cured and pains are alleviated, but because wounds and pains become openings or occasions for a new vision" (94). As Wheaton students learned, it is through the sharing of suffering and the giving of compassionate service that the experiences of the Camino begin to overcome the physical and emotional pain that pilgrims often experience during their journey.

Conclusion

For the Wheaton students who walked the Camino, their pilgrimage still continues. They continue to frequently eat together, speak of their shared Camino experiences, and continue to write in their reflective journals. While the present COVID-19 pandemic has led to an upsurge in online groups participating in virtual pilgrimages of the Camino de Santiago, with many forum sites actively encouraging pilgrims to seek the Camino spirit in their local spaces, the students know that there is no real substitute for being on the Camino, for it is through true physical sacrifice and experience that true reflection and *communitas* can take place. Indeed, for the students, the Camino journey afforded them a special opportunity to face physical and mental opposition while sojourning as strangers in a foreign land, and in doing so helped them to (1) develop strategies of resilience as they face pain and suffering in their own lives in the future, (2) understand that sacrifice provides growth and honesty about one's self that cannot be gained otherwise, and (3) acquire more sympathy and empathy for the struggles of strangers or others. In reality, by completing the Camino these evangelical students truly understand what it takes to "what is real, what is of God" (Cousineau 1998: xxx), and in doing so are better prepared to continue their life pilgrimage toward their eternal home.

Bibliography

Aziz, H. (2001) 'The journey: An overview of tourism and travel in the Arab/Islamic context', in D. Harrison (ed), *Tourism and the Less Developed World: Issues and Case Studies* (pp. 151–159). Wallingford, UK: CABI.

Bajc, V. (2007) 'Creating ritual through narrative, place and performance in Evangelical Protestant pilgrimage in the Holy Land', *Mobilities*, 2(3): 395–412. https://doi.org/10.1080/17450100701597319.

Bourke, V.J. (1958) *Saint Augustine: City of God*. New York: Crown Publishing.

Brodman, J. (2009) *Charity and Religion in Medieval Europe*. Washington, DC: The Catholic University of America Press.

Carr, D.M. (2010) *An Introduction to the Old Testament: Sacred Texts and Imperial Contexts of the Hebrew Bible*. Chichester, UK: John Wiley & Sons.

Challenger, D. (2014) 'Secularization and the Camino de Santiago pilgrimage', in L.D. Harman (ed), *A Sociology of Pilgrimage: Embodiment, Identity, Transformation* (pp. 128–145). New York: Ursus Press.

Chaucer, G. (1985) *The Canterbury Tales*. London and Boston, MA: G. Allen & Unwin.

Coelho, P. (1998) *The Pilgrimage*. New York: Harper Collins.

Coffey, T., Davidson, L., & Dunn, M. (1996) *The Miracles of St. James*. New York: Italica Press.

Coleman, S. (2014) 'Pilgrimage as trope for an anthropology of Christianity', *Current Anthropology*, 55(1): 281–291.

Cousineau, P. (1998) *Art of Pilgrimage*. San Francisco: Conari Press.

Davies, J.G. (1988) *Pilgrimage Yesterday and Today: Why? Where? How?* London: SCM Press Ltd.

de Cervantes Saavedra, M. (1998) *Don Quixote*. Trans T. Smollett. London: Bibliophile Books.

de Úbeda, F.L. (1968) *La pícara Justina*. Barcelona: Ediciones Zeus.

Di Giovine, M.A. (2011) 'Pilgrimage: Communitas and contestation, unity and difference—An introduction. *Tourism*, 59(3): 247–269.

Din, K.H. (1989) 'Islam and tourism: Patterns, issues, and options', *Annals of Tourism Research*, 16: 542–563.

Eade, J., & Albera, D. (2016) 'Pilgrimage studies in global perspective', in D. Albera & J. Eade (eds), *New Pathways in Pilgrimage Studies: Global Perspectives* (pp. 1–17). London: Routledge.

Eade, J., & Sallnow, M. (1991) *Contesting the Sacred. The Anthropology of Christian Pilgrimage*. Champagne: Illinois University Press.

Estevez, E. (2010) *The Way* [film]. Los Angeles.

Fields, D.H. (1995) 'Hospitality', in D.J. Atkinson & D.H. Fields (eds), *New Dictionary of Ethics & Pastoral Theology* (pp. 459–460). Downers Grove, IL: InterVarsity Press.

Fiore, R.L. (1984) *Lazarillo de Tormes*. Boston, MA: Twayne Publishers.

Foster, C. (2010) *The Sacred Journey*. Nashville, Thomas Nelson.

Frey, N.L. (1998) *Pilgrim Stories: On and Off the Road to Santiago*. Berkeley: University of California Press.

Gardner, S., Mentley, C., & Signori, L. (2016) 'Whose camino is it? (Re)defining Europe on the Camino de Santiago', in S. Sánchez y Sánchez & A. Hesp (eds), *The Camino de Santiago in the 21st Century: Interdisciplinary Perspectives and Global Views* (pp. 57–80). London: Routledge.

Goh, R.B.H. (2021) 'Protestant evangelical pilgrimages: Hagiography, supernatural influence, and spiritual mapping', *Journal of Cultural Geography*. 38(1): 1–27.

Howarth, S. (1982) *The Knights Templar*. New York: Barnes & Noble Publishing.

Janin, H. (2002) *Four Paths to Jerusalem: Jewish, Christian, Muslim, and Secular Pilgrimages, 1000 BCE to 2001 CE*. Jefferson, NC: McFarland & Company, Inc.

Jipp, J. (2017) *Saved by Faith and Hospitality*. Grand Rapids: Eerdmans.

Josan, I. (2009) 'Pilgrimage: A rudimentary form of tourism', *Geo Journal of Tourism and Geosites*, 2(4): 160–168.

Kaell, H. (2014) *Walking Where Jesus Walked: American Christians and Holy Land Pilgrimage*. New York: NYU Press.

Laansma, J. (2017) *The Letter to the* Hebrews. Eugene, OR: Cascade Books.

Larsen, T. (2007) 'Defining and locating Evangelism', in T. Larsen & D. Treier (eds), *The Cambridge Companion to Evangelical Theology* (pp. 1–14). Cambridge: Cambridge University Press.

MacLaine, S. (2000) *The Camino*. New York: Simon and Schuster.

McEwen, B., Gray, J., & Nasca, C. (2015) 'Recognizing resilience: Learning from the effects of stress on the brain', *Neurobiology of Stress*, 1: 1–11.

Melczer, W. (1993) *The Pilgrim's Guide to Santiago de Compostela*. New York: Lightening Source Incorporated.

Meyer, B. (2017) 'Catholicism and the study of religion', in K. Norget, V. Napolitano, & M. Mayblin (eds), *The Anthropology of Catholicism: A Reader* (pp. 305–315). Berkeley: University of California Press.

Mitchell-Lanham, J. (2015) *The Lore of the Camino de Santiago*. Minneapolis: Two Harbors Press.

Morinis, A. (1992) Introduction: The territory of the anthropology of pilgrimage', in A. Morinis (ed), *Sacred Journeys: The Anthropology of Pilgrimage* (pp. 1–28). Westport, CT: Greenwood Press.

Murray, M., & Graham, B. (1997) 'Exploring the dialectics of route-based tourism: The Camino de Santiago', *Tourism Management*, 18(8): 513–524.

Nicholson, H. (2001) *The Knights Hospitaller*. Woodbridge: Boydell Press.

Norman, A. (2009) 'The unexpected real: Negotiating fantasy and reality on the road to Santiago', *Literature and Aesthetics*, 19(2): 50–71.

Nouwen, H. (1979) *The Wounded Healer*. New York: Image Books.

O'Gorman, K.D. (2006) 'The legacy of monastic hospitality: The Rule of Benedict and rise of Western monastic hospitality', *Hospitality Review*, 8(3): 35–44.

Olsen, D.H. (2011) 'Towards a religious view of tourism: Negotiating faith perspectives on tourism', *Journal of Tourism, Culture and Communication*, 11(1): 17–30.

Olsen, D.H. (2019) 'Religion, spirituality, and pilgrimage in a globalizing world', in D.J. Timothy (ed), *Handbook of Globalisation and Tourism* (pp. 270–283). London: Edward Elgar.

Pohl, C. (1999) *Making Room. Recovering Hospitality as a Christian Tradition*. Grand Rapid: Eerdmans Publishing.

Rogers, S.S. (2011) *Inventing the Holy Land: American Protestant Pilgrimage to Palestine, 1865–1941*. Plymouth, UK: Lexington Books.

Ron, A.S., & Feldman, J. (2009) From spots to themed sites – The evolution of the Protestant Holy Land, *Journal of Heritage Tourism*, 4(3): 201–216.

Schmidt, W.S. (2012) *Walking with Stones: A Spiritual Odyssey on the Pilgrimage to Santiago*. Bloomington, IN: Trafford.

Turner, V. (1969) *The Ritual Process: Structure and Anti-Structure*. London: Routledge and K. Paul.

Turner, V., & Turner, E. (1978) *Image and Pilgrimage e in Christian Culture*. New York: Columbia University Press.

Tweed, T.A. (2000) 'John Wesley slept here: American shrines and American Methodists', *Numen*, 47(1): 41–68.

Webb-Mitchell, B. (2007) *The School of the Pilgrim. An Alternative Path to Christian Growth*. Louisville: Westminster John Knox Press.

Zimmer, D. (2018) 'Between art tourism and 'Protestant pilgrimage': Individual journeys to artworks in two churches of Reformed denomination near Bern, Switzerland', *Culture and Religion*, 19(4): 361–375.

19

GENDER AND PERFORMANCE IN THE CONTEXT OF RELIGIOUS AND SPIRITUAL TOURISM

Pilgrimage and sacred mobilities

Avril Maddrell

Introduction

The ways in which the social construction of gender has had a significant impact on the institutional organisation, regulation, power structures, practices, and performances of organised religion are well rehearsed. This includes gender-exclusive clergy and spaces, authority and roles based on gendered norms, and gender-defined codes of behaviour and dress within particular religious communities and/or spaces. As Morin and Guelke (2007: xix) note, "Most of the theology and practices of Christianity, Islam and Judaism were developed by men at times and places where women's inferior social (if not metaphysical) status seemed self-evident and where men placed premiums on women's sexual virtue". Even where religious teaching insists on equality in belief, the lived experience of that equality may be compromised, with social gendered norms shaping religious-spiritual beliefs, practices and mobilities. Indeed, "Women's and men's religious experience may differ significantly because religions often promote segregation in religious practices, congregation attendance, ordination, religious life, and religious identities" (Morin & Guelke 2007: xix). Historically, it is also significant that *across different faiths* gendered roles and beliefs have frequently been marked in differences between institutional and tactical religion, with women's beliefs and practices often constituting everyday tactical religion rather than loyalty to text-based authority (Ahmed 1999). It is also interesting to note that outside of institutionalised religion, women are the obvious majority in the sphere of (Western) holistic spirituality (Sointu & Woodhead 2008).

Many of these gendered relations have been highlighted and their uneven power relations are critiqued by feminist scholars (e.g., Jansen & Notermans 2012). However, if the study of religion has been gender-blind, so has gender studies often lacked sensitivity to the significance of religion, constituting a co-existing "double blindness" in this field (Gemzöe & Keinänen 2016). Attention to the mutual significance of gender and religion has prompted fruitful study of women's everyday religious experience (Ahmed 1999), including in majority and minority contexts (Aitchison, Hopkins & Kwan 2007) and in relation to different masculinities (Hopkins 2004) and femininities (Gemzöe 2005). This includes everyday spirituality (MacKian 2012), Queer spirituality (Browne, Munt & Yip 2010), and the significance of often gendered alternative spiritual beliefs-practices such as the belief in angels and Goddess worship (Rountree 2002; Utriainen 2016). At the heart of a number of these conceptual shifts is a focus on the ways in which beliefs intersect with and shape lived and embodied experience, often expressed through practice and performance—themes that are central to this chapter.

Despite these shifts and tourism's engagement with gender issues since the mid-1990s (Byrne Swain 1995; Apostolopoulos, Sönmez & Timothy 2001), gender typically remains marginal to the majority of studies of religious and spiritual travel, either ghettoised or omitted, with participant "voices" often being homogenised. While trends identified in recent religious tourism journal articles include greater diversity of beliefs and settings, the de-differentiation of religion and tourism, and transformative travel (e.g., Collins-Kreiner 2020), gender and associated intersectional identities are absent from current trends or efforts to signpost future agendas for the field. It is self-evident that if gender is significant to the study of both tourism and religion, it is significant to the study of religious-spiritual tourism. This chapter outlines the intellectual benefits of gender-sensitive research, drawing on recent scholarship and outlining an agenda for future research.

Each of the terms in this chapter's core theme of "religious and spiritual tourism" is loosely defined, the focus is on travel of any distance or duration centred on religious-spiritual sites, events, and rituals. Within the scholarship, the "religious" and "spiritual" are frequently dichotomised as mutually exclusive. However, then in fact overlap in the lived experience and practices of many (Olsen & Timothy 2006). Echoing the use of the term "pilgrim-tourist" (Vukonić 1996; Coleman & Eade 2004; Stausberg 2011), this blurred relationality is embodied in the conscious use of the hyphenated terms here, such as religious-spiritual, where appropriate. This does not deny the boundaries that institutions, individuals, and communities may choose to draw around their creeds and designations of sacred spaces and practices. Rather, the *analytical* insistence on blurred boundaries and relationality centres attention on their intersection and the spaces, practices, and experiences produced at and by those intersections rather than being unduly constrained by definitions and categories of who, what, or where is included or excluded. In response to arguments that the study of religious tourism is overly fragmented due to a focus on individual case studies (Collins-Kreiner 2020), this chapter highlights the value of attention to gender through thematic analysis of embodied participant-centred experience and meaning-making across *varied* case studies of organised and individual pilgrimages as a form of sacred mobility (religious-spiritual, spiritual and secular) (Maddrell, Terry & Gale 2015).

Gender and pilgrimage

Many pilgrimages stress the equality of pilgrims, and aspects of pilgrimage experience may cut through social and gendered norms, as reported by pilgrims in numerous contexts (see Hopkins, Stevenson, Shankar, Pandey, Khan & Tewari 2015 on the Magh Mela). Nonetheless, many forms of pilgrimage are highly gendered (Gemzöe 2012). For example, female pilgrims are not allowed to land on the Athos peninsula to visit the Orthodox monastic community, and Roman Catholic and Buddhist religious communities or retreats are commonly single sex spaces and practices. Worldwide, pilgrimage sites are typically controlled by male authorities and gatekeepers: "operating sacred sites is an almost exclusively male prerogative" (Shackley 2001: xv). However, case studies frequently demonstrate the predominance of women as pilgrims (on Christian sites, see Gemzöe 2009; Maddrell 2011; Jansen 2012). Indeed, it has been argued that the very feminisation of pilgrimage explains its marginalisation in the (Western) study of religion prior to the 2000s (Gemzöe 2005).

For many devout and/or impoverished women, pilgrimage is the only justification for travel and deprioritising material service to family, albeit they are enacting spiritual service for their families, as is evident in both Roman Catholic and Muslim women's narratives (McLoughlin 2015; Maddrell 2016; Notermans, Tuolla & Jansen 2016). This pilgrimage-in-service-of-family role has been evidenced in Catholic shrines where women overwhelmingly undertake the emotional spiritual labour of booking masses and collecting or buying commodities deemed to have sacred site-specific spiritual-therapeutic qualities such as holy water, oil, rosaries, and prayer cards (Shackley 2006; Maddrell 2016). However, in all but the most hasty of visits (as may be the case on tours), these gendered familial roles and practices do not preclude individual spiritual practice and renewal of wellbeing (Maddrell & della Dora 2013b). Just as non-institutionalised holistic spirituality practices provide a framework for women's gendered negotiation of selflessness and expressive selfhood (Sointu & Woodhead 2008), the holistic space-time of pilgrimage offers people of faith an opportunity to reconfigure the self spiritually-emotionally-physically-relationally (Maddrell 2013; Nilsson & Tesfahuney 2016). As Werbner (2015: 27) argues of Muslim pilgrimage, "the whole study of ritual embodiment, and of charisma as sacred embodiment, necessarily hinges on an understanding of symbolic movement as effecting both a metonymic and metaphoric transformation."

The work of Jill Dubisch has been pivotal in stimulating and shaping gender-informed pilgrimage studies. Using ethnographic methods, Dubisch's (1995) study of a Greek Orthodox pilgrimage church on the island of Tinos highlighted the ways in which pilgrimage devotions there are highly feminised and the ways in which such research is inflected by the gendered biography of the researcher. Other studies have shown Jewish women's devotion at the three Rachel shrines in Israel (the only women venerated in Jewish pilgrimage) (Sered 2005), and the predominance of women pilgrims at Marian shrines (Gemzöe 2009; Hermkens, Jansen & Notermans 2009), including where those shrines attract devotees from multiple faiths (Jansen 2009). While there is equality of status for women and men on the Hajj, women need to have a male chaperone and do not wear the simple white wrap worn by male pilgrims to symbolise both death and rebirth (McLoughlin 2015; Werbner 2015). Further, transgender pilgrims have been refused visas to travel to Mecca (Wasif 2017), which signals the diversity of gender issues significant to understanding the complexities of pilgrimage and other forms of religious tourism.

In the context of religious diversity and the parallel growth of secularisation in the West, it is important to note the polysemic nature of pilgrimage practices, particularly evident in "unregulated" or "open" pilgrimages (Maddrell & della Dora 2013a, 2013b). Also important to note is that not all pilgrimages are framed by religious or spiritual meaning or intent (Shackley 2001), secular pilgrimages to symbolic places or events are imbued with secular-sacred status, attributed with "magnetism" and evoke ritual performances, devotions, and embodied-emotional responses, which some describe in spiritual terms. In a later study, Dubisch (2004, 2005) participated in the annual motorcycle pilgrimages to "The Wall" in Washington, DC, commemorating US service personnel who died in the Vietnam War. This work evidenced the interstices of veteran and biker masculinities, communal remembering, and therapeutic spaces within this secular commemorative practice. A similar sense of collective identity and shared purpose is identified in other secular biking events, such as the Isle of Man TT races, which attract predominantly male spectators from around the world. Many of the 30,000 spectators who travel to the island for the annual TT festival with over 10,000 motorcycles, couch their commitment to attending the races as the compulsion to fulfil an annual commitment or a "once in a lifetime" secular pilgrimage, some explicitly using the vocabulary of pilgrimage. While on the island they perform various embodied rituals, including visiting famous-symbolic sites, meeting racers, and riding the road course. These rituals, in conjunction with clothing, particular forms of events/socialising, and bike and racecourse knowledge, coalesce around event-specific hegemonic masculinities (Maddrell, Terry, Gale & Arlidge 2015).

Within Christian pilgrimage, men tend to dominate numerically in military and nationalistic pilgrimages (Eade & Katić 2017) or in those pilgrimages that centre on physical endurance such as the Camino de Santiago de Compostela. In the latter, forms of hyper-masculinised and literal "muscular" Christianity are identified, centring variously on embodied speed and the ability to protect female pilgrims they perceive as "vulnerable", as well as authoritative knowledge (e.g. using GPS technology to master routes and blogs to establish pilgrimage narratives) (Gemzöe 2012). Rather than being vulnerable, many middle-aged Swedish women found themselves empowered by their experience of walking the Camino, captured in the eponymous quote from one respondent that she felt "big, strong and happy" (quoted in Gemzöe 2012: 37). However, it is important to note that both men walking the Camino and those attending the TT races demonstrate multiple masculinities rather than a singular monolithic masculinity. Likewise, military pilgrimages may be undertaken in the spirit of international peace-keeping as well as by those asserting nationalistic pride or territorial claims (Eade & Katić 2017). A recent study interrogating pilgrim motivations of Italian pilgrims visiting Medjugorje showed gendered differences, with men being more interested in "discovery", while "social" dimensions were more important for women (Scaffadi Abbate & Di Nuovo 2013). However, the social dimensions can also be important for some male pilgrims (Maddrell 2015), which underscores the need for nuanced gender-sensitive research design, methods, and analysis on both individual and macro scales.

Similarly, different expressions of femininities are experienced and expressed through different pilgrimages. Marian shrines are a focus for women's relation to the divine through the female personhood of Mary, with whom they identify, and see themselves and their lives. Hence, women's Marian devotion often centres on female embodied concerns such as fertility and giving birth, as well as the role and responsibilities of family and lifelong mothering (Gemzöe 2005; Hermkens et al. 2009; Jansen 2012), including diasporic family (Notermans 2012; Notermans et al. 2016). However, Mary is also a focus of intimate relationality, which can give those same women spiritual-social agency and authority, including the justification for religious travel as an expression of Marian and familial devotion. This agency and authority were notable at the Maltese national shrine of the Madonna Ta Pinu, on Gozo where, despite the institutional patriarchy of the Roman Catholic Church and male dominance of formal liturgy, older women were clearly the matriarchs, occupying the inner sanctuary of the Madonna, leading recitations of the rosary for public broadcast, and teaching spiritual practices to younger generations. Women of all ages were notably more active and expressive in their devotions at the shrine, some approaching the Madonna and the graves of venerated locals on their knees. When asked about this, the female shrine staff commented: "Men tend to give money to Our Lady rather than walk on their knees ... women are less ashamed to do these things" (Maddrell 2016: 232; also see Gemzöe 2005). Whereas masculine veneration of the Madonna, including that of visiting popes, is typically performed through financial or material gifting, feminine-relational devotion was evident in embodied performances of intimate respect-petition.

Agency and authority are also sought, and found, by women in New Age religions such as Neo-Paganism and alternative or self-spiritualities as experienced through practices such as Reiki massage and yoga retreats (Woodhead 2007). The Goddess movement has mobilised largely middle age and middle-class Western women to engage with their spirituality outside of patriarchal religious and social structures, thereby representing a political project as well as an opportunity to reimagine self and femininity through belief and ritual. This movement has generated both individual and themed group tours-pilgrimages, primarily to sites associated with ancient matriarchal societies and/or female deities, often encompassing more than one country and belief system. The Goddess pilgrim is frequently a tourist-pilgrim, described as a "postmodern figure, alienated from many of modern society's values; collecting a plethora of deities, myths, rituals and sacred sites from the world's religious traditions" (Rountree 2002: 478). Rountree's (2002) study demonstrated the ways in which embodied experience and expression is a key element of Goddess pilgrimage, whether through ritual acts or embodied resonance with site-specific "energies" through contact with the past through the materiality of place. In this she highlights the ways in which bodies simultaneously are sites of *embodied* experience and *embody* values and beliefs, as is common with other pilgrimage practices, and that in the specific case of Goddess pilgrims, they further "experience their female bodies as sacred, themselves as divine" (Rountree 2002: 494).

Beliefs, practices, and embodied performativity

Calls to engage with the psychosomatic aspects of pilgrimage (Morinis 1992) have coalesced with interest in the "spiritual magnetism" of places (Preston 1992), related notions of therapeutic spaces and practices (Winkelman & Dubisch 2005), feminist attention to gender as an analytical theme, and gender as scripted and performed identities (Butler 1990). These approaches have also served to focus attention and understanding of the *lived and embodied* nature of the human experience, including religious-spiritual beliefs and practices. Attention to embodiment, practice, and performance, combined with gender-sensitive methods and analyses, can provide valuable insights. As with gender, the study of religion in recent years has drawn much from its focus on embodiment, practice, and performance as a counterbalance to emphasis on beliefs, texts, and material artefacts, but in reality, these are often inseparable.

Religion and gender are both performative—not only represented by but also constituted in and through embodied performance (Butler 1990; Jansen 2009) as well as through mobilities that shape both religious-spiritually motivated travel and the rituals or activities enacted en route and at destination sacred sites. Mobilities, as performative embodied action and movement, is both a semantic field and metaphor (Coleman & Eade 2004); further, religious-spiritual travel and performances encompass movement and being moved, such as through embodied-material-emotional-spiritual-psychological mobilities (Hermkens, Jansen & Notermans 2009; Maddrell 2011).

This can be seen across case studies from different faiths and spiritual practices. For example, the embodied physicality of Goddess pilgrim spiritual practice was exemplified by one pilgrim at the Neolithic temple at Hagar Qim, Malta, who described lying against the ancient stones of the temple in order to connect with the "energy" of the sacred place and religious practice and practitioners of the past: "I lay down against the curved wall, fitted my body within their warm contours, and felt utterly connected to the past, present and future in the great cycle of being" (Rountree 2002: 488). For this pilgrim, her experience was an encounter with the female deity, reflected in the curved womb-like walls of the temple and the sense of personal-gendered self-acceptance and empowerment this evoked. Similar relational embodied-performative-material religious-spiritual experience through touch is also seen in other contexts. For example, Muslim pilgrims touching the Kaaba; Buddhists spinning prayer wheels; Orthodox Christian pilgrims kissing icons; and those experiencing a sense of temporal and spiritual liminality when encountering the remains of early medieval Celtic Christian chapels (Maddrell & della Dora 2013a; Maddrell 2015; Maddrell & Scriven 2016). Significantly, these heightened experiences through embodied performance are commonly reported across genders, as recorded in the experience of prayer walks in the Isle of Man:

> it was a profound experience to touch with Christians from centuries ago and with [other] Christians [....] as we prayed together
>
> *(Female, Roman Catholic)*

> Touching the stones and sitting on the walls gives me a great sense of connectedness to Christianity, to our ancestors and to this beautiful Island ... the Celtic crosses are reminders again from whence we have come. I love to come and just be in their presence ....
>
> *(Male, Methodist)*

Similar embodied-spiritual experiences have been reported in recent descriptions of the Hajj, where British pilgrim accounts, from both men and women typically reflected on bodily performances and related emotional responses:

> .... I see it as a very holistic experience and I think mainly it was the fact that I was walking in the footsteps of these great personalities that I had always heard about and spoken about and read about and thought about and imagined .... [at the Kaaba] You get this sort of rush through your body ... it just fills you up ... you just want to stand there forever, just not move from that one spot ... a lot of what you experience happens on such a spiritual level, almost like a molecular level, when you're vibrating with that anticipation ... (Aminah, 20s, London, Hajj 1997).
>
> *(Quoted in McLoughlin 2015: 49, 51)*

Recent participant-centred analysis of the Hajj and Umrah pilgrimages to Mecca, Saudi Arabia, evidence other embodied experiences and meaning-making. Women's accounts highlighted the risks associated with Hajj crowds and with particular phases and situated performative elements of the pilgrimage (e.g., stoning the pillars representing the devil), which are spaces of corporeal danger as well as of spiritual benefit. These observations were intermeshed with the gendered roles of motherhood, articulated by one mother who was conscious of her children left at home while she fulfilled her religious obligation to complete the Hajj: "that was one of the things that was upsetting me the most about leaving my children behind" (Asma, 40s, West Midlands, Hajj 2008) (quoted in McLoughlin 2015: 48). Another recent study analysing selfies from pilgrimages at Mecca (Hajj and Umrah) posted on social media by English-speaking pilgrims identified and analysed selfies as a performance of religious identity and belonging as well as religious travel. Analysis of these publicly available images showed that while both men and women pilgrims took and posted selfies, gender played a role in the framing and reception of the images. Men were overwhelmingly more likely to take *solo* selfies, compared to women, who more commonly posted group selfies. Female pilgrims were also more likely to be *criticised* for posting pictures of themselves, including by clergy from the global Muslim community. However, online community-solidarity between these women facilitated a collective response, for example assertively posting photos tagged #Selfie4Siauw after Felix Siauw, an Indonesian cleric, declared Muslim women's selfies showing their faces to be "shameless" and against Islamic law (Moumtaz 2015; Caidi, Beazley & Colomer Marquez 2018). As these brief vignettes highlight, the study of embodied sacred mobilities—religious-spiritual travel, situated and virtual practices and performances—highlights the need to interrogate the embodied experience "not least in terms of gender, ethnicity, class and empowerment. .... [and] the, as yet, underdeveloped analysis of the politics of pilgrimage" (Maddrell & Scriven 2016: 19) as part of the wider politics of gendered bodies and norms within religious travel practices.

Conclusion: setting an agenda for a critical gendered analysis of pilgrimage and spiritual tourism

Despite widespread rhetoric about the equality of pilgrims (gender, wealth, ethnicity-race, and health), many studies of pilgrimage travel and performance evidence gender (and other) differences. These differences are variously identified in degrees of inclusion/exclusion; the power relations of gendered institutional and liturgical roles; literal and normative permission to travel; gendered spaces, including gendered bodies; gendered practices and performances, including hyper-masculinity or -femininity; gendered norms of behaviour, etiquette, and dress; and circumscribed or empowered gendered agencies through religious-spiritual authority and performance. Clearly, there is a need for more systematic examinations of gender as an important aspect of intersectional identity that shapes access to, and the experience of, pilgrimage and other forms of religious-spiritual travel. Attention to gender requires reflection on the theoretical framing, research design, and methodologies of research (Maddrell 2020), as well as engagement with gendered critiques in relation to the hierarchies, discourse, opportunities, and agencies associated with empirical case studies. However, just as we need to be sensitive to the public patriarchal power relations that have shaped many formal religious institutions, pilgrimages are also mediated by their context and individual experience, evidenced in the accounts of those who find empowerment through their experience of sacred mobilities and matriarchal or gender-equality spiritual networks despite the gendered constraints of their religion or wider society. In addition to the character of the religion-spiritual practice and cultural context, socio-economic agency and age are also significant factors that intersect with gender, as is life-cycle stage, including parenthood, work commitments, retirement, elder status, and specific "significant" life events that are culturally defined as pivotal time-spaces for personal reflection and possible redirection.

The study of pilgrimage and religious-spiritual travel has been impoverished where gender has been omitted/silenced and/or monolithic (often masculine) accounts/perspectives have been presented as normative. While focusing on women's experience in pilgrimage is a priority, so too is moving beyond overly simplified gender stereotypes and binaries in studies of religious-spiritual experience. As noted above, gender needs to be treated as contingent and intersecting with other axes of identity—beliefs, institutional role, class, ethnicity, nationality, sexuality, and so on—in order to situate participant (and researcher) positionality. This in turn requires participant-centred research methods to be at the heart of such work, complemented by other methods as appropriate, including ethnographic methods and visual or textual discourse analysis.

References

Ahmed, L. (1999) *A Border Passage: from Cairo to America – A Woman's Journey*. New York: Penguin.

Aitchison, C., Hopkins, P., & Kwan, M. (eds.) (2007) *Geographies of Muslim Identities: Diaspora, Gender and Belonging*. Aldershot: Ashgate.

Apostolopoulos, Y., Sönmez, S., & Timothy, D.J. (eds.) (2001) *Women as Producers and Consumers of Tourism in Developing Regions*. Westport, CT: Praeger.

Browne, K., Munt, S.R., & Yip, A.K.T. (2010) *Queer Spiritual Spaces*. Farnham: Ashgate.

Butler, J. (1990) *Gender Trouble: Feminism and the Subversion of Identity.* New York: Routledge.

Byrne Swain, M (1995) 'Gender in tourism', *Annals of Tourism Research*, 2: 247–266.

Caidi, N., Beazley, S., & Colomer Marquez, L. (2018) 'Holy selfies: Performing pilgrimage in the age of social media', *The International Journal of Information, Diversity, & Inclusion*, 2(1–2): 8–31.

Coleman, S., & Eade, J. (eds) (2004) *Reframing Pilgrimage: Cultures in Motion.* London: Routledge.

Collins-Kreiner, N. (2020) 'A review of research into religion and tourism', *Annals of Tourism Research*, 82, 102892 (online first https://doi.org/10.1016/j.annals.2020.102892).

Dubisch, J. (1995) *In a Different Place: Pilgrimage, Gender and Politics of a Greek Island.* Princeton, NJ: Princeton University Press.

Dubisch, J. (2004) '"Heartland America": Memory, motion and the (re)construction of history on a motorcycle pilgrimage', in S. Coleman & J. Eade (eds.), *Reframing Pilgrimage: Cultures in Motion* (pp. 104–132). London: Routledge.

Dubisch J. (2005) 'Healing "the wounds that are not visible": A veteran's motorcycle pilgrimage', in J. Dubisch & M. Winkelman (eds.), *Pilgrimage and Healing* (pp. 135–154). Tucson: University of Arizona Press.

Eade, J., & Katić, M. (2017) *Military Pilgrimage and Battlefield Tourism: Commemorating the Dead.* London: Routledge.

Gemzöe, L. (2005) 'The feminization of healing in pilgrimage to Fatima', in J. Dubisch & M. Winkelman (eds.), *Pilgrimage and Healing* (pp. 25–48). Tucson: University of Arizona Press.

Gemzöe, L. (2009) 'Caring for others: Mary, death and the feminization of religion in Portugal', in A. Hermkens, W. Jansen, & C. Notermans (eds.), *Moved by Mary: The power of Pilgrimage in the Modern World* (pp. 149–164). Farnham: Ashgate.

Gemzöe, L. (2012) 'Big, strong and happy: Reimagining femininity on the way to Compostela', in W. Jansen & C. Notermans (eds.), *Gender, Nation and Religion in European Pilgrimage* (pp. 37–54). Farnham: Ashgate.

Gemzöe, L., & Keinänen, M. (2016) 'Contemporary encounters in gender and religion: Introduction', in L. Gemzöe, M. Keinänen, M., & A. Maddrell (eds.), *Contemporary Encounters in Gender and Religion: European Perspectives* (pp. 1–28). London: Palgrave Macmillan.

Hermkens, A.-K., Jansen, W., & Notermans, C. (2009) 'Introduction: The power of Marian pilgrimage', in A.-K. Hermkens, W. Jansen, & C. Notermans (eds.), *Moved by Mary: The Power of Pilgrimage in the Modern World* (pp. 1–16). Farnham: Ashgate.

Hopkins, P.E. (2004) 'Young Muslim men in Scotland: Inclusions and exclusions', *Children's Geographies*, 2(2): 257–272.

Hopkins, N., Stevenson, C., Shankar, S., Pandey, K., Khan, A., & Tewari, S. (2015) 'Being together at the Magh Mela: The social psychology of crowds and collectivity', in A. Maddrell, A. Terry, & T. Gale (eds.), *Sacred Mobilities* (pp. 19–40). Farnham: Ashgate.

Jansen, W. (2009) 'Marian images and religious identities in the Middle East', in A. Hermkens, W. Jansen, & C. Notermans (eds.), *Moved by Mary: The Power of Pilgrimage in the Modern World* (pp. 33–48). Farnham: Ashgate.

Jansen, W. (2012) 'Old routes, new journeys: Reshaping gender, nation and religion in European pilgrimage', in W. Jansen & C. Notermans (eds.), *Gender, Nation and Religion in European Pilgrimage* (pp. 1–8). Farnham: Ashgate.

Jansen, W., & Notermans, C. (eds.) (2012) *Gender, Nation and Religion in European Pilgrimage.* Ashgate, Farnham.

MacKian, S. (2012) *Everyday Spirituality: Social and Spatial Worlds of Enchantment.* London: Palgrave Macmillan.

Maddrell, A. (2011) '"Praying the keeills": Rhythm, meaning and experience on pilgrimage journeys in the Isle of Man', *Landabrefið*, 25: 15–29.

Maddrell, A. (2013) 'Moving and being moved: More-than-walking and talking on pilgrimage walks in the Manx landscape', *Journal of Culture and Religion*, 14(1): 63–77.

Maddrell, A. (2015) 'Mobilising the landscape and the body in search of the spiritual: Journeying, performance and community', in A. Maddrell, V. della Dora, A. Scafi, & H. Walton (eds.), *Christian Pilgrimage, Landscape and Heritage: Journeying to the Sacred* (pp. 150–170). London: Routledge.

Maddrell, A. (2016) 'Gendered spaces, practice, emotion and affect at the Marian Shrine of Ta Pinu, Malta', in L. Gemzöe, M. Keinänen, & A. Maddrell (eds.), *Contemporary Encounters in Gender and Religion. European Perspectives* (pp. 219–240). London: Palgrave Macmillan.

Maddrell, A. (2020) 'Mixed mobile methods for a mobile practice: Inclusive research on pilgrimage mobilities', in M. Büscher, M. Freudendal-Pedersen, S. Kesselring, & N. Grauslund Kristensen (eds.), *Handbook of Methods and Applications for Mobilities Research* (pp. 202–211). Cheltenham: Edward Elgar.

Maddrell, A., & della Dora, V. (2013a) 'Crossing surfaces in search of the holy: Landscape and liminality in contemporary Christian pilgrimage', *Environment and Planning A*, 45(5): 1104–1126.

Maddrell, A., & della Dora, V. (2013b) 'Editorial: Spaces of renewal', *Culture and Religion: An Interdisciplinary Journal*, 14(1): 1–7.

Maddrell, A., & Scriven, R. (2016) 'Celtic pilgrimage past and present: Liminal landscapes, liturgical practices and embodied mobilities', *Social and Cultural Geography*, 17(2): 300–321.

Maddrell, A., Terry, A., & Gale, T. (eds) (2015) *Sacred Mobilities*. Farnham: Ashgate.

Maddrell A., Terry, A., Gale, T., & Arlidge, S. (2015) 'At least once in a lifetime': Sports pilgrimage and constructions of the TT races as 'sacred journey'', in A. Maddrell, A. Terry, & T. Gale (eds.), *Sacred Mobilities* (pp. 7–91). Farnham: Ashgate.

McLoughlin, S. (2015) 'Pilgrimage, performativity, and British Muslims: Scripted and unscripted accounts of the Hajj and Umra', in M. Luitgard & M. Buitelaar (eds.), *Hajj: Global Interactions through Pilgrimage* (pp. 41–64). Leiden: Sidestone Press.

Morin, K.M., & Guelke, J.K. (eds.) (2007) *Women, Religion, & Space: Global Perspectives on Gender and Faith*. Syracuse, NY: Syracuse University Press.

Morinis, A. (1992) 'Introduction: The territory of the anthropology of pilgrimage', in A. Morinis (ed.), *Sacred Journeys: The Anthropology of Pilgrimage* (pp. 1–28). Westport, CT: Greenwood Press.

Moumtaz, N. (2015) 'Refiguring Islam', in S. Altorki (ed.), *A Companion to the Anthropology of the Middle East* (pp. 125–150). Hoboken, NJ: Wiley.

Nilsson, M., & Tesfahuney, M. (2016) 'Performing the "post-secular" in Santiago de Compostela', *Annals of Tourism Research*, 57: 18–30.

Notermans, C. (2012) 'Interconnected and gendered mobilities: African migrants on pilgrimage to Our Lady of Lourdes in France', in W. Jansen & C. Notermans (eds.), *Gender, Nation and Religion in European Pilgrimage* (pp. 19–36). Farnham: Ashgate.

Notermans, C., Tuolla, M., & Jansen, W. (2016) 'Caring and connecting: Reworking religion, gender and families in post-migration life', in L. Gemzöe, L. Keinänen, & A. Maddrell (eds.), *Contemporary Encounters in Gender and Religion: European Perspectives* (pp. 241–258). London: Palgrave Macmillan.

Olsen, D.H., & Timothy, D. (2006) 'Tourism and religious journeys', in D.J. Timothy & D.H. Olsen (eds.), *Tourism, Religion and Spiritual Journeys* (pp. 1–13). London: Routledge.

Preston, J. (1992) 'Spiritual magnetism: An organising principle for the study of pilgrimage', in A. Morinis (ed.), *Sacred Journeys: The Anthropology of Pilgrimage* (pp. 31–46). Westport, CT: Greenwood Press.

Rountree, K. (2002) 'Goddess pilgrims as tourists: Inscribing the body through sacred travel', *Sociology of Religion*, 63(4): 475–496.

Sered, S.S. (2005) 'Healing as resistance: Reflections upon new forms of American Jewish healing', in L.L. Barnes & S.S. Sered (eds), *Religion and Healing in America* (pp. 231–252). Oxford: Oxford University Press.

Scaffidi Abbate, C., & Di Nuovo, S. (2013) 'Motivation and personality traits for choosing religious tourism: A research on the case of Medjugorje', *Current Issues in Tourism*, 16(5): 501–506.

Shackley, M. (2001) *Managing Sacred Sites*. London: Continuum.

Shackley, M. (2006) 'Religious retailing at Ireland's national shrine', in D.J. Timothy & D.H. Olsen (eds.), *Tourism, Religion and Spiritual* Journeys (pp. 94–103). London: Routledge.

Sointu, E., & Woodhead, L. (2008) 'Spirituality, gender, and expressive selfhood', *Journal for the Scientific Study of Religion*, 47(2): 259–276.

Stausberg, M. (2011) *Religion and Tourism: Crossroads, Destinations and Encounters*. London: Routledge.

Terry, A., Maddrell, A., Gale, T., & Arlidge, S. (2015) 'Spectators' negotiations of risk, masculinity and performative mobilities at the TT Races', *Mobilities*, 10(4): 628–648.

Utriainen T. (2016) 'Desire for enchanted bodies: The case of women engaging in angel spirituality', in L. Gemzöe, M.L. Keinänen, & A. Maddrell (eds.), *Contemporary Encounters in Gender and Religion* (pp. 175–193). Cham: Palgrave Macmillan.

Vukonić, B. (1996) *Tourism and Religion*. Oxford: Elsevier.

Wasif, S. (2017) 'Transgender community barred from Hajj, Umrah', *Gale Onefile*, December 17, Online: https://go.gale.com/ps/i.do?p=STND&u=rdg&id=GALE|A519364123&v=2.1&it=r&sid=STND&asid=145f37d9. Accessed June 11, 2020.

Winkelman, M., & Dubisch, J. (2005) 'Introduction: The anthropology of pilgrimage', in J. Dubisch & M. Winkelman (eds) *Pilgrimage and Healing* (pp. ix–xxxvi). Tucson: University of Arizona Press.
Werbner, P. (2015) 'Sacrifice, purification and gender in the Hajj: Personhood, metonymy, and ritual transformation', in L. Mols & M. Buitelaar (eds.), *Hajj: Global Interactions through Pilgrimage* (pp. 27–39). Leiden: Sidestone Press.
Woodhead, L. (2007) 'Why so many women in holistic spirituality?' in K. Flanagan & P. Jupp (eds.), *The Sociology of Spirituality* (pp. 115–125). Farnham: Ashgate.

20

ISSUES OF AUTHENTICITY IN RELIGIOUS AND SPIRITUAL TOURISM

Maureen Griffiths and Maximiliano E. Korstanje

Introduction

Scholars and policy-makers agree that religious tourism is a vital force that not only reinvigorates local economies but is also one of the leading subsegments of tourism worldwide (Cochrane 2007; Olsen & Timothy 2006). The growth of religious tourism has coincided with the interests of researchers who are studying these complex matters. Religion and tourism are inextricably intertwined (Stausberg 2011; Vukonić 2002). As Olsen and Timothy (2006) note, even if religious tourism has not traditionally been seen as a phenomenon associated with economic development, today the segment has reached unparalleled levels of spending and economic growth for host destinations. In fact, much of the increasing academic interest in religious tourism has been driven by its economic potential.

This has raised serious concerns among some policy-makers for various reasons. First, sacred places are commoditized and sold to international heritage seekers who are attracted mainly by cultural and recreational opportunities. As a result, sacred sites, such as mosques, temples, and churches, have become secular tourist attractions. However, there is a dichotomy, or at best a dissociation, between religious tourism as a sacred rite and religious tourism as a secular form of gazing. To put it in another way, while religion was historically rooted in humankind's spiritual nature, which provided people's needed transcendence, tourism has been conceived traditionally as a secular activity oriented toward the maximization of pleasure. Such a misconception of tourism (pleasure-seeking is only one of many travel motivations) caused much debate around the belief that religious tourism leads to a dissociation from the 'authentic' experience. Some authors have emphasized the ideological nature of religious tourism and pilgrimage (Augé 1995; MacCannell 1976; Norman 2011; Weidenfeld & Ron 2008), whereas others contend that religious tourism, far from simply commoditizing the sacred, enhances social cohesion and group solidarity (Cohen 1979; Korstanje 2018; Raj & Griffin 2015a). A dichotomous classification between pleasure travelers and pilgrims was originally introduced by Cohen (1992), who distinguishes between two different types of travelers: those who depart from the 'center' to explore the periphery in a quest for pleasure, and others who are oriented towards understanding the 'Other'. To some extent, this position marked a conceptual differentiation between pilgrims who seek an authentic religious experience, and other travelers who may seek hedonic experiences and leisure-oriented services (Cohen 1992).

This chapter explores these two contrasting positions. In so doing, it discusses advances in the sociology of religion, from Durkheimian insights in the *L'Annee Sociologuique* to the present. Starting from the premise that religion plays a leading role in the configuration of culture, the chapter further interrogates the intersection of spirituality and society. Further, the morphology and functionality of religiously determined social institutions are examined.

Due to length constraints, this chapter is limited to considering the following questions in relation to the nature of authenticity: is religious tourism an activity that alienates pilgrims or simply a modern manifestation of the sacred pilgrimage? What are the motivations of this emerging subsegment? Is secularization theory enough to explain these types of complex issues? What are the main challenges religious tourism will face in the future?

The first section of this chapter reviews the main ideas of Max Weber (1864–1920) and Emile Durkheim (1858–1917), two senior sociologists who laid the foundations of the sociology of religion. Both were concerned about the advance of industrialization, which potentially erodes and disrupts social reciprocity. The future of religion for them is grim, whereas a depersonalized spirit of rationalization occupies a central position in all spheres of society. Durkheim was convinced that industrialization would not only destroy the authenticity of proper tribal life, but also social ties. For him, religion is nearly a form of authenticity, while secularization engenders closed and alienated forms of consumption. This critical position leads many scholars to accept the binary profane-sacred dichotomy. The second section focuses on the positives and negatives of religious tourism, as well as the socioeconomic impacts on the destination community. Although scholars have not reached a consensus about the different manifestations of religious tourism, they argue that it is time to reconsider the idea of sacredness. This led to an empirical case study, in the third section, which looks to the pilgrimage at Lujan, Argentina, to highlight certain elements of authenticity at a pilgrimage event. The views presented diverge sharply from previous publications, offering a fresh theoretical perspective regarding authenticity and religious tourism.

Religion and society

Sociologists and anthropologists were historically interested in the impact of religion on societal scaffolding. Emile Durkheim and Max Weber were pioneers in the study of religion and its sociological roots, and they emphasized the future of religion in modern society. They understood tribalism to reflect all that is repressed in contemporary society. The notion of authenticity occupied a central position in the work of Durkheim and Weber. Since he was neither an anthropologist nor a sociologist in the strictest sense, but rather a philosopher, Durkheim never conducted personal fieldwork to validate his hypotheses. In fact, his groundbreaking ideas came from previous ethnological studies conducted by others in the Pacific islands. In 1912, he published a seminal book titled *The Elementary Forms of Religious Life* (republished in 2008). He believed that religion and religiosity were social phenomena, wherein the community develops an emotional mechanism of adaptation in order for social cohesion to be successfully maintained. Vulnerability, as well as human fragility before a hostile environment, leads the community to create symbols to achieve a durable sentiment of security. Based on earlier studies of totemic groups in Australia, Durkheim maintains that the figure of animals (or plants) holds a sacred power that sublimates into the political hierarchy of the clan. In this way, animals express the role of gods as a much deeper system of beliefs and narratives that legitimate the morality of the community. In brief, this sacred origin, in Durkheim's (2008) thinking, is the common factor all religions ultimately share. He casts some doubt with respect to the future of industrial societies, suggesting that the secularization process will not only dismantle the influence of religiosity but also deteriorate social ties. In his view, industrialization may very well jeopardize social reciprocity (Durkheim 2008).

The same concern was expressed in the early works of Max Weber (1993). Unlike Durkheim, who supported his evidence with ethnographical studies conducted by other scholars, Weber ultimately adopted a historical exploration of religion. Weber conceptualized an all-encompassing model of many non-western religions. He shared the same concern as Durkheim regarding global secularization through the advance of an industrial ethos, but he aimed to explain how religion structures the socioeconomic background of society. In other words, Weber suggested that Calvinism and Protestantism gradually evolved towards the consolidation of capitalism. The capitalist system was based on the combination of a "sense of predestination" and the logic of instrumentality. In contrast to older economic forms, capitalism is the only system where the belief in magic is likely to decline (disenchantment with the world). Although Durkheim doubted the possibility that capitalism could be reversed, Weber was not so doubtful. Capitalism emerged as a consequence of professionalization, which was accompanied by the rationalization of social life. The depersonalization and the rise of social conflict are inevitable signs of the consolidation of such a process. Although Marx, too, exerted some influence in the sociology of religion, he will not be included in this discussion. He departed from a materialist tradition, which led him to avoid debates about religion.

Both Durkheim and Weber left a rich legacy to be followed by other historians and sociologists of religion (Mommsen & Osterhammel 2013; Stark, Doyle & Rushing 1983). Although Weber has not been widely incorporated in the field of tourism studies (see Olsen 2010), Durkheimian thinking can be seen in the early work of American anthropologist, Dean MacCannell, moved by understanding the power of religion, and of course tourism, in society. His innovative vision extends the anthropology of consumption to the social influence of religion, including his advances in linguistics and semiotics (structuralism). In essence, he is principally interested in describing, like Durkheim, how society remains united. With John Urry (1990, 2002), whose work has been widely debated, MacCannell was one of the pioneers of the sociology of tourism.

To understand MacCannell (1973, 1976) fully, a previous knowledge of Levi-Straussian structuralism or Durkheimian texts is required. MacCannell contends that the capitalist system has prospered not only by evolutionary means, where more sophisticated means of production replace tribal economies, as Durkheim observed but also due to a gap between two contrasting spheres: leisure and labor. While the latter determines the economic hierarchy of society, engendering a state of psychological frustration and resentment, the former refers to a recreational sphere where the resulting deprived worker is rejuvenated.

In his seminal work, *The Tourist: A New Theory of the Leisure Class*, MacCannell (1976) postulates, citing Durkheim, that sociology advanced due to the discovery of tribal totems as a key factor in the social cohesion of the tribal community. Likewise, there is an unquestionable divide between the sacred and the profane in industrial societies. Secularization has accelerated the decline of religion. In tribal organizations, the main source of power emanates from the figure of the totem. However, in industrial cultures, there is a gap, which MacCannell believes is filled by leisure consumption and tourism. To put it otherwise, tourism is a modern-day totem (MacCannell 1976). To some extent, this resolves part of the problem. It partially explains how society maintains cohesion. The production of capital is fixed by the combination of costs and prices. Marginal poverty is a direct consequence of capitalist production. This happens because capital owners earn profits by monopolizing the means of production, as well as a favorable legal framework. Workers are mere consumers who voluntarily transfer their earnings back to capital owners. While on holiday, workers become consumers who pass their earnings from paid wages back into the system (MacCannell 1976). Hence, tourism reportedly serves as an ideological platform that revitalizes the psychological frustrations occurring in the context of work. However, this leads to a more difficult question: what is the main motivation of modern tourists to visit fabricated landscapes?

In response, MacCannell (1976) coins the term 'staged-authenticity', in accordance with Goffman's (1959) work, to denote an analogy between theatre and social life. At a theatre, countless microinteractions among subjects take place. The staff carefully constructs a staged front to engage others (i.e. audience) in a play, while the backstage shows life as it really is. In the modern world, many tourists are obsessed with consuming 'authenticity', which, paradoxically, opens the door to 'staged authenticity' (Chhabra 2019; Chhabra, Healy & Sills 2003). The growth in consuming fabricated narratives and landscapes in relation to the decline of religious adherence, ushers society into needs-transforming theatricalization and authenticity-seeking, as the main reason to travel (MacCannell 1973, 1976; Ron & Timothy 2019).

As previously noted, tourism has evolved and grown rapidly, accompanied by higher levels of depersonalization, standardization, and predictability endemic to the notion of McDonaldization. Modern tourism has replicated not only the material asymmetries of capitalism but also the meta-discourse that has legitimized the authority of the ruling elites. As a result, cultures, cities, and even people are commoditized and exchanged in the marketplace where digital technologies have produced inauthentic empty spaces. The alterity is not significant to tourism consumption unless it can be adjusted to what tourists finally want.

Ethics plays a leading role in sanitizing the negative effects of modern tourism, as MacCannell (2001, 2011, 2012) concludes. Although he was widely criticized, like Durkheim, for offering a generalized and essentialized theory of tourism, which adjusts empirically only to modern destinations, and not to tourism as a social institution (Cohen 1987; King 2000; Knudsen & Rickly-Boyd 2012; Korstanje 2016), it is no less true that his insights have notably influenced religious tourism studies, for better or for worse. MacCannell provides a critical diagnosis of tourists' behavior and their quests for authenticity, which seems to be externally imposed by a superstructure that precedes it. Thus, methodologically speaking, what tourists feel or experience is not important as a source of information for the fieldworker. The answers lie in the meta-discourse, which is pre-fixed in the tourist consciousness, a type of mega-matrix composed of beliefs, stories, allegories, and signs carefully articulated to keep the hegemony of the status quo. In short, what tourists overwhelmingly seek is an authentic experience that is personalized and subjective (Wang 1999).

Religious tourism

Pilgrimage studies has a very long history. However, its close cousin, religious tourism, has increasingly attracted the interest of social scientists in recent years. It has also become a mainstream form of tourism in many sacred destinations (Butler & Suntikul 2018; Stausberg 2011; Timothy & Olsen 2006). In some pilgrimage destinations, local residents are reluctant to receive tourists when their intentions are not primarily associated with religion and faith (Collins-Kreiner, Shmueli, & Ben Gal 2015). This discussion begins with religious tourism as a sacred or profane activity. One of the aspects that characterize sacredness is its impermeability with the outside. Religious beliefs and rituals are locally practiced and, of course, what is seen as sacred by one group may be seen simply as culturally interesting by others

(Raj & Griffin 2017). This may be a source of conflict, as well as a mechanism that helps foster social bonds.

Raja and Raj (2017) emphasize the ability of religious tourism to enhance social cohesion and may even allow the convergence of different faiths and cultural values into the same community. Furthermore, there is no firm evidence to suggest that nonreligious tourism only embraces secularism or religious tourism only embraces a sense of sacredness. Olsen and Timothy (2006) acknowledge the reappearance of pilgrimage stemming from many combined factors, such as the rise of religious fundamentalism, the retreat of some traditional expressions to medieval spirituality, the expansion of transport and access, and other variables. People who may be subject to countless deprivations have a need to believe in a supreme being but also understand the deeper meaning of life. However, as Olsen and Timothy (2006) suggest, the notion of 'pilgrimage' may don secularized forms that range from nostalgic attractions to the homes and graves of celebrities.

> Likewise, many people travel to a widening variety of sacred sites not only for religious or spiritual purposes or to have an experience with the sacred in the traditional sense but also because they are marked and marketed as heritage or cultural attractions to be consumed. They may visit because they have an educational interest in learning more about the history of a site or understanding a particular religious faith and its culture and beliefs, rather than being motivated purely by pleasure-seeking or spiritual growth.
>
> *(Olsen & Timothy 2006: 5)*

As previously noted, Raj and Griffin (2015b) and Ron and Timothy (2019) lament that tourism is widely misconceived as the epitome of pleasure-driven postmodern consumerism. From its inception, tourism emulated a sacred journey-like quest for a lost paradise. This coincides with the reflections of Digance (2006), who premises that the notion of pilgrimage associates directly with the sacred journeys of the Middle Ages, even if now the term has a less spiritual connotation. Nonetheless, this dichotomous distinction between secular and sacred journeys rests on unsound foundations. As Digance notes, ancient travelers attending festivals or games should not be recognized as pilgrims. By the same token, these days, religious festivals and pilgrimages to holy cities may be considered significant nonpilgrimage tourist events. She believes that pilgrimages activate a much deeper and more philosophical search for meaning, above all concerning the mysteries of life. Hence, we have to consider two types of secular journeys: those that satisfy the spiritual needs within the parameters of traditional religion and those that emerge from people's non-religious desires for inner spirituality.

> All pilgrims share the common trait in that they are searching for, and expect also to be rewarded with, a mystical or magico-religious experience –a moment when they experience something out of the ordinary that marks a transition from the mundane secular humdrum world of our everyday existence to a special and sacred state.
>
> *(Digance 2006: 38)*

Rotherham (2015) notes that pilgrimage is one of the oldest forms of travel and human mobility and has been defined as such by scholars. The literature suggests the need to define pilgrimage, tourism, and religious tourism separately (Collins-Kreiner 2010a, 2010b). As a sacred journey, Rotherham believes that pilgrimage exhibits an act of devotion, which is rooted in religion. Tourism, however, denotes the act of moving out of the usual environment for recreational ends, although this is an outdated and misconceived view of tourism, for tourism also includes travel for health, religion and education. Thus, according to Rotherham, tourists and locals maintain religious and nonreligious beliefs that are interchanged in the same contested spaces. Visitors to religious sites may experience a range of motivations, some with no religious connections (Griffiths 2011), and the management of sacred sites must take these realities into consideration (Wiltshier & Griffiths 2016). Some agnostic visitors may experience a 'religious' or spiritual episode that generates a strong attachment to the place they visit. Equally important, the church or sacred site benefits economically from visiting secular tourists. Hence, the border between the sacred and the secular is blurred in the geographies of pilgrimage and religious tourism. In fact, religious tourism represents the most important economic sector in many sacred localities (Ron & Timothy 2019). Paradoxically, in Europe, although church-going is facing a serious decline, there is growing interest in Catholic churches and pilgrimage routes as heritage attractions. In the context of higher levels of uncertainty and anxiety, many people are abandoning the traditional institutions to enthusiastically embrace new forms of faith (Liutikas 2014; Timothy & Conover 2006; Zwissler 2011). This search for meaning can be seen in the rise of religious tourism as a major subsegment of the industry (Richards & Fernandes 2007), despite declining attendance. The tension between sacred and profane spaces was examined by Bremer (2006), who argues that places are culturally constructed and negotiated and even take different shapes through time. Places materialize reciprocal relations among peoples and interactions, which transcend the individuality of the subject.

> The importance of place in social relationships goes beyond its role as the site where people interact. In fact, place serves as an integral element in all social relations, both as a determinant of those relations and as a product of them as well. This is specially evident in places of religion. The special character of holy site endows its occupants with a degree of social prestige.
>
> *(Bremer 2006: 36)*

Religious practices, accompanied by the hierarchies they create, are far from being static but evolve over time. With this background in mind, Bremer acknowledges that religious places denote relational discourses that are articulated in a specific semiotic infrastructure. This point suggests two important connotations. On the one hand, the same place can sustain differently motivated tourists, for example, adherents and non-adherents. On the other hand, these visitors experience different religious interpretations and meanings, even though they visit the same places simultaneously. This reflects a strong sense of subjective authenticity, for what may be authentic to one visitor might not be authentic to another. What both forms of tourism, leisure tourism and religious tourism, have in common seems to be a quest for authentic experiences (Belhassen, Caton & Stewart 2008; Terzidou, Scarles & Saunders 2018). In this vein, the discourse of authenticity matches with religiosity in the same way that both leisure and religious tourists' need to consume authenticity as a mediated experience (Belhassen et al. 2008; Bremer 2006).

It appears that a universal definition of religious tourism remains elusive. Some authors have highlighted the negative impacts on the environment and local economy after thousands of pilgrims visit a sacred site (Shinde 2007). The maladies exacerbated by tourism such as physical degradation, cultural change and appropriation, depletion of resources, and social discord are also ever-present in religious destinations (Egresi, Kara & Bayram 2014; Karar 2010; Raj & Griffin 2017; Richards 2007; Shackley 1999). Not all pilgrimages are strictly linked to religiosity or interpersonal mediation. As the next section shows, pilgrimage philosophically interrogates the capacity of the self to undertake the journey (existential authenticity) rather than a question of faith. Examples of pilgrimages that are fraught with alcohol or drug abuse, as well as other secular practices oriented to entertainment, abound. The etymology of the term 'pilgrimage' comes from the Latin *peregrinare*, which means to wander or cross the land. The ancients used this word to express any touring that lacked comfort. These kinds of journeys were reserved for people, who for any reason, decided to travel outside well-established or civilized areas.

Pilgrimage to Lujan

On December 8 every year, the city of Lujan, in Buenos Aires province, Argentina, pays tribute to *Nuestra Señora de Lujan* (Our Lady of Lujan), a representation of the Blessed Virgin Mary. This event is the largest pilgrimage in Argentina and attracts thousands of participants who arrive from different points of Buenos Aires and neighboring counties at the Cathedral of Lujan. The 70-km walk generally takes one or two days. The priest Federico Grote organized the first pilgrimage on October 29, 1893. Originally, he was accompanied by approximately 400 worshippers who swore loyalty to *Nuestra Señora de Lujan* and fervently promised to return every year. The pilgrimage now attracts more than one million people every year.

The pilgrimage to Lujan takes place twice a year, so the timeframe for gathering first-hand data is limited. Fieldwork took place during the pilgrimage in 2014 and 2015, which involved covert observations and interviews with 30 pilgrims aged 20–50 years old, including 10 males and 20 females. Pilgrims were observed and the 30 participants were interviewed during the walk to Lujan city. Although the outcomes of the fieldwork described below are not exhaustive, they reveal at least a partial analysis of the motives and behaviors of pilgrims participating in the event. Based on the observations and interview data, three types of pilgrims were identified.

Secularized pilgrims

The people in this category are not motivated by religion or faith; rather, they seek new and interesting experiences. These are secular tourists who accept the challenge to walk more than one day to satisfy their own sense of self-actualization. In some cases, they are amateur athletes or students in martial arts, or some other extreme sport. Specifically, this segment represents the age cohort of 25–30 years. As an example, one of the interviewees (John, 28 years old) works as an insurance agent in Buenos Aires. He holds a BA degree in economics, is unmarried and has no children. His motivation is the adrenaline rush of walking without knowing whether he will reach the finish line (the cathedral). The cathedral is not a religious destination for him, but rather a goal to reach as a demonstration of endurance. This secular pilgrimage symbolizes not only hard work but also human nature. For John, his demonstrated athleticism and the adrenaline rush, not devotion, is the main reason behind the walk. He considers himself agnostic and holds a critical view of the Catholic Church. Not only does John not believe in the divinity of Christ, he also strongly agrees with the separation of powers between the state and the Catholic Church. Though he believes that religion has value because it educates people, when church hierarchy gets into politics, its goals are distorted. It is noteworthy that John is a supporter of left-wing feminism and pro-abortion laws. As he puts it,

> I do believe that Islam is a religion of peace and love, but what happens when the leaders become political? The expansion of radicalized cells comes from a distorted view

> of Islam, which ideologically promises a better place in the after-life, while promoting violence as a form of extortion. The case of terrorism in the Middle East and Europe seems to be a clear example. The same happens in Argentina when the Church opposes the right of free abortion. This law would save thousands of young lives if you ask me.
>
> *(John)*

In summary, secularized pilgrims are not only unsympathetic with religious motives, but they also participate in the pilgrimage as a form of testing their physical endurance. Like John, other interviewees in this category maintain a certain level of respect for the dogma of the Catholic Church. For example, abortion is a hot topic in Argentina and something that the church has strongly opposed, although it has recently received a lot of attention in the media and in parliamentary debates. In liberal circles, it has tainted the image of the church even further. Abortion is mentioned here because two secularized pilgrims demonstrated a strong anti-Catholic sentiment in relation to this issue. The two interviewees did not blame the faith but rather the perceived injustices perpetuated by the church.

For secularized-pilgrims, the event is not an objectively authentic faith-based experience, but an existential authentic (Wang 1999) opportunity to challenge their physical stamina. In general, they do not believe in God. They trust only their own capacity to achieve life's goals and believe the strongest people will survive. These participants focus on personal performance and individual achievements. The Lujan pilgrimage is only an event to participate in for personal gratification and testing one's physical strength. For them, it has nothing to do with religion.

Devoted pilgrims

This segment is on the opposite end of the spectrum and includes people with strong religious convictions. Some of them work directly for, or are otherwise linked to, the Catholic Church. For them, *Nuestra Señora de Lujan* is a unique event that reconfirms their faith, loyalties and trust in the Lord. Surprisingly, unlike secularized pilgrims, who tend to be more youthful, this segment demonstrated a wider age variation, from 18 to 25 and 45 to 65 years. One interviewee, Mary Anne, a grandmother and pensioner 67 years old, walked to Lujan because her granddaughter overcame cancer. She not only took advantage of this event to demonstrate her devotion but also to show gratitude to God. She was not motivated by testing her physical stamina. Instead, Mary Anne was on a quest for something more complex. She was motivated by a spiritual need. Mary Anne is a supporter of the church and an opponent of the abortion law. She goes on to say,

> We, Argentineans, have the honor of being the birth place of Pope Francisco. This proves that God protects us. Francisco, I believe, is doing it well...he came to the Church to guide God´s people through the lens of humility and charity... it is important the Church meddles in politics; if not, the left-wing and communist parties will cynically promote laws to destroy and erode the bases of the Catholic Church.
>
> *(Mary Anne)*

Although Mary Anne doubts she will reach Lujan by foot, because of her physical inability, she is confident she will get there, probably by taking a taxi or other transportation. This sojourner category is marked by a need to be in contact with deity and their spiritual selves while walking patiently towards an epitomous center. Lujan for them is the spiritual center of the nation, as well as the guiding star of the Argentinian Catholic community. For this group, the existential experience is special because it is a moment of communion between believers and deity, an extraordinary experience that affects the rest of their lives. The authenticity of the Lujan pilgrimage for this segment is not based only on a simple faith in God, but on the sacrifice, the hardship of walking more than 70 kilometers. In similitude of Christ's sacrifice on the cross for the salvation of humankind, the pilgrims' sacrifice helps them purify their souls. Unlike other subtypes, pleasure and leisure orientations are contrary to the genuine purpose of the pilgrimage. Devoted pilgrims can be divided into two subtypes: those who undertake the journey to demonstrate gratitude for a miracle in their lives and those who do it to supplicate God for blessings.

Devoted pilgrims strongly believe that authorities should be much stricter in proscribing secular practices such as alcohol consumption and drug use. These are considered inappropriate during the pilgrimage and are inauthentic practices. Members of this group feel that the pilgrimage should only be open to believers.

Recreational pilgrims

Unlike the other two segments, this one has unique characteristics. These people are not entirely motivated nor distracted by left or right politics. Instead, their motives aim to maximize enjoyment and achieve a novel experience. In contrast to the secular typology, this segment is not interested in facing challenges or overcoming obstacles. Rather, they want to be with friends while experiencing the event as a way to escape from their humdrum routines. Relaxation, recreation, and socializing are the main motivations. Some interviewees were affected by drugs or alcohol. Although drug use and some alcohol consumption are illegal, people consume these as a recreational pastime. Recreational pilgrims see the pilgrimage as a leisure event, a reason to be outdoors and to have fun. They have no intentions of developing spiritually or making it all the way to Lujan. Many of them choose to camp along the way. In their camps, these recreational 'pilgrims' drink alcohol, smoke marijuana, play music and sing, and play games. For them, the event is inauthentic from a religious perspective, but it is somewhat authentic in that it serves as a means of getting to know other people, gather with friends, and spend leisure time.

Members of this group manifest in different ways and their experiences are subjectively constructed. Some of them believe in God or some other force in the universe, but like secularized pilgrims, they have some distrust against the Catholic Church. When they were asked about the abortion law or the politics of the church, they maintained indifference and/or hostility towards any political party and in some cases even the church. Essentially, recreational pilgrims are not committed to the entire journey or arriving at the cathedral. Instead, their focus is 'hanging out', having a good time with their friends. Along the journey, they spend time in their tents playing the guitar or singing alongside the road. This cohort ranges from 18 to 25 years old. Alex (Alejandro) works as an urban artisan; he is well skilled in making necklaces and bracelets. He has neither a political affiliation nor interest in the abortion law. Though he believes in a supreme being who governs the destiny of humanity as a pantheist force, the event is a good opportunity to drink beer, listen to music and socialize with friends.

These three segments demonstrate different dynamics, behaviors, motivations, and authenticities. In contrast to the earlier secular-sacred discussion, this typology gives fresh insight into the motivations of certain pilgrims who walk to Lujan City. Some of them are fervently devoted to the Virgin Mary and Jesus, while others are detractors, or are indifferent to Catholicism. As with Rotherham's (2015) conclusion, the same event attracts differently-minded pilgrims who, from different viewpoints may try to reach the basilica. To this extent, the dichotomy of sacredness and secular logic, as it was formulated by Durkheim and MacCannell, should be revisited. In the pilgrimage of Lujan, religious and recreational pursuits coexist simultaneously and so do different authenticities.

Conclusion

This chapter provides empirical observations from a pilgrimage in Argentina and reflects an interesting point of view in the religious tourism debate. It is important not to lose sight of the fact that religious tourism and pilgrimage represent a growing segment of the tourism industry today. Besides the economic nature of tourism, religious tourism has captivated the attention of many scholars who are interested in revealing its impact on society. Originally, the sociological literature suggested that whereas tourists seek to maximize pleasure, pilgrims were in a quest for something more authentic. This axiomatic relationship is no longer clear-cut as the borders between the sacred and the profane have now blurred considerably. This chapter proposes a tripartite typology of pilgrimage participants at Lujan, Argentina. This includes secularized pilgrims, devoted pilgrims, and recreational pilgrims. Each subtype lives the notion of authenticity differently and demonstrates different ways of associating with the Catholic Church. The devoted pilgrims see the event as an auspicious opportunity to enhance their communion with God, whereas secularized and recreational pilgrims do not. Secularized pilgrims claim not to believe in God, whereas recreational participants maintain a modicum of faith, such as the idea of an afterlife, but they disagree with many of the church's doctrines such as its abortion stance. Still further, the secularized pilgrim is motivated primarily by the freedom and adventure of traversing the pilgrimage route, especially as it demonstrates physical stamina. They want to confront the uncertainties of the trail to show how strong human nature can be or that they have something to prove. Devoted pilgrims look for a real spiritual connection with God. They are moved by strong religious attachments with divinity. Lastly, recreational pilgrims desire escape from their daily routines and the disciplined structure of modern society. Though the model is not exhaustive nor statistically representative of the whole, it provides a new understanding of the interplay of tourism and religiosity.

Additional research is needed to understand the manifestations of authenticity in multiple faith contexts. The empirical part of this chapter focuses on a Roman Catholic example, but such situations also exist in other Christian denominations, as well as in Buddhism, Islam, Hinduism, Judaism, Sikhism, and other religions (Hung, Yang, Wassler, Wang, Lin & Liu 2017; Moufahim & Lichrou 2019). The meanings and values of authentic spiritual experiences will likely differ among religious traditions and within various geographical contexts. However, it is almost certain that religious adherents and leisure tourists both will seek their own authenticities during their visits to sacred heritage places.

References

Augé, M. (1995) *Non-lieux*. London: Verso.

Belhassen, Y., Caton, K., & Stewart, W.P. (2008) 'The search for authenticity in the pilgrim experience', *Annals of Tourism Research*, 35: 668–689.

Bremer, T. (2006) 'Sacred spaces and tourist places', in D.J. Timothy & D.H. Olsen (eds), *Tourism, Religion and Spiritual Journeys* (pp. 36–48). London: Routledge.

Butler, R.W., & Suntikul, W. (2018) *Tourism and Religion: Issues and Implications*. Bristol: Channel View Publications.

Chhabra, D. (2019) 'Authenticity and the authentication of heritage: Dialogical perceptiveness', *Journal of Heritage Tourism*, 14(5–6): 389–395.

Chhabra, D., Healy, R., & Sills, E. (2003) 'Staged authenticity and heritage tourism', *Annals of Tourism Research*, 30: 702–719.

Cochrane, J. (ed.) (2007) *Asian Tourism: Growth and Change*. London: Routledge.

Cohen, E. (1979) 'A phenomenology of tourist experiences', *Sociology*, 13(2): 179–201.

Cohen, E. (1987) '"Alternative tourism"—A critique', *Tourism Recreation Research*, 12(2): 13–18.

Cohen E. (1992) 'Pilgrimage centers: Concentric and eccentric', *Annals of Tourism Research*, 19: 33–50.

Collins-Kreiner, N. (2010a) 'Researching pilgrimage: Continuity and transformations', *Annals of Tourism Research*, 37: 440–456.

Collins-Kreiner, N. (2010b) 'The geography of pilgrimage and tourism: Transformations and implications for applied geography', *Applied Geography*, 30(1): 153–164.

Collins-Kreiner, N., Shmueli, D.F., & Ben Gal, M. (2015) 'Understanding conflicts at religious-tourism sites: The Baha'i World Center, Israel', *Tourism Management Perspectives*, 16: 228–236.

Digance, J. (2006) 'Religious and secular pilgrimage: Journeys redolent with meaning', in D.J. Timothy & D.H. Olsen (eds), *Tourism, Religion and Spiritual Journeys* (pp. 36–48). London: Routledge.

Durkheim, E. (2008) *The Elementary Forms of Religious Life*. New York: Courier Corporation.

Egresi, I., Kara, F., & Bayram, B. (2014) 'Economic impact of religious tourism in Mardin, Turkey', *Journal of Economics and Business Research*, 18(2): 7–22.

Goffman, E. (1959) *The Presentation of Self in Everyday Life*. Garden City, NY: Doubleday.

Griffiths, M. (2011) 'Those who come to pray and those who come to look: Interactions between visitors and congregations', *Journal of Heritage Tourism*, 6(1): 63–72.

Hung, K., Yang, X., Wassler, P., Wang, D., Lin, P., & Liu, Z. (2017) 'Contesting the commercialization and sanctity of religious tourism in the Shaolin Monastery, China', *International Journal of Tourism Research*, 19(2): 145–159.

Karar, A. (2010) 'Impact of pilgrim tourism at Haridwar', *The Anthropologist*, 12(2): 99–105.

King, C.R. (2000) 'The (mis)uses of cannibalism in contemporary cultural critique', *Diacritics*, 30(1): 106–123.

Knudsen, D.C., & Rickly-Boyd, J.M. (2012) 'Tourism sites as semiotic signs: A critique', *Annals of Tourism Research*, 39: 1252–1254.

Korstanje, M.E. (2016) 'The portrait of Dean MacCannell–towards an understanding of capitalism', *Anatolia*, 27(2): 298–304.

Korstanje, M.E. (2018) 'The lost paradise: The religious nature of tourism', in H. El-Gohary, D.J. Edwards, & R. Eid (eds), *Global Perspectives on Religious Tourism and Pilgrimage* (pp. 129–141). Hershey, PA: IGI Global.

Liutikas, D. (2014) 'Lithuanian valuistic journeys: Traditional and secular pilgrimage', *Journal of Heritage Tourism*, 9(4): 299–316.

MacCannell, D. (1973) 'Staged authenticity: Arrangements of social space in tourist settings', *American Journal of Sociology*, 79(3): 589–603.

MacCannell, D. (1976) *The Tourist: A New Theory of the Leisure Class*. Berkeley: University of California Press.

MacCannell, D. (2001) 'Tourist agency', *Tourist Studies*, 1(1): 23–37.

MacCannell, D. (2011) *The Ethics of Sightseeing*. Berkeley: University of California Press.

MacCannell, D. (2012) 'On the ethical stake in tourism research', *Tourism Geographies*, 14(1): 183–194.

Mommsen, W.J., & Osterhammel, J. (2013) *Max Weber and His Contempories*. London: Routledge.

Moufahim, M., & Lichrou, M. (2019) 'Pilgrimage, consumption and rituals: Spiritual authenticity in a Shia Muslim pilgrimage', *Tourism Management*, 70: 322–332.

Norman, A. (2011) *Spiritual Tourism: Travel and Religious Practice in Western Society*. London: Bloomsbury.

Olsen, D.H. (2010) 'Pilgrims, tourists and Max Weber's "ideal types"', *Annals of Tourism Research*, 37: 848–851.

Olsen, D.H., & Timothy, D.J. (2006) 'Tourism and religious journeys', in D.J. Timothy & D.H. Olsen (eds), *Tourism, Religion and Spiritual Journeys* (pp. 1–21). London: Routledge.

Raj, R., & Griffin, K.A. (eds) (2015a) *Religious Tourism and Pilgrimage Management: An International Perspective*. Wallingford: CABI.

Raj, R., & Griffin, K. (2015b) 'Introduction to sacred or secular journeys', in R. Raj, & K. Griffin (eds), *Religious Tourism and Pilgrimage Management: An International Perspective* (pp. 1–15). Wallingford: CABI.

Raj, R., & Griffin, K. (2017) 'Introduction to conflicts, religion and culture in tourism', in R. Raj & K. Griffin (eds), *Conflicts, Religion and Culture in Tourism* (pp. 1–10). Wallingford: CABI.

Raja, I., & Raj, R. (2017) 'The essence of community cohesion through religious tolerance', in R. Raj & K. Griffin (eds), *Conflicts, Religion and Culture in Tourism* (pp. 44–54). Wallingford: CABI.

Richards, G. (2007) *Cultural Tourism: Global and Local Perspectives*. London: Psychology Press.

Richards, G., & Fernandes, C. (2007) 'Religious tourism in northern Portugal', in G. Richards (ed.), *Cultural Tourism: Global and Local Perspectives* (pp. 215–238). New York: Haworth Hospitality Press.

Ron, A.S., & Timothy, D.J. (2019) *Contemporary Christian Travel: Pilgrimage, Practice and Place*. Bristol: Channel View Publications.

Rotherham, D.I. (2015) 'Sacred sites and the tourist: Sustaining tourism infrastructures for religious tourists and pilgrims–a UK perspective', in R. Raj & K. Griffin (eds), *Religious Tourism and Pilgrimage Management: An International Perspective* (pp. 37–56). Wallingford: CABI.

Shackley, M. (1999) 'Managing the cultural impacts of religious tourism in the Himalayas, Tibet and Nepal', in M. Robinson & P. Boniface (eds), *Tourism and Cultural Conflicts* (pp. 95–112). Wallingford: CABI.

Shinde, K.N. (2007) 'Pilgrimage and the environment: Challenges in a pilgrimage centre', *Current Issues in Tourism*, 10(4): 343–365.

Stark, R., Doyle, D.P., & Rushing, J.L. (1983) 'Beyond Durkheim: Religion and suicide', *Journal for the Scientific Study of Religion*, 22(2): 120–131.

Stausberg, M. (2011) *Religion and Tourism: Crossroads, Destinations and Encounters*. London: Routledge.

Terzidou, M., Scarles, C., & Saunders, M.N. (2018) 'The complexities of religious tourism motivations: Sacred places, vows and visions', *Annals of Tourism Research*, 70: 54–65.

Timothy, D.J., & Conover, P.J. (2006) 'Nature religion, self-spirituality and New Age tourism', in D.J. Timothy & D.H. Olsen (eds), *Tourism, Religion and Spiritual Journeys* (pp. 139–155). London: Routledge.

Timothy, D.J., & Olsen, D.H. (eds) (2006) *Tourism, Religion and Spiritual Journeys*. London: Routledge.

Urry, J. (1990) *The Tourist Gaze*. London: Sage.

Urry, J. (2002) *Consuming Places*. London: Routledge.

Vukonić, B. (2002) 'Religion, tourism and economics: A convenient symbiosis', *Tourism Recreation Research*, 27(2): 59–64.

Wang, N. (1999) 'Rethinking authenticity in tourism experience', *Annals of Tourism Research*, 26: 349–370.

Weber, M. (1993) *The Sociology of Religion*. New York: Beacon Press.

Weidenfeld, A., & Ron, A.S. (2008) 'Religious needs in the tourism industry', *Anatolia*, 19(2): 357–361.

Wiltshier, P., & Griffiths, M. (2016) 'Management practices for the development of religious tourism sacred sites: Managing expectations through sacred and secular aims in site development; report, store and access', *International Journal of Religious Tourism and Pilgrimage*, 4(7): 1–8.

Zwissler, L. (2011) 'Pagan pilgrimage: New religious movements research on sacred travel within Pagan and New Age communities', *Religion Compass*, 5(7): 326–342.

21

AFTERMATH

Calculating the effects of pilgrimage

Purna Roy and Hillary Kaell

Introduction

Are pilgrims changed when they return? Why are souvenir objects circulated? How are sites managed and maintained once visitors leave? Across the world, in very different locations, pilgrims must negotiate the trip's aftermath; returnees incorporate new actions and attitudes, arrange photos, and give gifts. These acts of memory-making often become integral to the experience of being at a sacred site.

Despite the shared importance of the return, comparatively little anthropological and sociological research has evaluated the topic in a sustained or systematic way.[1] This lacuna is sometimes ascribed to a Christian bias in pilgrimage studies. "While most Christian notions of pilgrimage start in a certain place and 'go-forth' to another place", writes one anthropologist of Islam, "Muslim spatial orientations imagine both the point of origin *and the act of return* as central to the experience of pilgrimage" (Kenny 2007, p. 364, Italics ours). Studies of Muslim pilgrimage have, in fact, offered the most robust engagement with returnees. Yet it was Roman Catholic sites that first drew ethnographic attention (Hertz 1913; Wolf 1958) and, in many ways, set the tone for the subfield. Victor and Edith Turner's (1978) study of Catholic pilgrimage is the most influential example. It encouraged a longstanding anthropological/sociological interest in how communities congregate and re-solidify social ties, as well as newer work on performativity and ritualization. In both respects, the Turners' example focused on the journey and the "center out there", in their well-known phrase (see also Loustau & DeConinck 2019).[2] While this model did not preclude the possibility of studying aftermaths, it tended to eclipse them.

Today, studies that do explore what happens after pilgrims return adopt a few major strategies. Many, if not most, draw inferences based on promotional materials, faith-based guidebooks, and interviews at the shrine site. A second group focuses only on the after effect, usually through focus groups and interviews, without working with pilgrims before or during their trips. This strategy is especially prevalent in studies of the *hajj*, a pilgrimage that is closed to non-Muslims and where former pilgrims may be identifiable by the addition of "hajji" or "hajja" to their names. Another "strategy," which has yielded some of the most evocative studies of aftermaths, is often more happenstance than design. Ethnographers who may not have set out to study pilgrimage but work in a particular location where people happen to take such journeys know pilgrims well before they depart and can follow up for a long period after the return (Delaney 1990; Gold 1988).

There are good reasons why comparatively few studies of pilgrimage explore aftermaths, and why those that do tend to fall into the categories above: it is costly and time-consuming for scholars, and especially for graduate students, to go on pilgrimage. If one does go, it is potentially even more costly and time-consuming to follow up with participants, who may hail from many places. Only a few scholars, such as Frey (1998, 2004),[3] have made it a point to systematically follow up with pilgrims in various countries over, in her case, a year-long period. Working on Spain's Camino de Santiago in the 1990s, she drew her inspiration from the (at the time) recent trend toward multi-sited ethnography. She notes the difficulty involved in maintaining contact with former pilgrims, as do other scholars who have used this method more recently (Fedele 2012; Kaell 2014).

This chapter draws on studies that have tracked pilgrimage aftermaths in order to highlight trends organized around three key themes: pilgrims, sites, and objects. The first section examines how scholars have evaluated the results for pilgrims at a personal and societal level. The second discusses the impact on sites and site management after pilgrims leave. The last section explores the circulation of souvenirs.

Pilgrims

At the heart of most studies of pilgrimage is the pilgrim. Is he changed? Did she attain her goal? And how do scholars evaluate such outcomes? All pilgrimage scholars are conversant with documents such as memoirs, guides, and handbooks that shape what pilgrims hope, or expect, to find on the trip. The difficulty lies in pinpointing how media about what one ought to accomplish interacts with, and may differ from, the journey's actual results. We have already noted a few key strategies for studying aftermaths. When it comes to observing changes within an individual, a fair number of studies rely on auto-ethnography, such as Michalowski and Dubisch's (2001) work on US motorcycle pilgrimages, Laksana's (2014) study of Catholic and Muslim pilgrimages in Java, and Hammoudi's (2005) work on the *hajj*. Laksana (2014, p. 223), for example, attests to how, as a Javanese Catholic, pilgrimage made him better understand the intersection of those two aspects of his identity and appreciate the integration of Islam into Javanese culture. Hammoudi's (2005, pp. 265–272) work details the shock of his journey home: falling out with a fellow pilgrim, dealing with corrupt petty officials, being solicited by a prostitute in Jeddah airport, feeling exhausted and disoriented after the evanescent happiness he had felt in Mecca (see also Frey 1998). Upon reflection, he writes, the pilgrimage did result in an unexpected form of clarity about life and its challenges.

Upon return, most pilgrims transmit aspects of their experience to others, effectively turning pilgrimage places into "storied spaces" (Feldman 2014) through the circulation of memories and memoirs (Coleman & Elsner 1995). Maurice Halbwachs' (1992, p. 196, cited in Feldman 2014) pioneering work on memory used the Holy Land to show how temporal and spatial distance from the sacred site itself is, in fact, fundamental to creating and preserving these collective memories. As Frey (1998, p. 184) points out in her study of the Camino de Santiago, however, contemporary pilgrims may not come from contexts in which their journey is understood or appreciated. She found that, compared to Spanish pilgrims, Americans and others (including non-Catholics) faced a "monumental task" narrating their experience and explaining the very act of pilgrimage itself, since it was unfamiliar to their audience. Today, post-trip testimonies also proliferate through social media, which has prompted new forms of analysis in which hundreds of post-trip tweets or Instagram photos can be cataloged. Though such studies cannot usually tell why pilgrims create particular images and who consumes them, bigger data sets may unearth interesting results, such as a recent article that suggests young hajjis post selfies to "create opportunities for self-representation and community building in a context of increasing Islamophobia" (Caidi, Beazley & Marquez 2018, p. 8, see also Aukland 2018).

Sociologists and political scientists have done the most to track and quantify post-trip narratives, generally by coding themes in focus group interviews and surveys. The quality of such studies varies; some rely on small samples and only vaguely contextualize responses (Toguslu 2017). Yet even the thinnest of studies may show significant patterns when considered in light of others. For example, surveys of returned hajjis in locations including Pakistan, the Caucasus, Belgium, and London have all confirmed that the trip enhances in-group identity and cohesion. They also suggest that returnees have a heightened sense of individuality and are more accepting of Christians and of diversity within Islam (Alexseev & Zhemukhov 2015; Clingingsmith, Khwaja & Kremer 2009; DeHanas 2013). Laksana (2014, p. 221) finds similar results in his study of the relations between Javanese Catholic and Muslim pilgrims, which he calls an improved "dynamic of mutual openness". Based on Toguslu's (2017) study of Belgian hajjis, in the context of minority groups, such as Muslims in Europe, the "home" feeling in Mecca may encourage a sense of out-of-placeness in their physical home. He found that returned hajjis actually nurtured the feeling that had bothered them before—feeling out of place as religious and racial minorities in Belgium—to encourage a commitment to Muslim practices, such as praying five times a day.

Muslim contexts are also a good example of how pilgrimage often raises a returnee's social standing. In Islam, this new authority is most obviously signaled in the title of "hajji" attached to one's name. Working in Kankan, Guinea, Kenny (2007, p. 371) notes that returnees may even greet their fellows differently as a result: with the hand turned downwards to avoid grasping the palm of those who the hajji now considers less ritually pure (see also Delaney 1990). Kenny also shows how returned pilgrims may come to be viewed as globally situated people, sometimes using the journey to make or cement business contacts that serve them upon return. In many contexts, they are also viewed as benefitting others at home through their prayers and blessings. In her work among Mbororo pastoralists in Cameroon, Virtanen (2014) shows how such patterns become more evident if scholars view pilgrimage holistically within a particular society. For example, the Mbororo view pilgrimage as part of Allah's moral/economic equilibrium: those wealthy enough to sell cows and go to Mecca must unstintingly share souvenirs and blessings with others upon return. In this way, the Mbororo spread the (spiritual) "wealth" of pilgrimage, while also developing a sense of their moral righteousness as a community over against their Muslim neighbors from other tribes, who they view as profiting *personally* from hajj, often by selling the goods they bring back.

In her work with US Christians, Kaell (2016, p. 400) also focuses on the circulation of prayers in the Holy Land and afterward, but notes that the people who choose to go on pilgrimage are usually already considered "spiritual experts". This status may be enhanced, but it is not radically changed upon return (see also Fedele 2012). Indeed, already being a devoted Christian is usually viewed as a prerequisite for undertaking the journey. Of course, gaining new or enhanced respect depends on local "scripts" for understanding pilgrimage. When these are absent, such as in journeys that are new age or more idiosyncratic, a returnee's reception may be more ambivalent; some pilgrims feel they are more respected, while others feel isolated because "no one really understands" their experience (Frey 1998, p. 187).

This work raises the question of failure: what if the trip does not accomplish hoped for ends? Prayers or blessings may not have the intended effect. One may have to grapple with disorienting or puzzling experiences in a sacred center where one is expected to feel belonging (Delaney 1990; Hammoudi 2005). Through pre- and post-trip interviews over a series of months, Kaell (2014, 2016) found that even pilgrims who were disappointed at first generally came to frame the trip as successful within a few months. A key component was how these pilgrims incorporated "home" rituals into what might be thought of as an extensible pilgrimage experience. Post-trip actions became linked to the journey itself and later were often narrated as such. Thus, an unanticipated outcome, such as improving one's marriage, became narrated as part of the "successful" pilgrimage, while pre-trip goals that were not accomplished were generally forgotten. Coleman (2014) observes something similar based on his conversations with returnees from Walsingham, England, where former pilgrims often narratively connect the journey to preexisting friendships or activities, such as nurses who drew a parallel between touching statues and pilgrims at the shrine and the comforting touch of a nurse in a hospital. Coleman (2014, p. 288) calls it a "mediating chain providing a bodily link between their place of pilgrimage...and places of home and work in another part of the country". In this way, a pilgrimage's aftereffects linger into an indistinct future and may be seen to have far-reaching effects on multiple people in a pilgrim's life.

Sites

As travel has become more affordable, the number of visitors has risen sharply at many sites, which are attracting a broader, global audience than their local retinue of devotees. Site managers are therefore confronted with "a multitude of visitor motivations and expectations, which increases the frequency and difficulty of management challenges and issues" (Olsen 2006, p. 107). At many sacred centers, these difficulties are compounded by the fact that professional or volunteer clergy are usually the main caretakers of the sacred centers. Their main priority might be to encourage worship and, even if there is a detailed management plan, they might not have the necessary training or resources to tackle the challenges that so often arise in the aftermath of pilgrimage (Olsen 2006; Shackley 2001).

When a large number of people visit a given site each year, one of the main concerns is waste production, which creates environmental hazards and physical degradation of the site. In Shinde's (2007) study of the Hindu sacred complex of Tirumala-Tirupati in south India, a site visited by approximately 1.2 million pilgrims every month, he makes an important observation about how particular rituals play a huge role in generating waste. He notes that over 20,000 coconuts[4] pile up each day, which the shrine management has to remove from the premises. While most of the items used in the worship of the presiding deity at any Hindu shrine are biodegradable (fruits, incense, etc.), their plastic and polythene packaging causes serious damage to the Tirupati-Tirumala hill ecosystem. Likewise, studies of the hajj have noted the piles of plastic, wrappers, and other refuse pilgrims throw along the roadways (Hammoudi 2005, p. 264). In Sikh pilgrimage shrines (Shinde 2007), waste management post-pilgrimage is an even more pressing need due to the presence of *langars* (free communal kitchen service that is part of every gurdwara) where the accumulation of non-biodegradable waste (plastic and glass cups, polythene, etc.) and polluted water due to inadequate sewerage facilities becomes a common problem. Shinde also observes that the majority of Hindu pilgrimage sites lack any institutional structure to manage the effects of religious tourism and the pilgrimage economy functions almost "exclusively through informal social networks" (Shinde 2012, p. 282).

However, issues of waste management and cleanliness are not universal to all mass pilgrimage and tourism sites. Reader (2005, p. 232) remarks that in the Zen temples of Japan "cleaning processes are strictly formalized and made into ritual practices with specific religious meanings that transcend the physical function of simply making place". Where the traditional Japanese aesthetic concept of *wabi-sabi* allows for sites of cultural or religious significance to decay, the concept of renewal is a well-established theme in Japanese religion—both philosophically and architecturally. In other words, both real and imagined decay are expected to set into sites which can then be periodically cleansed. One prominent example is the Ise Grand Shrine—the most important site in the Shinto religion—which is ritually deconstructed and rebuilt every twenty years to remind pilgrims of the transience of this world and the impermanence of all things. Of course, where cleanliness is concerned, it cannot be overlooked that Japan is a wealthy, developed country without the same economic constraints that shrines in countries like India face.

As Douglas (2003) argues, religious ideals of purity and secular conceptions of sanitation are not as different as scholars previously thought. In a tradition like Hinduism where ritual purity is a central religious ideal (particularly among upper-caste Hindus), "dirt" or "impure substances" in a sacred space could arguably raise concerns among devout traditional pilgrims about the site's religious purity and sanctity. After all, the "physical purity of all parts of the Temple is a precondition for their spiritual purity too. Materials used in worship, for example, cannot become fit objects for the gods unless their physical purity and that of their environment is maintained" (Fuller 1979, p. 473). Why, then, do so few scholars of Hinduism (cf. Alley 1994, 1998; Shinde 2007, 2012) talk about what happens to the items discarded around the shrine after ritual ceremonies?

Another aspect of site management concerns the street vendors who are a ubiquitous part of pilgrimages but can be perceived as detrimental to the aesthetic value of sacred shrines. This problem is particularly acute in developing countries, such as India, where the vendors are unregulated and often take up entire stretches of footpath, disrupting movements of pilgrims and traffic. In a shrine development act from 2007 pertaining to Bodhgaya,[5] a major sacred center and UNESCO designated site, the government of Bihar implemented measures to tackle and forcibly eliminate street vending. Without warning or plans for relocation, the state government sent in bulldozers and deployed police to remove the vendors, along with small business stands, and shanty homes in the vicinity of the shrine, as encroachments on government property.

The state government's act had two intentions: "development" of the state and "cleaning up the town" through restructuring urban spaces in order to strengthen the rising popularity of the Buddhist pilgrimage circuit within Bihar. Rodriguez (2017, p. 67) notes that this action plan was enacted in part as a response to the numerous visitor complaints about harassment by "aggressive street vendors" (a matter that is frequently mentioned in popular Indian tourist guide books to caution travelers). "Cleaning up" the pilgrimage center effectively (re)imagined Bodhgaya as a "serene, spiritual, 'authentically' Buddhist place", distanced from any materialistic transaction. According to Rodriguez, since Bodhgaya marks the location where Buddha "awakened" to realize that desire is the root of all suffering, many foreign Buddhist pilgrims supported the government's decision to remove street vendors perceived to be engaged in, what many considered, materialistic pursuits of accumulating wealth. Therefore, arguably, "cleaning up" Bodhgaya can be seen as a strategic move by the state government, guaranteeing proper maintenance and in turn encouraging pilgrim-tourists to return to the holy shrine.

Objects

Some of the most substantial work on pilgrimage aftermaths concerns the transfer of objects—and more particularly, gifts and souvenirs. Most studies in this vein frame their work through broader theoretical frameworks, notably anthropologist Marcel Mauss's (1990) concept of hau, or the "spirit of the gift", and sociologist Pierre Bourdieu's (1993) work on social capital. Also influential is a body of work on the "social life of things" (Appadurai 1986), which promotes research that follows an object's circulation.

Pilgrims bring back sacred statuary, prayer beads, and prayer rugs, along with souvenirs of no express religious value, such as bookmarks or clothing. Many return with little pieces of the place itself, such as Zamzam water (Mecca), olive wood (Israel/Palestine), or twigs, leafs, and rocks (Swanson & Timothy 2012). All pilgrim objects are generally viewed as potent because of their site of acquisition, regardless if they have been manufactured elsewhere, such as in China. In many traditions, an object's holiness is also amplified if a pilgrim prays with it or receives an authoritative blessing upon it at the shrine site. Sathya Sai Baba, an important South Asian guru, garnered widespread attention for his purported miraculous materializations of *vibhuti* (holy ash) and other objects, such as rings or necklaces, which he gave as gifts to the pilgrims who visited him. As Fedele (2012, p. 246) shows, spiritual or New Age tourists in France, who viewed objects as gaining power from the place itself rather than a holy person, transported objects from home to a particular site in order to "charge them with [its] energy" (see also Ron & Timothy 2019).

Regardless of how power, holy presence, or "energy" is transmitted, it is clear that across traditions most pilgrims give (and sometimes sell) these traveling objects to people at home. In Islam, giving gifts provides the opportunity for a pilgrim to confer blessings on others, which may occur publicly in festivities celebrating the hajjis' return (Alexseev & Zhemukhov 2015; Kenny 2007; Virtanen 2014). In Christianity, gifting objects is usually private—between family members and friends—although returnees may give a wide variety of smaller objects to colleagues and acquaintances in order to cement certain social relations. More informal and personal prayers often accompany these gifts, even in cases where the recipient is unaware of such intentions (Kaell 2012).

Objects are also important for pilgrims themselves. They arrange and display them in their homes, often in private spaces such as a bedroom or personal shrine. These objects may help focus their prayers or offer curative properties. Other times, the value of the pilgrimage souvenir may be purely aesthetic or symbolic of certain ideals (Morinis 1992). Modes of use and display differ depending on religious background and, often just as importantly, social class (Kaell 2014, p. 183). Studies of Sathya Sai Baba point to another important factor in traditions with pilgrimage to a holy individual. When recipients return home, the gifts initiate a "quasi contractual relationship" between devotee and guru. According to Kent (2004, p. 48), the gifts, which were mostly to be worn on the body or consumed, made Sai Baba's "imperceptible presence...physically contiguous with the recipient" (see also Srinivas 2012, p. 287). Through these acts, the former pilgrim "becomes obliged firstly to the person of Sai Baba, but through him to the redemptive mission and finally to the development of his own inner divinity" (Kent 2004, pp. 50–51).

Photographs are another set of common objects in contemporary pilgrimage. When posted online, photos become "intimate traces left behind...on social media platforms for multiple audiences to see" (Caidi, Beazley & Marquez 2018, p. 10). More commonly, scholars have explored how pilgrims arrange photos in albums and present these photos to friends. Many such studies codify patterns in the album and/or in the pilgrim's narration of events

(Schermerhorn & McEnaney 2017). Some try to trace the broader social life of photos as they are gifted and used by others, perhaps in ways the pilgrim might not expect. In his study of Hindu pilgrimage, Smith (1995) explores another kind of image: wallet-sized god posters of the deity that are ingested as a curative later by family members or framed in domestic shrines. The latter are decorated with fresh garlands to sanctify the object and open it to darśan—a ritual "sight", or reciprocal gaze between the devotee and the deity. For some family members, the image creates a feeling of closeness to divinity without physically traveling to sacred shrines. Smith also observes that Hindu sacred souvenirs, including pilgrims' maps, mythological charts and mystical geometrical diagrams (yantra) that guide pilgrims at the site, find second lives when they are re-used to "facilitate an inner, spiritual return" (Smith 1995, p. 45), allowing the pilgrim to re-experience the journey through a meditative, devotional state.

The last type of object "afterlife" generally operates on a much grander scale when individuals or institutions create replicas of sacred sites. This act turns foreign pilgrimage sites into local ones, usually relying on the idea that the sacred is moveable and replicable. Scholars of Catholic pilgrimage, who have traced this phenomenon in a number of locations, show how the builders of such sites are often former pilgrims who bring back presence-filled pieces of the original site—rocks, water, etc.—to integrate into the replica in a new landscape. These "surrogate pilgrimages" then use built environments as "stand-ins" for the original landscape (Barush 2016, cited in Karst 2017, p. 30). One of the most famous replicated sites is the Lourdes grotto in France, which is now multiplied many times around the world. In one case, Karst (2017) examines how a former pilgrim, Father Sorin, built a new grotto in South Bend, Indiana, in 1878 after he brought back holy water from the original site in France. Although these landscapes certainly derive their initial power from the presence of pilgrimage objects, as Karst points out, builders "may not always be concerned with creating authentic likenesses, but rather authentic spaces for devotion". In South Bend, for example, the "new" Lourdes grotto has different specialities than its French progenitor and more in keeping with its location on Notre Dame University's campus: instead of seeking physical cures, pilgrims "still whisper of miracles—of tests passed and degrees earned, of football games won, of relationships born or mended" (Karst 2017, pp. 30, 35).

Conclusion

In her classic anthropological account of Hindu pilgrimage, Gold (1988, p. 1) remarked on scholars' tendency to focus on the "journey's destination – the riverbank, the temple town, the lake or mountain shrine with little or no attention to its closure or return lap". More than a decade later, Frey (2004, p. 96) observed that studies still concentrated "on the pilgrim's journey and action at the goal...the return seems to be culturally constructed as unimportant, uninteresting, or simply unnoticed". Today, more scholars of contemporary pilgrimage are exploring the trip's aftermath—and there remains much to be said on the subject, especially in a comparative framework.

One factor hindering this work in anthropology and sociology is, as Frey (2004, p. 96) goes on to note, the methodological dilemma of trying to observe "a moving population that shares a common destination but [often] not a common home". Veteran researchers of pilgrimage, Coleman and Eade (2018), express some of the other, more conceptual, challenges involved:

> For our purposes, one of the most fascinating dimensions of the piece is precisely the difficulty...of determining not just the size but also the *location* of [a pilgrimage's]

> impact: should one focus on the site alone, the immediate locality, the region, the country, or the places from around the world that some pilgrims have come from? Such impact...can be seen as 'direct, indirect and induced' (Saayman et al. 2014: 410), pointing to the numerous and ramifying channels through which pilgrimage activities – and effects – flow.
>
> *(Coleman & Eade 2018, p. 9)*

Coleman and Eade (2018, p. 9) are sanguine about the possibilities, concluding "but the point is to remain open to where the pilgrimage assemblage seems to lead". Yet they also lament how many scholars still treat shrine sites as "bounded containers" and the concomitant failure to nurture wider conversations outside the subfield (Coleman & Eade 2018, p. 4, see also Singh 2013). While this problem has a few root causes, attending to aftermaths can certainly help. Doing so, scholars can unravel pilgrimage from a center "out there" to entangle it more fruitfully within a variety of societal dynamics and institutional frameworks operating in broader contexts.

The study of aftermaths also contributes to important new directions in pilgrimage studies. It may stimulate more work on social media and collective memory. It prompts questions about the legibility of travel narratives for people at home in an age of luxury travel, an issue especially relevant in traditions, such as Hinduism, where the scriptural texts privilege the journey's physical hardships. Studies of aftermaths also suggest that we benefit from framing pilgrimage as a temporally extended and "ritual-like" (Bell 2009, p. x) experience that contains more cohesive rituals within it and may therefore be reinterpreted or compel new experiences even long after the journey is done (Coleman 2014; Kaell 2014, 2016). This approach allows scholars to better trace how even a journey undertaken by one individual is often understood to benefit whole families and communities. Focusing on aftermaths may also enliven new work on "serial pilgrims" who return repeatedly to sites and therefore never really "end" the journey at all (Agnew 2019). Studies of local shrine sites is especially promising in this regard. Another key theme concerns ecology, sustainability, and waste management, which are a growing concern at many pilgrimage sites and have thus far been the purview mainly of studies of tourism management. Anthropologists and sociologists ought to start thinking more holistically about sites as also "experiencing" aftermaths, in parallel with pilgrims as they return home.

Exploring aftermaths in more depth is integral to scholars' ability to better evaluate the confident declarations—promoted by believers, religious doctrine, pop culture products, and tourism professionals—that the journey will have some kind of impact. It alerts us to how a journey's goals may undergo significant changes, result in unpredictable outcomes, or even remain unfulfilled.

Notes

1 Our chapter focuses on "fieldwork" studies, which broadly include scholarship in anthropology, sociology, cultural geography, tourism studies, and religious studies. By necessity, we have limited the discussion in a few ways. We do not consider metaphorical or virtual/online travel or scholarship that takes the term "pilgrimage" to refer more broadly to journeys home (e.g. Harman 2017) or diasporic migrations (Tweed 1997). We have also focused on scholarship and therefore omitted handbooks, guides, or memoirs by pilgrims and pilgrimage promoters (for more on these sources, see Eade and Mesaritou 2018).

2 As Durkheim's student, Hertz (1913) set the tone by couching pilgrimage as a ritual that knit a community together and recreated social order. We should note that the Turnerian model, as it

was adapted, is not precisely the same as the Turners' (1978, p. 22) own work, which at least gestured at how studies might join pilgrimage to wider societal institutions and concerns. By way of contrast to fieldwork studies, we might consider studies of pilgrimage by historians or psychologists; whether consciously or not, both fields focus on the aftermath because of their source materials (on a similar point, see Frey 2004, pp. 96–97).

3 Our title recalls Frey (2004). Another, much rarer, model is to study a pilgrimage that draws from a comparatively local catchment area over many years to observe diachronic change (e.g. Coleman 2014).

4 In south India, coconuts and coconut saplings are often used as gifts in ritual exchanges between deities and their devotees.

5 The Mahabodhi Temple is a 2000-year-old Buddhist temple in Bodhgaya popularly believed to be the location where the Buddha attained enlightenment. Shortly after India's independence, with the passing of the Bodh Gaya Temple Act of 1949, the management of the temple passed from the Hindu *mahant* (abbot) to the state government of Bihar, which established a Bodh Gaya Temple Management Committee (BTMC).

References

Agnew, M. (2019) '"This is a glimpse of Paradise": Encountering Lourdes through serial and multisited pilgrimage', *Journal of Global Catholicism*, 3(1), 26–63.

Alexseev, M.A., & Zhemukhov, S.N. (2015) 'From Mecca with tolerance: Religion, social recategorisation and social capital', *Religion, State and Society*, 43(4), 371–391.

Alley, K.D. (1994) 'Ganga and Gandagi: Interpretations of pollution and waste in Benaras', *Ethnology*, 33(2), 127–145.

Alley, K.D. (1998) 'Images of waste and purification on the banks of the Ganga', *City & Society*, 10(1), 167–182.

Appadurai, A. (1986) *The Social Life of Things: Commodities in Cultural Perspective.* Cambridge: Cambridge University Press.

Aukland, K. (2018) 'At the confluence of leisure and devotion: Hindu pilgrimage and domestic tourism in India', *International Journal of Religious Tourism and Pilgrimage*, 6(1), 18–33.

Barush, K.R. (2016) 'The root of the route: Phil's Camino Project and the Catholic tradition of surrogate pilgrimage', *Practical* Matters, 9, 70–80.

Bell, C. (2009) *Ritual: Perspectives and Dimensions.* Oxford: Oxford University Press.

Bourdieu, P. (1993 [1986]) 'The production of belief: Contribution to an economy of symbolic goods', In R. Johnson (ed.) *The Field of the Cultural Production: Essays in Art and Literature* (pp. 74–111). New York: Columbia University Press.

Caidi, N., Beazley, S., & Marquez, L.C. (2018) 'Holy selfies: Performing pilgrimage in the age of social media', *The International Journal of Information, Diversity, & Inclusion*, 2(1–2), 8–31.

Clingingsmith, D., Khwaja, A.I., & Kremer, M. (2009) 'Estimating the impact of the Hajj: Religion and tolerance in Islam's global gathering', *The Quarterly Journal of Economics*, 124(3), 1133–1170.

Coleman, S. (2014) 'Pilgrimage as trope for an anthropology of Christianity', *Current Anthropology*, 55(S10), 281–291.

Coleman, S., & Eade, J. (eds) (2018) *Pilgrimage and Political Economy: Translating the Sacred.* New York: Berghahn.

Coleman, S., & Elsner, J. (1995) *Pilgrimage: Past and Present in the World Religions.* Cambridge, MA: Harvard University Press.

DeHanas, D.N. (2013) 'Of Hajj and home: Roots visits to Mecca and Bangladesh in everyday belonging', *Ethnicities*, 13(4), 457–474.

Delaney, C. (1990) 'The Hajj: Sacred and secular', *American Ethnologist*, 17(3), 513–530.

Douglas, M. (2003) *Purity and Danger: An Analysis of Concepts of Pollution and Taboo.* London: Routledge.

Eade, J., & Mesaritou, E. (2018) 'Pilgrimage', In J.L. Jackson (ed.) *Oxford Bibliographies in Anthropology* (n.p., online resource). Oxford: Oxford University Press.

Fedele, A. (2012) *Looking for Mary Magdalene: Alternative Pilgrimage and Ritual Creativity at Catholic Shrines in France.* Oxford: Oxford University Press.

Feldman, J. (2014) 'Contested narratives of storied places—the Holy Lands', *Religion and Society*, 5, 106–127.

Frey, N. (1998) *Pilgrim Stories: On and Off the Road to Santiago, Journeys along an Ancient Way in Modern Spain*. Berkeley: University of California Press.

Frey, N. (2004) 'Stories of the return: Pilgrimage and its aftermaths', In E. Badone & S.R. Roseman (eds) *Intersecting Journeys: The Anthropology of Pilgrimage and Tourism* (pp. 89–109). Urbana: University of Illinois Press.

Fuller, C.J. (1979) 'Gods, priests and purity: On the relation between Hinduism and the caste system', *Man, New Series*, 14(3), 459–476.

Gold, A.G. (1988) *Fruitful Journeys: The Ways of Rajasthani Pilgrims*. Berkeley: University of California Press.

Halbwachs, M. (1992) 'The sacred topography of the gospels', In L.A. Coser (ed.) *Maurice Halbwachs on Collective Memory* (pp. 193–235). Chicago: University of Chicago Press.

Hammoudi, A. (2005) *A Season in Mecca: Narrative of a Pilgrimage* (Trans. Pascale Ghazaleh). Cambridge: Polity Press.

Harman, L.D. (2017) 'Journeying home: Toward a feminist perspective on pilgrimage', *International Journal of Religious Tourism and Pilgrimage*, 5(2), 29–34.

Hertz, R. (1913) 'Saint Besse: Étude d'un culte alpestre'. *Revue de l'Histoire des Religions*, 67, 115–180.

Kaell, H. (2012) 'Of gifts and grandchildren: American Holy Land souvenirs', *Journal of Material Culture*, 17(2), 133–151.

Kaell, H. (2014) *Walking Where Jesus Walked: American Christians and Holy Land Pilgrimage*. Albany: New York University Press.

Kaell, H. (2016) 'Can pilgrimage fail? Intent, efficacy, and evangelical trips to the Holy Land', *Journal of Contemporary Religion*, 31(3), 393–408.

Karst, L.A. (2017) 'A new creation: Translating Lourdes in America'. *Liturgy*, 32(3), 29–37.

Kenny, E. (2007) 'Gifting Mecca: Importing spiritual capital to West Africa', *Mobilities*, 2(3) 363–381.

Kent, A. (2004) 'Divinity, miracles and charity in the Sathya Sai Baba movement of Malaysia', *Journal of Anthropology*, 69(1), 43–62.

Laksana, A.B. (2014) *Muslim and Catholic Pilgrimage Practices: Explorations through Java*. Aldershot: Ashgate.

Loustau, M.R., & DeConinck, K. (2019) 'Editors' introduction', *Journal of Global Catholicism*, 3(1), 13–25.

Mauss, M. (1990 [1923]) *The Gift: the Form and Reason for Exchange in Archaic Societies* (translated by W.D. Halls). New York: W.W. Norton.

Michalowski, R.J., & Dubisch, J. (2001) *Run for the Wall: Remembering Vietnam on a Motorcycle Pilgrimage*. New Brunswick, NJ: Rutgers University Press.

Morinis, A. (1992) 'Introduction: The territory of the anthropology of pilgrimage', In A. Morinis (ed.) *Sacred Journeys: The Anthropology of Pilgrimage* (pp. 1–28). Westport, CT: Greenwood Press.

Olsen, D.H. (2006) 'Management issues for religious heritage attractions', In D.J. Timothy & D.H. Olsen (eds) *Tourism, Religion and Spiritual Journeys* (pp. 104–118). London: Routledge.

Reader, I. (2005) 'Cleaning floors and sweeping the mind: Cleaning as ritual process', In J. van Bremen & D.P. Martinez (eds) *Ceremony and Ritual in Japan: Religious Practices in an Industrialized Society* (pp. 227–245). London: Routledge.

Rodriguez, J. (2017) 'Cleaning up Bodhgaya: Conflicts over development and the worlding of Buddhism', *City & Society*, 29(1), 59–81.

Ron, A.S., & Timothy, D.J. (2019) *Contemporary Christian Travel: Pilgrimage, Practice and Place*. Bristol: Channel View Publications.

Saayman, A., Saayman, M., & Gyekye, A. (2014) 'Perspectives on the regional economic value of a pilgrimage', *International Journal of Tourism Research*, 16(4), 407–414.

Schermerhorn, S., & McEnaney, L. (2017) 'Through indigenous eyes: A comparison of two Tohono O'odham photographic collections documenting pilgrimages to Magdalena', *Religious Studies and Theology*, 36(1), 21–53.

Shackley, M. (2001) *Managing Sacred Sites: Service Provision and Visitor Experience*. London: Continuum.

Shinde, K. (2007) 'Pilgrimage and the environment: Challenges in a pilgrimage centre', *Current Issues in Tourism*, 10(4), 343–365.

Shinde, K. (2012) 'Policy, planning, and management for religious tourism in Indian pilgrimage sites', *Journal of Policy Research in Tourism, Leisure and Events*, 4(3), 277–301.

Singh, V. (2013) 'Work, performance, and the social ethic of global capitalism: Understanding religious practice in contemporary India', *Sociological Forum*, 28(2), 283–307.

Smith, H.D. (1995) 'Impact of "God Posters" on Hindus and their devotional traditions', In L.A. Babb & S.S. Wadley (eds) *Media and the Transformation of Religion in South Asia* (pp. 24–50). Delhi: Motilal Banarsidass Publishers.

Srinivas, T. (2012) 'Articles of faith: Material piety, devotional aesthetics and the construction of a moral economy in the transnational Sathya Sai movement', *Visual Anthropology*, 25(4), 270–302.

Swanson, K.K., & Timothy, D.J. (2012) 'Souvenirs: Icons of meaning, commercialization, and commoditization', *Tourism Management*, 33(3), 489–499.

Toguslu, E. (2017) 'The meaning of pilgrimage (Hajj): Re-shaping the pious identity of Belgian Turkish Muslims', *Islam and Christian–Muslim Relations*, 28(1), 19–32.

Turner, V., & Turner, E. (1978) *Image and Pilgrimage in Christian Culture*. New York: Columbia University Press.

Tweed, T. (1997) *Our Lady of the Exile: Diasporic Religion at a Cuban Catholic Shrine in Miami*. New York: Oxford University Press.

Virtanen, T. (2014) 'Transforming cattle into blessings: The moral economy of Mbororo pilgrimage', *Journal of Religion in Africa*, 44(1), 92–126.

Wolf, E.R. (1958) 'The Virgin of Guadalupe: A Mexican national symbol', *The Journal of American Folklore*, 71, 34–39.

SECTION IV

Managing religious and spiritual tourism

22

SOCIOPOLITICAL AND ECONOMIC IMPLICATIONS OF RELIGIOUS AND SPIRITUAL TOURISM

Dallen J. Timothy

Introduction

Every type of tourism has positive and negative consequences. Most of the negative impacts of tourism are associated with 'mass tourism', where large-scale visitation over long periods of time results in harmful environmental, sociocultural, and economic outcomes for the destination. These impacts have been well documented in thousands of research articles and a number of critical books since the 1970s that focus on understanding the diverse range of tourism's impacts and how best to manage them (Gursoy & Nunkoo 2019; Hall & Lew 2009; Mathieson & Wall 1982; Singh, Timothy & Dowling 2003).

In response to the detrimental effects of overtourism, or unbridled visitation, especially in less developed regions and crowded historic urban centers, 'alternative tourisms' and 'special interest tourisms' have been suggested and developed as a means of creating specialty niche markets that will ostensibly respect destinations better and decrease mass visitation to major urban destinations, such as Venice, Barcelona, and Prague, by raising the interest profile of alternative destinations and activities. Although the results of these movements since the 1990s have seen mixed results, the end result of alternative and special interest tourism has been the development of 'mass alternative tourism'. This has led to specialists once again calling for increased attention to overtourism and its impacts, suggesting that 'responsible tourism' may be part of the answer (Dodds & Butler 2019; Milano, Cheer & Novelli 2019; Pechlaner, Innerhofer & Erschbamer 2019).

This perspective suggests that tour operators and other service providers, destination management organizations, destination residents, and even the tourists themselves must take responsibility for their actions and behave accordingly. Efforts toward green tourism, small-scale tourism, and slow tourism reflect this changing mood in the industry. As part of their own responsible actions, many over-visited destinations, such as Venice, have started initiating programs that force responsibility upon external sellers (e.g., cruise companies) and the tourists themselves by charging extra landing fees, limiting the number of people who can visit each day, restricting where they can go, and levying entrance fees into the most crowded parts of the city (Timothy 2021).

Although much religious and spiritual tourism takes place on a small scale and is welcomed by local people, the largest tourist gatherings in the world are of a religious nature (e.g., Kumbh Mela and Hajj). Thus, religious tourism, pilgrimage, and many manifestations of spiritual tourism are key elements of the mass tourism phenomenon and are significant contributors to environmental, social, and economic problems and opportunities. For example, the overtourism problem in Barcelona directly affects the famous Sagrada Familia, and religious and spiritual crowding problems affect the churches of Venice, Amsterdam, and Rome, as well as the temples of Beijing and Kyoto.

This chapter describes some of the most pertinent sociopolitical and economic impacts of pilgrimage and religious and spiritual tourism. In particular, it examines a variety of sociopolitical implications and outcomes, including social distance, contestation and dissonance, and the commercialization of the sacred. The second part of the chapter then considers mixed positive and negative economic perspectives on religious tourism as an economic driver, unique employment considerations, the scale of tourism and religious heritage as a place brand, the implications of gentrification, and the effects of seasonality and other modifiers of tourism demand.

Sociopolitical implications of sacred tourism

The sociocultural impacts of tourism have been well documented. These include, among others, overcrowding, bad behavior on the part of tourists and many residents, increased crime and drug use, prostitution, tourists' disregard for local customs, tainted place image, and cultural demise. Host communities and outside commentators often see tourists in a negative, neocolonialist fashion, suggesting that tourism has many of the same characteristics as colonialism: exploitation of people and resources, external control, profiteering, and creating a deeper hierarchical divide between the haves and the have-nots (Hall & Tucker 2004). In a positive light, tourism is known in some cases to help develop social solidarity, improve communities' quality of life through jobs and the provision of public services, and to act as an impetus for the preservation of cultural traditions and built heritage (Mustafa 2014). Although these are important in the realm of religious tourism, this section examines only three specific interrelated social contexts: social distance, dissonance and contestation, and the commercialization of the sacred.

Social distance

Social distance refers to the level of intimacy, understanding, and cultural nearness between individuals and groups of people. This can apply to a wide range of cultural or demographic elements, including language, race, nationality, age, sexuality, socioeconomic status, or education level (Smith et al. 2014). In a faith context, social distance means the nearness or separateness of different sets of beliefs, rituals, and practices, and can influence religious compatibility in marriage or community living (Brinkerhoff & Mackie 1986; Cavan 1971). According to this theory, Catholics would have a relatively close social distance with Anglicans and Lutherans but a much wider social distance with Jains or Zoroastrians. Jews and Samaritans would share a much narrower social distance than that shared between Taoists and Druze.

This has interesting and important implications in the context of pilgrimage, religious tourism, and spiritual tourism. When the majority of local residents, the hosting community, belongs to the same faith as the pilgrims, there is a closer social distance between them (Terzidou et al. 2008). This can result both in benevolent and hospitable relationships as well as conflictual relations if the pilgrim visitors are seen to misbehave in some way beyond the norms of local culture. Just because they share a religion does not mean they share the same cultural behavioral norms. When the destination adheres largely to a different faith than that of the visitor, the same conflicted or benevolent relationships may develop. The Quran, for example, exhorts Muslims to be hospitable and caring towards non-*Ummah* (people not of the Muslim faith) guests. Christians too are instructed in the Bible to care for those in need, regardless of their beliefs. This translates into a well-known reputation of 'Arab hospitality', which is common in Muslim majority countries (Stephenson & Ali 2019). Although different religions shared by hosts and guests may create greater social distance, religious mandates and practices ostensibly draw them closer together.

In instances where the dominant religion hosts sacred sites of others, there is significant potential for a wide social distance. However, for the most part, in a religious tourism setting, relations remain positive. This can be seen clearly at sacred Jewish sites throughout Europe and North Africa in areas where relatively few Jews remain, but which are important destinations for Jews to experience their religious and diasporic heritage and undertake 'pilgrimages of nostalgia' (Ioannides & Ioannides 2002). In Morocco, for example, many synagogues have been well preserved and protected on behalf of the Jewish people by the Muslim communities that surround them (Soussi 2020). Similarly, many Jewish sites throughout Christian Europe continue to be commemorated and maintained by a non-Jewish majority (Krakover 2017; Russo & Romagosa 2010).

Social distances are widened when nonbeliever tourists visit the sacred sites of others and misbehave by photographing worshippers or liturgical services, speaking loudly, dressing inappropriately or disrespecting the faith in other ways. This frequently causes significant annoyance on the part of adherents and raises calls for precluding nonmembers from visiting sacred sites during worship or prayer times or even closing sacred sites to nonmembers permanently (Timothy 2021). New Age tourists and spiritual tourists in general face similar situations when, as part of their spiritual pathway, they visit sites that are deemed sacred to certain religions or Indigenous People. For example, in Sedona, Arizona, which is dubbed the 'New Age Capital of the World' by many new agers, spirit-seekers, and community marketing specialists, there are frequent clashes between the US National Forest Service (USNFS) and the New Age spiritualists in response to the pilgrims' propensity to build ritual alters on USNFS land and melt candles, burn fires, and scratch ritualistic symbols on delicate sandstone surfaces. The visitors have also clashed with Native Americans for their appropriation of Indigenous culture for use in their spiritual enlightenment (Olsen 2003; Timothy & Conover 2006). Another example is in Lumbini, Nepal, where the majority population is Muslim but the sacred sites there are Buddhist (the birthplace of Buddha). For the most part, Buddhist pilgrims and their Muslim hosts get along well. The local Muslim majority population are hospitable and help the local Buddhist inhabitants care for the sacred site and lead the area's commercial activities. Social distance can therefore help explain why adherents to different religions visit the sacred sites of others and why motivations to visit the sacred sites of others vary depending on visitors' own faith (Nyaupane et al. 2015).

Dissonance and contestation

Heritage, including religious heritage, is a hotly contested concept. Heritage dissonance entails a lack of harmony between groups that share at least part of a common history. Heritage contestation or dissonance is often heightened and enhanced in the context of religion. Timothy and Ron (2019) examine dissonance as manifested in denominational differences and conflicts in sacred space. The world's religions are largely comprised of subsectors or individual denominations whose histories diverge and doctrines vary from one sect to another. Doctrinal differences frequently manifest in religious tourism. For example, in the Holy Land, various Christian denominations are inclined to visit specific localities. The spot where the Church of the Holy Sepulchre stands is believed to be the location of Jesus' crucifixion and entombment among most Catholic and Orthodox adherents, whereas many Protestant groups believe the Garden Tomb outside the ancient walls of Jerusalem is the proper location of Jesus' tomb (Olsen & Ron 2013; Timothy & Ron 2019). Likewise, among different denominations of the Latter-day Saint movement (e.g., The Church of Jesus Christ of Latter-day Saints and the Community of Christ), there are overlapping and diverging versions of history, which manifest in how each denomination views its sacred spaces and the heritage sites it chooses to preserve and promote (Olsen & Timothy 2002).

Although these examples do not result in violence or animosity, they do illustrate the diverging nature of religious site management and visitation in sacred locales. Some inter-sectoral dissonance does result in power struggles and conflict in general terms as well as with regard to religious tourism (Olsen & Emmett 2021; Timothy & Emmett 2014). Perhaps the best example of this is the Sunni-Shiite divide in Islam, each of which has its own places of pilgrimage in various countries but also shares the sacrosanct sites of Mecca and Medina (Ziaee & Amiri 2019). The passionate conflict between the handful of Christian sects that own and manage small sectors of Jerusalem's Church of the Holy Sepulchre has at times resulted in the church being closed or barricaded (Bowman 2011; Timothy & Emmett 2014).

Dissonance is not limited to inner-religious sectoral disagreements but also between entirely different religions. For example, during the Bosnian War (1992–1995), the belligerents were divided along religious and ethnic lines. The Catholic Croats, Orthodox Serbs, and Muslim Bosniaks fought one another countrywide, with each group targeting the religious heritage and pilgrimage sites of the others. The Christians destroyed mosques and Muslims obliterated churches and shrines. Some estimates suggest that approximately 1,000 mosques, 483 Catholic churches and 400 Serbian Orthodox churches were destroyed or severely damaged (Mose 1996). Even in 2020, scars remain in the faithscapes of Bosnia and Herzegovina, with the wedges between the faithful in that country remaining raw and fresh. The religious dissonance and the war continue to underscore much of the tourism sector in that country with memorial tours and visits to sites of destruction continuing to comprise much of the tourism product there (Kudumovic 2020; Timothy 2021; Walasek et al. 2015).

Similar divisions have appeared in India in recent years. Hindus and Muslims have long contested the sacred locality in Ayodhya where the Babri Mosque stood until it was destroyed in 1992 by Hindu extremists. Hindus believed that the spot was the birthplace of the deity Rama and that the mosque had been built atop an ancient Hindu complex. The destruction and protests that followed pitted Indians of different faiths against one another throughout the country (Tahan 2020). The situation has been partially resolved by the Indian courts—a Hindu temple will be built on the site and a mosque will be built nearby. The locale has become a significant religious tourist attraction. Although these extreme examples are not necessarily a direct result of tourism, tourism is assuredly involved in the conflicts, particularly when pilgrimage-oriented sacred sites are targeted for destruction by extremist groups through dueling claims over sacred space.

Commercialization and commodification of the sacred

Although pilgrimage has traditionally been seen as the province of humility, penance, and in some cases poverty and destitution, with only a few exceptions, such is no longer the case.

Since the mid-twentieth century, the tides have turned to the point where religion, faith, and spirituality are among the most commoditized motives for travel. Billions of dollars are spent each year on pilgrimages or other religious journeys. Luxury retailscapes and holiday hotels and resorts, golf courses and amusement parks, and large estates of second homes now decorate the tourismscapes of many sacred sites and their environs throughout the world (Huang et al. 2017; Hung et al. 2017; Qurashi 2017; Ron & Timothy 2019; Shackley 2006b; Shinde 2020). Spiritual localities that were once destitute, lowly, and underprivileged because of their focus on faith-based travel, are now among some of the most overtouristified and over-commercialized destinations on the planet (Kouchi et al. 2018; Qurashi 2019). Perhaps the most vivid contemporary example of this is in Saudi Arabia where al-Saadi (2014) describes the 'Vegasization' of Mecca and its holy places. The Hajj's pinnacle position in Saudi Arabia's tourism economy has emboldened the kingdom to focus its commercialization efforts on the holy cities of Mecca and Medina (Bokhari 2019; Kouchi et al. 2018; Qurashi 2019).

Although such a high degree of commercialization has not yet occurred directly to shrines and their immediate vicinities in many other major religions, most of them are undergoing a commoditization process, albeit at a slower pace (Geary 2018; Hung et al. 2017; Nyaupane et al. 2015; Ron & Timothy 2019; Timothy 2018). Beyond the physical development of commerce and tourism in sacred places, other signs of commercialization have emerged and grown in recent years. First, theme parks have become more pervasive elements of sacredscapes, especially in the context of Christianity. Plethoric biblical theme parks or themed environments depicting certain events in the Bible are now common fixtures in industrial Christianity. The Creation Museum and the life-sized replica of Noah's Ark at the Ark Encounter in Kentucky, USA, and Noah's Ark Park in Hong Kong, as well as the Museum of the Bible in Washington, DC, are recent examples of extreme commodification of the Bible for tourism purposes (Bielo 2018; Paine 2019; Ron & Timothy 2019).

Religious cruises are another popular way of commercializing the sacred and are especially popular in Christian tourism. Some eastern Mediterranean cruises have a biblical theme related to the missionary movements of Jesus' early apostles, who criss-crossed the Mediterranean to spread the gospel (Timothy & Ron 2019). Likewise, many religious events have become commercial endeavors such as certain conferences and pageants. Religious mega-events include official pilgrimages, such as the Hajj and Kumbh Mela, as well as various holiday feasts and other celebrations in multitudes of faith traditions. In addition to these, however, are specially planned events, such as youth and women's conferences, pageants and plays, music concerts, Jubilee year celebrations, and other large gatherings. Although some of them are free and utilize volunteer labor and lodging in people's homes, many of them have commercial undertones (Olsen & Esplin 2020). Finally, the souvenir industry is a multi-billion dollar sector and is relevant to all types of tourism, including religiously or spiritually motivated travel (Houlihan 2000; Maldonado-Erazo et al. 2019; Shackley 2006b).

Olsen (2012) argues that over-commercialization of the sacred often leads to dissatisfying experiences and may ultimately result in conflict and contestation between the sacrum and the profanum in religious tourism spaces. Likewise, Sharpley (2009) notes that the hyperdevelopment of religious tourism may stifle the spirit of place and result in people having less satisfying spiritual experiences. Ron and Timothy (2019) point out that funding is required to operate and maintain sacred sites, so some level of commercialization is normal and expected (Shackley 2006a). This reflects the changing marketplace where religious travelers today live harried lives and are dependent upon modern technology, such as mobile phones, apps, the internet, comfortable lodging, and mass transportation. Despite all of this,

> because pilgrimage has been subsumed into the global tourism system, it does not mean that sacred sites and religious travel experiences are no longer genuine expressions of faith. On the contrary, such changes have functioned to democratize religious tourism in a way that enables more devotees to travel than ever before.
>
> *(Ron & Timothy 2019: 62)*

Although many faith travelers despise the overt commoditization of sacrosanct places, they are generally willing to overlook it because they realize the need for economic development in destination communities and funding to protect, maintain, and manage sacred sites.

Economic implications of religious tourism

Like all other types of tourism, religious and spiritual tourism have enormous economic consequences for sacred destinations (Bokhari 2019; Collins-Kreiner & Gatrell 2006; Din 1982; Nayak & Prabhu 2015; Nyikana 2017; Pourtaheri et al. 2012; Rizzello & Trono 2013; Shackley 2006a). Large-scale pilgrimage and religious tourism at places such as Jerusalem, Ranakpur, Bethlehem, Rome, Medjugorje, Lourdes, Lhasa, Amritsar, Medina, Fátima, Haifa, Kyoto, Santiago de Compostela, Knock, Salt Lake City, Sedona, Mecca, Varanasi, Montserrat, Bodh Gaya, and Lumbini, to name only a few, plays a significant role in the fiscal livelihoods of those destinations. The economic value of religious tourism is particularly notable in places that rely on pilgrimage for the majority of their tourism receipts. Despite religious tourism, pilgrimage, and spiritual travel having similar economic outcomes for destinations to other types of tourism, there are several unique perspectives that pertain directly to these specific markets.

Religious tourism as economic powerhouse

Religious tourism has until recently been seen as a less profitable form of travel than other types of tourism, largely owing to the currently invalid assumption that religious tourists spend far less than other tourists do (Egresi, Kara & Bayram 2012; Gupta & Raina 2008; Olsen 2013; Vukonić 1998, 2002). Although this may be true for the faithful traveling by foot on a pilgrimage route (Fernandes et al. 2012), it is typically not the case in the final destination. Lois-González and Santos' (2015) study finds that pilgrims to Santiago de Compostela have similar spending patterns to visitors overall, but they do tend to choose cheaper accommodations such as guesthouses and hostels. While non-spending or low-spending might have characterized most pilgrimages in the Middle Ages and even into the twentieth century for many pilgrims of different faiths, it is far from being true in the modern day. Even in ancient Jerusalem, the pilgrimage economy two millennia ago was massive for its time and scale (Goodman 2007).

Anciently, pilgrims usually traveled by foot or pack animal, lodged in tents or pilgrim hostels, and ate provisions they bring from home or foraged along the way. However, as noted above, normative pilgrimages today are far different from what they were a century or more ago. With only a few exceptions, such as numerous Hindus at the Kumbh Mela (Buzinde et al. 2014), this notion of poverty in pilgrimage is no longer the norm, even if religious institutions have tried to downplay pilgrimage economics (Nayak & Prabhu 2015; Vijayanand 2012). Even where pilgrims dress in simple sackcloth and overnight under the stars or in tents, pilgrimage is big business (Sujatha & Dadakalandar 2019; Vijayanand 2012; Vukonić 2002). Turner's (1973) original idea of *communitas* was partly based on the notion that everyone was the same during pilgrimage—that pilgrimage was a classless pursuit of a common goal. However, there has been a social stratification of wealthy pilgrims versus poor pilgrims in the past 50 years. Qurashi's (2017) insights explain how it is becoming increasingly difficult for poor pilgrims to undertake the Hajj because it now caters to high-end travelers, and pre-packaged Hajj tours are now the norm.

Whereas individual shrines, churches, and temples may officially eschew commercialization and avoid seeking to profit from pilgrim and tourist visitation, they typically need to cover their costs in their altruistic efforts to continue their ministries. The destinations where they are located, on the other hand, thrive on the commoditization of faith with a 'captive' audience of adulators. From a tourism perspective, even the most pious pilgrims have economic impacts, which was apparent in ancient and medieval times when vendors began earning a living selling food, drink, accommodations, souvenirs, and lodging to pilgrims.

Employment perspectives

One of the main benefits of every kind of tourism is job creation. Hundreds of thousands of people are employed directly or indirectly to service the needs of religious and spiritual pilgrims throughout the world (Gupta & Raina 2008; Pourtaheri, Rahmani & Ahmadi 2012). In addition to generating jobs, pilgrimage also produces regional income and tax revenue, just as all other types of tourism do (Jude et al. 2018; Vijayanand 2012).

In the realm of religious tourism, there are three main types of employment: jobs that are generated in the destination by religious tourism, paid jobs that are directly under the auspices of the religious establishment, and volunteer workers. The jobs generated through pilgrimage and religious tourism within the destination are essentially the same as the jobs stimulated by other types of tourism. These include direct, indirect, and induced employment in the lodging, transportation, food services, retail, tour operations, agriculture, fishing, public policy, entertainment, and construction industries.

Specialized positions employed by the holy site are also an important element of tourism's economic impact. Many religious attractions hire guest services coordinators or liaisons to oversee visitor management programs. Most faith organizations see themselves in the liturgy business, not the tourism business, so they hire managers to oversee the visitor management side of operations. This allows clergy and other site officials to concentrate on their worship responsibilities. Guides are another important hired position and are of two types: those that act as tour guides, and those that assist pilgrims in carrying out their rites and rituals during sacred events (Abdellah & Ibrahim 2013; Caidi 2019). Medical staff are also employed at many pilgrimage locales to deal with the health of large crowds of pilgrims. This is particularly important at shrines visited by the extreme elderly and ill who visit sacrosanct places for their healing properties, which includes much of the crowd at Lourdes (Baldacchino 2010). Pilgrim hostellers oversee pilgrim accommodations, including guest services and housekeeping (Wang 2015). Other jobs, such as ticket sales, retail sales, and grounds keeping, are similar to jobs in other types of establishments.

Somewhat unique to religious tourism is the predominance of volunteer workers at sacred sites or religious events (Gallarza, Fayos-Gardo & Arteaga-Moreno 2019). Although some holy places have paid administrators and support personnel, the majority of sacred sites in the majority of faiths are managed and staffed by volunteers. Religious site volunteers can be divided into two groups: those who live locally and are heavily involved in their faith practices, and those who could be classified as 'volunteer tourists' (Ron & Timothy 2019). Many nonlocal volunteers serve as short-term or long-term missionaries, as is the case of the Garden Tomb in Jerusalem. There, most personnel come from the UK, North America, and other areas to function as guides, gift shop salespeople, security staff, or maintenance workers for set periods of time. Volunteers at Nazareth Village also donate their time and means to labor in furthering the kingdom of God through tourism (Engberg 2017; Ron 2010; Ron & Feldman 2009). Similar situations exist at Protestant and Latter-day Saint historic sites in North America (Olsen & Timothy 2018). Volunteer workers are an important component of the personnel at pilgrimage localities throughout the world in many faiths. Many people spend weeks, months, or years volunteering, and although they do not receive monetary remuneration for their service, they have a salient economic impact. Most cover their own costs of living, economically affecting the region where they volunteer as they pay rent, purchase food, utilize local transportation, and spend money on sightseeing and recreational activities.

Scale and branding

Scale is an important consideration in the economics of pilgrimage and spiritual tourism. Religious attractions range from small, local shrines to world-class heritage attractions that draw visitors by the millions. Thousands of 'folk shrines' are scattered throughout Europe, North America, Latin America, Asia, and Africa. These are important local markers of spiritual events and sacred spaces, but do not draw visitors beyond a few resident or regional worshippers who stop by to pray and give offerings or who visit with clergy and local congregants in what is commonly termed 'folk religiosity' (Banica 2016; Klimova 2011; Nolan & Nolan 1989; Rodosthenous & Varvounis 2014; Ron & Timothy 2019).

These local or regional sacred localities have a far smaller catchment area and are typically associated with small-scale events or miracles of local renown, and therefore fewer visitors see them. As a result, their economic impact is negligible in most instances compared to religious sites of a national or international nature. While many local shrines have little to no direct economic impact, religious sites of national and worldwide notoriety are sometimes the pecuniary mainstay of destination regions or countries (Timothy 2021). Angkor Wat, Machu Picchu, the great temples of Greece and Turkey, the Pyramids of Egypt, the Prambanan and Borobudur complexes in Java, and the Buddhist landscape of Bagan are prime examples of some of the world's leading historic sites that appeal to religious tourists and general heritage enthusiasts. They serve as 'anchor attractions', drawing people from all corners of the globe and stimulating extremely heavy economic impacts for their home regions and the country as a whole.

Many such places become surrogate brands for the countries they belong to, and many of these spaces of faith become globally recognized iconic images that are synonymous with their locations. From a heritage perspective, what would Peru be on the global stage without Machu Picchu, Cambodia without Angkor Wat, Armenia and Georgia without their medieval churches and monasteries, or Japan without its classic temples and shrines? UNESCO's World Heritage List is awash with sites of faith and worship in every part of the world, also highlighting these sites' importance as global heritage brands and potential tourist attractions (Adie 2017; Jimura 2019; Metreveli & Timothy 2010).

Gentrification

One of the most important macro-economic and social outcomes of tourism in recent years is the notion of gentrification, which denotes urban redevelopment and its resulting inflated property values, which price residents out of the marketplace. Much urban gentrification is tied directly to tourism. Urban tourism is known to cause over-inflation of property values, usually in historic city centers (Pinkster & Boterman 2017; Timothy 2021). When tourism grows in historic urban areas, including sacred cities, costs of living increase enormously, and local residents are priced out of their original neighborhoods (Zaban 2020). In the past five years, the Airbnb phenomenon and overtourism, in general, have been heavily blamed for accelerating urban gentrification, causing property taxes and rents to swell to unaffordable levels, stimulating the destruction of historic parts of the city in favor of new tourism-oriented and high-end residential development, and overcrowding.

Religious tourism-induced gentrification is apparent in many holy cities. In Mecca and Medina, much of the Muslim heritage fabric has been razed to make way for more pilgrims and to develop additional tourism services. As noted previously, the historic areas around the great mosques are now reserved for high-end hotels and other expensive services and residences (Qurashi 2019). Not only does this form of gentrification destroy the Islamic heritage of these holy cities, it creates socioeconomic conditions where small enterprises have no chance of competing against the urban monopolization of companies such as Hilton, Hyatt, and Marriott. In the words of one urban development consultant in Saudi Arabia, "Mecca, to my regret, has become a site for real estate speculation. The name of the game is proximity to the Kaaba" (quoted in al-Saadi 2014: n.p.). Shinde (2020) also notes this similar phenomenon happening in many pilgrimage towns in India, where upscale luxury apartment blocks and second homes are being built to house wealthy devotees who seek to own a part of the sacred landscape.

Seasonality and demand shifters

Demand shifters—forces that cause unbalanced demand in consumption and may be political, natural, social, or economic—and seasonality are a major concern for tourism destinations, and most destinations have marked seasonal variations in demand. Low tourism demand means that tourism revenues in general decline, employment rates go down, and many small businesses either temporarily shut their doors during the slowest times of year or, in extreme cases, permanently. The high tourism season is marked by an increases in employment, regional income, and higher tax revenues. There are a variety of causes of seasonal variations that are both natural and institutional. Institutional seasonality includes religious holidays, and festivals, public holidays and school breaks, whereas natural seasonality is brought on by weather conditions and time of year. Beyond normative seasonal variations in demand, other conditions cause fluctuations in demand, including political crises, price hikes, security concerns, and pandemics.

Religious tourism experiences similar vicissitudes in the market, but in general, it is less elastic to demand shifters such as those noted above than other types of tourism are. Even during times of security crises, pilgrimages remain more resilient than leisure-oriented travel. During flare-ups in the Israeli-Palestinian conflict, for example, pilgrimage tourism to the Holy Land typically remains steady (Collins-Kreiner et al. 2006). Likewise, even when general tourism had essentially ceased during the Bosnian War (1992–1995), pilgrimages to Medjugorje carried on in the midst of a civil war (Jurkovich & Gesler 1997; Wiinikka-Lydon

2010). Indeed, even during economic recessions, pilgrimage is essentially 'recession-proof' (Olsen 2003).

There are obvious seasonalities at sites where pilgrimages are prescribed at specific times of year or during particular events (Dixit 2005), such as the Hajj and the Kumbh Mela. During those times, Mecca and the holy cities of Allahabad, Haridwar, Nashik, and Ujjain swarm with activity, hosting millions of pilgrims from all over the world. Yet at other times, visitation can be rather slow. While Umrah pilgrimages are not required in Islam and have no set dates, they continue to create demand for religious rituals and tourism service providers in Mecca and Medina all year long (Timothy & Iverson 2006; Zamani-Farahani et al. 2019). In religions that do not have specific dates for required mass pilgrimages, such journeys can be taken at any time of year, which translates into a generally steady flow of visitors at places such as Lourdes, Knock, and Medjugorje, or Kyoto and the Shinto shrines of the Kii Mountains in Japan. Nevertheless, religious holidays, such as Christmas and Easter, bring about momentous changes in demand for destinations such as Jerusalem, Bethlehem, Nazareth, and Rome (Olsen & Ron 2013; Ron & Timothy 2019). Unstipulated pilgrimages can take place at any time of year and often resemble traditional vacation patterns, such as when children are on summer holidays away from school, and many people are now combining religious travel with vacation travel.

Conclusion

Mass tourism, including religious tourism, creates both positive and negative sociocultural, political, and economic impacts at tourism destinations. Community solidarity, social engagement, identity, and depth of religious commitment may be positive manifestations of growing religious tourism. However, sacred destinations are also prone to certain sociopolitical conditions that are either brought on by high religious tourism demand (e.g., overcommoditization and overcommercialization) or exacerbated by its massive growth, such as the effects of social distance and religious dissonance. All three of these relationships between religion and tourism have salient implications for spiritual experiences and community development. Several religious tourism-specific economic perspectives were also examined in this chapter, including employment, scale and branding, gentrification, and the vagaries of seasonality and other demand shifters that cause ebbs and flows in the religious tourism marketplace.

From a socioeconomic vantage point, it makes little difference to the destination whether a guest is a pilgrim, a leisure tourist, or a spirit-seeker. For the destination community, all tourists, including religious tourists and pilgrims, create social impacts. This is especially the case in overtouristified destinations, where religious heritage sites are an important part of the tourism product and where major religious events take place. By the same token, for the providers who service the needs of visitors, it makes little difference what the motivation is for a visitor because it takes all visitors spending money to support local economic development, public services, and infrastructure.

Although religious tourism has its own unique social, political, and economic manifestations, mass religious and spiritual tourism is just as socioculturally damaging and promising as any other type of tourism, just as politically charged as any other type of tourism (sometimes more so), and equally economically important or devastating as any other type of tourism.

References

Abdellah, A., & Ibrahim, M. (2013) 'Towards developing a language course for hajj guides in Al-Madinah Al-Munawwarah: A needs assessment', *International Education Studies*, 6(3): 192–232.

Adie, B.A. (2017) 'Franchising our heritage: The UNESCO World Heritage brand', *Tourism Management Perspectives*, 24: 48–53.

al-Saadi, Y. (2014) 'Mecca's changing face: Rejuvenation or destruction?', *Taghrib News*, March 8, 2014. Online at: http://www.taghribnews.com/en/report/153929/mecca-s-changing-face-rejuvenation-or-destruction. Accessed September 3, 2020.

Baldacchino, D.R. (2010) 'Caring in Lourdes: An innovation in students' clinical placement', *British Journal of Nursing*, 19(6): 358–366.

Banica, M. (2016) 'Coach pilgrimage: Religion, pilgrimage, and tourism in contemporary Romania', *Tourist Studies*, 16(1): 74–87.

Bielo, J.S. (2018) *Ark Encounter: The Making of a Creationist Theme Park*. New York: New York University Press.

Bokhari, A.A.H. (2019) 'The economics of religious tourism (Hajj and Umrah) in Saudi Arabia', in J. Álvarez-García, M. de la Cruz del Río Rama, & M. Gómez-Ullate (Eds.), *Global Perspectives on Religious Tourism and Pilgrimage* (pp. 159–184). Hershey, PA: IGI Global.

Bowman, G. (2011) '"In dubious Battle on the Plains of Heav'n": The politics of possession in Jerusalem's Holy Sepulchre', *History and Anthropology*, 22(3): 371–399.

Brinkerhoff, M.B., & Mackie, M.M. (1986) 'The applicability of social distance for religious research: An exploration', *Review of Religious Research*, 28(2): 151–167.

Buzinde, C.N., Kalavar, J.M., Kohli, N., & Manuel-Navarrete, D. (2014) 'Emic understandings of Kumbh Mela pilgrimage experiences', *Annals of Tourism Research*, 49: 1–18.

Caidi, N. (2019) 'Pilgrimage to hajj: An information journey', *International Journal of Information, Diversity & Inclusion*, 3(1): 44–76.

Cavan, R.S. (1971) 'A dating-marriage scale of religious social distance', *Journal for the Scientific Study of Religion*, 10(2): 93–100.

Collins-Kreiner, N., & Gatrell, J.D. (2006) 'Tourism, heritage and pilgrimage: The case of Haifa's Baha'i Gardens', *Journal of Heritage Tourism*, 1(1): 32–50.

Collins-Kreiner, N., Kliot, N., Mansfeld, Y., & Sagi, K. (2006) *Christian Tourism to the Holy Land: Pilgrimage during Security Crisis*. Aldershot: Ashgate.

Din, A.K.H. (1982) 'Economic implications of Moslem pilgrimage from Malaysia', *Contemporary Southeast Asia*, 4(1): 58–75.

Dixit, S.K. (2005) 'Tourism pattern in Uttaranchal: Cure for seasonality syndrome', *Tourism Today*, 5: 79–90.

Dodds, R., & Butler, R. (Eds.) (2019) *Overtourism: Issues, Realities and Solutions*. Oldenbourg, Germany: De Gruyter.

Egresi, I., Kara, F., & Bayram, B. (2012) 'Economic impact of religious tourism in Mardin, Turkey', *Journal of Economics and Business Research*, 18(2): 7–22.

Engberg, A. (2017) 'Ambassadors for the kingdom: Evangelical volunteers in Israel as long-term pilgrims', in M. Leppäkari & K. Griffin (Eds.), *Pilgrimage and Tourism to Holy Cities* (pp. 156–170). Wallingford: CABI.

Fernandes, C., Pimenta, E., Gonçalves, F., & Rachão, S. (2012) 'A new research approach for religious tourism: The case study of the Portuguese Route to Santiago', *International Journal of Tourism Policy*, 4(2): 83–94.

Gallarza, M.G., Fayos-Gardo, T., & Arteaga-Moreno, F.J. (2019) 'Volunteering in religious events: A consumer behavior perspective through two case studies of Catholic mega-events', in J. Álvarez-García, M. de la Cruz del Río Rama, & M. Gómez-Ullate (Eds.), *Handbook of Research on Socio-Economic Impacts of Religious Tourism and Pilgrimage* (pp. 310–337). Hershey, PA: IGI Global.

Geary, D. (2018) 'India's Buddhist circuit(s): A growing investment market for a "rising" Asia', *International Journal of Religious Tourism and Pilgrimage*, 6(1): 47–57.

Goodman, M. (2007) *Judaism in the Roman World*. Leiden: Brill.

Gupta, S.K., & Raina, R. (2008) 'Economic impact of Vaishno Devi pilgrimage: An analytical study', *International Journal of Hospitality and Tourism Systems*, 1(1): 52–64.

Gursoy, D., & Nunkoo, R. (Eds.) (2019) *The Routledge Handbook of Tourism Impacts: Theoretical and Applied Perspectives*. London: Routledge.

Hall, C.M., & Lew, A.A. (2009) *Understanding and Managing Tourism Impacts: An Integrated Approach*. London: Routledge.

Hall, C.M., & Tucker, H. (2004) *Tourism and Postcolonialism: Contested Discourses, Identities and Representations*. London: Routledge.

Houlihan, M. (2000) 'Souvenirs with soul: 800 years of pilgrimage to Santiago de Compostela', in M. Hitchcock & K. Teague (Eds.), *Souvenirs: The Material Culture of Tourism* (pp. 18–24). Aldershot: Ashgate.

Huang, K., Pearce, P., & Wen, J. (2017) 'Tourists' attitudes toward religious commercialization', *Tourism Culture & Communication*, 17(4): 259–270.

Hung, K., Yang, X., Wassler, P., Wang, D., Lin, P., & Liu, Z. (2017) 'Contesting the commercialization and sanctity of religious tourism in the Shaolin Monastery, China', *International Journal of Tourism Research*, 19(2): 145–159.

Ioannides, D., & Ioannides, M.W.C. (2002) 'Pilgrimages of nostalgia: Patterns of Jewish travel in the United States', *Tourism Recreation Research*, 27(2): 17–25.

Jimura, T. (2019) *World Heritage Sites: Tourism, Local Communities and Conservation Activities*. Wallingford: CABI.

Jude, O.C., Uchenna, O., & Ngozi, E. (2018) 'Impact of religious tourism in host communities: The case of Awhum Monastery', *American Journal of Social Sciences*, 6(3): 39–47.

Jurkovich, J.M., & Gesler, W.M. (1997) 'Medjugorje: Finding peace at the heart of conflict', *Geographical Review*, 87(4), 447–467.

Klimova, J. (2011) 'Pilgrimages of Russian Orthodox Christians to the Greek Orthodox monastery in Arizona', *Tourism*, 59(3): 305–318.

Kouchi, A.N., Nezhad, M.Z., & Kiani, P. (2018) 'A study of the relationship between the growth in the number of Hajj pilgrims and economic growth in Saudi Arabia', *Journal of Hospitality and Tourism Management*, 36: 103–107.

Krakover, S. (2017) 'A heritage site development model: Jewish heritage product formation in south-central Europe', *Journal of Heritage Tourism*, 12(1): 81–101.

Kudumovic, L. (2020) 'The experience of post-war reconstruction: The case of built heritage in Bosnia', *Open House International*, 45(3): 231–248.

Lois-González, R.C., & Santos, X.M. (2015) 'Tourists and pilgrims on their way to Santiago: Motives, caminos and final destinations', *Journal of Tourism and Cultural Change*, 13(2): 149–164.

Maldonado-Erazo, C.P., Correa-Quezada, R., Viñán-Merecí, C., & Sarango-Lalangui, P. (2019) 'Characterization of the population segment dedicated to the retail trade of religious souvenirs', in J. Álvarez-García, M. de la Cruz del Río Rama, & M. Gómez-Ullate (Eds.), *Handbook of Research on Socio-Economic Impacts of Religious Tourism and Pilgrimage* (pp. 290–309). Hershey, PA: IGI Global.

Mathieson, A., & Wall, G. (1982) *Tourism: Economic, Physical and Social Impacts*. London: Longman.

Metreveli, M., & Timothy, D.J. (2010) 'Religious heritage and emerging tourism in the Republic of Georgia', *Journal of Heritage Tourism*, 5(3): 237–244.

Milano, C., Cheer, J.M., & Novelli, M. (2019) 'Introduction', in C. Milano, J.C. Cheer, & M. Novelli (Eds.), *Overtourism: Excesses, Discontents and Measures in Travel and Tourism* (pp. 1–17). Wallingford: CABI.

Mose, G.M. (1996) 'The destruction of churches and mosques in Bosnia Herzegovina: Seeking a rights-based approach to the protection of religious cultural property', *Buffalo Journal of International Law*, 3(1): 180–208.

Mustafa, M.H. (2014) 'Tourism development at the Baptism Site of Jesus Christ, Jordan: Residents' perspectives', *Journal of Heritage Tourism*, 9(1): 75–83.

Nayak, N., & Prabhu, N.B. (2015) 'Socio-economic impacts of pilgrimage tourism with reference to Udupi Sri Krishna Temple, Karnataka', *The International Journal of Business & Management*, 3(1): 41–50.

Nolan, M.L., & Nolan, S. (1989) *Christian Pilgrimage in Modern Western Europe*. Chapel Hill: University of North Carolina Press.

Nyaupane, G.P., Timothy, D.J., & Poudel, S. (2015) 'Understanding tourists in religious destinations: A social distance perspective', *Tourism Management*, 48: 343–353.

Nyikana, S. (2017) 'Religious tourism in South Africa: Preliminary analysis of a major festival in Limpopo', *African Journal of Hospitality, Tourism and Leisure*, 6(1): 1–8.

Olsen, D.H. (2003) 'Heritage, tourism, and the commodification of religion', *Tourism Recreation Research*, 28(3): 99–104.
Olsen, D.H. (2012) 'Negotiating identity at religious sites: A management perspective', *Journal of Heritage Tourism*, 7(4): 359–366.
Olsen, D.H. (2013) 'A scalar comparison of motivations and expectations of experience within the religious tourism market', *International Journal of Religious Tourism and Pilgrimage*, 1(1): 41–61.
Olsen, D.H., & Emmett, C.F. (2021) 'Contesting religious heritage in the Middle East', in C.M. Hall, & S. Seyfi (Eds.), *Cultural and Heritage Tourism in the Middle East and North Africa* (pp. 54–71). London: Routledge.
Olsen, D.H., & Esplin, S.C. (2020) 'The role of religious leaders in religious heritage tourism development: The case of The Church of Jesus Christ of Latter-day Saints', *Religions*, 11(5): 256.
Olsen, D.H., & Ron, A.S. (2013) 'Managing religious heritage attractions: The case of Jerusalem', in B. Garrod & A. Fyall (Eds.), *Contemporary Cases in Heritage: Volume 1* (pp. 51–78). Oxford: Goodfellow Publishers.
Olsen, D.H., & Timothy, D.J. (2002) 'Contested religious heritage: Differing views of Mormon heritage', *Tourism Recreation Research*, 27(2): 7–15.
Olsen, D.H., & Timothy, D.J. (2018) 'Tourism, Salt Lake City and the cultural heritage of Mormonism', in R. Butler & W. Suntikul (Eds.), *Tourism and Religion: Issues and Implications* (pp. 250–269). Bristol: Channel View Publications.
Paine, C. (2019) *Gods and Rollercoasters: Religion in Theme Parks Worldwide*. London: Bloomsbury.
Pechlaner, H., Innerhofer, E., & Erschbamer, G. (2019) *Overtourism: Tourism Management and Solutions*. London: Routledge.
Pinkster, F.M., & Boterman, W.R. (2017) 'When the spell is broken: Gentrification, urban tourism and privileged discontent in the Amsterdam canal district', *Cultural Geographies*, 24(3): 457–472.
Pourtaheri, M., Rahmani, K., & Ahmadi, H. (2012) 'Impacts of religious and pilgrimage tourism in rural areas: The case of Iran', *Journal of Geography and Geology*, 4(3): 122.
Qurashi, J. (2017) 'Commodification of Islamic religious tourism: From spiritual to touristic experience', *International Journal of Religious Tourism and Pilgrimage*, 5(1): 89–104.
Qurashi, J. (2019) 'Diminishing religious cultural heritage of holy Makkah and Medina due to commercialization of the sacred event', in R. Dowson, J. Yaqub, & R. Raj (Eds.), *Spiritual and Religious Tourism: Motivations and Management* (pp. 85–96). Wallingford: CABI.
Rizzello, K., & Trono, A. (2013) 'The pilgrimage to the San Nicola Shrine in Bari and its impact', *International Journal of Religious Tourism and Pilgrimage*, 1(1): 24–40.
Rodosthenous, N., & Varvounis, M. (2014) 'Contemporary parish and pilgrimage travel: Preconditions and targeting', *Tourism Today*, 14: 164–172.
Ron, A.S. (2010) 'Holy Land Protestant themed environments and the spiritual experience', in J. Schlehe, M. Uike-Bormann, C. Oesterle, & W. Hochbruck (Eds.), *Staging the Past: Themed Environments in Transcultural Perspectives* (pp. 111–133). Bielefeld, Germany: Verlag.
Ron, A.S., & Feldman, J. (2009) 'From spots to themed sites–the evolution of the Protestant Holy Land', *Journal of Heritage Tourism*, 4(3): 201–216.
Ron, A.S., & Timothy, D.J. (2019) *Contemporary Christian Travel: Pilgrimage, Practice and Place*. Bristol: Channel View Publications.
Russo, A.P., & Romagosa, F. (2010) 'The network of Spanish Jewries: In praise of connecting and sharing heritage', *Journal of Heritage Tourism*, 5(2): 141–156.
Shackley, M. (2006a) 'Costs and benefits: The impact of cathedral tourism in England', *Journal of Heritage Tourism*, 1(2): 133–141.
Shackley, M. (2006b) 'Empty bottles at sacred sites: Religious retailing at Ireland's national shrine', in D.J. Timothy & D.H. Olsen (Eds.), *Tourism Religion and Spiritual Journeys* (pp. 94–103). London: Routledge.
Sharpley, R. (2009) 'Tourism, religion and spirituality', in T. Jamal & M. Robinson (Eds.), *The Sage Handbook of Tourism Studies* (pp. 237–253). London: Sage.
Shinde, K.A. (2020) 'Managing the environment in religious tourism destination: A conceptual model', in K.A. Shinde & D.H. Olsen (Eds.), *Religious Tourism and the Environment* (pp. 42–59). Wallingford: CABI.
Singh, S., Timothy, D.J., & Dowling, R.K. (2003) *Tourism in Destination Communities*. Wallingford: CAB International.

Smith, J.A., McPherson, M., & Smith-Lovin, L. (2014) 'Social distance in the United States: Sex, race, religion, age, and education homophily among confidants, 1985 to 2004', *American Sociological Review*, 79(3): 432–456.

Soussi, H. (2020) 'Jewish heritage tourism in Morocco: Memories and visions', in D. Vanneste & W. Gruijthuijsen (Eds.), *Value of Heritage for Tourism, Proceedings of the 6th UNESCO UNITWIN Conference, 2019* (pp. 261–270). Leuven: University of Leuven.

Stephenson, M., & Ali, N. (2019) 'Deciphering 'Arab hospitality': Identifying key characteristics and concerns', in D.J. Timothy (Ed.), *Routledge Handbook on Tourism in the Middle East and North Africa* (pp. 71–82). London: Routledge.

Sujatha, P., & Dadakalandar, U. (2019) 'Socio-economic impact of pilgrim tourism in south India', *International Journal of Basic and Applied Research*, 9(5): 675–681.

Tahan, L.G. (2020) 'Archaeological destruction and tourism: Sites, sights, rituals and narratives', in D.J. Timothy & L.G. Tahan (Eds.), *Archaeology and Tourism: Touring the Past* (pp. 121–133). Bristol: Channel View Publications.

Terzidou, M., Stylidis, D., & Szivas, E.M. (2008) 'Residents' perceptions of religious tourism and its socio-economic impacts on the island of Tinos', *Tourism and Hospitality Planning & Development*, 5(2): 113–129.

Timothy, D.J. (2018) 'Cultural routes: Tourist destinations and tools for development', in D.H. Olsen & A. Trono (Eds.), *Religious Pilgrimage Routes and Trails: Sustainable Development and Management* (pp. 27–37). Wallingford: CABI.

Timothy, D.J. (2021) *Cultural Heritage and Tourism: An Introduction*, 2nd edn. Bristol: Channel View Publications.

Timothy, D.J., & Conover, P.J. (2006) 'Nature religion, self-spirituality and New Age tourism', in D.J. Timothy & D.H. Olsen (Eds.), *Tourism, Religion and Spiritual Journeys* (pp. 139–155). London: Routledge.

Timothy, D.J., & Emmett, C.F. (2014) 'Jerusalem, tourism, and the politics of heritage', in M. Adelman & M.F. Elman (Eds.), *Jerusalem: Conflict & Cooperation in a Contested City* (pp. 276–290). Syracuse, NY: Syracuse University Press.

Timothy, D.J., & Iverson, T. (2006) 'Tourism and Islam: Considerations of culture and duty', in D.J. Timothy & D.H. Olsen (Eds.), *Tourism, Religion and Spiritual Journeys* (pp. 186–205). London: Routledge.

Timothy, D.J., & Ron, A.S. (2019) 'Christian tourism in the Middle East: Holy Land and Mediterranean perspectives', in D.J. Timothy (Ed.), *Routledge Handbook on Tourism in the Middle East and North Africa* (pp. 147–159). London: Routledge.

Turner, V. (1973) 'The center out there: Pilgrim's goal', *History of Religions*, 12(3): 191–230.

Vijayanand, S. (2012) 'Socio-economic impacts in pilgrimage tourism', *International Journal of Multidisciplinary Research*, 2(1), 329–343.

Vukonić, B. (1998) 'Religious tourism: Economic value or an empty box?', *Zagreb International Review of Economics & Business*, 1(1): 83–94.

Vukonić, B. (2002) 'Religion, tourism and economics: A convenient symbiosis', *Tourism Recreation Research*, 27(2): 59–64.

Walasek, H., Hadžimuhamedović, A., Perry, V., & Wik, T. (2015) *Bosnia and the Destruction of Cultural Heritage*. London: Routledge.

Wang, K.Y. (2015) 'Live with the deity: Presence and significance of Taiwanese Taoist temple affiliated pilgrim accommodation', *Review of Religious Research*, 57(1): 157–158.

Wiinikka-Lydon, J. (2010) 'The ambivalence of Medjugorje: the dynamics of violence, peace, and nationalism at a Catholic pilgrimage site during the Bosnian war (1992–1995)', *Journal of Religion & Society*, 12: 1–18.

Zaban, H. (2020) 'The real estate foothold in the Holy Land: Transnational gentrification in Jerusalem', *Urban Studies*, 57(15): 3116–3134.

Zamani-Farahani, H., Carboni, M., Perelli, C., & Torabi Farsani, N. (2019) 'Islamic tourism in the Middle East', in D.J. Timothy (Ed.), *Routledge Handbook on Tourism in the Middle East and North Africa* (pp. 125–136). London: Routledge.

Ziaee, M., & Amiri, A. (2019) 'Islamic pilgrimage tourism in Iran: Challenges and perspectives', in S. Seyfi & C.M. Hall (Eds.), *Tourism in Iran: Challenges, Development and Issues* (pp. 69–83). London: Routledge.

23
THE ENVIRONMENTAL IMPACTS OF RELIGIOUS AND SPIRITUAL TOURISM

Kiran A. Shinde

Introduction

No matter what the form of religious tourism, the influx of pilgrims, religious tourists, and other visitors exacts significant environmental impacts in religious tourism destinations and at sacred sites of all religions and faiths. It is important to recognise that the term 'environment' in the context of sacred places is much more than physical and natural, and its use crosses over to the social, cultural, and religious spheres (Lochtefeld 2010; Shinde 2011; Terzidou, Stylidis & Szivas 2008). Hence, it is unsurprising that many studies have examined impacts from social, cultural, economic, and religious dimensions, as well as physical and natural environment (Bleie 2003; Choe & Hitchcock 2018; Della Dora 2012; Hung, Yang, Wassler, Wang, Lin & Liu 2017; Shinde 2012b, 2018). In recent years, there have been more focused investigations into the environmental impacts of religious tourism (Ahammad, Sreekrishnan, Hands, Knapp & Graham 2014; Alipour, Olya & Forouzan 2017; Aminian 2012; Hanandeh 2013; Henderson 2011; Malodia & Singla 2017; Qurashi 2020; Sati 2015; Shinde 2007). This chapter provides a critical review of the emerging literature that focuses on the physicality of sacred place and the impacts of visitation on the physical elements of the sacred environment.

From the extant literature on the environmental impacts of religious and spiritual tourism, it is possible to discern four themes: differences and commonalities of impacts across different sacrosanct environmental settings; the peculiarities of environmental behaviour in religious tourism that influence impacts on sacred sites; the nature of environmental management in religious tourism; and the distinguishing characteristics of spiritual tourism-environment interactions. The last theme requires a separate treatment because "spirituality and religiosity are not synonymous" (Timothy 2013: 36) and the practice of spiritual tourism intersects quite differently with sacred sites. While these themes provide the organising structure of this chapter, the concluding section suggests a future research agenda.

Environmental setting and impacts

Natural sacred sites

Natural formations such as mountain peaks, rocks, caves, lakes, and rivers, are often attributed a sacred value. Their sanctity is incorporated into religious and spiritual belief systems that offer ways of expressing reverence to such formations (Kiernan 2015). In organised religions and indigenous faiths, the natural landscape and its elements have long been revered and visited by human beings (Farra-Haddad 2020b; Olsen 2020). To reinforce their divine connections, natural landscapes are often featured in stories and mythologies of divine apparitions and deified presences in those landscapes. Nolan and Nolan (1989: 303) note that "between 33 and 42 percent of Europe's current shrines are associated with environmental features accorded a certain aura of sanctity". Natural sacred sites, which Timothy (2013: 33) calls "sacrosanct environments", present some unique challenges with respect to visitor influxes.

Mountains

Visiting mountains is one of the earliest expressions of pilgrimage and a deliberate way of "becoming closer to the spirits" (Cochrane 2009: 110) because pilgrims believed that the gods resided in the mountains. Mountain worship derived from a fear of the unknown and the need to appease the mountain-gods. Many popular pilgrimage places and sites of ascetic practices across different faiths are nestled within mountain ranges (Guichard-Anguis 2011; Timothy 2013, 2021b).

The fragile ecosystems of mountains are especially vulnerable to the impacts of visitors. Most mountainous regions are characterised by small settlements (often around monastic life) from which people traditionally walked into the mountains to worship (Nepal et al. 2020). As such, barring a few bare necessities, there traditionally was little development in mountains. However, in recent years, many formerly isolated mountainous regions have become accessible by road. Several pilgrimage studies in the Himalayas show that with the increasing influx of visitors, these settlements are expanding into dangerous zones of increased landslides and tectonic activity, and their built environment is becoming denser, stressing the mountain ecosystem's carrying capacity (Lochtefeld 2010; Pinkney 2013; Sati 2015; Singh 2004, 2005). In a study of the famous Hindu pilgrimage sites of *char-dham* in the Garhwal Himalayas, Sati (2015: 177) observes that "this region is highly prone to debris flow, flashfloods, landslides and mass movements" and that the situation is exacerbated due to "large-scale soil erosion" caused by the construction of settlements and roads while pollution is worsened by the "dumping of waste and litter in the open spaces and in the waterbodies".

Because of their physiographic characteristics, mountains are prone to natural calamities, which are accentuated by human activities, including pilgrimage and other forms of tourism. For instance, on June 16–17, 2013, in the Himalayan towns of Badrinath and Kedarnath, a "cloudburst [was] followed by debris-flow and flashfloods [which] killed more than 10,000 pilgrims and the local people" (Sati 2015: 177). In the Nepalese Himalayas, Bleie (2003) discusses the negative impacts of the opening of a cable car facility to enable access to the Manakamana Temple. Although the problems of "drinking water, energy, and timber crisis" were acute and the demand for "hotel facilities, hot food, and imported goods" led to problems of waste accumulation and management, more significant was the levelling of mountain slopes, which "destroyed the old drainage corridor there and caused the flooding and erosion" (Bleie 2003: 182). As sacred places in mountains are converted into "coveted recreational space" for tourists, vulnerable landforms become increasingly vulnerable (Singh 2005: 221).

Because of their challenging topography, mountains, to some extent, are still able to deter large numbers of visitors. In her study of Mount Athos, Greece, which is accessible only by ferry boat, Della Dora (2012: 960) argues that the limited numbers of monasteries strictly regulate access to Mount Athos where "a maximum of 120 Orthodox Christian visitors are allowed per day, whereas foreigners of other religious affiliations are limited to 10 per day (excluding pilgrims invited directly by the monasteries)". There are no hotels on the mountain peninsula. Instead, monasteries host visitors for a limited time and instruct them to follow the monastic routine during their visit. Thus, mountainscapes can impose certain forced patterns of visitation, which may have fewer negative impacts. This is something that Jimura (2016) also observed in the case of the Kii Mountains in Japan.

Caves

While caves are revered as sacred spaces in many faiths, mainly indigenous ones, Kiernan (2015: 187) argues, "they are perhaps less obvious targets for veneration" because of their "hidden" nature and poor visibility. A few studies have noted that when a temple or shrine is built near a cave, there is proliferation of religious activities (Cochrane 2009; Farra-Haddad 2020a; Kasim 2011; Paniandi, Albattat, Bijami, Alexander & Balekrisnan 2018). Kasim (2011) found that the celebration of the festival of Thaipusam attracted considerable religious tourism to Batu Caves in Malaysia. This festival involves a three-day procession where the "statue of Lord Muruga from the Sri Maha Mariamman temple is taken to the Caves". During this procession, devotees indulge in "masochistic acts of self-mutilation, body piercing, and heavy kavadi dragging [which] is seen as very cleansing to their soul" (Kasim 2011: 447). The natural setting of the caves seems to contribute heavily to the devotees being in "the state of being in a trance" (Kasim 2011: 447). While Kasim does not discuss the impacts on the setting, he refers to his informants being disturbed by the presence of non-adherent tourists. Similar findings were reported by Cochrane (2009) in a study of the volcanic crater, Mount Bromo in Java. She found that during the annual Tenggerese festival of the Kasodo, thousands of devotees walk in a procession from the temple at the foothill to the crater and offer "flowers, rice, and meat...to appease its resident spirits" (Cochrane 2009: 113). In this context, the main impacts were the noise and fumes from tourists' vehicles.

Waterbodies and wells

Water holds great ritualistic and religious significance in all faiths. It is believed to be purificatory, being able to wash away impurities and sins. For instance, Hindu pilgrimage thrives on the symbolic idea that crossing a ford of water is equivalent to crossing over from this world to heaven. Similarly, in some belief systems, bathing in a river is an important religious ritual.

In Christian sites such as Lourdes, holy water is believed to have miraculous powers. Across faiths, water-related rituals are central to pilgrimage practice. In India, the life-giving quality of water draws hordes of pilgrims to river sources in places like Yamunotri (headwaters of the Yamuna River), Gangotri (the origin of the Ganges River), and so on (Alley 2002; Eck 1981; Feldhaus 1995; Haberman 2006). Waterbodies are susceptible to the considerable negative impacts of pilgrimage, the most prominent being pollution from materials used in ritual offerings. Sites on riverbanks are used for funerary rites, such as burning dead bodies, which also causes severe pollution (Alley 2002). In developing countries, where sanitation is lacking, less-educated pilgrims from rural areas often resort to open defecation near riverbanks. A study of river pollution during the pilgrimage season (May–June) in Rishikesh-Haridwar, India, found that "500,000 additional visitors cause 20 times more pollution in the upper Ganges River compared to the rest of the year" (Ahammad, Sreekrishnan, Hands, Knapp & Graham 2014, cited in Hussein, Shahid, Basim & Chelliapan 2016: 3). The influx of pilgrims also generates considerable waste that finds its way into the sewers and eventually into the river.

Pilgrimage and religious tourism also damage the natural ecosystem of waterbodies. Riverbank erosion due to excessive crowding, presents imminent danger of severe flooding during the monsoon season. Sati (2015: 166) describes several instances where the rivers in places like Haridwar and Rishikesh "flow above danger marks [and] that lead to occurrences of severe disasters". In a study of Vrindavan, Shinde (2012b) notes that the use of riverbanks for recreational activities during fairs and festivals causes severe environmental degradation. The damage is exacerbated when boating services for tourists are present within the same river (Doron 2005). Like mountains, rivers also attract adventure activities (Malodia & Singla 2017). Many river-rafting spots have developed near popular pilgrimage sites, which is likely to worsen the ecological damage (Cooper 2009).

Landscapes

In pilgrimages, entire landscapes are considered sacred as miracle stories of deities and gods are imprinted throughout a given landscape. For instance, in north India, religious scriptures devoted to worshiping the Hindu god Krishna eulogise Braj as Krishna's playground, including the Yamuna River, Govardhan Mountain, 12 forests and numerous lakes and hamlets; each place is imbued with stories of his miracles (Entwistle 1987; Shinde 2012b). Pilgrimage landscapes—pilgrimagescapes—developed around stories of the divine and are found in all religions and faiths. For Christians, the Holy Land is a landscape infused with the life of Christ and the spread of Christianity, including many sites of immensely sacred miracles (Fleischer 2000; Liutikas 2015; Ron & Timothy 2019). Liutikas (2015: 16) observes that since the fifteenth century, "copies of the route that Christ had taken, started to appear in areas where natural features such as hills, valleys and streams resembled the relief of Jerusalem". He explains how this landscape was symbolised in Lithuania through various 'cavalries'. Similarly, a Buddhist pilgrimagescape developed around the places in Nepal and India where Gautam Buddha lived and laboured (Geary & Mason 2016).

The trails and routes that provide access to, and linkages between, these sacred places within and outside their sacrosanct landscape not only serve as physical representations of the boundaries of such sacred landscapes but also render sacred meaning through the performance of pilgrimage (Aukland 2017; Guichard-Anguis 2011; Jimura 2016). Frequently, the *terra sancta* of pilgrimage landscapes is defined by the very circumambulatory religious journeying that takes place within these landscapes.

Studies of spiritual landscapes suggest that the historical descriptions found in religious texts may not match their current status for several reasons. Some places become more popular than others and develop as centres of pilgrimage activities on their own because of certain social-spatial relations, while others struggle to maintain their identities (Haberman 1994). The practicalities of access and travel favour the developed centres over the natural sites, and over time, many natural sites almost disappear from popular itineraries or lose their naturalness through urbanisation and physical development. For instance, in the case of Braj, Shinde (2012b) found that the pilgrimagescape covered an area of about 2,500 km^2 and a 300-km long circuitous journey, but the sites continued to reduce over time. The eighteenth-century Vaishnava literature described hundreds of sites including forests, lakes, ponds, *kunds* and shrines. In 1885, there were 133 known sites, and the Braj circuit in the 1980s involved visits to 73 sacred places including 12 main forests and 36 *kunds* (waterbodies). The itinerary that Shinde participated in 2005 was limited to the sites most accessible by car, which included 25 sacred sites (5 forests, 3 *kunds*, and 17 temples).

Sacred groves

Sacred groves have been studied for their role in protecting biodiversity and conservation (Apffel-Marglin & Parajuli 2000; Chandran & Hughes 1997), but studies that discuss them as independent objects of pilgrimage are limited (Olsen 2020). Believed to be the abode of spirits or localities of angelic manifestations, such groves are often located outside villages and settlements and are mainly used for shamanistic practices. As such, these involve significant esoteric rituals and are usually visited only on certain occasions. Often these accessible only to those who live in the area or who adhere to the faith and thus have controlled visitor flows (Cochrane 2009; Pfaffenberger 1983). This means the temporality, scale, and nature of visitation may be why visitor impacts are less distinguishable on sacred groves.

From the above review, it is obvious that most natural sacred sites have a shrine, temple, or other commemorative structure, however small, which serves to demarcate and commemorate their sacred value. This tangible manifestation of sanctity eventually becomes the object of veneration and stimulates the growth of pilgrimage and other forms of religious tourism. However, a locality's remoteness and terrain typically determine the possibility of developing infrastructure for pilgrimage. With significant improvements in infrastructure technology recently, things are likely to change in the near future as places become more accessible to spiritual tourists.

Pilgrim towns

Pilgrim towns are settlements that evolve around sacred places and landscapes and have peculiar environmental characteristics. At their core is the material manifestation of the sacred embodying the 'spirit of the place'. This can be a natural sacred site, a site dedicated to a god or deity, or saints and eulogised through mythological legends and religious scriptures. The physical form and appearance of this core may change to accommodate the needs of people wanting to see, touch, and feel the source of holiness. Around this core, places of worship such as temples, tombs, and shrines are erected to venerate the sacred. Outside the core, religious infrastructure for devotees and others includes lodging and boarding services, food services, and souvenir vendors (Joseph 1994; Rinschede 1995; Shinde 2017). Cosmological and religious considerations are paramount in constructing a religious infrastructure that is built in devotion to deity and to satisfy the pious needs of pilgrims (Fraser 2015; Heitzman 1987; Mack 2002; Petrillo 2003). As such, most buildings employ the best materials, methods, and design principles, yielding rich tapestries with elaborate ornamentation (Fraser 2015; Garcia-Fuentes 2020; Shinde 2012a). Opulent religious buildings are some of the world's best known and most visited heritage attractions (Fernandes, Coelho & Brázio 2015; Nolan & Nolan 1992; Olsen 2019; Shackley 2001a; Timothy 2021a). The built environment in pilgrim towns often reflects the nature of the patronage at sacrosanct locales (Rinschede 1995; Shinde 2012a).

Several studies have documented the urban growth of pilgrim towns into popular religious tourism destinations (Alipour, Olya & Forouzan 2017; Ambrósio 2003; Aminian 2012; Hung et al. 2017; Rinschede 1986, 1995; Rodriguez 2017; Shinde 2012a, 2017). From these, three distinct processes in pilgrim town urbanisation can be identified. First is the conventional process of building temples and other faith-oriented infrastructure by religious institutions mainly for the purpose of providing services for devotees and pilgrims. These are dedicated to piety and divine services and include nunneries, old-age homes, convents, ashrams, hospices, and health services. Other supporting services and infrastructure, such as shops, residences for service providers, and transport facilities, follow in transforming the natural landscape into an urbanscape, generating concentric rings of development, usually radiating from the spiritual core (Rinschede 1986). The second process is driven by the desire of wealthy and resourceful devotees to have a second home in the vicinity of the object of their veneration (Jha 2007; Shinde 2012a; Shukla 2008). In the past, this saw the construction of mansions with beautiful architecture. Its contemporary manifestation involves more banal homes that correspond to the economic standing of patrons and take the form of real estate developments. What emerges is a housing development in an ordinary town but with a unique difference: absentee ownership and sporadic occupancy. The third process emanates from the other two. Corresponding to contemporary patterns of religious tourism, this process is distinctly characterised by increased demand for tourism infrastructure and support services including hotels, service stations, parking, shopping areas, and recreational facilities (Alipour et al. 2017; Lochtefeld 2010; Shinde 2017).

These processes have created distinct sacred and profane spaces in religious tourism destinations (Joseph 1994; Rinschede 1986; Ron & Timothy 2019). However, such a distinction may be unnecessary because in most pilgrimage towns, the boundary between the sacred and the profane is blurred, and these processes all lead to significant environmental changes that accompany the transformation of natural landscapes (e.g. forests and pastoral lands) into urban areas. For example, land-use changes sometimes occur with conventional agriculture being replaced by horticulture types (e.g. flowers and other products) for use in pilgrimage rituals (Ghosal & Maity 2010). Likewise, the addition of newer buildings and urbanisation threaten the built religious heritage. Thus, the very nature of religious activities inherent in pilgrimage and religious tourism may cause significant indirect impacts that are gradual over an extended timeframe in pilgrimage destinations.

In pilgrim towns, the most visible impacts derive directly from the influx of visitors. These are readily quantifiable and most visible in the form of traffic congestion, crowding, pollution, and solid-waste accumulation (Haberman 2006; Henderson 2011; Hussein et al. 2016; Sati 2015; Shackley 2001a). Tourism growth also stresses environmental services, such as water supplies, drainage and sanitation management, which not only pose challenges to carrying capacities but also lead to their physical deterioration.

Pilgrim trails

On pilgrim trails, the route, the landscape, and the journey are often of more paramount importance than the destination (Murray & Graham 1997; Olsen & Trono 2018; Wilkinson 2018). Places on the route derive sacred meaning and identity because of the route (Kim et al. 2019; Olsen & Wilkinson 2016; Timothy & Olsen 2018). Because of their scale, spatial and temporal characteristics, the impacts of pilgrimage on trails vary considerably (Bambi & Barbari 2015; Mason & Chung 2018; Rodrígueza et al. 2018; Trombino & Trono 2018). Where trails attract individual spiritual seekers, the impacts tend to be negligible (Geary & Mason 2016). However, in situations where pilgrim trails have become mass attractions, the impacts are more notable. Large processions on pilgrim routes resemble 'moving sacred towns' and significant impacts occur for the duration the pilgrimage (Shinde 2018). Shinde (2018) describes a walking pilgrimage called Palkhi in western India where pilgrims walk 210 kilometres with overnight stays at 14 places in towns and villages along the way. Although they erect tents for temporary accommodation, the entourage itself is comprised of trucks, bullock-carts, mobile shops, and clusters of latrines—all contributing to heavy traffic congestion and increased stress on the local infrastructure and environment. However, their impacts typically do not last more than two days, as the pilgrim mass moves on to its next destination.

More significant impacts on pilgrimage trails seem to result from increasing touristification and the presence of tourism activities (Mason & Chung 2018; Reader 2007). The development of accommodation facilities and tourism enterprises changes the socio-spatial fabric of some places en route that experienced only limited visitor numbers (Riguccio et al. 2015; Rodrígueza et al. 2018; Trombino & Trono 2018).

Festivals

During festivals, heavy influxes and intense activities of visitors have direct impacts that are concentrated over time and space (Gupta & Basak 2018; Ruback et al. 2008). The impacts of festivals have historically been dealt with using event management approaches (Maclean 2003; Raj & Griffin 2015; Shinde 2010). For instance, the Kumbh Mela, a Hindu festival that takes place every 12 years at one of the four holy sites in India, is attended by more than 20 million visitors. The state government provides more than 2,000 campsites for religious organisations for nearly two months and all facilities for the pilgrims are provided (Buzinde et al. 2014).

Smaller-scale annual or periodic festivals pose more significant challenges to the design and development of resilient environmental services in water supply, sanitation and solid waste management that can absorb strenuous impacts of fluctuating demands and scale of visitation. For instance, Hussien et al. (2016: 8) note the challenges in modelling the sewerage system in the holy city of Karbala in Iran as heavy sewer overflows during pilgrimage times, pose "a potential threat to the ecological and public health of the waterways that receive these overflows".

However, some festivals have positive benefits for the environment, particularly in indigenous faiths, where opportunities are provided for participants to maintain, upkeep, and rejuvenate the sacred natural landscapes (Cochrane 2009; Suntikul & Dorji 2016).

Religious tourism and environmental behaviour

The environmental behaviour of visitors in sacred places is highly contested. Several studies reinforce the stereotype of pilgrims as religiously motivated travellers who value the sacredness of the place and therefore behave in a manner that causes few impacts (Buzinde et al. 2014; Damari & Mansfeld 2016; Della Dora 2012; Guichard-Anguis 2011; Gupta & Basak 2018; Jimura 2016; Liutikas 2015; Moufahim & Lichrou 2019; Nepal et al. 2020; Olsen 2010; Pinkney & Whalen-Bridge 2018; Raj 2015; Terzidou et al. 2017; Wilkinson 2018). Their religious worldviews, attitudes and expressions of reverence through rituals seem to make them less responsible for the negative impacts to which they contribute (Bleie 2003; Shinde 2011). Non-pilgrim tourists, on the other hand, are censured for their hedonistic behaviour in sacred destinations. This is summed up succinctly by Singh (2004: 59) in the context of Himalayan pilgrimages: "Where pilgrims were quite happy drinking water from natural springs, streams and rivers, tourists and some modern pilgrims now expect or demand bottled water". The dissonance between worshippers and sightseers is discussed by Cochrane (2009: 116) at length in the Javanese case. In Tinos, Greece, Terzidou et al. (2008: 123) note that "residents' support of religious tourists is influenced by the respect they feel towards the pilgrims' spirituality and religiosity, whereas some irritations attributed to religious tourists arise from congestion, traffic and pollution that are characteristics of mass tourism". Several studies of Hindu sacred places have unequivocally found that tourists are blamed for all negative impacts and are seen as a threat, whereas pilgrims are hailed as innocent devotees who are necessary for a pilgrim town to survive (Aukland 2018; Lochtefeld 2010; Shinde 2011), despite the mounds of ritual litter left behind, the polluted water, and the inadequate sewerage systems that are overtaxed by their visits (Qurashi 2020; Timothy 2021a).

Environmental behaviour is driven by motivations. Visitors motivated by sightseeing and leisure may be more likely to consume the sacredscape in damaging ways, whereas pilgrims who seek the numinous are probably more willing to act according to instructions by clergy and the religious authorities who look after the object of veneration. Della Dora (2012: 971) found this distinction in her study of Meteora, Greece. She cites responses from a Greek monk and an Egyptian Coptic priest: "We are not here for the landscape; tourists are after landscape! We are here for the saints" and "[P]ilgrims are not bothered by other pilgrims. But tourists are by other tourists". Consternation is not only caused by tourists but also by pilgrims who behave like tourists. For instance, at Manakamana Temple in Nepal, Bleie (2003) notes that pilgrims who used the cable car to get to the main temple instead of the traditional pedestrian trail were violating religious norms. By patronising the cable car, they were party to the negative impacts on the fragile mountainscape. As noted above, often host communities believe that pilgrims are believers and followers of their own faith and therefore no different from themselves, so how can they be responsible for negatively impacting the environment (Alipour et al. 2017; Terzidou et al. 2008). Some studies have found that residents point to the scale of visitation as the culprit rather than individual pilgrim behaviours (Shinde 2011; Suntikul & Dorji 2016).

Environmental behaviour also depends on the value placed on environmental resources in a sacred place. Water, mountains, land, soil, and vegetation are all imbued with sacredness and are worshiped by pilgrims and residents. For their divine quality, these resources are widely used in purification and healing rituals. For instance, after visiting pilgrimage places on the banks of the Ganges River, Hindu pilgrims carry home some of its holy water bottles and jars as souvenirs. Pilgrims see the river as "a purifying goddess and an object of worship" and thus find it extremely difficult to see the pollution in the river even though ecologically it is "often filled with contaminants and pollution" (Lochtefeld 2010: 4). Similarly, pilgrims in Vrindavan collect soil and apply it to their foreheads as a blessing from Krishna (Shinde 2012b). Devotees believe that every physical element of sacrosanct places is a manifestation of the divine and should therefore be accepted, regardless of its condition; the environmental behaviour of visitors and residents in a sacred place is largely governed by their religious frame of reference.

Not all environmental behaviours of visitors in sacred places lead to negative impacts. Jimura (2016) argues that visitors appreciate the inscription of the Kii Mountains on UNESCO's World Heritage List, which has inspired increased efforts to protect the sacred environment of the mountains. The sacredness of a place also seems to inspire a conscious response to environmental problems (Lafortune-Bernard et al. 2020). There are several instances where pilgrim groups volunteer to mitigate problems as a "divine calling" (Singh 2005: 222). Singh (2004: 62) argues that the "traditional concepts of religious travel had sustainability built into it" but with infiltration of touristic characteristics in such travel, the behaviour of pilgrims and tourists are likely to change and lead to more complex outcomes with regard to environmental management.

Environmental management for religious tourism

The literature on sacred site management has focused primarily on visitor management and the challenges facing managers as they try to balance pilgrims' needs with those of the normative tourist gaze (Fadare & Benson 2015; Jimura 2016; Olsen 2019; Petrillo 2003; Presti & Petrillo 2010; Raj & Griffin 2015; Shackley 2001b; Timothy & Olsen 2006; Wong et al. 2016). Environmental management is one of the least researched topics in religious tourism studies, with only a handful of studies inquiring into the subject (González & Medina 2003; Henderson 2011; Shinde 2011). Studies dedicated exclusively to environmental management in sacred places localities note the complexities in addressing environmental problems that derive from pilgrimage and religious tourism (Shinde & Olsen 2020). At the heart of the issue is identifying who and what causes these impacts. As suggested above, individual visitors alone do not appear to be the crux of the problem but rather the masses of people who visit at the same time. Residents tend to be more concerned with the socio-cultural impacts than they are with the physical impacts. As a first step, the causes and nature of environmental impacts need to be clearly recognised and articulated for action. The next question is who responds and who should respond.

Historically, pilgrimage events encompassed large congregations of worshippers through which infectious diseases could easily spread and, as such, managers focused on ensuring hygienic conditions and maintaining law and order. Presently, a diverse range of institutional arrangements exists for managing sacred places (González & Medina 2003; Henderson 2011; Shackley 2001a). These can be seen on a spectrum (Shinde 2017) with one end being occupied by religious institutions that act as custodians of the sacred centre, which is the focus of the pilgrimage. These exist in many forms, including individual gurus, priests, owners of religious establishments, churches and faith organisations, clergy, ashrams, monasteries, charitable trusts, and faith societies that are directly involved in organising pilgrimage activities. Some of these may respond to the impacts of visitors, but their involvement is limited to their jurisdictions and restricted to providing services to pilgrims. Matters outside of their realms of reasonability are left for the government to manage (Alipour et al. 2017; Shinde 2011, 2017). The other end of the spectrum is anchored around the state authority that has the mandate to govern and administer the place (e.g. Mecca and the Hajj) (Henderson 2011; Raj 2015).

The challenges for environmental management in religious destinations are compounded for many reasons: the unique pattern of urbanisation driven by religious tourism; difficulties in assessing infrastructure needs due to fluctuating visitor demand; the informal economy of religious tourism driven by faith institutions operating outside the state domain, and other situations (Shinde & Olsen 2020). Based on their social and religious authority, religious actors use environmental resources to their benefit but hardly engage with wider debates about environmental protection, which they believe to be the responsibility of the state. Because religious institutions are driven by sacred mandates and state agencies operate within a secular framework, there are differences in the ways they articulate their environmental responsibility and participate (or not) in the mitigation of environmental impacts (Lafortune-Bernard et al. 2020; McGee 2000; Shinde 2011). This may not be the situation where religious authorities heavily influence state functions, such as in case of the Hajj (Hanandeh 2013; Henderson 2011; Qurashi 2020) and Chinese Buddhist sites (Wang et al. 2016; Wong et al. 2016). However, the exclusive focus on visitor needs may lead to a "spatial divide" that has serious implications for intensifying impacts. For instance, in the case of

Mashadad, Alipour et al. (2017: 177–178) note that "a form of dualism has developed because authorities have a spatial bias that manifests itself in the allocation of efforts toward areas near the shrine, to the detriment of the rest of the city" and "the further you are from the shrine, the fewer the resources that are devoted to environmental improvement".

Following the ethical-moral stand and inherent teachings of many religious traditions towards respect for nature, there are instances where religious institutions have begun to emphasise environmental stewardship and address environmental degradation in sacred places (Apffel-Marglin & Parajuli 2000; de Jong & Grit 2019; Guichard-Anguis 2011; Jimura 2016; Kiernan 2015; Sullivan 1998; Timothy 2013). The same ethic is also driving cooperation between religious institutions and state agencies towards environmental protection. One example is the Vrindavan Conservation Project in India. Vrindavan is a pilgrim town dedicated to the Hindu god Krishna and is believed to be Krishna's recreational playground. As part of this project, religious gurus have taken the lead to restore the forests in the name of Krishna and encourage other stakeholders to be more environmentally responsible (Nash 2012).

The institutions involved in managing natural sacred sites are structurally different from those in urban areas. Many natural sites come under the ambit of natural resource agencies and are faced with issues such as biodiversity conservation and community livelihoods as they intersect with tourism impacts (Bleie 2003; Sati 2015; Singh 2004, 2005). Since traditional systems of governance continue to be practiced in these areas, pilgrimage activities should be managed in ways that minimise impacts (Choe & Hitchcock 2018; Cochrane 2009).

Environmental management is a political activity and its success or failure depends on the worldviews, mandates, vested interests, and engagement of stakeholders (Tanner & Mitchell 2002). Based on a study of six pilgrimage sites in India, Shinde (2020) suggests a conceptual model for a holistic understanding of environmental management related to religious tourism. This model comprises three analytical categories: environmental processes, institutional responsibility, and place attachment. Environmental processes are contingent upon factors such scale, scope, the geography of the place (its resilience) and influx of visitors. Institutional responsibility refers to the "collective ways of articulating and defining environmental problems and addressing them through moral, ethical, and legal authority" by different actors, both religious and non-religious. How different actors respond depends on place attachment, which is fostered through the framework of religious practices involved in pilgrimage.

Spiritual tourism and environmental impacts

Although spirituality is a "close relative" of religion (Kraft 2007: 234), the intersections between spiritual tourism and the environment are somewhat different compared to the interaction with normative religious tourism and pilgrimage from the perspective of spiritual tourists' motivations and behaviours and the scale of their visitation (Timothy 2013). In spiritual tourism, the focus is on self-actualisation and individual quests to discover one's existential place in the world. While spirituality can be accessed through many different means (Di Giovine & Choe 2019), religious sites and sacred places provide ideal conditions for spirit seekers (Huang et al. 2020; Timothy 2013). Spiritual experiences encompass emotions such as restorative energy, spiritual nourishment and renewal, reflectivity and reflexivity, finding meaning in life, enlightenment, universal solidarity, reverence for the earth, expressing gratitude, and a general oneness with nature (Bond et al. 2015; Moufahim & Lichrou 2019; Moufakkir & Selmi 2018; Norman 2011; Sharpley 2009; Singleton 2017). Due to its connotations with human-nature and human-earth connectivity, spiritual tourism is more likely to generate sympathetic environmental behaviour.

The environmental impacts of spiritual tourism can be better understood by grouping spiritual tourists into two categories. One is those who believe in some of the teachings of oriental religions and participate in some spiritual 'products' (in the sense that spirituality can be bought) through traditional practices such as meditation, yoga and chanting with gurus in religious centres to achieve enlightenment (Kraft 2007). These kinds of spiritual tourist dabble in, or 'check out' certain spiritual philosophies and practices are often foreigners experimenting with elements of other faith traditions. They stay longer in the spiritual destination, but they do not become permanent members or followers (Kraft 2007; Sharpley & Sundaram 2005). Kraft (2007) argues that such seekers are fewer in number and frequently travel as backpackers on frugal budgets. Because of their low numbers, spiritual tourists are thought to causes far fewer impacts on the environment, although the prolonged presence of foreigners at religious sites may have notable impacts on the local socio-cultural fabric. For instance, some foreigners who first came to Vrindavan to immerse themselves in Krishna worship, began buying real estate and opening shops to cater to foreign visitors (Shinde 2012a).

The other category is comprised of New Age spiritual followers. These self-professed pilgrims travel to 'spirit centres' such as Glastonbury, England, and Sedona, USA (Reader 2007: 213), as well as to other places held sacrosanct by indigenous people, in search of spiritual enlightenment. Although New Agers may align with environmental paradigms such as "deep ecology, ecofeminism, environmental conservation and nature adulation", their "movement is one of the most environmentally controversial" (Timothy 2013: 39). Timothy (2013) argues that since there are no original sites dedicated solely for New Age worship, these seekers venerate the sacred spaces of others, including indigenous people and other religions. They are frequently blamed for cultural appropriation, misbehaviour, and violating the sanctity of other people's sacred as by discarding 'ritual litter', taking away objects as souvenirs, and vandalising local natural and cultural heritage (Timothy 2013: 39).

Conclusion

Religious and spiritual tourism scholars have grappled with defining pilgrimage, religious tourism and spiritual tourism, as well as understanding religious tourists' motivations, questions of authenticity, phenomenological experiences, and to some extent management challenges. Given the enormous scale of global religious tourism today, it is prudent to expand our understanding of environmental concerns related to this massive phenomenon. This chapter takes us that direction and presents an overview of the intersection between spirituality, religion, and tourism from an environmental impact perspective.

The review has revealed some common impacts of religious tourism on sacred environments, including congestion, depletion of resources, strains on services, and an overall deterioration of the physical and socio-cultural environment. However, these impacts depend on many factors such as the geography of the place, the intensity and frequency of visitor influxes, the types and extent of infrastructure *in situ*, and resilience of the locality. Because of their environmental settings, the four types of religious destinations described here, namely natural sacred sites, pilgrim towns, pilgrimage trails, and festivals, exhibited certain differences in the impacts they experience. The current literature is rich in observations of environmental impacts, but more rigorous measurements are necessary.

A few areas for further research can be identified. The equating of nature with the divine inspires spiritually inclined visitors, but the evidence is not conclusive on whether and how such inspiration translates into action towards protecting biodiversity and other elements of the environment. Similarly, it seems that because most religions advocate environmental stewardship (however implicit that may be), it would motivate religious leadership and believers to be more conscious of their actions and actively participate in mitigating and managing environmental impacts. Again, some studies report that this is not always the case. It would be worthwhile to investigate the reasons why "believers have often failed fully to follow out such concepts of stewardship" (Tanner & Mitchell 2002: 208) in sacred destinations.

Another area for exploration is environmental governance, which is much more than the simpler notion of management. Here, studies need to explore decision-making in highly charged environments where religious sanctions, social acceptance, and moral authorities divide and polarise action between and across religious and non-religious actors and institutions. Scholars need to investigate factors that affect actions or inactions and how they generate discourses about environmental protection and conservation and the overall sustainability of sacred sites.

The 'spirit of the place' must also be addressed. The physicality of a sacred site is anchored in its geography and landscape, so how cab authenticity be ensured given that religious tourism tends to alter the very landscape it is meant to venerate? Cochrane (2009: 117) frames this problem with an example: "if a focal point – a cave, an ancient temple – is so restored or developed that it loses its historical connection or special atmosphere, will visitors be able to access the spiritual gratification which they seek". There are also concepts in tourism-environment research, namely carrying capacity and the Tourism Area Life Cycle, which can be further developed in the context of sacrosanct destinations.

Finally, it is important to note that not all impacts are negative. Religious tourism may also influence positive change in environmental conditions. For instance, as the value of religious heritage is realised, more organisations and institutions may come together to preserve it. More examples and case studies are required to demonstrate that religion, religious travel, and spiritual pilgrimages indeed can be sustainable.

References

Ahammad, Z., Sreekrishnan, T., Hands, C., Knapp, C., & Graham, D. (2014) 'Increased waterborne bla NDM-1 resistance gene abundances associated with seasonal human pilgrimages to the upper Ganges River', *Environmental Science & Technology*, 48(5): 3014–3020.

Alipour, H., Olya, H.G.T., & Forouzan, I. (2017) 'Environmental impacts of mass religious tourism: From residents' perspectives', *Tourism Analysis*, 22: 167–183.

Alley, K.D. (2002) *On the Banks of the Ganga: When Waste Water Meets a Sacred River.* Ann Arbor: University of Michigan Press.

Ambrósio, V. (2003) 'Religious tourism: Territorial impacts in the EU Marian sanctuary towns', in C. Fernandes, F. Mcgettigan, & J. Edwards (eds.), *Proceedings of the 1st Expert Meeting of ATLAS-Religious Tourism and Pilgrimage Research Group* (pp. 43–56). Fatima: Tourism Board of Leiria.

Aminian, A. (2012) 'Environmental performance measurement of tourism accommodations in the pilgrimage urban areas: The case of the Holy City of Mashhad, Iran', *Procedia-Social and Behavioral Sciences*, 35: 514–522.

Apffel-Marglin, F., & Parajuli, P. (2000) '"Sacred grove" and ecology: Ritual and science', in C.K. Chapple & M.E. Tucker (eds.), *Hinduism and Ecology: The Intersection of Earth, Sky and Water* (pp. 291–312). Cambridge, MA: Harvard University Press.

Aukland, K. (2017) 'Pilgrimage expansion through tourism in contemporary India: The development and promotion of a Hindu pilgrimage circuit', *Journal of Contemporary Religion*, 32(2): 283–298.

Aukland, K. (2018) 'Repackaging India's sacred geography: Travel agencies and pilgrimage-related travel. *Numen*, 65(2–3): 289–318.

Bambi, G., & Barbari, M. (eds.) (2015) *The European Pilgrimage Routes for Promoting Sustainable and Quality Tourism in Rural Areas.* Florence: Firenze University Press.

Bleie, T. (2003) 'Pilgrim tourism in the Central Himalayas', *Mountain Research and Development*, 23(2): 177–185.
Bond, N., Packer, J., & Ballantyne, R. (2015) 'Exploring visitor experiences, activities and benefits at three religious tourism sites', *International Journal of Tourism Research*, 17: 471–481.
Buzinde, C.N., Kalavar, J.M., Kohli, N., & Manuel-Navarrete, D. (2014) 'Emic understandings of Kumbh Mela pilgrimage experiences', *Annals of Tourism Research*, 49: 1–18.
Chandran, M.S., & Hughes, J.D. (1997) 'The sacred groves of south India: Ecology, traditional communities and religious change. *Social Compass*, 44(3): 413–427.
Choe, J., & Hitchcock, M. (2018) 'Pilgrimage to Mount Bromo, Indonesia', in D.H. Olsen & A. Trono (eds.), *Religious Pilgrimage Routes and Trails: Sustainable Development and Management* (pp. 180–195). Wallingford: CABI.
Cochrane, J. (2009) 'Spirits, nature and pilgrimage: The "other" dimension in Javanese domestic tourism', *Journal of Management, Spirituality and Religion*, 6(2): 107–119.
Cooper, M. (2009) 'River tourism in the South Asian subcontinent', in B. Prideaux & M. Cooper (eds.), *River Tourism* (pp. 23–40). Wallingford: CABI.
Damari, C., & Mansfeld, Y. (2016) 'Reflections on pilgrims' identity, role and interplay with the pilgrimage environment', *Current Issues in Tourism*, 19(3): 199–222.
de Jong, M., & Grit, A. (2019) 'Implications for managed visitor experiences at Muktinath Temple (Chumig Gyatsa) in Nepal: A netnography', in M. Griffiths & P. Wiltshier (eds.), *Managing Religious Tourism* (pp. 135–143). Wallingford: CABI.
Della Dora, V. (2012) 'Setting and blurring boundaries: Pilgrims, tourists, and landscape in Mount Athos and Meteora', *Annals of Tourism Research*, 39: 951–974.
Di Giovine, M.A., & Choe, J. (2019) 'Geographies of religion and spirituality: Pilgrimage beyond the 'officially'sacred', *Tourism Geographies*, 21(3): 361–383.
Doron, A. (2005) 'Encountering the 'other': Pilgrims, tourists and boatmen in the city of Varanasi', *The Australian Journal of Anthropology*, 16(2): 157–179.
Eck, D. (1981) 'India's tirthas: 'Crossing' in sacred geography', *History of Religions*, 20(2): 323–344.
Entwistle, A. (1987) *Braj: Centre of Krishna Pilgrimage*. Groningen: Egbert Forsten.
Fadare, S., & Benson, E. (2015) 'The consumption and management of religious tourism sites in Africa', in R. Raj & K. Griffin (eds.), *Religious Tourism and Pilgrimage Management: An International Perspective* (pp. 218–233). Wallingford: CABI.
Farra-Haddad, N. (2020a) 'Archaeology and religious tourism: Sacred sites, rituals, sharing the baraka and tourism development', in D.J. Timothy & L.G. Tahan (eds.), *Archaeology and Tourism: Touring the Past* (pp. 106–120). Bristol: Channel View Publications.
Farra-Haddad, N. (2020b) 'Interreligious dialogue: Trees, stones, water, and interfaith itual experiences in Lebanon', in K.A. Shinde & D.H. Olsen (eds.), *Religious Tourism and the Environment* (pp. 95–104). Wallingford: CABI.
Feldhaus, A. (1995) *Water and Womanhood: Religious Meanings of Rivers in Maharashtra*. Oxford: Oxford University Press.
Fernandes, C., Coelho, J., & Brázio, M. (2015) 'Revisiting religious tourism in northern Portugal', in R. Raj & K. Griffin (eds.), *Religious Tourism and Pilgrimage Management: An International Perspective* (pp. 254–266). Wallingford: CABI.
Fleischer, A. (2000) 'The tourist behind the pilgrim in the Holy Land', *Hospitality Management*, 19: 311–326.
Fraser, V. (2015) 'Accommodating religious tourism: The case of the Basilica of the Virgin of Guadalupe in Mexico', *Interiors*, 6(3): 329–350.
Garcia-Fuentes, J.-M. (2020) 'Mimicking mountains: Antoni Gaudi's Sagrada Familia and the Shrine of Montserrat', in K.A. Shinde & D.H. Olsen (eds.), *Religious Tourism and the Environment* (pp. 105–115). Wallingford: CABI.
Geary, D., & Mason, D. (2016) 'Walking in the Valley of the Buddha: Buddhist revival and tourism development in Bihar', *Journal of Global Buddhism*, 17: 57–63.
Ghosal, S., & Maity, T. (2010) 'Development and sustenance of Shirdi as a centre for religious tourism in India', in R.P.B. Singh (ed.), *Holy Places and Pilgrimages: Essays on India* (pp. 159–173). New Delhi: Shubhi Publications.
González, R., & Medina, J. (2003) 'Cultural tourism and urban management in northwestern Spain: The pilgrimage to Santiago de Compostela', *Tourism Geographies*, 5(4): 446–460.

Guichard-Anguis, S. (2011) 'Walking through a World Heritage forest in Japan: The Kumano pilgrimage', *Journal of Heritage Tourism*, 6(4): 285–295.

Gupta, S., & Basak, B. (2018) 'Exploring pilgrim satisfaction on facilities for religious events: A case of Ratha Yatra at Puri', *Asia Pacific Journal of Tourism Research*, 23(8): 765–779.

Haberman, D.L. (1994) *Journey through the Twelve Forests: An Encounter with Krishna*. New York: Oxford University Press.

Haberman, D.L. (2006) *River of Love in an Age of Pollution: The Yamuna River of Northern India*. Berkeley: University of California Press.

Hanandeh, A.E. (2013) 'Quantifying the carbon footprint of religious tourism: The case of Hajj', *Journal of Cleaner Production*, 52: 53–60.

Heitzman, J. (1987) 'Temple urbanism in Medieval south India', *The Journal of Asian Studies*, 46(4): 791–827.

Henderson, J.C. (2011) 'Religious tourism and its management: The Hajj in Saudi Arabia', *International Journal of Tourism Research*, 13(6): 541–552.

Huang, K., Pearce, P., Guo, Q., & Shen, S. (2020) 'Visitors' spiritual values and relevant influencing factors in religious tourism destinations', *International Journal of Tourism Research*, 22(3): 314–324.

Hung, K., Yang, X., Wassler, P., Wang, D., Lin, P., & Liu, Z. (2017) 'Contesting the commercialization and sanctity of religious tourism in the Shaolin Monastery, China', *International Journal of Tourism Research*, 19(2): 145–159.

Hussein, A., Shahid, S., Basim, K., & Chelliapan, S. (2016) 'Modeling sewer flow in a pilgrimage city', *Journal of Environmental Engineering*, 142(12), 05016005.

Jha, S.K. (2007) 'Realty ride on divine living', *Reality Plus: The Real Estate Review*. Online: http://www.realtyplusmag.com/spotlight_fullstory.asp?spotlight_id=30 [Accessed 12/10/2007].

Jimura, T. (2016) 'World heritage site management: A case study of sacred sites and pilgrimage outes in the Kii Mountain Range, Japan. *Journal of Heritage Tourism*, 11(4): 382–394.

Joseph, C.A. (1994) *Temples, Tourists and the Politics of Exclusion: The Articulation of Sacred Space at the Hindu Pilgrimage Centre of Pushkar, India*. Unpublished PhD thesis, University of Rochester.

Kasim, A. (2011) 'Balancing tourism and religious experience: Understanding devotees' perspectives on Thaipusam in Batu Caves, Selangor, Malaysia', *Journal of Hospitality Marketing Management*, 20(3–4): 441–456.

Kiernan, K. (2015) 'Landforms as sacred places: Implications for geodiversity and geoheritage', *Geoheritage*, 7(2): 177–193.

Kim, H., Yilmaz, S., & Ahn, S. (2019) 'Motivational landscape and evolving identity of a route-based religious tourism space: A case of Camino de Santiago', *Sustainability*, 11(13): 1–20.

Kraft, S.E. (2007) 'Religion and spirituality in Lonely Planet's *India*', *Religion*, 37: 230–242.

Lafortune-Bernard, A., Rajendra, N.S., Weise, K., & Coningham, R.A.E. (2020) 'Religious tourism and environmental conservation in Lumbini, the Birthplace of Lord Buddha, World Heritage Site, Nepal', in K.A. Shinde & D.H. Olsen (eds.), *Religious Tourism and the Environment* (pp. 83–94). Wallingford: CABI.

Liutikas, D. (2015) 'Religious landscape and ecological ethics: Pilgrimage to the Lithuanian Calvaries', *International Journal of Religious Tourism and Pilgrimage*, 3(1): 12–24.

Lochtefeld, J. (2010) *God's Gateway: Identity and Meaning in a Hindu Pilgrimage Place*. New York: Oxford University Press.

Mack, A. (2002) *Spiritual Journey, Imperial City: Pilgrimage to the Temples of Vijayanagara*. New Delhi: Vedams.

Maclean, K.K. (2003) *Power and Pilgrimage: The Kumbha Mela in Allahabad, 1765–1954*. Unpublished Ph.D. thesis, La Trobe University.

Malodia, S., & Singla, H. (2017) 'Using HOLSAT to evaluate satisfaction of religious tourist at sacred destinations: The case of religious travelers visiting sacred destinations in the Himalayas, India', *International Journal of Culture, Tourism and Hospitality Research*, 11(2): 255–270.

Mason, D., & Chung, M.-H. (2018) 'The burgeoning of the Baekdu-daegan Trail into a new religious-pilgrimage tourism asset of South Korea', *Journal of Tourism and Leisure Research*, 20(4): 425–441.

McGee, M. (2000) 'State responsibility for environmental management: Perspectives from Hindu texts on polity', in C.K. Chapple & M.E. Tucker (eds.), *Hinduism and Ecology: The Intersection of Earth, Sky and Water* (pp. 59–86). Cambridge, MA: Harvard University Press.

Moufahim, M., & Lichrou, M. (2019) 'Pilgrimage, consumption and rituals: Spiritual authenticity in a Shia Muslim pilgrimage', *Tourism Management*, 70: 322–332.
Moufakkir, O., & Selmi, N. (2018) 'Examining the spirituality of spiritual tourists: A Sahara Desert experience', *Annals of Tourism Research*, 70: 108–119.
Murray, M., & Graham, B. (1997) 'Exploring the dialectics of route-based tourism: The Camino de Santiago', *Tourism Management*, 18(8): 513–524.
Nash, J. (2012) 'Re-examining ecological aspects of Vrindavan Pilgrimage', in L. Manderson, W. Smith, & M. Tomlinson (eds.), *Flows of Faith: Religious Reach and Community in Asia and the Pacific* (pp. 105–121). Dordrecht: Springer.
Nepal, S.K., Mu, Y., & Lai, P.-H. (2020) 'The Beyul: Sherpa perspectives on landscapes characteristics and tourism development in Khumbu (Everest), Nepal', in K.A. Shinde & D.H. Olsen (eds.), *Religious Tourism and the Environment* (pp. 70–82). Wallingford: CABI.
Nolan, M.L., & Nolan, S. (1989) *Christian Pilgrimage in Modern Western Europe*. Chapel Hill: University of North Carolina Press.
Nolan, M.L., & Nolan, S. (1992) 'Religious sites as tourism attractions in Europe', *Annals of Tourism Research*, 19: 68–78.
Norman, A. (2011) *Spiritual Tourism: Travel and Religious Practice in Western Society*. London: Continuum.
Olsen, D.H. (2010) 'Pilgrims, tourists and Max Weber's 'ideal types'', *Annals of Tourism Research*, 37: 848–851.
Olsen, D.H. (2019) 'Best practice and sacred site management: The case of Temple Square in Salt Lake City, Utah', in M. Griffiths & P. Wiltshier (eds.), *Managing Religious Tourism* (pp. 65–78). Wallingford: CABI.
Olsen, D.H. (2020) 'Pilgrimage, religious tourism, biodiversity, and natural sacred sites', in K.A. Shinde & D.H. Olsen (eds.), *Religious Tourism and the Environment* (pp. 23–41). Wallingford: CABI.
Olsen, D.H., & Trono, A. (2018) *Religious Pilgrimage Routes and Trails: Sustainable Development and Management*. Wallingford: CABI.
Olsen, D.H., & Wilkinson, G. (2016) 'Are fast pilgrims true pilgrims? The Shikoku pilgrimage', *Annals of Tourism Research*, 61: 213–267.
Paniandi, T.A., Albattat, A.R., Bijami, M., Alexander, A., & Balekrisnan, V. (2018) 'Marketing mix and destination image, case study: Batu Caves as a religious destination', *Almatourism*, 9(17): 165–186.
Petrillo, C.S. (2003) 'Management of churches and eligious sites', in C. Fernandes, F. Mcgettigan, & J. Edwards (eds.), *Proceedings of the Religious Tourism and Pilgrimage: ATLAS - Special Interest Group 1st Expert Meeting* (pp. 71–86). Fatima: Tourism Board of Leiria.
Pfaffenberger, B. (1983) 'Serious pilgrims and frivolous tourists: The chimera of tourism in the pilgrimages of Sri Lanka', *Annals of Tourism Research*, 10: 57–74.
Pinkney, A.M. (2013) 'Very present history in the land of the gods: Contemporary Māhātmya writing on Uttarākhand', *International Journal of Hindu Studies*, 17(3): 229–260.
Pinkney, A.M., & Whalen-Bridge, J. (2018) *Religious Journeys in India: Pilgrims, Tourists, and Travelers*. Albany: State University of New York Press.
Presti, O.L., & Petrillo, C.S. (2010) 'Co-management of religious heritage: An Italian case-study', *Tourism*, 58(3): 301–311.
Qurashi, J. (2020) 'The natural environment and waste management at the Hajj', in K.A. Shinde & D.H. Olsen (eds.), *Religious Tourism and the Environment* (pp. 133–143). Wallingford: CABI.
Raj, R. (2015) 'Pilgrimage experience and consumption of travel to the city of Makkah for the Hajj ritual', in R. Raj & K.A. Griffin (eds.), *Religious Tourism and Pilgrimage Management: An International Perspective* (pp. 173–190). Wallingford: CABI.
Raj, R., & Griffin, K.A. (2015) *Religious Tourism and Pilgrimage Management: An International Perspective*. Wallingford: CABI.
Reader, I. (2007) 'Pilgrimage growth in the modern world: Meanings and implications. *Religion*, 37: 210–229.
Riguccio, L., Russo, P., Carullo, L., Lanteri, P., & Tomaselli, G. (2015) 'The transformation of the landscape along the pilgrimage routes in the province of Catania - Sicily', in G. Bambi & M. Barbari (eds.), *The European Pilgrimage Routes for Promoting Sustainable and Quality Tourism in Rural Areas* (pp. 165–182). Florence: Firenze University Press.
Rinschede, G. (1986) 'The pilgrimage town of Lourdes', *Journal of Cultural Geography*, 7(1): 21–34.

Rinschede, G. (1995) 'The pilgrimage centre of Loreto, Italy: A geographical study', D.P. Dubey (ed.), *Pilgrimage Studies: Sacred Places, Sacred Traditions* (pp. 157–178). Allahabad: Society of Pilgrimage Studies.

Rodriguez, J. (2017) 'Cleaning up Bodhgaya: Conflicts over development and the worlding of Buddhism', *City & Society*, 29(1): 59–81.

Rodrígueza, X.A., Martínez-Rogetb, F., & González-Muriasa, P. (2018) 'Length of stay: Evidence from Santiago de Compostela', *Annals of Tourism Research*, 68: 9–19.

Ron, A.S., & Timothy, D.J. (2019) *Contemporary Christian Travel: Pilgrimage, Practice and Place*. Bristol: Channel View Publications.

Ruback, R.B., Pandey, J., & Kohli, N. (2008) 'Evaluations of a sacred place: Role and religious belief at the *Magh Mela*', *Journal of Environmental Pshychology*, 28: 174–184.

Sati, V.P. (2015) Pilgrimage tourism in mountain egions: Socio-economic and environmental implications in the Garhwal Himalaya', *South Asian Journal of Tourism and Heritage*, 8(2): 164–182.

Shackley, M. (2001a) *Managing Sacred Sites: Service Provision and Visitor Experience*. London: Continuum.

Shackley, M. (2001b) 'Sacred World Heritage Sites: Balancing meaning with management', *Tourism Recreation Research*, 26(1): 5–10.

Sharpley, R. (2009) 'Tourism, religion and spirituality', in T. Jamal & M. Robinson (eds.) *The SAGE Handbook of Tourism Studies* (pp. 237–253). London: Sage.

Sharpley, R., & Sundaram, P. (2005) 'Tourism: A sacred Journey? The case of ashram tourism, India', *International Journal of Tourism Research*, 7: 161–171.

Shinde, K.A. (2007) 'Pilgrimage and the environment: Challenges in a pilgrimage centre', *Current Issues in Tourism*, 10(4): 343–365.

Shinde, K.A. (2010) 'Managing Hindu festivals in pilgrimage sites: Emerging trends, opportunities and challenges', *Event Management*, 14(1): 53–69.

Shinde, K.A. (2011) '"This is religious environment": Environmental discourses and environmental behaviour in the Hindu pilgrimage site of Vrindavan', *Space and Culture*, 14(4): 448–463.

Shinde, K.A. (2012a) 'Place-making and environmental change in a Hindu pilgrimage site in India', *GeoForum*, 43(1): 116–127.

Shinde, K.A. (2012b) 'Shifting pilgrim trails and temple-towns in India: Problems and prospects', in T. Winter & P. Daley (eds.), *Routledge Handbook on Heritage in Asia* (pp. 328–338). London: Routledge.

Shinde, K.A. (2017) 'Planning for urbanization in religious tourism destinations: Insights from Shirdi, India', *Planning Practice & Research*, 32(2): 132–151.

Shinde, K.A. (2018) 'Palkhi: A moving sacred town', in D.H. Olsen & A. Trono (eds.), *Religious Pilgrimage Routes and Trails: Sustainable Development and Management* (pp. 150–166). Wallingford: CABI.

Shinde, K.A. (2020) 'Managing the environment in religious tourism destinations: A conceptual model', in K.A. Shinde & D.H. Olsen (eds.) *Religious Tourism and the Environment* (pp. 42–59). Wallingford: CABI.

Shinde, K.A., & Olsen, D.H. (eds.) (2020) *Religious Tourism and the Environment*. Wallingford: CABI.

Shukla, A. (2008) 'Spirituality drives India's realty industry *NERVE India*. Online: http://www.nerve.in/news:253500135576 [Accessed 20/08/2008].

Singh, Sagar (2004) 'Religion, heritage and travel: Case references from the Indian Himalayas', *Current Issues in Tourism*, 7(1): 44–65.

Singh, Shalini (2005) 'Secular pilgrimages and sacred tourism in the Indian Himalayas', *GeoJournal*, 64(3): 215–223.

Singleton, A. (2017) 'The summer of the spirits: Spiritual tourism to America's foremost village of spirit mediums', *Annals of Tourism Research*, 67: 48–57.

Sullivan, B.M. (1998) 'Theology and ecology at the cirthplace of Krsna', in L.E. Nelson (ed.), *Purifying the Earthly Body of God: Religion and Ecology in Hindu India* (pp. 247–268). Albany: State University of New York Press.

Suntikul, W., & Dorji, U. (2016) 'Local perspectives on the impact of tourism on religious festivals in Bhutan', *Asia Pacific Journal of Tourism Research*, 21(7): 741–762.

Tanner, R., & Mitchell, C. (2002) *Religion and the Environment*. New York: Palgrave.

Terzidou, M., Scarles, C., & Saunders, M.N. (2017) 'Religiousness as tourist performances: A case study of Greek Orthodox pilgrimage', *Annals of Tourism Research*, 66: 116–129.

Terzidou, M., Stylidis, D., & Szivas, E.M. (2008) 'Residents' perceptions of religious tourism and its socio-economic impacts on the island of Tinos', *Tourism and Hospitality Planning & Development*, 5(2): 113–129.

Timothy, D.J. (2013) 'Religious views of the environment: Sanctification of nature and implications for tourism', in A. Holden & D.A. Fennell (eds.), *The Routledge Handbook of Tourism and the Environment* (pp. 53–64). London: Routledge.

Timothy, D.J. (2021a) *Cultural Heritage and Tourism: An Introduction*, 2nd edn. Bristol: Channel View Publications.

Timothy, D.J. (2021b) 'Tourism and the multi-faith heritage of the Middle East and North Africa: A resource perspective', in S. Seyfi & C.M. Hall (eds.), *Cultural and Heritage Tourism in the Middle East and North Africa: Complexities, Management and Practices* (pp. 34–53). London: Routledge.

Timothy, D.J., & Olsen, D.H. (eds.) (2006) *Tourism, Religion and Spiritual Journeys*. London: Routledge.

Timothy, D.J., & Olsen, D.H. (2018) 'Religious routes, pilgrim trails: Spiritual pathways as tourism resources', in R. Butler & W. Suntikul (eds.), *Tourism and Religion: Issues and Implications* (pp. 220–235). Bristol: Channel View Publications.

Trombino, G., & Trono, A. (2018) 'Environment and sustainability as related to religious pilgrimage routes and trails', in D.H. Olsen & A. Trono (eds.), *Religious Pilgrimage Routes and Trails: Sustainable Development and Management* (pp. 49–60). Wallingford: CABI.

Wang, W., Chen, J.S., & Huang, K. (2016) 'Religious tourist motivation in Buddhist Mountain: The case from China', *Asia Pacific Journal of Tourism Research*, 21(1): 57–72.

Wilkinson, G. (2018) 'The Shikoku pilgrimage: Popularity and the pilgrim's transactions', in D.H. Olsen & A. Trono (eds.), *Religious Pilgrimage Routes and Trails: Sustainable Development and Management* (pp. 196–209). Wallingford: CABI.

Wong, C.U.I., McIntosh, A., & Ryan, C. (2016) Visitor management at a Buddhist sacred site', *Journal of Travel Research*, 55(5): 675–687.

24
MARKETING RELIGIOUS AND SPIRITUAL TOURISM EXPERIENCES

Farooq Haq

Introduction

People's interest in religion and spirituality seems to have awakened from the horrible events of 9–11. The subsequent wars, known as wars on terror, and various global crises have further motivated others to join religious organizations or otherwise enrich their spiritual lives by visiting sacred sites and locales. These journeys of spiritual development increased the recognition of religious and spiritual tourism by academic researchers (Moal-Ulvoas & Taylor 2014; Norman 2014; Olsen 2013; Timothy & Olsen 2018). Intellectuals from diverse backgrounds have meditated, studied, and delivered on spiritually motivated travel for centuries, although it seems to have entered the commercial travel glossary more recently (Andriotis 2009; Haq & Medhekar 2019; Mitroff & Denton 1999). Beyond its expansive research coverage in the social sciences, spirituality and religiosity have recently become an increasingly important research subject in business studies and travel industry management (Coats 2008; Cochrane 2009; Norman 2014; Olsen 2013; Pesut 2003). Although an important part of business management studies, marketing has received considerably less research attention in the spiritual and religious tourism arena (Eid 2012; Štefko, Kiráľová & Mudrík 2015).

This chapter examines the marketing of religious and spiritual tourism experiences and may be useful for tourism service providers, destination managers, religious groups, and government agencies. A thorough conceptual comparison of religion and spirituality is beyond the scope of this chapter and has already been discussed at the outset of the book; however, it will be examined briefly in the next section. Moreover, based on what other authors have already concluded, this chapter considers religion and spirituality to be similar concepts with respect to tourism marketing strategies, and although the author is aware of conceptual and practical differences, the two terms will be used interchangeably throughout this chapter.

Religion has inexplicitly influenced spiritual tourism, and the religious tourism literature has focused descriptively on pilgrims, religious tourists, and other people visiting sacred destinations, shrines, sacrosanct attractions, and religious-themed festivals. Most of the literature has focused on religious tourists who are affiliated with a specific faith or religion. There has been a much leaner analysis to develop theory in the realm of religious tourism. However, three recurring themes appeared during the process of reviewing the literature for this chapter. The dominant theme is descriptions of travel to specific destinations. The secondary theme focuses on the impact of significant people associated with a religion. Third, there is much written about the organization and management of religious events.

Although some consider marketing a necessary evil in the commercialization of any product or service, business success depends on the effective design and implementation of a marketing-oriented roadmap. Some passionate spiritual tourists engaged by the author, when inquired about marketing spiritual tourism, were offended, considering such action to be taboo, as if it were a sin to talk about the marketability of their sacred journeys. Likewise, some Hajjis in Makkah considered it oxymoronic to connect spiritual tourism with marketing. The fact remains that the billions of dollars the Saudi government earns from the Hajj every year is the outcome of cleverly crafted and implemented marketing strategies, just as it is in many pilgrimage destinations. This chapter evaluates various strategic options for marketing religious and spiritual tourism, ranging from a conventional marketing mix approach to a pragmatic relationship marketing strategy.

Understanding religious and spiritual tourism

Tourism is a service industry driven by many consumer motivations and comprising many services (Cohen 1979; Olsen 2013; Shackley 2002). The elements of all of these motives (demand) and services (supply) are not more important than the other (Andriotis 2009; Haq 2014; Jauhari & Sanjeev 2010). An example of the interconnection of tourism elements can be demonstrated by a tourist traveling to the tomb of Sufi Ali Hujveri, known by devotees as Daata Sahib, in Lahore (Qureshi 1995). In this case, religion, spirituality, destination qualities, food, and music all play a part in determining the religious tourist's level of satisfaction. Many other examples from diverse faiths illustrate the importance of marketing to the success of religious tourism and to creating memorable and satisfying pilgrimage experiences (Eid 2012; Haq 2013; Ron & Timothy 2019; Štefko, Kiráľová & Mudrík 2015).

The desire or need to find meaning and purpose in life has important business implications, but accepting that religiosity includes a belief in a Supreme Power who controls the entire universe does not fit well with organizational management theory (Coats 2008; Mitroff & Denton 1999; Taylor 1989). A non-religious definition of spirituality describes the phenomenon as an unfolding mystery, harmonious interconnectedness, and inner strength (Pesut 2003). A multi-religious definition explains spirituality as an inner truth that is the focal point of all religions (Haq & Medhekar 2019). Ibn Al-Arabi, the spiritual teacher of the founders of the Ottoman Empire, produced a multi-religious definition of spirituality in the twelfth century:

> my heart has become capable of every form: it is the pasture of gazelles and a covenant of Christian monks and a temple of idols and the pilgrim's Ka'ba (Makkah) and the tables of the Tora and the book of Quran. I follow the religion of love: whatever way love's camel takes, that is my religion and faith.
>
> *(Nicholson 1978, p. 67)*

Tourists' levels of satisfaction are connected with their values, beliefs, habits, and cultural norms (Rinschede 1992; Shackley 2002). Tourism research has provided much business insight for tourism marketers, developing both theoretical and applied perspectives (Buhalis 2000; Haq & Jackson 2009; Riege & Perry 2000). Compared with other essential components of culture, religion has molded people's beliefs and life habits. Even individuals who claim to be non-religious, have been influenced by religion. Religion has also been seen as a dominant element in cultural tourism (Coats 2008; Raj & Morpeth 2007; Shackley 2002).

Religion inspires religious tourism and spiritual tourism, both of which are sub-segments of cultural tourism (Haq & Jackson 2009; Norman 2014; Timothy 2021). Spirituality has been defined as the essence of religion and an inspirational quality that even non-religious people seek in their lives (Mitroff & Denton 1999; Smith 1992; Timothy & Olsen 2018). Conceptually, spiritual tourism, religious tourism, and pilgrimage are closely intertwined, since they are among the oldest forms of travel and are all linked to religion, belief systems, and personal improvement. They offer hope to travelers and an experience beyond the self-directed orientation of most other forms of tourism (Raj & Morpeth 2007; Rinschede 1992; Vukonić 1996). The conceptual boundary between spirituality and religion is illustrated by Mitroff (2003), who rejects the notion that religion and spirituality are the same. Some researchers have analyzed the spiritual tourism phenomenon separately. For example, Coats (2008) examines spiritual tourism in the context of eco-conscious New Age tourists in Sedona, Arizona, for whom self-actualization is important.

The lack of clarity about religious and spiritual tourism may be the outcome of scholarly debate based on different theoretical and managerial perspectives. Norman (2014) claims that spiritual and religious tourism reflect the individual's self-discovery and preservation of well-being. Taylor (1989) argues that to specify the location of the self in the social universe requires a central drive for individuality. There are congruencies between different kinds of spiritual tourism and other forms of tourism, including religious tourism, in a range of contexts (Taylor 1989). The similarity of these concepts is also noted by Tilson (2005), who considers religious-spiritual tourism a single term in his article on the Camino de Santiago. Likewise, Jauhari and Sanjeev (2010) note the travel and business opportunities associated with religious, cultural, and spiritual tourism in India, and highlight situations that are both religious and spiritual: holy festivals, sacred sites, and saints' days.

To clarify further, this chapter looks at pilgrimage being described as institutionalized religious tourism, which includes tour packages comprised of different services and offerings (Clingingsmith, Kwaja & Kremer 2008; Collins-Kreiner & Gatrell 2006; Raj & Morpeth 2007; Ron & Timothy 2019). Cohen (1992) polarizes pilgrimage centers into two types: popular and formal. Within popular pilgrimage centers, "the ludic and folksy activities are of greater importance and may even take precedence over the more serious and sublime activities" (Cohen 1992, p. 36). A formal pilgrimage center is one in which "the serious and sublime religious activities are primarily emphasized" and "the pilgrim's principle motive for the journey…is to perform a fundamental religious obligation" (Cohen 1992, p. 36). Based on the preceding discussion, it is important to emphasize that in this chapter, the differences between spiritual tourism, religious tourism and pilgrimage is not emphasized and will be used interchangeably.

Religious tourism, as it sounds, has a religious code-based motivation, though people also travel to religious destinations and visit sacred attractions with a different purpose, such as showing respect, gaining knowledge and understanding, and appreciating culture (Andriotis 2009; Cochrane 2009; Lupu, Brochado & Stoleriu 2019; Rodrigues & McIntosh 2014). Some researchers classify these more recreational travelers as religious tourists (Collins-Kreiner & Gatrell 2006; Cohen 1979; Olsen 2012). These are the people who are attracted to a holy place because of its architecture, artistic beauty, prominence and popularity, or historical value (Cochrane 2009; Egresi & Kara 2018; Haq & Medhekar 2019). Travelers without religious convictions should not be identified as pilgrims, religious tourists, or spiritual tourists. By contrast, religious tourism and formal pilgrimage, with their focus on inspiring faith or fulfilling religious duties, can be seen as spiritual, so they can be regarded as categories of spiritual tourism.

This idea is supported by the research of Clingingsmith et al. (2008), Topik (1999), and Haq (2014), which rejects the old theory that religion and business are adversarial or incompatible. As illustrated earlier, religion and its commercial connections are evidence of the significant economic impact of the Hajj on Makkah, where locals have nicknamed pilgrims their "crops" (Topik 1999). This commercialization of pilgrimage is a form of commodification (Hernandez-Ramdwar 2013; Olsen 2013; Raj & Morpeth 2007; Qurashi 2017; Ron & Timothy 2019), but this discussion is not within the scope of this chapter. Considering that Hajj involves approximately three million people undertaking the annual pilgrimage to Makkah, being selected from a much bigger pool of hopeful applicants, and spending an average of US$10,000 per head, the pilgrimage's commercial success must involve some element of marketing (Haq & Jackson 2009). The whole three-day ritual process injects huge sums of capital into the local economy and into the Saudi national economy as the Hajjis purchase travel necessities, transportation, food, accommodation, and ritual guides. They also pay government fees and taxes. The billions of dollars earned each year by the Saudi Arabian government presents Hajj as a combined business activity, social exercise and classic example of pilgrimage, religious tourism, and spiritual tourism (Clingingsmith et al. 2008; Haq & Jackson 2009; Timothy & Iverson 2006).

Religious and spiritual tourism experiences

Research during the 1960s shifted from general human behavior to a "humanistic, existential, and phenomenological perspective", which covered behavior and consumer experience (Privette 1983, p. 1361). Based on Privette's (1983) pioneering study, several researchers developed their thoughts on consumption and experience leading to the landmark study by Pine and Gilmore (1999), which declared the beginning of the experience economy. The new focus on the experience economy triggered many studies in tourism, hospitality, and services marketing. Walls, Okumus, Wang and Kwun (2011) presented three research directions: crafting a category of experiences; examining causes of, or clarifying, an experience; comparing relationships between experiences and other concepts. This chapter agrees with other researchers who stress the need to understand experiences as a basis for managing or marketing tourism products and services (Carù & Cova 2003; Cohen 1979; Gnoth 1997; Norman 2014; Walls et al. 2011). Walls et al. (2011) contributed to understanding experiential tourism from four perspectives: ordinary to extraordinary, cognitive to emotive experiences, physical experience and human interactions, and individual characteristics and situational factors.

Ordinary to extraordinary

The first perspective represents the range of experiences from ordinary to extraordinary. Consumer experiences in hospitality and tourism happen outside of the daily routine of home. At the highest level, they are peak or transformative experiences (Cohen 1979; Smith 1992). Carù and Cova (2003) differentiate between ordinary and extraordinary experiences with the latter being the most desired goal of tourists. Ordinary experiences are based on routine events, daily life and regular trips, while extraordinary encounters consist of unexpected outcomes with total immersion or flow experience (Carù & Cova 2003; Norman 2014; Walls et al. 2011). Most tourism marketers today claim that all religious and spiritual tourism delivers extraordinary travel experiences, more so than many other types of tourism.

Cognitive (objective) to emotive (subjective) experiences

This factor signifies that individuals can initiate the process in which an experience may occur. Walls et al. (2011) suggest that an experience can occur by coincidence or it can be self-generated. That the consumer can opt to have the experience or not, implying that everyone is not equally affected by every consumer experience. Tourism marketers must be sharp with their segmentation, as religious or spiritual tourists could be seeking cognitive and emotive experiences. The study by Haq and Jackson (2009) shows that Pakistanis undertaking the Hajj from a Muslim majority country enjoyed the cognitive experience, while Australians traveling from a Muslim minority country found their Hajj experience more emotive. Marketing strategies can be crafted in line with the segmentation based on expected experiences.

Physical experience and human interactions

Travel consumers generally assess their experiences based on physical and human interactions. The physical elements include the tangible part of the experience including the hotel room, the buildings, transportation and food service, while the human interaction entails the intangibles (e.g. spiritual growth and sense of community) and that make the experience more memorable (Walls et al. 2011). Marketing religious and spiritual tourism experiences should consider the physical experience for tourism packaging and pricing. The outcomes measured as tourists' delight and loyalty could be mapped with the human interaction before, during, and at the end of the journey.

Individual characteristics and situational factors

Situational factors and individual features are generally considered to be outside the control of business managers due to the variance in each consumer's individual characteristics and life situation (Carù & Cova 2003). Situational factors in tourism include the purpose of the trip, travel companions, and the nature of the destination. Individual characteristics include, among others, personality type and sensitivity to the environment (Walls et al. 2011). For religious and spiritual tourists, situation factors might include the events, the destination, and timing—for example, attending Christmas Mass in the Vatican and visiting Makkah during Ramadan. The individual characteristics of religious and spiritual tourists will link back with the segmenting challenge depending upon the religious conviction and depth of one's faith, which is almost impossible to measure, and extensive qualitative research is required.

The Tourism Experience Model (TEM) (Gnoth 2003) provides insight into experiences in tourism to identify and study tourists' behaviors. This model may also be useful in guiding destinations toward better market positioning and planning for memorable tourist experiences. Gnoth (2003) presents the model with two axes—activities and consciousness, which are secured by the poles of exploration and recreation, based on role authenticity and existential authenticity. The role authentic person side is described by expectations of social roles, while the existential authentic side is marked by feelings that define the person. Recreation is branded by self-reflexive and self-recreational activities, while exploration indicates learning and development.

The model consists of two axes: consciousness, and activity. Consciousness relates to the style of how tourists receive their destination experience. Following sociological/structuralist insights, one end of the consciousness-axis relates to the socially constructed worldview that forms the individual person whose authenticity is reflected in socially accepted role performances (Gnoth & Matteucci 2014). Consciousness means gaining experiences as guided by role-expectations. The more role-conforming one is, the more role authentic the person is. The other end of the consciousness dimension relates to the human being and the state of being. Existentialism is defined by Gnoth (1997) as self-discovery by freeing one's self from society and becoming closer to one's own true self. The other axis consists of types of activities that are either recreational or exploratory. Recreational activities are experienced regularly and relate to habits and daily routines, while exploratory activities entail searching for new insights and understandings.

There are several well-recognized tourist experience models that relate to multiple types of spiritual tourist experiences (Norman 2014). These experiences also identify forms of tourism that are not immediately "spiritual" or "religious". Norman (2014) suggests five types of tourism experiences related to religion and spirituality: spiritual tourism as healing, spiritual tourism as experiment, spiritual tourism as quest, spiritual tourism as retreat, and spiritual tourism as collective.

In the first category, healing-based tourism is concerned with practices that improve people's problematic daily life. This may include individuals testing the strength and value of their personal relationships. In the second category, experiment-based experiences involve people trying different options when routine life feels like a problem. This may require an overall review of one's self-awareness. Spiritual tourism as quest is the most relevant form of spiritual experience in which people undertake a quest for personal discovery, wisdom or truth (Smith 1992). The quest also includes the process of discovering a sense of spiritual reality. The retreat category can be understood as an escape from humdrum daily life to appreciate sacred time or ritual renewal. Tourists also connect it with well-being, nature, and spiritual isolation (Cochrane 2009). Spiritual tourism as collective may be the opposite of retreat where the tourist seeks human interaction and intragroup belonging (Collins-Kreiner & Gatrell 2006). This may be more connected with pilgrimage, such as the Catholic faithful walking on the Camino de Santiago, or Muslims visiting Makkah (Raj & Morpeth 2007; Tilson 2005).

Experiential segmentation for religious and spiritual tourism

Norman's (2014) five spiritual tourism experiences can be aligned with the tourism experience model (TEM) of Gnoth (2003) and the four elements of experiential tourism from Walls et al. (2011) to derive a comprehensive classification of religious and spiritual tourism experiences. The classifications can be mapped with a clear and effective relationship marketing strategy. A synthesis and cross-mapping of the three models with various dimensions of experiential tourism and religious and spiritual tourism experiences can lead to sharply narrowed segmentation, as illustrated in Table 24.1.

Further analysis of Table 24.1 implies that Norman's (2014) five types of spiritual tourism experiences can be correlated with the type of authenticity and nature of the experience, which can be further aligned with the four elements of experiential tourism proposed by Walls et al. (2011). The five types of experiences illustrate the layers or segments of religious and spiritual tourists that can be targeted or managed separately or collectively by applying various elements of relationship marketing. The concept, application, and structure of relationship marketing for religious and spiritual tourism in discussed in the following section.

Table 24.1 Religious and spiritual tourism experiences

Spiritual Experience	*Type of authenticity*	*Nature of Experience*	*Element of experiential tourism*
Healing	Role authenticity	Exploratory	Extraordinary—Individual
Experiment	Role authenticity	Exploratory	Individual—Cognitive
Quest	Existential authenticity	Recreational	Individual—Emotive
Retreat	Existential authenticity	Recreational	Individual—Physical experience
Collective	Role authenticity	Exploratory	Cognitive—Human interaction

After Norman (2014), Gnoth (2003) and Walls et al. (2011).

Marketing religious and spiritual tourism experiences

The literature identifies three strategic approaches to tourism marketing: consumer-oriented, competitor-oriented, and trade-oriented (Buhalis 2000; Riege & Perry 2000). The consumer-oriented approach concentrates on individuals or groups of tourists, their behaviors, and attitudes. The competitor-oriented approach accommodates competitive forces that can affect business success. The trade-oriented approach focuses on intermediaries. Riege and Perry (2000) consider basic tourism products and do not focus on applications of special tourism products. Likewise, the role of tourism product orientation has been largely ignored. To cover this weakness, the fourth approach of product-oriented marketing was suggested by several authors (e.g. Buckley 2018; Chen & Tseng 2005; Kavaratzis & Ashworth 2008). The tourism product orientation also highlights three vital issues for marketing in tourism: the tourist's emotional state at the place, prior expectations of the destination, and the tourist's satisfaction with the destination (Buhalis 2000; Kavaratzis & Ashworth 2008; Riege & Perry 2000).

To achieve fit between the proposed marketing strategies and the marketing activities selected to implement each strategy, the discussion of spiritual tourism marketing for specific groups adopted the relationship marketing approach (Buhalis 2000; Gronroos 1997; Wang 2008). The classical framework of the marketing mix (see McCarthy 1964) has been criticized for being too simplistic, unidimensional, and more theoretical than practical (Gronroos 1989; Haq 2014; Kavaratzis & Ashworth 2008). Critics have rejected the adoption of the marketing mix concept and suggested a paradigm shift toward relationship marketing that is orientated toward relationships rather than transaction-based marketing.

The literature on tourism marketing follows the paradigm shift from Morrison's (2002) tourism marketing mix to the application of relationship marketing in tourism (Buhalis 2000; Gronroos 1997; Kavaratzis & Ashworth 2008). The marketing mix was judged as a static, rigid, and linear framework, and hence relationship marketing was suggested as a realistic, flexible, and robust alternative. Unlike relationship marketing, the marketing mix approach did not accomplish the essentials of the marketing concept but offered a production-oriented marketing rather than customer-oriented marketing (Gronroos 1989; Wang 2008). Relationship marketing is based on trust, promise fulfillment, exchange, and communication among all partners (Buhalis 2000; Gronroos 1997; Wang 2008). Adopting these essentials of relationship marketing, this chapter suggests marketing religious and spiritual tourism experiences based on product (or service) offered, people involved, and collaboration between all players. The application of these elements (product, people, and collaboration) for successful relationship marketing of religious and spiritual tourism experiences is discussed below.

The product

The product consists of the destination assets accessible to tourists, including their features and benefits (Morrison 2002). This includes the tangible elements of the sacrosanct place and the enlightening experiences spiritual and religious tourists seek by interacting with those spaces. The spiritual tourism product is not new. It has received considerable research attention and commercial analysis, and with the recent upsurge in spirituality-related businesses, it seems to be on the cusp of a major growth phase (Buckley 2018; Kavaratzis & Ashworth 2008). A product-based relationship marketing strategy includes moving toward strategic segmentation; relating product performance to customer needs, modifying the product if necessary; intense distribution; and building efficiencies in production and marketing (Buckley 2018; Wang 2008). The investment in product development and improvement to build trust and loyalty with customers is a key to relationship marketing (Chen & Tseng 2005; Wang 2008).

The people

People play an important role in relationship marketing linked to a service product (Buckley 2018; Gronroos 1997). In spiritual tourism, several people interact with the tourists at various places during the trip, ranging from travel agents to destination guides and religious leaders who may be able to transform transactions into relationships (Kavaratzis & Ashworth 2008). A good example of the "people" at the destination is that at all Sufi shrines in Pakistan promoted as spiritual tourism destinations the visitors get a chance to listen to "*qawwali*" performance, which is "an authentic spiritual song" (Qureshi 1995, p. 1) written by a Sufi praising God and the Prophet Muhammad. However, the *qawwali* has experienced a massive makeover for folk, pop, mixed, and movie themes in India and Pakistan. The performers of this artistic tradition play an important role in relationship building and relationship marketing as they contribute to the sacredness of the place and the rewards of visiting.

The people working to create relationships need to be selected and trained so that they meet tourists' expectations. The category of "people" in this context runs the entire gamut, including ritual guides, entertainers and artisans, religious advisers and prayer leaders, hospitality providers, and security personnel in some cases. Relationship marketing designed with the interaction between providers and customers, together with discounted pricing of packages to repeat customers, should benefit both the customer and the tourism operator as satisfying experiences are co-created between the destination's people and the religious or spiritual consumers who visit.

Collaboration

Collaboration drives relationship marketing to build operating efficiencies among various stakeholders involved in the spiritual tourism industry. For collaboration to be successful, all partners on the demand side and supply side must collaborate and benefit (Augustyn & Knowles 2000; Haq 2013; Tilson 2005). The tourism industry typically includes both vertical and horizontal partnerships with both forward and backward integration. Vertical partnerships and integration include the expansion of the tourism product or service into a related service, for example, a hotel also offering sightseeing tours (Weidenfeld 2018). Horizontal partnerships refer to tourism companies entering into collaboration or integration with related business, for example, a hotel buying or entering a strategic alliance with an airline (Augustyn & Knowles 2000; Weidenfeld 2018).

Horizontal partnerships identify the amount of cooperation with public and private operators and providers of spiritual tourism products and services (Augustyn & Knowles 2000). Vertical partnerships identify the degree of collaboration between operators and various stakeholders, including transport companies, hotels, media channels, insurance companies, destination management, and financial institutions such as banks and credit card companies (Augustyn & Knowles 2000; Chen & Tseng 2005). This chapter suggests that relationship marketing for religious and spiritual tourism needs collaboration in relation to their position on both vertical and horizontal axes.

The depth and width of any collaboration will indicate the quantity and variety of stakeholders. The scope of this chapter does not deliver recommendations for selecting partners, but it does present basic collaboration issues and the criteria that could be considered by tourism marketers. Researchers have proposed certain parameters for establishing successful collaborative efforts in the tourism industry (Augustyn & Knowles 2000; Tilson 2005; Weidenfeld 2018). Adopting the parameters that have been proposed in the literature, it is suggested that collaboration in spiritual tourism requires reciprocal grounds, mutual goals, expert participation, developmental structure, and sustainability.

Conclusion

Recent publications in special interest tourism associated with faith, spirituality, and religion have awarded generous space to discussions on spirituality versus religion. This chapter goes a step further and suggests that spirituality and religion may be near equivalent terms and hence spiritual and religious tourism could be marketed together, especially as most religious people see themselves as spiritual, even if many spiritual people do not necessarily see themselves as religious. Although some people might consider the notions of spiritual travel and marketing to be incompatible, it remains a fact that even the most dedicated pilgrimage destinations undertake significant marketing actions, whether these are overt and visible to consumers or hidden in the backstage reaches of destination managers. Marketing takes many forms and is now considered a necessary tool for creating positive and satisfying experiences for all kinds of tourists, including spiritual tourists and pilgrims.

To achieve holistic spiritual and religious tourism marketing, market segmentation needs to be sharpened where relationship marketing is suggested as the palpable option. Segmenting spiritual and religious tourists based on experiences was carried out by selecting relevant methods from experts such as Gnoth and Matteucci (2014), Norman (2014) and Walls et al. (2011). A cross-examination of their segmentation based on tourist' attitudes and behaviors led this study to design five segments as illustrated in Table 24.1 and Figure 24.1.

With the purpose of crafting an effective relationship marketing strategy, various marketing elements were assessed. The study concludes that products, people, and collaboration are the key elements of the relationship marketing strategy to be mapped with the five specific segments of spiritual and religious tourists. Tourism planners and marketers can consider these five segments of spiritual and religious tourists and check where their own target

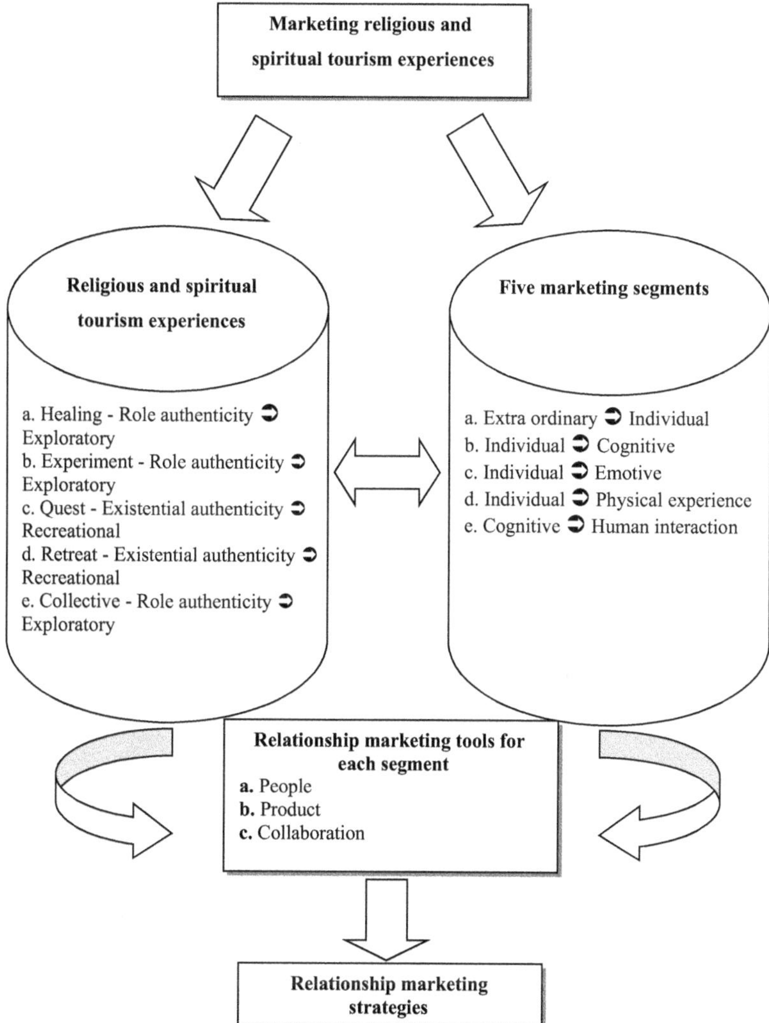

Figure 24.1 Religious and spiritual tourism segments and relationship marketing

customers fit in, based on their attitudes and behaviors. The three elements of relationship marketing could be considered a key focus to build an effective marketing framework. Empirical work is needed to verify the five segments with the tourists and assess relationship marketing with tourism providers.

References

Andriotis, K. (2009) 'Sacred site experience: A phenomenological study', *Annals of Tourism Research*, 36: 64–84.

Augustyn, M.M., & Knowles, T. (2000) 'Performance of tourism partnerships: A focus on York', *Tourism Management*, 21(4): 341–351.

Buckley, R. (2018) 'Tourism and natural World Heritage: A complicated relationship', *Journal of Travel Research*, 57(5): 563–578.

Buhalis, D. (2000) 'Marketing the competitive destination of the future', *Tourism Management*, 21(1): 97–116.

Carù, A., & Cova, B. (2003) 'Revisiting consumption experience: A more humble but complete view of the concept', *Marketing Theory*, 3(2): 267–286.

Chen, H., & Tseng, C. (2005) 'The performance of marketing alliances between the tourism industry and credit card issuing banks in Taiwan', *Tourism Management*, 26(1): 15–24.

Clingingsmith, D., Kwaja, A.I., & Kremer, M. (2008) *Estimating the Impact of the Hajj: Religion and Tolerance in Islam's Global Gathering.* Cambridge, MA: Harvard University, Kennedy School.

Coats, C. (2008) 'Is the womb barren? A located study of spiritual tourism in Sedona, Arizona, and its possible effects on eco-consciousness,' *Journal for the Study of Religion, Nature and Culture*, 2(4): 483–507.

Cochrane, J. (2009) 'Spirits, nature and pilgrimage: The other dimension in Javanese domestic tourism', *Journal of Management, Spirituality & Religion*, 6(2): 107–120.

Cohen, E. (1979) 'A phenomenology of tourist experiences', *Sociology*, 13(2): 179–201.

Cohen, E. (1992) 'Pilgrimage centre: Concentric and excentric', *Annals of Tourism Research*, 18: 33–50.

Collins-Kreiner, N., & Gatrell, J. (2006) 'Tourism, heritage and pilgrimage: The case of Haifa's Baha'i Gardens,' *Journal of Heritage Tourism*, 1(1): 32–50.

Egresi, I.O., & Kara, F. (2018) 'Residents' attitudes to tourists visiting their mosques: A case study from Istanbul, Turkey', *Journal of Tourism and Cultural Change*, 16(1): 1–21.

Eid, R. (2012) 'Towards a high-quality religious tourism marketing: The case of Hajj service in Saudi Arabia', *Tourism Analysis*, 17(4): 509–522.

Gnoth, J. (1997) 'Tourism motivation and expectation formation', *Annals of Tourism Research*, 24: 283–304.

Gnoth, J. (2003) 'Developing the tourism experience model', Paper presented at the Australian and New Zealand Marketing Academy. Otago, New Zealand. Retrieved from: https://www.researchgate.net/publication/256118025_Developing_the_tourism_experience_model

Gnoth, J., & Matteucci, X. (2014) 'A phenomenological view of the behavioural tourism research literature', *International Journal of Culture, Tourism and Hospitality Research*, 8(1): 3–21.

Gronroos, C. (1989) 'Defining marketing: A market-oriented approach', *European Journal of Marketing*, 23(1): 52–60.

Gronroos, C. (1997) 'From marketing mix to relationship marketing - towards a paradigm shift in marketing', *Management Decision*, 35(4): 322–339.

Haq, F. (2013) 'Islamic spiritual tourism: An innovative marketing framework', *International Journal of Social Entrepreneurship and Innovation*, 2(5): 438–447.

Haq, F. (2014) 'The significance of partnership as a marketing strategy for Islamic spiritual tourism', *Journal of Islamic Marketing*, 5(2): 258–272.

Haq, F., & Jackson, J. (2009) 'Spiritual journey to Hajj: Australian and Pakistani experience and expectations', *Journal of Management, Spirituality and Religion*, 6(2): 141–156.

Haq, F., & Medhekar, A. (2019) 'Is spiritual tourism a peace vehicle for social transformation and economic prosperity in India and Pakistan', in A-M. Nedelea & M-O. Nedelea (eds), *Marketing Peace for Social Transformation and Global Prosperity* (pp. 189–211). Hershey, PA: IGI Global.

Hernandez-Ramdwar, C. (2013) 'African traditional religions in the Caribbean and Brazil: Models of religious tourism and impacts of commodification', *Journal of Heritage Tourism*, 8(1): 81–88.

Jauhari, V., & Sanjeev, G.M. (2010) 'Managing customer experience for spiritual and cultural tourism: An overview,' *Worldwide Hospitality and Tourism Themes*, 2(5): 467–476.

Kavaratzis, M., & Ashworth, G.J. (2008) 'Place marketing: How did we get here and where are we going?' *Journal of Place Management and Development*, 1(2): 150–159.

Lupu, C., Brochado, A., & Stoleriu, O. (2019) 'Visitor experiences at UNESCO monasteries in northeast Romania', *Journal of Heritage Tourism*, 14(2): 150–165.

McCarthy, E.J. (1964) *Basic Marketing: A Managerial Approach.* Homewood: Richard D. Irwin.

Mitroff, I. (2003) 'Do not promote religion under the guise of spirituality', *Organization*, 10(2): 375–380.

Mitroff, I., & Denton, E.E. (1999) *A Spiritual Audit of Corporate America: A Hard Look at Spirituality, Religion, and Values in the Workplace.* San Francisco: Jossey-Bass Inc.

Moal-Ulvoas, G., & Taylor, V.A. (2014) 'The spiritual benefits of travel for senior tourists', *Journal of Consumer Behaviour*, 13(6): 453–462.

Morrison, A.M. (2002) *Hospitality and Travel Marketing*, 3rd edn. New York: Delmar.

Nicholson, R.A. (1978) *The Tarjuman al-Ashwaq: A Collection of Mystical Odes by Ibn Al-Arabi,* translated from original. London: Theosophical Publishing House.
Norman, A. (2014) 'The varieties of the spiritual tourist experience', *Literature & Aesthetics*, 22(1): 20–37.
Olsen, D.H. (2012) 'Negotiating identity at religious sites: A management perspective', *Journal of Heritage Tourism*, 7(4): 359–366.
Olsen, D.H. (2013) 'A scalar comparison of motivations and expectations of experience within the religious tourism market', *International Journal of Religious Tourism and Pilgrimage*, 1(1): 41–56.
Pesut, B. (2003) 'Developing spirituality in the curriculum: Worldviews, intrapersonal connectedness, interpersonal connectedness', *Nursing and Health Care Perspectives*, 24(6): 290–294.
Pine, B.J., & Gilmore, J.H. (1999) *The Experience Economy: Work Is Theatre & Every Business a Stage.* Boston, MA: Harvard Business School Press.
Privette, G. (1983) 'Peak experience, peak performance and flow: A comparative analysis of positive human experiences,' *Journal of Personality and Social Psychology*, 45(1): 1361–1368.
Qurashi, J. (2017) 'Commodification of Islamic religious tourism: From spiritual to touristic experience', *International Journal of Religious Tourism and Pilgrimage*, 5(1): 89–104.
Qureshi, R.B. (1995) *Sufi Music of India and Pakistan: Sound, Context and Meaning in Qawwali.* Chicago: University of Chicago Press.
Raj, R., & Morpeth, N.D. (2007) *Religious Tourism and Pilgrimage Festivals Management: An International Perspective.* Wallingford: CABI.
Riege, A.M., & Perry, C. (2000) 'National marketing strategies in international travel and tourism', *European Journal of Marketing*, 34: 1290–1305.
Rinschede, G. (1992) 'Forms of religious tourism', *Annals of Tourism Research*, 19: 51–67.
Rodrigues, S., & McIntosh, A. (2014) 'Motivations, experiences and perceived impacts of visitation at a Catholic monastery in New Zealand', *Journal of Heritage Tourism*, 9(4): 271–284.
Ron, A.S., & Timothy, D.J. (2019) *Contemporary Christian Travel: Pilgrimage, Practice and Place.* Bristol: Channel View Publications.
Shackley, M. (2002) 'Space, sanctity and service: The English cathedral as heterotopia', *International Journal of Tourism Research*, 4(5): 345–352.
Smith, V.L. (1992) 'Introduction: The quest in guest', *Annals of Tourism Research*, 19: 1–17.
Štefko, R., Kiráľová, A., & Mudrík, M. (2015) 'Strategic marketing communication in pilgrimage tourism', *Procedia-Social and Behavioral Sciences*, 175(12): 423–430.
Taylor, C. (1989) *Sources of the Self: The Making of the Modern Identity.* Cambridge, MA: Harvard University Press.
Tilson, D.J. (2005) 'Religious-spiritual tourism and promotional campaigning: A church-state partnership for St. James and Spain', *Journal of Hospitality & Leisure Marketing*, 12(1/2), 9–40.
Timothy, D.J. (2021) *Cultural Heritage and Tourism: An Introduction*, 2nd edn. Bristol: Channel View Publications.
Timothy, D.J., & Iverson, T. (2006) 'Tourism and Islam: Considerations of culture and duty', in D.J. Timothy & D.H. Olsen (eds), *Tourism, Religion and Spiritual Journeys* (pp. 186–205). London: Routledge.
Timothy, D.J., & Olsen, D.H. (2018) 'Religious routes, pilgrim trails: Spiritual pathways as tourism resources', in R. Butler & W. Suntikul (eds), *Tourism and Religion: Issues and Implications* (pp. 220–235). Bristol: Channel View Publications.
Topik, S. (1999) 'The bottom line between heaven and hell', *World Trade*, 12(3): 92–95.
Vukonić, B. (1996) *Tourism and Religion.* Oxford: Pergamon Press.
Walls, A.R., Okumus, F., Wang, Y.R., & Kwun, D.J.W. (2011) 'An epistemological view of consumer experiences', *International Journal of Hospitality Management*, 30(1): 10–21.
Wang, Y. (2008) 'Collaborative destination marketing: Understanding the dynamic process', *Journal of Travel Research*, 47(2): 151–166.
Weidenfeld, A. (2018) 'Tourism diversification and its implications for smart specialisation', *Sustainability*, 10(2): 319.

25

MANAGING RELIGIOUS AND SPIRITUAL TOURISM SITES

Simon Coleman and Daniel H. Olsen

Introduction

Travel for religious purposes has occurred for millennia, and people from all cultures have traveled to sacred sites, whether it be for curiosity, initiatory or cleansing rites, to fulfill vows, or to worship (Morinis 1992). However, since World War II, there has been a sharp growth in the number of people who journey to religious and spiritual sacred locations owing to the "democratization of travel", where increases in transportation and telecommunications technologies, along with greater disposable income in many regions of the world, have provided more opportunities to travel (Olsen 2019a). Religious and spiritual tourism is now a powerful social, cultural, and economic force in the contemporary world, with over 300 million people a year engaging in "national and international religious journeys" to visit religious sites around the world (UNWTO 2011; Tomljenović & Dukić 2017: 1). Many governments and tourism promoters use religious heritage to encourage tourist visitation to their destinations (Olsen 2003).

In addition to the increasing number of visitors to sacred sites, the composition of visitors to these sites has also become more varied to include pilgrims, tourists, and spiritual seekers. Today, in many cases, tourists outnumber conventional worshipers to the point that in some instances regular religious functions are almost impossible to perform (Griffiths 2011; Clancy 2020; Gowreesunkar & Thanh 2020). Of these visitors, many are religious tourists who consciously and deliberately travel in order to satisfy their devotional needs, while others are spiritual tourists or seekers who engage in secular forms of journeying in search of powerful mystical or transcendent experiences (Tomljenović and Dukić 2017: 2). Such groups have varying motivations and expectations (Olsen 2013).

While it is often useful to highlight the degree of premeditation involved in undertaking a journey, for the purposes of this chapter religion and spirituality are more usefully distinguished in terms of their different orientations toward official expressions of faith. Religious practices usually reinforce conventional forms of devotion and identity, while spiritual engagements are typically individualized and sometimes actively resistant to rituals and spaces managed by religious authorities. For example, in her account of new forms of spirituality associated with Mary Magdalene and the Sacred Feminine, Fedele (2013) discusses how pilgrims visiting Roman Catholic shrines in Europe distance themselves from formal shrine rituals, which they regard as patriarchal and alienating. However, in other cases, such as most visitors to the Christian shrines at Walsingham in England, sharp distinctions between religious and spiritual attitudes are harder to discern, since many who come express multiple or unclear reasons for their trips. Visitors to the sites may not always be fully aware of their motivations, or may change degree, direction, and formality of ritual commitment during a single visit (Coleman 2018).

While the drawing of firm distinctions between religious pilgrims and secular tourists is of great importance to some managers of sites, such divisions are increasingly difficult to maintain in a "world of postmodern travel" (Badone & Roseman 2004: 2; see Collins-Kreiner 2010; Olsen 2010). The meaning of the term "sacred" has been subject to much debate (see Olsen 2019b), especially given the Durkheimian tradition of regarding it as symbolizing social collectivity (Badone & Roseman 2004: 2). For some scholars, religious or spiritual tourism is constituted by attending to a sporting event or cemetery quite as much as visiting the shrine of a saint (see also Reader & Walter 1993; Gammon 2004; Knox & Hannam 2014; Olsen 2014, 2019a; Maddrell & Terry 2015; Pliberšek, Basle & Lebe 2020). Ideas about appropriate behavior in relation to the sacred also vary considerably across cultures and religious traditions. For example, Reader (2014: 6–7) describes how in 2008 priests from various pilgrimage shrines created miniature replicas of several famous regional pilgrimage routes, including soil from these sites, at a shopping mall at Centrair, the international airport outside Nagoya in Japan. The pilgrimage priests reasoned that since more people were visiting shipping malls than religious sites it made sense to take the sacred to the shopping mall. Meanwhile, Orland and Bellafiore (1990) note how a part of the pilgrimage culture in India is the toleration of people who beg pilgrims for money and food, such as occurs at Bodh Gaya, the site where tradition holds Gautama Buddha achieved enlightenment. However, with the site becoming an important tourist attraction, government officials have removed these beggars from the area so as not to offend the sensibilities of Western tourists.

Increasing numbers of visitors to religious heritage sites have led to a growing concern among some scholars over the amplification of management issues when tourism space overlaps with religious space (Bremer 2006). This development has raised substantial questions regarding how destinations and religious sites are organized and maintained in order to attract, welcome, and even withstand these visitors while also protecting the fabric and aesthetics of sites. The management of holy place offers both challenges and opportunities for administrators, clergy, and both regional and national governments. Research related to sacred site management tends to focus on human-built sacred sites because of their importance as sites of ritual performance and their links to rising numbers of tourists (Shackley 2001b; Woodward 2004; Olsen 2006; Wiltshier & Griffiths 2016; Griffiths & Wiltshier 2019). In this chapter, the authors explore the huge variation of religious sites around the world—human as well as natural—alongside the wide range of management strategies used to sustain and operate them.

Types of religious and spiritual sites

In her seminal work on the management of sacred locations, Shackley (2001a: 2) suggested a typology that ranged from human-built to sacred natural sites, "earth energy" sites, sacred mountains and islands, and sites of secular pilgrimage. Many of the places listed by Shackley are *human-built*—places on the earth's surface that are viewed as sacred because of a hierophanic or significant religious event that occurred there or because they relate to the lives of religious persons. These places have been ritually set apart from profane space, given boundaries, marked by a religious building, and managed and maintained as such (Kong 2001). Different types of human-built sites include:

- Single nodal sacral sites (e.g., Canterbury, UK; Hagia Sophia, Turkey);
- Archaeological sites (e.g., Machu Picchu, Peru; Chichen Itza, Mexico);
- Temple complexes (e.g., Laliblea, Ethiopia; St Katherine's Monastery, Egypt);
- Cities (Rome, Italy; Varanasi, India; Jerusalem);
- Burial sites (e.g., Catacombs, Rome; Arlington Cemetary, Washington, DC); and
- Secular sacred sites (e.g., Robbin Island, South Africa; Holocaust sites) (Shackley 2001a).

In some cases, secular sites are related to "dark tourism" destinations memorializing death and suffering (Stone et al., 2018; Olsen & Korstanje 2020); "commemorative pilgrimages" to places related to secular religion, such as Gettysburg or the Lincoln Memorial (compare Campo 1998; West 2008); and "reminiscence pilgrimages" related to people returning to their homelands (Uusihakala 2011; Wright 2020). Most of these human-built sites are in areas, whether urban or rural, that are easily accessible by pilgrims and tourists.

Sacred sites located in wilderness areas are generally referred to as *sacred natural sites* (SNSs). SNSs are "areas of land or water having special spiritual significance to people and communities" (Wild & McLeod 2008: xi). Oviedo and Jeanrenaud (2007: 77) define SNSs as "natural areas recognized as sacred by indigenous and traditional peoples, as well as natural areas recognized by institutionalized religions or faiths as places for worship and remembrance". There are several different types of SNS:

- Mountains and volcanoes (e.g., Mt Bromo, Indonesia; San Francisco Peaks, USA; Mt Kailash, Tibet);
- Rivers and larger water bodies (e.g., Ganges River, India; Lake Titicaca, Bolivia, and Peru; Basum Lake, Tibet);
- Forests and groves (e.g., Osun Sacred Forest Grove, Nigeria; Buoyem Sacred Grove, Ghana; Lumbini Grove, Nepal);
- Islands (e.g., Rapa Nui, Chile; Iona Island, Scotland; The Holy Island of Lindisfarne, UK);
- Plants (e.g., the lotus flower [*nelumbo nucifera*] in Hinduism and Buddhism; the Mediterranean cypress [*cupressus sempervirens*], Turkey); and
- Animals (e.g., cows in Hinduism; monkeys in Nepal; lions in China; lizards in New Zealand), and larger natural landscapes (Shackley 2001a; Timothy 2012; Olsen 2020; Chapter 23 this volume).

Caves and grottoes are also included in this category, many of which have been re-engineered by religious groups to include religious statues, wall art, artificial lighting, and seating. Prominent examples include the Mogao Grottoes, Longman Caves, and the Cave Temple of Dunhuang in China; St Michael's Cave in Italy, the Dambulla Cave Complex in Sri Lanka; and the Salt Cathedral of Zipaquira in Colombia. Many sacred natural sites are uninhabited because they are marked as off-limits by religious or spiritual taboos as they are considered the abode of the gods. Other sacred natural sites, however, are multi-functional and act as contexts where sacral practices take place (Aitpaeva 2013), the source of material resources for community livelihoods, and biodiversity conservation areas (Verschuuren et al. 2010; Samakov & Berkes 2017; Olsen 2020). Earth energy sites—also termed "power places"

(Ivakhiv 2003)—are locations within the earth's "planetary [energy] grid system" where divine power is manifest (Ivakhiv 2007: 265). This manifestation of power can be *linear*, as in the case of "ley lines" or lines of positive energy within in the earth, or *nodal*, such as Stonehenge and Glastonbury Tor, UK, and Sedona, Arizona, from which ley lines originate, or at which they meet. Travel to these earth energy locations is an integral part of New Age movements, many of which focus on hybrids of human-built and sacred natural sites. Many religious communities, such as the Roman Catholic Church, have "appropriated" several earth energy sites and built cathedrals over them, and so these appropriated places become the focus of New Age pilgrims who seek to reclaim them (Fedele 2014).

One type of sacred location not listed by Shackley (2001a) is that of *indigenous or aboriginal sacred sites*. Most indigenous or traditional sites are SNSs, in part because of belief systems that often hold the natural environment to be inherently sacred. Indeed, indigenous groups have held SNSs to be sacred long before institutionalized religions developed theologies of the environment and spiritual connections with nature (Oviedo, Jeanrenaud & Otegui 2005). Indigenous sacred sites are found around the world, including the Arctic (Heinämäki & Herrmann 2017). While sacred sites are vital to the perpetuation of indigenous culture (Butzier & Stevenson 2014), such communities have often been disenfranchised or removed from their sacred geographies as a result of colonization, so that many indigenous communities have come to view their sacred sites as "diasporic space" (Coleman 2016). In some cases, these indigenous sacred locations are found within urban spaces (Jackson & Ormsby 2017) or inside protected natural areas (Verschuuren et al. 2010). Indigenous claims to traditional sacred sites are being increasingly recognized by national and international law, international governing bodies, and nongovernmental organizations (Butzier & Stevenson 2014; Liljeblad & Verschuuren 2019)—Uluru being a recent example of this (Bickersteth, West & Wallis 2020). More and more, they are being managed in consultation with indigenous communities (Bakht & Collins 2017; Tran, Ban & Bhattacharyya 2020).

Many of these human-built and SNSs are presently included on UNESCO's World Heritage List. Stemming from the *Convention Concerning the Protection of the World Cultural and Natural Heritage* in 1972, the World Heritage List contains over 1100 human, natural, and mixed heritage sites deemed to be of universal value to humanity and therefore worthy of preservation for future generations (UNESCO 2020a). Examples of sacred sites on the World Heritage List include the Rock-hewn Churches of Lalibela in Ethiopia; Ephesus in Turkey; the San Antonio Missions in the USA; and the Temple and Cemetery of Confucius in China (UNESCO 2020b; see Olsen & Timothy 2021). Some sacred sites are in the "mixed" category on the World Heritage List because they have both cultural and natural value. Examples include the Lapponian Area in Sweden; the sacred Kii Mountain Range in Japan; and Mount Athos in Greece (UNESCO 2020b). UNESCO's Representative List of the Intangible Cultural Heritage of Humanity also contains several examples of intangible religious heritage related to pilgrimage and celebrations that need assistance to keep this inheritance alive. Examples include the Círio de Nazaré in the City of Belém, Brazil; Holy Week processions in Popayán, Columbia; and Qālišuyān Rituals of Mašhad-e Ardehāl in Kāšān, Iran (UNESCO 2020c; see Olsen & Timothy 2021).

Pilgrimage routes and trails make up some of the oldest transportation networks in the world, and in many cases have served as the "building-blocks" for modern systems of conveyance (Olsen 2019a). In addition to linking numerous sacred sites together, they may represent a linear geography that becomes sacred in its own right. The Camino de Santiago de Compostela, with its various routes that converge on the remains of the apostle St James, has become revitalized as a religious pilgrimage route for Roman Catholics in the modern era but is also seen as a spiritual pathway by agnostics and people from non-Catholic religious traditions (Lois González 2013; Farias et al. 2019). The Hajj in Saudi Arabia is one of the larger pilgrimages in the world, with around three million people participating each year, and academic debate still exists as to the role of Mecca in pre-Islamic trade (Bukharin 2009). The Appalachian Trail is a 2,200 mile (3,540 km) long route in the United States that is viewed by many as a spiritual trail for those who wish to commune with nature (Redick 2009, 2018). Other famous pilgrimage trails include the Shikoku and Kumano Kodō pilgrimage trails in Japan, the Mount Kailash pilgrimage in Tibet, Machu Picchu in Peru, and pilgrimages to sacred places such as Croagh Patrick in Ireland, Adam's Peak in Sri Lanka, and the Char Dham pilgrimage sites in India.

Management challenges

The immense material and spatial range of locations mentioned above suggests that the character and scope of their management will vary, influenced by the physical characteristics of the site—its geographical location and geographic extent (e.g., point, line, or area), accessibility, and vulnerability to various forms of erosion. Equally important are the social and cultural contexts of site use and the various external "claims" made on the religious and spiritual resources within these sites. Managers need to balance popularity against vulnerability and potential profit against preservation.

The management of sacred sites is therefore a multifaceted task, considering the multitude of issues related to the internal and external management of these sites. Internally, the "core business" of religious heritage site management is to maintain a "sense of place", or to provide an atmosphere that encourages visitors to have enhanced religious or spiritual experiences (Shackley 2001a, 2001b, 2002). However, maintaining this "sense of place" has become more difficult with increasing visitors, particularly in urban areas or places with political and social instability, which can heighten urban management problems, disrupt visitor flows, and make it difficult to enhance existing visitor management systems or expand religious tourism development (Shackley 2001a, pp. 7–8). This situation is also the case with SNSs, where site managers struggle to ensure that SNSs are not destroyed by the sheer weight of numbers of people.

Scholars have expressed particular concern over how increased visitation can damage the "sense of place" and aesthetics of these sites. Physical impacts include theft, vandalism, and accidental damage, different types of pollution, microclimatic change, and overcrowding (see Table 25.1). Scholars have also discussed several other internal management issues related to the effective interpretation of religious heritage sites to visitors (Poria et al. 2009; Olsen 2012; Hughes et al. 2013; Ballantyne et al. 2016; Thouki 2019); the enhancing of visitor experience (Voase 2007; Williams 2007; Francis et al. 2008; Francis et al. 2010; Bond 2015); the ethics over "pay[ing] to pray" (Shackley 2001a; Woodward 2004; Qurashi 2017); the commodification of religious sites and relics (Winter & Gasson 1996; Askew 2008; Shinde 2010b; Wiltshier & Clarke 2012; Hung et al. 2017); how to best control visitor flows and behavior (Shackley 2001a; Wong et al. 2016); the best marketing strategies related to religious heritage sites (Shackley 2001a; Pavicic, Alfirevic & Batarelo 2007; Eid 2012; Huang & Pearce 2019); managing religious festivals and events (Shinde 2010a; Raj, Griffin & Blackwell 2015); and the use of social media, including the taking of "holy selfies", at these sites (Prats, Aulet & Vidal 2015; Caidi, Beazley & Colomer Marquez 2018; see Chapter 27, this volume).

Table 25.1 Types of physical tourism impacts at religious heritage tourism sites

Type of impact	*Change*	*Examples*
Theft of artifacts	Loss of resource	Removing tiles from Taj Mahal (Shackley 2001a, 2006); Illegal digging in Palestinian Territories (Al-Houdalieh 2010).
Vandalism/graffiti	Damage to resource	Uluru (Ayers Rock), Australia (Shackley 2001a); Islamic monuments in Bukhara and Samarkand, Central Asia (Airey & Shackley 1998); Mnajdra megalithic temple, Malta (Shackley 2003); Rock-cut tombs in Beth Shearim National Park, Israel (Merhav & Killebrew 1998).
Accidental damage/ wear and Tear	Damage to resource	English cathedrals (Shackley 2002; Rotherham 2007); Rock-cut churches in Lalibela, Ethiopia (Shackley 2001a; Carlisle 2009); St. Katherine's monastery, Mount Sanai, Egypt (Shackley 1998); Masada, Israel (Merhav & Killebrew 1998).
Pollution (air, noise, fouling, litter)	Reduced attractiveness, loss of atmosphere	Ganges River, India (Alley 1992, 2002); Sacred landscapes, Bhutan (Allison 2014); Hajj (Qurashi 2020)
Microclimatic change	Damage to resource	Egyptian tombs (Shackley 2001a); Frescoes in cathedrals (Bernardi et al. 2000); Travelling religious relics (della Dora 2009)
Crowding	Increase in other impacts occurring	Notre Dame Cathedral (Boyer 2000; Shackley 2002; Ashworth 2009).

After Shackley (2001a) and Olsen and Ron (2013).

In addition to these internal issues, some scholars have commented on several external challenges that can affect management strategies at religious heritage sites. For example, religious heritage sites are affected by the politics and social trends of the area in which they are located, and national, regional, and local political and social instability can heighten administrative problems. As Shackley (2001a: 7) notes, even if visitor numbers are high, the easiest religious heritage sites to run are those where socio-political stability is also high. For example, it is difficult to manage sacred sites properly when multiple religious groups compete over the ownership, maintenance, and interpretation of sites deemed sacred (Olsen & Timothy 2002). Examples include the Middle East, where Christians, Muslims, and Jews compete over ownership and interpretation of religious sites throughout this region (Collins-Kreiner et al. 2006; Olsen & Ron 2013), and Ayodhya, India, where Hindus tore down the sixteenth-century Babri Mosque in 1992 to reclaim Hindu sacred space (Shaw 2000).

Another key area of external concern relates to the multiple stakeholders, such as government and tourism officials, urban and regional planners, local business enterprises, travel agents, tour guides, local tourism enterprises, chambers of commerce, local residents, tourists, religious congregations, and historical societies, who have varied interests regarding how religious heritage sites are managed, including issues related to accessibility, visitor characteristics, promotion, and maintenance (Olsen 2006; Petreas 2011). Indeed, the management of sacred sites is rarely just a matter of administrative or logistical competence: it involves delicate and sometimes passionate theological, cultural, and social negotiations between different interests and stakeholders. This balancing act involves weighing the relative merits and demands of working with external stakeholders in marketing to either a mass tourism market, which Throsby (2009: 13) refers to as "high volume/low yield", or the more niche market of cultural tourism audience, which is "low volume/high yield". According to Olsen (2006) and Olsen and Esplin (2020), for religious heritage site managers, the most important stakeholders are the ecclesiastical leaders within their own religious group, who establish the organizational structure and management practices of sacred sites based on the views they hold of tourism and the core theological goals they wish to see fulfilled through their sacred spaces (see also Olsen 2011, 2012). However, from a more secular tourism perspective, governments would likely be viewed as the most important stakeholder, considering their role in the development of policies within which tourism functions (Jeffries 2001; Kunkel 2010) and in cases where governments control religious functions at sacred sites (see Bozonelos, Chapter 3, this volume). From this secular perspective, tourism businesses would be considered the next most important stakeholder, followed by tourists themselves.

These internal and external issues play an important role in the spatial development and organization of sacred sites. For example, in recent decades the increase in pilgrims coming to Saudi Arabia to perform the Hajj has been accompanied by fundamental alterations to the urban fabric of Mecca, including the creation of a much more commercially oriented space that contains a giant clock tower and a hotel that dwarfs the Grand Mosque. The moral and material justifications for such changes have been fiercely debated across the Muslim world, leading Bianchi (2013) to observe that

> The indictment is not that Saudi Arabia has planned poorly, but that it has planned destructively by willfully despoiling an irreplaceable resource that—unlike their oil wealth—belongs not to them, but to all Muslims and to all humanity, including generations not yet born.
>
> *(p. 18)*

Increasing visitation to sacred sites can also lead to the creation of temporal and spatial divisions between classes of visitors in order to preserve the sacrality of the place. Coleman (2018) notes that managers of many English cathedrals reinforce such boundaries in performing liturgical services during no-peak visitor times, with cathedral staff roping off space near the altar to demarcate a special area for the congregation. During these times the cathedral may close to visitors who do not identify themselves as worshipers.

These internal and external management issues also lead to clashes in values that center on what types of activities should be allowed in particular spaces. Such debates are manifest in the case of Uluru noted above, where some tourists and government officials have been at odds with the Anangu people over whether tourists should be able to climb the rock (Digance 2003; James 2007; Hueneke & Baker 2009). Other areas of contention include the appropriateness of charging for access to a sacred space—raising questions of visitor motivation (Shackley 2001a; Woodward 2004; Olsen 2006). Some English cathedrals, for instance, charge tourists a fee to enter while permitting permit pilgrims free entrance. Sites may also choose to avoid directly charging for admission, but instead ask for donations in an attempt to avoid the appearance of commercializing the site even though the amount of money made from donations is often not enough to cover maintenance costs (Olsen 2006; Levi & Kocher 2009). This concern over commercialism can lead to seemingly mundane decisions related to how close a gift store should be located to a saint's tomb or even if a gift shop should be allowed on the premises (Hung et al. 2017). The striking of an appropriate balance between fees and donations and commodification can however be difficult, as Di Giovine (2012) demonstrates in his account of the cult of Padre Pio based in San Giovanni Rotondo, Italy, where shrine guardians are often accused by the faithful of inauthenticity and commodification in their presentation and marketing of the saint. This attitude is also evident in Hung et al.'s (2017) research on monasteries in China where monks are accused of being more interested in selling religious items than catering to the religious and spiritual needs of visitors.

Contestation can also occur in instances where sacred sites serve as multi-purpose buildings, catering to different constituencies and staging a wide variety of events, or where they are owned and controlled by multiple entities. While mixed economies with multiple stakeholders grow over time around sacred sites (Valenta & Strabac 2016), the co-existence of different models of management is often highly problematic. English cathedrals are again revealing in this context, given that older churches have become increasingly important parts of the heritage landscape of many cities and have therefore become subject to secular, professional norms of administration and accountability. Coleman (2018: 128) notes the contrasts in style and objective between ecclesiastical and tourist managers. Distinctions between different roles focusing on religious and tourism aspects of site management are indicated in the titles of posts, ranging from "Dean", "Canon", and "Verger" to "Visitor Operations Manager" and "Director of Marketing". In practice, these two managerial regimes rely on each other: the former to provide spiritual and cultural capital, and the latter to accumulate the economic capital that enables the site to survive as a living institution. In similar fashion, Sánchez-Carretero (2013: 146–147) discusses how the development of heritage typically brings together two "logics" that may exist in mutual tension: that of the market and that of the politics of identity. Bowman (2011, 2014) has also written on the politics of ownership and joint management at the Church of the Holy Sepulchre in Jerusalem, noting how difficult it has been to make any management decisions because of the view that heritage management is a "zero-sum game", where "an advantage to one stakeholder is simultaneously seen to disadvantage another" (Olsen & Guelke 2004: 503).

Management techniques and opportunities

While sacred site managers face several internal and external management issues, there are several ways in which these challenges can be turned into positive cultural and economic opportunities. Stausberg (2011: 92) proposes various strategies or techniques to manage and mitigate negative management challenges, including "entrance fees, pay perimeters, transport and vehicle management, queue controls, temporary closures, [and] forcing people [to participate in] guided tours". MacCannell (1976) suggests that signage provides a useful way to direct visitors to and through sites, as well as tourist or religious guides who provide information, discreetly move visitors to places favored by managers, and act as mediators between the requirements of administrators and the desires of consumers (Feldman 2016). Guides are often hired by local site managers, but may also be used by external tourist agencies who attempt to anticipate the needs and interests of different traveling constituencies. A wide variety of guides with different informational and linguistic backgrounds can cater to the interpretational needs and wants of different kinds of visitors. While the management strategies and measures present by Stausberg may appear "defensive", managers of sites and numerous other constituencies are also presented with positive cultural and economic opportunities by the presence of large, diverse constituencies of people. Candea and da Col (2012) note that offering hospitality to strangers at these sites is an exercise of both welcome and control, where the institution or tried and proven management techniques provide visitors with access to valued sites, while encouraging—or disciplining—these visitors into being "responsible" guests and consumers.

Tourism, including its religious and spiritual manifestations, provides economic opportunities not only for managers of sacred sites but also for regions that may lack other means of economic development. Smith (2003: 10) refers to the fostering of heritage-oriented travel as supplying "developing" countries with important sources of income as well as cultural recognition. In practice, similar principles apply across the world. Rotherham (2007: 64), for example, argues that visits to religious sites in the United Kingdom have promoted both economic regeneration and community sustainability because these sites appeal to visitors at multiple levels of belonging and identity. As Rotherham notes, "Cultural roots may be embedded in places or images regardless of whether or not the contemporary tourist is a believer" (p. 65). Well-known sacred sites also play significant geopolitical roles. Geary (2018), for example, has shown how the well-known Buddhist pilgrimage complex of Bodh Gaya in north-east India has become a "soft power" heritage resource for strengthening partnerships and creating formal bilateral aid arrangements across Asia, including collaborations with the Japanese government.

If sacred sites are to have positive impacts that reach beyond the confines of any given location or religious organization, their management teams must address the wider physical, cultural, and social contexts in which such locations operate. Assessing the possibilities of creating sustainable local development through heritage tourism, Throsby (2009: 19ff) refers to certain "golden rules", including the measurement of "nonmarket effects" such as the "aesthetic, spiritual, social, historical, symbolic, [and] authenticity value" of what is being offered. Throsby looks not only to how sites benefit managers and visitors, but also to the ways in which they promote "liveability" in the environments where they are situated—possibly helping to promote tangible benefits such as transport links, and intangible benefits such as senses of place and identity, among local inhabitants.

At the same time, religious and spiritual sites offer important possibilities for the coexistence of different stakeholder entities. Though firm boundary-marking and separation is the norm at many multi-use and -owned sacred sites, other sites are organized so that positive awareness of cultural and religious others is maintained while frictional interactions are reduced as much as possible. Bowman (2016) refers to such interactions as "giving ground", which involves recognizing "the right of the Other to be in the same place as oneself as well as committing to the rites of negotiating her presence" (p. 260). While the idea of "giving ground" can operate as a useful metaphor, it can also take literal physical form as sites offer opportunities for religious co-presence or even merging. Albera (2019) traces numerous examples of Marian shrines around the Mediterranean region where a common ritual syntax of visiting tombs of saints and lighting candles permits the sharing of a commonly revered environment. Creating the possibilities for such coexistence of religious and spiritual diversity is a significant responsibility for site managers who must decide whether or not to intervene in orchestrating the enactment of religious identities.

Conclusion

Scholars who write about the popularity of sacred sites often focus on the aesthetic qualities of place, the benefits to be gained from devotion to a given deity, or the pressures to visit exerted by family or religious leaders. This chapter has demonstrated the significance of management practices in the promotion and preservation of sacred sites and their "sense of place". Site administrators and managers form an important "human infrastructure" at and around sanctified locations, and are constantly making choices and decisions that affect the physical and spiritual fabric of these sites. Effective site management is likely to entail gaining a good understanding of the physical boundaries and affordances of a site, an appreciation of local, national, and possibly global political and managerial contexts, and a sensitive awareness of the needs and expectations of visitors. Such virtues are both tested and demonstrated at the time of writing (summer 2020) when the presence of the COVID-19 virus around the world has rendered physical visits to sites temporarily difficult in many cases (see Olsen and Timothy 2020). Although the challenges remain significant, imaginative administrators have demonstrated an ability to turn current circumstances to their advantage by extending the spatiality and sacrality of their sites. Sheklian (2020), for example, has documented how Armenian Orthodox Christians held Easter Celebrations during the COVID-19 pandemic by staging an Internet service run by a resident deacon and priest in a New York seminary, which drew in people located in numerous continents.

While much of the literature on sacred site management comes from the Anglophone tourism literature and is written through Western, Christian, and post-Christian lenses, increasing attention needs to be paid to management practices conducted by other cultural and ethnic populations around the world. For example, Jafari and Scott (2014: 1) refer to the "Muslim world and its tourisms", suggesting that both scholars and administrators need to recognize ways in which religious and secular practices are closely intertwined in such contexts, and how Western concepts of management may not always be the most salient.

References

Airey, D., & Shackley, M. (1998) 'Bukhara (Uzbekistan): A former oasis town on the Silk Road', in M. Shackley (ed), *Visitor Management: Case Studies from World Heritage Sites* (pp. 10–25). Oxford, UK: Butterworth-Heinemann.

Aitpaeva, G. (2013) 'Introduction', in G. Aitpaeva (ed), *Sacred Sites of the Southern Kyrgyzstan: Nature, Manas, Islam* (pp. 6–10). Bishkek, Kyrgyzstan: Aigine Cultural Research Center.

Albera, Dionigi (2019) 'Ritual mixing and interrituality at Marian shrines', in M. Moyaert (ed), *Interreligious Relations and the Negotiation of Ritual Boundaries. Interreligious Studies in Theory and Practice* (pp. 137–154). New York: Palgrave Macmillan.

Al-Houdalieh, S.H. (2010) 'Archaeological heritage and related institutions in the Palestinian National Territories 16 years after signing the Oslo Accords', *Present Pasts*, 2(1): 31–53.

Alley, K.D. (1992) 'On the banks of the Ganga', *Annals of Tourism Research*, 19(1): 125–127.

Alley, K.D. (2002) *On the Banks of the Gaṅgā: When Wastewater Meets a Sacred River*. Ann Arbor: University of Michigan Press.

Allison, E. (2014) 'Waste and worldviews: Garbage and pollution challenges in Bhutan', *Journal for the Study of Religion, Nature and Culture*, 8(4): 405–428.

Ashworth, G.J. (2009) 'Do tourists destroy the heritage they have come to experience?' *Tourism Recreation Research*, 34(1): 79–83.

Askew, M. (2008) 'Materializing merit: The symbolic economy of religious monuments and tourist-pilgrimage in contemporary Thailand', in P. Kitiarsa (ed), *Religious Commodifications in Asia: Marketing Gods* (pp. 89–119). London: Routledge.

Badone, E., & Roseman, S. (2004) 'Approaches to the anthropology of pilgrimage and tourism' in E. Badone & S. Roseman (eds), *Intersecting Journeys: The Anthropology of Pilgrimage and Tourism* (pp. 1–23). Champaign: University of Illinois Press.

Bakht, N., & Collins, L. (2017) '"The Earth is our Mother": Freedom of religion and the preservation of Indigenous sacred sites in Canada', *McGill Law Journal/Revue de droit de McGill*, 62(3): 777–812.

Ballantyne, R., Hughes, K., & Bond, N. (2016) 'Using a delphi approach to identify managers' preferences for visitor interpretation at Canterbury Cathedral World Heritage Site', *Tourism Management*, 54: 72–80.

Bernardi, A., Todorov, V., & Hiristova, J. (2000) 'Microclimatic analysis in St. Stephan's church, Nessebar, Bulgaria after interventions for the conservation of frescoes', *Journal of Cultural Heritage*, 1(3): 281–286.

Bianchi, R. (2013) *Islamic Globalization: Pilgrimage, Capitalism, Democracy, and Diplomacy*. Singapore: World Scientific.

Bickersteth, J., West, D., & Wallis, D. (2020) 'Returning Uluru', *Studies in Conservation*, 65(1): 9–17.

Bond, N. (2015) 'Exploring pilgrimage and religious heritage tourism experiences', in R. Raj & K. Griffin (eds), *Religious Tourism and Pilgrimage Management: An International Perspective* (2nd ed., pp. 118–129). Wallingford, UK: CABI.

Bowman, G. (2011) '"In dubious battle on the plains of heav'n": The politics of possession in Jerusalem's Holy Sepulchre', *History and Anthropology*, 22(3): 371–399.

Bowman, G. (2014) 'The politics of ownership: State, governance and the status quo in the Church of the Anastasis (Holy Sepulchre)', in E. Barkan & K. Barkey (eds), *Choreographies of Shared Sacred Sites: Religion and Conflict* Resolution (pp. 199–234). New York: Columbia University Press.

Bowman, G. (2016) 'Grounds for sharing—occasions for conflict: An inquiry into the social foundations of cohabitation and antagonism', in R. Bryant (ed), *Post-Ottoman Coexistence Book: Sharing Space in the Shadow of Conflict Book* (pp. 258–275). Oxford, UK: Berghahn.

Boyer, J.M. (2000) 'Impact of the public on monuments', in J.M. Ballester (ed), *Sustained Care of the Cultural Heritage Against Pollution* (pp. 159–171). Strausbourg, Cedex, France: Council of Europe Publishing.

Bremer, T.S. (2006) 'Sacred spaces and tourist places', in D.J. Timothy & D.H. Olsen (eds), *Tourism, Religion and Spiritual Journeys* (pp. 41–51). London: Routledge.

Bukharin, M. (2009) 'Mecca on the caravan routes in pre-Islamic antiquity', in *The Qur'ān in Context: Historical and Literary Investigations into the Qur'ānic Milieu* (pp. 115–134). Leiden: Brill.

Butzier, S.R., & Stevenson, S.M. (2014) 'Indigenous peoples' rights to sacred sites and traditional cultural properties and the role of consultation and free, prior and informed consent', *Journal of Energy & Natural Resources Law*, 32(3): 297–334.

Caidi, N., Beazley, S., & Colomer Marquez, L. (2018) 'Holy selfies: Performing pilgrimage in the age of social media', *The International Journal of Information, Diversity, & Inclusion*, 2(1–2): 8–31.

Campo, J.E. (1998) 'American pilgrimage landscapes', *The Annals of the American Academy of Political and Social Science*, 558(1): 40–56.

Candea, M., & da Col, G. (2012) 'The return to hospitality', *Journal of the Royal Anthropological Institute*, 18: 1–19.

Carlisle, S. (2009) 'Lalibela, Ethiopia', in M. Shackley (ed), *Visitor Management: Case Studies from World Heritage Sites* (pp. 139–160). Oxford, UK: Butterworth-Heinemann.

Clancy, M. (2020) 'Overtourism and resistance: Today's anti-tourist movement in context', in H. Pechlander, E. Innerhofer, & G. Erschbamer (eds), *Overtourism: Tourism Management and Solutions* (pp. 14–24). London: Routledge.

Coleman, D. (2016) 'Indigenous place and diaspora space: Of literalism and abstraction', *Settler Colonial Studies*, 6(1): 61–76.

Coleman, S. (2018) 'On praying in an old country: Ritual, replication, heritage, and powers of adjacency in English cathedrals', *Religion*, 49(1): 120–141.

Collins-Kreiner, N. (2010) 'The geography of pilgrimage and tourism: Transformations and implications for applied geography', *Applied Geography*, 30(1): 153–164.

Collins-Kreiner, N., Kilot, N., Mansfeld, Y., & Saig, K. (2006) *Christian Tourism to the Holy Land: Pilgrimage during Security Crisis*. Aldershot, UK: Ashgate.

della Dora, V. (2009) 'Taking sacred space out of place: From Mount Sinai to Mount Getty through travelling icons', *Mobilities*, 4(2): 225–248.

Digance, J. (2003) 'Pilgrimage at contested sites', *Annals of Tourism Research*, 30(1): 143–159.

Di Giovine, M. (2012) 'Padre Pio for sale: Souvenirs, relics, or identity markers?' *International Journal of Tourism Anthropology*, 2(2): 108–127.

Eid, R. (2012) 'Towards a high-quality religious tourism marketing: The case of Hajj service in Saudi Arabia', *Tourism Analysis*, 17(4): 509–522.

Farias, M., Coleman III, T.J., Bartlett, J.E., Oviedo, L., Soares, P., Santos, T., & Bas, M.D.C. (2019) 'Atheists on the Santiago Way: Examining motivations to go on pilgrimage', *Sociology of Religion*, 80(1): 28–44.

Fedele, A. (2013) *Looking for Mary Magdalene: Alternative Pilgrimage and Ritual Creativity at Catholic Shrines in France*. New York: Oxford University Press.
Fedele, A. (2014) 'Energy and transformation in alternative pilgrimages to Catholic shrines: Deconstructing the tourist/pilgrim divide', *Journal of Tourism and Cultural Change*, 12(2): 150–165.
Feldman, J. (2016) *A Jewish Guide in the Holy Land: How Christian Pilgrims Made Me Israeli*. Champaign: Illinois University Press.
Francis, L.J., Mansfield, S., Williams, E., & Village, A. (2010) 'Applying psychological type theory to cathedral visitors: A case study of two cathedrals in England and Wales', *Visitor Studies*, 13(2): 175–186.
Francis, L.J., Williams, E., Annis, J., & Robbins, M. (2008) 'Understanding cathedral visitors: Psychological type and individual differences in experience and appreciation', *Tourism Analysis*, 13(1): 71–80.
Gammon, S. (2004) 'Secular pilgrimage and sport tourism', in B.W. Ritchie & D. Adair (eds), *Sport Tourism: Interrelationships, Impacts and Issues* (pp. 30–45). Clevedon, UK: Channel View Publications.
Geary, D. (2018) 'Transnational courting through Shakyamuni Buddha: Japanese pilgrimage and geographical dowries in North India', in S. Coleman & J. Eade (eds), *Pilgrimage and Political Economy: Translating the Sacred* (pp. 40–58). Oxford, UK: Berghahn.
Gowreesunkar, V.G., & Thanh, T.V. (2020) 'Between overtourism and under-tourism: Impacts, implications, and probable solutions', in *Overtourism: Causes, Implications and Solutions* (pp. 45–68). Cham, Switzerland: Springer.
Griffiths, M. (2011) 'Those who come to pray and those who come to look: Interactions between visitors and congregations', *Journal of Heritage Tourism*, 6(1): 63–72.
Griffiths, M., & Wiltshier, P. (eds) (2019) *Managing Religious Tourism*. Wallingford, UK: CABI.
Heinämäki, L., & Herrmann, T.M. (eds) (2017) *Experiencing and Protecting Sacred Natural Sites of Sámi and Other Indigenous Peoples: The Sacred Arctic*. Cham, Switzerland: Springer.
Huang, K., & Pearce, P. (2019) 'Visitors' perceptions of religious tourism destinations', *Journal of Destination Marketing & Management*, 14: 100371. https://doi.org/10.1016/j.jdmm.2019.100371.
Hueneke, H., & Baker, R. (2009) 'Tourist behaviour, local values, and interpretation at Uluru: 'The sacred deed at Australia's mighty heart'', *GeoJournal*, 74(5): 477–490.
Hughes, K., Bond, N., & Ballantyne, R. (2013) 'Designing and managing interpretive experiences at religious sites: Visitors' perceptions of Canterbury Cathedral', *Tourism Management*, 36: 210–220.
Hung, K., Yang, X., Wassler, P., Wang, D., Lin, P., & Liu, Z. (2017) 'Contesting the commercialization and sanctity of religious tourism in the Shaolin Monastery, China', *International Journal of Tourism Research*, 19(2): 145–159.
Ivakhiv, A. (2003) 'Nature and self in new age pilgrimage', *Culture and Religion*, 4(1): 93–118.
Ivakhiv, A. (2007) 'Power trips: Making sacred space through new age pilgrimage', in D. Kemp & J.R. Lewis (eds), *Handbook of New Age* (pp. 263–286). Leiden, The Netherlands: Brill.
Jackson, W., & Ormsby, A. (2017) 'Urban sacred natural sites – A call for research,' *Urban Ecosystems*, 20: 675–681.
Jafari, J., & Scott, N. (2014) 'Muslim world and its tourisms', *Annals of Tourism Research* 44(1): 1–19.
James, S. (2007) 'Constructing the climb: Visitor decision-making at Uluru', *Geographical Research*, 45(4): 398–407.
Jeffries, D.J. (2001) *Governments and Tourism*. London: Routledge.
Knox, D., & Hannam, K. (2014) 'The secular pilgrim: Are we flogging a dead metaphor?', *Tourism Recreation Research*, 39(2): 236–242.
Kong, L. (2001) 'Mapping 'new' geographies of religion: Politics and poetics in modernity', *Progress in Human Geography*, 25(2): 211–233.
Kunkel, L.M. (2010) *International Tourism Policy and the Role of Governments in Tourism in the Context of Sustainability*. Norderstedt, Germany: GRIN Verlag.
Levi, D., & Kocher, S. (2009) 'Understanding tourism at heritage religious sites', *Focus: Journal of the City and Regional Planning Department*, 6(1): 17–21.
Liljeblad, J., & Verschuuren, B. (2019) *Indigenous Perspectives on Sacred Natural Sites: Culture, Governance and Conservation*. London: Routledge.
Lois González, R.N.C. (2013) 'The Camino de Santiago and its contemporary renewal: Pilgrims, tourists and territorial identities', *Culture and Religion*, 14(1): 8–22.
MacCannell, D. (1976) *The Tourist: A New Theory of the Leisure Class*. New York: Schocken Books.

Maddrell, A., & Terry, A. (2015) '"At least once in a lifetime": Sports pilgrimage and constructions of the TT Races as "sacred" journey', in Madrell, A. Terry & T. Gale (eds), *Sacred Mobilities: Journeys of Belief and Belonging* (pp. 87–108). Abingdon: Ashgate.

Merhav, R., & Killebrew, A.E. (1998) 'Public exposure: For better and for worse', *Museum International*, 50(4): 15–20.

Morinis, A. (1992) 'Introduction: The territory of the anthropology of pilgrimage', in A. Morinis (ed), *Sacred Journeys: The Anthropology of Pilgrimage* (pp. 1–28). Westport, CT: Greenwood Press.

Olsen, D.H. (2003) 'Heritage, tourism, and the commodification of religion', *Tourism Recreation Research*, 28(3): 99–104.

Olsen, D.H. (2006) 'Management issues for religious heritage attractions', in D.J. Timothy & D.H. Olsen (eds), *Tourism, Religion and Spiritual Journeys* (pp. 104–118). London: Routledge.

Olsen, D.H. (2010) 'Pilgrims, tourists, and Weber's "Ideal Types"', *Annals of Tourism Research*, 37(3): 848–851.

Olsen, D.H. (2011) 'Towards a religious view of tourism: Negotiating faith perspectives on tourism,' *Journal of Tourism, Culture and Communication*, 11(1): 17–30.

Olsen, D.H. (2012) 'Teaching truth in "third space": The use of religious history as a pedagogical instrument at Temple Square in Salt Lake City, Utah', *Tourism Recreation Research*, 37(3): 227–237.

Olsen, D.H. (2013) 'A scalar comparison of motivations and expectations of experience within the religious tourism market', *International Journal of Religious Tourism and Pilgrimage*, 1(1): 41–61.

Olsen, D.H. (2014) 'Metaphors, typologies, secularization, and pilgrim as hedonist: A response', *Tourism Recreation Research*, 39(2): 248–258.

Olsen, D.H. (2019a) 'Religion, spirituality, and pilgrimage in a globalizing world', in D.J. Timothy (ed), *Handbook of Globalisation and Tourism* (pp. 270–283). Cheltenham, UK: Edward Elgar.

Olsen, D.H. (2019b) 'The symbolism of sacred space', in N. Crous-Costa, S. Aulet, & D. Vidal-Casellas (eds), *Interpreting Sacred Stories: Religious Tourism, Pilgrimage and Intercultural Dialogue* (pp. 29–42). Wallingford, UK: CABI.

Olsen, D.H. (2020) 'Pilgrimage, religious tourism, biodiversity, and natural sacred sites', in K.A. Shinde & D.H. Olsen (eds), *Religious Tourism and the Environment* (pp. 23–41). Wallingford, UK: CABI.

Olsen, D.H., & Esplin, S.C. (2020) 'The role of religious leaders in religious heritage tourism development: The case of The Church of Jesus Christ of Latter-day Saints', *Religions*, 11(5): 256. https://doi.org/10.3390/rel11050256.

Olsen, D.H., & Guelke, J.K. (2004) 'Spatial transgression and the BYU Jerusalem Center controversy', *The Professional Geographer*, 56(4): 503–515.

Olsen, D.H., & Korstanje, M. (eds) (2020) *Dark Tourism and Pilgrimage*. Wallingford, UK: CABI.

Olsen, D.H., & Ron, A.S. (2013) 'Managing religious heritage attractions: The case of Jerusalem', in B. Garrod & A. Fyall (eds), *Contemporary Cases in Heritage: Volume 1* (pp. 51–78). Oxford, UK: Goodfellow Publishers.

Olsen, D.H., & Timothy, D.J. (2002) 'Contested religious heritage: Differing views of Mormon heritage', *Tourism Recreation Research*, 27(2): 7–15.

Olsen, D.H., & Timothy, D.J. (2020) 'The COVID-19 pandemic and religious travel: Present and future directions', *International Journal of Religious Tourism and Pilgrimage*, 8(7): 170–188

Olsen, D.H., & Timothy, D.J. (2021) 'Contemporary perspectives of pilgrimage', in D. Liutikas (ed), *Pilgrims: Values and Identities* (pp. 224–238). Wallingford, UK: CABI.

Orland, B., & Bellafiore, V.J. (1990) 'Development directions for a sacred site in India', *Landscape and Urban Planning*, 19(2): 181–196.

Oviedo, G., & Jeanrenaud, S. (2007) 'Protecting sacred natural sites of indigenous and traditional peoples', in J.-M. Mallarch & T. Papayannis (eds), *Protested Areas and Spirituality: Proceedings of the First Workshop of the Delos Initiative—Montserrat 2006* (pp. 77–99). Gland, Switzerland: The World Conservation Union (IUCN).

Oviedo, G., Jeanrenaud, S., & Otegui, M. (2005) *Protecting Sacred Natural Sites of Indigenous and Traditional Peoples: An IUCN Perspective*. Gland, Switzerland: The World Conservation Union (IUCN).

Pavicic, J., Alfirevic, N., & Batarelo, V.J. (2007) 'The management and marketing of religious sites, pilgrimage and religious events: Challenges for Roman Catholic pilgrimages in Croatia', in R. Raj & N.D Morpeth (eds), *Religious Tourism and Pilgrimage Management: An International Perspective* (1st ed., pp. 48–63). Wallingford, UK: CABI.

Petreas, C. (2011) 'The conflicting interactions among stakeholders in religious and pilgrimage sites', in K. Andriotis, A. Theocharous, & F. Kosti (eds), *Proceedings of the International Conference on Tourism: Tourism in an Era of Uncertainty.* http://iatour.net/wp-content/uploads/2014/12/Proceedings_ICOT2011.pdf (accessed 2 November 2017).

Pliberšek, L., Basle, N., & Lebe, S.S. (2020) 'From burial spaces to pilgrimage places: The changing role of European cemeteries', in D.H. Olsen & M.E. Korstanje (eds), *Dark Tourism and Pilgrimage* (pp. 75–84). Wallingford, UK: CABI.

Poria, Y., Biran, A., & Reichel, A. (2009) 'Visitors' preferences for interpretation at heritage sites', *Journal of Travel Research*, 48(1): 92–105.

Prats, L., Aulet, S., & Vidal, D. (2015) 'Social network tools as guides to religious sites', in R. Raj & K. Griffin (eds), *Religious Tourism and Pilgrimage Management: An International Perspective* (2nd ed., pp. 146–159). Wallingford, UK: CABI.

Qurashi, J. (2017) 'Commodification of Islamic religious tourism: From spiritual to touristic experience', *International Journal of Religious Tourism and Pilgrimage*, 5(1): 89–104.

Qurashi, J. (2020) 'The natural environment and waste management at the Hajj', in K.A. Shinde & D.H. Olsen (eds), *Religious Tourism and the Environment* (pp. 133–143). Wallingford, UK: CABI.

Raj, R., Griffin, K., & Blackwell, R. (2015) 'Motivations for religious tourism, pilgrimage, festivals and events', in R. Raj & K. Griffin (eds), *Religious Tourism and Pilgrimage Management: An International Perspective* (2nd ed., pp. 103–117). Wallingford, UK: CABI.

Reader, I. (2014) *Pilgrimage in the Marketplace.* London: Routledge.

Reader, I., & Walter, T. (1993) *Pilgrimage in Popular Culture.* New York: Palgrave Macmillan.

Redick, K. (2009) 'Wilderness as axis mundi: Spiritual journeys on the Appalachian Trail', in G. Backhaus & J. Murungi (eds), *Symbolic Landscapes* (pp. 65–90). Dordrecht, The Netherlands: Springer.

Redick, K. (2018) 'Interpreting contemporary pilgrimage as spiritual journey or aesthetic tourism along the Appalachian Trail', *International Journal of Religious Tourism and Pilgrimage*, 6(2): 78–88.

Rotherham, I. (2007) 'Sustaining tourism infrastructures for religious tourists and pilgrims within the UK', in R. Raj & N.D. Morpeth (eds), *Religious Tourism and Pilgrimage Management* (pp. 64–77). Wallingford, UK: CABI.

Samakov, A., & Berkes, F. (2017) 'Spiritual commons: Sacred sites as core of community-conserved areas in Kyrgyzstan', *International Journal of the Commons*, 11(1): 422–444.

Sánchez-Carretero, C. (ed) (2013) *Heritage, Pilgrimage and the Camino to Finisterre: Walking to the End of the Earth.* Berlin: Springer.

Shackley, M. (1998) 'A gold calf in sacred space? The future of St. Katherine's monastery, Mount Sinai (Egypt)', *International Journal of Heritage Studies*, 4(3–4): 124–134.

Shackley, M. (2001a) *Managing Sacred Sites: Service Provision and Visitor Experience.* London: Continuum.

Shackley, M. (2001b) 'Sacred world heritage sites: Balancing meaning with management', *Tourism Recreation Research*, 26(1): 5–10.

Shackley, M. (2002) 'Space, sanctity and service: The English cathedral as hetertopia', *International Journal of Tourism Research*, 4: 345–352.

Shackley, M. (2003) 'Management challenges for religion-based attractions', in A. Fyall, B. Garrod & A. Leask (eds), *Managing Visitor Attractions: New Directions* (pp. 159–170). Oxford, UK: Butterworth-Heinemann.

Shackley, M. (2006) 'Visitor management at world heritage sites', in A. Leask & A. Fyall (eds), *Managing World Heritage Sites* (pp. 83–93). Oxford, UK: Butterworth-Heinemann.

Shaw, J. (2000) 'Ayodhya's sacred landscape: Ritual memory, politics and archaeological 'fact'', *Antiquity*, 74(285): 693–700.

Sheklian, C. (2020) 'Communion in quarantine: How liturgical Christian churches celebrated Easter', *Cultural Anthropology* 'Editors' Forum: Covid-19' https://culanth.org/fieldsights/communion-in-quarantine

Shinde, K.A. (2010a) 'Managing Hindu festivals in pilgrimage sites: Emerging trends, opportunities, and challenges', *Event Management*, 14(1): 53–67.

Shinde, K.A. (2010b) 'Entrepreneurship and indigenous enterpreneurs in religious tourism in India', *International Journal of Tourism Research*, 12(5): 523–535.

Smith, M. (2003) *Issues in Cultural Tourism Studies.* London: Routledge.

Stausberg, M. (2011) *Religion and Tourism: Crossroads, Destinations, and Encounters.* London: Routledge.

Stone, P.R., Hartmann, R., Seaton, A.V., Sharpley, R., & White, L. (eds) (2018) *The Palgrave Handbook of Dark Tourism Studies.* London: Palgrave Macmillan.

Thouki, A. (2019) 'The role of ontology in religious tourism education—Exploring the application of the postmodern cultural paradigm in European religious sites', *Religions*, 10: 649. https://www.mdpi.com/2077-1444/10/12/649.

Throsby, D. (2009) 'Tourism, heritage and cultural sustainability: Three "golden rules"', in L. Fusco Girard & P. Nijkamp (eds), *Cultural Tourism and Sustainable Local Development* (pp. 13–29). Farnham: Ashgate.

Timothy, D.J. (2012) 'Religious views of the environment: Sanctification of nature and implications for tourism', in A. Holden & D.A. Fennell (eds), *The Routledge Handbook of Tourism and the Environment* (pp. 53–64). London: Routledge.

Tomljenović, R., & Dukić, L. (2017) 'Religious tourism—from a tourism product to an agent of societal transformation', in M. Stanišić (ed), *Religious Tourism and the Contemporary Tourism Market* (pp. 1–8). Belgrade: Singidunum University.

Tran, T.C., Ban, N.C., & Bhattacharyya, J. (2020) 'A review of successes, challenges, and lessons from Indigenous protected and conserved areas', *Biological Conservation*, 241: 108271.

UNESCO. (2020a) World Heritage List Statistics. Available at: https://whc.unesco.org/en/list/stat (accessed 7 August 2020).

UNESCO. (2020b) World Heritage List. Available at: https://whc.unesco.org/en/list/ (accessed 7 August 2020).

UNESCO. (2020c) UNESCO. (2020a) Representative List of the Intangible Cultural Heritage of Humanity. Available at: https://ich.unesco.org/en/lists (accessed 7 August 2020).

UNWTO. (2011) Religious Tourism in Asia and the Pacific. World Tourism Organization, Madrid.

Uusihakala, K. (2011) 'Reminiscence tours and pilgrimage sites', *Suomen Antropologi: Journal of the Finnish Anthropological Society*, 36(1): 57–64.

Valenta, M., & Strabac, Z. (2016) 'The dramaturgical nexus of ethno-religious, tourist and transnational frames of pilgrimages in post-conflict societies: The Bosnian and Herzegovinian experience', *Tourist Studies*, 16(1): 57–73.

Verschuuren, B., Wild, R., McNeely, J.A., & Oviedo, G. (2010) *Sacred Natural Sites: Conserving Nature & Culture*. London: Earthscan.

Voase, R. (2007) 'Visiting a cathedral: The consumer psychology of a 'rich experience'', *International Journal of Heritage Studies*, 13(1): 41–55.

West, B. (2008) 'Enchanting pasts: The role of international civil religious pilgrimage in reimagining national collective memory', *Sociological Theory*, 26(3): 258–270.

Wild, R., & McLeod, C. (2008) *Sacred Natural Sites: Guidelines for Protected Area Managers*. Gland, Switzerland: IUCN.

Williams, T.M. 2007. *Evaluating the Visitor Experience: The Case of Chester Cathedral*. Unpublished master's thesis. University of Chester, United Kingdom.

Wiltshier, P., & Clarke, A. (2012) 'Tourism to religious sites, case studies from Hungary and England: Exploring paradoxical views on tourism, commodification and cost-benefits', *International Journal of Tourism Policy*, 4(2): 132–145.

Wiltshier, P., & Griffiths, M. (2016) 'Management practices for the development of religious tourism sacred sites: Managing expectations through sacred and secular aims in site development; report, store and access', *International Journal of Religious Tourism and Pilgrimage*, 4(7): 1–8.

Winter, M., & Gasson, R. (1996) 'Pilgrimage and tourism: Cathedral visiting in contemporary England', *International Journal of Heritage Studies*, 2(3): 172–182.

Wong, C.U.I., McIntosh, A., & Ryan, C. (2016) 'Visitor management at a Buddhist sacred site', *Journal of Travel Research*, 55(5): 675–687.

Woodward, S.C. (2004) 'Faith and tourism: Planning tourism in relation to places of worship', *Tourism and Hospitality Planning & Development*, 1(2): 173–186.

Wright, K., (2020) 'Finding roots: Pop culture pilgrimage and the affective geographies of Kunta Kinteh Island', in D.H. Olsen & M.E. Korstanje (eds), *Dark Tourism and Pilgrimage* (pp. 185–196). Wallingford, UK: CABI.

26

MANAGING COMPLEX ISSUES IN RELIGIOUS AND SPIRITUAL EVENTS, FESTIVALS AND CELEBRATIONS

Rev. Ruth Dowson

Introduction

This chapter undertakes a critical review of the current state of research into the complex issues faced by those who organise religious and spiritual events, festivals and celebrations. The current literature emanates from a range of multidisciplinary sources, which have formerly been located in distinct, traditional disciplinary silos (Dowson, 2019a). The chapter also investigates research into the contemporary context of the dynamic between religious and spiritual aspects and origins of festivals, celebrations and associated events within the background of the global experience economy. It notes the impact of culture on religious and spiritual festivals, celebrations and events, considering the differences between appropriation and appreciation of other cultures as they coalesce. In identifying the complex issues that arise from these circumstances and perspectives, the chapter assesses the future development of conceptual and theoretical approaches to religious and spiritual festivals, celebrations and events.

The chapter will identify and assess management issues that face the organisers of religious and spiritual events, festivals and celebrations both now and in the future. These management issues include sector-specific concerns, as well as those that apply across all events and festivals. This chapter incorporates a range of relevant examples, from both religious and spiritual perspectives, and attempts to critically consider the different types of issues arising from events, festivals and celebrations that involve or result in religious and spiritual tourism, and which require the identification and management of specific issues that arise from the religious or spiritual nature of the events.

Event contexts

In an era dominated by the experience economy (Pine & Gilmore, 2011), in which the acquisition of physical goods is increasingly subordinated to the pursuit of experiences that result in memories (Wood, 2009, 2017; Wood & Moss, 2015; Wood & Slater, 2015), events often provide the means of gaining such experiences. Although tourism emerged from its roots in religious travel (Timothy & Olsen, 2006), there continues to be a considerable growth and profusion of different types of religious tourism, and now also of spiritual tourism, across the globe. Such tourism today is often embedded within the context of an event, or located around a series of events. The life we live in the twenty-first century is increasingly based and set within an environment where we engage with others through the medium of events. This trend has become known as 'Eventization', and is situated within the 'Critical Events' turn, and elsewhere (Becci, Burchardt & Casanova, 2013; Dowson, 2016, 2019b; Lamond & Platt, 2016; Maasø, 2018; Pfadenhauer, 2010; Spracklen, Richter & Spracklen, 2013). The religious world is not exempt from this development, as faith organisations embrace the structure of events to promote their messages, to build community, strengthen relationships, and raise funds, amongst other purposes (Ron & Timothy, 2019).

Concept of eventization of faith

Roche's (2000, p. 3) declaration that events are 'sociologically important in characterising and understanding modern societies' is vital to understanding the importance and role of events in the creation and growth of flourishing religious and spiritual communities, as well as influencing individual lives. Wood (2009) suggests that experiential events are utilised in marketing to effectively generate immediate choices and consequences, while embedding enduring transformation, for example, through the ongoing purchasing of a product. Pfadenhauer (2010) suggested that faith could be such a product, identifying eventization as a concept and applying it to the Roman Catholic Church. Pfadenhauer (2010) builds on an interpretation of the use of events as a marketing tool in promoting religion on a large scale. She studied a very large Roman Catholic Church event, World Youth Day, which brings together young people from around the world, combining the traditional forms of Roman Catholic festivals within a contemporary (modern) event structure. According to Pfadenhauer, this is a similar activity to any brand activation or experiential marketing event in the secular world, but with the overall purpose of recruiting or evangelising new members for the church. Pfadenhauer labelled this activity 'eventization of faith'. Becci, Burchardt and Casanova (2013) drew their conclusions from religious topography, acknowledging the religious innovations that have followed secularisation. They identify traditional religious celebrations as becoming united with the general move towards the eventization of leisure, even inhabiting secular public spaces. Becci et al (2013) interpret the burgeoning extension of events in the religious space as branding and marketing, supporting Pfadenhauer's (2010) conclusions.

From a secular perspective, and explicitly within the context of the turn to critical events studies, Spracklen, Richter and Spracklen (2013) considered the revolutionary nature of the changes that have commodified leisure, eventizing and monetising public leisure spaces for new forms of shared activity that often become ritualised, developing shared meaning for participants. This trend equally influences the delivery of religious and sacred events and festivals, as accepted societal norms evolve, whereby social activities that draw people together increasingly take the form of organised events, whether in public or private spaces—and often require payment to hire venue space. Indeed, we have observed the increased use of secular spaces for religious events (Dowson, 2016). This move sees venues such as the Leeds First Direct Arena or Butlin's seaside holiday camps host religious events. Such events often include elements of the spectacular (Debord, 2012), drawing into question their authenticity.

Critical event studies aim to ground the academic study of events beyond the mere logistical, no longer limited to typological analyses. Lamond and Platt (2016) welcome the growing number of university modules that consider cultural, social, political and even religious perspectives on the neo-liberal world of events. This critical approach is developed further by Spracklen and Lamond (2016) in a self-acknowledged polemical research monograph, aimed to encourage critical thinking in events students, in place of the standard logistics as taught on many undergraduate and postgraduate university events courses. Both Lamond and Platt, and Spracklen and Lamond, recognise the importance of context—social, political and cultural—in understanding the nature of events—and here I would add the existence of a layer of religious and spiritual considerations to the mix. All too often, the theological nature of a religious or spiritual event is forgotten, to the detriment not only of meaning but practical delivery, and even organisational reputation.

No consideration of the impact of religious and spiritual events would be complete without a consideration of Falassi's (1967) pioneering typological model of ritual. Falassi's study of festivals identified a range of nine ritual aspects that can productively be applied to the context of religious and spiritual celebrations, festivals and events, construing sacred time, in contrast to the ordinary time of everyday life. Table 26.1 considers these different rites and provides relevant examples.

Table 26.1 Falassi's typology of ritual applied to religious and spiritual events, festivals and celebrations

Rites	*Religious and Spiritual rituals in events, festivals and celebrations*
Rites of purification, cleansing, chasing away evil by fire/holy water/sacred relics/ symbols	• Anointing with holy oil • Smudging to cleanse a space with burning sage
Rites of passage, marking transition from one stage of life to another, e.g. initiations	• Baptism, confirmation, Bar/Bat mitzvah • Moon cycle circle group
Rites of reversal—symbolic inversion, e.g. (masks &) costumes; gender misidentification; role confusion; using sacred places for profane activities	• Costumes—robed choir, robed clergy • Samhain festival marking the return of winter with bonfires and wearing disguising costumes to visit neighbours
Rites of conspicuous display—objects of high symbolic value put on display, touched or worshipped; used in processions; guardians and social/ political/ religious elite display their powers	• Statues of Mary in processions on 15 August • Beltane celebrations at Glastonbury Tor or Stonehenge
Rites of conspicuous consumption—feasts, food & drink; gifts showered on guests; sacred (Holy) communion	• Holy Communion (Eucharist) services using contemporary or traditional language • Shared meals—bring and share vegan food, 'breaking bread together' • Antithetical—the opposite of conspicuous consumption
Ritual dramas, retelling of myths and legends or historical re-enactment	• Passover Meal—re-telling the story of the Hebrew people leaving Egypt for the Promised Land • Coronation of Danish kings at Stonehenge
Rites of exchange from commerce—buying and selling—to gift exchanges and charitable donations	• Resources shop or stall selling clothing, music, DVDs and books, CDs of talks, publications, Fair Trade goods • Energy exchange—engagement with the process or event, in place of financial contribution

(*Continued*)

Rites	*Religious and Spiritual rituals in events, festivals and celebrations*
Rites of competition—games, sports, contests of all kinds either highly unpredictable and merit-based or ritualised and predictable	• Family or all-age services, 'Messy church' with games and food • 'Spiritual marketplace' as different spiritualities and religions offer choice
De-valorization rites—take place at the end of the event; Restoring normal time and space, closing ceremonies; Formal/informal farewells	• Hajj—pilgrims change back into their own clothes from white Irham dress • Spiritual circle is broken, participants leave the site

The power of ritual in encouraging and even enforcing action should not be overlooked, as there may be serious implications that arise as a result, impacting on logistics, health and safety, as well as other aspects of the event.

Event logistics

Events, festivals and other celebrations are not as simple to manage as might be thought by the non-professional, as health and safety considerations combine with (often unexpected) external impacts to require the adjustment of even well-thought-through plans (Dowson & Bassett, 2018, Theodore, 2018). However, the complexities involved in managing religious and spiritual events, festivals and celebrations transcend (Turner, 1982) those of standard event management, precisely because they include an additional layer of meaning—which is the religious or spiritual element (Dowson, 2015). Such a perspective can have an impact on the event logistics in many ways; the following section considers the event management issues that can arise in the development and delivery of religious and spiritual events, festivals and celebrations.

Event management issues

The section identifies and assesses the issues that face the organisers of religious and spiritual events, festivals and celebrations, both now and in the future. These issues include sector-specific event management concerns as well as those that apply across all events and festivals.

Low levels of events professionalism

Many religious organisations have professional religious officer-holders and leaders, such as imams, rabbis, monks, priests or ministers, and many of these are employed or paid by their organisation. However, it is common practice that such organisations generally rely on members or adherents to undertake the vast array of voluntary activities that promote the organisation and enable it to function. For spiritual (non-religious) activities, there are fewer structured arrangements in place, and less centrally-coordinated organisational activity overall, even where such organisations do exist. As a result, activities such as planning and delivering events are often undertaken or at least supported by ordinary attendees; the church, temple, synagogue or mosque with a professional event manager within the ranks of its regular membership is indeed blessed to have such input, but on the whole, those people who are active in supporting their local place of worship, are (hopefully enthusiastic) volunteers, with little if any training in events management or processes. It is rare to find even large churches that have complex events programmes employing qualified events professionals. They are much more likely to incorporate volunteers into the event team—gap year students from home or abroad, or semi-retired and retired members with time on their hands. Consequently, despite their enthusiasm and even years of experience, those who develop and deliver religious and spiritual events are often unacquainted with basic knowledge of event management processes or relevant health and safety regulations.

The pulsating nature of the events workforce from the build-up construction phase of events to the post-event break-down means that fewer workers are required in planning stages than at the event itself. This is reflected in Alvin Toffler's (1990) concept of the 'pulsating organisation' that morphs in size, in response to circumstances. As a result, religious and spiritual event organisers may struggle to recruit adequate numbers of event staff, whether paid or volunteers, and most will certainly not include trained, qualified or experienced event professionals.

The levels of event professionalism in religious and spiritual events management and delivery are much lower than in other sectors of the events industry, such as business or sports events management. However, religious organisations run the risk of event failure and serious reputational damage by allowing (or encouraging, and even coercing) 'enthusiasts' or staff untrained in events management to play major roles in event strategic planning and operational delivery, without consideration of the cost to the organisation and to the individuals of such actions. An emerging challenge is to acknowledge the complexities of event management and acquire the professional event management skills needed to ensure safe delivery (Dowson & Bassett, 2018).

The event planning process requires strategic consideration of who does what—agreeing in advance the roles that are required, who will fill those roles, as well as how they will be recruited, managed and trained, and what tasks they will undertake. Although some religious organisations are beginning to acknowledge the need for training, thus far there are few that encourage professional education (for example, qualifying through a university degree in events management), or training (such as one-day workshops for volunteers involved in leading event programmes in their place of worship) or the encouragement to use practical resources, such as event planning and management textbooks. All too often, those who support their religious organisations in events are doing so because no-one else showed up to help (or the usual suspects appear), and even when they are enthusiasts, little is done by the organisation leadership to ensure the safe delivery of an event.

High reliance on volunteers in the sector

There is a heavy reliance in events management generally, but more so in events run by religious institutions and charitable organisations, on the use of volunteers to undertake key event tasks and roles, perhaps supported by a small core team that often includes religious leaders and lay employees. The pulsating nature of the events workforce (Toffler, 1984; Toffler, 1990; Van der Wagen, 2007) requires a much larger team to deliver an event on site than it does to plan the event in advance, and many organisations—and especially in the religious and spiritual sector—do not have the financial resources to pay for such staffing levels. Thus, they are dependent on volunteers completing operational tasks and even assuming strategic roles. A key issue in the use of such volunteer labour in events is the lack of professionalism in selecting, training and managing those volunteers, while the over-use of untrained volunteers risks event failure as well as the organisation's reputation within the local community and beyond, with wider implications when problems emerge in national and local press, which is all the more common in the era of social media where almost everyone has an opinion and a camera-phone.

Lack of clarity or forethought of the purpose of an event

Many religious and spiritual events may be timed on a seasonal basis or connected with celebrating lifecycle occasions. But any religious or spiritual event, whether intended for internal or external audiences, will have an ultimate purpose. Such intentions may be explicit or implicit, or even hidden or covert. In particular, the overwhelmingly paramount consideration of mission or evangelism is almost certainly present in many religious or spiritual events, which may not be expressed or even recognised by the organisers (Dowson, 2015). In Table 26.2, the top row shows the resulting groupings from research into church event purposes; the subsequent rows in each grouping show that there are different types of event purposes, which are explained later.

To give examples, with the governance grouping (column 1, row 1), the purpose is corporate governance (column 1, row 2). In a traditional church context, this governance could include the Annual Parochial Council Meeting (APCM), and meetings of the Parish Church Council (PCC). The type of event involved in this purpose would be meetings, but a PCC meeting is easier to organise than an APCM. Some event church purposes involve a spiritual

Table 26.2 Groupings of identified event purposes

Governance	*Spiritual church activity*	*Internally-driven events*	*Community focus*	*External organisations hiring facilities for events*
Corporate governance	Catechesis Discipleship Holiness Initiation Life-cycle Liturgical Ritual Seasons Worship	Fundraising Networking and growing sustainable networks Social justice		Commercial activity Community-based activity
	Change and transformation Education Evangelism and Mission Forming group identity Learning Nurturing Pastoral Relationship-building within the church Ritual Reputation Teaching		Ecumenism Interfaith Relationship-building within wider community Civic events	

Source: Dowson (2015, p. 179).

dimension, such as learning about the Christian faith, learning how to live as a Christian, and worshipping God, but there are also vital spiritual dimensions in baptisms, confirmations and ordinations. The life-cycle purpose includes events around birth, marriage and death. Meanwhile, there are other events that are not restricted to the people who attend church, and these events might fall within the 'spiritual' grouping, although there are times when they are events that fall outside the purely spiritual, for example, an Alpha course (evangelistic seeker programme) (Holy Trinity Brompton, 2018) or a social event. The community focus grouping might be a fundraising event, or working to support the local community through 'family fun days'. Civic events also have a community focus, from Remembrance Day to annual civic services. Many churches are being encouraged to consider hiring out their spaces to external organisations, whether on a commercial basis or for the local community, so for example the local camera club might meet in the church hall, or parents who attend the church toddler group might hire the hall for a birthday party.

The key to understanding the purpose of religious events, festivals and celebrations is that there is complexity in marrying the two aspects—of event purpose, and event form (i.e. what type of event). There are many varied event purposes for church events, but it is not always possible to match these up with a specific event type; a fundraiser may take many different event forms, but the purpose is fundraising. Religious and spiritual event organisers often fail to consider what the purpose of a specific event is. This is complicated by the fact that events can have multiple purposes. This is especially true of evangelism and mission, which are at times implicit, rather than explicit, and may even have covert or hidden purposes. It is vital for event organisers to consider and agree on the purpose of any event before they start planning (Bowdin, Allen, O'Toole, Harris, & McDonnell, 2011; Dowson & Bassett , 2018).

There are aspects of repetition of religious and spiritual events, festivals and celebrations, as shown in Figure 26.1, indicating the frequency of the event. The frequency of an event carries specific management issues, as those which are delivered more often are easier to remember in terms of processes and activities required to deliver a safe and successful event. Any new or one-off event will bring with it new challenges and will consume more energy in preparation and planning, recognising, responding to and dealing with new issues as they

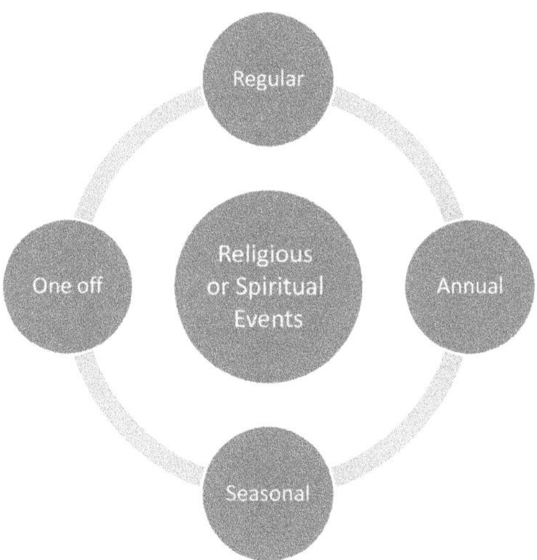

Figure 26.1 Frequency of spiritual and religious events

arise. Regular events may bring their own challenges as event teams become complacent and forget to be vigilant. For all events, it is important to develop a manual that includes event purpose and details, health and safety aspects—especially continuously updated rolling risk assessments, budgets, operational planning, contingencies, staffing, marketing and communications, legal requirements and evaluations, and capturing improvements for the next event. It should be noted that for some events, 'regular' might mean every 10, 12 or even 20 years; for example, the Lambeth Conference of Anglican bishops from around the world is held every 10 years (Lambeth Conference, 2019), while the Hindu Kumbh Mela festival takes place four times over the course of 12 years, attracting tens of millions of participants (Prayagraj Mela Pradhikaran, 2019), spanning the Hindu religion, as well as the science of astronomy, astrology, spirituality, traditional rituals, and socio-cultural customs and practices.

Lack of consideration of theological implications

The theological concerns that impact on the delivery and planning of religious and spiritual events are rarely considered by those involved, and yet they influence the construction, design and execution of such an event, and should therefore form part of the planning process. This aspect of planning for religious and spiritual events, festivals and celebrations is one that has rarely been addressed by religious organisations, although in 2018–2019, some UK churches and cathedrals have experienced the negative impacts of introducing events into their sacred spaces that are deemed out of kilter with the nature of the building. This aspect links into the earlier consideration of the purpose of the event, as well as the type of event format that is deemed appropriate for the sacred surroundings of a place of worship.

Types of events

One of the complexities that emerges from studying and understanding religious and spiritual events is that there are many different types of events, festivals and celebrations in this sector. It is worth considering the implications of some of these differences, and their evental origins, as they impact on, and contribute to, some of the complex issues that require management. Existing events, festivals and celebrations may have been handed down from generation to generation, perhaps with explicit meaning and purpose. Alternatively, traditional events, festivals and celebrations may have continued, having lost their original meanings in the mists of time, or having been adopted, adapted, appropriated or appreciated by and in other cultures and contexts. In contemporary society, the creation, emergence and evolution of new types of events, festivals and celebrations continue to expand the breadth of spiritual and religious events, as secular event designers develop new experiences.

Technical aspects

Technical aspects of events affect the way participants receive the event content. For example, sound quality can make the difference between a poor experience and a truly connected event, influencing and transforming the lives of those present and facilitating connection with the numinous. The primary function of speaker systems in churches is the spoken word, but many churches with contemporary worship styles have applied sound systems and audio technology to enhance the musical worship experience, making it more immersive (Heinze, 2019). Such systems localise where the sound is coming from, providing congregations with a less distracting listening experience, as the audio speakers are situated closer to the people, enhancing the connection between worship leaders and worshippers and resulting in more engaging immersive veneration. Such technical developments have transferred from the secular music world, where there is an expectation of high-quality audio-visual experience at an event, to address the issues of poor intelligibility in reverberant spaces through acoustic treatment, and high-quality sound reinforcement systems with directional loudspeakers. Outdoor spiritual and religious celebrations and festivals benefit equally from such technological advances, and can also include sound and noise-limiting equipment, protecting neighbouring areas from unwanted noise pollution.

Cultural risks

In every event, risks need to be addressed. In professional event management, the planning process includes undertaking a detailed risk assessment, to identify potential risks at the event, and to enhance and protect the event and public safety. An event organiser is responsible for many event aspects, but none more vitally important than meeting the obligations to provide a safe event. There will always be risks associated with events, and importantly the event organiser must put the risk into some sort of context. A risk assessment will enable the event organiser to identify potential risks and decide what measures need to be put in place or what actions need to be taken to reduce the risk.

There are three basic steps involved in a risk assessment, which is considered a 'living' document as it continues through pre-event planning, the event itself and beyond. These steps are:

1 Look for hazards (i.e. anything that can cause harm). For large events it will be useful to divide the site into smaller sectors. A secular outdoor festival site, for example, could be divided into the following sectors: ticket office, wristband exchange, car parking, main stage, main arena, arena entrance, individual marquees, and camping field.
2 Decide who might be harmed and how. Examples of people at risk at a secular festival site would include festival-goers, employees, temporary workers, volunteers, members of the public, contractors, and suppliers. Event managers should always pay particular attention to vulnerable people, such as young children, people with disabilities or inexperienced staff. Examples of how people could be harmed on a festival site include cuts and bruises, sprains and strains, broken and dislocated bones, absorbing substances, noise injuries, burns and scalds, crushing or trapping injuries, and flying or falling objects.
3 Evaluate the risks and decide what needs to be done first (if anything). To help evaluate the risks, a scoring system is used. Numerical scores are given to the Probability (P) and Severity (S) of risks, and these scores are multiplied to get a rating for the risk. The risk factor is calculated by multiplying the Probability score by the Severity score.

Theodore (2018) developed a specialist approach to risk assessment for the religious event space, which identifies the risk factors in logistics, operations and event administration. Theodore suggests that these are general risk factors. In addition to the usual risk considerations around people, property, finances, systems, environment and image, religious events and religious sites should also include an assessment of cultural, historical and religious values. A well-prepared and executed risk assessment document prepared by a competent person with expertise in health and safety matters will not prevent accidents from taking place at a religious or secular event, festival or celebration, but it will enable the event manager to address health and safety issues as and when they arise, with a well-considered plan in place. Korstanje, Raj and Griffin (2018) have focused in depth on the health, safety and risk challenges that face events and tourism in the religious context.

Cultural risk assessment

An innovative Cultural Risk Assessment model was developed by Dowson and Albert, which supports the identification and management of cultural risks at an event. This is especially important in the context of religious and spiritual events, festivals and celebrations. The role of the event manager is to identify and acknowledge any issues concerning values, rituals, behaviours and diverse practices, and respect them, working towards providing and supporting an inclusive environment in which the event festival or celebration takes place (Table 26.3).

Table 26.3 Cultural risk assessment

Issue	*Details/examples*
General Guest Protocols	High profile/VIP guests
Specific Religious Protocols and Religious Considerations	Muslim Protocols Prayer Times Direction of Prayer Location of Prayer Halal Options including crockery and cutlery Alcohol Catholic Protocols Sunday Mass 7th Day Adventist Practices Specific Religious Traditions Specific Dietary Requirements, including for different levels of religious practice and religiosity, e.g. Glatt kosher Entertainment—consider suitability for audience
Demographic context	Review attendee lists for differences in demographics and make relevant recommendations for content, style and structure of the event
Location issues	Including relevant food dishes in the menu, e.g. local cuisine, or dishes that reflect the event host culture
Language	Language barriers
Accessibility	Ensure venue is accessible for those with physical disabilities
Event Timing	Avoid certain times and seasons, e.g. no multi-faith events on Friday afternoon/evening
Dietary Requirements	Includes guests having different tastes in food, allergies, religious and non-religious restrictions and requirements, e.g. Halal, Kosher, and Vegan All dishes must be labelled with ingredients, label Vegetarian, Vegan and Halal dishes clearly so guests will be confident in knowing they are eating the correct food Some religions look poorly on food waste, which could cause offense to some guests if there is a notable amount of food waste

Issue	*Details/examples*
Alcohol	Consider sensibilities of guests who do not drink for religious or personal reasons. All drinks and menus should have clear labelling as to what is non-alcoholic, and bartenders must be clear when serving drinks
Etiquette	Could include wearing modest dress
Weather	Check climate and weather seasonal trends and potential issues, advise guests on appropriate dress and behaviour
Anxiety and mental health	Some guests may suffer with anxiety or mental health; being in social situations they are not used to may trigger attacks. Ensure there are quiet places to which people can go to in case of panic attacks or stressful moments. If someone is known to have anxiety, ensure they are introduced to people carefully so they feel comfortable in knowing that they are welcomed and have familiar faces
Personality	Diverse backgrounds will include extrovert and introvert. It is important that everyone understands each other's boundaries to ensure no one is uncomfortable or offended. For a multicultural wedding, having an event before the actual day will enable people to adjust to one another
Gender	Ensure designated spaces for men and women to pray separately if needed. Show people where prayer spaces are located.
Sexuality	Homosexuality views in a diverse context; whilst it may not be appropriate or possible to change perspectives towards different lifestyles, it may be possible to remind all guests why they are there, e.g. to celebrate the couple at a wedding
Understanding Different Cultures; New Experience	Could suggest shared activities to engage in other's cultural contexts

Source: Dowson and Albert (2019).

The Cultural Risk Assessment suggests specific areas to consider in planning a religious or spiritual event, proposing actions to manage the risks. The outcome of undertaking such an assessment as part of the ongoing event planning process is to enhance the event experience for participants and to reduce the potential for damage or negative issues that might arise.

Conclusion

Although there are many complex issues that arise from the planning and delivery of religious and spiritual events, festivals and celebrations, there are many gaps in the theoretical literature and practical academic research that specifically relate to this context. Religious and spiritual events, festivals and celebrations are all governed by the same processes and logistical management structures that relate to any other event. However, it is clear that the religious or spiritual dimension of these events adds an additional layer of complexity, which provides potential for specific research in the future. The growth of the global experience economy in general influences the growth of events in the religious and spiritual sector, encouraging expansion that mirrors the developments in secular events.

The future conceptual and theoretical development of understanding religious and spiritual festivals, celebrations and events should include work in several specific areas. First, little is known about the detailed event management and logistical aspects of religions and spiritual events. These should become a central focus of research because, as noted above, these types of events differ in their roles and contexts. Second, training religious and spiritual event volunteers is also crucial for the sector, as they typically do not have the same level of expertise that exists in other event realms. Third, the impact of the adoption, adaptation, appropriation and appreciation of religious and spiritual rituals in secular event contexts, and the role of rituals in spiritual and religious events are worthy of additional research, as these elements overlap considerably, such as in certain community festivals, founders celebrations and harvest festivals. Fourth, we likewise know little about the theological perspectives on religious and spiritual events. A deeper understanding of this would benefit religious representatives, faith organisations, as well as event managers, and provide deeper meaning, authenticity and identity in the context of religious and spiritual events. Fifth, the eventization of faith is a new area of research but one that has a lot of potential for future work. Finally, understanding the use of secular spaces for religious events and sacrosanct spaces for secular events has the potential to create conflict or enhance harmonious relations. These issues and many more stand at the forefront of future research on religious and spiritual events. There remains much work to be done to understand the relationships between religion, spirituality and events.

References

Becci, I., Burchardt, M., & Casanova, J. (2013) *Topographies of Faith: Religion in Urban Spaces.* Boston, MA: Brill.

Bowdin, G., Allen, J., O'Toole, W., Harris, R., & McDonnell, I. (2011) *Events Management*, 3rd edn. London: Butterworth-Heinemann.

Debord, G. (2012) *Society of the Spectacle*, 2nd edn. Eastbourne: Soul Bay Press.

Dowson, R. (2015) 'Religion, community and events', in A. Jepson & A. Clarke (eds), *Exploring Community Festivals and Events* (pp. 169–186). London: Routledge.

Dowson, R. (2016) 'Event as spiritual pilgrimage: A case study of the 'Cherish' Christian Women's Conference', *International Journal of Religious Tourism and Pilgrimage*, 4(2): 12–28.

Dowson, R. (2019a) 'Religious and spiritual festivals and events', in J. Mair (ed.), *The Routledge Handbook of Festivals* (pp. 313–322). London: Routledge.

Dowson, R. (2019b) 'Motivations to visit sacred historical objects: The Lindisfarne Gospels' visit to Durham 2013 – a sacred journey?', in R. Dowson, J. Yaqub & R. Raj (eds), *Spiritual and Religious Tourism: Management and Motivations* (pp. 109–121). Wallingford: CABI.

Dowson, R., & Albert, B. (2019) 'Celebration, ritual & culture', Unpublished lecture notes, module content, UK Centre for Events Management, Leeds Beckett University, Leeds, UK.

Dowson, R., & Bassett, D. (2018) *Event Planning and Management: Principles, Planning and Practice*, 2nd edn. London: Kogan Page.

Falassi, A. (1967) 'Festival: Definition and morphology', in A. Falassi (ed.), *Time Out of Time: Essays on the Festival* (pp. 1–10). Albuquerque: University of New Mexico Press.

Heinze, C. (2019) 'Sound all around: A look at immersive sound technologies', *Church Production.* March 4, 2019. Available online: https://www.churchproduction.com/education/sound-all-around-a-look-at-immersive%C2%A0sound-technologies/. Accessed 10 February 2019.

Holy Trinity Brompton (2018) Alpha course. Online at: https://www.htb.org/alpha. Accessed 26 September 2019.

Korstanje, M., Raj, R., & Griffin, K. (eds) (2018) *Risk and Safety Challenges for Religious Tourism and Events.* Wallingford: CABI

Lambeth Conference (2019) About the Lambeth Conference. Available online: https://www.lambethconference.org/about/. Accessed 10 February 2019.

Lamond, I.R., & Platt, L. (eds) (2016) *Critical Event Studies: Approaches to Research.* London: Palgrave Macmillan.

Maasø, A. (2018) 'Music streaming, festivals, and the eventization of music', *Popular Music and Society*, 41(2): 154–175.

Pfadenhauer, M. (2010) 'The eventization of faith as a marketing strategy: World Youth Day as an innovative response of the Catholic Church to pluralization', *International Journal of Nonprofit and Voluntary Sector Marketing*, 15(4): 382–394.

Pine, B.J., & Gilmore, J.H. (2011) *The Experience Economy*. Boston, MA: Harvard Business Review Press.

Prayagraj Mela Pradhikaran (2019) 'Introduction: Discover Kumbh, discover India, discover yourself', Available online: http://kumbh.gov.in/en/about-kumbh. Accessed 10 February 2019.

Roche, M. (2000) *Mega-Events and Modernity: Olympics and Expos in the Growth of Global Culture*. London: Routledge.

Ron, A.S., & Timothy, D.J. (2019) *Contemporary Christian Travel: Pilgrimage, Practice and Place*. Bristol: Channel View Publications.

Spracklen, K., & Lamond, I.R. (2016) *Critical Event Studies*. London: Routledge.

Spracklen, K., Richter, A., & Spracklen, B. (2013) 'The eventization of leisure and the strange death of alternative Leeds', *City*, 17(2): 164–178.

Theodore, B. (2018) 'Risk assessing and the religious event space: The context for the risk assessment', in M. Korstanje, R. Raj & K. Griffin (eds), *Risk and Safety Challenges for Religious Tourism and Events* (pp. 77–88). Wallingford: CABI.

Timothy, D.J., & Olsen, D.H. (eds) (2006) *Tourism, Religion and Spiritual Journeys*. London: Routledge.

Toffler, A. (1984) *Future Shock*. New York: Bantam Books.

Toffler, A. (1990) *Power Shift: Knowledge, Wealth, and Violence at the Edge of the 21st Century*. New York: Bantam Books.

Turner, V. (ed.) (1982) *Celebration: Studies in Festivity and Ritual*. Washington, DC: Smithsonian Institution.

Van der Wagen, L. (2007) *Human Resource Management for Events: Managing the Event Workforce*. Oxford: Butterworth-Heinemann.

Wood, E.H. (2009) 'Evaluating event marketing: Experience or outcome?', *Journal of Promotion Management*, 15(1): 247–268.

Wood, E.H. (2017) 'The value of events and festivals in an age of austerity', in E. Lunberg, J. Armbrecht, T.D. Andersson & D. Getz (eds), *The Value of Events* (pp. 10–35). London: Routledge.

Wood, E.H., & Moss, J. (2015) 'Capturing emotions: experience sampling at live music events'. *Arts and the Market*, 5(1): 45–72.

Wood, E.H., & Slater, A. (eds) (2015) *The Festival and Event Experience*. Colchester: Leisure Studies Association.

27

THE USE OF INFORMATION AND COMMUNICATION TECHNOLOGIES IN RELIGIOUS TOURISM

Suzanne Amaro, Cristina Barroco, and Paula Fonseca

Introduction

It is well known that information and communication technologies (ICTs) have completely transformed the tourism industry (Bethapudi 2013; Buhalis & Law 2008; Buhalis & O'Connor 2005; Ukpabi & Karjaluoto 2017), creating unprecedented opportunities for tourism and hospitality businesses (Law, Buhalis & Cobanoglu 2014). Not only have they changed the way firms are managed, but they have also changed consumer behavior in all phases of the decision-making process (Law et al. 2014).

These transformations have attracted researchers for many years, driving many studies on the use of ICTs in tourism (Bethapudi 2013; Buhalis & Law 2008; Hausmann & Weuster 2018; Neuhofer, Buhalis & Ladkin 2015). These studies stress how new technologies enhance the tourism experience, provide tourist information, improve service quality, and contribute to greater traveler satisfaction. However, most of these studies have focused on general tourism. There are tourism niches, such as religious tourism, where one may argue that new technologies are less important. Indeed, this argument seems irrefutable in the religious context considering that secularization theory suggests that "strong religious affiliation will be negatively related to all forms of mass media use because a vast majority of media content does not reflect traditional religious values" (Armfield & Holbert 2003, p. 30) and that, with modernization, institutionalized religions will become more irrelevant (Kluver & Cheong 2007).

According to this line of thought, it is interesting to note that almost 20 years ago Armfield and Holbert (2003) found that the more religious people were, the less likely they were to use the Internet. However, this is unlikely to be the case today. Recently, the Pope unveiled the "Click to Pray" app, calling for people to download the app and pray with him (Lapin 2019). He pointed out that the Internet was a way "to stay in touch with others, to share values and projects and to express the desire to form a community." This shows how times have changed and that perhaps religion and ICTs can go hand in hand. Indeed, ICTs are now widely used for religious purposes (Ramos, Henriques & Lanquar 2016) and are being increasingly used by all players: religious site managers, religious tourists, faith tour organizers, and other interested parties. Hence, technology is changing religious tourism, just as it has changed other areas (Rashid 2018), such as heritage tourism, more generally.

Nowadays, managing religious sites presents major challenges. On the one hand, visitors are diverse (Hughes, Bond & Ballantyne 2013), ranging from devoted pilgrims to tourists with secular motivations, both of which may include people who travel alone or with families and people with disabilities (Gassiot, Prats & Coromina 2015). On the other hand, overtourism, preservation of heritage sites and sustainable tourism practices are topics that have recently received much attention (de Jong & Varley 2018; Gurira & Ngulube 2016). Furthermore, ICTs have contributed to a more informed tourist that has consequently become more demanding.

This chapter provides an overview of the use of ICTs in the religious tourism industry, evidencing how useful they can be in this specific tourism context, as they are in the tourism industry in general. The chapter also examines how ICTs can be used to manage religious sites, events, and pilgrimages more effectively in such a challenging environment.

Religious tourism

Religious tourism, considered to be one of the oldest types of tourism (Rinschede 1992), has been growing in popularity worldwide in recent years (Bond, Packer & Ballantyne 2015; Griffin & Raj 2017). Travelling to sacred places was, and continues to be, an inherent aspect of almost every culture and today worldwide, more and more people are travelling to sacred sites (Blackwell 2007).

The World Tourism Organization estimates that over 300 million tourists visit the world's key religious sites every year, with approximately 600 million national and international religious trips taking place throughout the world (Griffin & Raj 2017). Moreover, religion is increasingly seen as an important driver of tourism (deAscaniis & Cantoni 2016). For instance, countries that are home to religious buildings or sites have the potential to become popular pilgrimage tourism destinations (Bideci & Albayrak 2016).

Religious tourism benefits several stakeholders and can have significant impacts. As an example, Egresi, Kara and Bayram (2014) concluded that the effect of religious tourists in the province of Mardin, Turkey, was clearly positive. There were higher occupancy rates in the hotels, new hotels had been built, visitors ate at local restaurants, and new jobs had been created. Durán-Sánchez, Álvarez-García, del Río-Rama and Oliveira (2018) also argue that religious tourism benefits all its stakeholders, including shrines and holy sites through increased donations and the hospitality sector through increased income. The *hajj* pilgrimage to Mecca illustrates the positive economic benefits of pilgrimage. It is estimated to bring in 16.5 billion USD annually, around 3% of Saudi Arabia's gross domestic product (D'Ignoti 2016). Given its economic importance, it is not surprising that religious tourism has captured the attention not only of academics but also practitioners, such as destination marketing organizations, planners, and even moviemakers (Choe & O'Regan 2015).

Defining religious tourism has been a major challenge over the years, and there is no consensus on a standard definition. One of the earliest definitions was proposed by Rinschede (1992), who argues that it is a form of tourism whose participants are motivated partially or exclusively by religious reasons. In contrast, Geybels (2014) suggests that religious tourism includes people on holiday choosing to visit religious sites or festivals without religious or existential engagement. Religious tourism can include excursions to pilgrimage shrines, religious conferences, religious festivities, sacred heritage sites, and walking religious routes, among other activities (Barroco, Antunes & Dias 2017). Studies have clearly evidenced that visits to religious sites and pilgrimages may not all be motivated by faith. For instance, grand cathedrals attract visitors with cultural motivations (Bond et al. 2015), whereas pilgrims on the Way of Saint James (*Camino de Santiago*) are mostly motivated by spiritual reasons (Amaro, Antunes & Henriques 2018). Other studies have shown that the Camino is also becoming an increasingly popular heritage route for non-devotees (Ron & Timothy 2019; Timothy & Olsen 2018). Several other studies have shown that there are many other reasons to visit sacred sites besides religion (Bideci & Albayrak 2016; Cerutti & Piva 2016; Drule, Chiş, Băcilă & Ciornea 2012).

Deconstructing whether religious tourism is conducted for purely religious motives or not is not the aim of this chapter. Instead, for the purposes of this chapter, religious tourism is a type of tourism that includes a religious site as a destination or a pilgrimage to a religious site or religious event, regardless of the traveler's motivation.

Religious tourism and ICTs

ICTs are being increasingly used for religious purposes (Ramos et al. 2016) and, as a consequence, are changing the face of religious tourism (Rashid 2018; Tripko & Dragan 2011). The subsequent four sections provide an overview of the various media through which ICTs are being used for religious tourism purposes: webpages, social media, augmented and virtual reality, and mobile devices and applications.

Webpages

Religious tourists, like other tourists, frequently use the Internet to plan their trips. According to Akbulut and Ekin (2017), webpages may be the first source of information about a sacred site or destination and have an inevitable impact during the destination selection process. Indeed, the Internet has become a popular tool for gathering information about sacred sites, as well as providing images and other required details related to travel planning (Barroco, Antunes & Amaro 2018).

Gupta, Editor and Gulla (2010) conducted a study targeting people who had already visited the Vaishno Devi Shrine in India, to examine visitors' use of the shrine's official website. Most of the respondents (65%) were aware of, or had used, the shrine's webpage. Sánchez-Amboage, Ludeña-Reyes and Viñán-Merecí (2017) also found that the Internet (Google and Facebook) was the online information source most used on the pilgrimage to the Virgin of El Cisne, one of the largest religious events in Ecuador.

Websites are not only used to promote religious sites but also religious events. The Catholic World Youth Day is an example of such a worldwide religious event that has a website to promote the encounter with the Pope every three years in a different country. Narbona and Arasa (2019) conducted a study on users' visits to the website during the 31st World Youth Day in Krakow. Their findings suggest that official websites are the most appropriate channels to communicate practical information about such mass faith events, especially since the main target is young people who commonly use new technologies in their daily lives.

Despite the importance of webpages for religious tourists, many religious destinations lack an Internet presence. Akbulut and Ekin (2017) found that only 64% of the sacred sites in Europe had a website, and they suggest that these sacred localities should improve search results by highlighting more relevant search engine keywords. In a study conducted about the presence of Portuguese Marian Shrines on the Internet and the quality of their websites, Barroco, Antunes and Amaro (2018) found that only 8 of the 39 sanctuaries had their own webpage. Rashid (2012) also found that the official site for the Blue Mosque in Istanbul was disappointing, since it was poorly designed, not useful and did not consider tourists' needs, arguing that the webpage needs to be customer-focused to enhance the visitor experience.

In addition to providing information, websites can also be used to create virtual pilgrimages, a new way of doing pilgrimages that allows an individual to use technology and the Internet to create a "mythscape", to experience the divine, using interactive tools to entertain the traveler in a unique form and finally to provide people the opportunity to do this type of activity in the comfort of their own home (MacWilliams 2002). Hill-Smith (2009), who uses the term "cyberpilgrimage", highlights other benefits: 24-hour availability of pilgrimage sites; everyone is welcomed, regardless of faith; the visit is free of charge; shrines and artifacts can be seen without physical restrictions; and finally, safety is ensured since many holy locations are targets for man-made and natural disasters.

Social media

Social media has provided tourism companies with unprecedented opportunities to understand tourists better (Leung, Law, Van Hoof & Buhalis 2013), engage with them and promote their businesses. Religious tourism can also gain from the use of social media as it is an efficient support in promoting religious destinations and religious programs (Yesil 2013). Prats, Aulet and Vidal (2015) note that many religious sites have a Facebook page, both with religious and touristic content. Religious sites use other social media as well. For instance, the Vatican News, besides having a Facebook page, also utilizes Twitter, Instagram and YouTube. The YouTube channel has a webcam that allows users to view the Vatican 24 hours a day. These social media platforms are useful to provide tourists with relevant information and to engage with them, while at the same time promoting the religious sites. The Abbey of Montserrat, in Catalonia, Spain, is a notable example of how social media can be used to promote religious tourism, by promoting a contest, #SelfieMontserrat, on Instagram, Facebook and Twitter.

Sánchez-Amboage, Ludeña-Reyes and Viñán-Merecí (2017) carried out a survey among visitors at the Virgin of el Cisne shrine in Ecuador and found that social media played an important role in promoting the pilgrimage and the sanctuary. The study found that the pilgrimage's Facebook pages have high levels of engagement, with Santuario de la Virgen de El Cisne having the highest (54.4%). This is a high level of engagement, considering that Facebook social engagement for national tourism organizations does not usually reach even 1% (Mariani, Mura & Di Felice 2018).

Tourism-specific social media websites can also be used to study religious tourists' needs and satisfaction. For instance, Cerutti and Piva (2016) assessed how the Italian religious destinations of Sacred Mount of Oropa and Sacred Mount of Orta were perceived by tourists by analyzing TripAdvisor reviews. They assert that understanding tourists' feedback on destinations through these types of platforms is a useful tool for local management in promoting their destinations better and producing more satisfied guests. Reviews of religious sites on TripAdvisor were also analyzed in a case study of the Lalibela Rock-Hewn Churches in Ethiopia (Ndivo & Cantoni 2016). Many reviews indicated a lack of tourist satisfaction and provided usable information about why this was the case. These examples show that religious site reviews online can contribute to a better understanding of visitors' real needs and, consequently, local management can better plan strategies to improve the quality of future visitor experiences.

Many studies have found that recommendations and reviews on social media have a salient impact on tourists' travel planning and decision making (Fotis, Buhalis & Rossides

2011; Hudson & Thal 2013). In the religious tourism context, social media is also an important source of information for travelers and provides an electronic tool to disseminate word of mouth (e-WOM) about sacred places and events. Religious tourists are similar to other tourists regarding their engagement in e-WOM during the information search phases of trip planning and during the journey itself (Khan & Khan 2015). Moreover, religious tourists tend to depend a great deal on the recommendations of those who have gone before them. E-WOM is therefore an important planning tool, in particular for those visiting a religious site for the first time (Iriobe & Abiola-Oke 2019).

Augmented and virtual reality

The use of augmented reality (AR) and virtual reality (VR) technology is another widespread trend within the tourism sector, with one of the main aims being to improve the tourist experience (Kounavis, Kasimati & Zamani 2012). Both technologies integrate virtual and real-world elements (Marr 2019). Several studies have provided evidence and discussed how VR and AR can enhance the travel experience (Guttentag 2010; Han, tom Dieck & Jung 2018; Kounavis et al. 2012; Neuburger & Egger 2017; tom Dieck, Jung & Rauschnabel 2018). Ramos et al. (2016) argue that these technologies can also enhance tourists' personal and spiritual experiences in a religious context, such as pilgrimages. Other studies have shown that all visitors to religious heritage sites can benefit from the use of AR and VR technologies, especially as they help reconstruct the ruins of churches and monasteries virtually to enable visitors to experience how they might have looked and felt centuries ago (Fusté-Forné 2020; Rueda-Esteban 2019).

With VR, users put on a head-mounted display or a VR headset and are immersed in a computer-generated "reality", sensing that they are moving among virtual objects on a screen (Marr 2019). An example of VR for religious tourism is the Experience Mecca Virtual Reality. This virtual tour takes people around the major landmarks of Mecca's holy places, offering a gravitating and immersive look at the *hajj* rituals. It also allows users to interact with other worshippers, or "go back in time" to see how Muhammad and other prophets lived (Budgor 2014).

AR is understood as the virtual world coexisting within the real (physical) world (Craig 2013). It can be very useful since it provides access to location-based information, relevant to tourists' immediate surroundings, content is timely and updated, and it provides interactive annotations that are integrated with map-based services and additional information (Yovcheva, Buhalis & Gatzidis 2012). Another benefit of the use of AR applications is that they can contribute to preserve heritage and religious sites (Fusté-Forné 2020). On the one hand, they provide an alternative to access localities that are being threatened by overtourism, reducing the number of visitors (Ramos et al. 2016). On the other hand, since many sacred sites restrict the use of information signs, AR can provide relevant information to tourists without negatively impacting the environment (Jung & Han 2014).

Some examples of AR for religious tourism include experiencing medieval churches as they existed hundreds of years ago (Fusté-Forné 2020; Rueda-Esteban 2019) or seeing what the Sagrada Família in Barcelona will look like when construction is eventually completed. Zaibon, Pendit and Abu Bakar (2015) developed an AR application to help visitors experience their visit to Melaka, Malaysia, a historical city that is home to many mosques, temples, churches and other cultural heritage sites that together form a UNESCO World Heritage Site. Zaibon et al.'s (2015) study found that users preferred AR over traditional media and that they would like to use it again in the future.

Mobile devices and applications

The use of smartphones and mobile applications (apps) is another major trend in the use of ICTs in religious tourism. For instance, regarding pilgrimages, Nickerson, Austreich and Eng (2014) found that 77% of the respondents who had walked the Way of Saint James (*Camino de Santiago*), had taken mobile devices, and 38% had used a Camino-specific app. Pilgrims are interested in using apps during the pilgrimage, as they are useful guides and allow them to save time (Antunes & Amaro 2016). Research has shown that pilgrimage participants will be more willing to use an app for a pilgrimage if it has general information about the route and less information about the religious features (Amaro, Duarte & Antunes 2019).

There seem to be contrasting opinions regarding the use of technologies by pilgrims. Nickerson et al. (2014) discovered that views about the use of technology on the *Camino de Santiago* can be quite strong, both positively and negatively. Qurashi and Sharpley (2018) also found that new technologies have both positive and undesirable implications for pilgrims' spiritual experiences. Considering some of the least positive aspects, pilgrims reported that new technologies negatively affected their spiritual experience because they were a distraction. Feeling the spirit demands concentration on worship and the non-material elements of pilgrimage rituals (Qurashi & Sharpley 2017). One respondent even suggested that using new technologies on the hajj was "nothing but a sin" (Qurashi & Sharpley 2018, p. 43), and it is likely that many pilgrims feel the same. When pilgrims who were using Camino apps were asked if they were better than paper-based guides, only 12% agreed or strongly agreed (Nickerson et al. 2014). Many pilgrims feel that mobile technology can distract them from the Camino experience because of the constant need to charge their devices (Nickerson & Eng 2017).

However, on a more positive note, some people believe that ICTs can enhance the religious tourism experience. For example, 57% of the respondents who had carried mobile devices on the Camino de Santiago believed that the devices had enhanced their Camino experience (Nickerson et al. 2014). Alshattnawi (2012) highlights the benefits of new technologies in pilgrimages, by proposing the use of Quick Response (QR) codes for pilgrims on the journey to Mecca. A QR code is a two-dimensional form of a barcode that can be read with a smartphone or other device, connecting directly to websites, phone numbers and much more (QRme 2019). Traditionally, pilgrims on the way to Mecca resort to reading brochures that explain what they should do to follow the steps according to the prescribed rituals and also the locations they may visit during their stay in Mecca (Alshattnawi 2012). With QR codes posted at different locations along the pilgrimage route, pilgrims can access information with their mobile devices at specific locations, helping them make their journey more efficient (Alshattnawi 2012).

In a different religious tourism context, Narbona and Arasa (2016) reflect on the importance of mobile apps and instant messaging during large religious events, using the beatification of the Catholic bishop Álvaro del Portillo as a case study. This ceremony was held in Madrid in September 2014, with more than 250,000 people from over 70 countries (Narbona & Arasa 2016). The Álvaro del Portillo app was created for this event to provide useful information and support for attendees, such as news, spiritual texts and logistical information (e.g. how to travel from the airport to the event, schedules and maps) (Narbona & Arasa 2016). The app also included a map that allowed pilgrims to visit places linked to the life of the bishop, bringing pilgrims to other places in the city, such as Museo del Prado, the Paseo de la Castella and the Plaza Mayor. The organization saved money that would have been spent on booklets for the Mass, since the app contained a ceremony book, with the texts needed to follow the religious ceremony. Álvaro del Portillo was also available on WhatsApp. Messages were sent to those who joined, promoting the website and Twitter, with invitations to download the ceremony books, messages with useful contacts and texts requesting donations. The messages sent via WhatsApp, helped increase visits to the site. This example illustrates how an app created for a religious event can serve as an excellent form of communication, collect donations, save money, promote touristic sites and maintain contact with participants.

Apps can also be important in making religion accessible to all. An example to demonstrate this is the MyEyes app of the Shrine of Fatima in Portugal (one of the most important Catholic shrines in the world dedicated to the Virgin Mary). This technology allows blind pilgrims to experience the shrine with greater autonomy. Tourists with visual impairments can receive relevant information about the shrine: a description of the location, practical directions to help them get around and the location of useful facilities (e.g. information desk and public restrooms) and the Pilgrim Itinerary. Another example is a mobile application launched in June 2019 in Jerusalem by Israel's Center for the Blind (Lesley 2019). This new app makes the city's holy sites accessible to the visually impaired by catering to their specific needs. It identifies people's location within the city and connects them to a database that describes the place where they currently are.

Apps are also emerging in religious tourism to attract and entertain different types of travelers, such as children, who usually travel with their families to holy sites but are not normally the main audiences for this type of tourism. These types of apps can educate, amuse and even captivate younger tourists to want to visit the religious destination. The Shrine of Fatima created two such apps for children. One of these, *Fatima, a Story Filled with Light*, is an interactive book designed for children between the ages of 3 and 8. This app encourages children to learn about the story of Fatima by interacting with the main characters using text, auditory prompts and purposeful touch. The other app, *The Little Shepherds Game*, is a game for children over the age of 4.

Tracking technologies

To better plan and manage tourist areas, it is necessary to comprehend how tourists move through space and time (Lew & McKercher 2006; Shoval & Ahas 2016). With the use of tracking technologies, such as Global Positioning Systems (GPS), Bluetooth technology, Radio-Frequency Identification (RFID), smartphones and other mobile devices with embedded sensors, researchers are better able to measure tourists' movements and explore new paths that will allow tourism research to develop further.

Shoval and Isaacson (2007) conducted a study using digitally based methods to see if these were feasible for collecting data about the spatial behaviors of tourists. They performed three different experiments, which included three types of tracking technologies: GPS, land-based antennas, and a hybrid solution. Two of the three experiments were linked to religious destinations. One involved the Old City of Jerusalem and the other the biblical historic city of Nazareth. They concluded that tracking devices are effective tools for analyzing tourists' movements in space and time.

In a longer study taking place over four years, involving 1,030 tourists, Cohen-Hattab and Shoval (2015, p. 12) collected data using "highly accurate GPS devices" to study time-space patterns in the Old City of Jerusalem. This sample contained a large number of Jewish groups (900) and 40 Christian groups. This digitally based method allowed the researchers to conclude that the tourist areas visited within the Old City are highly segmented.

To understand crowd dynamics during the annual hajj in Mecca, Jamil, Basalamah, Lbath and Youssef (2015) used tiny, wearable Bluetooth Low Energy (BLE) tags and some smartphones containing GPS to collect mobility data among pilgrim groups containing some 732 volunteer pilgrims from three different countries. The results allowed them to analyze different behaviors in group dynamics when related to entry/exit times, length of stay at the event, group cohesion and hotspots for crowd gathering.

Not only are these tracking technologies important for examining religious tourists' behavior, but they also benefit religious tourists. Indeed, Saudi Arabia gave out GPS-enabled electronic bracelets to Mecca pilgrims in 2016, following a major trampling tragedy in which more than 750 people died and 900 were injured. The bracelets included personal and medical information to help authorities in case of an emergency, as well as other information for worshippers, such as prayer times and a multi-lingual helpdesk to guide non-Arabic speakers (BBC News 2016). The pilgrims pointed out that the bracelet allowed them to access important information and provided safety and security because it helped locate lost devotees and delivered help to those in need (Qurashi & Sharpley 2018).

Conclusion

The symbiosis of ICTs and the tourism industry is undeniable, and religious tourism as a niche form of tourism is no exception. Tourists today seek to learn more and discover the world using ICTs, which has led to an increase in tourist activity at religious sites and a challenge for those who manage these sites (Hughes et al. 2013). This chapter highlighted several uses of ICTs in religious tourism contexts and how religious sites, events and pilgrimages can be more effectively managed. Despite some people not approving of the use of new technologies in religious tourism, it is undeniably a powerful tool with enormous potential. There is still a lot of research to be done in this area given that few studies examine the use of technologies in religious tourism contexts, a similar conclusion that Amaro et al. (2019) noted. Nevertheless, this chapter has shown several developments in the use of ICTs in religious tourism and suggests six major areas where ICTs can intervene in religious tourism:

1 Enhance tourists' experience
2 Promote religious tourism
3 Facilitate accessible tourism
4 Ensure sustainable tourism
5 Provide safety
6 Advance religious tourism research

Regarding the first area, many studies have shown that ICTs can enhance tourists' experience in several ways. For instance, they can provide information before and during a religious journey. This is a benefit that visitors to religious sites and events value, increasing their satisfaction and perhaps their intent to return. ICTs can also enhance the religious tourist's experience with the use of AR, VR, QR codes and apps, allowing them, for instance, to interact with other worshippers, access information in their own language, access relevant information or even to "go back in time". Moreover, analyzing online reviews of religious sites and pilgrimages can help managers better understand religious tourists and, thus, offer services adapted to their needs, thereby enhancing their experience.

The second major area of ICT intervention is to promote religious tourism. New technologies, including websites, social media and other digital sources, are the least expensive marketing media available and are powerful tools in planning and promoting religious events as they aid in heightening pilgrims' experiences, leading to a higher likelihood of success of the event (de la Cierva, Black & O'Reilly 2016). ICTs can help attract more visitors and different types of religious tourists, such as families with small children, non-religious tourists or devoted pilgrims from different religions with secular motivations. Furthermore, ICTs can even contribute to obtaining donations that are much needed in several cases to support religious site maintenance.

The third major area of ICT intervention is to facilitate accessible tourism. People with disabilities are restricted in the number of places they can visit because of their limitations or because of the lack of accessible infrastructure. People with special access needs visit religious sites and feel more attracted to those where they feel independent and dignified than others (Gassiot et al. 2015). This chapter has shown how ICTs can be a powerful instrument for tourists with special needs by providing, for example, apps for blind people or by allowing people to virtually visit a religious site or undertake a virtual pilgrimage.

ICTs can contribute to a more sustainable tourism, the fourth major area identified, in at least two ways. First, since many religious sites restrict the use of information signs, ICTs can provide information and experiences without creating signage pollution. Likewise, technology also helps destinations and attractions avoid printing brochures and pilgrimage maps, which reduces reliance on paper materials. Second, with virtual pilgrimages, virtual tours, 360° photographs, and videos online, religious tourists can now visit a sacred site without having to physically travel to the location, reducing the impacts of mass footfall, a major problem for some religious sites that are currently threatened by overtourism.

ICTs also help address safety concerns, the fifth major area, related to holy locations that are many times targets for man-made and natural disasters. "Cyberpilgrimage" is an example of how religious tourists may be safer not physically being at a holy location that is considered dangerous. The use of electronic bracelets by pilgrims travelling to Mecca with personal and medical information can help authorities in case of an emergency and in locating lost pilgrims. Yamin (2019) provides many suggestions on how technology can be used to avoid catastrophes at crowded religious events.

Finally, ICTs can contribute to advancing religious tourism research. Indeed, tools such as tracking devices can help researchers and managers gain more comprehensive knowledge of religious tourists' behaviors regarding, for example, entry/exit times, length of stay and hotspots for crowd gathering. From a practical point of view, this can help religious site managers better design religious attractions, improve infrastructure and other services unique to this type of tourism in order to better accommodate all types of religious tourists (Shoval & Ahas 2016). Moreover, tracking technology can be useful in forecasting positive and negative events and prevent overcrowding (De Cantis, Ferrante & Shoval 2018). The use of such technology, however, raises ethical concerns for such research efforts that might infringe on the privacy of participants (Raun, Ahas & Tiru 2016; Shoval & Ahas 2016).

Religious heritage sites are used by governments and destinations to attract tourists (Olsen 2006). To manage these sites more effectively, it is important to identify aspects that will contribute to a better tourism experience and find ways to appeal to a more diverse and demanding range of visitors (Hughes et al. 2013). In this context, religious site managers should consider the benefits of ICTs and develop strategies to incorporate them in religious tourists' touchpoints to increase innovation and differentiation from other competitors. These strategies will not only benefit religious sites, but all stakeholders, in particular tourists and local businesses.

References

Akbulut, O., & Ekin, Y. (2017) 'How visible are sacred sites online? Availability of European sacred site websites', *International Journal of Religious Tourism and Pilgrimage*, *5*(1), 105–124.

Alshattnawi, S. (2012) *Effective Use of QR Codes in Religious Tourism.* Paper presented at the International Conference on Advanced Computer Science Applications and Technologies (ACSAT), Kuala Lumpur, Malaysia.

Amaro, S., Antunes, A., & Henriques, C. (2018) 'A closer look at Santiago de Compostela's pilgrims through the lens of motivations', *Tourism Management*, *64*, 271–280.

Amaro, S., Duarte, P. A. O., & Antunes, A. (2019) 'Determinants of intentions to use a pilgrimage app: A cross-cultural comparison', *International Journal of Religious Tourism and Pilgrimage*, 7(2), 19–30.

Antunes, A., & Amaro, S. (2016) 'Pilgrims' acceptance of a mobile app for the Camino de Santiago', In A. Inversini & R. Schegg (Eds.) *Information and Communication Technologies in Tourism 2016* (pp. 509–521). Cham: Springer.

Armfield, G. G., & Holbert, R. L. (2003) 'The relationship between religiosity and internet use', *Journal of Media and Religion*, *2*(3), 129–144.

Barroco, C., Antunes, J., & Amaro, S. (2018) *The Internet as an Important Tool in Developing the Marian Shrine Route in Portugal.* Paper presented at the 10th Annual International Religious Tourism and Pilgrimage Conference, Santiago de Compostela, Spain.

Barroco, C., Antunes, J., & Dias, H. (2017) *Motivations for Visiting Sacred Sites: The Case of Senhora da Lapa Schrine.* Paper presented at the 9th Annual International Religious Tourism and Pilgrimage Conference, Armeno, Italy.

BBC News. (2016) 'Hajj pilgrims to be given e-bracelets', Retrieved from https://www.bbc.com/news/technology-36675180

Bethapudi, A. (2013) 'The role of ICT in tourism industry', *Journal of Applied Economics and Business*, *1*(4), 67–79.

Bideci, M., & Albayrak, T. (2016) 'Motivations of the Russian and German tourists visiting pilgrimage site of Saint Nicholas Church', *Tourism Management Perspectives*, *18*, 10–13.

Blackwell, R. (2007) 'Motivations for religious tourism, pilgrimage, festivals and events', In R. Raj & D. Morpeth (Eds.) *Religious Tourism and Pilgrimage Festivals Management: An International Perspective* (pp. 35–47). Wallingford: CABI.

Bond, N., Packer, J., & Ballantyne, R. (2015) 'Exploring visitor experiences, activities and benefits at three religious tourism sites', *International Journal of Tourism Research*, *17*(5), 471–481.

Budgor, A. (2014) 'Mecca 3D allows Muslims to make a pilgrimage using VR', Retrieved from https://killscreen.com/articles/mecca-3d-aims-bring-us-all-closer-god/

Buhalis, D., & Law, R. (2008) 'Progress in information technology and tourism management: 20 years on and 10 years after the Internet—The state of eTourism research', *Tourism Management*, *29*(4), 609–623.

Buhalis, D., & O'Connor, P. (2005) 'Information communication technology revolutionizing tourism', *Tourism Recreation Research*, *30*(3), 7–16.

Cerutti, S., & Piva, E. (2016) 'The role of tourists' feedback in the enhancement of religious tourism destinations', *International Journal of Religious Tourism and Pilgrimage*, *4*(3), 6–16.

Choe, J., & O'Regan, M. (2015) 'Religious tourism experiences in South East Asia', In R. Raj & K. Griffin (Eds.) *Religious Tourism and Pilgrimage Management: An International Perspective* (2nd ed., pp. 191–204). Wallingford: CABI.

Cohen-Hattab, K., & Shoval, N. (2015) *Tourism, Religion, and Pilgrimage in Jerusalem.* London: Routledge.

Craig, A. B. (2013) *Understanding Augmented Reality: Concepts and Applications.* Waltham, MA: Elsevier.

De Cantis, S., Ferrante, M., & Shoval, N. (2018) 'Walking tourism in urban destinations: Measurement opportunities, challenges and case studies', Paper presented at the international conference, Challenges for European City Tourism, May 1–5, Jerusalem.

de Jong, A., & Varley, P. (2018) 'Foraging tourism: Critical moments in sustainable consumption', *Journal of Sustainable Tourism*, *26*(4), 685–701.

de la Cierva, Y., Black, J. L., & O'Reilly, C. (2016) 'Managing communications for large Church events: Best practices for Krakow 2016 and Dublin 2018', *Church, Communication and Culture*, *1*(1), 110–134.

deAscaniis, S., & Cantoni, L. (2016) 'Pilgrims in the Digital Age: a research manifesto', *International Journal of Religious Tourism and Pilgrimage*, *4*(3), 1–5.

D'Ignoti, S. (2016) 'Religious tourism is on the rise, research says', Online at: https://www.lastampa.it/vatican-insider/en/2016/07/12/news/religious-tourism-is-on-the-rise-research-says-1.34834112

Drule, A. M., Chiş, A., Băcilă, M. F., & Ciornea, R. (2012) 'A new perspective of non-religious motivations of visitors to sacred sites: Evidence from Romania', *Procedia - Social and Behavioral Sciences*, *62*, 431–435.

Durán-Sánchez, A., Álvarez-García, J., del Río-Rama, M., & Oliveira, C. (2018) 'Religious tourism and pilgrimage: Bibliometric overview', *Religions*, *9*, 1–15.

Egresi, I., Kara, F., & Bayram, B. (2014) 'Economic impact of religious tourism in Mardin, Turkey', *Journal of Economics and Business Research*, *18*(2), 7–22.

Fotis, J., Buhalis, D., & Rossides, N. (2011) 'Social media impact on holiday travel planning: The case of the Russian and the FSU markets', *International Journal of Online Marketing*, *1*(4), 1–19.

Fusté-Forné, F. (2020) 'Mapping heritage digitally for tourism: An example of Vall de Boí, Catalonia, Spain', *Journal of Heritage Tourism*, 15(5), 580–590.

Gassiot, A., Prats, L., & Coromina, L. (2015) 'Analysing accessible tourism in religious destinations: The case of Lourdes, France', *International Journal of Religious Tourism and Pilgrimage*, *3*(2), 47–56.

Geybels, H. (2014) 'Religious common culture and religion tourism', *Yearbook for Ritual and Liturgical Studies/Jaarboek voor Liturgie-onderzoek*, *30*, 39–50.

Griffin, K., & Raj, R. (2017) 'The importance of religious tourism and pilgrimage: Reflecting on definitions, motives and data', *International Journal of Religious Tourism and Pilgrimage*, *5*(3), ii–ix.

Gupta, K., Editor, T., & Gulla, A. (2010)' Internet deployment in the spiritual tourism industry: The case of Vaishno Devi Shrine', *Worldwide Hospitality and Tourism Themes*, *2*(5), 507–519.

Gurira, N. A., & Ngulube, P. (2016) 'Using contingency valuation approaches to assess sustainable cultural heritage tourism use and conservation of the outstanding universal values (OUV) at Great Zimbabwe World Heritage Site in Zimbabwe', *Procedia - Social and Behavioral Sciences*, *225*, 291–302.

Guttentag, D. A. (2010) 'Virtual reality: Applications and implications for tourism', *Tourism Management*, *31*(5), 637–651.

Han, D.-I., tom Dieck, M. C., & Jung, T. (2018) 'User experience model for augmented reality applications in urban heritage tourism', *Journal of Heritage Tourism*, 13(1), 46–61.

Hausmann, A., & Weuster, L. (2018) 'Possible marketing tools for heritage tourism: The potential of implementing information and communication technology', *Journal of Heritage Tourism*, 13(3), 273–284.

Hill-Smith, C. (2009) 'Cyberpilgrimage: A study of authenticity, presence and meaning in online pilgrimage experiences', *The Journal of Religion and Popular Culture*, *21*(2), 6.

Hudson, S., & Thal, K. (2013) 'The impact of social media on the consumer decision process: Implications for tourism marketing', *Journal of Travel & Tourism Marketing*, *30*(1–2), 156–160.

Hughes, K., Bond, N., & Ballantyne, R. (2013) 'Designing and managing interpretive experiences at religious sites: Visitors' perceptions of Canterbury Cathedral', *Tourism Management*, *36*, 210–220.

Iriobe, O., & Abiola-Oke, E. (2019) 'Moderating effect of the use of eWOM on subjective norms, behavioural control and religious tourist revisit intention', *International Journal of Religious Tourism and Pilgrimage*, 7(3), 38–47.

Jamil, S., Basalamah, A., Lbath, A., & Youssef, M. (2015) 'Hybrid participatory sensing for analyzing group dynamics in the largest annual religious gathering', Paper presented at the UbiComp '15: Proceedings of the 2015 ACM International Joint Conference on Pervasive and Ubiquitous Computing.

Jung, T., & Han, D. (2014) 'Augmented reality (AR) in urban heritage tourism', *e-Review of Tourism Research*, *5*, 1–5.

Khan, G., & Khan, F. (2015) 'Motivations to engage in eWom among Muslim tourists: A study of inbound Muslim tourists to Malaysia', *International Journal of Islamic Marketing and Branding*, *1*(1), 69–80.

Kluver, R., & Cheong, P. H. (2007) 'Technological modernization, the internet, and religion in Singapore', *Journal of Computer-Mediated Communication*, *12*(3), 1122–1142.

Kounavis, C. D., Kasimati, A. E., & Zamani, E. D. (2012) 'Enhancing the tourism experience through mobile augmented reality: Challenges and prospects', *International Journal of Engineering Business Management*, *4*, 1–6.

Lapin, T. (2019) 'Pope Francis unveils new "Click to Pray" app', *New York Post*, Online at: https://nypost.com/2019/01/21/pope-francis-unveils-new-click-to-pray-app/

Law, R., Buhalis, D., & Cobanoglu, C. (2014) 'Progress on information and communication technologies in hospitality and tourism", *International Journal of Contemporary Hospitality Management, 26*(5), 727–750.

Lesley, P. B. (2019) 'Jerusalem's holy sites app for the blind', *World Religion News*. Online at: https://www.worldreligionnews.com/?p=62122

Leung, D., Law, R., Van Hoof, H., & Buhalis, D. (2013) 'Social media in tourism and hospitality: A literature review', *Journal of Travel & Tourism Marketing, 30*(1–2), 3–22.

Lew, A., & McKercher, B. (2006) 'Modeling tourist movements: A local destination analysis', *Annals of Tourism Research*, 33(2), 403–423.

MacWilliams, M. W. (2002) 'Virtual pilgrimages on the internet', *Religion, 32*(4), 315–335.

Mariani, M. M., Mura, M., & Di Felice, M. (2018) 'The determinants of Facebook social engagement for national tourism organizations' Facebook pages: A quantitative approach', *Journal of Destination Marketing & Management, 8*, 312–325.

Marr, B. (2019) 'The important difference between virtual reality, augmented reality and mixed reality', *Forbes*. Online at: https://www.forbes.com/sites/bernardmarr/2019/07/19/the-important-difference-between-virtual-reality-augmented-reality-and-mixed-reality/#326d95be35d3

Narbona, J., & Arasa, D. (2016) 'The role and usage of apps and instant messaging in religious mass events', *International Journal of Religious Tourism and Pilgrimage, 4*(3), 29–42.

Narbona, J., & Arasa, D. (2019) 'Mass religious events as opportunities for tourism promotion: An analysis of users' visits to the website of World Youth Day 2016 in Krakow', *Church, Communication and Culture, 3*(3), 379–388.

Ndivo, R., & Cantoni, L. (2016) 'The efficacy of heritage interpretation at the Lalibela rock-hewn churches in Ethiopia: Exploring the need for integrating ICT-mediation', *International Journal of Religious Tourism and Pilgrimage, 4*(3), 17–28.

Neuburger, L., & Egger, R. (2017) 'Augmented reality: Providing a different dimension for museum visitors', In R. Schegg & B. Stangl (Eds.) *Information and Communication Technologies in Tourism 2017* (pp. 65–77). Cham: Springer.

Neuhofer, B., Buhalis, D., & Ladkin, A. (2015) 'Technology as a catalyst of change: Enablers and barriers of the tourist experience and their consequences', In I. Tussyadiah & A. Inversini (Eds.) *Information and Communication Technologies in Tourism 2015* (pp. 789–802). Cham: Springer.

Nickerson, R. C., Austreich, M., & Eng, J. (2014) 'Mobile technology and smartphone innovation analysis', Paper presented at the Twentieth Americas Conference on Information Systems, Savannah, Georgia, USA.

Nickerson, R. C., & Eng, J. (2017) 'Use of mobile technology and smartphone apps on the Camino De Santiago: A comparison of American and European pilgrims', Paper presented at the Conference of the Portuguese Association for Information Systems.

Olsen, D. H. (2006) 'Management issues for religious heritage attractions', In D. J. Timothy & D. H. Olsen (Eds.) *Tourism, Religion and Spiritual Journeys* (pp. 104–118). London: Routledge.

Prats, L., Aulet, S., & Vidal, D. (2015) 'Social network tools as guides to religious sites', In R. Raj & K. Griffin (Eds.) *Religious Tourism and Pilgrimage Management: An International Perspective* (2nd ed., pp. 146–159). Wallingford: CABI.

QRme. (2019) 'What is the QR Code?' Online at: https://www.qrme.co.uk/#qr-codes

Qurashi, J., & Sharpley, R. A. (2017) 'SMART media technologies impact on the spiritual experience of hajj pilgrims', Paper presented at the 9th Annual International Religious Tourism and Pilgrimage Conference.

Qurashi, J., & Sharpley, R. A. (2018) 'The impact of SMART media technologies (SMT) on the spiritual experience of hajj pilgrims', *International Journal of Religious Tourism and Pilgrimage, 6*(4), 37–48.

Ramos, C. M., Henriques, C., & Lanquar, R. (2016) 'Augmented reality for smart tourism in religious heritage itineraries: Tourism experiences in the technological age', In J. Rodrigues, P. Cardoso, J. Monteiro & M. Figueiredo (Eds.) *Handbook of Research on Human-Computer Interfaces, Developments, and Applications* (pp. 245–272). Hershey, PA: IGI Global.

Rashid, A. G. (2018) 'Religious tourism – a review of the literature', *Journal of Hospitality and Tourism Insights, 1*(2), 150–167.

Rashid, T. (2012) 'Web-based customer-centric strategies: New ways of attracting religious tourist to religious sites', *International Journal of Tourism Policy, 4*(2), 146–156.

Raun, J., Ahas, R., & Tiru, M. (2016) 'Measuring tourism destinations using mobile tracking data', *Tourism Management*, 57, 202–212.

Rinschede, G. (1992) 'Forms of religious tourism', *Annals of Tourism Research, 19*(1), 51–67.

Ron, A. S., & Timothy, D. J. (2019) *Contemporary Christian Travel: Pilgrimage, Practice, and Place.* Bristol: Channel View Publications.

Rueda-Esteban, N. R. (2019) 'Technology as a tool to rebuild heritage sites: The second life of the Abbey of Cluny', *Journal of Heritage Tourism*, 14(2), 101–116.

Sánchez-Amboage, E., Ludeña-Reyes, A.-P., & Viñán-Merecí, C. (2017) 'Impact of religious rourism in aocial media in the Andean eegion of Ecuador: The xase of the pilgrimage of the Virgin of El Cisne and the trade fair of Loja', In F. Campos Freire, X. Rúas-Araújo, Martínez-Fernández, & L.-G. V.-A., X. (Eds.) *Media and Metamedia Management* (pp. 303–308). Cham: Springer.

Shoval, N., & Ahas, R. (2016) 'The use of tracking technologies in tourism research: The first decade', *Tourism Geographies*, 18(5), 587–606.

Shoval, N., & Isaacson, M. (2007) 'Tracking tourists in the digital age', *Annals of Tourism Research*, 34(1), 141–159.

Timothy, D. J., & Olsen, D. H. (2018) 'Religious routes, pilgrim trails: Spiritual pathways as tourism resources', In R. Butler & W. Suntikul (Eds) *Tourism and Religion: Issues and Implications* (pp. 220–235). Bristol: Channel View Publications.

tom Dieck, M. C., Jung, T. H., & Rauschnabel, P. A. (2018) 'Determining visitor engagement through augmented reality at science festivals: An experience economy perspective', *Computers in Human Behavior, 82*, 44–53.

Tripko, D., & Dragan, R. (2011) 'Impact of IT and other technologies to religious tourism', *International Journal of Economics and Law, 1*(2), 31–40.

Ukpabi, D. C., & Karjaluoto, H. (2017) 'Consumers' acceptance of information and communications technology in tourism: A review', *Telematics and Informatics, 34*(5), 618–644.

Yamin, M. (2019) 'Managing crowds with technology: Cases of hajj and Kumbh Mela', *International Journal of Information Technology*, 11(2), 229–237.

Yesil, M. M. (2013) 'The social media factor in the development and promotion of religious tourism', *Journal of Turkish Studies, 8*(7), 733–733.

Yovcheva, Z., Buhalis, D., & Gatzidis, C. (2012) 'Smartphone augmented reality applications for tourism', *E-review of Tourism Research, 10*(2), 63–66.

Zaibon, S. B., Pendit, U. C., & Abu Bakar, J. A. (2015) 'Applicability of mobile augmented reality usage at Melaka cultural heritage sites', Paper presented at the International Conference on Computing and Informatics, ICOCI 2015.

28

INTERPRETING RELIGIOUS AND SPIRITUAL TOURISM DESTINATIONS/SITES

Tomasz Duda

Introduction

The space of religious tourism, including pilgrimage, is created by delicate and extremely complicated relationships between the sacred and the profane heritage of a place. This results in the need to understand the place and experience, and how these affect the experiential recipient in very individual and unique ways. These complex relations are most fully manifested when contact occurs between the characteristics of sacred spaces and the characteristics of the visitors. What influences do such encounters have on the meanings and perceptions of a sacred site? Do the progression of globalization and the universality of narrative techniques have negative impacts on the originality and uniqueness of the *genius loci*? How should sacred places be interpreted to preserve their identity and spiritual significance, while allowing them to engage the recipient in learning about the site's other values? Contemporary interpretation is an extremely important element of the dialogue between visitors and attractions. It must therefore reconcile the growing consumption of heritage and the demand for alternative forms of tourism in a balanced way while maintaining the elemental nature of the sacred site and its spiritual significance.

From the earliest times, pilgrimage—travel undertaken for religious or spiritual reasons—satisfied important social needs and was one of the most salient short- and long-distance movement of people. Many authors even claim that tourism, like earlier forms of human mobility, was born of pilgrimage travel (e.g. Burns & Holden 1995; Różycki 2016; Vukonić 1996). However, while once the dominant motive was strictly religious engagement, multitudinous motivations have emerged since the Middle Ages, including those of an educational or recreational nature (e.g. Jackowski 2003; Olsen & Timothy 2006; Olsen, Trono & Fidgeon 2018; Vukonić 2006). Nowadays, however, a more frequent departure from the classic superficial 'sightseeing' includes engaging in deeper, more authentic experiences in connection with sacred space (Duda & Doburzyński 2019). Through proper destination or site interpretation, people should be able to sense the authenticity and spiritual depth of the sacred places they visit without excessive banalization or touristification. This chapter aims to characterize, classify and evaluate interpretation and narratives in religious tourism, as well as to indicate trends and future development directions.

Interpretation in religious and spiritual tourism

Tourism focuses largely on consuming products, impressions, pleasures, experiences and more (Vukonić 1996). Apart from knowledge, there is a strong need for feelings: deepened, moving and often calming experiences. Humans naturally seek these sensations in spiritual contexts and sacred space. Religious tourism, spiritual tourism and pilgrimage have become partly a response to these deeper senses. However, many people visit holy sites in the broadest sense not only for religious or spiritual reasons but also because these places represent valued cultural heritage and function simultaneously as tourist attractions where people can learn about the history of the place, its architecture and the religious denomination it represents.

As early as the mid-1950s, Tilden (1957) noted that the connection between a site or destination and its visitors is an appropriate narrative and interpretive message. The course of these processes is based on an intuitive understanding of the content being communicated. Tilden even speaks of 'information-based revelation', in which interpretation ought to provoke the recipient to think independently and create a meaningful image. Attempts to interpret sacred sites in this way were already made in ancient times, such as in ancient Egypt, where pilgrims visiting temples were accompanied by priests who explained the meaning of symbolisms and the course of everyday rituals (Lansangan-Cruz 2008; Mikos von Rohrscheidt 2019a) or in ancient Greece, where professional guides (perigetai—ταί περιεχει) and explanatory staff (exegetai—ἐξηγηταί) were present. One of the first tourist guides was created in Greece. It explained, among other things, the meaning of objects of worship, their symbolisms and history. This was called '*Peregesis tes Hellados*'.

The interpretation of sacred objects in those times, although unequivocal and narrowly catered to individual, local sites of worship was an extremely modern concept far ahead of its time. It used original narratives and aimed to help visitors understand messages. The ancient interpretation was normally an oral message and most often an uncritical narrative. Mikos von Rohrscheidt (2019a) defines this type of interpretation as a 'single-dimensional' one, given its limited scope of interaction.

A significant increase in the importance of sacred space, the development of symbolism, as well as a supra-local and supra-regional transmission of religious ideas occurred in the Middle Ages. Regardless of location, culture or religious circle, the influence of these times has strongly molded the religious dimensions of the social and political life of most nations. The development of sites of worship, spiritual centers and pilgrimage resulted in the intense need for interpreters who could guide and explain the meanings of the most important destinations. The narrative increasingly concerned not only individual sites of worship or spiritual objects (i.e. churches, chapels, monasteries, temples, springs and groves), but also linear systems, such as emerging pilgrim routes. Sacred space was interpreted through the use of various narrative techniques such as lectures, sermons, stories related to the figures of saints or sages, processions and biblical staging. At that time, detailed guidebooks for pilgrims were also created, which not only described the route and the religious heritage passed along the way, but also indicated the best accommodations and inns. They also mentioned the times and dates when it is best to set off on a journey of faith (Robinson 1997).

> In Canterbury, tears of regret and groans of gratitude mingle with howls and screams of the sick. Crowds are pushing ahead to check each new miracle after it is announced, and priests must barge themselves through to examine the patient personally... At the

> back of the church there are the best places for a tiny bit of sleep, despite the noise and closeness of the heat from thousands of candles.
>
> *(Sumption 1975, p. 93)*

> Another place where a visitor can easily become a participant in spiritual experiences is the barren desert east of Jerusalem and Bethlehem. Few tours leave such an indelible impression as the one at sunset, leading from the Dead Sea to Jerusalem.
>
> *(Sox 1985, p. 213)*

It is easy to see that in earlier times, the interpretation of holy sites was quite naturally combined with experiences of a mystical nature. Explaining the meaning of sacred space was done in a manner that integrated spiritual space and geographical space, or landscape and surrounding nature. During this period, numerous itineraries were created, which constituted descriptions of the path leading to known and universally recognized sacred sites. In the first half of the twelfth century, the text *Mirabilia Urbis Romae* was written. It was essentially a guide to the Eternal City prepared by a canon of the basilica of St. Peter in the Vatican and was intended for pilgrims coming to Rome. Some descriptions of pan-European pilgrim routes were penned even earlier, including the description of *Via Francigena* by the pilgrim bishop, Sigerica, or the famous *Codex Calixtinus* describing the route to Santiago de Compostela (Ron & Timothy 2019). Large portions of those descriptions were devoted to natural phenomena and the landscape. The interpretation of nature as a context for culture and spiritual experiences often begins with a specific phenomenon (e.g. beauty and misunderstanding of the genesis of natural phenomena, unusual events or the culmination of the relief), which recipients must discover for themselves, hence the key importance of implementing the narrative *in situ* and in circumstances facilitating active contact with the surrounding environment. This type of multidimensional interpretation, which marries the relations between the recipient, the sacral, and geographical space can be described as 'integrated interpretation'.

Modern times are a period of dynamic social, cultural, political and religious changes. In the context of migration and increasing human mobilities, it is also a period of intensive tourism growth. At the end of the eighteenth century and throughout the nineteenth century, various forms of leisure emerged. Along with this, active and leisure tourism appeared, including the trend of comprehensive sightseeing, such as the so-called Grand Tour. Within Christianity, largely because of the Reformation, pilgrimage motivated strictly by religious purposes ceased to be the dominant form of travel. The border between *the sacrum* and *the profanum* became increasingly blurred, and the pilgrimage phenomenon began to be perceived as shared geographical, historical, cultural and sacral spaces (Adler 2002; Ambrósio 2007; Burns & Holden 1995; Cohen 1998; Olsen & Timothy 2006).

Changes in the social structure of traveling translated into changes in perceptions of sacred sites and sacred spaces, as well as the interpretation thereof. The nineteenth and twentieth centuries were also a time for becoming aware of one's own heritage, identifying with it and creating a characteristic narrative around it. Thus, sacred space was increasingly treated as a combination of the *sacrum* and elements of cultural, religious and historical heritage. The visitor not only needed experiences of a spiritual nature, direct contact with holiness but also desired information on the genesis of the site and its past, as well as its relationship with the present. The visitor, regardless of his or her main motivation for reaching a holy site, also wanted to understand the significance of the place for its broader cultural and spiritual importance, as well as the timeless message that accompanied it. Therefore, the interpretation of the sacrosanct was associated with local heritage and other assets and was carried out by independent interpreters. Through this process, many of the pilgrimage centers known to date have evolved into large tourist destinations of regional and global importance. Examples are the shrines of Fatima, Lourdes, Rome and Guadalupe, large pilgrimage centers, including Częstochowa, Santiago de Compostela, Mecca and Varanasi, or smaller centers of religious worship throughout the world. In addition, the popularity of guidebooks contributed to the broadening of narration and encouraged visitors to independently, yet supported by interpretation, focus on the educational and cultural values of religious tourism.

Since the second half of the nineteenth century, interpretation comprehensively covered the interconnectedness between past, present and future. Due to the nature and broad spectrum of impact of this approach, it can be described as 'extended interpretation'. It includes not only educational and narrative on-site activity but also in the space surrounding the sacred site, such as on the pilgrimage route or in places more closely connected to the sacred central point (Kanaan-Amat, Crous-Costa & Aulet 2019).

Mikos von Rohrscheidt's (2019b) publication on the history of heritage interpretation in tourism noted that in the second half of the twentieth century, a departure from the classical educational form was discernable in favor of 'balanced and holistic interpretation'. This reflects a significant expansion of areas subject to interpretation and changes in the form of communication and participant involvement. Alternative narratives and new forms of interpretation emerged, also in religious tourism contexts, including places of worship and spiritual experiences. According to Tilden's (1957) thinking, the main goals of interpretation are to stimulate engagement in the process of cultural self-development, to broaden visitors' horizons and to convey a clear message and historical truths. The interpretation process is additionally subject to constant change (Kanaan-Amat et al. 2019), reflecting a response to the rapidly changing expectations of tourists, including pilgrims. Sites that traditionally emitted a 'spiritual magnetism' and attracted mainly religiously motivated people (Olsen 2019; Preston 1992) have become a locus of intercultural and inter-religious exchange, as well as a place of heritage (Duda 2019). Changes in perceptions of sacred sites, as well as the global trend of individualized tourist experiences, have contributed to significant changes in the interpretation of religious and spiritual tourism destinations. From the end of the twentieth century, 'participatory interpretation' has become popular. This is based largely on the interactive involvement of recipients in the educational and narrative process (Hodges 2020).

Special examples of this type of interpretation include engaged sightseeing, combined with the experience of the hardships of the pilgrimage route, stories about ancient pilgrims and accommodations in former monasteries, such as along the Way of St. James, the Way of St. Olaf, the Route of the Templars and the old routes to Jerusalem in several European countries. Participatory interpretation, at least in its initial form, did not take into account the notion of authenticity, which so many tourists seek. The most important were personal impressions affected by appealing interpretation (Beck & Cable 1998; Uzzell 1998). At sites of a religious or spiritual nature, the *genius loci* also matters. Interpretation of the sacred must be done more carefully than in other situations and with more appropriate media to avoid distracting from pilgrims' personal experience (Kanaan-Amat et al. 2019; Morales Miranda 2001). Duda (2019) similarly concluded many sacred site visitors do not need additional interpretation, for their encounters are extremely personal and individual: 'personalized interpretation'.

Forms and rules of interpretation in religious tourism

The tourist's encounter with the sacred, irrespective of his or her origin, cultural context or spiritual involvement, is incomplete and devoid of authenticity if it does not include an appreciation of the heritage of the region and its people (Mikos von Rohrscheidt 2019b). On the other hand, the vast majority of religious heritage includes non-material spiritual elements, which are difficult to interpret and are characterized by distinct features of individuality. Everyone perceives holiness differently, and therefore, religious motives and manifestations can be comprehended differently (Kaelber 2002, 2006; Olsen et al. 2018; Różycki 2016)

Despite certain conceptual differences, religious tourism is frequently identified with pilgrimage. In this sense, the *sacrum* zone increasingly penetrates the *profanum*. The pilgrim undertaking a trip for strictly religious purposes is no longer only involved in ceremonies, worship services or rituals, but also a keen consumer of the destination locality with a desire to broaden her or his knowledge of faith, religion, culture and history (Duda & Doburzyński 2019; Mikos von Rohrscheidt 2016; Olsen 2011). The study by Duda and Doburzyński (2019) indicates that there are differences in the perceptions of sacral space among people traveling to consecrated destinations for strictly religious purposes and those who, aside from the religious significance, also sightsee the other historical, cultural and social values of the destination. Therefore, the types and forms of interpretation, as well as the narrative media used, are closely related to the needs and expectations of the recipient.

Castells (2001) identifies three basic spheres of influence of interpretation in the broader heritage context:

- The *emotional (or sensual) sphere* refers to the multi-sensory presentation of a site being an attractive space and affecting the recipient's senses, which translates into the perception of the site by the recipient with its aesthetic values. Using appropriate narrative tools and forms, the interpreter helps give the site a specific atmosphere, refers to the character of the site and emphasizes its cultural, aesthetic and spiritual values;
- The *ideological sphere* refers to the nature of the site and the conditions of the functioning thereof in a social, economic, political, cultural or religious context;
- The *instrumental sphere* uses a variety of tools (including schematics, engravings, photographs, models, animations and modern multimedia) and narrative forms (reconstructions, fictionalized narratives, storytelling and games).

In the case of sacral spaces, however, more spheres of influence are identifiable. The list of tools used by the interpreter at this type of site will also be different and slightly more selective. In holy places, there are different restrictions than what other heritage localities might face, such as the inviolability of the *sacrum* zone, which highlights the specificity of interpretation in religious and spiritual tourism. Considering these unique and delicate conditions, sacred space includes the spheres distinguished by Castells (2001) as follows:

- The *symbolic sphere* includes the interpretation of sites particularly sensitive to specific religious or denominational groups, sites full of symbols and meanings, which, on the one hand, require some explanation and proper narrative supporting their comprehension (in particular for visitors belonging to different religious groups, non-believers or people from different cultural and social domains). On the other hand, they constitute a

kind of closed semantic circle reserved for a given religious group and enter into direct relations between the believer and the *sacrum* surrounding him/her;
- The *spiritual* sphere is the individual sphere of the recipient, which translates into difficulties in its identification and proper interpretation; it is worth emphasizing that the spiritual sphere does not refer only to the internal experiences of believers, but reflects the whole spiritual heritage of the site and its universal message identified with the given cultural circle of the site. The spiritual sphere is also an integral part of the site's authenticity the process known as 'searching self-identity through travel', described by Olsen (2008, 2012) and Collins-Kreiner (2010) as the so-called 'third space'.

To comprehend better the ideas that guide the development of sacred space as a tourism asset, one should be aware of the fact that the vast majority of cases are not purposely planned to be tourist attractions. As Millar (1999) indicates, such sites originated as social spaces where individual religious groups expressed their faith, cultivated spirituality and met the needs of communion with the *sacrum*. Over time, these relationships have created features, symbols and material and non-material characteristics that identify them with the specific cultural and religious heritage of the surrounding societies. In addition, they reflect current social, political or cultural and religious conditions (Olsen 2019). Examples from around the world show that sacred space can be a venue for contestation and inter-sectoral or inter-faith disputes between different religious groups (e.g. Jerusalem or Ayodhya) (Tahan 2020; Timothy & Emmett 2014), as well as a place where personal identity, solidarity or opposition related to certain political, social or religious groups manifest (Hassner 2003; Jackowski & Sołjan 2000). Olsen (2019) suggests the notion of 'sacred spaces in motion', in which pilgrims or tourists may look for alternative spaces to avoid conflicts or even displays of aggression in politically charged or socially divisive sacred sites.

Tourism focused on the heritage of faith and broadly conceived spiritual values, treats non-material resources (e.g. ideas, symbols, universal messages) as basic elements of place. The more universal their meaning and the more widespread they are outside their place of origin, the more diverse their tourist audiences will be. Therefore, they are interpreted in the context of the social and cultural religious conditions of the era. Interpretation often uses an extensive inventory of methods and techniques that can be attractive and accessible to a wide audience. Their use largely depends on the nature and spatial assumption of the sacred place or object, as well as the profile of the recipients themselves. Interpretation at sacred places or other localities of a spiritual nature should follow the main principles and goals as defined by Tilden (1957) and Beck and Cable (1998). Due to the nature of sacred space, however, interpreters should take into account the individual needs of the recipients, particularly with regard to maintaining their private perception of the site and personal relationships with holiness. The interpretation of religious and spiritual attractions should not interfere too much with the deep meaning of the *genius loci* and the authenticity of the site. It should take into account the specific sensitivities of religiously involved people. The approach to presenting elements of spiritual heritage allows visitors to learn and understand the universal message that is shared with visitors. Hopefully, it also translates into better protection and preservation, but also to show certain processes involving identifying one's relationship with the *sacrum*, the development of beliefs, as well as shaping attitudes and building close relationships with the surrounding environment.

The specific, individual and often unique features of sacrosanct places necessitate a very precise and well-planned interpretive program. The diversity of techniques and tools and the technological capabilities depend both on external factors (e.g. size and significance of the site, trail or object being interpreted, its development, architectural value and even financial condition) as well as internal factors (e.g. religious and cultural influence, heritage transfer, the relation of *sacrum* to the *profanum* or the authenticity and universality of the transmitted messages). The elements connecting these factors are the tourists/pilgrims themselves and their preparation to receive the interpreted content—religious, spiritual or cultural involvement, willingness to accept the narrative addressed to them, individual perception of the site, as well as their interpersonal relations and the ability to use new technologies.

Due to the large diversity of religious and spiritual tourism destinations, as well as the developing profiles of visitors, there are many standard techniques and tools used in interpreting sacred sites. Some of them are interpersonal, relating to direct narration between the interpreter and the recipient. Others use modern technology and narrative multimedia tools. They are mainly used by individuals who engage with interpretive media by themselves based on prepared narratives. The scope of interpretation and its physical location may be temporarily or permanently limited because of the presence of the *sacrum* zone, which is the physical area of a pilgrim's personal relations with holiness (Duda 2019; Duda & Doburzyński 2019). This mainly applies to guided narrations, wherein guides explain the meaning of individual sites in the sacred space in a professional manner.

By compiling all of the above elements that shape the appropriate narration and create an interpretation of the sacral destination, the following are the most important forms, tools and goals of interpretation in religious and spiritual tourism.

Interpersonal interpretation

Heritage places have long utilized a wide range of interpersonal interpretive tools to educate visitors and help them create experiences. Such tools involve live people acting as guides, information brokers and re-enactors. All of these approaches to heritage tourism have been effective and are now widely used in religious tourism settings (Timothy 2021).

Oral communication

Often comprised of fictionalized legends, storytelling and thematic narratives, oral communication techniques focus on a specific topic related to works of art or stories about the genesis of the place, its development and meaning, or the character (e.g. saints, former spiritual masters, spiritual guides). Oral transmission is usually carried out *in situ* by a guide and refers to a specific place, environment or element of the destination. It is usually used for a larger group of recipients, most often for those not directly related to the place or even a given religious, cultural or national group. Due to the ease of falling into a routine during the so-called systematic or repetitive interpretation—common in the case of the guide's work, oral communication should be conducted so that it is easy to read and understand and be respectful of the worldviews of recipients. Due to the nature of the destination, oral communication cannot be delivered everywhere and at any time even during the site's opening hours. Restrictions apply to, among others, the presence of the *sacrum* zone and during religious practices and worship services.

Lectures, sermons, hagiographic narratives

These include thematic speeches focused on a specific topic related to the place, its symbolism and/or its physical characteristics. The narrative, which is much longer than in the case of traditional oral communication, aims not only to provoke thinking but also to increase interest in issues directly or indirectly related to the place being interpreted. This type of interpretation is addressed to certain recipients who, because of their own interests, decide to participate in the narrative. Some kinds of this type of narration are fictionalized meetings, including staged characters, where the interpreter plays the role of a character from a specific period and actively presents information about the issue, for example, a visit to the family home of Pope John Paul II in Wadowice, Poland. Actors dressed in period costumes, such as at Nazareth Village, narrate stories about events and living during the time of Christ. Another example is multimedia lectures, in which the interpreter's narration is complemented by films and other such media.

Participatory interpretation

Participatory interpretation emphasizes the interaction between the interpreter and the recipient (pilgrim, tourist). The idea of the participatory narrative is to draw the recipients into the interpretation by jointly discovering its secrets, searching for elements of the narration in the place, participating in thematic workshops, competitions or discussions. This type of interpretation is present, among others, on pilgrim routes or so-called prayer paths, where spiritual guides engage the pilgrim or other tourist in participating in religious practices, explaining at the same time their meanings and symbolisms. This type of interpretation includes, for instance, organized meetings referring to old religious traditions (e.g. Feast of Tabernacles—Sukkot in Israel). A special type of participatory interpretation is the thematic workshops often organized at large monastic premises, sanctuaries or other religious places. Examples include icon writing workshops in Cistercian monasteries and illuminating incunabulas in Benedictine monasteries. A specific form of interpretation participation is multi-day stays of an educational, contemplative or formative nature in the area of a working temple or monastery, or among a religious order. Such stays include not only educational components but also creative exercises of spiritual or contemplative values (e.g. the Taizè community in France).

Intercultural narratives

These narratives are found in places where various religious or cultural-religious groups dominate, as well as in objects that constituted an important historical spiritual center. Many such places are found in the Holy Land, where the three monotheistic religions—Judaism, Christianity and Islam—dominate a small area. Some places in Germany, Scandinavia, Poland and the Czech Republic exemplify these. In areas that were once the regional domain of Protestant, Orthodox or other Christian sects but which are currently managed by the Catholic Church, a double interpreter has been introduced. In this instance, two guides may be involved in explaining the meanings of church architecture, decor and symbolism from an inter-denominational perspective. Interpretation from each of these faiths or denominations allows the recipient to become acquainted with diverging views of the same heritage. Together these perspectives provide a more holistic picture of the place as a spiritual center of societies that inhabited the region through different periods of time.

Staging and stylized narrative

These are less common in religious tourism than in other forms of heritage tourism. They rely on the staging of an event, a person's life history, a legend or an historical event related to the place or object being interpreted. Events partly staged in religious tourism include various types of church fairs, processions or sacralized representations of past events. Examples include La Festa dei Ceri (Gubbio, Italy), Misteri d'Elx (Elx, Spain), El Cant de Sibilla (Mallorca, Spain) and the annual passion play in Kalwaria Zebrzydowska, Poland (Sawicka 2007).

Extended interpretation

These types of methods go beyond the interpersonal ones and are more dependent on the visitors' own use of certain media. These include printed material, placards, displays and technologies that assist visitors in understanding the cultural and religious significance of the place they are visiting.

Educational information boards

Interpretive plaques or information boards display facts, knowledge and information in an accessible manner. This often includes pictures, diagrams, explanations and stories in textual form. The most accessible interpretive panels include braille and audio options for visually impaired visitors. In religious tourism, this medium is most often used for objects and sites that have significant architectural or aesthetic values, or displays in a museum setting. The advantages of this medium are low cost, no additional fees, easy maintenance, the personalization of the message and access to wider audiences, regardless of the time or number of visitors. The main disadvantages of this tool, however, are the difficulty and expense of making changes when the information needs to be updated and the limited surface space for typesetting.

Multimedia exhibits

These include various types of dioramas, information kiosks, touch screens or interactive games and videos. In faith tourism and in broader religious heritage contexts, such tools are often used in cultural and sacral heritage objects or temples, which, in addition to sacral functions, also have museum functions. New technologies provide multidimensional presentations, as well as multi-sensory and interactive learning. Augmented reality is an important component of this category, which can benefit users by virtually rebuilding parts of churches and temples that have disappeared over centuries or millennia. Such technologies allow visitors to see what certain structures, features and artworks might have looked like during their heyday or when they were first constructed, such as at the Abby of Cluny and in the Romanesque Churches of Catalonia (Fusté-Forné 2020; Rueda-Esteban 2019). Thanks to new audio and video technology, visitors can create their own sightseeing paths, supported by stationary or mobile interpretation devices. In some facilities such as the Basilica of St. Francis of Assisi, Italy, triggered illumination of the main features of the church (e.g. Giotto's frescoes from the thirteenth century) combined with a simultaneous explanation of their history and message, adds considerable interpretive appeal and value.

Personal digital interpretation

Recent advances in information and communication technology (ICT) have significantly influenced the interpretation and management of cultural heritage (Hausmann & Weuster 2018; Timothy 2021). ICT has spurred the development of many different technology-driven interpretive media as noted in the section above, but much of the change has resulted in highly personalized tools that utilize new forms of narration and modern technology. GPS-powered mobile phone applications are the best current example of this. Phone apps provide many personalized interpretive media, such as site guides which the user controls freely, QR code readers, and virtual and individualized augmented reality technologies (Bohlin & Brandt 2014; Solima & Izzo 2018). An example would be the Benedictine Abbey of Montserrat in Catalonia, Spain. Modern digital interpretation also allows for learning about places and objects that have been destroyed over the centuries and have not survived to modern times, or for various reasons visiting them is not possible (e.g. some Aboriginal sacred sites in Australia or the unapproachable hermits of the abbots in the Apennines). Despite the many advantages of modern technology and its educational capabilities, it contains a number of disadvantages that are often unacceptable for religious tourism and pilgrimage. The main concern is the potential for ICT to shallow the message, or overshadow the message. It may also help create mass narratives devoid of deeper reflection on the symbolisms of the sacred locale, reproducing stereotypes and blurring the authenticity of the site. Nevertheless, this type of interpretation is frequently used at religious centers for its cognitive value and to function as an additional attraction.

Fictionalized experiential interpretation

This relatively new form of narration in religious tourism has many features in common with participatory interpretation, encouraging the visitor to participate in co-creating the experience and knowledge of a region's religious heritage. It also uses tools characteristic of multi-threaded fictional narratives (e.g. printed descriptions, mobile apps, riddles, games and competitions) (Champion 2008; Ćosović & Brkić 2019). For example, to learn about the history of the Cistercian order in France, visitors can utilize 'fictionalized touring' (so-called 'questing') of the former monastery in Citeaux (Burgundy) and the Abbey of Chorin (Germany). There, visits are staged and involve the creation of a medieval atmosphere through choral music, burning torches and tasting monastery beers.

Conclusion: the future of interpretation in religious tourism

With the growth in the need for more educational and more experiential immersion, the interpretation of cultural heritage, including religious and spiritual heritage, is one of the most important characteristics of tourism development throughout the world. On the other hand, sacred space is one of the most important elements of religious destinations and religious landscapes, constituting the subject of interest for tourists and pilgrims. Its functioning, shape and boundaries result from the influence of both the elements of the *sacrum* and the broadly understood *profanum* (Cohen 2006; Duda & Doburzyński 2019; Puşcaşu 2015). One of the bridges linking these two elements is an adequate and appropriate interpretative program that will engage participants and explain the place's religious and spiritual values, and its cultural heritage role in the region. In many cases, the full understanding of the *sacrum* involves the use of an appropriate narrative, the origins of which derive from the tradition of the *profanum*. The elements of the *profanum* have therefore impacted the character of sacral spaces, as well as the manner of understanding and perceiving the *sacrum* itself. This impact does not necessarily reflect negatively or raise conflict between these seemingly two different worlds. Research by the author among sacred site visitors has even pointed to the interpenetration of these profane elements (Duda 2019).

One of the basic aims of interpretation, as indicated by Tilden (1957), is the universality of the message and the reference to the whole site, strongly related to its surroundings and larger region. Interpretation should also be addressed to everyone, regardless of their cultural backgrounds. However, there are some barriers or limitations that must be overcome by using more universal narrative techniques and methods. An example of such a barrier is cross-border religious heritage. In this situation, religious heritage and faithscapes are divided by a political boundary, which in many parts of the world strongly deters transfrontier collaboration, including in tourism. So what should the interpretation of cross-border sacredscapes look like in light of the barriers raised by political boundaries? One solution might be to use a common intercultural narrative to present a fuller picture of the site's diversity in terms of nationality and culture. Unfortunately, in many cases, the lack of cross-border cooperation becomes an insurmountable barrier. Another question is how we should interpret sacred places and objects in conflict, such as in the case of a community that currently occupies in area that is sacred for another group. In the literature, this phenomenon is described as 'difficult heritage', 'dissonant heritage' and 'contested heritage' and applies to sites where one community (culture, religion) does not recognize, or is in conflict with, another (e.g. Mikos von Rohrscheidt 2019b; Olsen & Timothy 2002; Timothy & Emmett 2014; Tunbridge & Ashworth 1996). Well-organized interpretation can help mitigate interfaith conflict and may lead to greater understanding and ultimately mutual tolerance and common ground. On the other hand, the implementation of a common narrative may be difficult because of the overlapping identities associated with the site.

It is difficult to predict the future of interpretation in religious and spiritual tourism contexts. At many sites, interpretation is so unique to a heritage of faith, that it lacks equivalents in other areas of heritage and tourism. Authenticity is also understood somewhat differently in such places. In the case of holy localities, authenticity is more often reflected in non-material values (e.g. communion with deity and the spirit of the place). In addition, it is important to remember that the *sacrum* zone in many cases constitutes an inviolable membrane between different interpretations of faith heritage. It is the very personal zone where visitors interact with holiness. Interference in this space and state of being may be perceived as interference in people's most private and deepest emotional states. It is this understanding of sacred space that should determine the extent, content and media of interpretation. The overall values shaping sacred sites and their messages as part of a region's cultural heritage derived from the original religious significance of that locale. This should underscore the look of interpretation at places of faith, religion and spirituality.

References

Adler, J. (2002) 'The holy man as travel attraction: Early Christian Asceticism and the moral problematic of modernity', in W.H. Swatos, & L. Tomasi (eds), *From Medieval Pilgrimage to Religious Tourism: The Social and Cultural Economics of Piety* (pp. 25–50). Westport, CT: Praeger.

Ambrósio, V. (2007) 'Sacred pilgrimage and tourism as secular pilgrimage', in R. Raj & N.D. Morpeth (eds), *Religious Tourism and Pilgrimage Management: An International Perspective* (pp. 78–88). Wallingford: CABI.

Beck, L., & Cable, T.T. (1998) *Interpretation for the 21st Century: Fifteen Guiding Principles for Interpreting Nature and Culture*. Urbana, IL: Sagamore.

Bohlin, M., & Brandt, D. (2014) 'Creating tourist experiences by interpreting places using digital guides', *Journal of Heritage Tourism*, 9(1): 1–17.

Burns P., & Holden P. (1995) *Tourism: Towards the 21st Century*. Hemel Hempstead: Prentice-Hall.

Castells, M. (2001) *La era de la Información: Economia, Sociedad y Cultura, Volumen II: el Poder de la Identidad*. Mèxico City: Siglo XXI Editore.

Champion, E.M. (2008) 'Otherness of place: Game-based interaction and learning in virtual heritage projects', *International Journal of Heritage Studies*, 14(3): 210–228.
Cohen, E. (1998) 'Tourism and religion: A comparative perspective', *Pacific Tourism Review*, 2(1): 1–10.
Cohen, E.H. (2006) 'Religious tourism as an educational experience', in D.H. Olsen & D.J. Timothy (eds), *Tourism, Religion and Spiritual Journeys* (pp. 78–93). London: Routledge.
Collins-Kreiner, N. (2010) 'Researching pilgrimage: Continuity and transformations', *Annals of Tourism Research*, 37(2): 440–456.
Ćosović, M., & Brkić, B.R. (2019) 'Game-based learning in museums—cultural heritage applications', *Information*, 11(1): 22–34.
Duda, T. (2019) 'Does a religious tourist need a guide? Interpretation and storytelling in sacred places', in D. Vidal-Casellas, N. Creus-Costa & S. Aulet (eds), *Tourism, Pilgrimage and Intercultural Dialogue: Interpreting Sacred Stories* (pp. 105–114). Wallingford: CABI.
Duda, T., & Doburzyński, D. (2019) 'Religious tourism vs. sacred space experience: Conflict or complementary interaction?' *International Journal of Religious Tourism and Pilgrimage*. 17(5): 1–10.
Fusté-Forné, F. (2020) 'Mapping heritage digitally for tourism: An example of Vall de Boí, Catalonia, Spain', *Journal of Heritage Tourism*, 15(5): 580–590.
Hassner, R.E. (2003) 'To halve and to hold: Conflicts over sacred space and the problem of indivisibility', *Security Studies*, 12(4): 1–33.
Hausmann, A., & Weuster, L. (2018) Possible marketing tools for heritage tourism: The potential of implementing information and communication technology. *Journal of Heritage Tourism*, *13*(3), 273–284.
Hodges, S. (2020) 'Interpreting the past: Telling the archaeological story to visitors', in D.J. Timothy & L.G. Tahan (eds), *Archaeology and Tourism: Touring the Past* (pp. 167–185). Bristol: Channel View Publications.
Jackowski, A. (2003) *Sacred Space of the World: The Basics of the Geography of Religion* (in Polish). Kraków: Wydawnictwo Naukowe UJ.
Jackowski, A., & Sołjan, I. (2000) Millenium polskiego pielgrzymowania. *Peregrinus Cracoviensis*, 8: 9–38.
Kaelber, L. (2002) 'The sociology of medieval pilgrimage: Contested views and shifting boundaries', in W.H. Swatos & L. Tomasi (eds), *From Medieval Pilgrimage to Religious Tourism: The Social and Cultural Economics of Piety* (pp. 51–74). Westport, CT: Praeger.
Kaelber, L. (2006) 'Paradigms of travel: From medieval pilgrimage to the postmodern virtual tour', in D.J. Timothy & D.H. Olsen (eds), *Tourism, Religion and Spiritual Journeys* (pp. 49–63). London: Routledge.
Kanaan-Amat, M., Crous-Costa N., & Aulet S. (2019) 'Interpretation tools for religious heritage', in D. Vidal-Casellas, N. Creus-Costa, & S. Aulet (eds), *Tourism, Pilgrimage and Intercultural Dialogue: Interpreting Sacred Stories* (pp. 85–95). Wallingford: CABI.
Lansangan-Cruz, Z. (2008) *Principles and Ethics of Tour Guiding*. Quezon City: Rex Books.
Mikos von Rohrscheidt, A. (2016) 'Analysis of the programs of tourist trips as the basis for a typology of participants in religious tourism', (in Polish), *Folia Turistica*, 39: 67–88.
Mikos von Rohrscheidt, A. (2019a) 'An outline of the history of heritage interpretation in tourism and the proposal of its systematics', (in Polish), *Turystyka Kulturowa*, 3: 7–32.
Mikos von Rohrscheidt, A. (2019b) 'Object of heritage interpretation in contemporary cultural tourism – the main types of interpreted resources and leading contents', (in Polish), *Turystyka Kulturowa*, 4: 65–91.
Millar, S. (1999) 'An overview of the sector', in A. Leask & I. Yeoman (eds), *Heritage Visitor Attractions: An Operations Management Perspective* (pp. 1–21). New York: Cassell.
Morales Miranda, J. (2001) *Guía pára la interpretación del patrimonio: El arte de acercar el legado natural y cultural al público visitante*, 2nd edn. Sevilla: Junta de Andalucía.
Olsen, D.H. (2008) *Contesting Identity, Space and Sacred Site Management at Temple Square in Salt Lake City, Utah*. Unpublished doctoral dissertation, University of Waterloo, Ontario, Canada.
Olsen, D.H. (2011) 'Towards a religious view of tourism: Negotiating faith perspectives on tourism', *Tourism Culture & Communication*, 11(1): 17–30.
Olsen, D.H. (2012) 'Teaching truth in third space: The use of history as a pedagogical instrument at Temple Square in Salt Lake City, Utah', *Tourism Recreation Research*, 37(3): 227–237.

Olsen, D.H. (2019) 'The symbolism of sacred space', in D. Vidal-Casellas, N. Creus-Costa, & S. Aulet (eds), *Tourism, Pilgrimage and Intercultural Dialogue: Interpreting Sacred Stories* (pp. 29–42). Wallingford: CABI.

Olsen, D.H., & Timothy, D.J. (2002) 'Contested religious heritage: Differing views of Mormon heritage', *Tourism Recreation Research*, 27(2): 7–15.

Olsen, D.H., & Timothy, D.J. (2006) 'Tourism and religious journeys', in D.J. Timothy & D.H. Olsen (eds), *Tourism, Religion and Spiritual Journeys* (pp. 1–21). London: Routledge.

Olsen, D.H., Trono, A., & Fidgeon, P.R. (2018) 'Pilgrimage trails and routes: The journey from the past to the present', in D.H. Olsen & A. Trono (eds), *Religious Pilgrimage Routes and Trails. Sustainable Development and Management* (pp. 1–13). Wallingford: CABI.

Preston, J.J. (1992) 'Spiritual magnetism: An organizing principle for the study of pilgrimage', in A. Morinis (ed), *Sacred Journeys: The Anthropology of Pilgrimage* (pp. 31–46). Westport, CT: Praeger.

Puşcaşu, V. (2015) 'Religious tourism or pilgrimage? *European Journal of Science and Theology*, 11(3): 131–142.

Robinson, M. (1997) *Sacred Places, Pilgrim Paths: An Anthology of Pilgrimage*. New York: Harper Collins.

Ron, A.S., & Timothy, D.J. (2019) *Contemporary Christian Travel: Pilgrimage, Practice and Place*. Bristol: Channel View Publications.

Różycki, P. (2016) *Turystyka a Pielgrzymowanie, Wyd.* Kraków: WAM.

Rueda-Esteban, N.R. (2019) 'Technology as a tool to rebuild heritage sites: The second life of the Abbey of Cluny', *Journal of Heritage Tourism*, 14(2), 101–116.

Sawicka, A. (2007) *Camins i cruïlles de la cultura catalana* (in Polish). Kraków: Ksiegarnia Akademicka.

Solima, L., & Izzo, F. (2018) 'QR Codes in cultural heritage tourism: new communications technologies and future prospects in Naples and Warsaw. *Journal of Heritage Tourism*, *13*(2), 115–127.

Sox, D. (1985) *Relics and Shrines*. London: Allen & Unwin.

Sumption, J. (1975) *Pilgrimage: An Image of Medieval Religion*. London: Faber & Faber.

Tahan, L.G. (2020) 'Archaeological destruction and tourism: Sites, sights, rituals and narratives', in D.J. Timothy & L.G. Tahan (eds), *Archaeology and Tourism: Touring the Past* (pp. 121–133). Bristol: Channel View Publications.

Tilden, F. (1957) *Interpreting Our Heritage*. Chapel Hill: University of North Carolina Press.

Timothy, D.J. (2021) *Cultural Heritage and Tourism: An Introduction*, 2nd edn. Bristol: Channel View Publications.

Timothy, D.J., & Emmett, C.F. (2014) 'Jerusalem, tourism, and the politics of heritage', in M. Adelman & M.F. Elman (eds), *Jerusalem: Conflict & Cooperation in a Contested City* (pp. 276–290). Syracuse, NY: Syracuse University Press.

Tunbridge, J., & Ashworth, G. (1996) *Dissonant Heritage: The Management of the Past as a Resource in Conflict*. Chichester: Wiley.

Uzzell, D.L. (1998) 'Interpreting our heritage: A theoretical interpretation', in D.L. Uzzell & R. Ballantyne (eds), *Contemporary Issues in Heritage and Environmental Interpretation: Problems and Prospects* (pp. 11–25). London: The Stationery Office.

Vukonić, B. (1996) *Tourism and Religion*. New York: Pergamon.

Vukonić, B. (2006) 'Sacred places and tourism in the Roman Catholic tradition', in D.J. Timothy & D.H. Olsen (eds), *Tourism, Religion and Spiritual Journeys* (pp. 237–253). London: Routledge.

29

COEXISTENCE BETWEEN TOURISTS AND MONKS

Managing temple-stay tourism at Koyasan, Japan

Kaori Yanata and Richard Sharpley

Introduction

Long acknowledged to be one of the earliest manifestations of tourism (Kaelber, 2006; Raj & Griffin, 2015), travel for religious purposes has evolved into a major sector of the contemporary global tourism market (McKelvie, 2005). To some extent, this reflects the remarkable growth in tourism – both domestic and international – in general; as ever-increasing numbers of people enjoy the means and ability to travel, it is perhaps inevitable that participation in religious tourism, in particular, is also growing. At the same time, however, both the scope and meaning of religious tourism have expanded significantly in recent decades. Not only is the variety of religious sites and events that are developed and promoted for touristic purposes becoming increasingly diverse but also, reflecting the broadening and blurring of the concepts of religion and spirituality, the very concept of religious tourism is becoming more widely interpreted (Olsen & Timothy, 2006; Sharpley, 2009). Both religion and religious tourism are being rebranded (Carrette & King, 2005) whilst supporting Suntikul and Butler's (2018: 1) argument that 'the common ground between tourism and religion has moved beyond what has been termed "religious tourism"', new touristic experiences, such as 'new age' (Attix, 2002), holistic (Smith & Kelly, 2006) or wellness tourism (Voigt & Pforr, 2013), have in recent years become increasingly popular and have spiritual connotations.

One such experience is retreat tourism (Heintzman, 2013), or so-called temple-stay tourism, a more specific form of retreat tourism most usually referring to visitors staying in religious buildings or complexes in Asian countries such as China, Thailand or Korea (Son & Xu, 2013). The provision of accommodation at religious sites in general is not, of course, a new phenomenon; monasteries and other religious places have long welcomed travellers in need of overnight facilities, typically in return for a donation (Chun, Roh & Spralls, 2017; Ron & Timothy, 2019). However, as Shackley (2004) observes, even at a time when participation in formal religious institutions and practices is on the wane, a commensurate growth in the search for spiritual fulfilment through other means (Heelas & Woodhead, 2005; Wuthnow, 1998) has manifested in increasing demand for retreat tourism 'by people in search of peace, quiet and spiritual input' (Shackley, 2004: 227). Around the world, numerous retreat houses, sometimes utilising secular locations such as historic cultural properties or purpose-built facilities but more typically in places associated with a particular religious tradition, offer visitors accommodation, food, spaces for quiet contemplation or 'doing nothing' (Shackley, 2004: 228) and, dependent on their markets, varying degrees of organised religious or spiritual activity, from attending workshops or classes to participating in formal worship.

Significantly, such contemporary retreat or temple-stay tourist experiences are usually provided on a commercial basis, often representing a vital source of income for monasteries, convents, temples and other religious places and institutions seeking to maintain both their physical and spiritual fabric. This is not surprising. As religious tourism more generally continues to grow in scale and scope, so too has its economic potential become increasingly recognised and exploited (Griffin & Raj, 2017; Ron & Timothy, 2019). In other words, there has always been a commercial aspect to religion; for example, not only have religious artefacts long been produced for sale to worshippers (Shackley, 2006; Shi, 2011) but also the construction of places of worship has also long been dependent on donations from devotees (Ward, 2006). Indeed, it is argued that the world's major religions have always competed within a global religious marketplace (Einstein, 2008). However, religious sites, attractions and destinations are demonstrating an increasing propensity to be developed and marketed as tourist attractions (Triantafillidou et al., 2010; Vukonić, 2002) and, as a consequence of this increasingly commercial focus, are facing the challenges of commodification (Olsen, 2003).

Surprisingly, despite the growth in popularity of temple-stay tourism in particular and the increasing academic attention paid to it, few if any attempts have been made to consider the potential impacts of this commodification process on the lives of those at the religious locations who interact with tourists. In other words, a number of studies have focused on the perceived consequences of the development of monasteries as tourist destinations, such as Drule, Chis and Ciornea's (2012) research in Transylvania, Ryan and McKenzie's (2003) study of the transformation of a Benedictine settlement in Western Australia into a tourist destination, or Wong, McIntosh and Ryan's (2013) study of how monks and nuns at Pu-Tuo-Shan pilgrimage site in China cope with the challenges of tourism (see also Rodrigues & McIntosh, 2014). However, by its very nature, temple-stay tourism allows tourists to penetrate and share, albeit temporarily, the religious lives of the residents of the places where they stay yet, as discussed shortly, most studies of temple-stay tourism focus on the tourists and their experiences rather than on how the residents manage the commodification of their religious practices or, more succinctly, how they coexist with tourists.

The purpose of this chapter is to address this gap in the literature. Specifically, drawing on research undertaken at Koyasan, a mountain temple complex located in the north-west of Wakayama Prefecture in Japan, it explores how those with the responsibility for maintaining the religious sanctity of the complex manage the demands of tourists. In so doing, it seeks to offer a new dimension to knowledge and understanding of the phenomenon of temple-stay tourism, albeit within the specific context of Koyasan. To provide a conceptual framework for the research, the chapter commences with a necessarily brief review of temple-stay tourism and the potential consequences of commodification before going on to introduce the empirical example and discussing the research and its outcomes.

Temple-stay tourism: the commodification of religious practices?

As noted above, temple-stay tourism has in recent years enjoyed growing popularity as a niche manifestation of retreat tourism, the latter embracing elements of spiritual, religious, holistic and wellness tourism. Typically associated with the provision of accommodation and other services at Buddhist religious sites (Chun et al., 2017 Wang, 2011), although not unique to that religious tradition – Hindu ashrams in India have, for example, long been popular amongst Western tourists (Sharpley & Sundaram, 2005) – temple-stay tourism and the development of Buddhist heritage sites as tourist attractions more generally has been in increasing evidence in countries such as Myanmar, Thailand, Bhutan, China and Nepal (Kaplan, 2010). Typically, this occurs on an ad-hoc basis, although one country noted for its formal temple-stay programme is South Korea which, perhaps unsurprisingly, has also become the focus of much academic research (for example, Chun et al., 2017; Kaplan, 2010; OECD, 2009; Son & Xu, 2013; Song et al., 2015; Wang, 2011).

The Korean temple-stay programme originated in 2002, coinciding with the country's hosting of the FIFA World Cup, the catalyst being the government's recognition of the need for additional accommodation facilities to meet the needs of the large number of tourists attending the tournament. Initially, the idea was met with some resistance on the part of the religious authorities. However, following the use of temples as accommodation for attendees at subsequent sports events, a more formal relationship evolved in 2004 when the Jogye Buddhist Order (the representative order of traditional Korean Buddhism) established its Cultural Corps of Korean Buddhism, partly funded by the Ministry of Culture and Tourism (Kaplan, 2010; Wang, 2011), the purpose being to formally promote temple stays as a necessary source of income, as well as offering visitors the opportunity to experience traditional Korean culture and to propagate the teachings of Korean Buddhism. The programme has since evolved into a successful sector of tourism in Korea. In 2017, almost 71,000 international tourists participated in temple stays whilst, since 2002, the programme has attracted a total of 495,000 international and 3.07 million domestic tourists (Korea Bizwire, 2018). Research reveals that experiencing traditional Korean Buddhist culture is a major motivating factor among visitors, while education/learning about Buddhism, escape, self-reflection/self-growth and being in nature are also significant motivators (Chun et al., 2017; OECD, 2009; Song et al., 2015). However, few if any studies explore the impacts of temple-stay tourism from the perspective of the hosts, both at Korean temples and elsewhere, although some have considered the more general impacts of tourism on monasteries and monastic life (Rodrigues & McIntosh, 2014; Shackley, 1998).

The temple-stay programme in South Korea is unique in terms of both its scale and organisation. Nevertheless, temple-stay tourism in any context is susceptible to the challenges arising from the commodification of religious heritage for tourist consumption. Commodification in general is the process by which anything – a good, a service, a cultural ritual or practice – is accorded an exchange value (Appadurai, 1986); it becomes a commodity that is produced, marketed and exchanged, usually for a financial payment. Hence, tourism as a contemporary social and economic phenomenon is a commodified activity, based as it is primarily on the purchase of goods, services and experiences. Thus, as Olsen (2003: 101) argues, one way in which 'religion and religious built heritage are commodified [for tourism] is where religious groups commodify their doctrines, customs and beliefs for economic gain'.

Within the tourism literature, the consequences of such commodification have long attracted academic scrutiny, typically, though not exclusively, in the context of the authenticity of otherwise tourist experiences. Specifically, following MacCannell's (1973) early thesis that tourist experiences are purposefully staged by the host community and, hence, such staging or commodification of tourism inevitably thwarts tourists in their presumed quest for authentic experiences, numerous commentators have debated the relationship between commodification and authenticity (Cohen, 1988; Olsen, 2002; Pearce & Moscardo, 1986; Steiner & Reisinger, 2006; Wang, 1999). A consideration of these debates is beyond the scope of this chapter, but it is important to note that more limited attention has been paid in general to the effects of commodification on the experiences of those producing or performing tourist experiences. In addition to the very presence of large numbers of tourists, particularly in religious settings, which may disrupt the day-to-day lives of the local community, the significance or meaning (authenticity) of the production of cultural artefacts, rituals or performances may be diminished as they become commodified for touristic consumption. For example, in an early and widely-cited study, Greenwood (1989), describes how the Alarde, a public ritual that celebrates the northern Spanish town's victory over the French in the seventeenth century, became a commodified tourist attraction, not least because it came to be performed twice on the day to meet tourist demand. He later acknowledges that it has regained some significance to its participants, though as a contemporary political statement. Other studies suggest that performances for tourists may in fact reinforce local cultural identity (McKean, 1989; Xie, 2003), although more generally the potential exists for cultural practices to lose their meaning for performers, and what might be referred to as 'coping strategies' may be adopted to mitigate such negative consequences. For example, Wong (2017, 2018) discusses how monks and nuns at Pu-Tuo-Shan in China adapt to large numbers of visitors. This chapter now turns to the example of temple-stay tourism at Koyasan, Japan, to explore if and how such coping strategies are adopted to enable the monks to coexist with tourists.

The development of temple-stay tourism at Koyasan

Koyasan is a large temple settlement in north Wakayama Prefecture, Japan, and is located on an alpine plain some 800 meters above sea level, surrounded by eight peaks. These collectively are said to represent a lotus plant and, hence, Koyasan is considered to be the 'pure land' (Matsunaga, 2014). Its history can be traced back more than 12 centuries to 816 when the High Priest Kukai (774–835), founder of the Shingon (or Esoteric) school of Buddhism, was granted permission by then-reigning Emperor Saga to establish a monastic training centre there. Kongobuji Temple, which Kukai (posthumously known as Kobo Daishi) founded, subsequently became the head temple of the Shingon sect but, over the centuries, the Koyasan complex expanded significantly in line with its increasing importance as a pilgrimage centre. Initially, only nobles and Samurai visited in an early manifestation of religious tourism in Japan but, as Koya-Hijiri monks were dispatched around the country to preach Buddhism, greater numbers of people began to visit Koyasan (see Funck & Cooper, 2015). As a consequence, not only were more temples built but also from the late fourteenth century onwards, temple lodgings were established to accommodate pilgrims. By the early eighteenth century, Koyasan had become a popular religious tourism destination among the wider population, with many people visiting to hold memorial services to their ancestors, and had developed into a small town. A 1793 map details about 750 buildings, including temples, lodgings, shops and houses (Miyasaka & Sato, 1984). By the end of the Edo period, there were more than 680 temples although, primarily as a result of two major fires but also because of consolidations and the closure of many temples (see below), this number had declined to 137 by 1891. Nowadays, there are more than 100 temples in Koyasan, including the well-known Okunoin Temple, which not only houses the mausoleum of Kobo Daishi, but is also the site of Japan's largest cemetery, the resting place of more than 200,000 souls (Tables 29.1 and 29.2). There are also 52 temple lodgings, or *shukobo*, accommodating more than 84,000 international tourists in 2017, and the town offers shops, tea rooms and other facilities to meet the needs of well over one million day visitors (Tourist Bureau, Department of Commerce, Industry, Tourism and Labor in Wakayama Prefecture, 1959–2017).

Of particular relevance to this chapter, until the mid-nineteenth century Koyasan was a strictly religious town; its population was limited to those engaged in temple-related activities, both religious and commercial (Matsunaga, 2014), and tourists visited only for religious purposes. However, a number of factors underpinned its commodification and transformation into the popular tourist destination it is today. Initially, during the Meiji Restoration, the declaration of Shintoism as the state religion undermined the influence of Buddhism, leading to a decline in Koyasan's fortunes with many temples being closed and, hence, the need to secure alternative sources of income other than religious donations. However, continuing improvements in access were also influential. Between January and June 1884, some 40,000 people made the climb on foot up to Koyasan (Miyasaka & Sato, 1984). Following the opening of the railway from Osaka in 1925, approximately 500,000 visitors were recorded that year, and numbers increased further following the construction of a cable car up to the town in 1930. As a consequence, Koyasan became increasingly attractive as a tourist destination, and in response, temple lodgings were refurbished and improved to appeal to this new tourist market. Soon after the end of the Second World War, they then became subject to the Inns and Hotels Act and were obliged to charge fixed prices (Akiyama, 2015). Following subsequent collaboration with tour agencies, temple lodgings began to welcome organised groups seeking rest and relaxation (Akiyama, 2018).

The construction of road links in the 1960s, however, firmly established Koyasan as a domestic mass tourism destination. In 1965, for example, more than a million people travelled there to celebrate the temples' 1150-year anniversary (Ikeda, 2015). Yet, this ease of access had a fundamental impact on tourism to Koyasan. Day visits became more popular leading to a decline in overnight stays, and accommodation was again transformed to meet the needs of mainly independent secular visitors. The popularity of temple stays was then re-established in the early 2000s, particularly among international tourists, as a result of both Japan's policy to increase inbound tourism and the inclusion of Koyasan in the World Heritage Site inscription of the 'Sacred Sites and Pilgrim Routes in the Kii Mountain Range' in 2004. Hence, Koyasan now caters for large numbers of domestic and international tourists both religious and secular, wishing to experience traditional Buddhist culture. In 2015, more than 440,000 temple-stay tourists were welcomed to *shukobo* in Koyasan, the number of international visitors in particular having tripled since 2013 (Japan Times, 2018). It is with these tourists that the monks must negotiate a coexistence. How they do so is the focus of the research now discussed.

The research: how monks manage temple-stay tourism at Koyasan

Research method

Given the overall purpose of the research, to identify how monks at Koyasan manage temple-stay tourism to minimise the potential negative consequences of the commodification of their religious lives or, more precisely, how they coexist with the tourists who temporarily share their religious world, qualitative data were deemed the most appropriate. Specifically, the first author of this chapter spent a number of days working as an intern in one temple lodging, during which time she engaged in participant observation. Subsequently, semi-structured interviews were conducted with head monks, subordinate monks and an office manager (not a monk) at the Kongobuji Temple (as noted above, the head temple at Koyasan) and nine other temples. These were selected with the support of the Koyasan Temple Lodging Association, which fulfils an administrative role for all lodgings or *shukobo* at Koyasan, on the basis of their different approaches to the provision of accommodation and

Table 29.1 Temple lodging interviews at Koyasan

Believers only and Japanese only	*Believers/Tourists and Japanese only*
Temple Lodging A (Head monk)	Temple Lodging B (Head monk) Temple Lodging C (Subordinate monk)
Tourists and Japanese/Foreigners	*Believers/Tourists and Japanese/Foreigners*
None	Temple Lodging D (Head monk) Temple Lodging E (Office manager) Temple Lodging F (Head monk) Temple Lodging G (Subordinate monk) Temple Lodging H (Subordinate monk) Temple Lodging I (Vice head monk)

other services. In particular, two sets of variables were employed to select lodgings: Those whose guests are 'believers only' (religious) or are also 'tourists' (secular) and those whose guests are 'Japanese only' (domestic visitors) or are also 'foreigners' (international tourists). Table 29.1 details the interviewees based upon these variables. As discussed shortly, these variables are also evidence of a principal manner in which different temples at Koyasan respond to the challenges of offering temple-stay experiences to tourists.

Research outcomes

To contextualise the more specific questions regarding specific means of achieving a balance between commodification for economic gain and maintaining the sanctity of the temples and the lives of their residents, the research first sought to establish the monks' overall attitudes towards temple-stay tourism. Broadly, those interviewed held positive attitudes, for two reasons. First, it was accepted as a necessary way of addressing the financial difficulties faced by the Shingon sect, not only at Koyasan itself but also at 'branch' temples elsewhere in Japan. On the one hand, the number of monks and practicing Buddhists visiting the temples has long been in decline; on the other hand, the number of tourists visiting Koyasan continued to demonstrate an upward trend. Hence, admission fees and other spending represented a vital source of income to sustain the religious activities of the temples. As one respondent remarked:

> Kongobuji Temple is the head temple of the Koyason Shingon sect, with about 3600 branch temples all around Japan. Therefore, Kongobuji has two kinds of income... on the one hand, money from monks and from branch temples, but this income is declining. On the other hand, the main income at the head temple is from admission fees, offering charms, amulets and prayers, and from donations. We are thankful that many people visit Koyasan, regardless of whether they are believers or tourists, because they pay admission fees, buy charms and offer donations.

Second, the monks also perceive the opening up of Koyasan to tourists to be an opportunity to fulfil their mission of sharing the teachings of Buddhism as widely as possible, supporting a similar finding from other studies (Wong et al., 2013). This was clearly articulated by one monk (Temple C):

> I don't care about individual motivations to visit Koyasan... However, Koyasan is a sect of Mahayana Buddhism [which] teaches that everyone can take a large boat together,

Table 29.2 A typology of temple-stay lodgings at Koyasan

	Specialist Faith	*Faith and Stay*	*Experiential*
Core service	Faith: religious services, prayers	Faith (prayers, etc. and accommodation/food)	Temple stay Buddhist training
Facilitating services	Accommodation/food Postcards Personal consultation/ counselling	Multi-language service Luggage storage Reservation services	Multi-language service Luggage storage Reservation services
Supporting services	Buddhist training	Buddhist training	Food
Guests	Believers/Japanese only (Temple A)	Believers/Tourists and Japanese/International (Temples B, D, E, F, and H)	Believers/Tourists and Japanese/International (Temples C, G, and I)

> whilst Theravada Buddhism teaches that individuals need to take a small boat by themselves. I would like to let many visitors take this large boat together through teaching the Esoteric Buddhism developed by the Great Monk Kukai.

Nevertheless, despite this overall positive perspective, the nature of the provision of temple-stays at Koyasan varies between temples dependent upon both the financial needs and the position adopted by the head monks of individual temples with regard to the extent to which tourism is embraced. The research revealed three distinctive approaches to (or levels of) commodification of the temples for touristic consumption which, for convenience, can be framed by Gronroos' (2000) widely-cited concept of service products comprising three elements: core, facilitating and supporting services. Table 29.2 summaries these three approaches, defined here as 'Specialist Faith', 'Faith and Stay' and 'Experiential'. These are discussed in more detail in the following sections.

Specialist faith

The core focus of this form of temple stay is on religious practices, particularly praying for ancestors, and is available only to Japanese 'believers' who actively engage in Buddhism. The provision of accommodation and food (the same food that monks eat) is hence one of a number of facilitating services, such as mailing postcards or counselling guests on personal or religious matters from a Buddhist perspective, whilst the opportunity exists for guests to participate in training/education in Buddhism as an additional supporting service. For the temples, the benefit of this form of temple-stay tourism is that it is, in a sense, 'low impact-high value'; guests pay relatively high prices but also develop a relationship with the temple that continues from one generation to another. Moreover, numbers are restricted to one or two small groups at a time, so the monks are able to focus on the spiritual needs/ experiences of individual guests who are able to concentrate on their prayers in a quiet and calm atmosphere. Thus, an authentic experience is maintained for both monks and guests.

Faith and stay

This is similar to the 'specialist faith' stay inasmuch as religious services are offered as a core 'product'. However, accommodation and food are also offered as core services and, significantly, guests, who may be domestic or international and practicing /non-practicing

Buddhists, and include both individuals and organised groups, are not required to attend religious services. Hence, this temple-stay can be thought of as 'ordinary' accommodation but within a religious complex, and the spiritual element may be limited to attending training in Buddhist activities such as meditation or Sutra (Buddhist scripture) copying, considered to be a devotional practice in Buddhism. Significantly, the monks do not make concessions to the potential needs of international/non-Buddhist guests (for example, all guests are served the same vegetarian food). That is, all guests are received and treated, as far as possible, in an 'authentic' religious manner, although written instructions on how to behave within the temple are provided in English for international visitors. In addition, these temple-stays are not promoted through agencies, again limiting the degree of commodification:

> Koyasan was originally founded as a training site, so I think offering a temple-stay service is enough if it earns money for running the temple. So, I do not refuse or select guests, but I don't contract with specific tourist agencies. And I keep time to pray in the daytime.
>
> *(Temple B)*

Experiential

Three temples in this research (C, G and I) offer a temple-stay 'package' that perhaps most closely aligns with the broader concept of retreat tourism. The core product is a temple-stay experience that offers visitors, the great majority of whom are international tourists not familiar with Buddhism, the opportunity to experience prayer/religious services as well as training in Buddhist practices. Moreover, it is also the most commodified form of temple-stay tourism at Koyasan; the temples have their own Internet and social media pages to attract and communicate with potential guests, and information is provided in English to explain day-to-day life and practices at the temples. In addition, one temple (Temple C) has created a package specifically targeting female guests, including special vegetarian menus and sutra copywriting:

> We have made a package called the 'maiden plan', collaborating with a tourist agency. For a long time, Koyasan prohibited entry to women... however, we would like even women on their own to feel free to visit Koyasan...
>
> *(Temple C)*

Another temple created an additional experience to complement the temple-stay package, namely, a night tour of Okunoin cemetery. Initially created unofficially by monks for groups of visitors, it became transformed into an organised tour available also to non-temple-stay guests:

> We started the night tour six or seven years ago... the reason that we started it was that we saw many guests with nothing to do [after mediation and dinner]... We visited Okunoin and prayed at night time on the monthly date anniversary of Kobo Daishi's death, and we decided to do it in English. On average, we had 20 to 40 participants. We found they did not even know the name of Kobo Daishi; that is why we decided to talk about him and his teachings on the night tour.
>
> *(Temple I)*

Another temple (G) occasionally organises special stays based on 'purifying mind and body'. Outside specialists offer training in yoga and vegetarian cooking, while the traditional activities of meditation and sutra copying are supplemented with other activities such as group running. Such commodified experiences compete with traditional notions of the Buddhist faith, yet the monks in these temples consider it a legitimate means of sharing Buddhist teachings:

> We would like to talk about kobo Daishi's teaching... we are sure we can spread Buddhist teachings, even to international guests, as long as we work hard.
>
> *(Temple I)*

Generally, then, the segmentation of temple-stay tourism based on different visitor demands, the financial needs of specific temples and the varying attitudes of monks appears to be a pragmatic means of achieving coexistence between monks and visitors. Nevertheless, as the following section discusses, the research revealed a number of challenges that still exist in managing temple-stay tourism at Koyasan.

Managing temple-stay tourism at Koyasan

Although the temples in this research distinguished between 'believers' and 'tourists' (or non-believers), a major issue to arise from the research was that some monks expressed concern that even those who visit Koyasan to pray for their ancestors are losing faith:

> It seems that believers now don't hesitate to watch TV and drink alcohol after the ceremonies.
>
> *(Temple I)*

> Most Japanese visitors are day trippers; they now have less desire to pray for their ancestors or visit graveyards... More and more guests visit Koyasan to enjoy the cool air in summer, to experience forest therapy and to eat vegetarian food. Moreover, some misunderstand temple lodgings; they think they are general accommodation, and ask if they have lunch there.
>
> *(Temple E)*

Perhaps as a consequence, many temples face a number of challenges in managing the behaviour of visitors, both domestic and international. Respondents in the research lamented that many guests do not observe the times for meals, prayers and training sessions, thereby disrupting these activities and the lives of the monks (Temple G), while other guests often talk loudly, disturbing the calm atmosphere of the temples (Temple C). International tourists in particular do not follow traditional modes of behaviour. For example, they fail to take off their shoes, they lean on sliding paper doors and damage them, and they wear swimming costumes in the baths (Temple I).

To an extent, such problems are addressed by the provision of instructions on arrival, a guide book in guests' rooms and signage around the temples (Temples C, E, G and I). Monks also warn visitors verbally if their behaviour is inappropriate (Temples C and G). Nevertheless, the growing number of guests that some temples are now attracting makes it more difficult for monks to meet all their needs, particularly when catering for both traditional believers and for experience-seeking tourists. For example, during the research, it was observed that two Japanese women at a ceremony (at Temple G) attended primarily by international tourists were confused when, following prayers to ancestors, rather than giving a sermon as conventionally expected, the attendant monks commenced a question and answer session on Buddhism in English with the international visitors. Nor is it only visitors who are impacted; growing demands for training in sutra copying and other activities is imposing on the lives and workloads of the monks. For example, Temple G has a monk who can instruct on meditation and two on sutra copying. One of those is the same monk who instructs on meditation; he once lost his chance to have dinner because of the time it took for these sessions. Interestingly, a recent newspaper article (Japan Times, 2018) reports that one monk has filed a lawsuit for damages and unpaid wages against his temple, claiming that he was suffering depression from overwork.

In addition, although the temples at Koyasan seek to maintain their sanctity and traditions as far as possible, there is a need for some, particularly those attracting international visitors, to respond to the needs of their guests (often revealed in on-line evaluations of their experiences) to remain competitive. For example, one temple decided to provide Wi-Fi, coffee lounges and morning baths following repeated requests from guests (Temple I), while another offered alternative vegetarian food to suit international tastes (Temple G). Temple C found it necessary to redesign its interior to improve accessibility for its ageing, predominantly domestic, clientele. In other words, as the temples of Koyasan become increasingly dependent on tourism as a source of income, they are having to adapt and commodify the services they provide and the overall temple experience in order to meet the varying and evolving needs of their guests.

Conclusion

As observed in the introduction to this chapter, temple-stay tourism is becoming an increasingly sought-after experience (Son & Xu, 2013). Yet, as with all forms of religious tourism, the provision of temple-stay experiences involves some degree of commodification of religious places and practices, potentially impacting upon their sanctity and on the authenticity of the religious lives of the temples' residents. Hence, the purpose of this chapter was to explore how the monks of Koyasan in Japan, as a specific case study, seek to manage temple-stay tourism in order to maintain as far as possible their traditional religious existence or, in other words, to coexist with tourists and tourists.

Temple-stay tourism at Koyasan has developed primarily out of financial need; without the income from tourists, the future of the temples would be uncertain. Yet, the commodification of their religion (Mahayana Buddhism) for touristic consumption is justified by a central tenet of that religion – to spread the teachings of Buddhism as widely as possible. Hence, the provision of temple-stay experiences not only contributes necessary financial resources but also enables the monks to fulfil, in part, their religious duties.

Nevertheless, a variety of temple-stay 'products' are offered by different temples at Koyasan, essentially on a continuum from traditional, religiously-focused stays to a broader temple experience for secular visitors. This enables the individual temples to manage tourism according to both financial need and the extent to which they wish to minimise the commodification of their religious practices, although as this chapter has revealed, a number of more specific measures are also necessary to ensure that the behaviour of visitors not only conforms with the traditions of temples but also does not disrupt the day-to-day lives of the monks.

However, as this chapter has also suggested, these practical arrangements may not be sufficient in the longer term to maintain coexistence between the monks and tourists. More specifically, those temples or *shukobo* offering temple-stay experiences may be required to respond to transformations in the markets for tourism to Koyasan, not least the increasing numbers of international non-Buddhist tourists and the evident increasing secularisation of the domestic Japanese market. As a consequence, it may become more difficult for the temples and the monks not to compromise their traditional religious practices; that is, it may become increasingly necessary for them to adapt to the needs of tourists (to commodify further the temple-stay) and, in so doing, perhaps seek means of distinguishing between the provision of temple-stay experiences to tourists and their own religious practices. In other words, it may become necessary to formalise the temple-stay 'business' with, for example, defined tasks and roles for the monks, separate from but supporting the continuation of their traditional activities.

References

Akiyama, F. (2015) 高野山における宿坊の変容と観光 [Transformation of temple lodgings and tourism in Koyasan] (master's thesis). Wakayama University, Japan.

Akiyama, F. (2018) 高野山の宿坊 [Temple lodgings in Koyasan]. In K. Kanda, Y. Ohura & K. Kato (Eds), *大学的和歌山ガイド：こだわりの歩き方 [Wakayama guide from university's specialist perspective: Selected ways of stroll]*. Kyoto: Showado, pp. 25–41.

Appadurai, A. (1986) *The Social Life of Things: Commodities in Social Perspective*. Cambridge: Cambridge University Press.

Attix, S. (2002) New age-oriented special interest travel: An exploratory study. *Tourism Recreation Research*, 27(2), 51–58.

Carrette, J., & King, R. (2005) *Selling Spirituality: The Silent Take-over of Religion*. Abingdon: Routledge.

Chun, B., Roh, E., & Spralls, S. (2017) Living like a monk: Motivations and experiences of international participants in templestay. *International Journal of Religious Tourism and Pilgrimage*, 5(1), 39–55.

Cohen, E. (1988) Authenticity and commoditization in tourism. *Annals of Tourism Research*, 15(3), 371–386.

Drule, A.M., Chis, A., & Ciornea, R. (2012) The spiritual, ethical and economical impact of religious tourism: The case of Transylvanian monasteries. In *The Proceedings of the International Conference 'Marketing: From Information to Decision*. Cluj-Napoca: Babes Bolyai University, pp. 89–100.

Einstein, M. (2008) *Brands of Faith: Marketing Religion in a Commercial Age*. Abingdon: Routledge.

Funck, C., & Cooper, M. (2015) *Japanese Tourism: Spaces, Places and Structures*. Oxford: Berghahn Books.

Greenwood, D. (1989) Culture by the Pound: An anthropological perspective on tourism as cultural commoditization. In V. Smith (Ed.), *Hosts and Guests: The Anthropology of Tourism*, 2nd edn. Philadelphia: University of Pennsylvania Press, pp. 171–185.

Griffin, K., & Raj, R. (2017) The importance of religious tourism and pilgrimage: Reflecting on definitions, motives and data. *International Journal of Religious Tourism and Pilgrimage*, 5(3), 2–9.

Gronroos, C. (2000) *Service Management and Marketing: A Customer Relationship Management Approach*, 2nd edn. Hoboken, NY: John Wiley & Sons.

Heelas, P., & Woodhead, L. (2005) *The Spiritual Revolution: Why Religion Is Giving Way to Spirituality*. Oxford: Blackwell.

Heintzman, P. (2013) Retreat tourism as transformational tourism. In Y. Reisinger (Ed.), *Transformational Tourism: Tourist Perspectives*. Wallingford: CABI, pp. 68–81.

Ikeda, K. (2015) 聖地の観光地化とマス・ツーリズム：高野山における交通の発達に伴う聖地空間の再編と役割の変化 [The Cultural Influence of the Development of Infrastructure in a Religious Site Koyasan]. *The Tourism Studies*, *26*(2), 61–72.

Japan Times (2018) Monk sues temple at Mount Koya World Heritage site over heavy workload. Available at: https://www.japantimes.co.jp/news/2018/05/17/national/crime-legal/monk-sues-temple-mt-koya-world-heritage-site-heavy-workload/#.XSVJfj_7SUk (Accessed 23 July 2019).

Kaelber, L. (2006) Paradigms of travel: From medieval pilgrimage to the postmodern virtual tour. In D.J. Timothy & D.H. Olsen (Eds), *Tourism, Religion and Spiritual Journeys*. Abingdon: Routledge, pp. 49–63.

Kaplan, U. (2010) Images of monasticism: The temple stay program and the re-branding of Korean Buddhist temples. *Korean Studies*, 34, 127–146.

Korea Bizwire (2018) Over 70,000 foreigners experienced temple stay programs last year. Available at: http://koreabizwire.com/over-70000-foreigners-experienced-temple-stay-programs-last-year/114189 (Accessed 5 July 2019).

MacCannell, D. (1973) Staged authenticity: Arrangements of social space in tourist settings. *American Journal of Sociology*, 79(3), 589–603.

Matsunaga, Y. (2014) *高野山 [Koyasan]*. Tokyo: Iwanami Shinsyo.

McKean, P. (1989) Towards a theoretical analysis of tourism: Economic dualism and cultural involution in Bali. In V. Smith (Ed.), *Hosts and Guests: The Anthropology of Tourism*, 2nd edn. Philadelphia: University of Pennsylvania Press, pp. 119–138.

McKelvie, (2005) Religious tourism. *Travel and Tourism Analyst* 4, 1–17.

Miyasaka, Y., & Sato, T. (1984) *高野山史 [History of Koyasan]* Tokyo: Shinkousya.

OECD (2009) Temple stay programme: Republic of Korea. In *The Impact of Culture on Tourism*, OECD, pp. 115–128. Available at: http://www.mlit.go.jp/kankocho/naratourismstatisticsweek/statistical/pdf/2009_The_Impact.pdf (Accessed 5 July 2019).

Olsen, D.H. (2003) Heritage, tourism, and the commodification of religion. *Tourism Recreation Research*, 28(3), 99–104.

Olsen, D.H., & Timothy, D.J. (2006) Tourism and religious journeys. In D.J. Timothy & D.H. Olsen (Eds), *Tourism, Religion and Spiritual Journeys*. Abingdon: Routledge, pp. 1–22.

Olsen, K. (2002) Authenticity as a concept in tourism research: The social organization of the experience of authenticity. *Tourist Studies*, 2(2), 159–182.

Pearce, P., & Moscardo, G. (1986) The concept of authenticity in tourist experience. *The Australian and New Zealand Journal of Sociology*, 22(1), 121–132.

Raj, R., & Griffin, K. (Eds) (2015) *Religious Tourism and Pilgrimage Management: An International Perspective*, 2nd edn. Wallingford: CABI.

Rodrigues, S., & McIntosh, A. (2014) Motivations, experiences and perceived impacts of visitation at a Catholic monastery in New Zealand. *Journal of Heritage Tourism*, 9(4), 271–284.

Ron, A.S., & Timothy, D.J. (2019) *Contemporary Christian Travel: Pilgrimage, Practice and Place*. Bristol: Channel View Publications.

Ryan, M., & McKenzie, F. (2003) A monastic tourist experience: The packaging of a place. *Tourism Geographies*, 5(1), 54–70.

Shackley, M. (1998) A golden calf in sacred space: The future of St. Katherine's Monastery, Mount Sinai (Egypt), *International Journal of Heritage Studies*, 4, 123–134.

Shackley, M (2004) Accommodating the spiritual tourist: The case of religious retreat houses. In R. Thomas (Ed.), *Small Firms in Tourism: International Perspectives*. Oxford: Elsevier, pp. 225–237.

Shackley, M. (2006) Empty bottles at sacred sites: Religious retailing at Ireland's National Shrine. In D.J. Timothy & D.H. Olsen (Eds), *Tourism, Religion and Spiritual Journeys*. Abingdon: Routledge, pp. 94–103.

Sharpley, R. (2009) Tourism, religion and spirituality. In T. Jamal & M.K. Robinson (Eds), *The Sage Handbook of Tourism Studies*. London: Sage Publications, pp. 237–253.

Sharpley, R., & Sundaram, P. (2005) Tourism: A sacred journey? The case of ashram tourism, India. *International Journal of Tourism Research*, 7(3), 161–171.

Shi, F. (2011) Business at religious sites: Bless or sin? In K. Andriotis, A. Theocarous & F. Kotsi (Eds), *Proceedings of Tourism in an Era of Uncertainty Conference*. Cyprus: International Association for Tourism Policy, pp. 207–215.

Smith, M., & Kelly, C. (2006) Holistic tourism: Journeys of the self? *Tourism Recreation Research*, 31(1), 15–24.

Son, A., & Xu, H. (2013) Religious food as a tourism attraction: The roles of Buddhist temple food in Western tourist experience. *Journal of Heritage Tourism*, 8(2/3), 248–258.

Song, H.J., Lee, C.K., Park, J.A., Hwang, Y.H., & Reisinger, Y. (2015) The influence of tourist experience on perceived value and satisfaction with temple stays: The experience economy theory. *Journal of Travel & Tourism Marketing*, 32(4), 401–415.

Steiner, C., & Reisinger, Y. (2006) Understanding existential authenticity. *Annals of Tourism Research*, 33(2), 299–318.

Suntikul, W., & Butler, R. (2018) Tourism and religion: Origins, interactions and issues. In R. Butler & W. Suntikul (Eds), *Tourism and Religion: Issues and Implications*. Bristol: Channel View Publications, pp. 1–13.

Tourist Bureau, Department of Commerce, Industry, Tourism and Labor in Wakayama Prefecture (1959–2017) *和歌山県観光客動態調査報告書 [Wakayama tourist movement investigation report]*.

Triantafillidou, A., Koritos, C., Chatzipanagiotou, K., & Vassilikopoulou, A. (2010) Pilgrimages: The 'promised land' for travel agents? *International Journal of Contemporary Hospitality Management*, 22(3), 382–398.

Voigt, C., & Pforr, C. (Eds) (2013) *Wellness tourism: A Destination Perspective*. Abingdon: Routledge.

Vukonić, B. (2002) Religion, tourism and economics: A convenient symbiosis. *Tourism Recreation Research*, 27(2), 59–64.

Wang, N. (1999) Rethinking authenticity in tourism experience. *Annals of Tourism Research*, 26(2), 349–370.

Wang, W. (2011) *Explore the Phenomenon of Buddhist Temple Stay in South Korea for Tourists*. UNLV Theses, Dissertations, Professional Papers, and Capstones. 1074. Available athttps://digitalscholarship.unlv.edu/thesesdissertations/1074 (Accessed 2 July 2019).

Ward, G. (2006) The future of religion. *Journal of the American Academy of Religion*, 74(1), 179–186.

Wong, C. (2017) Being a monk for a day. Really? The case of Pu-Tuo-Shan. Critical Tourism Studies Proceedings, 17, Article 145. Available at: http://digitalcommons.library.tru.ca/cts-proceedings/vol2017/iss1/145 (Accessed 16 July 2019).

Wong, C. (2018) The monks and nuns of Pu-Tuo as custodians of their scare Buddhist site. In R. Butler & W. Suntikul (Eds), *Tourism and Religion: Issues and Implications*. Bristol: Channel View Publications, pp. 99–114.

Wong, C, McIntosh, A., & Ryan, C. (2013) Buddhism and tourism: Perceptions of the monastic community at Pu-Tuo-Shan, China. *Annals of Tourism Research*, 40, 213–234.

Wuthnow, R. (1998) *After Heaven: Spirituality in America Since the 1950s*. Berkeley: University of California Press.

Xie, P. (2003) The bamboo-beating dance in Hainan, China: Authenticity and commodification. *Journal of Sustainable Tourism*, 11(1), 5–16.

30

FOOD AND RELIGION

Tourism perspectives

Silvia Aulet, Carlos Fernandes, and Dallen J. Timothy

Introduction

Eating is essential to sustain life, but beyond satisfying physiological needs, food has close connections to other aspects of life. People's alimentary behaviours are influenced by many sociocultural and economic factors, including religion and nationality. Although it can be a source of tension within a community, food is a social connector; it helps create pleasant experiences, underscores social well-being, and is fodder for familial traditions and memory-making (Carolan 2012; Contreras & Gracia 2005; Gofton 1989; Poulain 2017; Ward, Coveney & Henderson 2010).

Many scholars have drawn clear connections between food and tourism, not only with regard to food services as part of the hospitality system but also gastronomy and cuisine as a crucial part of an area's cultural traditions and its production methods as tourism assets (Hall & Gössling 2016; Hjalager & Richards 2002; Richards 2012). The linkages between food and tourism also provide a foundation for local economic development, regional tourism branding, and a platform for destination marketing, as well as a mechanism for building social solidarity in protecting local culture (Bessière 2013; Richards 2012).

Much contemporary discourse about food focuses on sustainability, security, and culinary traditions (Pinstrup-Andersen 2009; Timothy 2016). Part of the latter notion is the expanding work on culinary heritage and the role of food in national or regional identity formation (Frost & Laing 2016; Ramshaw 2016). The World Tourism Organization (2012, 2019) has repeatedly emphasized the importance of culinary heritage and gastronomic traditions in the development of regional tourism and in the formation of regional identity.

As already noted, food is an extremely important part of a region's heritage. It commemorates important happenings and underscores the identity of peoples and places. Gastronomy is celebrated with major cultural events, and food is a core part of every society's celebrations of birth, death, life stages, and faith. In fact, food and religion are inextricably connected in many ways. Indeed, numerous religious and spiritual practices have direct or indirect connections to food. As religion is, like food, a key part of social identity and cultural heritage, the juxtaposition of food, faith, and tourism is important for developing a deeper understanding between religion, human heritage, and religious and spiritual practice. This chapter examines food as heritage and religious practice within the context of religious and spiritual tourism. It is divided into three sections, namely food and heritage as cultural practice, the role of food in religion, and food, religion, and tourism.

Food, heritage, and culture

Food and eating have important social and cultural dimensions (Bessière 2013; Jolliffe 2010; Timothy 2016, 2021). Every society and cultural group has its own food preparation techniques, ingredients, and eating traditions that are important and symbolic. The cultural values and social codes where people grow up determine cooking and food consumption practices. Thus, alimentary traditions frequently differ between cultures and enable individuals to express their cultural identity (González Turmo 2007; Nunes dos Santos 2007).

Food is a crucible of cultural knowledge and a manifestation of history. A region's history of wealth or poverty, abundance or scarcity, climatic conditions, and humankind's struggles with nature is evident in the culinary traditions that dominate a country or region. Culinary traditions are among the longest-surviving elements of culture. They reinforce social identity, support feelings of nationalism or patriotism and mirror lifestyles, religious beliefs, habits, and customs. They tell tales of colonialism and slavery, Indigeneity and subjugation, survival and death, and triumph and defeat (Timothy 2016). Thus, food consumption behaviours denote an expression of social class and resource availability (Carolan 2012; Poulain 2017). In the words of Timothy and Ron (2013a: 275),

> food, cuisine and culinary traditions are among the most foundational elements of culture. While there is a long tradition of identifying many places with their traditional foods, cuisine is becoming an ever more important part of the contemporary cultural heritage of regions and countries.

Gastronomy and cuisine refer to a set of knowledge and practices related to culinary art, recipes, ingredients, techniques, and methods. They entail historical evolution, cultural meanings, and the natural environment from which food sources are obtained and the way in which they are used, as well as the social and cultural aspects that intervene in the relationship. Gastronomy and cuisine are also included as part of intangible heritage—the practices, representations, expressions, knowledge, and techniques that provide communities, groups, and individuals with a sense of identity and continuity according to the UNESCO Convention of 2011 (World Tourism Organization 2012).

With the globalization of people's tastes, and as people have become increasingly willing to widen their gustatory spaces, there has been a rapid growth in culinary tourism (Everett 2016; Hall & Sharples 2003; Sormaz et al. 2015; Timothy & Ron 2013b). Given that local cuisine is so fully imbued with a place's intangible heritage, tourists can gain a more authentic or holistic cultural experience by partaking of local specialty foods (Okumus, Okumus & McKercher 2007).

The Committee on Culture and Education of the European Parliament has recognized the importance of food and gastronomy as artistic and cultural expressions and fundamental pillars of family and social relationships (European Parliament 2014). Many of the heritage expressions on UNESCO's Intangible Heritage list are gastronomic traditions and food items. UNESCO's inscription requirements emphasize the importance of food rites, preparation methods, group knowledge, traditions, and symbols related to the act of preparing and eating specific foods (UNESCO 2020). Recognizing the importance of food and gastronomy in human heritage, including religion, it would be remiss not to examine the relationships between food and religion as well as the ties between food, religion, and tourism.

Food and religion

Food and drink play a pivotal role in religion and often lie at the core of doctrines, dogmas, rituals, and practices. As Anderson (2014: 189) notes, food 'is central to religion as a symbol, as a subject of prayers, as a maker of sharing and as communion'.

Religion is a defining feature of human cultures and has existed in various forms since prehistoric times. Religion and spirituality are studied from many different angles within the social sciences (Aulet & Sureda 2019), especially from a social phenomenological perspective, which is the focus of this chapter. Religion, faith, and spirituality underscore the human experience and have long provided a moral and ethical compass and social framework for living for most of humankind throughout the Anthropocene (Ron & Timothy 2019; Velasco 1982)

Religion plays several important social roles. From a food or gastronomic perspective, two of these deserve particular mention (Díez de Velasco 1998). First, religious ecology deals with how faith determines people's interactions with their host ecosystems. Ecological protective measures that minimize anthropic impacts on vital resources are an example, as is certain religion-based taboos against eating specific animal species (Harris 1980; Jenkins et al. 2017). Second, religion provides a framework for human co-existence and behavioural norms within society. This has implications for food and meals, such as the *langar*, or community dining in Sikhism where food is offered to all attendants as a way of showing charity (Kaur 2020; Popli 2010), as well as the rising phenomenon of providing food for the poor as an act of faith-filled service (Denning 2021; Sack 2000).

All religions encompass rites, rituals and spiritual performances that are unique to each one. These procedures demonstrate individual or collective devotion to deity and typically encompass prayers, offerings, and actions that educe transcendent experiences (Gaarder et al. 2009). Rituals vary between religions and even between individual sects within religions, but they almost all share certain common elements, such as communion with deity in sacred places and times and leadership of revered individuals. They also possess high symbolic values and perpetuate the religious narrative, follow prescribed guidelines and actions, and often include music, aromatics, or food (Aulet 2012). Related to this last point, Timothy and Ron (2016: 104) suggest that "food and religion are in many ways inseparable. Food and drink form part of the physical manifestation of beliefs and spiritual traditions, and food is frequently the centre of people's spiritual universe". In addition to food's clear connections to religious rituals, many people's gustatory choices and behaviours outside of ritualistic circumstances in everyday life are determined by the faith they follow.

The Bible frequently mentions food as a manifestation of miracles, as a gift from God, as a medium of sustenance, as a sign of godly devotion, and as a substance to give to the poor and needy (Ron & Timothy 2013, 2019; Timothy & Ron 2016). The Quran also mentions eating and drinking frequently, with "sustenance being one of the chief signs of God's existence and concern for which humans should give thanks" (Feeley-Harnik 1995: 567). Muslims are exhorted to be aware of what they eat and to eat according to tradition, for a healthy body corresponds to a healthy soul (Rouse & Hoskins 2004). Sacred scriptures and dogmas in most religions discuss the importance of food—what to eat, what to avoid, how to prepare or eat a meal, the importance of celebratory feasts, and encouraging adherents to feed the needy. Some of these concepts are elaborated on below.

Food prohibitions

Most culinary prohibitions regard the animal kingdom. While Christianity is quite lax on this issue, religions such as Islam, Hinduism, and Judaism have stricter codes related to consuming animals. Animals such as pigs, wild boars, dogs, snakes, monkeys, carnivorous animals with claws and tusks, clawed birds of prey, harmful animals (e.g., rats), animals generally considered repulsive (e.g., flies), mules and donkeys, poisonous animals, and some aquatic animals are proscribed for Muslims and deemed *haram* (forbidden) (Ribagorda Calasanz 1999). Animals that are not slaughtered according to Islamic law are also *haram*. Likewise, alcoholic beverages are forbidden (Regenstein et al. 2003; Riaz & Chaudry 2004; Timothy & Ron 2016).

Judaism's *kosher* laws pertain to foods that meet strict requirements for production and consumption. For example, blood cannot be consumed, and meat and milk products cannot be served together in the same dish or even in the same meal. The Torah is quite clear regarding which animals can be consumed and which cannot. Terrestrial animals that have cloven hooves and ruminates (i.e., cows, sheep, goats, and deer) are kosher, while those that do not meet these two conditions are not. The law forbids consuming pigs, rabbits, squirrels, dogs, cats, camels, horses, and so forth. The Torah also lists impure birds and insects that may not be consumed. Kosher marine animals must simultaneously have fins and scales, which preclude shellfish. Logically, all products derived from these animals are also prohibited, except for a curious exception: honey, since it is considered a product of flowers rather than bees (Masoudi 1993; Regenstein & Regenstein 1991; Regenstein et al. 2003; Timothy & Ron 2016).

Hinduism does not require a vegetarian diet, although it is recommended because of *ahimsa*—non-violence against all life forms, including animals. Many Hindus interpret holy writ condemning violence as a justification for a vegetarian diet, including the Sanskrit epic Mahabharata's maxim that non-violence is the highest duty and the highest teaching. There is also a common precept that non-vegetarian diets are detrimental to the mind and spirit (Ghassem-Fachandi 2006; Kaza 2005; Subramaniam 2011). However, within Hinduism, there is a great diversity of gustatory traditions that range from strictly vegan diets to vegetarian diets that include ovo-lacto foods, as well as meat and fish-eating traditions. Beef is rarely eaten, as the cow is considered a sacred animal and symbol of life (Khare 1992).

Other religious traditions may have recommendations or restrictions but are not as strict as the ones already mentioned. For example, Buddhism does not have set dietary laws, but adherents are encouraged to follow a vegetarian regimen whenever possible (Khare 1992; Rosen 2011). Monks and some very conservative orders of Buddhism in East Asia are prohibited from eating 'pungent' vegetables, such as onions, leeks, and garlic, and are discouraged from using added flavours, such as salt, chilies, and certain spices (Son & Xu 2013).

Christianity is the least regulated religion with regard to culinaria, although certain Christian denominations do have dietary restrictions. Many protestant sects discourage excessive alcohol consumption. Based on biblical interpretations, Jehovah's Witnesses refrain from consuming blood or blood-based products. They may eat meat as long as the animal is properly bled out. The strictest Seventh-day Adventists follow a vegetarian diet, although some adherents eat meat that is low in fat and 'clean', or meats that are described as kosher in the biblical Book of Leviticus. Most Adventists also refrain from using alcohol, tobacco, drugs, and caffeine. Most members of the Church of Jesus Christ of Latter-day Saints also adhere to a strict policy of eating meat in moderation, avoiding the consumption of alcohol and hot caffeinated beverages (e.g., tea and coffee), and avoiding tobacco and non-prescribed narcotics (Timothy & Ron 2016). In some Christian monastic societies, food rules prohibit the eating of four-legged animals (Just 2007).

In short, many religions forbid eating certain foods because the foods are impure (e.g., Judaism and Islam) because people should do no harm to other living creatures (e.g., Hinduism and Buddhism) or because they are unhealthy (e.g., Buddhism and certain Christian sects). Consuming fruits, vegetables, and grains is encouraged by almost all religions, with only a few exceptions with prohibitions against eating certain grains or consuming certain foods at specific times of the year (Timothy & Ron 2016).

Food rituals

Besides regulating the content of food, some religious mores also prescribe how to handle and prepare food. In Islam, *halal* (permissible) animals must be slaughtered ritually according to Islamic law, which dictates that the slaughterer must be a Muslim who knows correct Islamic procedures. Kosher law requires similar rituals. Animals that die naturally cannot be eaten and the ritual slaughter must be carried out in a very specific manner (Popovsky 2010; Regenstein & Regenstein 1991; Regenstein et al. 2003). In Hinduism, there are specific rules for preparing and eating food, such as ritual washings of hands, feet, and mouth before a meal (Khare 1992; Timothy & Ron 2016). Eating is usually done with the fingers of the right hand in Hinduism and Islam (Regenstein et al. 2003).

Hindu traditions consider cooking a sacred rite. Since ancient times, food has been used to venerate gods and comfort the human soul and body. In Indian temple kitchens, food is prepared for deities and devotees. Temple food was traditionally considered to have healing properties. Historically, offerings (*bhog*) to Hindu gods were prepared only with traditional and natural local ingredients (nothing processed or artificial) and could only be cooked or warmed with naturally occurring elements such as coconut husks or logs from certain trees. According to Hindu tradition, food is impregnated with the essence of the fuel sources, which help cleanse suffering when eaten, so only certain types of natural fuels should be used (Appadurai 1981; Contreras 2007; Srinivas 2006).

Religious connections to food are also manifested during festivals and other celebratory times. Certain comestibles are either required or encouraged during pilgrimages, religious services, or on holidays. Religious feasts are an important manifestation of godly adoration and fulfilment of mortal duty. Almost every religion on earth celebrates certain events through formal feasts. Thanksgiving is the largest annual feast in the United States and Canada. Although its origins have been questioned by historians but believed to have originated as a harvest celebration, it became romanticized in American folklore as a sacred holiday—a time to express gratitude to God for bounteous blessings (Baker 2009). Today, although the holiday is becoming increasingly secular, it is celebrated each November (October in Canada), and for many faiths and families, Thanksgiving continues to represent a day of feasting and prayer, as a symbolic representation of gratitude for the blessings of heaven. Throughout Christendom, Christmas and Easter celebrations are marked by large feasts, with themes of hope, renewal, and rebirth. Religious feasts abound in Judaism, Islam, and other major world religions. Jews commemorate seven important historical events each year with feasts. Among the best known of these are Passover, Yom Kippur, and the Feast of the Tabernacles (Sukkot) (Ioannides & Ioannides 2006). Eid al-Fitr marks the end of the month-long fast in Islam and is celebrated with special meals. The annual Eid al-Adha, or Feast of the Sacrifice, commemorates Abraham's willingness to sacrifice his son, Ismael, in obedience to the will of God. This holiday is also marked with special feasts (Winchester 1999).

Whereas ritual feasting is encouraged on certain days of the year, moderation is generally encouraged at other times. Moderation is a core religious principle related to eating that deals with both health and self-control. Buddhism and Hinduism call for austerity and moderation. Muhammad instructed Muslims not to "kill your hearts (soul) by eating a lot and drinking a lot, since the heart is like a land of agriculture that if you pour a lot of water on it, it will be damaged" (quoted in Di Marzo 2016: n.p.). To exercise self-restraint, monastic communities of several religious traditions encourage moderation as well (Guzmán Parejo 2018; Just 2007; Thera 1994).

Connected to the larger phenomenon of moderation, fasting is a common practice in many faiths. Abstinence and fasting have a clear spiritual purpose: to promote self-discipline, to symbolize penance, and to facilitate spiritual communion with a higher power. Timothy and Ron (2016) note that in most religions, fasting can be done whenever one feels the need, but there are some prescriptions or requirements. Fasting during the month of Ramadan is the standard for all Muslims and one of the basic pillars of Islam, which has both ascetic and social connotations. During Ramadan, Muslims fast from sunrise to sunset. According to the Quran, fasting is required during Ramadan and the pilgrimage to Mecca, as well as in everyday life when needed as a sign of penance for wrongdoing (Grimm 2002; Mujtaba 2016; Vajda 2017).

In Judaism, fasting on Yom Kippur and Tish B'av is prescribed. Buddhist monks do not eat after the midday meal, and in other Asian religions, including Jainism and Hinduism, certain fasts are observed at certain times (Neusner & Sonn 1999; Smit 2014; Timothy & Ron 2016; Vajda 2017). In some sects of Christianity, Lent is considered a period devoted to fasting, abstinence, and penitence. However, fasting practices differ by denomination. In Catholicism, meat is forbidden on Fridays between Ash Wednesday and Holy Saturday. In the Orthodox churches, meat and milk are forbidden during the whole week of Lent and for several weeks afterwards (Akakios 1996; Grumett & Muers 2010). Other Christian denominations encourage periodic fasting as a means of increasing individual spirituality and drawing closer to God. Members of the Church of Jesus Christ of Latter-day Saints fast for 24 hours (or two meals) on the first Sunday of each month and donate to charity the money not spent on the meals they missed. The purpose of Latter-day Saint fasting is to humble themselves, to supplicate God for blessings, and to show devotion.

In most traditional societies, food was a link between humans and gods and formed the basis for ritual offerings (Timothy & Ron 2013a). For example, the religion of pre-contact Hawaii was a complex belief system wherein almost the entire material world was imbued with sacred and symbolic qualities (Abbott 1992). The gods were believed to inhabit natural features (e.g., volcanoes), which were considered sacred to Native Hawaiians (Handy & Handy 1972). Animals were also deified and used as food offerings in ritual sacrifices in temples dedicated to nature gods. Pigs, the most valued food in Polynesia, were considered especially sacred and highly valued by the gods, as were coconuts, bananas, chickens, and certain fish—often red ones—because red was a sacred colour (Titcomb 1972). Food offerings were literally intended to feed the gods, as a means of increasing their power (O'Connor 2008). Death-day celebrations around the tombs of saints included huge feasts of sacred food attended by as many as a hundred people (Feeley-Harnik 1995). Today, food offerings continue to be a salient part of people's lived religious life. Daily food donations to deity are a common practice in many religious cultures throughout the world and are a routine practice in daily life. Beyond actions at home, however, people participate in food-related rituals and activities during holidays or other travel opportunities. This is especially true during pilgrimages.

Food, religion, and tourism

Increased research attention has been devoted to the crossover between food, religion, and tourism in recent years (Aulet, Mundet & Vidal 2017; Aulet, Vidal & Crous 2015; Hall & Prayag 2020; Lee & Wall 2020; Ron & Timothy 2013, 2019; Son & Xu 2013; Timothy & Ron 2016). As previously noted, religion and gastronomy are inextricably connected with one another, and both are important (and inseparable) elements of cultural heritage. When tourism is brought into the mix, additional perspectives are manifested. Ron and Timothy (2013) conceptualized a typology of relationships between food and religious tourism (see Figure 30.1). As this typology suggests, there are many ways in which the relationships between food, religion, and tourism intersect (Ron & Timothy 2013: 236–237), some of which were noted in the previous section. The first relationship entails godly offerings. During official pilgrimages and even on informal visits to temples, shrines, or other sacred sites, many of the faithful 'sacrifice' food items as a sign of devotion. Such actions may be seen as more altruistic or auspicious during journeys away from home. Second, bestowing money or food to the needy is an important part of many pilgrimages. The meat of sacrificial animals is often donated to the poor during holidays and feasts, and other food items (e.g., fruit and bread) are donated to organizations to help feed the disadvantaged in the destination. The third relationship is the role of religiously proscribed food items and preparation methods. These come to bear on religious adherents' travel experiences, whether they travel for religious purposes or on holiday. Fourth, there are also prescribed foods that should be consumed during pilgrimages or other types of travel, such as certain sacraments and sacrificial animals. The ways in which food and eating can unite the faithful in a sense of solidarity is the fifth relationship. Solidarity between fellow travellers can be enhanced through mealtime socialization, even when the meal is not religious in nature. Sixth, foodies, or serious culinary tourists, have been likened to 'secular pilgrims' who travel to seek personal fulfilment by immersing themselves in the cultural foodscapes of the destination, just as other

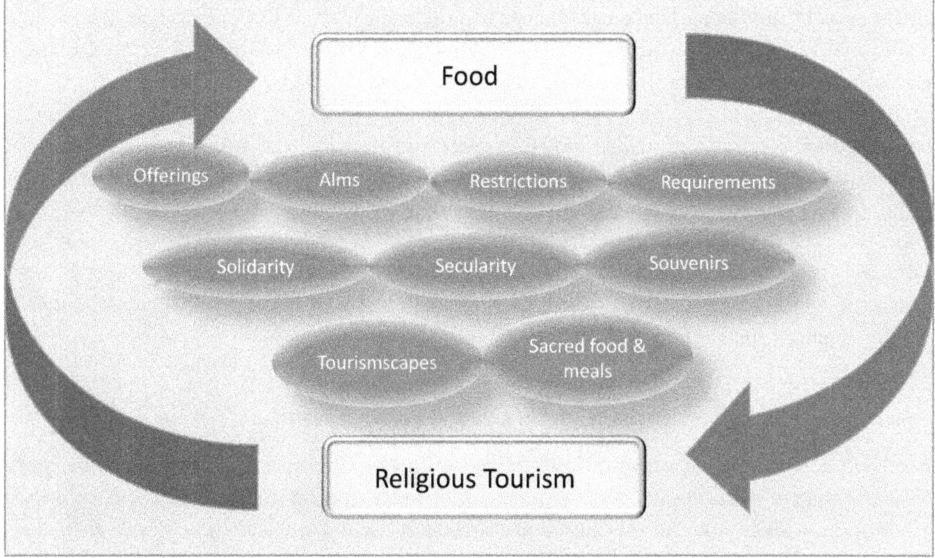

Figure 30.1 Relationships between religion and food in the tourism/pilgrimage context
Source: adapted from Ron and Timothy (2013).

secular pilgrims might do in other visitor contexts. For some travellers, culinary immersion may be a spiritual experience. The seventh association is the emergence, production, and consumption of religious food souvenirs. The eighth connection is the development of religious food-oriented tourismscapes, which is common in several sacred localities, such as Jerusalem, Bethlehem, and Nazareth. There, stimulated by mass religious tourism in the Holy Land, biblical foodscapes have emerged, emphasizing the importance of aliment in the Old and New Testaments. Examples include restaurant names, menus, signs, souvenirs, and the prevalence of biblically themed eateries and meal services. Finally, related to the previous point, sacred meals and food items are a salient part of many people's pilgrimage journeys. Eating a re-created 'Last Supper' in Israel, or a snack of manna or milk and honey with dates in Egypt or Israel, enhances one's Holy Land experience and may draw one closer to God.

Owing to space constraints, not all of these relationships between religion, tourism, and food will be reviewed in this chapter. Instead, we focus on religious food as a cultural attraction, sacred meals as a pilgrimage event, monastic and temple food, food-based souvenirs, and the religious requirements in hospitality services to illustrate some of the tourism connections between religion and culinaria.

Religious food as cultural attraction

In most cases, pilgrims, as well as certain other religious tourists, can be compared to 'serious cultural tourists', who systematically pursue an interest—in this case, learning about and experiencing heritage destinations (McKercher 2002; Stebbins 1996; Timothy 2021). The primary goal of pilgrims is to encounter the sacred, develop themselves spiritually, or demonstrate penance and humility. For most pilgrims, the journey is very sacred and serious as they visit holy heritage locales and have inspiring and transcendent spiritual experiences. Meanwhile, other elements of cultural heritage enter the pilgrimage and spiritual tourism equation to enhance the travel experience even further. For 'serious religious tourists', or pilgrims, the focus of their pilgrimage is the main religious attractions—the spiritual centre or shrine—and the cultural landscapes that surround them. Part of this cultural milieu includes traditional foods and local dishes, many of which have been modified or created through religious heritage traditions, as will be discussed below.

Poulain (2017) suggests that the act of eating is an essential part of a journey that provides first-hand contact with autochthonous cultures and people, who in many cases share the faith of the visitors. This is an important part of the cultural and solidarity experience of religious tourism, just as it is in other tourism contexts (Okumus & Cetin 2018). In Cerutti and Piva's (2016) study, the majority of visitors to the Sacred Mount of Oropa, Italy, were attracted foremost by the Marian sanctuary there, but 23% of their study participants also noted the appeal of the local gastronomic heritage, including the local dish *polenta concia*, in positively enriching their sacred journeys.

Sacred meals as pilgrimage event

A related concept is the consumption of religious meals as spiritual or religious experience. One of the best examples of this is when religious tourists and pilgrims to Israel exhibit an emerging curiosity and fascination about ancient biblical food and their desire to partake of authentic New Testament-era meals. Similarly, representations of Biblical cuisine have appeared more prominently in the tourism landscapes of Jerusalem and other parts of Israel, also as an important part of the pilgrimage experience by reinforcing the scriptural association with food. As food-based religious tourists place importance on authentic ancient comestibles, local establishments reconstruct the Last Supper and serve food which would have been eaten in New Testament times. To walk literally where Jesus walked, or to eat the food he ate, are powerful motives for Christians to travel to Israel. Thus, the more objectively authentic an experience can be for Christian pilgrims, the closer they feel to Jesus and the more faith promoting their experience will be. Authentic Biblical food is an important part of this experience for many tourists and pilgrims and may draw them closer to deity (Ron & Timothy 2013, 2019).

Outside the Holy Land, historic meals of religious importance can be found in other Christian destinations. For example, the municipality of Manresa, near Barcelona, collaborated with other agencies and foundations to develop a pilgrimage food focus on the Camino Ignaciano, the trail of Saint Ignatius of Loiola, which starts at his birthplace in Spain's Basque Country and ends in Manresa where he experienced many miracles and much learning before leaving Spain for the Holy Land (Abad Galzacorta & Guereño Ómil 2016; Compañia de Jesús 2020). St. Ignatius considered food to be a gift from God which helped people on their spiritual pathways and drew them closer to Christ (Dalmau 2018). In line with St. Ignacio's thinking, historical research was undertaken to discover common food items in the region during the sixteenth century. Culinary items were discovered and are offered to pilgrims on arrival in Manresa, making their experience deeper and more authentic. For example, one of the main dishes is the 'Pilgrim's Broth', 'a light and purifying broth, one hundred percent vegetable and made with homemade products, ideal for everyone that arrives at Manresa following the path of Saint Ignatius' (Fundacio Alicia 2018: 5). These sorts of sacred meals, as with those in the Holy Land, accentuate the pilgrimage experience and can heal the soul.

Monastic and temple food

A study by Aulet, Mundet and Vidal (2017) examines the culinary legacies of monasteries. Because monks and nuns traditionally lived isolated lives, were self-sufficient, and prepared humble meals for their fellow devotees, certain foods developed with specific recipes and dining customs. In the Medieval Ages, European monasteries grew their own food to meet the needs of their monastic community, but they also attended to the needs of pilgrims. Through their simplicity, self-reliant agriculture, and unique culinary traditions, many monasteries developed practices of making jam and sweets to use up any surplus fruit and vegetables or to give as gifts. Today, many of these products, as well as baked goods and other agricultural products, are still a key source of funding for those monasteries (Aulet 2012).

Monasteries have also traditionally played an important role in the production of wine and oil, as both were used for rituals and religious rites. Still today in Europe, many monasteries continue to produce wine and oil. Some of them have become attractions in enotourism (wine tourism) and several participate as visitor nodes on long-distance wine routes. One example is the Abbey of Saint Hildegard, in Germany, where one sister formally studied enology and now manages wine production (Beltrán Peralta 2018). The Cistercian Abbey of Santa Maria de la Oliva in Navarra, Spain, is another example. It is one of the last monasteries in Spain still producing its own wine (Aulet et al. 2017).

Montserrat Monastery in Catalonia, Spain, is a prominent example of monastery food that has become part of the attraction of this famous international destination. The Black Madonna of Montserrat is the patron saint of Catalonia and is guarded by a monk community. The site receives around three million visitors per year (including local, national and international visitors) (Aulet et al. 2019). Historically, certain foods were prepared by the monks at Montserrat, and these culinary items have become important attractions for pilgrims and other religious and cultural tourists. Popular among domestic visitors are three food products: *Aromes de Montserrat*, *coca* and *mató* (Abadia de Montserrat 2015). *Aromes de Montserrat* is a distilled herbal liqueur made with a traditional monastery recipe. Historically, Montserrat's monks used to collect herbs from the mountain to prepare the liqueur. It is no longer prepared in the monastery, but its historic origins are still celebrated locally, and it continues to be produced by a local company using the original recipe. *Coca* is a sweet cake, which is a popular snack in Catalonia. These cakes are still produced in the monastery's bakery using the original monk recipe. *Mató* is a fresh, soft Catalonian cheese which is normally eaten with honey as dessert. Although it did not necessarily originate in the monastery, it is closely associated with it today. The cheese is produced in nearby villages and sold in street stalls outside the monastery. In the nineteenth century, many of the families in the villages around Montserrat lived on wine production, but a phylloxera grapevine infestation destroyed many vineyards, ruining the livelihoods of many locals. As a result of this tragedy, Montserrat's monks authorized villagers to sell their other farm products, including cheese, to visitors and pilgrims. This tradition continues today and is seen as an important part of local heritage and lore (Mulet 2019).

Like many of Europe's monasteries, numerous temples in East Asia, particularly in South Korea, have begun opening their doors to commercial activities. Spiritual retreats have become especially popular among New Age spiritualists, as well as Buddhists, Hindus, Sikhs, Jains, and adherents to other religions from around the world (Lee & Wall 2020; Son & Xu 2013). Much of the focus of these retreats is educational in nature, including learning about healthy eating, vegetarian diets, detox eating, and meditation (Heintzman 2013; Ouellette et al. 2005). As an extension of this, Buddhist temple food, which is known for its freshness and simplicity, has become far more popular beyond the walls of temples and monasteries, and there are now Buddhist temple food restaurants throughout Asia and even in North America and Europe. Thus, the co-mingling of gastronomy and spirituality is exceptionally evident in this growing new trend (Kim, Lee & Ryu 2018; Moon 2008).

Souvenirs

Souvenirs come in all shapes and sizes, and some are even intangible, such as memories of eating and gustatory flavours. Food souvenirs are becoming increasingly popular in destinations that are known for specific food items (Swanson & Timothy 2012), and in the context of faith travel and general visits to sacred sites, food souvenirs are increasingly considered a source of revenue as well as a memory holder for visitors. Food items that are deemed sacred, or which are connected to people, places and events of faith through a common heritage, such as the monastic wine and oils noted above, play an important commercial role in religious tourism. Most sacred site managers within the world's major religions officially eschew commercialization, even if the destinations where they are located do not (Raj & Griffin 2017; Shackley 2006). Most shrines, churches, temples, and other sacred sites sell small-scale souvenirs such as postcards, booklets, prayer beads, candles, holy water, and icons. Increasingly, however, these sites are selling food items that are either directly related to the religious rituals of the locality or indirectly linked through a heritage connection.

In the Holy Land, food souvenirs are very popular. Olives and olive oil from Israel, Palestine, and Jordan, dates from Egypt and Israel, wine from the village of Cana, unleavened bread from Bethlehem and Nazareth, and spices from Jerusalem's spice markets are popular take-home souvenirs. These and many others are directly linked to stories and events in the Bible. Timothy and Ron (2016) suggest that religious souvenirs may become ostensibly tastier at home and more meaningful when purchased from a sacred site, as they may be symbolically imbued with 'the spirit' and bring back pleasant memories of a journey.

As noted above, Buddhist temple foods, such as various kimchis in Korea, as well as beer from Belgian and German monasteries, may be more meaningful and valuable when purchased from a sacred site and during liminal times. The monastery of Montserrat has created its own product brand, 'Gastronomia de Montserrat' which, apart from the products already described, also includes locally produced goods such as wine, olive oil, cookies, and chocolate. Bringing home a food souvenir has become a salient part of the pilgrimage ritual at Montserrat.

Religious hospitality requirements

Because culture and religion so frequently determine people's food choices at home and in everyday life, it also determines their culinary experiences while traveling. Although some people may be more lenient with their diets while traveling away from home, most devotees are aware of the need to maintain their religious alimentary practices. According to Minkus-McKenna (2007), some 70% of surveyed Muslim travellers follow their religious food rules strictly and consume only halal food. For some people, the puritanical bonds of home may be considered a limitation to enjoying the full range of culinary experiences while traveling (Cohen & Avieli 2004; Mak et al. 2012; Moira et al. 2017).

Different levels of hygiene, different eating habits, and language communication barriers can be important constraints for religionists who want to understand ingredients and cooking procedures. Thus, choosing a meal in countries where the primary religion(s) differs from the traveller's faith may be risky or difficult (Cohen & Avieli 2004; Mohd Nawawi et al. 2019). The food restrictions of Jews and Muslims have led hospitality services and tourism destinations to cater to these groups' specific religious needs and develop labels and brands such as 'halal tourism' and 'kosher tourism' (Moira et al. 2017; Moufakkir, Reisinger & AlSaleh 2019). This adaptation, which the tourism industry has undergone for the past decade or so, has resulted in Muslims and Jews being able to holiday abroad with an assurance of familiarity, safety, and religious adherence, which enables them to enjoy their experiences better. Although halal tourism goes far beyond food alone, including such things as separate male and female recreation spaces, food and its preparation methods are indeed an important element of the phenomenon (Weidenfeld & Ron 2008).

The differences between people who belong to a certain faith traveling to sacred places or on a pilgrimage versus the same people traveling on a relaxing vacation are vast. For the most part, Hindus and Buddhists can find plenty of vegetarian restaurants in pilgrimage locales, Jews can easily locate kosher meals in destinations important in Judaism, and nearly all food services in Muslim countries where pilgrimages typically take place are halal. Hindus, Jews, and Muslims on holiday outside such areas, however, may find it difficult to locate suitable meals (Moufakkir et al. 2019; Timothy & Ron 2016). The Network of Spanish Jewries (*Red de Juderías de España*) was developed in many locations with a former Jewish presence throughout Spain to promote the country's Jewish heritage and identify restaurants, accommodations, other services and cultural offerings (Russo & Romagosa 2010). This network is increasingly popular among Jewish tourists, but although many restaurants in the system claim to offer 'Sephardic' meals, they are not guaranteed to be kosher (Red de Juderías de España 2020).

These unique situations notwithstanding, it is becoming increasingly easy to find halal or kosher foods in urban areas throughout the world. This is because of growing immigrant populations, increasing local expertise about the need to satisfy foreign tourists, and the fact that more restaurants and packaged foods are being labelled kosher or halal in non-Jewish and non-Muslim majority countries (El-Gohary 2016; Havinga 2010; Mohsin & Ryan 2019).

Conclusion

Although the number of studies about the interface of tourism and gastronomy has grown in recent years, there is still a notable gap in our understanding of the relationships between food, tourism, and religion. The aim of this chapter is to narrow that gap by consolidating much of what is known in this tripartite relationship. However, this review is not exhaustive, and many gaps remain.

Although recent research has shown that in the West there appears to be a secularization in society, indicating a decline in religious adherence. Paradoxically, however, there is a notable increase in interest in visits to religious sites and pilgrimage (Blackwell 2007; Collins-Kreiner 2019; Olsen & Trono 2018), which partly indicates a growing interest in spirituality over religiosity (Liutikas 2017; Ron & Timothy 2019). Concomitantly, there has been a growth in special interest food tourism and gastronomy (de Albuquerque Meneguel et al. 2019; Everett 2016). Trends in food consumption, including among foodie travellers, show an increasing emphasis on healthy diets, local and fresh food, slow food, heritage cuisines, authentic aliment, and greater stress on sustainable agriculture and production (Calicioglu et al. 2019; Libcralo et al. 2020; Ramankutty et al. 2018; Schmitt et al. 2017). Much of this aligns with the growth in monastic and temple foods in Europe and Asia. Religious gastronomy and the diverse relationships between faith and food play an important role in these culinary trends.

Besides sharing an increasing interest in contemporary societies, religion and food have traditionally played a role as social connectors. Both of them are a critical part of cultural identity formation, living culture, and heritage meaning-making. McGettigan (2003) explores the relations between cultural tourism and religious sites and proposed a way of seeing the intercourse between religious tourism, spiritual tourism and cultural tourism (Figure 30.2). Such a diagram also applies to the crossover between food and religious tourism/pilgrimage, with spirituality, religiosity, and cultural heritage being interdependent and inseparable with regard to food production, consumption, and use in sacred contexts.

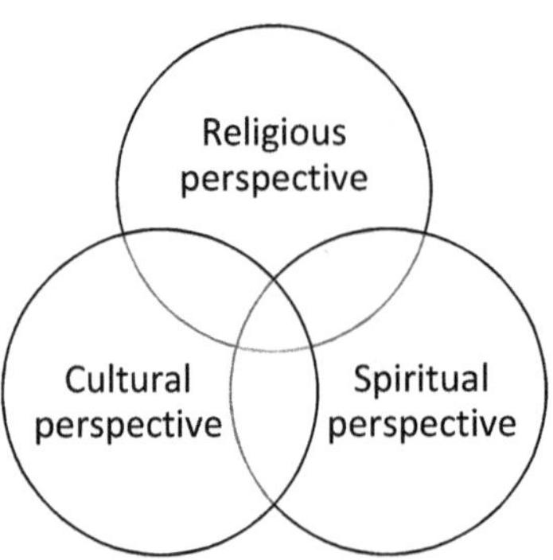

Figure 30.2 Interrelations between religious tourists/pilgrims and food
Source: adapted from McGettigan (2003).

Although many authors have explored the characteristics that distinguish pilgrims from other religious travellers (e.g., Aulet & Vidal 2018; Collins-Kreiner 2010, 2019; Greenia 2014; Fernandes et al. 2003; Reader 2007), this chapter treats them interchangeably in the context of food. Food and gastronomy are indelibly connected with religion. Food is a crucial part of every region's cultural heritage, and religion is one of the clearest manifestations of culture (Figure 30.2). Likewise, in almost every case, food is inseparably connected to religious ritual and practice, prohibitions, and praise. Aliment-related religious dogma and proscriptions are both travel constraints and attractions, although not usually simultaneously for the same people.

More than half of the world's population is religiously constrained by food restrictions at home and during travel. It is incumbent upon destinations that desire to attract large market segments of the Muslim, Jewish, Buddhist, and Hindu communities to understand the culinary needs and restrictions of these lucrative markets.

Food and food-related rituals are tourist attractions, especially when there is a strong element of alimentary heritage involved with monasteries, shrines, and temples, as well as authentic meals and foods that might have been eaten in ancient times. To consume a meal that Jesus, Mohammed, or Buddha might have eaten lends considerable spiritual and cultural appeal to the experience. Food is also often seen as lending added appeal to a pilgrimage, which can help visitors have an enhanced religious experience. Food is an important part of daily faith and its value intensifies during times of pilgrimage or other visits to sacred sites.

Culinary heritage and religion will forever be intertwined. The chapter aims to reconceptualise some of the current thinking in this arena. However, more research is needed to understand the deeper connections among these realities, especially regarding visitors' culinary experiences and how supply can meet religious demand.

References

Abad Galzacorta, M., & Guereño Ómil, B. (2016) 'Las necesidades del peregrino Ignaciano: Percepciones de una experiencia', *International Journal of Scientific Management and Tourism*, 2: 9–25.

Abadia de Montserrat (2015) *Alimentación—La Botiga de Montserrat*. Online at: https://botiga.montserratvisita.com/es/10-alimentacion. Accessed July 30, 2020.

Abbott, I.A. (1992) *La'au Hawaii: Traditional Hawaiian Uses of Plants*. Honolulu: Bishop Museum Press.

Akakios, A. (1996) *Fasting in the Orthodox Church*. Etna, CA: Center for Traditionalist Orthodox Studies.

Anderson, E.N. (2014) *Everyone Eats: Understanding Food and Culture*, 2nd edn. New York: New York University Press.

Appadurai, A. (1981) 'Gastro-politics in Hindu South Asia', *American Ethnologist*, 8(3): 494–511.

Aulet, S. (2012) *Competitivitat del Turisme Religiós en el Marc Contemporani: Els Espais Sagrats i el Turisme*. Unpublished doctoral theesis, University of Girona, Spain.

Aulet, S., Altayó, J., & Vidal-Casellas, D. (2019) 'Managing visitors at sacred sites: The case of Montserrat', in P. Wiltshier & M. Griffin (Eds.), *Managing Religious Tourism* (pp. 47–64). Wallingford: CABI.

Aulet, S., Mundet, L., & Vidal, D. (2017) 'Monasteries and tourism: Interpreting sacred landscape through gastronomy', *Revista Brasileria de Pesquisa En Turismo*, 11(1): 175–196.

Aulet, S., & Sureda, M. (2019) 'The semantics of the sacred: A tool for interreligious dialogue', in D. Vidal-Casellas, S. Aulet, & N. Crous-Costa (Eds.), *Tourism, Pilgrimage and Intercultural Dialogue. Interpreting sacred stories* (pp. 14–28). Wallingford: CABI.

Aulet, S., & Vidal, D. (2018) 'Tourism and religion: Ssacred spaces as transmitters of heritage values', *Church, Communication and Culture*, 3(3): 237–259.

Aulet, S., Vidal, D., & Crous, N. (2015) 'Religious and spiritual tourism as an opportunity for rural tourism: The case of Girona', in G. Bambi & M. Barbari (Eds.), *The European Pilgrimage Routes for Promoting Sustainable and Quality Tourism in Rural Areas* (pp. 703–714). Florence: Florence University Press.

Baker, J.W. (2009) *Thanksgiving: A Biography of an American Holiday*. Lebanon, NH: University of New Hampshire Press.

Beltrán Peralta, N. (2018) *Vino y Religión: Estudio de los Monasterios Benedictinos en Europa*. Unpublished master's thesis, University of Girona, Spain.

Bessière, J. (2013) "Heritagisation', a challenge for tourism promotion and regional development: An example of food heritage', *Journal of Heritage Tourism*, 8(4): 275–291.

Blackwell, R. (2007). Motivations for religious tourism, pilgrimage, festivals and events. In R. Raj and N.D. Morpeth (Eds.), *Religious tourism and pilgrimage festivals management: An international perspective* (pp. 35–47). Wallingford, UK: CABI.

Calicioglu, O., Flammini, A., Bracco, S., Bellù, L., & Sims, R. (2019) 'The future challenges of food and agriculture: An integrated analysis of trends and solutions', *Sustainability*, 11(1): 222–243.

Carolan, M. (2012) *The Sociology of Food and Agriculture*. London: Routledge.

Cerutti, S., & Piva, E. (2016) 'The ole of tourists' feedback in the enhancement of religious tourism destinations', *International Journal of Religious Tourism and Pilgrimage*, 4(3): 6–16.

Cohen, E., & Avieli, N. (2004) 'Food in tourism: Attraction and impediment', *Annals of Tourism Research*, 31: 755–788.

Collins-Kreiner, N. (2010) 'The geography of pilgrimage and tourism: Transformations and implications for applied geography', *Applied Geography*, 30(1): 153–164.

Collins-Kreiner, N. (2019) Pilgrimage tourism—past, present and future rejuvenation: A perspective article. *Tourism Review*, 75(1), 145–148.

Compañia de Jesús (2020) *Camino Ignaciano*. Online at: www.caminoignaciano.org. Accessed June 3, 2020.

Contreras, J. (2007) 'Alimentación y religión', *HUMANITAS Humanidades Médicas*, 16: 1–22.

Contreras, J., & Gracia, M. (2005) *Alimentación y Cultura: Perspectiva Antropologicas*. Barcelona: Ariel.

Dalmau, B. (2018) *Alimentacio i vida espiritual*. Montserrat: Publicacions de l'Abadia de Montserrat.

de Albuquerque Meneguel, C.R. Mundet, L., & Aulet, S. (2019) 'The role of a high-quality restaurant in stimulating the creation and development of gastronomy tourism', *International Journal of Hospitality Management*, 83: 220–228.

Denning, S. (2021) 'Religious faith, effort and enthusiasm: Motivations to volunteer in response to holiday hunger', *Cultural Geographies*, 28(1): 57–71.

Díez de Velasco, F. (1998) *Introducción a la historia de las religiones: hombres, ritos, dioses*, 2nd edn. Madrid: Trotta.

Di Marzo, S. (2016, June 16) *La mística de la alimentación musulmana*. Bienmesabe. Online at: https://elestimulo.com/bienmesabe/la-mistica-de-la-alimentacion-musulmana/. Accessed June 8, 2020.

El-Gohary, H. (2016) 'Halal tourism, is it really Halal?', *Tourism Management Perspectives*, 19: 124–130.

European Parliament (2014) *European Parliament resolution of 12 March 2014 on the Europpean gastronomic heritage: cultural and educational aspects*. Online at: https://www.europarl.europa.eu/doceo/document/TA-7-2014-0211_EN.html?redirect. Accessed July 3, 2020.

Everett, S. (2016) *Food and Drink Tourism: Principles and Practice*. London: Sage.

Feeley-Harnik, G. (1995) 'Religion and food: An anthropological perspective', *Journal of the American Academy of Religion, 63*(3): 565–582.

Fernandes, C., McGettigan, F., & Edwards, J. (Eds.) (2003) *Religious Tourism and Pilgrimage: ATLAS Special Interest Group, 1st Expert meeting*. Leiria: Tourism Board of Leiria.

Frost, W., & Laing, J. (2016) 'Cuisine, migratiton, colonialism and diasporic identities', in D.J. Timothy (Ed.), *Heritage Cuisines: Traditions, Identities and Tourism* (pp. 37–52). London: Routledge.

Fundacio Alicia (2018) *La Cuina del Pelegrí Guia gastronòmica de la Manresa ignasiana i el Bages*. Ajuntament de Manresa. Online at: https://www.manresa.cat/docs/docsArticle/9771/2018-05-31_la_cuina_del_pelegri.pdf. Accessed July 1, 2020.

Gaarder, J., Hellern, V., Notaker, H., Baggethun, K., & Lorenzo Torres, M. (2009) *El Libro de las Religiones*. Madrid: Siruela.

Ghassem-Fachandi, P. (2006) *Sacrifice, Ahimsa, and Vegetarianism: Pogrom at the Deep End of Non-Violence*. Unpublished doctoral dissertation, Cornell University, Ithaca, New York.

Gofton, L. (1989) 'Sociology and food consumption', *British Food Journal*, 91(1): 25–31.

González Turmo, I. (2007) 'Patrimonio gastronómico, cultura y turismo: El caso de la España del siglo XX en el contexto mediterrâneo', in J. Tresserras & X. Medina (Eds.), *Patrimonio Gastronómico y Turismo Cultural en el Mediterráneo* (pp. 197–216). Barcelona: Ibertur-Universidad de Barcelona/Instituto Europeo del Mediterráneo.

Greenia, G. (2014) 'What is pilgrimage? The religious origins of pilgrimage studies', in L. Harman (Ed.), *A Sociology of Pilgrimage: Embodiment, Identity, Transformation* (pp. 8–18). Nashville: Ursus Press.

Grimm, V. (2002) *From Feasting to Fasting: The Evolution of a Sin*. London: Routledge.

Grumett, D., & Muers, R. (2010) *Theology on the Menu: Asceticism, Meat and Christian Diet*. London: Routledge.

Guzmán Parejo, J. (2018) Diversidad alimenticia según las prescripciones religiosas: Cuestiones jurídicas. *Ilu, Revista de Ciencias de Las Religiones*, 23: 191–216.

Hall, C.M., & Gössling, S. (Eds.) (2016) *Food Tourism and Regional Development: Networks, Products and Trajectories*. London: Routledge.

Hall, C.M., & Prayag, G. (Eds.) (2020) *The Routledge Handbook of Halal Hospitality and Islamic Tourism*. London: Routledge.

Hall, C.M., & Sharples, L. (2003) 'The consumption of experiences or the experience of consumption? An introduction to the tourism of taste', in C.M. Hall, L. Sharples, R. Mitchell, N. Macionis, & B. Cambourne (Eds.), *Food Tourism around the World: Development, Management and Markets* (pp. 13–36). Oxford: Butterworth-Heinemann.

Handy, E.S.C., & Handy, E.G. (1972) *Native Planters in Old Hawaii: Their Life, Lore and Environment*. Honolulu, HI: Bernice P. Bishop Museum.

Harris, M. (1980) *Vacas, cerdos, guerras y brujas: Los enigmas de la cultura*. Madrid: Alianza.

Havinga, T. (2010) 'Regulating halal and kosher foods: Different arrangements between state, industry and religious actors', *Erasmus Law Review*, 3(4): 241–255.

Heintzman, P. (2013) 'Retreat tourism and a form of transformational tourism', in Y. Reisinger (Ed.), *Transformational Tourism: Tourist Perspectives* (pp. 68–81). Wallingford: CABI.

Hjalager, A.-M., & Richards, G. (2002) *Tourism and Gastronomy*. London: Routledge.

Ioannides, M.W.C., & Ioannides, D. (2006) 'Global Jewish tourism: Pilgrimages and remembrance', in D.J. Timothy & D.H. Olsen (Eds.), *Tourism, Religion and Spiritual Journeys* (pp. 156–171). London: Routledge.

Jenkins, W., Tucker, M.E., & Grim, J. (Eds.) (2017) *Routledge Handbook of Religion and Ecology*. London: Routledge.

Jolliffe, L. (ed.) (2010) *Coffee Culture, Destinations and Tourism*. Bristol: Channel View Publications.

Just, C. (2007) *Regla de Sant Benet*, 2nd edn. Montserrat: Publicacions de l'Abadia de Montserrat.

Kaur, I.N. (2020) 'Sikhism', in D.B. Yaden, Y. Zhao, K. Peng, & A.B. Newberg (Eds.), *Rituals and Practices in World Religions: Cross-Cultural Scholarship to Inform Research and Clinical Contexts* (pp. 151–166). Cham, Switzerland: Springer.

Kaza, S. (2005) 'Western Buddhist motivations for vegetarianism', *Worldviews: Environment, Culture, Religion*, 9(3): 385–411.

Khare, R.S. (1992) *The Eternal Food: Gastronomic Ideas and Experiences of Hindus and* Buddhists. Albany: State University of New York Press.

Kim, K.S., Lee, T.J., & Ryu, K. (2018) 'Alternative healthy food choice for tourists: Developing Buddhist temple cuisine', *International Journal of Tourism Research*, 20(3): 267–276.

Lee, A., & Wall, G. (2020) 'Temple food as a sustainable tourism attraction: Ecogastronomic Buddhist heritage and regional development in South Korea', *Journal of Gastronomy and Tourism*, 4(4): 209–222.

Libcralo, P., Mendes, T., Mendes, T., & Liberato, D. (2020) 'Culinary tourism and food trends', in A. Rocha (Ed.), *Smart Innovation, Systems and Technologies* (pp. 517–526). Cham, Switzerland: Springer.

Liutikas, D. (2017) 'The manifestation of values and identity in travelling: The social engagement of pilgrimage', *Tourism Management Perspectives*, 24: 217–224.

Mak, A.H., Lumbers, M., & Eves, A. (2012) 'Globalisation and food consumption in tourism', *Annals of Tourism Research*, 39(1): 171–196.

Masoudi, G.F. (1993) 'Kosher food regulation and the religion clauses of the first amendment', *The University of Chicago Law Review*, 60(2): 667–696.

McGettigan, F. (2003) 'An analysis of cultural tourism and its relationship with religious sites', in C. Fernandes, F. McGettigan, & J. Edwards (Eds.), *Religious Tourism and Pilgrimage: ATLAS Special Interest Group 1st Expert Meeting* (pp. 13–26). Leiria: Tourism Board of Leiria.

McKercher, B. (2002) 'Towards a classification of cultural tourists', *International Journal of Tourism Research*, 4(1): 29–38.

Minkus-McKenna, D. (2007) 'The pursuit of halal', *Progressive Grocer*, 86(17): 42.

Mohd Nawawi, M.S.A., Abu-Hussin, M.F., Faid, M.S., Pauzi, N., Man, S., & Mohd Sabri, N. (2019) 'The emergence of halal food industry in non-Muslim countries: A case study of Thailand', *Journal of Islamic Marketing*, 11(4): 917–931.

Mohsin, A., & Ryan, C. (2019) 'Halal tourism: A growing market on a global stage', in D.J. Timothy (Ed.), *Routledge Handbook on Tourism in the Middle East and North Africa* (pp. 309–318). London: Routledge.

Moira, P., Mylonopoulos, D., & Vasilopoulou, P. (2017) Kosher tourism: A case study from Greece', in R. Raj & K. Griffin (Eds.), *Conflicts, Religion and Culture in Tourism* (pp. 144–154). Wallingford: CABI.

Moon, S.S. (2008) 'Buddhist temple food in South Korea', *Korea Journal*, 48(4): 147–180.

Moufakkir, O., Reisinger, Y., & AlSaleh, D. (2019) 'Much ado about halal tourism: Religion, religiosity or none of the above?', in D.J. Timothy (Ed.), *Routledge Handbook on Tourism in the Middle East and North Africa* (pp. 319–329). London: Routledge.

Mujtaba, U. (2016) 'Ramadan: The month of fasting for Muslims, and tourism studies—Mapping the unexplored connection', *Tourism Management Perspectives*, 19: 1070–1177.

Mulet, M. (2019) 'Historia de vida sobre la vida agraria y el patrimonio alimentario en los pueblos colindantes al Monasterio de Montserrat', Paper presented ata the *XXX Jornades De Recerca De L'observatori De L'Alimentació (Odela)* conference, 13 February 2019.

Neusner, J., & Sonn, T. (1999) *Comparing Religions through Law: Judaism and Islam*. London: Routledge.

Nunes dos Santos, C. (2007) 'Somos lo que comemos: Identidad cultural, hábitos alimenticios y turismo', *Estudios y Perspectivas En Turismo*, 16: 234–242.

O'Connor, K. (2008) 'The Hawaiian Luau', *Food, Culture & Society*, 11(2): 149–172.

Okumus, B., & Cetin, G. (2018) 'Marketing Istanbul as a culinary destination', *Journal of Destination Marketing & Management*, 9: 340–346.

Okumus, B., Okumus, F., & McKercher, B. (2007) 'Incorporating local and international cuisines in the marketing of tourism destinations: The cases of Hong Kong and Turkey', *Tourism Management*, 28: 253–261.

Olsen, D.H., & Trono, A. (Eds.) (2018) *Religious Pilgrimage Routes and Trails: Sustainable Development and Management*. Wallingford: CABI.

Ouellette, P., Kaplan, R., & Kaplan, S. (2005) 'The monastery as a restorative environment', *Journal of Environmental Psychology*, 25(2): 175–188.

Pinstrup-Andersen, P. (2009) 'Food security: definition and measurement', *Food Security*, 1(1): 5–7.

Popli, U.K. (2010) 'Sikhism and social work', in T. Gracious (Ed.), *Origin and Development of Social Work in India* (pp. 182–201). New Delhi: Indira Gandhi National Open University.

Popovsky, M. (2010) 'The constitutional complexity of kosher food laws', *Columbia Journal of Law and Social Problems*, 44: 75–107.

Poulain, J. (2017) *The Sociology of Food: Eating and the Place of Food in Society*. New York: Bloomsbury.

Raj, R., & Griffin, K. (Eds.) (2017) *Conflicts, Religion and Culture in Tourism*. Wallingford: CABI.

Ramankutty, N., Mehrabi, Z., Waha, K., Jarvis, L., Kremen, C., Herrero, M., & Rieseberg, L.H. (2018) 'Trends in global agricultural land use: Implications for environmental health and food security', *Annual Review of Plant Biology*, 69(1): 789–815.

Ramshaw, G. (2016) 'Food, heritage and nationalism', in D.J. Timothy (Ed.), *Heritage Cuisines: Traditions, Identities and Tourism* (pp. 53–64). London: Routledge.

Reader, I. (2007) 'Pilgrimage growth in the modern world: Meanings and implications', *Religion*, 37(3): 210–229.

Red de Juderías de España (2020) Caminos de Sefarad: Red de Juderías de España. Online at: https://redjuderias.org/rasgo-2/v. Accessed June 3, 2020.

Regenstein, J.M., Chaudry, M.M., & Regenstein, C.E. (2003) 'The kosher and halal food laws', *Comprehensive Reviews in Food Science and Food Safety*, 2(3): 111–127.

Regenstein, J.M., & Regenstein, C.E. (1991) 'Current issues in kosher foods', *Trends in Food Science & Technology*, 2: 50–54.

Riaz, M., & Chaudry, M. (2004) *Halal Food Production*. Boca Raton, FL: CRC Press.

Ribagorda Calasanz, A. (1999) 'Los animales en los textos sagrados del Islam', *Espacio Tiempo y Forma: Serie III, Historia Medieval*, 12: 101–138.

Richards, G. (2012) 'An overview of food and tourism trends and policies', in OECD (Ed.), *Food and the Tourism Experience: The OECD-Korea Workshop* (pp. 13–46). Paris: The OECD Tourism Studies.

Ron, A.S., & Timothy, D.J. (2013) 'The land of milk and honey: Biblical foods, heritage and Holy Land tourism', *Journal of Heritage Tourism*, 8(2–3): 234–247.

Ron, A.S., & Timothy, D.J. (2019) *Contemporary Christian Travel: Pilgrimage, Practice and Place*. Bristol: Channel View Publications.

Rosen, S. (Ed.) (2011) *Food for the Soul: Vegetarianism and Yoga Traditions*. Santa Barbara, CA: Praeger.

Rouse, C., & Hoskins, J. (2004) 'Purity, soul food, and Sunni Islam: Explorations at the intersection of consumption and resistance', *Cultural Anthropology*, 19(2): 226–249.

Russo, A.P., & Romagosa, F. (2010) 'The Network of Spanish Jewries: In praise of connecting and sharing heritage', *Journal of Heritage Tourism*, 5(2): 141–156.

Sack, D. (2000) *Whitebread Protestants: Food and Religion in American Culture*. New York: Palgrave.

Schmitt, E., Galli, F., Menozzi, D., Maye, D., Touzard, J.M., Marescotti, A., Six, J., & Brunori, G. (2017) 'Comparing the sustainability of local and global food products in Europe', *Journal of Cleaner Production*, 165: 346–359.

Shackley, M. (2006) 'Empty bottles at sacred sites: Religious retailing at Ireland's National Shrine', in D.J. Timothy & D.H. Olsen (Eds.), *Tourism, Religion and Spiritual Journeys* (pp. 94–103). London: Routledge.

Smit, P.B. (2014) 'Reaching for the Tree of Life: The role of eating, drinking, fasting, and symbolic foodstuffs in 4 Ezra', *Journal for the Study of Judaism*, 45(3): 366–387.

Son, A., & Xu, H. (2013) 'Religious food as a tourism attraction: The roles of Buddhist temple food in Western tourist experience', *Journal of Heritage Tourism*, 8(2/3): 248–258.

Sormaz, U., Akmese, H., Gunes, E., & Aras, S. (2015) 'Gastronomy in tourism', *Procedia Economics and Finance*, 39: 725–730.

Srinivas, T. (2006) 'Divine enterprise: Hindu priests and ritual change in neighbourhood Hindu temples in Bangalore', *South Asia: Journal of South Asia Studies*, 29(3): 321–343.

Stebbins, R.A. (1996) 'Cultural tourism as serious leisure', *Annals of Tourism Research*, 23: 945–948.

Subramaniam, M. (2011) 'An introduction to the concept of vegetarianism among Hindus in Malaysia', *Jurnal Pengajian India*, 10(1): 51–60.

Swanson, K.K., & Timothy, D.J. (2012) 'Souvenirs: Icons of meaning, commercialization and commoditization', *Tourism Management*, 33(3): 489–499.

Thera, N. (1994) *Dhammapada: La enseñanza de* Buda. Madrid: Editorial EDAF.

Timothy, D.J. (2016) 'Heritage cuisines, foodways and culinary traditions', in D.J. Timothy (Ed.), *Heritage Cuisines: Traditions, Identities and Tourism* (pp. 1–24). London: Routledge.

Timothy, D.J. (2021) *Cultural Heritage and Tourism: An Introduction*, 2nd edn. Bristol: Channel View Publications.

Timothy, D.J., & Ron, A.S. (2013a) 'Heritage cuisines, regional identity and sustainable tourism', in C.M. Hall & S. Gössling (Eds.), *Sustainaible Culinary Systems: Local Foods, Innovation, Tourism and Hospitality* (pp. 275–290). London: Routledge.

Timothy, D.J., & Ron, A.S. (2013b) 'Understanding heritage cuisines and tourism: Identity, image, authenticity and change', *Journal of Heritage Tourism*, 8(2/3): 99–104.

Timothy, D.J., & Ron, A.S. (2016) 'Religious heritage, spiritual aliment and food for the soul', in D.J. Timothy (Ed.), *Heritage Cuisines: Traditions, Identities and Tourism* (pp. 105–118). London: Routledge.

Titcomb, M. (1972) *Native Use of Fish in Hawaii*. Honolulu: University of Hawaii Press.

UNESCO (2020) *Intangible Cultural Heritage*. Online at: https://ich.unesco.org/en/lists. Accessed May 30, 2020.

Vajda, G. (2017) 'Fasting in Islam and Judaism', in J. Hawting (Ed.), *The Development of Islamic Ritual* (pp. 133–149). London: Routledge.

Velasco, M. (1982) *Introducción a la Fenomenología de la Religión*. Madrid: Cristiandad.

Ward, P., Coveney, J., & Henderson, J. (2010) 'A sociology of food and eating: Why now?', *Journal of Sociology*, 46(4): 347–351.

Weidenfeld, A., & Ron, A.S. (2008) 'Religious needs in the tourism industry', *Anatolia*, 19(2): 357–361.

Winchester, F. (1999) *Muslim Holidays*. Mankato, MN: Bridgestone Books.

World Tourism Organization (2012) *Tourism and Intangible Cultural Heritage*. Madrid: UNWTO.

World Tourism Organization (2019) *Guidelines for the Development of Gastronomy Tourism*. Madrid: UNWTO.

31
SAFETY, FEAR, RISK, AND TERRORISM IN THE CONTEXT OF RELIGIOUS TOURISM

Maximiliano E. Korstanje and Babu George

Introduction

In what has been considered the bloodiest attack on American soil after Pearl Harbor, September 11, 2001, represented a turning point for security-related specialists and policy makers throughout the world (Dalby 2003; Hall, Timothy & Duval 2003). This event saw the division of the world into secure and insecure nations. Importantly, this event marked the first time that terrorists employed mass transportation as real weapons against civilian targets (Korstanje & Olsen 2011). Since this event, multiple studies have focused on the effects of terrorism on local economies as well as the global tourism and hospitality industry (Bonham, Edmonds & Mak 2006; Korstanje & Clayton 2012; Pappas 2010; Raine 2013; Saha & Yap 2014; Yan et al. 2016). Although the notion of risk perception underscores models that predict terrorist attacks in leisure hotspots (Floyd et al. 2004; Fuchs et al. 2013; Reisinger & Mavondo 2005), the post-9/11 security dilemma opened the doors to new, unanswered questions. As Bianchi (2006) notes, the excessive public attention on the struggles of tourist destinations to strengthen security has become problematic, as the fear of terrorism, together with more general global anxiety over violence, vulnerability, and uncertainty, has paved the way for a new security climate in which, despite the efforts and material resources to make destinations safer, reinforces the tenets of terrorism.

While religious tourism is lauded as a means of developing peripheral regions, or places which historically were exploited by imperial powers, certain radical discourses have played a crucial role in generating resentment against Western tourists. As a result, religious festivals and mega-events are often targeted in order to grab the attention of government officials. As such, it is particularly important not to lose sight of the fact that as ambassadors of their home countries, tourists count on being protected by the host country during their visits (Korstanje, Raj & Griffin 2018). Within this context, this is organized into four parts. The first section deals conceptually with the terms risk, security, and fear so that readers will have a better understanding of the broader context. The chapter then examines the extent to which religious events and religious tourism in general are fertile grounds for terrorism and local violence. Since religiosity does not disappear but rather adapts to specific contexts, the authors suggest that religious tourism not only revitalizes the psychological frustrations of citizens in their day-to-day life but also boosts social trust, which trust is necessary for a society to function properly. As such, religious mega-events and tourism involve rituals that must be performed safely and carefully. Third, the chapter enumerates the most significant security considerations in planning religious tourism events. Finally, the last section examines the security challenges facing the religious tourism industry.

Fear, safety, and risk

In an older publication, Roehl and Fesenmaier (1992) summarized 30 years of psychological research regarding travel examining the correlation between risk perception and travelers' demographic features. In their analysis of this literature, they highlighted three dimensions of risk according to consumers' demographic characteristics: physical-equipment risk, vacation risk, and destination risk. Each of these risk dimensions varied in intensity and longevity. They found that far from being static, risk perception is contextual, varying between different people and their travel and personality types. In this vein, Plog (1991) developed a theoretical model to explain why some personalities experience an extreme fear of flying while others do not. Plog concluded that motivations, fears, and phobias come from people's unique and individual life experiences. From a travel experience perspective, Plog believed there to be two types of travelers: those who grow up in a climate of respect and solidarity, and those who have been socialized in contexts of conflict and fear. The former travelers are more inclined to seek novelty and contact with others, whereas the latter travelers tend to be more afraid of strangers, experiences, and places that are outside their normal comfort zones. From this analysis, Plog developed the well-known model of tourist behavior with three types of tourists: allocentrics (who are open to new and un-ordinary experiences), psychocentrics (who are fearful or distrusting of the "Other" and seek well-established tourist destinations), and mid-centrics (who exhibit aspects of the other two types of tourists). Although widely criticized for its simplicity, Plog was a pioneer devoted to understanding the intersections of travel, pleasure, and fear (Korstanje 2009).

One of the limitations of Plog's theory was its inability to provide an adequate understanding of risk. This was rectified in part by German sociologist Ulrich Beck (1992), who argued that the 1986 Chernobyl accident was the touchstone for the rise of a new period in late capitalism, which period he dubbed "the risk society"—the beginnings of modern society's organization and reaction to risk. Paradoxically, the very technology employed to improve this world became the potential recipe for its demise. This paradox represented one of the dichotomies of late capitalism, which placed all classes of people into egalitarian conditions before risk (Beck 1992). Luhmann (2017), however, criticized Beck for excessive alarmism. As Luhmann argued, risks as described by Beck are different from threats, where the former are direct results of decision-making of people in positions of power while the latter corresponds with external events that cannot be avoided or controlled. Luhmann observed that people who generate risks rarely deal with the consequences of their decisions and actions. For example, passengers who end up in an airplane accident are not responsible for this event. Rather, responsibility rests on managers who chose to lower the costs of flying by cutting corners on maintenance and security. As such, while the passengers do not create risks, they are the ones who face potential external dangers (Luhmann 2017).

In the context of tourism, risk is understood as any danger, whether real or imagined, that jeopardizes the functioning of the tourism system (Sönmez & Graefe 1998). The notion of 'real or imagined' risks has been hotly debated around the academic literature regarding risk. For example, Larsen (2007) argued that since risk is rooted in people's life experiences and their access to information, and is therefore cognitive in nature, there are other, much deeper sentiments that have been overlooked by rationalist psychologists. To further this argument, Larsen differentiates between "risks" and "worries". While risks or potential dangers are recognized or ignored according to individual perceptions and choices, tourist worries are associated with the emotional disposition of an individual traveling or not traveling to a certain destination as related to concerns of potential future events that may have a negative outcome (Larsen 2007; Larsen, Brun & Øgaard 2009). Far from being rational agents, tourists are motivated by internal emotions, which emotions derive in large part from previous experiences. This explains, in part, why some travelers are risk-seekers while others are risk-avoiders (Wolff & Larsen 2014). As such, increasing fear by tourists of the unknown or potential future negative events forces policy makers to address their concerns over security at tourism destinations (Adams 2001; Bianchi 2006).

However, as Bianchi (2006) suggests, fear is paradoxical—as a reaction to unknown or potential negative events, it is difficult to engage in communication or marketing campaigns that will fully alleviate people's fear. For example, several studies reveal that there are cultural differences regarding travel risk perceptions (Reisinger & Mavondo 2005; Sharifpour, Walters & Ritchie 2014). More specifically, tourists from English-speaking countries and Protestant Christians feel less secure while traveling abroad than tourists from other nationalities and faiths (Fuchs & Reichel 2004; He, Park & Roehl 2013; Lepp & Gibson 2003; Quintal, Lee & Soutar 2010), in part because citizens of states involved in the "War on Terror" tend to be more reluctant to visit Muslim countries than citizens from other nationalities (Korstanje 2009).

Fear may be defined as the rise of unpleasant emotions caused by external threats. Not only do these emotions jeopardize the ontological security of the person, but as a basic emotion, ensures human survival by orchestrating different responses with both an individual and a collective actions to preserve life. However, at times, risk can paralyze and individual or group, keeping them from making the best decisions in the moment (Bauman 2013). Although biological at their roots, fears have adapted to cultures and times (Douglas & Wildavsky 1983). Unfortunately, fear has not received adequate research attention commensurate with its importance within tourism studies. As Bianchi (2006) noted, discourses on security emphasize governance as tug-of-war that supports democracies in fighting terrorism. The neoliberalization of the world, which underpins globalization, has enhanced the human right of mobility among western tourists (Hall, Amore & Arvanitis 2019; Wearing et al., 2019). As such, the ideological essence of modern tourism is therefore intertwined with liberty and consumption. This process of democratization, which is supported and epitomized by globalization and tourism (Timothy 2019), however, flies in the face of totalitarian states, often known as "rogue states". Nevertheless, the neoliberal discourse is trapped in a conceptual gridlock, where "The capacity-freedom presupposes liberty, but liberty does not presuppose capacity freedom" (Bianchi 2006: 67).

Since its genesis, the modern state has enthusiastically embraced people's right to travel and the freedom of mobility as two related and mainstream cultural values as a means of shoring up their sovereign power. However, as noted above, the events of September 11, 2001 inaugurated a new period in which fear became a global commodity, creating or stimulating new consumer products and modes of surveillance (Bianchi 2006). However, although humanity is more secure today than preceding generations, the excess of freedom in today's world in most societies engenders a climate of uncertainty, which is fertile ground for fomenting fears and anxieties (Bianchi 2006; Bianchi & Stephenson 2013). As such, the question must be asked: Will risk and fear transform tourism, and if so, in what ways?

Partly in answer to this question, Korstanje and Clayton (2012) suggest that tourism and terrorism share many common features, including the use of mass transportation and media management, albeit for many different goals. Their premise is that western rationality introduces extortion as an economic and cultural relationship. Terrorism, or at least the threat of terrorism, leads to changes in both technological innovations (i.e., the logic of destructive creation) and the cost of consumer products related to travel and risk. In this case, security becomes a commodity that is exchanged in different economic circuits, with peripheral economies that cannot afford to make the appropriate changes falling behind more developed economies. As well this gap between peripheral and developed economies is created through their capacity (or lack of) to create economic value (Lash, Urry & Urry 1993). In a globalized economy, personal and media perception divides the world into safe and unsafe tourism destinations. As such, there has been a growth of a network of security and risk experts, which growth has occurred proportionally to the increasing perception of fear and risk as legitimated by modern science since 9/11 (Beck 1992; Luhmann 2017). Countries that have access to these experts are generally perceived as being safer. However, as Mansfeld and Pizam (2006) note, one of the limitations within the field of tourism security is the difficulty in demonstrating how theory and practice can be effectively combined. This in part because of the complexity of the tourism system, with its multiple stakeholders, all of whom pursue their own interests, making holistic planning attempts difficult at best.

Society and religious tourism

Emile Durkheim, one of the pioneers of modern sociology, stressed in his early essays the importance of religion in society. Based on his work with aboriginal tribes, Durkheim believed that the power and authority invested in individuals in a particular society emanated from a divine source he called a "totem". This totem, whether based on animals, plants, or human-made icons, became an object around which social cohesion and a sense of social identity were developed. In the case of religion, Durkheim argued that religion would eventually disappear in industrial societies because industrialization would undermine the social ties of faith or the totem (Durkheim 2008). MacCannell (1973) argued the same but also suggested that the new "totem" in industrialized societies would be leisure and tourism. For MacCannell, tourism provided relief from the frustrations and daily grind of everyday life through the channels of leisure consumption and hedonism. While MacCannell hinted that the spheres of the "sacred" and the "profane" should be seen as separate entities in modern society, he believed that people's need to experience some form of "reality" and novelty—something unfamiliar and far from the humdrum conditions of home—would lead to the acceptance of "staged authenticity", which would fill the gap left by the decline of religion in daily life (MacCannell 1973, 1999, 2002).

However, Raj and Griffin (2015) and Olsen (2019) recognize that far from disappearing from the public sphere, religion remains alive within post-secular societies. Since many forms of tourism are motivated by maximizing pleasure, some scholars suggest that pilgrimage is somehow different from tourism (Collins-Kreiner 2010; see Cohen 1992). Indeed, pilgrims and tourists not only vary in their psychological motivations but also in the levels of anxieties and their religious attachment to the divinity (Collins-Kreiner 2010; Korstanje & Seraphin 2017), without mentioning the divorce between the "sacred" and the "profane" (Korstanje & George 2012). Ample evidence shows that some forms of pilgrimage and religious tourism are very much motivated by pleasure, enjoyment, and leisure, in addition to religious obligations, such as religious festivals and holy day celebrations (Olsen 2010). As such, this distinction between pilgrimage and tourism is blurry at best (Kaelber 2006), and precise borders cannot reliably be drawn between pilgrimage and tourism.

For example, Korstanje and George (2012: 162) note that the Falkland Islands (Islas Malvinas) off the coast of Argentina are "a mythical archetype enrooted in nationalism, heroism, human rights, and political tolerance" despite its national "sacredness" for the Argentinian state and its citizens, most Argentinians do not visit the Islas Malvinas/Falkland Islands for religious or even for nationalistic pilgrimages. Indeed, tourism is not promoted to the islands in an effort to maintain the mythical and sacrosanct spaces of the Islands through excluding humans. This differs from MacCannell's view that spaces become "sanctified" through mass tourism. Critiquing the sacred/profane dichotomy, Korstanje and George (2012) coined the terms "sacred-sacred" and "sacred-profane", with the former term referring to sacredness regardless of a human presence or not, while the latter term connotes the human labeling and symbolization of the divine. It is not otiose to assume that tourism operates in the dimension of the "sacred-profane", while the "sacred-sacred" is reserved for the domain of religion (Korstanje & George 2012). In this vein, Stausberg (2011) introduces the term "*homo turisticus religious*" to express the impossibility of fully dissociating tourism from religion and pilgrimage.

Religious events or festivals are key elements in enhancing social cohesion, narrowing the distance between gods and humankind. Because of this, the orchestration of rites and rituals not only should be carefully monitored by religious authorities to avoid conflict between multiple religious groups or within a religious group (Dayan & Katz 1994; Olsen & Timothy 2002), but policy makers should also devote considerable effort and material resources to avoiding conflict or any manifestation of violence in pilgrimage and religious festivals to prevent a broader climate of political instability (Henderson 2003). At the same time, because pilgrimages are conceived as liminal experiences for people who leave the comfort of home to travel to a spiritual center, faith communities develop their own ontological conceptions of security and risk (Collins-Kreiner, Kliot, Mansfeld & Sagi 2017; Nikjoo, Razavizadeh & Di Giovine 2020). In contrast with tourists, many pilgrims and religious leaders tolerate higher levels of uncertainty, discomfort, and risk in their pilgrimages since their journey a quires a redemptive value, particularly if death occurs during their travels (Belhassen, Caton & Stewart 2008).

Ultimately, as noted above, it is important to view limitations in the development of security systems beyond objective reality. As Sunstein (2005) puts it, people fear things that are sometimes imagined, and their emotional dispositions regarding those imagined things often distort external risks that are otherwise regulated by government experts in a "populist climate" of dissatisfaction. Regular citizens, due to media influences, for example, frequently exaggerate risks that are relatively minor while mysteriously ignoring larger, more eminent risks (Sunstein 2005). As Clarke (2018) notes, ideology plays an important role in determining why humans loathe some frightening objects but love others. This is, as also noted above, culturally contingent, as the collective memory of a society is formed through discourses, anxieties, myths, narratives, and texts that determine social and individual behavior. People's misgivings about terrorism or other risks are engendered by previous experiences and by what they are socially conditioned to see as good or bad. All this suggests that the world of ideas (ideology) should be taken more seriously when developing security-related programs, particularly those related to pilgrimage and religious tourism.

Understanding religious tourism security

The effects of risk in public spectacles depend on at least three variables. The first variable is the threshold of control, or the degree of control society has over the security of an event. This involves surveillance and the ability to predict the time and place of a potential terrorist attack. Hypothetically, for example, a bombing can be more easily and visually monitored than food contamination by microorganisms. This leads to increased surveillance and concern over potential attacks against iconic heritage or political sites or important urban destinations that resonate deeply in public opinion compared to attacks that take place in peripheral or less popular regions. This explains in part why international terrorism targets tourist destinations and attractions or popular urban areas.

The second variable is the probability that a negative event could be repeated in the short term. When specialists identify a risk as having the potential to occur again, resources should be mobilized to prevent it from happening again; or, if this is not possible, to lessen its effects. Meanwhile, once a threat is repeated (e.g., terrorism), a desensitization may occur, and a general sense fatalism among both the locals and potential tourists may build. As specialists become more familiar with terrorism, perpetrators must devise new innovative and crueler forms of violence to grab public attention.

The third variable involves the status of victims of any security crises. While any casualty that comes about from violent acts is unacceptable, society seems to esteem different population segments differently. For example, society feels more threatened and vulnerable when women and children become casualties of violent acts. Likewise, high-status individuals who are affected by attacks or other crises draw more public attention than when ordinary people are affected. This also goes for foreign tourists. When Western tourists are affected by dangerous situations and terrorist attacks, it is more newsworthy and shocking than when locals are impacted. Terrorists know this, and therefore frequently attack tourist targets for their widespread impact and notoriety.

Terrorist attacks and other events can have an immediate effect on tourist arrivals. This is especially true in less-developed countries, where there is a stronger dependency between tourism and economic development. However, it is important to note that not all risks follow the same dynamic because they are caused by different phenomena. As noted earlier, while people's perceptions influence their travel behavior and views of a destination, under some conditions these perceptions obscure more than they clarify—with people focusing on either past momentous events or the potential of future events (even if they are of a low probability) while at the same time trivializing other major risks, whether they be viruses and pandemics, food insecurity, terror attacks, airplane and road accidents, natural disasters, dangerous climatic of topographical conditions, crimes against tourists, and political instability (Sunstein 2005).

Out of these risks, each of which generates different reactions in the minds of tourists, terrorism is widely esteemed as a global risk, including in the context of pilgrimage and religious tourism. The risk cannot be understood without a social imaginary that constitutes the existence of risk. In this vein, tourists choose travel opportunities according to their level of risk awareness, which is usually developed by their own experiences and outside sources of information. This means that while some risks (e.g., terrorism) are deeply feared and avoided, others (e.g., local crime) are often overlooked, possibly because these types of risks are widely accepted as a common possibility (e.g. car accidents and pickpocketing).

In support of the above, Achcar (2015) contends that not all victims have the same social value. Western tourists are often a much sought-after tourist market segment, even more than domestic tourists or international tourists from non-Western countries because of their higher expenditures. Because of the global political power of the countries from which these Western tourists come from, terrorist organizations seem to prioritize tourists from the Global North over tourists from the Global South in terms of targeting. When people from developing societies support and sympathize with victims from the developed world, they feel a part of a "global empire" in a mega-structure that systematically rejects them as lower-order dwellers in a globalized world. However, since many terrorist networks originate in countries with underdeveloped economies, they cannot compete with global powers and remain largely outside the economic structure of the world (Korstanje 2016; Tzanelli 2016).

Much of the motivations of terrorists are related to religion, whether based on apocalyptic prophesy, the establishment of theocratic political spaces, or creating a pure religious state (Chowdhury et al. 2017). In this context, pilgrimage and religious tourism sites are prime targets for attacks that aim to create political instability, as disturbing sacred rituals and mega-events and damaging religious sites is an efficient way of undermining the authority of the state and religious leaders. For example, in their examination of pilgrimage to Shri Mata Vaishno Devi Shrine in India, Khajuria and Khanna (2014) found that while safety and security lead to a decline in tourists. The religious diversity associated to ethnic disputes can engender potential tension between pilgrimages and locals. This is aggravated by the fact many religious destinations are situated in isolated areas, with roads that do not allow a rapid intervention of security forces. For that reason, pilgrimages' perception of risk is vital to understand in what ways destinations are valorized as safe or unsafe. While Khajuria and Khanna also found that safety and security issues were not concerns among the majority of pilgrims, there were some differences in risk perception among different ages of pilgrims. Still further, leisure tourists and pilgrims develop different expectancies respecting to risk. For example, accidents and risk of unhygienic food were pondered as major threats. Unlike classic tourists who are very demanding, pilgrims compensate for the bad experiences derived from the lack of infrastructure with their faith.

Religious principles for crisis leadership

Religion is one way in which people organize themselves and create leadership structures. Founding leaders of religions develop both the leadership structure based on their personal leadership ideas (Worden 2005) as well as the eternal rewards that come to the faithful (Frunza 2017). Yet the question is often asked: Why are people attracted to religion? Religious adherence is often a response to a sense of loss or the lack of meaning or purpose that all humans go through at some point in their lives. Religion provides answers to some of life's biggest questions and establishes a framework for better living. However, the success of a religion depends upon its ability to provide solutions to the human condition. Meaning and purpose are time-transcendent promises, and many religions require people to delay personal gratification and sacrifice present pleasures for greater eternal rewards in the life to come. This same reasoning, however, may also lead to the development of terrorist activities, especially when otherworldly rewards are combined with violent acts in the here and now.

This caveat aside, religion teaches positive values that may help individuals rise above the challenges of crises and become leaders at different scales. Religious leadership comes in many forms, including "leader as teacher" and "leader as seer". As well, religious leaders may encourage adherents to incorporate religious virtues in their daily and work leadership, as well as take on a servant-leadership role in their dealings with other people. These views of leadership, however, can differ between different religions. For instance, some strands of Hinduism and Buddhism highlight that the individual is their own leader. In the Adwaita tradition of Hinduism, this idea is articulated as *Tat Tvam Asi* ("Thou Art That"). Abrahamic religions, on the other hand, clearly distinguish leaders from followers, although in Christianity, there is much emphasis on emulating Christ and his attributes among both leaders and followers.

A central idea that resonates with most religions is the servant-leadership theory, in which leaders should serve those they lead. There is no better situation to put this into practice in times of crisis. While Christianity claims this philosophy as its own, the idea of servant-leadership is evident in Judaism, Islam, Buddhism, Hinduism, as well as other faiths (Shirin 2014). In the Abrahamic religions, ancient leaders such as Moses and Abraham are revered as examples of transformational servant-leaders. In Buddhism, Buddha is seen as the ultimate example of servant-leader. Hinduism exhorts individuals to transform into servant-leaders, although it also teaches that people are already transformed beings but do not realize it due to *Maya* (i.e., illusion or magic), and can only come to a realization of such through interactions with strong servant-leaders such as gurus.

Charismatic leadership theory is another common thread among the world's religions, where the founders of charismatic faiths gained a considerable following as a result of their charism, whether through healing or sheer personal character (Barnes 1978). Charismatic leaders can play a fundamental role during crises, when people look for guidance and instruction. Religious leaders can also play an important role in developing people's attitudes towards a range of things in the secular realm (Emerson & Mckinney 2010), including avoiding or engaging in risky situations, as demonstrated during the COVID-19 pandemic, where some religious leaders encouraged adherents to follow government and health official's guidelines while other religious leaders wanted adherents to leave things in the hands of God (Olsen and Timothy, 2020).

Conclusion

One of the conceptual problems when studying perceptions of risks is that scholars overemphasize what tourists perceive of as risk factors as based on questionnaires and interviews. Methodologically, this is somewhat risky, as study participants may either lie to protect their own interests or simply refrain from expressing their innermost feelings. In many cultures, men are culturally conditioned to reject fear as a real emotion since it can diminish their masculinity. As such, many studies often do not provide deeper insight into the difference between risks and threats (Korstanje 2009). Therefore, more studies are needed to interrogate the intersection of language in the formation of risk. The best anecdote to illustrate the functioning of risk derives from an ethnography done in an aboriginal community living alongside an active volcano in southern Argentina, while the first author was a student. Many volcanologists and other scientists were asked by government officials to evacuate the aboriginal community. They were deeply surprised when the locals not only rejected evacuation attempts but also pointed out that the worrisome eruptions were a direct result of God's rage. In fact, the technology introduced to predict the volcanic activity in the region was considered by the local population to be the main reason for the disaster. This illustrates the power of spiritual beliefs and religious perceptions in people's ontological construction of security (Korstanje & Baker 2018).

Religion is like fire—it can soothe or boost aggression in individuals and groups. During religious travel, people go through a liminal transition from one psychological state to another. To ensure this transition is more meaningful, purposeful, and impactful, spiritual leadership is needed. In the event of terrorism during a religious gathering, religious leaders should act as crisis managers, with crisis leadership being based upon religious principles. As Korstanje and George (2020) note, religious tourism has economic potential in some peripheral regions, but run the risk of creating radicalized discourses which direct a much deeper sentiment of resentment against foreigner tourists, particularly in the COVID-19 era where pilgrims are spreaders of the virus (Korstanje 2020; Olsen & Timothy, 2020). As ambassadors of their respective nations, religious tourists are easy targets. This chapter stresses the importance of managing crimes in religious tourism and how leadership principles from the teachings of religions themselves can help improve management practices within the context of risks. The promotion of security-related programs should include the voice and opinion of religious leaders who placate the historical discontent or the dormant inter-ethnic conflicts of some groups. Religious leaders not only guide pilgrims in interpreting the sacred text in the correct way, but they also manage human emotions. Hate and resentment are powerful emotions, which, unless they are successfully regulated by religious leaders, may lead very well to a process of radicalization (Korstanje & George 2012). Unfortunately, there is not much literature on this topic. However, societies should continue to provide avenues for spiritual emancipation, with religious tourism being an important part of this. Therefore, it is somewhat ironic that events that are held to uplift individuals and societies are often targeted for terrorist attacks or other major crimes. Likewise, religiously motivated crimes sometimes target festivities that are considered profane by those perpetrating the attacks.

References

Achcar, G. (2015) *Clash of Barbarisms: The Making of the New World Disorder.* London: Routledge.

Adams, K.M. (2001) 'Danger-zone tourism: Prospects and problems for tourism in tumultuous times', in P. Teo, T.C. Chang, & K.C. Ho (eds.), *Interconnected Worlds: Tourism in Southeast Asia* (pp. 265–281). Amsterdam: Pergamon.

Barnes, D.F. (1978) 'Charisma and religious leadership: An historical analysis', *Journal for the Scientific Study of Religion*, 19(1): 1–18.

Bauman, Z. (2013) *Liquid Fear.* New York: Wiley.

Beck, U. (1992) *Risk Society: Towards a New Modernity.* London: Sage.

Belhassen, Y., Caton, K., & Stewart, W.P. (2008) 'The search for authenticity in the pilgrim experience', *Annals of Tourism Research*, 35: 668–689.

Bianchi, R. (2006) 'Tourism and the globalisation of fear: Analysing the politics of risk and (in)security in global travel', *Tourism and Hospitality Research*, 7(1): 64–74.

Bianchi, R., & Stephenson, M.L. (2013) 'Deciphering tourism and citizenship in a globalized world', *Tourism Management*, 39: 10–20.

Bonham, C., Edmonds, C., & Mak, J. (2006) 'The impact of 9/11 and other terrible global events on tourism in the United States and Hawaii', *Journal of Travel Research*, 45(1): 99–110.

Chowdhury, A., Raj, R., Griffin, K., & Clarke, A. (2017) 'Terrorism, tourism and religious travellers', *International Journal of Religious Tourism and Pilgrimage*, 5(1): Article 3. DOI: 10.21427/D7BD8G.

Clarke, A. (2018) 'Religion, ideology and terrorism', in M. Korstanje, R. Raj, & K. Griffin (eds.), *Risk and Safety Challenges for Religious Tourism and Events* (pp. 8–17). Wallingford: CABI.

Cohen, E. (1992) 'Pilgrimage and tourism: Convergence and divergence', in A. Morinis (ed.), *Sacred Journeys: The Anthropology of Pilgrimage* (pp. 47–61). Westport, CT: Greenwood Press.

Collins-Kreiner, N. (2010) 'Researching pilgrimage: Continuity and transformations', *Annals of Tourism Research*, 37: 440–456.

Collins-Kreiner, N., Kliot, N., Mansfeld, Y., & Sagi, K. (2017) *Christian Tourism to the Holy Land: Pilgrimage during Security Crisis.* London: Routledge.

Dalby, S. (2003) 'Calling 911: Geopolitics, security and America's new war', *Geopolitics*, 8(3): 61–86.

Dayan, D., & Katz, E. (1994) *Media Events.* Cambridge, MA: Harvard University Press.

Douglas, M., & Wildavsky, A. (1983) *Risk and Culture: An Essay on the Selection of Technological and Environmental Dangers.* Berkeley: University of California Press.

Durkheim, E. (2008) *The Elementary Forms of the Religious Life.* New York: Courier Corporation.

Emerson, T.L.N., & Mckinney, J.A. (2010) 'Importance of religious beliefs to ethical attitudes in business', *Journal of Religion and Business Ethics*, 1(2): 5–9.

Floyd, M.F., Gibson, H., Pennington-Gray, L., & Thapa, B. (2004) 'The effect of risk perceptions on intentions to travel in the aftermath of September 11, 2001', *Journal of Travel & Tourism Marketing*, 15(2–3): 19–38.

Frunza, S. (2017) 'Ethical leadership, religion and personal development in the context of global crisis', *Journal for the Study of Religions and Ideologies*, 16(46): 3–16.

Fuchs, G., & Reichel, A. (2004) Cultural differences in tourist destination risk perception: An exploratory study', *Tourism*, 52(1): 21–37.

Fuchs, G., Uriely, N., Reichel, A., & Maoz, D. (2013) 'Vacationing in a terror-stricken destination: Tourists' risk perceptions and rationalizations', *Journal of Travel Research*, 52(2): 182–191.

Hall, C.M., Amore, A., & Arvanitis, P. (2019) 'The globalising force of human mobilities', in D.J. Timothy (ed.), *Handbook of Globalisation and Tourism* (pp. 55–65). Cheltenham: Edward Elgar.

Hall, C.M., Timothy, D.J., & Duval, D.T. (2003) 'Security and tourism: Towards a new understanding', *Journal of Travel and Tourism Marketing*, 15(4): 1–18.

He, L., Park, K., & Roehl, W.S. (2013) 'Religion and perceived travel risks', *Journal of Travel & Tourism Marketing*, 30(8): 839–857.

Henderson, J.C. (2003) 'Managing tourism and Islam in peninsular Malaysia', *Tourism Management*, 24(4): 447–456.

Kaelber, L. (2006) 'Paradigms of travel: From medieval pilgrimage to the postmodern virtual tour', in D.J. Timothy & D.H. Olsen (eds.), *Tourism, Religion and Spiritual Journeys* (pp. 65–79). London and New York: Routledge.

Khajuria, S., & Khanna, S. (2014) Tourism risks and crimes at pilgrimage destinations–a case study of Shri Mata Vaishno Devi. *International Journal of Event Management Research*, 8(1): 77–93.

Korstanje, M.E. (2009) 'Revisiting risk perception theory in the context of travel', *E-Review of Tourism Research*, 7(4): 68–80.

Korstanje, M.E. (2016) *The Rise of Thana Capitalism and Tourism*. Abingdon: Routledge.

Korstanje, M.E. (2020) Tourism and the War against a Virus? Available at: https://northernnotes.leeds.ac.uk/tourism-and-the-war-against-a-virus-also-in-spanish/ [accessed 6 July 2020].

Korstanje, M.E., & Baker, D. (2018) 'Politics of dark tourism: The case of Cromañón and ESMA, Buenos Aires, Argentina', in Stone P. et al (eds.), *The Palgrave Handbook of Dark Tourism Studies* (pp. 533–552). London: Palgrave Macmillan.

Korstanje, M.E., & Clayton, A. (2012) 'Tourism and terrorism: Conflicts and commonalities', *Worldwide Hospitality and Tourism Themes*, 4(1): 8–25.

Korstanje, M.E., & George, B.P. (2012) 'Falklands/Malvinas: A re-examination of the relationship between sacralisation and tourism development', *Current Issues in Tourism*, 15(3): 153–165.

Korstanje, M E., & Olsen, D.H. (2011) 'The discourse of risk in horror movies post 9/11: Hospitality and hostility in perspective', *International Journal of Tourism Anthropology*, 1(3–4): 304–317.

Korstanje, M.E., Raj, R., & Griffin, K. (eds.) (2018) *Risk and Safety Challenges for Religious Tourism and Events*. Wallingford: CABI.

Korstanje, M.E., & Seraphin, H. (2017) 'Revisiting the sociology of consumption in tourism', in S.K. Dixit (ed.), *Routledge Handbook of Consumer Behaviour in Tourism and Hospitality* (pp. 16–25). London: Routledge.

Korstanje, M.E., & George, B. (2020) Safety, fear, risk and terrorism in the context of religious tourism. In M.E. Korstanje & H. Seraphin (eds.), *Tourism, Terrorism and Security: Tourism Security-safety and Post Conflict Destinations* (pp. 89–102). Bingley: Emerald.

Larsen, S. (2007) 'Aspects of a psychology of the tourist experience', *Scandinavian Journal of Hospitality and Tourism*, 7(1): 7–18.

Larsen, S., Brun, W., & Øgaard, T. (2009) What tourists worry about–Construction of a scale measuring tourist worries. *Tourism Management*, 30(2), 260–265.

Lash, S.M., Urry, S.L.J., & Urry, J. (1993) *Economies of Signs and Space*. Thousand Oaks, CA: Sage.

Lepp, A., & Gibson, H. (2003) Tourist roles, perceived risk and international tourism', *Annals of Tourism Research*, 30(3): 606–624.

Luhmann, N. (2017) *Risk: A Sociological Theory*. London: Routledge.

MacCannell, D. (1973) 'Staged authenticity: Arrangements of social space in tourist settings', *American Journal of Sociology*, 79(3): 589–603.

MacCannell, D. (1999) *The Tourist: A New Theory of the Leisure Class*. Berkeley: University of California Press.

MacCannell, D. (2002) *Empty Meeting Grounds: The Tourist Papers*. London: Routledge.

Mansfeld, Y., & Pizam, A. (eds.) (2006) *Tourism, Security and Safety*. London: Routledge.

Nikjoo, A., Razavizadeh, N., & Di Giovine, M.A. (2020) What draws Shia Muslims to an insecure pilgrimage? The Iranian journey to Arbaeen, Iraq during the presence of ISIS. *Journal of Tourism and Cultural Change*. DOI: 10.1080/14766825.2020.1797062.

Olsen, D.H. (2010) 'Pilgrims, tourists, and Weber's "ideal types"', *Annals of Tourism Research*, 37(3): 848–851.

Olsen, D.H. (2019) 'Religion, spirituality, and pilgrimage in a globalizing world', in D.J. Timothy (ed.), *Handbook of Globalisation and Tourism* (pp. 270–283). London: Edward Elgar.

Olsen, D.H., & Timothy, D.J. (2002) 'Contested religious heritage: Differing views of Mormon heritage', *Tourism Recreation Research*, 27(2): 7–15.

Olsen, D.H., & Timothy, D.J. (2020) 'The COVID-19 pandemic and religious travel: Present and future directions', *International Journal of Religious Tourism and Pilgrimage*, 8(7): 170–188.

Pappas, N. (2010) 'Terrorism and tourism: The way travelers select airlines and destinations', *Journal of Air Transport Studies*, 1(2): 76–96.

Plog, S. (1991) *Leisure Travel: Making it a Growth Market Again!* New York: Wiley.

Quintal, V.A., Lee, J.A., & Soutar, G.N. (2010) 'Risk, uncertainty and the theory of planned behavior: A tourism example', *Tourism Management*, 31(6): 797–805.

Raine, R. (2013) 'A dark tourist spectrum', *International Journal of Culture, Tourism and Hospitality Research*, 7(3): 242–256.

Raj, R., & Griffin, K.A. (2015) 'Introduction to sacred or secular journeys', in R. Raj & K.A. Griffin (eds.), *Religious Tourism and Pilgrimage Management: An International Perspective* (2nd ed., pp. 1–15). Wallingford: CABI.

Reisinger, Y., & Mavondo, F. (2005) 'Travel anxiety and intentions to travel internationally: Implications of travel risk perception', *Journal of Travel Research*, 43(3): 212–225.

Roehl, W.S., & Fesenmaier, D.R. (1992) 'Risk perceptions and pleasure travel: An exploratory analysis', *Journal of Travel Research*, 30(4): 17–26.

Saha, S., & Yap, G. (2014) 'The moderation effects of political instability and terrorism on tourism development: A cross-country panel analysis', *Journal of Travel Research*, 53(4): 509–521.

Sharifpour, M., Walters, G., & Ritchie, B.W. (2014) Risk perception, prior knowledge, and willingness to travel: Investigating the Australian tourist market's risk perceptions towards the Middle East. *Journal of Vacation Marketing*, 20(2): 111–123.

Shirin, A. (2014) 'Is servant leadership inherently Christian?' *Journal of Religion and Business Ethics*, 3(1): 1–25.

Sönmez, S.F., & Graefe, A.R. (1998) 'Influence of terrorism risk on foreign tourism decisions', *Annals of Tourism Research*, 25: 112–144.

Stausberg, M. (2011) *Religion and Tourism: Crossroads, Destinations and Encounters.* London: Routledge.

Sunstein, C.R. (2005) *Laws of Fear: Beyond the Precautionary Principle.* Cambridge: Cambridge University Press.

Timothy, D.J. (2019) 'Globalization: The shrinking world of tourism', in D.J. Timothy (ed.), *Handbook of Globalisation and Tourism* (pp. 323–332). Cheltenham: Edward Elgar.

Tzanelli, R. (2016) *Thanatourism and Cinematic Representations of Risk: Screening the End of Tourism.* Abingdon: Routledge.

Wearing, S., McDonald, M., Taylor, G., & Ronen, T. (2019) 'Neoliberalism and global tourism', in D.J. Timothy (ed.), *Handbook of Globalisation and Tourism* (pp. 27–43). Cheltenham: Edward Elgar.

Wolff, K., & Larsen, S. (2014) 'Can terrorism make us feel safer? Risk perceptions and worries before and after the July 22nd attacks', *Annals of Tourism Research*, 44: 200–209.

Worden, S. (2005) 'Religion in strategic leadership: A positivistic, normative/theological, and strategic analysis', *Journal of Business Ethics*, 57(3): 221–239.

Yan, B.J., Zhang, J., Zhang, H.L., Lu, S.J., & Guo, Y.R. (2016) 'Investigating the motivation–experience relationship in a dark tourism space: A case study of the Beichuan earthquake relics, China', *Tourism Management*, 53: 108–121.

32

RELIGIOUS AND SPIRITUAL TOURISM

Sustainable development perspectives

Anna Trono

Introduction

Travel undertaken for religious purposes is currently enjoying a resurgence, although the purely religious aspect is less important than it was in the past, partly because modern societies are less dominated by organised religion. In numerical terms, however, it is substantial, as evident in the millions of tourists and pilgrims who annually visit famous places and events of worship such as Kumbh Mela, Mecca, the Holy Land, Lourdes, Fatima and Rome. In addition to individual places of faith, religious routes and trails draw a huge number of people who retread historic pilgrimage trails with extraordinary symbolic and spiritual values. In many cases, these are promoted and organised by associations and public or private bodies that invest in religious and spiritual tourism. Prime examples here are the cultural routes of the Council of Europe, which since 1987 has recognised the importance of the various religious 'ways' as key cultural and spiritual vehicles. Santiago de Compostela has seen an increase in popularity that has made the Galician city the biggest religious tourism destination in Europe (Pérez Guilarte & Lois González 2018).

Cultural and religious itineraries traditionally have represented a form of tourism understood as a slow journey of discovery that gradually reveals the local landscape and culture by means of ancient routes whose value is now recognised or in the process of being recognised (Timothy & Boyd 2015). Such routes enable travellers to rediscover themselves while drawing their attention towards the social dimension of travel as well as contributing to the sustainable development of the regions involved (Timothy 2018). The paths can provide profound experiences, spiritual growth and self-actualization and the visits to places of faith (e.g. sanctuaries, cathedrals and mosques) represent growth engines for the local economy, promoting the characteristic assets of the region with positive effects on employment. However, they also exhibit points of weakness that affect the development of the regions involved and their search for sustainability. Indeed, places of pilgrimage and religious routes frequently neglect sustainability, understood in terms of respect for the environment, but also social and cultural values and aspects such as accessibility, hospitality and the presence of structures providing assistance and/or services for specific types of market (Scidurlo & Callegari 2015). Finally, yet importantly, they frequently tend towards a barely concealed use of religious practices for economic ends, reflecting a tendency to see religious devotion from a purely economic point of view. Indeed, a disconnect frequently arises between pilgrimage to holy places (and spiritual tourism more generally) and the motives and sociocultural processes that originally gave rise to it and continue to sustain it. This produces a clash of principles that reflects a tension between tradition and modernity,[1] showing that the spiritual motive is much more than a simple ritual.

Given this premise, the objective of this chapter is to reflect on the sustainable development of religious and spiritual tourism, on the possible measures to adopt and on the forms of support that could help ensure sustainability in environmental, economic and sociocultural terms. To this end, the study that follows is structured into three sections. The first section, starting with a brief preliminary analysis of the concept of sustainable tourism in line with the recommendations of *Agenda 2030 for Sustainable Development,* looks at the factors that either favour or undermine sustainability.

After a brief overview of the literature on religious tourism, today more appropriately named spiritual tourism, the second section presents a diachronic analysis of the notions of journey and pilgrimage understood on one hand as an ascetic and mystic experience and a way to rediscover human, cultural, environmental and landscape values, and on the other hand as an experience of sharing with others and getting to know the local community. We are thus speaking of a wide-ranging experience, of a rediscovered equilibrium that restores respect and love for the environment, nature and culture. The section considers the factors that influence the sustainable development of the places visited and describes initiatives to generate and develop sustainable regional synergism in the management of cultural and religious heritage. It illustrates other influences in the secular and religious spheres that encourage the management of holy places in accordance with a pragmatic and utilitarian logic.

The third section analyses a few case studies in an attempt to determine whether the blend of tourism in general and a desire for culture and religious sentiment can be reconciled not only with the search for spirituality and contact with natural and uncontaminated landscapes but also with a focus on the sustainable development of the places visited. It considers the need for such tourism to avoid conflict with commercial imperatives, creating new market niches, new types of supply and demand, new circuits, new entrepreneurial figures and new wealth, acting as a dynamo of economic development of religious places and routes.

The conclusion to the chapter discusses the need for an eco-compatible religious and spiritual tourism that not only improves the quality of life of those who live along routes or in places of spiritual or religious interest but also upholds the values of solidarity, brotherhood and intercultural dialogue.

Sustainable tourism

Sustainable tourism has become a significant theme in the last few decades among academics and professionals, on which there now exists an extensive literature dealing with both its definitions and its applications and setting out the need to reduce the negative effects of tourism in accordance with an approach that is publicly and politically acceptable (Butowski 2019; Chirilă, Chirilă & Sîrbulescu 2018; Connell, Page & Bentley 2009; Farrell & Hart 1998; Sachs 2015).

There has been much debate on the meaning of the term 'development' and on the principles underpinning sustainability, starting with the foundation of the World Tourism Organization (UNWTO) in 1974, the formulation of *Agenda 21 for Tourism* and the drawing up of the *Lanzarote Charter for Sustainable Tourism* (1995). The latter has been followed by many other international charters and conventions, which recommend and promote new approaches to tourism, increasingly attentive to its social, regional, environmental and cultural components (Tinacci Mossello 2014). More recently, the theme has been revisited by *Agenda 2030*, approved by the United Nations in September 2015, in which the principles and recommendations for sustainable development are restated. Agenda 2030 sets out 17 Sustainable Development Goals (SDGs) to be reached by 2030, representing a substantial shift with regard to current development models (United Nations 2019). Specifically, Agenda 2030 highlights the urgent need to integrate the environmental question with the social and economic dimensions of development, inviting all countries (developed, emerging and developing) to contribute to the task of putting the world on the path to sustainability. All countries are obliged to draw up their own strategy for sustainable development, to state the results achieved as part of a process coordinated by the UN and, last but not least, to involve all components of society: from companies to the public sector, from civil society to philanthropic institutions, from universities and research centres to operators in the fields of information and culture.

Agenda 2030 recognises the potential of sustainable tourism to create jobs, promote culture and local products (SDG 8.9) and yield economic benefits, particularly for small island states and less developed countries. However, it also stresses the need for continuous monitoring and optimal management by means of suitable policies, considering the progressive increase in the number of tourists on an international level, which in the near future is likely to have a serious impact on the environment, the quality of the visitor experience and the quality of life in the destination.

Indeed, sustainable tourism means a lot more than providing development and its attendant economic and social benefits on an inter-generational level. It entails reducing and/or mitigating all undesired impacts on the natural, historic, cultural and social environments. It means intervening on an economic level, guaranteeing equitable and efficient economic development that can bring growth to future generations. On the sociocultural side, it means operating in ways that are compatible with local culture, values and identity, and on the environmental side, it means guaranteeing development that is compatible with the preservation of biological diversity and biological resources. In order to minimise its impact on the local environment and culture, therefore, it is necessary to balance the needs of tourists with those of the destination community. These objectives are all easy to list but hard to implement, considering the high degree of uncertainty regarding the field of application, the priorities to consider when seeking to make tourism more sustainable and the ways of putting this into practice.

While the positive aspects of tourism have frequently been stressed and emphasised, the negative implications have not always been highlighted because they often conflict with local and global socioeconomic interests. Some years ago, the UNWTO, with the financial support of the Directorate-General for International Cooperation and Development of the European Commission (CE) (World Tourism Organization 2013), began stressing the significant and growing influence of tourism on the local environment, society and economy. Tourism clearly affects local markets, limits access to resources and generates exploitative labour conditions, increases crime rates and sexual exploitation, and threatens traditions and social and cultural values.

Tourism undoubtedly plays a role in climate change, accounting for about 8% of global CO_2 emissions, mainly generated by transport (especially aircraft) but also produced by hotels and restaurants, infrastructure, retailing and the consumption of food and drink linked to tourism. In some areas, it contributes to water and soil pollution caused by inadequate treatment of solid and liquid waste produced by tourism companies and the activities of the tourists themselves. Hotels and resorts are often big consumers of precious and non-renewable resources, such as soil, energy and water.

Promoting environmentally sustainable tourism is an essential task, especially in regions where tourist pressure is acute and in those where the effects of climate change are most evident and rapid, with all that this implies in terms of depletion of drinking water reserves and the loss of biodiversity (Gössling, Hall & Scott 2015; Kliot 2019). As a consequence, the management of water resources, pollution and waste is one of the main challenges facing tourism. Inappropriate tourism activities in sensitive areas can be highly damaging for biodiversity, and poor management of visitors can have negative impacts on cultural heritage (Gray & Winton 2009). A growing threat in environmental terms is mass tourism, which tends to reduce the region being visited to a series of banal stereotypes, with the loss of its specific properties and environmental quality and generating pollution on a systemic level (Séraphin, Gladkikh & Thanh 2020; Visentin & Bertocchi 2019; Wall 2019).

To be sustainable, therefore, tourism needs to adapt to, and share the lifestyles of, the host communities, embracing sobriety and technological creativity, ensuring that relationships and participation are not lost. What is needed is a healthy approach to every asset, every item of beauty and every positive relationship that nature and people have to offer in the region being visited, together with close cooperation among all local stakeholders. 'Alternative tourism' is "compatible with natural, social and cultural values. It enables both hosts and visitors to benefit from the interaction generated by the experience of the visit" (Smith & Eadington 1992, quoted in Bimonte & Pagni 2003, p. 11).

Alternative tourisms are currently growing include slow tourism and nature tourism. Ensuring the safeguarding of all the resources of the environment with which the traveller comes into contact during their stay, these are an expression of a cultural shift in the tourism market. Another manifestation of tourism associated with slow mobility and sustainability is cultural/spiritual route-based tourism, which represents the new frontier of cultural tourism. Activities are often based on traditional routes of faith and culture.

Routes of the soul, places of worship and sustainability

Extremely popular are trails linked to one form or another of religious tourism or tourism of the soul, aimed at a more spiritual way of practising one's faith: a tourism which despite the financial crisis of the last few years and the security concerns generated by the rise of 'fundamentalist' movements, above all in certain developing and/or politically unstable countries, now plays an important economic and social role (Ron & Timothy 2019; Timothy 2018).

The literature on religious tourism is wide-ranging and rich in ideas, thanks partly to the authoritative knowledge it displays of the places and regions visited, full of history and culture, which make them highly attractive and appealing for contemporary visitors and tourists. Historic documentation indicates that in the past, journeys were undertaken for many reasons: a desire for adventure, a pastime, cultural interests and a search for the exotic and religious observance, all of which are recognisable in the modern world. The current model of travel, particularly that of a spiritual nature, differs in many ways, although not radically, from the forms of religious journey of the past. This is a theme that has been widely debated (Ambrosio 2007; Barber 1993; Blackwell 2007; Carbone, Corinto & Malek 2016; Cohen 1974, 1992, 1998, 2008; Dallari, Trono & Zabbini 2009; Damari & Mansfeld 2014; Graburn 1989; Griffin 2007; Hitrec 1990; Smith 1992; Stoddard 1997; Timothy & Olsen 2006; Trono 2014) to the point of exhaustion (Collins-Kreiner 2009, 2010), making every comparison between the ancient pilgrim and the modern tourist—who operates in quite different temporal and cultural contexts—quite futile (Olsen 2010). Indeed, the behaviours and requirements of those who undertake journeys of religious interest have changed. The journeys themselves meanwhile are today more akin to regular tourism, whose negative aspects they share (Gray & Winton 2009; Herrero, Sanz & Devesa 2009; Murray & Graham 1997; Raj & Morpeth 2007; Shackley 2001; Trono 2016). Indeed, the term tourism is increasingly associated with a sharply negative meaning and the word 'journey' is preferred, with its suggestion of discovery, learning something new and doing something different, as opposed to a model of tourism based on standardised and uniform products (Corna Pellegrini 2000; Toscano 1996; Visentini 1996). Thus, in contrast to modern tourism, often understood as the expression of consumerism and alienation, governed by a purely economic logic, there is the image of the traveller of the past, who is open to experience, ready to marvel at the spectacle of the world, to share knowledge and emotions and to seek a dimension that transcends everyday life: a form of travel that recalls the extraordinary labours and tribulations, material and spiritual, that were for centuries the lot of most pilgrims. Today, the journey towards a religious place or destination is seen as providing an opportunity to discover ourselves and the world around us. The path, which represented the bitterest and most tiring aspect for the medieval pilgrim, has now acquired a social and cultural value, offering the opportunity to meet other people and try new things, to marvel at cultural and natural landscapes and to conduct research or seek personal growth mediated by introspective experience. In many cases, the routes of the ancient pilgrims, once full of dangers, have become consolidated trails that make it possible to have an emotional, educational, social and participatory experience. Specifically, religious journeys and visits to sanctuaries offer the opportunity to socialise and compare one's own values and culture with those of others.

Today's pilgrim is a traveller who reflects, who is aware and attentive to the behaviour they must adopt on their journey: responsible and respectful towards the environment and traditional cultures of the local populations and eager to discover products that are authentic expressions of local culture. Religious and spiritual tourism is sustainable because it is respectful of the ethical and aesthetic values of the landscape but also and above all because it is centred on the values of community, in the sense of a shared memory of a local past, possibly religious, which is rich in echoes and meanings attributed to places and memories (Mcmillan & Chavis 1986). The sustainability of routes and places of faith also entails the meeting of people and different cultures, hospitality, conviviality, kindness and awareness of contributing to the social, cultural and economic development of the regions being visited. It is a form of tourism that promotes and brings together the full range of cultural resources and local products, providing new opportunities for enterprise and infrastructural development and an antidote to the abandonment of inland regions. It activates processes of regional renewal, facilitating the development of complex networks that are able to go beyond local boundaries, conferring on the region-specific and important functions.

Tourism of routes of the soul is centred on the beauty of silence, of physical and spiritual well-being, but also on that of nature and the tranquillity of the countryside, rural villages and urban monuments, re-establishing the links between the memories of places and recovering cultural and landscape heritage. As long ago as 1987, with the Declaration of Santiago and the recognition of the Way of St James as a European cultural route, the Council of Europe had confirmed its value in terms of both its cultural, social and symbolic capital and its capacity for promoting intercultural dialogue and responsible tourism. Tourism associated with the European Cultural Routes is aware and more sustainable than mass tourism because it is based on content that respects the environment and nature, and it is able to restore a sense of the common good and respect for local values, as well as dialogue and listening.

In addition, it has recovered the value of doing things slowly, particularly important for those who wish to travel without haste to their final destination (Davolio & Meriani 2011; Trono & Olsen 2018).

Tourism linked to routes of the soul can even contribute to the creation of value and sustainable development in difficult contexts such as Aspromonte in Calabria, a land with marvellous landscapes and a people known for their love, passion and hospitality, but also a "beautiful and bitter" land, as it was described by its bishops in the most recent document on organised crime. In this region, the creation of a route with a religious function and a historic destination in the form of the Sanctuary of the Madonna della Montagna in Polsi can enable

> the re-appropriation by the local communities of places that have for too long been contaminated and subjugated by the presence of the 'ndrangheta, facilitating virtuous processes of economic and social legality. Itineraries of faith and hope [facilitate] the recovery and conversion to viable assets of the local resources, restoring to the local communities often forgotten testimony of history and shared memory, fragments of life, literature and art, and serving as the catalyst for the development of enterprise, an equitable economy and responsible tourism.
>
> *(Di Gregorio, Chiodo & Nicolosi 2018, p. 17)*

While the features of a route are important, even more so are the motives that drive people to travel it and seek the final destination, which is the expression of a precise search, a reflection that incorporates the faith motive but also the appreciation of art and history, knowledge of one's own and other people's place of origin, and the desire to rediscover the meaning of life, which in their daily existence has been lost. A growing number of people are thus seeking mystical places that meet their needs for personal growth: places of creativity, where religious motives and/or the simple need for serenity and introspection combine with the beauty of the landscapes and art.

Well known and very popular for example in northern Italy are the *Marian Sanctuaries* of *Eastern Piedmont*, rich in history and precious works of art, *which make up the evocative* cultural and religious route *of the Black* Madonnas *of the* Holy Mountains (Afferni & Mangano 2009). An integral part of the Sacri Monti of Piedmont and Lombardy UNESCO World Heritage Site and a meeting point for worshippers and art lovers alike, the seven Holy Mountains of Piedmont are famous for the enchanting routes that rise to elevations of up to 1,200 metres, the 160 chapels built between 1,500 and 1,600, the 2,500 statues and the 12,000 painted figures, immersed in mature forests and large gardens. In an austerely beautiful Alpine setting, not far from Biella, stands the Sacro Monte di Oropa, steeped in the profound religiosity of its 12 chapels dedicated to the life of the Virgin Mary and the vast sanctuary of Oropa, which "reflects the practicality and impressive solidity of the people who created it and expanded it" (Gribaudi 1966, p. 409). The original project envisaged a complex of 20 chapels, which were supposed to project a wide-ranging narrative including significant episodes in Mary's life taken from the Holy Scriptures and the Apocrypha.

A resting place for travellers passing through the Valle d'Aosta since ancient times, Oropa has for centuries been a destination for pilgrims placing their faith in the Black Madonna (*Beata Maria Virgo Oropae in montibus Bugellae*). It is currently considered the most important Marian place of worship in the Alps: a magnet for visitors who wish to refresh their spirit in peace and silence and prayer. The important inscription of the Sacri Monti of Piedmont and Lombardy on UNESCO's World Heritage List in 2003 increased interest in the sanctuary

Figure 32.1 Sanctuary and the Sacro Monte di Oropa
Photo by Chiara Barbieri.

and the Sacro Monte di Oropa has become an important point of reference for national and international operators in religious and spiritual tourism (Figure 32.1). Committed to a management that is aware of the cultural values underpinning its integrity and authenticity and respects the views of those who seek to discover places of faith and spirituality, the sanctuary and the Sacro Monte di Oropa are the expression of a type of tourism that can satisfy both the needs of the spirit and the love of art and nature. This also involves bringing together providers of tourism and cultural services, a task that has been greatly facilitated by the presence of the Borsa di Oropa dei Percorsi Devozionali e Culturali, whose efforts are financed and organised by Piedmont Regional Administration.

The needs that arise in this context are not irreconcilable, and the dynamics of the relationship between worship and culture are simple and comprehensible. Churches, chapels and sanctuaries are visited by a growing number of worshippers, tourists and scholars because they are not only important places of faith but also because they constitute famous monuments, in some cases, the most famous cultural heritage attraction in an entire city. In Rome, for example, the Pantheon, a key building in western culture occupying a symbolic place in Italian history, is in the perception of many tourists above all or even solely an ancient temple, but it is also Santa Maria ad Martyres, a Christian place of worship. This is also the case with the church of Nôtre Dame in Paris, recently hit by a devastating fire. With over 12 million visitors a year, it is the second most visited monument in the French capital after the Eiffel Tower, above the Louvre, and the second most visited church in Europe after St Peter's Basilica in Rome. It is considered a masterpiece of Gothic art and is on UNESCO's World Heritage List.

Overcrowding, exemplified by the long queues to enter these monuments of faith and culture, generates safety and sustainability problems, to which the response is to charge a fee for entry, in an awkward equilibrium of religious, cultural and economic considerations. As Valerio Pennasso (Beltrami 2017), director of the national office for cultural ecclesiastical heritage and religious buildings of the Italian Bishops' Conference (CEI), points out, the perceptive shift from churches to museums (with or without tickets) is partly a result of social and economic changes in progress in the secular field. Above all, however, it reflects the impact of mass tourism, which devours sacred places in the same way it devours everything else it sees as a product for consumption. This is what legitimises the policies of selling admission tickets to control the flow of tourists and reserving some parts of the church for silence and prayer. The rules of the Canonical Code and measures by the ecclesiastical authorities aim to clarify, limit and discourage this practice (Azzimonti 2016; Feliciani 2010). This upholds the pre-eminence of the religious motive, which in the judgement of many religious authorities, however, conflicts with the need for management, conservation and restoration of cultural heritage, often carried out with the help of public funding from national and international sources, giving rise to much perplexity.

The problem of how to manage the financial flows associated with holy places and the promotion of journeys of the soul is clearly complex and fuels debates and controversies. The demand for visits to holy places is increasing, as is the range of goods and services offered by a vast and variegated body of operators: a universe of companies in tune with the latest developments in the international tourism market, characterised by an exceptional degree of competitiveness and capacity for forming strategic alliances with the main players in the sector, both public and private. Indeed, while the modern reasons for visiting holy places are often of a religious and spiritual nature, the ends of those who organise and manage them are frequently more profane and prosaic.

Sustainable religious tourism is predicated on adequate preparation, necessary to be able to truly appreciate the religious site and any associated cultural activities, rejecting pseudo-religious and pseudo-cultural speculation as well as the degeneration of visits to holy places typical of mass tourism (Battisti 1996). It makes a careful distinction between liturgical celebration and collateral activities posing as folklore, which draw an undifferentiated mass of tourists interested in traditional music and dance, in open conflict with the religious motive. This generates confusion and excessive anthropic pressure, along with problems of safety and environmental damage, which needs to be avoided and/or carefully monitored and analysed with an integrated multidisciplinary approach. The aim of such monitoring is to identify serious problems caused by tourism that may arise in the short and long term, but also to assess strategies that enable the economic development of the area while safeguarding the quality of the environment in accordance with environmental quality standards (EQSs).

A careful study of the pressure caused by pilgrims and tourists in holy places based on environmental indicators and the Driver-Pressure-State-Impact-Response (DPSIR) model was recently conducted for two case studies of an urban district and a protected area. The former considered the Sanctuary of Pompei, in Italy, a site of Marian pilgrimage visited by four million worshippers a year and thus exposed to strong anthropic pressures, while the latter looked at Kalakad Mundthurai Tiger Reserve in southern India, where the impact on the local fauna, caused by the increase in vehicular traffic linked to religious tourism, is a major environmental problem. Analysis of the increased water and energy consumption, waste production, discharges from water treatment plants and traffic emissions (public and private transport) generated by flows of worshippers has identified the key elements exerting pressure on the environment. In addition, it can provide those responsible for managing the arrival of the pilgrims with data that are useful for determining priorities and ensuring that changes or activities are in line with the principles of sustainability indicated by the World Commission on Environmental Development in the Brundtland Report of 1987 (Trombino & Trono 2018).

Sustainability also means encouraging a culture of sobriety and solidarity. Sobriety not just as an individual choice but as a collective orientation that becomes part of the culture and customs of the group, to be learned and taught to younger generations as an essential characteristic of places of worship. Journeys of faith and visits to sanctuaries originally arose from explicitly religious or spiritual needs, but in addition they have always been occasions for socialising and learning about others. However, sobriety and composure when visiting religious places and attending religious ceremonies are often accompanied by dangerous forms of fanaticism or subtle infiltration by outsiders with suspicious appeals to behaviours that have been repeatedly condemned by religious organisations, mayors and local prefects.

The entrepreneurial component also fuels controversy over the sustainability of religious journeys, as the attitude of those who organise religious journeys and provide services in holy places frequently tends to obscure the sacred motive and subordinate it to their economic interests (Shackley 2001, p. xvii). The problem is that the 'discovery' of religious tourism, with the flow of tourists and public funding that is often used to promote and support it, fuels expectations not just in the religious world but also in the secular sphere, composed of travel agencies and cultural associations, but also numerous other operators who are skilled at managing the religious aspect from a pragmatic and utilitarian perspective, exploiting holy places and images in a bad-taste blend of religion and marketing and seeing sacred places in purely business terms with very little that might be considered spiritual (Trono 2016).

Sustainability of the places and products of tourism of the soul: case studies

Cultural tourism associated with religious motives is known to generate economic dynamism and development in locations and areas that were often ignored by traditional tourism, and this alone justifies a large number of initiatives and events linked to the religious aspects (Campolo 2016; Varnavas, Rodosthenous & Vogazianos 2018). Traditions that celebrate patron saints, perceived as a new sign of local identity, have been revived. Tourism often becomes excursionism, appealing to the faithful and the curious alike, associating religious celebrations (rituals and processions) with interest in folklore and festivals (street lights, concerts, music and dance), in which the religious motif, in an evident evolution towards secularism, takes on a recreational and leisure dimension, with an emphasis on fun and conviviality.

In some countries, the situation is more complex; the rigorous observance of the rites of established religions has been declining at varying rates for many years and has been impacted especially by the development of alternative religions. The latter, with their focus on mystic experiences of spiritual rebirth, blend religious beliefs, expectations of regeneration, elements of eastern philosophy and magical or esoteric practices in a new-found equilibrium between human beings and nature.

Completely removed from such New Age religious experiences and in any case aimed at finding a symbiosis between nature and culture, a way of fulfilling the potential of the region to provide an authentic and immersive experience, are, as we have said, the European Cultural Routes, which follow the ancient pilgrimage routes and are travelled today by a multitude of persons regardless of their religious beliefs (Timothy 2018; Timothy & Boyd 2015). These travellers seek a new type of holiday, immersed in natural and evocative landscapes, often undertaken in solitude. This is the identifying motif of the Via Francigena, one of the most ancient roads in Europe (Ron & Timothy 2019). Linking the northern continent to Rome, it was travelled in the Middle Ages by hundreds of pilgrims to reach the 'Holy City'. The Francigena is currently the object of great interest on the part of private operators and the public, not particularly religious and highly secular, above all seeking a tourism that is slow, green and spiritual. It was not originally and is not today a single road. It is an itinerary composed of many alternative paths and routes, which pilgrims can follow in accordance with their spiritual direction and the time of year, enabling a form of tourism that is practised by well-informed travellers with an appreciation of the rural landscape and cultural heritage and is therefore sustainable. This kind of tourism now characterises some stretches of the Via Francigena in Tuscany, Italy, which has set up a regional system managed and developed in such a way as to obtain ecological and historico-cultural benefits. It has created a greenway with the important function of connecting cities with the countryside, combining religious heritage and cultural products, creating synergism between existing regional resources (Bambi & Barbari 2015; Bambi & Iacobelli 2017; Timothy & Boyd 2015).

It is a success story that policy-makers would like to see replicated in the Italian Mezzogiorno, and in particular in the southern Salento, the heel of Italy, and the terminal part of the Puglia region, through which runs the southern Via Francigena. Recently recognised by the regional authorities, the international Via Francigena system and the Council of Europe, it replicates the Via Traiana (a variant of the Via Appia), which was the main pilgrimage route between Rome and Jerusalem according to ancient geographers, chronicles, travel diaries and toponomastic data (Stopani 1992, 2008). The publication of numerous travel diaries, particularly those written from the twelfth century onwards, containing more explicit data on the state of the places and descriptions of towns and architecture, has made it possible to reconstruct with greater reliability one of the ancient pilgrimage routes, based on the Via Appia-Traiana, and to propose it as the continuation of the Via Francigena.

The Francigena Salentina, which in the last few years has drawn the attention of scholars (Trono 2012), politicians and local stakeholders, runs from Egnatia (in farmland near Fasano, in the province of Brindisi) to the southern cape of Puglia, retreading the Via Traiana, along the eastern side of southern Puglia as far as Santa Maria di Leuca, the site of the sanctuary of Santa Maria de Finibus Terrae, the objective of devotional routes from the Late Middle Ages (Cavaliere 2014).

The route enables the rediscovery of the rural landscape of the Salento: an inland area, without crowds, which still today fascinates visitors with the quality of its natural, cultural and landscape resources, fortunately spared the uncontrolled urbanisation of the coastal strip (Figure 32.2).

Along the route lie ancient *masseria* farmhouses, centres of life and production, today transformed into agritourism establishments, hotels or modern farms, resuming the economic role of times past. The landscape is enriched with woods and protected areas, but also dry-stone walls and other structures, creating unexpected effects that liven up the sun-baked

Figure 32.2 Via Francigena Salentina
Photo by Fabio Mitrotti.

and solitary landscapes with their unusual shapes. The small and very small towns have reinvented local traditions by opening new restaurants and eateries, but have also reactivated their ancient hospitality, creating B&Bs or small guest houses in historic homes. The Salento stretch of the Francigena thus has the right conditions for developing slow and spiritual tourism and is able to promote the traditional economy, linked to the products of its land and culture, while respecting its landscape and local traditions. Not all of the Via Francigena Salentina qualifies as sustainable, however. The morphological and pedological conditions of this route, with stretches of unmetalled road, rocky or loose terrain and steep slopes, pose accessibility problems for people who are not self-sufficient or suffer from motor, sensory or cognitive disabilities. These are the route's gravest defects, pointed out by the walkers themselves, who have also called for the creation of a system of accommodation providers, services for walkers and the presence of qualified personnel able to tackle a variety of needs. These are important suggestions, to which the *Associazione Comunità ospitante degli itinerari delle Vie Francigene della Puglia meridionale* is seeking to respond. Active since 2016, it works with a network of 12 local associations and 21 municipalities, all involved in a sort of 'regional laboratory', which promotes the Francigena route and raises awareness among the local population of the importance of the trails in cultural, natural and spiritual terms and of the need to attract sustainable tourism to enhance regional development.

Among the forms of religious tourism that unfold in marginal regions or urban areas affected by environmental and social blight, worthy of attention are the rites of the Holy Week.

The Easter celebrations constitute one of the most important events in the Christian liturgical calendar and are held in cities small and large, their organisation and management being handled by a combination of secular and religious interests. Over time, they have become the consolidated expression of authentic faith but also a means of social aggregation, of high anthropological value, not just because of the forms of exaltation and expressions of penitence among participants, but also their associative function. The presence of persons playing specific roles in the rituals and the participation of the public strengthens the bonds within the group by revitalising in each of its members the strong sense of belonging to the faith that binds them together. The religious ceremonies also play a socioeconomic role, understood as an opportunity to improve the attractiveness of the town, boost local development and quality of life (Brunet, Bauer, De Lacy & Tshering 2001). Attending religious celebrations means sharing in the emotions of an enthusiastic crowd enjoying a collective religious experience or more simply participating in a centuries-old rite that reflects the traditions of a people. The rites of Holy Week are an important form of cultural heritage that attracts tourists by offering them a route of faith and culture, full of emotions and other regional qualities.

In Italy, Spain, Portugal and Greece, Holy Week is characterised by intense expressions of devotion linked to the celebration of religious rites, customs and popular traditions, which attract a considerable number of worshippers, tourists and simply curious. The rites of Easter represent a significant part of Italy's traditional heritage, from Piedmont to Sicily, with a powerful social meaning that expresses the ties with the community.

In Puglia, as in many parts of southern Italy, the rites of Holy Week are celebrated in a profoundly moving way for the participants. They constitute an act of collective mourning, characterised by the intense but subdued involvement of the crowd. In the north of the region, on the mountains of the Gargano and in the cultural-religious fabric of the towns of Troia, Ascoli Satriano, Candela and Biccari on the Daunian hills, the ancient tradition of processions and evocative representations that transmit the mystery and the pathos of the Easter rites are a factor of tourist attraction and vivacity in towns at risk of isolation and decline. The local population, with its strong sense of connection to place, offer tourists a chance to share in the traditions and emotions associated with participation in the rites of Holy Week, together with a rich artistic, cultural and natural heritage that includes food and wine.

The rites of Holy Week are also very popular in Taranto, a city that suffers from low quality of life, health and jobs, but which continues to maintain interest in its local cultural traditions. This includes religious events such as the rites of Holy Week, in which it participates with intensity and passion, attracting increasing numbers of people. The success and the distinctive character of these rites lie in the slow pace of the hooded penitents, the customs of the confraternities and above all in the musicality of their ritual songs (sung in the local dialect), used in specific moments to mark the symbolic phases of the processions. The rites unfold in the old town of Taranto and lend themselves to intense emotions, especially during the nocturnal representation of the Passion, in itself fascinating and mysterious, the religious event thereby combining with the city's highly specific social circumstances. During the processions, faith becomes history and history becomes spectacle, arousing the curiosity and interest of an increasing number of worshippers and tourists. Holy Week in the old town of Taranto is without doubt the richest religious event in terms of celebrations, in which the liturgy intertwines with manifestations of popular piety (Via Crucis, processions, sacred representations), which unfold in a dramatic and theatrical fashion in accordance with a language suited to popular sensitivity that involves the emotional sphere (Figure 32.3).

Figure 32.3 Holy Week in Taranto
Photo by Stefania Stallo.

For the citizens of Taranto, it is a spectacle of faith and tradition that combines the fascination of a compelling and mysterious ritual, annually renewed, with a repertoire of cultural initiatives of interest to tourists (e.g. concerts and photographic shows). The rites of Holy Week in Taranto represent a decidedly positive phenomenon in emotional, social and cultural terms, but they are unsustainable in an old town of such architectural, historic and artistic value, which for years has suffered from a severe degree of urban and environmental neglect. The city lacks the services necessary to maintain this interdependence with the past and is also ill-prepared for the future and to attract sustainable quality tourism.

Quality tourism of this kind is also the objective of the Municipality of Bari with regard to the flow of Orthodox pilgrims who for nearly a thousand years have been coming from Russia to honour St Nicholas the Wonderworker of Myra, whose tomb is housed in the large Romanesque basilica. There were about 33,000 Russian overnight stays in 2019, and their number is growing exponentially, the commemorations of the Saint's birthday (6th December) and the transfer of his remains to Bari (7–9 May) being particularly busy times (Figure 32.4).

They mainly arrive by coach in organised groups accompanied by a guide, causing much congestion in the old town and producing little benefit for the local economy (Rizzello & Trono 2013; Trono & Rizzello 2014). The city, which is the regional capital, a metropolitan city and an important Mediterranean port for both cargo and passenger traffic, also regularly receives a large number of cruise ship passengers. The municipal administration is seeking to convert the flow of excursionists into religious tourists, whose regularity could enable economies of scale leading to city-wide repurposing. The Municipality of Bari and Puglia Regional

Figure 32.4 Orthodox pilgrims at the Church of San Nicola. Bari
Photo by Svetlana Alexeevna Kouzmenko.

Administration understand the value of these tourists and, as well as confirming the strategic importance for the economy of Puglia of commercial relations with Russia, they are pressing ahead with cultural exchanges and flights, which now arrive from Russia on a weekly basis. They believe it is a good idea to invest in this potentially substantial slice of the tourism market, in accordance with the principles of sustainability. This entails making Bari a more welcoming city, able to offer specific high-quality services for guests with religious motives.

Conclusions

One of the uses of spiritual and religious cultural heritage is tourism, which deserves particular attention because it recognises the values of the place being visited in terms of both devotion and its legacy, but also because it sees the sociocultural features of the destination as assets and an opportunity to promote local products. This can produce synergistic effects deriving from the close complementarity of religious, cultural and tourist services and the associated economic activities.

However, the excessive use of places and routes of faith and the massive participation in religious events produces anthropic pressure and poses problems for the environment, transport, accessibility and hospitality—in other words, sustainability. There is a need to plan and manage this type of tourism carefully, respecting the value of the religious and cultural site and the faith and culture of the visitor, as well as the meaning of the cultural and religious heritage for those who administer the site and the host community.

Having said this, perhaps it would be useful to consider rethinking how to communicate the site's identity and image without modifying the value of its content or purposes. For example, concerning the use of the term '*sagra*', which refers to both the religious and the profane festivities, the recreational and folklore aspects should not be allowed to prejudice, obscure or erase the more profound content of the religious celebration, characterised by acts and moments of profound devotion.

Concerning the cultural routes, it would be useful to place greater attention on accessibility. Many disabled persons feel a particularly strong inner need to undertake a journey, driven by a profound desire to experience travel and share their experiences on their return. For this to happen, it is necessary to provide the means to follow the route in the most self-sufficient way. The disabled do not expect the entire route, including hotels and restaurants, to be perfectly suitable and accessible, but it is legitimate to ask that efforts be made to eliminate some of the barriers and provide them with information and well-designed websites.

Route-based tourism needs to be controlled, disciplined and supported by making sure it has the right infrastructure (properly equipped green spaces, cycle paths and roads reserved for walkers) and services (including food and accommodation) that can be used in every season of the year (Timothy & Boyd 2015). Particular care needs to be taken when addressing the diffidence and uncooperative attitude of the local communities, sometimes unwilling or unable to interact with visitors who have needs that differ from those of the traditional tourist. Nor should the frequent and sometimes unhelpful influence of politics be underestimated, leaders often being neglectful of the need to involve local communities, indifferent to the social and cultural functions of the routes and the motives of those who follow them.

The compatibility and sustainability of the routes, as well as the consistency of the measures with respect to the context in which they are applied, depends on the horizontal integration of the regional system (environment, landscape and socioproductive systems) and the collaboration of all the actors involved (institutional and otherwise).

The problem is that the commitment of all parties and the way in which the item of religious or spiritual interest is managed depend on the awareness of its cultural and religious meaning. Unfortunately however, routes and sites of religious interest are targeted by a diverse range of interests, not always in accordance with the motives of those who set off on the journey to visit the holy place, and not always respectful of the place and the culture of the local community they visit.

To be successful, religious tourism requires first of all the capacity to manage demand and supply with regard to sacred spaces and secondly, planning.

The management of religious tourism is a relatively new paradigm and, as with any new model, it is still evolving and seems to be developing in stages that start with the study and recognition of the value of the religious attraction, subsequently involving the stakeholders and then proceeding to draw up a programme for its safeguarding and promotion.

Essential, therefore, are the cooperation and involvement of all institutions, both religious and secular. It is necessary to construct a network of alliances at several levels of government, with public and private subjects, from the secular and religious worlds. Such networks enable good governance of religious tourism while respecting the values of sustainability. Planning entails drawing up good projects in which studies of the region and the dynamics of tourism, together with the use of innovative technologies, are combined with knowledge of the market, the commitment and capacity of the enterprises involved, and the entire system of promoting religious/spiritual tourism. Promotion should develop in such a way as to embrace not just the cultural and religious element, but also to seek rational and shared approaches that lead to improved behaviours and a better quality of life, producing results that are sustainable for the communities and the region as a whole. The process of management requires control and monitoring of sacrosanct places, but professionalism and respect for ethical and cultural values are also needed. What makes sustainable religious/spiritual tourism desirable is the value of the sacred place or object to the tourist and the local community.

Note

1 Concerning the concepts of modernity, tradition and religious beliefs (and the relationships between them), see Gallino (2016).

References

Afferni, R., & Mangano, S. (2009) 'The sacred mounts of Piedmonte and Lombardia as alternative and sustainable experiences for religious tourism', in A. Trono (ed.), *Proceedings of the International Conference Tourism, Religion & Culture: Regional Development through Meaningful Tourism Experiences, Lecce, Poggiardo 27th-29th October 2009* (pp. 483–499). Galatina: Mario Congedo.

Ambrosio, V. (2007) 'Sacred pilgrimage and tourism as secular pilgrimage', in R. Raj & N.D. Morpeth (eds.), *Religious Tourism and Pilgrimage Festivals Management: An International Perspective* (pp. 78–88). Wallingford: CABI.

Azzimonti, C. (2016) 'I beni culturali ecclesiali in Italia', *Quaderni di diritto ecclesiale*, 29: 347–378.

Bambi, G., & Barbari, M. (Eds.) (2015) 'The European pilgrimage routes for promoting sustainable and quality tourism in rural areas'. *International Conference Proceedings 4–6 December 2014*. Florence: Firenze University Press.

Bambi G., & Iacobelli, S. (2017) 'Study and monitoring of itinerant tourism along the Francigena Route, by camera trapping system', *Almatourism*, 8(6): 144–164.

Barber, R. (ed.) (1993) *Pilgrimages*. London: Boydell Press.

Battisti, F.M. (1996) 'Comportamenti di massa e turismo religioso nel Giubileo', in E. Nocifora (ed.), *Il Viaggio. Da "Grand Tour" al turismo post-industriale* (pp. 277–293). Naples: Edizioni Magma.

Beltrami, A. (2017) 'Dibattito: Chiese col ticket, questione aperta', Online at: https://www.avvenire.it/agora/pagine/ticket. Accessed February 10, 2020.

Bimonte, S., & Pagni, R. (2003) *Protezione, fruizione e sviluppo locale: aree protette e turismo in Toscana, IRPET*. Florence: Regione Toscana.

Blackwell, R. (2007) 'Motivations for religious tourism, pilgrimage, festivals and events', in R. Raj & N.D. Morpeth (eds.), *Religious Tourism and Pilgrimage Festivals Management: An International Perspective* (pp. 35–47). Wallingford: CABI.

Brunet, S., Bauer, J., De Lacy, T., & Tshering, K. (2001) 'Tourism development in Bhutan: Tensions between tradition and modernity', *Journal of Sustainable Tourism*, 9(3): 243–256.

Butowski, L. (2019) 'Tourist sustainability of destination as a measure of its development', *Current Issues in Tourism*, 22(9): 1043–1061.

Campolo, D. (2016) 'The sustainable development of inland areas through cultural landscape and cultural routes', *Environment, Energy, Landscape*, 12: 80–84.

Carbone, F, Corinto, G., & Malek, A. (2016) 'New trends of pilgrimage: Religion and tourism, authenticity and innovation, development and intercultural dialogue: Notes from the diary of a pilgrim of Santiago', *AIMS Geosciences*, 2(2): 152–165.

Cavaliere, F. (2014) 'Arte e devozione nei santuari mariani del Salento meridionale lungo il camino per Santa Maria de Finibus Terrae', in A. Trono, Leo Imperiale, & G. Marella (eds.), *Walking Towards Jerusalem: Cultures, Economies and Territories* (pp. 231–249). Galatina: Mario Congedo.

Chirilă, D., Chirilă, M., & Sîrbulescu, C. (2018) 'Study regarding manifestation forms of sustainable tourism', *Scientific Papers: Animal Science and Biotechnologies*, 51(1): 191–198.

Cohen, E. (1974) 'Who is a tourist? A conceptual clarification', *Sociological Review*, 22(4): 527–555.

Cohen, E. (1992) 'Pilgrimage and tourism: Convergence and divergence', in A. Morinis (ed.), *Sacred Journeys: The Anthropology of Pilgrimage* (pp. 47–61). Westport, CT: Greenwood Press.

Cohen, E. (1998) 'Tourism and religion: A comparative perspective', *Pacific Tourism Review*, 2: 1–10.

Cohen, E. (2008) 'The changing faces of contemporary tourism', *Society*, 45(4): 330–333.

Collins-Kreiner, N. (2009) 'Researching pilgrimage: Continuity and transformations', *Annals of Tourism Research*, 37: 440–456.

Collins-Kreiner, N. (2010) 'The geography of pilgrimage and tourism: Transformations and implications for applied geography', *Applied Geography*, 30(1): 153–164.

Connell, J., Page S.J., & Bentley, T. (2009) 'Towards sustainable tourism planning in New Zealand: Monitoring local government planning under the Resource Management Act', *Tourism Management*, 30: 867–877.

Corna Pellegrini, G. (2000) *Turisti viaggiatori: per una geografia del Turismo Sostenibile*. Milan: Tramontana.

Dallari, F., Trono, A., & Zabbini E. (eds.) (2009) *I viaggi dell'Anima: Cultura e territorio Potenzialità e problemi dello Sviluppo del Turismo religioso*. Bologna: Patron.

Damari, C., & Mansfeld, Y. (2014) 'Reflections on pilgrims' identity, role and interplay with the pilgrimage environment', *Current Issues in Tourism*, 19: 199–222.

Davolio, M., & Meriani, C. (2011) *Turismo Responsabile. Che cos'é. Come si fa*. Milan: Touring.

Di Gregorio, D., Chiodo, A.P., & Nicolosi, A. (2018) 'Pellegrinaggio religioso in Aspromonte', *Labor-Est*, 17: 15–20.

Farrell, A., & Hart, M. (1998) 'What does sustainability really mean? The search for useful indicators', *Environment*, 40(9): 4–31.

Feliciani, G. (2010) 'La questione del ticket d'accesso alle chiese', *Aedon*, 3. Online at: www.aedon.mulino.it/archivio/2010/3/feliciani.htm. Accessed February 11, 2020.

Gallino, L. (2016) 'La modernizzazione mancata', *Quaderni di Sociologia*, 70–71: 53–69.

Gössling, S., Hall, C.M., & Scott, D. (2015) *Tourism and Water*. Bristol: Channel View Publications.

Graburn, N. (1989) ''Tourism: The sacred journey', in V. Smith (ed.), *Hosts and Guests: The Anthropology of Tourism* (pp. 17–32). Philadelphia: University of Pennsylvania Press.

Gray, M., & Winton, J. (2009) 'The effect of religious tourism on host communities', in A. Trono (ed.), *Proceedings of the International Conference on Tourism, Religion & Culture: Regional Development through Meaningful Tourism Experiences, Lecce, Poggiardo 27t -29th October 2009* (pp. 551–561). Galatina: Mario Congedo.

Gribaudi, D. (1966) *Piemonte e Val D'Aosta*. Turin: UTET.

Griffin, K.A. (2007) 'The globalization of pilgrimage tourism? Some thoughts from Ireland', in R. Raj & N.D. Morpeth (eds.), *Religious Tourism and Pilgrimage Festivals Management: An International Perspective* (pp. 15–34). Wallingford: CABI.

Herrero, L.C., Sanz, J.A., & Devesa, M. (2009) 'Who pays more for a cultural religious festival? A case study in Santiago De Compostela', in A. Trono (ed.), *Proceedings of the International Conference Tourism, Religion & Culture: Regional Development through Meaningful Tourism Experiences, Lecce, Poggiardo 27th-29th October 2009* (pp. 443–464). Galatina: Mario Congedo.

Hitrec, T. (1990) 'Religious tourism: Development - characteristics – perspectives', *Acta Turistica*, 2(1): 9–49.

Kliot, N. (2019) 'MENA as a critical meeting point between tourism and water resources', in D.J. Timothy (ed.), *Routledge Handbook on Tourism in the Middle East and North Africa* (pp. 163–177). London: Routledge.

McMillan, D.W., & Chavis, D.M. (1986) 'Sense of community: A definition and theory', *Journal of Community Psychology*, 14: 6–23.

Murray, M., & Graham, B. (1997) 'Exploring the dialectics of route-based tourism: The Camino de Santiago', *Tourism Management*, 18: 513–524.

Olsen, D.H. (2010) 'Pilgrims, tourists and Max Weber's "ideal types"', *Annals of Tourism Research*, 37: 848–851.

Pérez Guilarte, Y., & Lois González, R.C. (2018) 'Sustainability and visitor management in tourist historic cities: The case of Santiago de Compostela, Spain', *Journal of Heritage Tourism*, 13(6): 489–505.

Raj, R., & Morpeth, N.D. (eds.) (2007) *Religious Tourism and Pilgrimage Festivals Management: An International Perspective*. Wallingford: CABI.

Rizzello, K., & Trono, A. (2013) 'The pilgrimage to San Nicola Shrine in Bari and its impact', *International Journal of Religious Tourism and Pilgrimage*, 1(2): 24–40.

Ron, A.S., & Timothy, D.J. (2019) *Contemporary Christian Travel: Pilgrimage, Practice and Place*. Bristol: Channel View Publications.

Sachs, J.D. (2015) *The Age of Sustainable Development*. New York: Columbia University Press.

Scidurlo, P., & Callegari, L. (2015) *Guida al Cammino di Santiago per tutti*. Milan: Terre di Mezzo.

Séraphin, H., Gladkikh, T., & Thanh, T.V. (eds.) (2020) *Overtourism: Causes, Implications and Solutions*. Cham: Springer Nature.

Shackley, M. (2001) *Managing Sacred Sites: Service Provision and Visitor Experience*. London: Thompson.

Smith, V.L. (1992) 'Introduction: The quest in guest', *Annals of Tourism Research*, 19: 1–17.

Stopani, R. (1992) *La Via Francigena del Sud: l'Appia Traiana nel Medioevo*. Florence: Le Lettere.

Smith, V.L., & Eadington, W. (1992) *Tourism Alternatives: Potentials and Problems in the Development of Tourism*. Philadelphia: University of Pennsylvania Press.

Stopani, R. (2008) 'La Via Appia Traiana nel medioevo', in Banco di Napoli, Finmeccanica (ed.), *Roma-Gerusalemme: Lungo le Vie Francigene del Sud* (pp. 1–15). Rome: Civita, Tipografia Ostiense.

Stoddard, R H. (1997) 'Defining and classifying pilgrimages', in R.H. Stoddard & A. Morinis (eds.), *Sacred Places, Sacred Spaces: The Geography of Pilgrimages* (pp. 41–60). Baton Rouge: Louisiana State University.

Timothy, D.J. (2018) 'Cultural routes: Tourist destinations and tools for development', in D.H. Olsen, & A. Trono (eds.), *Religious Pilgrimage Routes and Trails: Sustainable Development and Management* (pp. 27–37). Wallingford: CABI.

Timothy, D.J., & Boyd, S.W. (2015) *Tourism and Trails: Cultural, Ecological and Management Issues*. Bristol: Channel View Publications.

Timothy, D.J., & Olsen, D.H. (eds.) (2006) *Tourism, Religion and Spiritual Journeys*. London: Routledge.

Tinacci Mossello, M. (2014) 'Prospettive di sviluppo del turismo sostenibile', in A. Trono, M.L. Imperiale, & G. Marella (eds.), *In Viaggio Verso Gerusalemme: Culture, Economie e Territori* (pp. 273–281). Galatina: Mario Congedo.

Toscano, M.A. (1996) 'Per una sociologia del viaggio note metodologiche', in E. Nocifora (ed.), *Il Viaggio: Da "Grand Tour" al turismo post-industriale* (pp. 9–21). Naples: Edizioni Magma.

Trombino, G., & Trono, A. (2018) 'Environment and sustainability as related to religious pilgrimage routes and trails', in D.H. Olsen & A. Trono (eds.), *Religious Pilgrimage Routes and Trails* (pp. 49–60). Wallingford: CABI.

Trono, A. (ed.) (2012) *Via Francigena: Cammini di Fede e Turismo Culturale*. Galtina: Mario Congedo.

Trono, A. (2014) 'Cultural and religious routes: A new opportunity for regional development', in R.C. Lois-González, X.M. Santos-Solla, & Taboada-De-Zúñiga, P. (eds.), *New Tourism in the 21st Century: Culture, the City, Nature and Spirituality* (pp. 5–25). Cambridge: Cambridge Scholars Publishing.

Trono, A. (2016) 'Logistics at holy sites', in M. Leppäkari, K. Griffin & R. Raj (eds.), *Pilgrimage and Tourism to Holy Cities: Ideological Perspectives and Practical Management* (pp. 117–132). Wallingford: CABI.

Trono, A., & Olsen, D.H. (2018) 'Pilgrimage trails and routes: Journeys from the present to the future', in D.H. Olsen & A. Trono (eds.), *Religious Pilgrimage Routes and Trails* (pp. 247–254). Wallingford: CABI.

Trono, A., & Rizzello, K. (2014) 'Ethical management and socio-cultural impacts of the Rites of Holy Week: A case of study in the Puglia region (Italy)', in G. Richards, L. Marques, & K. Mein (eds.), *Event Design: Social Perspectives and Practices* (pp. 92–108). London: Routledge.

United Nations (2019) *The Sustainable Development Goals Report*. Online at: https://unstats.un.org/sdgs/report/2019/The-Sustainable-Development-Goals-Report-2019.pdf. Accessed February 12, 2020.

Varnavas, A., Rodosthenous, N., & Vogazianos, P. (2018) 'Religious tourism as a tool for sustainability: The case of Cyprus', *Journal of Social Sciences Research*, 4: 285–293.

Visentin, F., & Bertocchi, D. (2019) 'Venice: An analysis of tourism excesses in an overtourism icon', in C. Milano, J.M. Cheer, & M. Novelli (eds.), *Overtourism: Excesses, Discontents and Measures in Travel and Tourism* (pp. 18–38). Wallingford: CABI.

Visentini, C. (1996) 'Il viaggio perduto? Dal "viaggio dei moderni" alla "fine dei viaggi"', in E. Nocifora (ed.), *Il Viaggio: Da "Grand Tour" al turismo post-industriale* (pp. 215–227). Naples: Edizioni Magma.

Wall, G. (2019) 'Perspectives on the environment and overtourism', in R. Dodds & R.W. Butler (eds.), *Overtourism: Issues, Realities and Solutions* (pp. 27–45). Berlin: De Gruyter.

World Tourism Organization (2013) *Sustainable Tourism for Development Guidebook: Enhancing Capacities for Sustainable Tourism for Development in Developing Countries*. Madrid: UNWTO.

SECTION V

Emerging and future directions

33

RELIGION, SPIRITUALITY, AND TOURISM

Emerging and future directions

Dallen J. Timothy and Daniel H. Olsen

Introduction

This handbook is about faith, religion, spirituality, and the sacred and the secular and how these intersect with the moderating variable of tourism. Through the current academic literature, much is known regarding people's motivations for undertaking pilgrimages and seeking spiritual experiences through travel. These include religious obligations, penance and forgiveness, promotion of faith and testimony, increasing proximity to deity, socializing with family and other likeminded devotees, and seeking an existential understanding of self. Indeed, much research has been done on the marketing and management of sacrosanct places and the economic, social, and environmental costs and benefits of mega and small-scale religious events and religious festivals, as well as religious and spiritual tourism more generally. Much of this knowledge has been eruditely reviewed in this book, but we have also tried to go beyond normative familiarity to identify and address lesser-known elements of religious tourism and spiritual travel.

This handbook has discussed a wide range of perspectives on many traditional elements of religious tourism and spiritual travel, with the chapters herein presenting many unique viewpoints that are not yet well documented in the literature. These include, for example, popular culture, political economy, religious theme parks, the UNESCO branding of religious heritage, gender, and authenticity. Although these subjects are well established in the social sciences and tourism studies more generally, until now they have not been systematically examined in the context of religious and spiritual tourism. At the same time, many avenues of research regarding the intersections of religion, spirituality, and tourism have not been covered in this handbook. As such, this concluding chapter suggests areas that are in particular need of additional research attention in the near future. In particular, the chapter focuses on six areas of potentially fruitful research: unusual spaces and places of religious tourism, dark tourism and pilgrimage, slow and religious tourism, religious hospitality, emerging travel markets and motives, and the effects of globalization on religious tourism.

Spaces of religious tourism

The contributors in this book have discussed pilgrimage, religious tourism, and spiritual travel in many different contexts and venues. The most common venues for these forms of tourism are, of course, temples, mosques, churches, shrines, archaeological sites, grottos and other natural areas, visitor centers, and various themed environments in both urban and rural settings. Yet, there are multitudes of other types of spaces, places, and venues that have not been discussed widely in the scholarly literature, even though they may have different impacts on the visitor experience and facilitate religious tourism differently. This dearth of academic attention derives from the relative newness of such alternative spaces in modern-day religious tourism, as well as the dominance of scholarship that focuses on prescribed holy places (e.g., the Masjid al-Haram in Mecca, the Western Wall in Jerusalem, and the Ghats and Ganges River in Varanasi) or proscribed places to avoid in mainstream pilgrimage traditions (Gürey 2017). Although alternative pilgrimage and religious tourism spaces could fill volumes, given our space limitations, this section is limited to the role of museums, trails, and cruise ships and resorts.

Museums

Museums are a prominent asset for heritage tourism, although people do not always think of them in the context of religious tourism (Timothy 2021b). There are essentially three types of religious museums. The first is comprised of religious archaeological sites and buildings, such as churches, temples, or mosques, which are classified as museums for the purpose of conserving artifacts and educating the public (Farra-Haddad 2020; Koren-Lawrence & Collins-Kreiner 2019). While they were not built with tourists in mind, many of these sites are now managed and marketed to cater to this tourism niche market. One of the best and most controversial examples of this is the Hagia Sophia in Istanbul, the fourth-century cathedral and later fifteenth-century mosque, which for many years has been Turkey's most visited tourist attraction. In 1931, the mosque was decommissioned and closed, but it opened in 1935 as a museum and functioned as such until July 2020 when, by presidential decree, it reverted to a functioning mosque and lost its museum status (Olsen & Emmett 2021). In Europe, it is common for cathedrals or other functioning houses of worship to contain sections cordoned off and managed specifically as museums (Shackley 2002). Likewise, the Mediterranean Basin and the Levant are full of religious archaeological sites that are major visitor attractions and functioning museums.

The second type of religious museum is comprised of buildings or staging areas that were built intentionally as archival stores to protect historical relics associated with a certain faith or several faiths within a region. These represent normative purpose-built museums where people go to learn about religious history by participating in interpretive programs and viewing artifacts and hearing narrations of history. The Coptic Museum in Cairo, Egypt, the Digambar Jain Sangahalay in Ujjain, India, the Confucius Museum in Qufu, China, and the International Buddhist Museum in Kandy, Sri Lanka, are highly-regarded examples of this type of attraction.

Third, religious museums may also be classified as 'themed environments' or 'theme parks' (Paine 2019; Ron & Timothy 2019; Shinde 2021; see Chapter 13) where religious subjects are on display or acted out by interpreter-actors. These religious theme parks serve three purposes: to spread a message of faith, to interpret religious heritage, and to entertain.

Famous examples of this category include the Ark Encounter in Kentucky, USA, the Holy Land Experience in Florida, USA, and Nazareth Village outdoor living museum in Israel (Bielo 2018; Chmielewska-Szlajfer 2017; Ron & Feldman 2009; Shoval 2000).

Trails and routes

Much has been written about pilgrims' experiences on the way to the center of faith and how the route or journey was the most arduous and impactful part of the experience (e.g., Buzinde et al. 2014; Damari & Mansfeld 2016; Maddrell 2011; Wu, Chang & Wu 2019). However, relatively little is known about today's trail-based experiences, except that researchers have distinguished between the recreation and religious users of various pilgrim trails (Murray & Graham 1997; Øian 2019; Olsen & Trono 2018).

Timothy and Boyd (2015) distinguish between purposive and organic tourism trails, organic trails being those that evolved naturally or which were not built specifically to be trails, but now function as such. This includes most pilgrimage routes, trade routes, railway lines, canal towpaths, and similar trails. Purposive trails are those that are put together for use in recreation or tourism. They are designed to link attractions and places that share a common historical theme. Historical pilgrim pathways fit the definition of organic trails, which many pilgrims and non-pilgrims use to this day. Many traditional pilgrimage routes, such as the Camino de Santiago, St Olaf's Way and Via Francigena, went into disuse during the Reformation but have now been resurrected as outdoor linear corridors for both pilgrims and other tourists. These are now marketed as some of Europe's most scenic hiking trails, and the majority of their users are non-religious, although many claim to have had 'spiritual' experiences while utilizing the trail (Lois-González, Santos & Romero 2018; Ron & Timothy 2019). Israel has recently developed a couple of long-distance Christian trails: the Jesus Trail and the Gospel Trail, both of which highlight localities that are mentioned in the New Testament, being part of the ministry of Christ (Collins-Kreiner & Kliot 2016). There are also several major purposive routes in Europe with religious themes, which the Council of Europe has been instrumental in developing as a tool for tourism development and cultural conservation. Some of these include the European Route of Jewish Heritage, the European Route of Cistercian Abbeys, and the Routes of the Reformation (Timothy & Olsen 2018).

Cruise ships and resorts

A uniquely modern manifestation of religious tourism is faith-based cruises and resort stays. Cruise ships and resorts are not typical or traditional venues for pilgrimage or spiritual experiences, and yet they are increasingly important for their roles in religious tourism. Jewish cruises are gaining popularity in the Mediterranean and other parts of Europe. Ports of call include cities that were once, or continue to be, important centers of Jewish faith and culture. These Jewish 'heritage cruises' take place on the Rhine and Danube Rivers through Germany, Austria, Hungary, and the Czech Republic, as well as in the Mediterranean with key stopovers in Egypt and Israel (Silverstein 2020). The second type of Jewish cruise is aimed at a more religious market—kosher cruises. While these cruises also visit important historic locales, their focus is on ensuring that those who follow Jewish dietary restrictions can also enjoy a broader range of vacation experiences knowing that all food onboard is kosher. Passover cruises are becoming increasingly popular and include organizers taking care of all the cooking and cleaning requirements affiliated with the Passover holiday (Silverstein 2020). Several other companies have recently begun offering Muslim cruises, or 'halal cruises', which are especially popular in Turkey and Malaysia. They focus on themes of interest to Muslim travelers and offer halal dining, lodging, and recreational experiences (Battour & Ismail 2016).

Cruise products are especially popular among various Christian denominations and serve two primary purposes (Ron & Timothy 2019). First, certain Mediterranean cruises are being sold as 'Christian cruises' or 'in the Footsteps of the Apostles' cruises and aim to trace the ancient sea journeys of Jesus' apostles. These sailings call at ports in Italy, Greece, Turkey, Cyprus, Israel, and Egypt, all of which were a part of the scriptural geography of the Old and New Testament and important locations in spreading the gospel following the death and resurrection of Jesus. The land portions of these cruises focus on sites of the Apostle Paul's missionary efforts and the conversion of ancient Mediterranean peoples to Christianity, while onboard entertainment includes many guest speakers or 'headliners' who are experts on biblical archaeology and history as well as well-known religious leaders.

The second type of Christian cruise generally has little to do with the destination and more to do with the onboard fellowship of like-minded devotees. Diverse Christian faiths program their own itineraries within existing cruise products in an effort to build a sense of community and fellowship and to strengthen faith through the stories of others and invited guest speakers. That these cruises visit ports that are not typically associated with Christian history or doctrine is irrelevant. Rather, it is the social solidarity element that is most important in this context. Such cruises are extremely popular in the Caribbean during cold northern winters as well as in China, where many groups' shore packages include 'solidarity visits' to Chinese Christian communities to show support and fellowship to their brothers and sisters who lack religious freedom in an official atheist state (Ron & Timothy 2019).

Similar to this second type of cruise product are Christian retreats and resort stays. Christian ranch-stays and camps are increasingly popular family vacation destinations. Likewise, several resorts in the Caribbean and Mexico are devoting specific areas of their properties to Christian-friendly holidays, including guest entertainers and speakers. Like some cruises, these ranches and resorts are geared toward creating a sense of *communitas* among Christian holidaymakers and providing family-friendly activities in a comfortable Christian environment without the influence of heavy partying, alcohol, and drugs (Wright 2008).

Dark tourism and religion

Dark tourism has received a great deal of academic awareness commensurate with its attention by the industry (e.g., Biran et al. 2011; Foley & Lennon 2000; Hartmann 2014; Ivanova & Light 2018; Kerr & Price 2016; Stone et al. 2018; Stone & Sharpley 2008). Dark tourism is a form of heritage tourism that focuses on aspects of human suffering, death, and other events and places of darkness. Its products and resources include cemeteries and battlefields and sites of human-induced or natural disasters, incarceration, slavery, torture and abuse, tragic accidents, terrorism, and museums of morbidity, among many others. There are many different shades of darkness, with some sites being especially gruesome (e.g., mass murder or terror attacks), while others are somewhat less morose but nonetheless sullen in their own right (e.g., historic war sites or Body Worlds exhibits) (Stone 2006; Timothy 2021a).

The parallels between dark tourism and pilgrimage have been highlighted by a small handful of scholars (e.g., Collins-Kreiner 2016; Olsen & Korstanje 2020), suggesting that people's behavior towards, and motivations for, dark tourism resemble many of the same motives and behaviors in relation to pilgrimage. Deep emotions are often involved in both types of travel experiences, and both dark tourists and pilgrims are frequently motivated by personal identity seeking, self-awareness, and a sense of their own mortality (Ivanova & Light 2018). Korstanje and Olsen (2020) suggest a number of analogous relations between dark tourism and pilgrimage, including that both entail forms of resilience, exhibit strong heritage values, and provide a means of coming to terms with death and eschatology.

Likewise, the notion that many types of travel, including dark tourism, may be seen as a type of secular pilgrimage, or a journey to reach a desired 'center of otherness', has long been a mantra in the religious tourism and pilgrimage literature (e.g., Turner & Turner 1978). These centers may be places of national or filial pride, locations associated with celebrities or personal heroes, or sites of society's collective anguish. Such journeys and places can induce a strong sense of inner connectedness between one's own spiritual self and a larger force beyond the self. Visits at the National September 11 Memorial and Museum in New York City or the Flight 93 National Memorial near Shanksville, Pennsylvania, can produce a combination of emotions, including a deep sense of loss, sadness, solidarity, and somber adulation (Croom et al. 2018; Kerr & Price 2018; Sather-Wagstaff 2016).

A third perspective is that many elements of religious heritage have connotations of darkness and suffering (Cohen 2011; Esplin & Olsen 2020). Spain's recent actions to protect and promote the country's Jewish heritage has occurred as the country has begun to come to terms with the state's forced conversions of Jews to Catholicism or their expulsion from Spain in the fifteenth century (Dulska in press; Russo & Romagosa 2010). The heritages commemorated and the interpretive narratives of many Latter-day Saint historic sites also memorialize a history of mass persecution, execution, and expulsion, which nineteenth-century members of the Church of Jesus Christ of Latter-day Saints withstood because many of their beliefs did not align with mainstream Christianity (Esplin & Olsen 2020; Olsen 2006b, 2013; Ron & Timothy 2019).

In truth, much of the world's religious history is fraught with darkness: abuse of power, persecution of peoples because of their faith, forced expulsions and migrations, wars, and other tragedies in the name of religion. There is, therefore, a clear connection between religious tourism and dark tourism, but we do not yet know the extent of these connections. There is room for additional research on the parallels between dark tourism and pilgrimage or other forms of religious tourism that go beyond only motivations and emotive experiences. Resource and product overlap, and the ways in which destinations are attempting to either connect the two or ignore them entirely, are areas we know very little about but which could enhance scholarly knowledge about the crossovers between dark tourism and pilgrimage, as well as other types of tourism and pilgrimage. Certainly, there are parallels that researchers have yet to identify.

Diaspora/roots tourism

Roots tourism and dark tourism have much in common. Many ethnic diasporas have a dark history, including slavery, starvation, and war. Roots/diaspora tourism also has much in common with religious tourism. For example, 'forced diasporas' often result from contestation or conflict between cultures, races, and faiths. The Jewish diaspora is the best known and largest forced diaspora, but the Arab diaspora, which continues to grow, is largely rooted in ongoing religiously motived wars and conflicts. The Armenian diaspora was in part a result of religious discord between the Armenian Christians and their Ottoman overlords through different periods of time. Many diasporas and forced migrations have occurred based on religious differences or nationality with immigrants representing several religions (e.g., Indians with a majority being Hindus, Sikhs, or Muslims). Other mass and individual migrations have taken place throughout history for many reasons, including employment opportunities, famines and starvation, escaping autocratic regimes, family solidarity, and colonial networking. Although not necessarily related to religion, diasporas of every kind create the potential for 'return travel' or 'roots tourism' (Coles & Timothy 2004).

Faith-based diasporas have spread religious adherents throughout the world. Many descendants of old and recent diasporas continue to practice the religions of their forebears. This results in many people traveling to their ancestral homelands to participate in pilgrimages, pilgrimage-like celebrations and to visit near and distant relatives (Olsen 2019). Travel by American Muslims to participate in the Hajj and Canadian Indians traveling to India to participate in Hindu or Sikh pilgrimages or religious holidays are prime manifestations of this religion-roots tourism relationship.

Among recent diasporic groups, there is a strong sense of connection to the homeland. Forced diasporic people tend to feel most disconnected and suffer most from a cultural 'identity crisis' (Coles & Timothy 2004), but even people of volunteer diasporas tend to have a common sense of identity and respect for their ancestors. This often results in 'roots tourism', 'genealogy tourism', 'personal heritage tourism', 'return travel' or 'diaspora tourism', all of which entail people traveling to visit their ancestral lands, visit close and distant relatives, undertake genealogical or family history research, and participate in activities their ancestors might have participated in (e.g. farming, gardening, hunting, attending church) (Alexander et al. 2017; Higginbotham 2012; Huang et al. 2013; Timothy 2008).

For many roots tourists, especially people of African descent whose ancestors were part of the slave trade, visits to slavery heritage sites in their own countries or in Africa can have deep and transcendent outcomes. Many African-Americans who visit slave sites in Ghana or other West African countries often report a deep sense of loss, emotional solidarity with their ancestors, and a spiritual connection with Africa and those who suffered the abuse and indignity of slavery (Timothy & Teye 2004; Wright 2020). For many African-Americans and other nationalities of African descent, such a deeply personal diasporic journey to the lands of their forebears becomes a true pilgrimage and something that either helps solidify their faith in humanity and God or results in increased bitterness (Timothy & Teye 2004).

Although other diasporas may not evoke the depth of spirit and emotion as that found within the African diaspora experience, they do indeed have a spiritual role to play. Visits to ancestral homelands 'have long been the focus of strong emotional attachments by various socio-religious groups' (Olsen 2006a: 114). In a world that is confusing, conflicted, and ungrounded, undertaking a genealogical journey can be very therapeutic; it can help heal the troubled soul (Timothy 2008). By understanding our ancestors and their lived experiences, people begin to partly understand their own spiritual needs today and how they came to be who they are (Kurzwell 1995; Mindell 2010). These emotive responses to visiting ancestral lands 'lead some genealogy travelers to...adopt the persona of "pilgrim", for they see themselves...as pilgrims on a sacred quest for spiritual enlightenment, peace, and identity confirmation' (Timothy 2008: 122).

The connections between genealogy, family history, spirituality, and travel are clear, but little empirical research has been done to take these relationships further. Understanding the spiritual connectedness between different diasporas, their ethnic homelands, the reasons for their dispersal, and their level of connectedness in their adopted land would go far in providing additional knowledge about the 'spiritual' or existential longing that many hyphenated and displaced migrant groups feel in their traditional homelands.

Religious tourism and slow tourism

Slow tourism stems from the 'slow food' movement, which began in the 1980s with the aim of protecting local foods and food cultures, growing food sustainably, encouraging healthy eating, avoiding waste, understanding food sources, supporting local culinary enterprises, and countering the rapid popularization of fast food. It also emphasizes the quality of food and dining experiences over quantity. This movement has since grown into a vast international organization, 'Slow Food', which has branch offices in many countries to help propagate its principles of healthy living and healthy environment. 'Slow Food' has several parallels with the kosher and halal movements, including an emphasis by the latter movements on pure food, whole food, and certain gustatory prohibitions (Atalan-Helicke 2015; Boyd 2016; Heiman et al. 2019).

Slow tourism encourages deeper and more meaningful travel experiences. Rather than rushing through many attractions in a day or destinations in a week, the way most mass-produced tour packages are organized today, slow tourism urges people to decelerate, enjoy the destination, become immersed in local cultures and foods, undertake healthy activities and engage more with nature (Dickinson & Lumsdon 2010; Fullagar et al. 2012). In doing this, tourists can have healthier, more restful, and more authentic holiday experiences. Most analysts argue that slow tourism is a more sustainable way of 'doing' tourism because travelers use local transportation, support local businesses (e.g., food services, farmers, and guest houses), spend more money in non-corporate establishments, and are more culturally in tune with local inhabitants (e.g., Dickinson & Lumsdon 2010; Fullagar et al. 2012; Shang, Qiao & Chen 2020).

Because of its emphasis on wellness and slow mobility, including cycling and walking, slow tourism has many parallels with traditional pilgrimage. As noted in other parts of this book, historical and traditional European pilgrimages entailed a lot of walking. After a nearly 400-year lull in some instances, the idea of walking pilgrimage trails has picked up pace again since the 1980s and is once again becoming a salient part of many people's religious travel experiences in Europe (Ashley & Deegan 2009; Olsen & Trono 2018; Ron & Timothy 2019; Timothy & Boyd 2015).

The mountain pilgrimages of Japan have existed since ancient times but are becoming more culture and wellness tourism-oriented beyond the sole domain of Shinto practitioners (Jimura 2016; Kato & Progano 2017). The Japanese pilgrim routes are analogous to the notion of slow tourism and have in fact been adopted by the slow tourism movement, as they entail walking, immersion in nature and place-based culture, and encourage wellness through healthy eating, yoga, and thermal baths (Kato 2017). They also encourage rural development and 'back to basics' modes of travel and enjoyment. Das and Islam (2017) and Kato (2017) draw many parallels between walking tourism in general and pilgrimage because both support increased mindfulness, place authenticity, wellness and is a lower-impact form of tourism (Redick 2018). At the same time, the ideal of pilgrimage as being a slow phenomenon is culturally and geographically contingent. As Olsen and Wilkinson (2016) note, some pilgrimages in Japan encourage pilgrims to complete their pilgrimage circuits as quickly as possible, as the more times one completes a pilgrimage circuit, the more merit one receives.

Hospitality and religion

There has been a lot written about Arab or Muslim hospitality (e.g., Sobh et al. 2013; Stephenson 2014; Stephenson & Ali 2019), yet Islam is not the only religion mandated to serve others and to demonstrate hospitality and kindness towards strangers (Kirillova et al. 2014;

Martin 2014; Pohl 1999). Outside of documented historical accounts, we know little about how travelers are welcomed and their needs satisfied according to holy writ and religious traditions in other faiths. According to the Quran and Hadiths, Muslims are required to demonstrate kindness and generosity to strangers. The Bible exhorts Christians and Jews to do the same, yet these groups and others have received much less attention in the study of hospitality and tourism (e.g., Sorensen 2005).

Part of the religion-hospitality intersection is the growing phenomena of halal tourism and kosher tourism, which were alluded to earlier in the cruise discussion. Increasing numbers of worldwide destinations are beginning to cater more overtly to the Muslim and Jewish markets. Following the events of September 11, 2001, many Middle Easterners curtailed their travel to the West for fear of encountering Islamophobia. Many Muslims reoriented their vacation choices to other Muslim-majority countries, which saw massive Middle Eastern arrivals in places such as Malaysia, Indonesia, and Turkey throughout the early 2000s. Since that time, there have been many efforts in non-Muslim majority countries to entice back the lucrative Muslim market by developing halal tourism (Battour & Ismail 2016; Hall & Prayag 2020; Mohsin & Ryan 2019; Moufakkir et al. 2019). Halal tourism is far more than an issue of food, although gastronomy and food services are an important part of the halal tourism phenomenon. It also includes extra services that make Muslim travelers feel welcomed and safe. A handful of halal ranking systems have been developed in Southeast Asia, Turkey, and other places to help Muslim travelers understand how halal a particular service provider is. For example, many resorts and hotels provide a schedule of prayer times; a qiblat (Mecca directional marker); Quran and prayer rug in the room or common prayer area; gender-segregated recreational facilities (e.g., swimming pools); and alcohol-free and halal-certified restaurants that will have a higher crescent rating than a property that only has a halal menu option (Timothy & Ron 2016; Weidenfeld & Ron 2008). There are similar kashrut and glatt kosher certifications for Jewish consumers in Israel and abroad. These are growing in areas that are popular among Israeli tourists who prefer to adhere to their dietary restrictions (Moira et al. 2017; Timothy & Ron 2016).

Within the many different faiths on the planet, there are diverging views of service and hospitality that intersect with tourism in and around their communities. While there is only a small literature on this topic (e.g., Hall & Prayag 2020; Islam & Kirillova 2020; Kirillova, Gilmetdinova & Lehto 2014; Lashley 2015; Weidenfeld 2006), more additional research is needed to gain a broader perspective on variations in travel needs and hospitality services between religions, especially among some of and alcohol the lesser researched faiths.

The religious travel markets and motives

People travel for a wide range of reasons. In fact, there are probably as many motivations for undertaking a journey as there are journeys, each one having its own unique set of purposes, interests, or obligations. Although there has been a great deal of research on religious travel motives, many of which have been explained in this volume, we still know relatively little about people's motives for religious and spiritual travel beyond the normative explanations of religious obligations, desire to commune with deity, to seek blessings, and visiting a famous shrine.

Understanding the market for pilgrimage, religious tourism, and spiritual travel is critical for destinations and service providers to offer satisfying faith or spiritually transformative experiences that will touch the lives of those who visit, increase their economic footprint, cause them to return, and result in positive word of mouth. Traditionally, the pilgrimage market has been accepted as an undifferentiated cohort of travelers who are easy to please because of their obligation to visit. Service quality was deemed somewhat irrelevant to whether or not a pilgrim would re-visit a destination or even recommend it to others; most pilgrimage destinations were populated by a 'captive audience' of worshippers who had little flexibility in their choice of destination.

Although this is still somewhat true in instances where pilgrimages are required at designated times and localities, the broader notion of religious tourism and the evolving nature of pilgrimage in many faiths means that faith travelers have an increasingly wide range of pilgrimages to choose from. With the exception of the required Hajj, most religious people can choose whether or not to go on a pilgrimage as a means of improving their spiritual selves, and much of the time they can choose the destination. Thus, the pilgrimage and religious tourism sector today within mainstream Christianity, Hinduism, Buddhism, Jainism, Sikhism, Taoism, Confucianism, Islam (outside the Hajj), and in most other religions, is increasingly cognizant that there are many pilgrimage destinations vying for people's religious travel plans (Heydari Chianeh et al., 2018; Ron & Timothy 2019).

Within the pilgrimage market, there are many types and motives. Markets may be understood in many ways. For example, pilgrims are distinguishable by their religious affiliation. This variable means that some people will only visit sacred sites designated holy by their own faith; adherents will avoid the sacred sites of others. Even within Christianity, members of certain churches would not likely visit places considered sacrosanct by other denominations (Ron & Timothy 2019). Likewise, many pilgrims are motivated by strict obligation. For this segment, attending a pilgrimage is a requirement of their faith and they have little choice in when they undertake pilgrimage and where they wish to go. Also, in some pilgrimage contexts, service quality has traditionally been seen as less important, but nowadays in a more competitive marketplace, it has become a higher priority for destinations (Eid 2012; Handriana et al. 2019).

As part of the growing service-mindedness in the religious tourism sector, people with disabilities are increasingly being facilitated at sacred destinations. Certain pilgrimage centers, such as Lourdes, have long catered to the special needs of the ill, the aged, and those with fragile health (Gesler 1996). More recently, however, many faith destinations are now seeking to improve people's experiences by making sacred sites, visitor centers, shrines, and places of worship more accessible to the ill or less mobile segments of society (Gassiot-Meilan et al. 2019).

Other new niches are continuing to appear in the marketplace. As standards of living continue to improve in developing countries, and with the spread of certain religions in key parts of the world, there has been a remarkable growth of religious and spiritual tourism in non-traditional markets. Christian pilgrims from many countries of Africa regularly visit Israel and the broader Holy Land by the millions. Greater numbers of Chinese travelers are seeking to experience the religious heritage of Europe, and even the underground growth of Christianity in China is causing many converts to seek pilgrimage experiences within and outside of China as a means of building their faith networks and socializing with other Christians in more open settings in the Western world.

Globalization

Globalization refers to the various processes that figuratively make the world smaller. Globalization connects people, places, businesses, and governments through trade, education, political alliances, transnational corporatization, social media, popular culture, and tourism. Tourism is both a product *of* globalization and a force *for* globalization (Timothy 2019). It manifests some of the most impactful elements of globalization, such as human mobility and international collaboration, and embraces much of the technology that has brought the world together. As tourism continues on its upward global trajectory and as it eventually recovers from the COVID-19 pandemic, religious and spiritual tourism will join the ranks of the fastest-growing travel sectors as the World Tourism Organization (2011, 2015) has predicted.

However, technology is perhaps one of the most influential forces of globalization in religious and spiritual tourism contexts (Olsen 2019). Especially popular in the past 15 years is the emergence of virtual pilgrimages, wherein the faithful who are unable to undertake a pilgrimage for whatever reason can do so virtually with mobile phone apps and online through virtual reality software or webcams. Most major religions today have apps that enable virtual experiences on pilgrim routes and tours, or performing rituals in sacred locales (McFarlan Miller 2020). Such technologies might allow non-adherents to explore and experience sacred places and events they are not permitted to attend in person (e.g., Mecca and the Hajj for non-Muslims). It also enables people to undertake an arduous pilgrimage that might be beyond their physical capabilities. Virtual technology may engender an interest in people to visit a new spiritual destination, just as it has in other tourism settings (Bogicevic et al. 2019), and help those who have lost their faith to regain it.

Other aspects of globalization, especially political globalization, have the potential to grow spiritual and religious tourism considerably. Most notably is the process of cross-border cooperation and supranationalism in which sovereign states unite for a common cause, usually economic development, military alliances and human mobility. The most progressive example of this is the European Union and its Schengen Treaty, both of which have functionally reduced most of Europe's interior borders to symbolic lines of division. The debordering of Europe through supranationalism has opened up a huge region with a vast array of religious heritages. This has stimulated increased cross-border pilgrimages (e.g., Buyskykh 2019), enabled religious tourism to thrive in certain formerly isolated regions, and is pushing small, regional shrines and sacred sites to the forefront of global tourism. This means that many folk shrines of local renown now have the potential to become destinations of more international acclaim.

The effects of globalization on religious adherence are still unknown compared to what we know about globalized economics, politics, and popular culture. This is especially true with regard to scale of religious heritage and pilgrimage shrines. While Olsen (2019) has suggested that globalization has led to the transplantation of religious and spiritual traditions due to increased mobility, the democratization of travel and increasing visitation to sacred sites, and increasing commodification of religion and spirituality, there is scope for much more research on many aspects of religious and spiritual tourism and their connections to globalization processes.

Final remarks

The chapters in this handbook in many ways have done what they were supposed to do: they answered many questions and raised many more. We have tried to be as comprehensive in covering the intersections between religion, spirituality, and tourism as possible, but there are inevitable deficiencies. These deficiencies, however, should not be viewed as weaknesses, but rather opportunities to delve even deeper into understanding the shared and individual nuances between each manifestation of faith, whether part of an organized religion or simply one's own quest for knowledge, spiritual connectedness with nature, or secular humanist pathway to a greater understanding of life. This concluding chapter highlights several areas that need more research attention. Volumes could be written about what we do not know, but space constraints do not allow us to delve into every missing connection. These are for future volumes of work.

As noted in the introductory chapter and in many chapters throughout this handbook, religious and spiritual tourism are not synonymous. Not all religious people are spiritual, and not all spiritual people are religious, and as such the travel experiences of religious and spiritual tourists may vary considerably. We must learn more about the differences and similarities between the two, but what we do know is that both can, and frequently do, manifest as pilgrimages that may be deeply meaningful to those who participate. As Ron and Timothy (2019: 124) note,

> Many outdoor enthusiasts, even those who would not describe themselves as 'religious', often report having 'spiritual' experiences in wilderness settings...Something about being immersed in nature, alone with one's thoughts and undistracted from the harried lifestyles of modern society, allows individuals to develop connections with an otherworldly force beyond themselves.

The varying doctrines and practices of the world's religions—what they share in common and what differentiates them—have the potential to interact with tourism or influence how it unfolds. There are in fact thousands of individual religions, faiths, sects, and denominations, numbering from a small handful of followers to having devotees in the millions or billions. Each one of these faiths and sects has different worldviews and expresses different ways of dealing with deity, society, and the afterlife. For some, faith-based travel is not expected. For others, it is required, and yet for some, it is prohibited or strongly discouraged (Timothy & Olsen 2006). We know a lot about religious travel of all kinds in Islam, Christianity, Judaism, Buddhism, and Hinduism but much less about the travel practices of other faith communities. Additional research is required so that we can better understand the unique practices and spiritual nuances associated with the religious or spiritual travel behavior of the multitudes of other faiths or those who seek meaning outside of institutionalized religion, which drive and inspire the people of the world.

References

Alexander, M., Bryce, D., & Murdy, S. (2017) 'Delivering the past: Providing personalized ancestral tourism experiences', *Journal of Travel Research*, 56(4): 543–555.

Ashley, K., & Deegan, M. (2009) *Being a Pilgrim: Art and Ritual on the Medieval Routes to Santiago*. London: Lund Humphries.

Atalan-Helicke, N. (2015) 'The halal paradox: Negotiating identity, religious values, and genetically engineered food in Turkey', *Agriculture and Human Values*, 32(4): 663–674.

Battour, M., & Ismail, M.N. (2016) 'Halal tourism: Concepts, practices, challenges and future', *Tourism Management Perspectives*, 19: 150–154.

Bielo, J.S. (2018) *Ark Encounter: The Making of a Creationist Theme Park*. New York: New York University Press.

Biran, A., Poria, Y., & Oren, G. (2011) 'Sought experiences at (dark) heritage sites', *Annals of Tourism Research*, 38(3): 820–841.

Bogicevic, V., Seo, S., Kandampully, J.A., Liu, S.Q., & Rudd, N.A. (2019) 'Virtual reality presence as a preamble of tourism experience: The role of mental imagery', *Tourism Management*, 74: 55–64.

Boyd, S.W. (2016) 'Reflections on slow food: From 'movement' to an emergent research field', in D.J. Timothy (Ed.), *Heritage Cuisines: Traditions, Identities and Tourism* (pp. 166–179). London: Routledge.

Buyskykh, I. (2019) 'In pursuit of healing and memories: Cross-border Ukrainian pilgrimage to a polish shrine', *Journal of Global Catholicism*, 3(1): 64–98.

Buzinde, C.N., Kalavar, J.M., Kohli, N., & Manuel-Navarrete, D. (2014) 'Emic understandings of Kumbh Mela pilgrimage experiences', *Annals of Tourism Research*, 49: 1–18.

Chmielewska-Szlajfer, H. (2017) '"Authentic experience" and manufactured entertainment: Holy Land Experience religious theme park', *Polish Sociological Review*, 200(4): 545–558.

Cohen, E.H. (2011) 'Educational dark tourism at an in populo site: The Holocaust Museum in Jerusalem', *Annals of Tourism Research*, 38(1): 193–209.

Coles, T., & Timothy, D.J. (2004) ''My field is the world': Conceptualizing diasporas, travel and tourism', in T. Coles & D.J. Timothy (Eds.), *Tourism, Diasporas and Space* (pp. 1–29). London: Routledge.

Collins-Kreiner, N. (2016) 'Dark tourism as/is pilgrimage', *Current Issues in Tourism*, 19(12): 1185–1189.

Collins-Kreiner, N., & Kliot, N. (2016) 'Particularism vs. universalism in hiking tourism', *Annals of Tourism Research*, 56(3): 132–137.

Croom, A.R., Squitiero, C., & Kerr, M.M. (2018) 'Something so sad can be so beautiful: A qualitative study of adolescent experiences at a 9/11 memorial', *Visitor Studies*, 21(2): 157–174.

Damari, C., & Mansfeld, Y. (2016) 'Reflections on pilgrims' identity, role and interplay with the pilgrimage environment', *Current Issues in Tourism*, 19(3): 199–222.

Das, S., & Islam, M. (2017) 'Hindu pilgrimage in India and walkability: Theory and praxis', in C.M. Hall, Y. Ram, & N. Shoval (Eds.), *The Routledge International Handbook of Walking* (pp. 242–250). London: Routledge.

Dickinson, J., & Lumsdon, L. (2010) *Slow Travel and Tourism*. London: Earthscan.

Dulska, A.K. (in press) 'Urban Jewish heritage and its (in)visibility: Perspectives from Navarre, Spain', *Journal of Heritage Tourism*. DOI: 10.1080/1743873X.2020.1810693.

Eid, R. (2012) 'Towards a high-quality religious tourism marketing: The case of Hajj service in Saudi Arabia', *Tourism Analysis*, 17(4): 509–522.

Esplin, S.C., & Olsen, D.H. (2020) 'Martyrdom and dark tourism in Carthage, Illinois', in D.H. Olsen & M.E. Korstanje (Eds.), *Dark Tourism and Pilgrimage* (pp. 99–110). Wallingford: CABI

Farra-Haddad, N. (2020) 'Archaeology and religious tourism: Sacred sites, rituals, sharing the Baraka and tourism development', in D.J. Timothy & L.G. Tahan (Eds.), *Archaeology and Tourism* (pp. 106–120). Bristol: Channel View.

Foley, M., & Lennon, J.J. (2000) *Dark Tourism*. London: Continuum.

Fullagar, S., Markwell, K., & Wilson, E. (Eds.) (2012) *Slow Tourism: Experiences and Mobilities*. Bristol: Channel View Publications.

Gassiot-Meilan, A., Coromina-Soler, L., & Planagumà, L.P. (2019) 'Accessible tourism in religious destinations: Comparing the motivations of people with and without special access needs', in J. Álvarez-García, M. de la Cruz del Río Rama, & M. Gómez-Ullate (Eds.), *Handbook of Research on Socio-Economic Impacts of Religious Tourism and Pilgrimage* (pp. 357–374). Hershey, PA: IGI Global.

Gesler, W. (1996) 'Lourdes: healing in a place of pilgrimage', *Health & Place*, 2(2): 95–105.

Gürey, N. (2017) 'Visiting graves, tombs and shrines in Islamic law', in R. Raj & K. Griffin (Eds.), *Conflicts, Religion and Culture in Tourism* (pp. 107–116). Wallingford: CABI

Hall, C.M., & Prayag, G. (Eds.) (2020) *The Routledge Handbook of Halal Hospitality and Islamic Tourism*. London: Routledge.

Handriana, T., Yulianti, P., & Kurniawati, M. (2019) 'Exploration of pilgrimage tourism in Indonesia', *Journal of Islamic Marketing*, 11(3): 783–795.

Hartmann, R. (2014) 'Dark tourism, thanatourism, and dissonance in heritage tourism management: New directions in contemporary tourism research', *Journal of Heritage Tourism*, 9(2): 166–182.

Heiman, A., Gordon, B., & Zilberman, D. (2019) 'Food beliefs and food supply chains: The impact of religion and religiosity in Israel', *Food Policy*, 83: 363–369.

Heydari Chianeh, R., Del Chiappa, G., & Ghasemi, V. (2018) 'Cultural and religious tourism development in Iran: Prospects and challenges', *Anatolia*, 29(2): 204–214.

Higginbotham, G. (2012) 'Seeking roots and tracing lineages: Constructing a framework of reference for roots and genealogical tourism', *Journal of Heritage Tourism*, 7(3): 189–203.

Huang, W.J., Haller, W.J., & Ramshaw, G.P. (2013) 'Diaspora tourism and homeland attachment: An exploratory analysis', *Tourism Analysis*, 18(3): 285–296.

Islam, M.S., & Kirillova, K. (2020) 'Non-verbal communication in hospitality: At the intersection of religion and gender', *International Journal of Hospitality Management*, 84: 102326.

Ivanova, P., & Light, D. (2018) 'It's not that we like death or anything': Exploring the motivations and experiences of visitors to a lighter dark tourism attraction', *Journal of Heritage Tourism*, 13(4): 356–369.

Jimura, T. (2016) 'World Heritage Site management: A case study of sacred sites and pilgrimage routes in the Kii Mountain range, Japan', *Journal of Heritage Tourism*, 11(4): 382–394.

Kato, K. (2017) 'Pilgrimage as a slow tourism development—Kumano-kodo pilgrimage, Wakayama, Japan', in C.M. Hall, Y. Ram, & N. Shoval (Eds.), *The Routledge International Handbook of Walking* (pp. 232–241). London: Routledge.

Kato, K., & Progano, R.N. (2017) 'Spiritual (walking) tourism as a foundation for sustainable destination development: Kumano-kodo pilgrimage, Wakayama, Japan', *Tourism Management Perspectives*, 24: 243–251.

Kerr, M.M., & Price, R.H. (2016) 'Overlooked encounters: Young tourists' experiences at dark sites', *Journal of Heritage Tourism*, 11(2): 177–185.

Kerr, M.M., & Price, R.H. (2018) '"I know the plane crashed": Children's perspectives in dark tourism,' in P. Stone, R. Hartmann, T. Seaton, R. Sharpley, & L. White (Eds.), *The Palgrave Handbook of Dark Tourism Studies* (pp. 553–583). London: Palgrave Macmillan.

Kirillova, K., Gilmetdinova, A., & Lehto, X. (2014) 'Interpretation of hospitality across religions', *International Journal of Hospitality Management*, 43: 23–34.

Koren-Lawrence, N., & Collins-Kreiner, N. (2019) 'Visitors with their 'backs to the archaeology': Religious tourism and archaeology', *Journal of Heritage Tourism*, 14(2): 138–149.

Korstanje, M.E., & Olsen, D.H. (2020) 'Negotiating the intersection between dark tourism and pilgrimage', in D.H. Olsen & M.E. Korstanje (Eds.), *Dark Tourism and Pilgrimage* (pp. 1–15). Wallingford: CABI.

Kurzwell, A. (1995) 'Genealogy as a spiritual pilgrimage', *Avotaynu*, 11(3): 15–20.

Lashley, C. (2015) 'Hospitality and hospitableness', *Research in Hospitality Management*, 5(1): 1–7.

Lois-González, R.C., Santos, X.M., & Romero, P.T.D.Z. (2018) 'The Camino de Santiago de Compostela: The most important historic pilgrimage way in Europe', in D.H. Olsen & A. Trono (Eds.), *Religious Pilgrimage Routes and Trails: Sustainable Development and Management* (pp. 72–87). Wallingford: CABI.

Maddrell, A. (2011) '"Praying the Keeills': Rhythm, meaning and experience on pilgrimage journeys in the Isle of Man', *Landabrefid*, 25: 15–29.

Martin, L.R. (2014) 'Old Testament foundations for Christian hospitality', *Verbum et Ecclesia*, 35(1): 01–09.

McFarlan Miller, E. (2020) 'Digital pilgrimages allow the faithful to travel the world from their couches', *National Catholic Reporter*, May 2, 2020. Online at: https://www.ncronline.org/news/people/digital-pilgrimages-allow-faithful-travel-world-their-couches. Accessed November 16, 2020.

Mindell, C. (2010) 'Jewish genealogy as a spiritual pursuit: Conversation with Arthur Kurzweil', *CT Jewish Ledger*, Online at: http://www.jewishledger.com/2010/04/jewish-genealogy-as-a-spiritual-pursuit/. Accessed June 30, 2020.

Mohsin, A., & Ryan, C. (2019) 'Halal tourism: A growing market on a global stage', in D.J. Timothy (Ed.), *Routledge Handbook on Tourism in the Middle East and North Africa* (pp. 309–318). London: Routledge.

Moira, P., Mylonopoulos, D., & Vasilopoulou, P. (2017) 'Kosher tourism: A case study from Greece', in R. Raj & K. Griffin (Eds.), *Conflicts, Religion and Culture in Tourism* (pp. 144–154). Wallingford: CABI.

Moufakkir, O., Reisinger, Y., & AlSaleh, D. (2019) 'Much ado about halal tourism: Religion, religiosity or none of the above?' in D.J. Timothy (Ed.), *Routledge Handbook on Tourism in the Middle East and North Africa* (pp. 319–329). London: Routledge.

Murray, M., & Graham, B. (1997) 'Exploring the dialectics of route-based tourism: The Camino de Santiago', *Tourism Management*, 18(8): 513–524.

Øian, H. (2019) 'Pilgrim routes as contested spaces in Norway', *Tourism Geographies*, 21(3): 422–441.

Olsen, D.H. (2006a) 'Management issues for religious heritage attractions', in D.J. Timothy & D.H. Olsen (Eds.), *Tourism, Religion and Spiritual Journeys* (pp. 104–118). London: Routledge.

Olsen, D.H. (2006b) 'Tourism and informal pilgrimage among the Latter-day Saints', in D.J. Timothy & D.H. Olsen (Eds.), *Tourism, Religion and Spiritual Journeys* (pp. 254–270). London: Routledge.

Olsen, D.H. (2013) 'Touring sacred history: The Latter-day Saints and their historical sites', in Hunter, J.M. (Ed.), *Mormons and American Popular Culture: The Global Influence of an American Phenomenon, Vol. 2* (pp. 225–242). Santa Barbara, CA: Praeger Publishers.

Olsen, D.H. (2019) 'Religion, spirituality, and pilgrimage in a globalizing world', in D.J. Timothy (Ed.), *Handbook of Globalisation and Tourism* (pp. 270–283). London: Edward Elgar.

Olsen, D.H., & Emmett, C.F. (2021) 'Contesting religious heritage in the MENA region', in C.M. Hall & S. Seyfi (Eds.), *Cultural and Heritage Tourism in the Middle East and North Africa* (pp. 54–71). London and New York: Routledge.

Olsen, D.H., & Korstanje, M.E. (Eds.) (2020) *Dark Tourism and Pilgrimage*. Wallingford: CABI.

Olsen, D.H., & Trono, A. (Eds.) (2018) *Religious Pilgrimage Routes and Trails: Sustainable Development and Management*. Wallingford: CABI.

Olsen, D.H., & Wilkinson, G. (2016) 'Fast pilgrimage? The case of the Shikoku pilgrimage', *Annals of Tourism Research*, 61: 228–230.

Paine, C. (2019) *Gods and Rollercoasters: Religion in Theme Parks Worldwide*. New York: Bloomsbury Publishing.

Pohl, C.D. (1999) *Making Room: Recovering Hospitality as a Christian Tradition*. Grand Rapids, MI: Wm. B. Eerdmans Publishing.

Redick, K. (2018) 'Interpreting contemporary pilgrimage as spiritual journey or aesthetic tourism along the Appalachian Trail', *International Journal of Religious Tourism and Pilgrimage*, 6(2): 78–88.

Ron, A.S., & Feldman, J. (2009) 'From spots to themed sites–the evolution of the Protestant Holy Land', *Journal of Heritage Tourism*, 4(3): 201–216.

Ron, A.S., & Timothy, D.J. (2019) *Contemporary Christian Travel: Pilgrimage, Practice and Place*. Bristol: Channel View Publications.

Russo, A.P., & Romagosa, F. (2010) 'The network of Spanish Jewries: In praise of connecting and sharing heritage', *Journal of Heritage Tourism*, 5(2): 141–156.

Sather-Wagstaff, J. (2016) *Heritage that Hurts: Tourists in the Memoryscapes of September 11*. London: Routledge.

Shackley, M. (2002) 'Space, sanctity and service: The English cathedral as heterotopia', *International Journal of Tourism Research*, 4(5): 345–352.

Shang, W., Qiao, G., & Chen, N. (2020) 'Tourist experience of slow tourism: From authenticity to place attachment–a mixed-method study based on the case of slow city in China', *Asia Pacific Journal of Tourism Research*, 25(2): 170–188.

Shinde, K.A. (2021) 'Religious theme parks as tourist attraction systems', *Journal of Heritage Tourism*, 16(3): 281–299. DOI: 10.1080/1743873X.2020.1791887.

Shoval, N. (2000) 'Commodification and theming of the sacred: Changing patterns of tourist consumption in the 'Holy Land'', in M. Gottdiener (Ed.), *New Forms of Consumption: Consumers, Culture, and Commodification* (pp. 251–263). Lanham, MD: Rowman & Littlefield.

Silverstein, E. (2020) 'Faith-based theme cruises', *CruiseCritic*, February 6, 2020. Online at: https://www.cruisecritic.com/articles.cfm?ID=1025. Accessed November 3, 2020.

Sobh, R., Belk, R.W., & Wilson, J.A. (2013) 'Islamic Arab hospitality and multiculturalism', *Marketing Theory*, 13(4): 443–463.

Sorensen, P.J. (2005) 'The lost commandment: The sacred rites of hospitality', *Brigham Young University Studies*, 44(1): 5–32.

Stephenson, M.L. (2014) 'Deciphering 'Islamic hospitality': Developments, challenges and opportunities', *Tourism Management*, 40: 155–164.

Stephenson, M.L., & Ali, N. (2019) 'Deciphering 'Arab hospitality': Identifying key characteristics and concerns', in D.J. Timothy (Ed.), *Routledge Handbook on Tourism in the Middle East and North Africa* (pp. 71–82). London: Routledge.

Stone, P.R. (2006) 'A dark tourism spectrum: Towards a typology of death and macabre related tourist sites, attractions and exhibitions', *Tourism*, 54(2): 145–160.

Stone, P., & Sharpley, R. (2008) 'Consuming dark tourism: A thanatological perspective', *Annals of Tourism Research*, 35(2): 574–595.

Stone, P.R., Hartmann, R., Seaton, A.V., Sharpley, R., & White, L. (Eds.) (2018) *The Palgrave Handbook of Dark Tourism Studies* (pp. 335–354). London: Palgrave Macmillan.

Timothy, D.J. (2008) 'Genealogical mobility: Tourism and the search for a personal past', in D.J. Timothy & J. Kay Guelke (Eds.), *Geography and Genealogy: Locating Personal Pasts* (pp. 115–135). Aldershot: Ashgate.

Timothy, D.J. (2019) 'Globalisation: The shrinking world of tourism', in D.J. Timothy (Ed.), *Handbook of Globalisation and Tourism* (pp. 323–332). Cheltenham: Edward Elgar.

Timothy, D.J. (2021a) *Cultural Heritage and Tourism: An Introduction*, 2nd edn. Bristol: Channel View Publications.

Timothy, D.J. (2021b) 'Tourism and the multi-faith heritage of the Middle East and North Africa: A resource perspective', in C.M. Hall & S. Seyfi (Eds.), *Cultural and Heritage Tourism in the Middle East and North Africa: Complexities, Management and Practices* (pp. 34–53). London: Routledge.

Timothy, D.J., & Boyd, S.W. (2015) *Tourism and Trails: Cultural, Ecological and Management Issues*. Bristol: Channel View Publications.

Timothy, D.J., & Olsen, D.H. (2006) 'Conclusion: Whither religious tourism?' in D.J. Timothy & D.H. Olsen (Eds.), *Tourism, Religion and Spiritual Journeys* (pp. 271–278). London: Routledge.

Timothy, D.J., & Olsen, D.H. (2018) 'Religious routes, pilgrim trails: Spiritual pathways as tourism resources', in R. Butler & W. Suntikul (Eds.), *Tourism and Religion: Issues and Implications* (pp. 220–235). Bristol: Channel View Publications.

Timothy, D.J., & Ron, A.S. (2016) 'Religious heritage, spiritual aliment and food for the soul', in D.J. Timothy (Ed.), *Heritage Cuisines: Traditions, Identities and Tourism* (pp. 104–118). London: Routledge.

Timothy, D.J., & Teye, V.B. (2004) 'American children of the African diaspora: Journeys to the motherland', in T. Coles & D.J. Timothy (Eds.), *Tourism, Diasporas and Space* (pp. 111–123). London: Routledge.

Turner, V., & Turner, E. (1978) *Image and Pilgrimage in Christian Culture: Anthropological Perspectives*. Oxford: Basil Blackwell.

Weidenfeld, A. (2006) 'Religious needs in the hospitality industry', *Tourism and Hospitality Research*, 6(2): 143–159.

Weidenfeld, A., & Ron, A.S. (2008) 'Religious needs in the tourism industry', *Anatolia*, 19(2): 357–361.

World Tourism Organization (2011) *Religious Tourism in Asia and the Pacific*. Madrid: UNWTO.

World Tourism Organization (2015) *Proceedings of the International Congress on Religious Heritage and Tourism: Types, Trends and Challenges*. Madrid: UNWTO.

Wright, K. (2020) 'Finding *Roots*: Pop culture pilgrimage and the affective geographies of Kunta Kinteh Island', in D.H. Olsen & M.E. Korstanje (Eds.), *Dark Tourism and Pilgrimage* (pp. 185–196). Wallingford: CABI.

Wright, K.J. (2008) *The Christian Travel Planner*. Nashville: Thomas Nelson, Inc.

Wu, H.C., Chang, Y.Y., & Wu, T.P. (2019) 'Pilgrimage: What drives pilgrim experiential supportive intentions?', *Journal of Hospitality and Tourism Management*, 38: 66–81.

INDEX

Note: **Bold** page numbers refer to tables and *italic* page numbers refer to figures and page number followed by "n" refer to end notes.